NEW TRENDS IN PROCESS CONTROL AND PRODUCTION MANAGEMENT

PROCEEDINGS OF THE INTERNATIONAL CONFERENCE ON MARKETING MANAGEMENT, TRADE, FINANCIAL AND SOCIAL ASPECTS OF BUSINESS (MTS 2017), 18–20 MAY 2017, KOŠICE, SLOVAK REPUBLIC AND TARNOBRZEG, POLAND

New Trends in Process Control and Production Management

Editors

Lenka Štofová & Petra Szaryszová

Department of Management, Faculty of Business Economics with seat in Košice, University of Economics in Bratislava, Slovak Republic

CRC Press
Taylor & Francis Group
Boca Raton London New York

CRC Press is an imprint of the
Taylor & Francis Group, an **informa** business

A BALKEMA BOOK

Published by:
CRC Press/Balkema
P.O. Box 447, 2300 AK Leiden, The Netherlands
e-mail: Pub.NL@taylorandfrancis.com
www.crcpress.com – www.taylorandfrancis.com

First issued in paperback 2020

Typeset by V Publishing Solutions Pvt Ltd., Chennai, India

ISBN 13: 978-0-367-73572-2 (pbk)
ISBN 13: 978-1-138-05885-9 (hbk)

Visit the Taylor & Francis Web site at
http://www.taylorandfrancis.com

and the CRC Press Web site at
http://www.crcpress.com

Table of contents

Preface

The goal of each business is to meet the customer's needs to secure the success of the business and the profit of the businessman. This goal can only be achieved with the help of appropriate management and marketing which is also a factor in the successful social development of the country. Global strategic development documents and implementation of the best available techniques and technologies clearly induces the development trends of research and development, modernization of scientific infrastructure and improvement of overall conditions for research and development.

The conference proceeding "New Trends in Process Control and Production Management" is a collection of scientific papers by researchers from several countries and presents the latest knowledge in the area of process, financial and production management, marketing, economy, environment and the social sphere of human society. Each of the research papers was presented at the 5th international scientific conference Marketing Management, Trade, Financial and Social Aspects of Business (MTS 2017) in relevant sections from 18 to 19 May 2017 in Košice (Slovak Republic) and on 20 May 2017 in Tarnobrzeg (Republic of Poland). It dealt with problems such as: strategic management, production and crisis management, process and financial management, taxes, auditing, direct foreign investments, convenient operations in foreign trade, marketing, networking, business intelligence, environmental and social aspects of business etc.

Commitment to provide quality products while saving the environment, care for the health and safety of employees is becoming a very important part of the strategy and image of enterprises looking for long-term success. Our special thanks go to the Chairman of the International Scientific Committee of the MTS 2017 conference, prof. Ing. Bohuslava Mihalčová, PhD. & PhD. and equally to all members of the Program and Organizing Committee for their responsible work and assistance in the realisation of the conference, that resulted also in this proceeding. We believe that this publication provides a great pledge of interesting knowledge for not only a scientific, but also the general public.

Editors

International scientific committee members

INTERNATIONAL SCIENTIFIC COMMITTEE

Prof. Ing. Bohuslava MIHALČOVÁ, PhD & PhD—*Chairman, University of Economics in Bratislava, SK*
Prof. Ing. Ferdinand DAŇO, PhD—*Rector of the University of Economics in Bratislava, SK*
Doc. RNDr. Zuzana HAJDUOVÁ, PhD—*Dean of the Faculty of Business Economics of the University of Economics in Bratislava with seat in Košice, SK*
Dr. h. c. prof. RNDr. Michal TKÁČ, CSc.—*University of Economics in Bratislava, SK*
Doc. Ing. Mgr. Zuzana JUHÁSZOVÁ, PhD.—*University of Economics in Bratislava, SK*
Prof. h. c. doc. Ing. Vladimír NĚMEC, PhD, mim. prof.—*Czech Technical University in Prague, CZ*
Prof. dr hab. Andrzej SZROMNIK—*Cracow University of Economics, PL*
Prof. Dr. Zoltán SZAKÁLY—*University of Debrecen, HU*
Prof. Ing. Kani KABDI, PhD.—*L. N. Gumilyov Eurasian National University Astana, KZ*
Prof. Marina KHAYRULLINA, D. Sc.—*Novosibirsk State Technical University, RU*
Prof. Ing. Adriana CSIKÓSOVÁ, CSc.—*Technical University in Košice, SK*
Prof. Ing. Jaroslav DVOŘÁČEK, CSc.—*VŠB—Technical University of Ostrava, CZ*
Dr. h. c. doc. Ing. Stanislav SZABO, PhD., MBA, LL.M—*Czech Technical University in Prague, CZ*
Doc. Ing. Elena ŠÚBERTOVÁ, PhD.—*University of Economics in Bratislava, SK*
Assoc. Prof.dr Igor BUDAK—*University of Novi Sad, RS*
Dr. hab. Inż. Magdalena RZEMIENIAK, prof. PL—*Lublin University of Technology, PL*
Dr. hab. Inż. Artur PAŹDZIOR, prof. PL—*Lublin University of Technology, PL*
Dr. hab. Anna SZYLAR—*State Higher Vocational School Memorial of Prof. Stanislaw Tarnowski in Tarnobrzeg, PL*
Dr. hab. Paweł MARZEC, prof. KUL—*The John Paul II Catholic University of Lublin, PL*
Dr. hab. Inżpil. Jarosław KOZUBA, Profesornzw.—*Polish Air Force Academy, PL*
Vitaliy SERZHANOV, PhD, Associate Professor—*Faculty of Economics of Uzhhorod National University, UA*
Ing. Attila TÓTH, PhD—*Novitech Partner s.r.o., SK*

PROGRAM COMMITTEE

Ing. Magdaléna FREŇÁKOVÁ, PhD—*Chairman, University of Economics in Bratislava, SK*
Doc. PhDr. Mária Ria JANOŠKOVÁ, PhD—*Vice-Chairman, Technical University in Košice, SK*
Prof. Ing. Michal PRUŽINSKÝ, CSc.—*University of Economics in Bratislava, SK*
Prof. dr hab. Zbigniew MAKIEŁA—*Jagiellonian University in Kraków, PL*
Prof. dr hab. inż. Stanisław SKOWRON—*Lublin University of Technology, PL*
Prof. h. c. Ing. Martin BOSÁK, PhD—*University of Economics in Bratislava, SK*
Doc. PhDr. Alena NOVOTNÁ, PhD—*Catholic University in Ružomberok, SK*
Ing. Alexander TARČA, PhD—*University of Economics in Bratislava, SK*
Dr. Magdalena MACIASCZYK—*Lublin University of Technology, PL*
Dr. Pawel MACIASCZYK—*State Higher Vocational School Memorial of Prof. Stanislaw Tarnowski in Tarnobrzeg, PL*
Ing. Peter ČEKAN, PhD—*Technical University in Košice, SK*
Ing. Jakub KRAUS, PhD—*Czech Technical University in Prague, CZ*
Ing. Radoslav ŠULEJ, PhD—*Technical University in Košice, SK*
Dr. Zsolt POLERECZKI—*University of Debrecen, HU*
Ing. Marianna KICOVÁ, PhD—*University of Economics in Bratislava, SK*

Ing. Jozef BALUN—*Novitech a.s., SK*
Mgr. Gabriela SANČIOVÁ—*University of Economics in Bratislava, SK*
Ing. Jozef LUKÁČ—*University of Economics in Bratislava, SK*

ORGANIZING COMMITTEE

Ing. Petra SZARYSZOVÁ, PhD—*Chairman, University of Economics in Bratislava, SK*
Ing. Lenka ŠTOFOVÁ, PhD—*Vice-Chairman, University of Economics in Bratislava, SK*
Doc. Ing. Jana NAŠČÁKOVÁ, PhD—*University of Economics in Bratislava, SK*
Ing. Cecília OLEXOVÁ, PhD—*University of Economics in Bratislava, SK*
Drinż. Joanna WYRWISZ—*Lublin University of Technology, PL*
Dr. Agnieszka RZEPKA—*Lublin University of Technology, PL*
Ing. Luboš SOCHA, PhD & PhD—*Technical University in Košice, SK*
RNDr. Ján SABOL, PhD—*Slovenská asociácia procesného riadenia, SK*
Ing. Róbert ROZENBERG, PhD—*Technical University in Košice, SK*
Ing. Vladimír SOCHA, PhD—*Czech Technical University in Prague, CZ*
Ing. Peter POÓR, PhD.—*Technical University in Košice, SK*
Ing. Jaroslav DUGAS, PhD—*University of Economics in Bratislava, SK*
Ing. František HURNÝ—*University of Economics in Bratislava, SK*
Ing. Martin MUCHA—*University of Economics in Bratislava, SK*

New Trends in Process Control and Production Management – Štofová & Szaryszová (Eds)
© 2018 Taylor & Francis Group, London, ISBN 978-1-138-05885-9

Options for the use of energy crops in the process of renewable energy production

P. Adamišin, E. Huttmanová & J. Chovancová
Department of Environmental Management, Faculty of Management, The University of Prešov,
Prešov, Slovak Republic

ABSTRACT: Growing energy crops is one of the ways to produce energy from renewable sources. By cultivating energy crops it is possible to ensure revitalization of damaged soils as well as maintenance of productive and non-productive soil properties. A suitable alternative to the use of agricultural land, which cannot be used for food production for various reasons, is the production of biomass for energy production. Biomass production is important as a source of energy but also in the socio-economic sphere. Especially in the countryside, it has the opportunity to create a range of new job opportunities while ensuring the maintenance and cultivation of the landscape. One option for achieving the EU's main objective by 2020 (to obtain 20% of energy from renewable sources) there is use of energy crops. The aim of the paper is to evaluate the possibilities of producing bioenergy from energy crops under conditions of Slovakia.

1 INTRODUCTION

The security of mankind's energy needs and ecological problems are among the basic problems of humanity (Trenčiansky et al. 2007, p. 87).

The evolution of energy consumption is exponentially growing. Fossil fuel supplies are not so large that humanity can afford to rely entirely on it (Fáber et al. 2012, p. 5). Primary energy production, however, produces substances that are most polluting the environment (Trenčiansky et al. 2007, p. 87).

In addition, the uneven distribution of fossil energy sources and the need for long-range transport do not contribute to ensuring the sustainable development. For this reason, it seems most advantageous to use the energy directly at the site of its origin. Renewable energy sources are one of the paths that need to be taken to ensure greater diversification and distribution of energy sources (Fáber et al. 2012, p. 5).

In general, an important source of such energy is agriculture and related energy from plant production. Slovakia is a country with limited natural resources and it is therefore necessary to look for new sources of energy so that dependence on the import of fossil fuels is reduced. From this point of view, growing energy crops represent an important issue, which aims to produce biomass for prime combustion, gasification, anaerobic digestion, biofuel production and for further energy and industrial use. The importance of biomass is not only that it is a new source of energy, the

more significant are the ecological aspect that is manifested in the landscape-forming function of the targeted biomass, it contributes to the efficient use of land also in the agriculturally less-favoured areas and also provides and creates new job opportunities (Porvaz et al. 2009, p. 66).

Slovakia is among the countries with a significant potential for renewable energy sources (RES). Their utilization has changed significantly over the last decade when in 2002 the share of renewable energy sources was about 1.6% of the total primary energy consumption (electricity, heat, cold and transport) currently over 10%. (Fáber et al. 2012, p.5).

In 2006, production of electricity from renewable energy sources and cogeneration of electricity and heat was supported by Act No. 309/2009 (Bosák et al. 2016, Majerník et al. 2015), thanks to this law, there was an increase in the building of RES equipment in Slovakia especially photovoltaics and biomass (Fáber et al. 2012, p.5).

The EU's current objective is to achieve a higher share of energy from renewable sources (by 2020, the renewable energy production must represent 20% of final energy consumption). Increasing the share of energy from renewable sources is also one of the basic priorities of the Slovak Energy Policy (Ministry of the Environment SR).

In Slovakia, the largest share of the total technically feasible potential of accused sources of energy – 35% is biomass with a value of 40 453 TJ/year (11237 GWh/year). In our conditions, it is realistic to use forest biomass energy crops, agricultural biomass, wood waste and food industry waste, and biomass waste from the industrial and communal sphere for energy purposes. (Fáber et al. 2012, p. 7). The opportunities for renewable energy production are further discussed by Andrejovský et al. 2013; Demoet al. 2013; Horbaj 2006; Jamriška 2007 and others, highlighting their positive contribution to the revitalization of damaged soils (Vilček et al. 2014), as well as the socio-economic area. Current trends, however, are rather directed towards the use of phytobiomas (Bejda et al. 2012); and the use of phytobiomas for energy purposes is directed towards the use of cereals and oilseeds (oilseed rape, sunflower), despite the fact that other alternative and economically more efficient sources are being sought (e.g. grasses - Festucaarundinacea, Arrhenatherumelatius, Phragmites australis etc.) (Jamriška 2007).

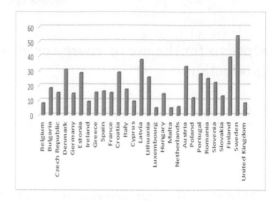

Figure 1. Share of energy from renewable sources in total energy consumption in %.
Source: Own processing; Data: Eurostat.

2 OBJECTIVE, MATERIAL AND METHODS

The aim of the paper is to evaluate the possibilities of bioenergy production from energy crops in the conditions of Slovakia.

The assessment of energy crop production will be carried out using a correlation and regression analysis where data from Eurostat were used as input data.

Through selected methods of statistical analysis we model the possible future development of RES production in Slovakia, we compare it with the average of the EU countries, as well as analyze the links between the economic development of the countries and the level of RES support. We measure the level of RES support at the level of individual countries through the indicator of the share of RES in total energy production. To quantify economic developments, we used the GDP per capita indicator, but for higher objectification of cross-reference in PPS units.

3 RESULTS AND DISCUSSION

European Union countries currently account for almost 17% of the share of energy from renewable sources in total energy consumption (Fig. 1). There are, however, significant differences between countries. The most positive values (the highest share of energy from renewable sources) are Sweden - almost 54%, Finland (almost 40%) and Latvia (37%), which have already achieved their national targets for this share.

By contrast, Luxembourg (5%), Malta (5%) and the United Kingdom (8.2%) have the lowest share of energy from renewable sources in total energy

consumption, whereby the national targets of these countries are two to three times higher.

By eliminating the partial results so far presented, we can assume that the individual EU countries are encouraging the development of RES. One factor that we are able to quantify (as opposed to qualitative as, for example, environmental awareness of society) is the economic maturity of the country.

We can assume that a country that is economically more advanced will put more emphasis on promoting long-term sustainability, also on the growth of RES production. Based on this axiom, we will identify the relationship between the share of RES and GDP per capita expressed in purchasing power parity. Results are visualized in a 2D (Fig. 2).

It is clear that it is not possible to identify a statistically significant relationship between the share of RES and the economic development of the country. Even expanding the regression band analysis with 95% reliability does not account for most countries within the specified interval.

The verification of the results ascertained by the visual assessment was carried out through correlation analysis processes.

The analysis was performed using both Pearson and Spearman correlation analysis. The results of the correlation analysis are shown in Table 1.

Neither of the correlation analyzes does not indicate the existence of a statistically significant link between the share of energy acquired by the country from RES and the economic development of the country.

Countries that are increasingly supportive of RES are governed by other principles and the economic development of the country is not a significant factor.

In the last decade, the Slovak Republic has increased the share of energy from renewable

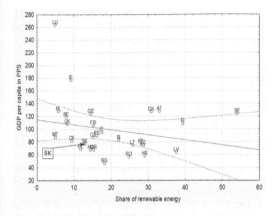

Figure 2. Relationship between the share of RES and GDP per capita based on PPP.
Legend: AT Austria, BE Belgium, BG Bulgaria, CY Cyprus, CZ Czech Republic, DE Germany, DK Denmark, EE Estonia, ES Spain, FI Finland, FR France, GR Greece, HR Croatia, HU Hungary, IE Ireland, IT Italy, LT Lithuania, LU Luxembourg, LV Latvia, MT Malta, NL Netherlands, PL Poland, PT Portugal, RO Romania, SE Sweden, SI Slovenia, SK Slovakia, UK United Kingdom.

Table 1. Results of the correlation analysis - verification of the relationship between the share of RES and GDP per capita.

	Pearson correlation		Spearman correlation	
	share _RE	GDP in PPS	share _RE	GDP in PPS
share_RE	1.0000	−0.2157	1,0000	−0.2487
p-value	–	0,27	–	0.21
GDP in PPS	−0.2157	1.0000	−0.2487	1.0000
p-value	0.27	–	0.21	–

Source: Own processing; Data: Eurostat.

sources, which is positive, when compared to 2004, the consumption of energy produced in this way doubled and currently reaches almost 13% (Fig. 3), but it is still below the EU-28 average.

It is clear from Chart 2 that the production of renewable energy sources, measured by the share of total energy consumption, is increasing both in Slovakia and in the EU.

Based on historical developments, we have modeled possible future developments through a regression analysis. In both cases, a linear regression model was chosen, as the model results were sufficiently significant in the case of linear models.

The regression model for the average EU countries is as follows:

$$y^{++} = 7.416^{++} + 0.773^{+}x \tag{1}$$

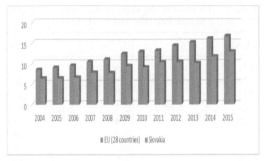

Figure 3. Share of energy from renewable sources in total energy consumption in % in Slovakia and at the average of EU countries.
Source: Own processing; Data: Eurostat.

The regression model for the SR is as follows:

$$y^{++} = 5.335^{++} + 0.574^{+}x \tag{2}$$

Based on the Europe 2020 objectives, the EU target for the share of energy from renewable sources was set at 20% for 2020. The projection of future developments based on historical data using this significant model for the year 2020 assumes a value of 20.56%. Based on this, we can assume that the targets set at the EU average will be achieved by 2020. The 95% confidence interval predicts the target value of the RES share in 2020 ranging from 20.07% to 21.04%.

In Slovakia, we can foresee the share of energy from renewable sources of 15.09% in 2020 (again based on the predictions from the regression model). It is a share that exceeds the target set for the SR (14% in 2020). With 95% confidence, the target value of the RES share in 2020 can be assumed to be based on the model, ranging from 14.01% to 16.18%.

Based on these assumptions we can conclude that meeting the share of energy produced from renewable sources should be achievable. Although this result is satisfactory from a given point of view, it creates risks, the gradual deepening of the difference between the share of energy produced in Slovakia and the EU average. This disparity is also evident from the value of parameter b1 in both regression models: while in the EU countries, the share of production from RES is 0.773% on average per year, in Slovakia it is only 0.574% on average. The projection of further development of the share of RES production in total energy production for the next 10 years after 2020 is presented in Table 2.

Although on the basis of historical results, we could expect a continuous and systematic increase in the share of energy production from RES, the

3

Table 2. Forecast of the share of RES production after 2020 in the Slovak Republic and at EU level.

YEAR	EÚ	SR
2021	21.3	15.7
2022	22.1	16.2
2023	22.9	16.8
2024	23.7	17.4
2025	24.4	18.0
2026	25.2	18.5
2027	26.0	19.1
2028	26.7	19.7
2029	27.5	20.3
2030	28.3	20.8
2031	29.1	21.4
2032	29.8	22.0
2033	30.6	22.6
2034	31.4	23.1
2035	32.2	23.7
2036	32.9	24.3

increase in the EU will be higher than in Slovakia. Assuming the preservation of historical trends, the difference between the share of energy use from RES in Slovakia and the EU will be exacerbated.

4 CONCLUSIONS

The above assessments show that at EU average, the targets will be met by 2020 and that EU countries will be able to produce more than 20% of renewable energy in the given period (subject to otherwise unchanged conditions). Despite this relatively positive assumption, interregional differences need to be taken into account in achieving that goal. The Slovak Republic should produce almost 15% of renewable energy in 2020, fulfilling the national target, but not reaching the EU target. Based on this, it can be concluded that not all EU countries are equally involved in achieving the set goals. However, a more detailed assessment of developments in individual EU countries requires more detailed research and it will be the subject of our further research interest, in spite of the finding that the statistically significant link between the economic performance of countries and the share of renewable energy has not been confirmed.

ACKNOWLEDGEMENT

This study has been created thanks to support under the Research and Development Operational Program, for the project: University Science Park TECHNICOM for Innovative Applications with the Support of Knowledge Technologies – II. Phase, project code: 313011D232, co-financed by the European Regional Development Fund.

REFERENCES

Andrejovský, P. et al. 2013. *Energy efficiency and economic support of regional energy.* In: International Multidisciplinary Scientific Geoconference Surveying Geology and Mining Ecology Management, SGEM, Volume 2, 2013, p. 103-110. Isbn 978-619710504-9.

Bejda, J. et al. 2002. *Laboratórny výskum briketovateľnosti bioodpadu.* In. Technikaochranyživotnéhoprostredia (TOP) 2002, STU Bratislava, s. 229–234.

Bosák, M., et al. 2016. *Application of new environmental recycling technology.* In: International Multidisciplinary Scientific GeoConference Surveying Geology and Mining Ecology Management, SGEM, Volume 2, 2016, p. 77–84. ISBN 978-619710556-8.

Demo, M., et al. 2013. *Production and energy potential of different hybrids of poplar in the soil and climatic conditions of southwestern Slovakia.* In: Wood Research. Volume 58, issue 3, 2013, p. 439–450. ISSN 1336-4561.

Fáber, A. et al. 2012. *Atlas obnoviteľných zdrojov energie na Slovensku.* Bratislava: Energické centrum Bratislava, 2012. 317 s. ISBN 978-80-969646-2-8. Available on: http://ecb.sk/fileadmin/user_upload/editors/documents/Kniha_OZE_A5_def_web.pdf.

Horbaj, P. 2006. *Možnosti využívania biomasy v SR.* In.: Acta Montanistica Slovaca, roč.11, 2006, s. 258–263.

Jamriška, P. 2007. *Rastlinná výroba – zdroj obnoviteľnej energie.* In Predpoklady využívania poľnohospodárskej a lesníckej biomasy na energetické a biotechnické využitie. Nitra: SAPV s. 20–27. ISBN 978-80-89162-32-1.

Kardis, K. & Kardis, M. 2011. *Wartosci a globalizacja na tle przemian spoleczno-gospodarczych wspólplczesnego swiata.* In Rodzina wobec wyzwan współczesnosci. Tom II. Krosno: Krosnienska Oficyna Wydawnicza Sp. z o.o., 2011. p. 146-148. ISBN 978-83-62843-12-1.

Majerník, M. et al. 2015. *Innovative model of integrated energy management in companies.* In: Quality Innovation Prosperity. Volume 19, Issue 1, 2015, p. 22–32. ISSN 1335-1745.

Porvaz, P. et al. 2009. *"Poľné plodiny ako zdroj biomasy na energetické využitie v podmienkach Slovenska."* In Inovatívne technologie pre efektívne využitie biomasy v energetike. s. 66–75. ISBN 978-80-225-2962-4. Available on: http://enersupply.euke.sk/wp-content/uploads/66-75_porvaz-nascakova- kotorova-kovac.pdf.

Tej, J. et al. 2014. *Crisis awarness of the municipal district sresidents: Implication for crisis management at the local government level.* In: Quality. Innovation. Prosperity. Vol. 18, No. 2 (2014), p. 1–14. ISSN 1335-1745.

Trenčiansky, M. et al. 2007. *Energetické zhodnotenia biomasy.* Zvolen: NLC. 2007. 147 s. ISBN 978-80-8093-050-9.

Vilček, J. et al. 2014. *Risk elements in soils of burdened areas of eastern Slovakia.* In: Polish journal of environmental studies. Volume 21, issue 5, 2012, p. 1429–1436. ISSN 1230-1485.

New Trends in Process Control and Production Management – Štofová & Szaryszová (Eds)
© 2018 Taylor & Francis Group, London, ISBN 978-1-138-05885-9

Ecological aspects of manufacturing enterprises activity: Analysis methods

B.A. Amanzholova & E.V. Khomenko
Faculty of Business, Novosibirsk State Technical University, Novosibirsk, Russian Federation

ABSTRACT: Disclosure of information on ecological activities in the reporting of manufacturing enterprises is very important issue in conditions of the Russian national model of sustainable development which doesn't have today sufficient experience and the developed regulatory framework, and also because of nature, intensity and scales of impact of the entities on the environment. It is obvious that under uncertainty of social responsibility parameters and objectively existing misbalance between the economic, social and ecological aspects disclosed in the reporting new approaches to the analysis of disclosure of the information on ecological activities are necessary. Authors offered specific methods and procedures that provides complex assessment of corporate social responsibility on the basis of the public reporting parameters variability. Approbation of the method provided systematization of the statutorily prescribed finance indicators of ecological activities, provisions and indirect liabilities according to accounting standards, and also voluntarily incurred liabilities in nature protection activities.

Disclosure of information on ecological activity is one of the factors making positive impact on sustainable development of the Russian manufacturing enterprises. However its action is limited for the reasons of uncertainty of parameters of social responsibility and lack of national standards of corporate social responsibility. High-quality disclosure of information promotes management decisions based on information on ecological costs, assets and liabilities, and provides realization of strategic approach in management of economic, ecological and social effectiveness of manufacturing enterprises activity. Informational content of the reporting reflecting the ecological liabilities which are legislatively fixed and voluntarily incurred by the enterprises increases investment prospects of manufacturing enterprises.

Today information on ecological aspects of activity is presented by the Russian manufacturing enterprises in the statistical-and-ecological reporting forms which aren't public, and also in the public reporting, including financial statements. However the range of the disclosed indicators significantly differs. The most important indicators of statistical survey are: payments for negative impact on the environment, and also investments on environmental protection. The system of payments for negative impact on the environment is the mechanism of stimulation of nature protection and environmentally safe investments. The main share in payments is made by payments for placement of production and consumption wastes (Table 1) (Ministry of Natural Resources and Environmental Protection of the Russian Federation 2016, p. 199).

It is evident than in the total amount of the considered payments, as shown in Table 1, at one times there was an excess of payments for negative impact within the allowable limits, at other times—payments for excess negative impact.

The most considerable volumes of nature protection and environmentally safe investments are made by the enterprises of such types of economic activity as "the manufacturing activity", "production and distribution of the electric power, gas and water" and "extraction of minerals" (Table 2) (Ministry of Natural Resources and Environmental Protection

Table 1. Payments for negative impact on the environment in Russia, billion rubles.

Payment type	2013	2014	2015
Total	24.7	23.2	27.9
including: negative impact on atmosphere	10.4	8.2	7.6
negative impact on water	3.8	4.3	5.0
placement of production and consumption waste	10.5	10.7	15.3
From the total amount of payments it is paid for: negative impact within the allowable limits	11.8	12.3	13.5
excess negative impact	12.9	10.9	14.4

Table 2. Investments into fixed capital on environmental protection by types of economic activity, million rubles.

Types of economic activity	2014	2015
Manufacturing activity	62,908	63,816
Production and distribution of the electric power, gas and water	25,548	28,693
Extraction of minerals	22,354	24,044
Transport and communication	2815	2478
Agriculture, hunting and forestry	752.5	597.3
Providing other municipal, social and personal services	561	1095.2

of the Russian Federation 2016, p. 199). Over three quarters of total annual volume of investments are the share of these types of economic activity.

The state initiative of transition to the new system of ecological regulation based on the principles of the best available techniques and technologies (BATT) has to promote fuller disclosure of information on ecological aspects of activity of the industrial enterprises. It is planned that 300 enterprises made about 60% of gross pollution will begin implementation of new system of ecological regulation in 2019.

The analysis of the existing practice of disclosure of information on ecological activity in the public reporting of the manufacturing enterprises making the greatest negative impact on the environment confirms lack of the detailed information even in financial statements. For example, environmental protection costs and recognized ecological liabilities are respectively included in assets or the total expenses and short-term liabilities and aren't shown separately. Indirect liabilities and provisions, arising in ecological activity are disclosed only in case of importance of their influence on a financial state, results of activity and cash flow.

In annual reports for shareholders and sustainable development reports the information is disclosed to the fullest extent possible. It contains financial and non-financial indicators of ecological activity, and also the main directions of environmental policy of the company (Amanzholova 2016).

It is obvious that current situation in the sphere of information support of the manufacturing enterprises management influence on the analytical potential of the reporting that significantly reduces quality of management decisions in the ecological sphere. "Content analysis is used to determine the extent of biodiversity disclosures found in the integrated and sustainability reports of companies included in the JSE's mining and food producer and retail sectors. This method is frequently used to study non-statutory disclosures in corporate reports and involves codification of information found in these reports into predefined categories

in order to highlight trends and make inferences" (Mansoor & Maroun 2016, p. 599).

One of the difficult questions of ecological activity management that is urgent for both internal and external users of information is the question of a object of the ecological accounting and a number of objects for the analysis. Authors hold a view of Sokolov (2010) and consider the economic event as an object of the ecological accounting. Recognition of the economic event as an object of the ecological accounting allows to identify the moments of economic process changing or confirming structure of assets or their sources (assets and sources at the same time), connected with ecological activity. Authors consider that such single moments of economic process can be grouped on nature protection costs and ecological liabilities.

Specific nature of ecological activity predetermines classification of costs in dependence on their purposes, but it has no significant influence on the accounting procedure determined by national accounting standards. So, equipment rebuilding costs made to decrease negative impact on the environment refers to capital costs, and finally, will increase the fixed assets value. The capital costs on creation of the new technologies minimizing impact on the environment can be recognized as intangible assets. The given examples demonstrate a high level of a regulation of recognition and registration in accounting of capital and current costs. However this conclusion doesn't extend to ecological liabilities for a number of reasons.

The first reason is a specific base for the origin of liabilities. Ecological liabilities may arise from legislative rules, contracts and delict at once. In turn, the base for the origin defines date of their recognition, an assessment method, documentary confirmation and other aspects which are defined by the organization independently.

Secondly, recognition of the direct liabilities requires some conditions to be satisfied whereas for recognition of the indirect liabilities and provisions professional judgment is required in order to assess the probability of corporate economic benefits reduction that may have significant impact on reliability and a representativeness of the information on liabilities.

Thirdly, "exit" of the economic events out of the reporting period is typical for ecological activity. It is about events after the reporting date which consequences may be disclosed in the reporting of the previous period using a special procedure or revealed as the economic events of a new period. The choice of recognition method is also a subject of professional judgment, and it may influence not only on reliability of reporting indicators, but also on their comparability between the reporting periods.

Thus, costs and liabilities are the objects of the analysis that are the most important for manufacturing enterprises having negative impact on the environment (Fig. 1).

It is obvious that the objects given on the Figure 1 are inter-related as in accounting system they characterize assets and their sources. However for recognition of assets and liabilities in accordance with the civil laws it is very important to declare possession, use and disposition rights or their combination. Besides, it is rather difficult to identify spheres of corporate social responsibility and legal backgrounds of reporting preparation. Authors selected legislatively fixed liabilities of the company and voluntarily assumed liabilities within agreements with shareholders as such legal backgrounds.

Speaking about corporate social responsibility, authors follow traditional approach when the economic, ecological and social aspects are considered. "The overall focus of sustainability accounting and assurance is risk assessment, which is based on sustainability from the perspective of all stakeholders-financial, social, environmental, and technological-and management of the risk. Assumptions of sustainability accounting are developed from postulates of financial accounting and include: Continuity. Reports assume business continuity sufficient to meet sustainability objectives and requirements, including product disposal and environmental cleanup" (Fagerström et. al. 2017).

The research has shown that under the conditions of uncertainty of parameters of social responsibility for the Russian manufacturing enterprises there is an imbalance between economic, social and ecological activities. So, a number of indicators were calculated for the company which has negative impact on the environment. Indicators were chosen to characterize sustainable development in economic, ecological and social activities (Table 3).

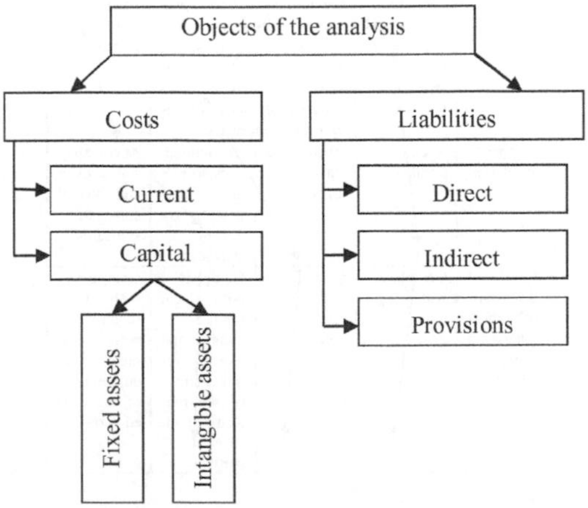

It should be mentioned that in Table 3 only a part of the system of indicators linking factors of sustainability and their influence is presented, however the provided figures reveal some imbalance in activities. So, the indicator "The created value added" is used for an integrated assessment of the sustainable development of the company and it shows significant increase in value added during the analyzed period. At the same time the economic sustainability indicators also tend to growth, but the same is not appropriate for social and ecological spheres.

The information from the accounting (financial) reporting; annual report; reporting on the sustainable development was used for calculations. So, in the conditions of information sources variety and non-mandatory provisions of the significant information disclosure the new approaches to the analysis of disclosure of the information on ecological activity are required (Fagerström et. al. 2007, Brown & Dillard 2014., Burchell et. al. 1985).

Authors have offered the method for analysis of representation and disclosure of the information on ecological activity in reporting (Fig. 2).

This method provides a complex assessment of corporate social responsibility in the sphere of nature protection activity taking into account variability of the parameters disclosed in the public reporting. The difference of the author's approach from existing ones is in focus of analyst's attention on the identification of the specific economic events and the analysis of influence of their consequences on ecological activity indicators. Such specific facts are events after the reporting date and contingencies.

Figure 1. Objects of the analysis specifically for manufacturing enterprises carrying out ecological activity.

Table 3. Dynamics of indicators of economic, ecological and social sustainability, %.

Indicators	2014	2015	Changes
Rates of change of the created value added	173.3	162.5	−10.8
Economic sustainability			
Rates of change of return on equity	104.2	132.7	28.5
Rates of change of capital productivity	109.9	119.3	9.4
Social sustainability			
Rates of change of labor productivity	123.4	86.5	−36.9
Ecological sustainability			
Rates of change of a share of payments for pollution of the environment in the created value added	53.8	66.9	13.1

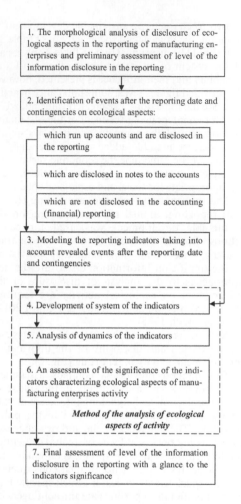

1. The morphological analysis of disclosure of ecological aspects in the reporting of manufacturing enterprises and preliminary assessment of level of the information disclosure in the reporting

2. Identification of events after the reporting date and contingencies on ecological aspects:

which run up accounts and are disclosed in the reporting

which are disclosed in notes to the accounts

which are not disclosed in the accounting (financial) reporting

3. Modeling the reporting indicators taking into account revealed events after the reporting date and contingencies

4. Development of system of the indicators

5. Analysis of dynamics of the indicators

6. An assessment of the significance of the indicators characterizing ecological aspects of manufacturing enterprises activity

Method of the analysis of ecological aspects of activity

7. Final assessment of level of the information disclosure in the reporting with a glance to the indicators significance

Figure 2. Proposed method of analysis of the ecological aspects of manufacturing enterprises activity.

The production enterprises which activity damages the environment make reserves on emergency response, land restoration and other similar reserves. In reporting for closed financial year these reserves reduce profit (or increase losses), but in the next period the sum of a reserve can be used on definite purposes, and can be restored. Both cases have significant effect on performance indicators, including ecological ones.

However the facts of reserves restoration need special attention as they are capable to misinform reporting users due to failure of the indicators comparability in the reporting. For this reason authors have included in the analysis method some procedures, directed to identification of events after the reporting date and contingencies on ecological aspects, and modeling the reporting indicators taking into account consequences of such events and facts.

On the first stage it is recommended to use a morphological analysis method for identifying the most significant for the analysis purposes indicators of the manufacturing enterprises public reporting. As it was mentioned above, variability of the public reporting parameters is caused by types of the reporting, legal backgrounds and spheres of corporate social responsibility of the enterprises.

In Figure 3 results of preliminary assessment of level of the information disclosure on ecological activity in different types of the public reporting (accounting (financial) reporting, annual report, sustainable development reporting) are presented.

Legal backgrounds of the reporting and spheres of corporate social responsibility also participate in preliminary assessment. Combination of these factors predetermines content of financial and

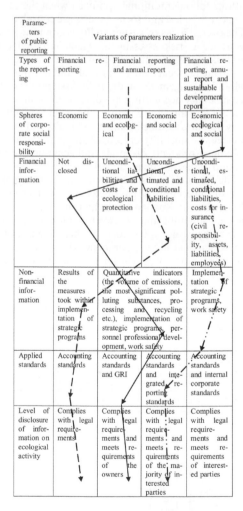

Figure 3. Morphological matrix of disclosure of information on ecological activity in the public reporting of the enterprises.

non-financial information on ecological activity of the manufacturing enterprises.

Verification of this hypothesis allowed estimating the level of disclosure of information on ecological activity in the public reporting of the enterprises for ensuring their sustainable development. During the analysis some situations typical for most manufacturing enterprises are revealed. It is disclosure in accounting reporting the economic indicators on ecological activity, legislatively provided and obligatory for the enterprises according to accounting standards. Facts of disclosure of not only financial, but also non-financial information according to accounting standards, sustainable development standards and corporate standards are revealed (Amanzholova et. al. 2016).

By results of preliminary assessment of disclosure of information on ecological activity the structure of the financial and non-financial information on objects of the analysis is identified. So, if the ecological liabilities are included in debit debt and credit debt, then on date of the analysis it is necessary to estimate their dynamics between the reporting date and date of the reporting submission to the interested parties. The procedures of the qualitative analysis of events after the reporting date and contingencies offered by authors are shown in Figure 4.

For the demonstration of proposed analysis method an example of the lawsuit abortive at the reporting date, which subject is quality of services provided within the nature protection activity is used. If works are performed and documented correctly, then their value is included in capital or current costs recognized in accounting and reporting. In this case credit debt is accounted. However, if the lawsuit is completed at the date of submission of the reporting to users, then the reasons for correction the data on assets and obligations arises. But for making the decision on such correction it is necessary to take into account materiality or potential materiality of consequences for the financial state, financial results and cash flow. That's why modeling of the reporting indicators taking into account revealed events after the reporting date and contingencies is a separate procedure in proposed analysis method. What concerns contingencies, the most important stage of modeling process is the analysis of consequences under uncertainty (Khayrullina et al. 2015).

So, if to come back to the example of a lawsuit, then the procedures shown in the lower part in Figure 4 are applied if it is abortive lawsuit at the date of submission of the reporting to the interested parties. Making a decision on correction of the ecological activity indicators it is necessary to proceed from assessment of the materiality of events after the reporting date and contingencies consequences.

Such correction may be done in different ways. First, data on assets, liabilities, capital, income and

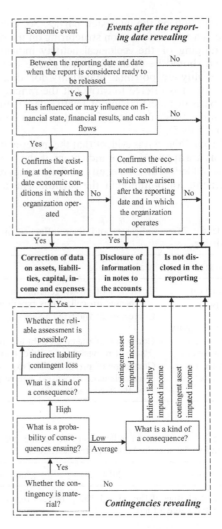

Figure 4. The procedures of the qualitative analysis of events after the reporting date and contingencies.

costs are ascertained, and the analysis is carried out taking into account a reliable assessment of the corrected indicators. Secondly, notes to the accounts are used for making corrections. In fact indicators correction may be done differently. According to Russian accounting standards, recognized contingencies should be included in general expenses (option 1), miscellaneous expenses (option 2) or asset value (option 3) depending on type of the contingence. In modeling process it is necessary to pay attention to the links, determined by the content of indicators and mechanism of their influence.

Identification of the performance indicators sensitive to events after the reporting date and to contingencies provides consolidation of information and cut in expenses for obtaining of useful information by the interested parties. So, there is a need for the information structuring on the basis of ranging of the ecological activity indicators, which should be disclosed with a glance to their materiality or potential materiality for decision making. It is offered to carry out the ranging on the basis of an estimation of relative departure of the indicators calculated with and without corrections (Tab. 5).

From Table 5 it is evident that the material deviations are appropriate for indicators of the created value added, return on equity, and also a share of payments for pollution of the environment in the created value added. So, the information about consequences of the events after the reporting date should be disclosed in reporting.

Thus, the proposed method of the ecological aspects analysis provides a reliable information basis for drafting the notes to accounts, contents of the annual report and the report on sustainable development. Focus on meeting the requirements of the interested parties is the main advantage of such approach. The disclosure of the data on ecological aspects and information interrelations of the sustainable development indicators in the reporting provides increase in validity of conclusions and decisions, and also a reliable assessment of social responsibility of manufacturing enterprises.

Table 4. Main indicators and their influence on the reporting.

Indicators	Main influence
Return on equity	Correction of profit (general expenses, miscellaneous expenses)
Capital productivity	Correction of the fixed assets value
Credit debt turnover	Correction of the credit debt for the sum of provision or indirect liability
Returns on assets	Correction of profit (general expenses, miscellaneous expenses), correction of the fixed assets value

Table 5. Dynamics of indicators of economic, ecological and social sustainability,% (after the modeling).

Indicators	2014	2015		Changes	
Rates of change of the created value added	173.3	162.5	145.2	−10.8	−28.1
Economic sustainability					
Rates of change of return on equity	104.2	132.7	112.9	28.5	8.7
Rates of change of capital productivity	109.9	119.3	119.3	9.4	9.4
Social sustainability					
Rates of change of labor productivity	123.4	86.5	86.5	−36.9	−36.9
Ecological sustainability					
Rates of change of a share of payments for pollution of the environment in the created value added	53.8	66.9	73.6	13.1	19.8

REFERENCES

Ministry of Natural Resources and Environmental Protection of the Russian Federation. 2016. The governmental report "About a state and protection of the environment in Russian Federation in 2015", *The basic ecological indicators of industries*: 196–199. Moscow: NIA-Priroda.

Amanzholova, B.A. et. al. 2016 Disclosure of information on ecological activity as a factor of sustainable development of manufacturing enterprises. *13th International Scientific-Technical Conference on Actual Problems of Electronics Instrument Engineering (APEIE)* Vol. 3: 202–206.

Mansoor, H. & Maroun, W. 2016. An initial review of biodiversity reporting by South African corporates: The case of the food and mining sectors (Review). *South African Journal of Economic and Management Sciences* Vol. 19, Issue 4: 592–614.

Sokolov, Ya.V. 2010. *Accounting as sum of economic events*. Moscow: Magister & Infra-M.

Fagerström, A. et. al. 2007. Compliance with consolidation (group) accounting standards – the vertical adjustment issue: a survey of Swedish multinationals. *Journal for Global Business Advancement* 1 (1): 37–48.

Fagerström, A. et. al. 2017. Accounting and Auditing of Sustainability: Sustainable Indicator Accounting (SIA). *Sustainability (United States)* 10 (1): 45–52.

Brown, J. & Dillard, J. 2014. Integrated reporting: On the need for broadening out and opening up. *Accounting, Auditing and Accountability Journal* 27 (7): 1120–1156.

Burchell, S. et. al. 1985. Accounting in its social context: Towards a history of value added in the United Kingdom. *Accounting, Organizations and Society* 10 (4): 381–413.

Khayrullina, M. et al. 2015. Production systems continuous improvement modelling. *Quality Innovation Prosperity* 19(2): 73–86.

Identification of the risks of PPP projects and their valuation

E. Augustínová
Faculty of Mining, Ecology, Process Control and Geotechnology, Technical University of Košice, Košice, Slovak Republic

ABSTRACT: Projects of water infrastructure (drinking water supply and drainage and sewage treatment), can be at various risks which may affect their feasibility. Therefore the important aspect of the project implementation in the water sector can be an assessment of the risk which may affect the project at various stages. When deciding on implementation of a project using conventional (PSC) model or PPP model, there is necessary to consider the possibility of effective risk management and risk allocation between the public and private sector. Risk is an important factor which affects the impact of the successful completion of a project in financial, temporal and qualitative perspectives. This article deals with identifying and quantifying potential risks on the side of private as well as public partner using the DBFO model. The Design Build Finance Operate (DBFO) model is a form of Public-Private Partnership (PPP).

1 INTRODUCTION

Water is a vital resource for life, and for our economy. However, the aquatic environment faces many serious challenges such as water scarcity, pollution and ecosystem degradation. These pressures will increase in Europe and globally due to climate change and the increasing global population. Therefore, urgent action is needed to tackle these challenges.

There are significant challenges in Europe with regard to water quality and quantity. As reflected by the Blueprint(COM/2012/0673) to safeguard Europe's water resources, almost half of Europe's freshwaters are at risk of not achieving good ecological status, the main objective of the EU Water Framework Directive, with adverse effects on biodiversity and public health and hampering the provision of ecosystem services. Water scarcity and droughts already affect one third of the EU territory across different latitudes, while floods cause deaths, displacements and large economic losses all over Europe. Significant investments are needed to build, operate, maintain and adapt water infrastructures in order to face these challenges inside Europe and in other developed countries. Many developing countries are still struggling with the provision of basic needs like adequate water supply and sanitation, which are a prerequisite to fight poverty and promote economic development.

These challenges are projected to increase due to climate change, socio economic developments and increasing water demand in agriculture to support essential ecosystem services as food and develop-ment of a bio-based economy. Recent studies show that by 2050, 2.3 billion more people than today are projected to live in river basins experiencing severe water stress.

Additional pressures will exacerbate competition between water users, putting irrigated agriculture, ecosystems, cities, industries and, in general, economic development at risks in several regions of the world (OECD 2010).

Public-Private Partnership (PPP) is a long term contract between a private party and a government entity for the provision of public service and development of public infrastructure in which responsibilities and ravers are shared.

The DFBO model is one of the PPP alternatives of the partnership between the public and private sectors. PPP in general is a type of cooperation between the state administrative bodies and business sector with the purpose to provide funding, building, renovation, administration or maintenance of public infrastructure or providing of public service.[1]

_____

1. *Public Services—The public services in Slovak Republic are the services created, organized, or regulated by the public administrative body, which should secure that the service will be provided in the way necessary to satisfy the social needs while respecting a principle of subsidiary, principle according to which all measures and authorities are executed on the lowest level of administration, which allows their execution or performance. In the area of technical environmental infrastructure the public services usually include the electricity and gas supplying, drinking water supplying and municipal wastewater disposal and treatment.*

Public discussions on suitability of PPP usually do not always give a clear answer to the question what we exactly understand under the term PPP with regard to above-mentioned definition, but after long-time discussions the experts gradually agreed on the clear definable term of PPP. The core elements of effective and true PPP models were defined by (Pfnür et al. 2010) as we can see them on the Figure 1.

2 PPP COOPERATION MODELS—DBFO

Each PPP model has its strengths and weaknesses which must be recognized and integrated.

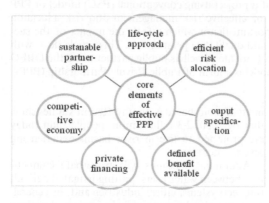

Figure 1. Core elements of effective PPP model.

Table 1. SWOT analysis—DBFO mode.

PPP does not provide a "quick fix" and should be applied only where suitable and when clear benefits and advantages can be demonstrated.

PPP structures must be adapted to sector and project context. Desired impacts and benefits will influence PPP selection and design (Kavas 2012) states the strengths and weaknesses of the DFBO model, which are summed up in the following Table 1.

3 MODELS OF PUBLIC-PRIVATE PARTNERSHIPS IN THE MUNICIPAL SEWAGE DISPOSAL

The most common types of PPP models created to secure the public services in the area of municipal water supplying are DBFO models or their variants DB, DBO, etc. According to (Renda et al. 2006) „*In Design-Build- Finance-Operate (DBFO) schemes, the private partner designs the service or the asset according to the requirements set by the public entity, ensures and finances the construction/ implementation of the asset/service following the design phase, and finally operates the facility. At the end of the PPP contract, the service or asset can be granted back to the public sector under the terms of the original PPP agreement; in alternative, the agreement is renegotiated. DBFO is the most complex type of PPP, since itguarantees all the implementation and operational efficiencies of the previous models, but also provides for new sources of*

Main Features	Application
• *Contract with a private party to design, build, operate and finance a facility for defined period, after which the facility reverts to the public sector* • *The facility is owned by the private sector for the contract period and it recovers costs through public subvention* • *Key driver is the utilization of private finance and transfer of design, construction & operating risk* • *Variant forms involve different combinations of the principle responsibilities*	• *Suited to projects that involve significant operating content* • *Particularly suited to roads, water and wastewater projects*

Strengths	Weaknesses
• *Transfer of design, construction and operating risk* • *Potential to accelerate construction* • *Risk transfer provides incentive for adoption of whole life costing approach* • *Promotes private sector innovation and improved value for money* • *Improved quality of operation and maintenance* • *Contracts can be holistic* • *Government is able to focus on core public sector responsibilities* • *Attracts private sector finance* • *Attracts debt finance discipline* • *Delivers more predictable and consistent cost profile* • *Greater potential for accelerated construction program; and* • *Increased risk transfer provides greater incentive for private sector contractor to adopt a whole life costing approach to design*	• *Possible conflict between planning and environmental considerations* • *Contracts can be more complex and tendering process can take longer than for BOT* • *Contract management and performance monitoring systems required* • *Cost of re-entering the business if operator proves unsatisfactory* • *Funding guarantees may be required* • *Change of management system required*

capital. The most common model is the DBFO con-cessionwhere the private investor designs, finances, constructs and operates a revenue-generating infra-structure in exchange for the right to collect the revenues for a specified period of time, generally for 25–30 years. Ownership of the asset remains with the public sector. This model is particularly suited for roads, water and waste projects and generally for services where user charges can be applied."

4 RISK ALLOCATION

In building infrastructure through public-private partnerships, these risks include the provision of infrastructure, operational risks, and the demand for services, financial risks and political risks.

Agreements on whom and what the risks involve both sides of the partnership are implicitly left to system users and are a key part of PPP contracts.

The PPP project has a number of risks because of its complexity. The importance of optimal dis-tribution of risks between the public and private sector should not be underestimated, because only in this way can we achieve an optimal price-per-formance ratio.

Therefore, the principle of optimal risk spread-ing is as follows: The risk is borne by the party that can best assess and minimize it. The risk distribu-tion of the project is therefore based on partner-oriented managerial skills. The prerequisite for the proper distribution of the risks involved is first to identify them.

Using of quantitative methods as support for decision process is one of the possibilities how to eliminate risk during decision. Many times man-agers solve problems, that they know not how to evaluate from the view of expert skills resp. from the view of practical experiences and therefore they use also other tools, as for example statis-tics application, operation analysis, mathematical modelling. (Teplická et al. 2012)

PUBLIC PRIVATE PARTNERSHIP (PPP) allows the public sector to get better value for money in the delivery of the public services. More-over, by switching its role from a provider to a buyer of services, the Government can focus on its core responsibilities of policy-making and regula-tion. Through closer partnership with the private sector, efficiency gains and other benefits can be reaped, particularly from the following sources:

1. *Private Sector Expertise and Competitive Advantage.* PPP allows Government to tap on to the private sectors expertise, innovation and competitive advantages in the delivery of pub-lic goods and services. This could raise quality and improve cost effectiveness through innova-tive designs or business tie-ups. In addition, the

Government can also tap on the private sectors" networks to maximise asset utilisation and com-mercial potential.

2. *Cost Efficiency through Lifecycle Optimisation.* By combining design, build, maintain and oper-ate functions in the same PPP contract with the provider, it gives the provider a strong incentive ensure the project design takes into consideration operational and other lifecycle costs. PPP also allows for private sector scrutiny of design specifi-cations and the business model, with the possibil-ity of achieving the same outcomes at lower cost.

3. Optimal Sharing of Risks.
In a PPP project, Government and the private sector share the risks of delivering a service. The risks may be allocated according to each party's expertise in managing and mitigating the risks in the service delivery process. Typical risks that are allocated to the private sector include design, construction and financing risks. On the other hand, the public sector may take on politi-cal and regulatory risks, while other risks such as demand / revenue risks will be assigned to whichever party is best able to bear it. By trans-ferring the financial risks to the private sector, there will also be greater certainty over Govern-ments future cash flows.

FOR THE PRIVATE SECTOR:

a. PPP offers more business opportunities to the private sector. The private sector will be engaged to deliver a full suite of services (e.g. design, construction, operations and maintenance) which were traditionally performed in-house by public agencies or performed by multiple private companies. The main business sense is not just maximizing profits at the cost of cost reduction, but also creating value and trying to reach customers by meeting their requirements and needs (Teplická et al. 2015).

b. PPP also allows the private sector to move from just constructing assets according to clearly specified designs, to designing and delivering innovative solutions. The private sector has more room to innovate and offer efficient solu-tions for public services. In addition, the private sector can also use its expertise and network to maximise asset utilisation and the commercial potential of the project.

c. The involvement of private sector players in PPP projects may also give companies valuable expertise and experience to spur their develop-ment in the PPP arena and position them to win overseas contracts.

FOR MEMBERS OF THE PUBLIC:

aa. PPP brings together the expertise of the Gov-ernment and the private sector to meet the

Table 2. Risks for PPP projects.

Risk	Description
Planning	
Changes in planning	*Delay and cost increase due to changes in planning entities*
The quality of planning	*Increasing the cost of the repayment schedule is necessary due to the poor planning quality*
Authorizations	*Additional costs as a result of granting a postponement authorization only /*
	In connection with increased requirements
Construction	
Acquisition of land	*Delays in procurement, construction, higher acquisition costs*
Basic construction, the choice of building materials	*Complications due to the geological characteristics of the country, archaeological sites, contamination, etc.*
Construction costs	*Cost overrun due to incorrect calculations / planning, poor project management*
Construction time	*Extensions due to misconduct, poor project management, bad weather*
Operation and use	
Management of buildings	*Incorrect calculation, in particular maintenance and repair costs, e.g. due to the insufficient quality of construction, deviations from performance standards*
Renovation and reinvestment	*Skip the necessary investments into the future at a higher price*
Usage or demand	*Increased/under-utilization leads to an additional increase or decrease costs.*
Vandalism	*Costs due to the arbitrary destruction of buildings and facilities that are accessible, particularly in public buildings, schools and the criminal sector*
Increase in prices	*Higher costs due to increased prices of basic resources (electricity, water, personnel costs)*
Aging	*The costs of prematurely obsolete technology and, where necessary, the renewal of equipment to meet performance standards (in particular information technology)*
Funding	
Loans	*The financing project will be late or will not happen.*
Variable funding	*Funding is more expensive than expected due to changes in interest rates, margin, rate, etc.*
Exceeding the financial framework	*Additional costs lead to insufficient funding or compel them to use it reserve lines at the bank.*
Refinancing	*Late/inadequate commissioning may jeopardize sufficient coverage of the debt service.*
Assessment	
Residual value	*The residual market value of the asset/state of the structure at the end of the life of the project*
Superior risks	
Force majeure	*Unforeseen expenses due to natural disasters*
Political force majeure	*Cost/loss of earnings due to the strike, civil war or war conflicts*
Risk of insolvency	*The insolvency of the general contractor, subcontractors or project company*
Interfaces risk	*Negative consequences of the coexistence of private and public services (From public officials)*
Changes in laws, regulations and standards	*Increased costs due to changes in building and operating standards, work practices and safety regulations, health, environmental protection standards, taxes, etc.*

needs of the public effectively and efficiently. When structured appropriately, PPP will deliver public services that can better meet the needs of the public without compromising public policy goals and needs.

bb. Government will also ensure that public interest is protected in all PPP projects and that service delivery will meet public needs at the best value for money when the private sector is brought in to provide government services.

In particular, the public sector will ensure that:

a. The private providers can meet the public needs effectively;
b. There is clear accountability when services are delivered by the private sector and the public knows who to approach for service queries and feedback;
c. Public security, health and safety will not be compromised in the PPP projects; and
d. Confidentiality of information will be observed. We will protect the personal data and information of the public agency's customers.

5 CONCLUSION

Risk allocation in PPPs is straightforward in principle-risks must be allocated to the party best able to manage them (at the lowest cost)—but challenging in implementation. Generic applications of this principle have resulted in more or less standardized notions of how risks should be allocated between public and private parties, which have reduced VFM.

Effective risk allocation requires creative and innovative thinking, customized to the unique characteristics of the project. It also requires additional guiding principles, including considering which party has the greatest incentives to undertake preventative risk management and to minimize the financial consequences of a risk. Partially transferring risks that are typically fully retained by the public sector may also create incentives for the private party to opt for more cost-efficient solutions. Because risks are continually evolving throughout the life of the project, general guiding principles or "rules of the game" should be devised in order to create predictability in the management of unexpected or new risks. Finally, throughout the risk allocation process, parties should avoid pursuing overly sophisticated risk management strategies that result in high monitoring, transaction and management costs, which can erode VFM.

ACKNOWLEDGEMENT

This contribution is part of project VEGA 1/0741/16 Innovation controlling of industry companies for maintenance and improvement of competiveness and KEGA 002TUKE-4/2014 Innovative didactic methods of education process at university and their importance in increasing education master of teachers and development of students competences.

REFERENCES

COM/2012/0673,http://ec.europa.eu/environment/water/blueprint/index_en.htm>

Kavas, D. 2012. Possible PPP models for cooperation in the Municipality of Ljubljana/*Institute for Economic Research,—Working paper* No. 68, ISSN 1581-80631.

OECD. Water Outlook to 2050, 2010: *The OECD calls for early and strategic action*, OECD Publishing, Paris.

Pfnür, A. et al. 2010:*Risiko- management bei Public Private Partnerships*. Springer Verlag Berlin Heidelberg, ISBN 978-3-642-01073-6.

Public Private Partnership: *Erfahrungen aus der Praxis*, http://www.weka.de/oldmediadb/000000812.pdf

Renda, A. & Schrefler, L. 2006. Public—Private Partnerships, *Models and Trends in the European Union* (IP/A/IMCO/SC/2005-161), IP/A/IMCO/NT/2006-3, 10p.

Teplická, K. et al. 2012. *Using of operation analysis models in selected industrial firm*. Acta Montanistica Slovaca, 17(3): 151–157.

Teplická, K. et al. 2015. Analysis of causal relationships between selected factors in process of performance management in *Industrial companies in Slovakia*. Economic journal, 63(5): 504–523.

New Trends in Process Control and Production Management – Štofová & Szaryszová (Eds)
© 2018 Taylor & Francis Group, London, ISBN 978-1-138-05885-9

Comparison of the development of waste generation and waste management in the EU

M. Bačová & M. Stričík
Faculty of Business Economics with seat in Košice, University of Economics in Bratislava,
Košice, Slovak Republic

ABSTRACT: It is important to monitor the development of the waste generated and of the waste treatment. Our contribution is focused on the development of the total waste generated by households and businesses by economic activity (according to NACE Rev. 2 and year) and the analyze of the development of the development of the municipal waste generation and treatment, by type of treatment method in EU-28, in EU-15 and in ten countries that joined the EU in 2004 with an emphasis on Slovakia. The main objective of the contribution is to assess the development and current state of waste generate and waste management in the EU.

1 INTRODUCTION

Waste is generated in connection with each human activity. It is a by-product of the production and provision of the service (Sivák et al. 2011). The by-product may arise in the form of undesirable output of economic activity as non-useful by-product, e.g. emissions, i.e. as outputs with negative market value. Also, products produced and intended for single-use, including packaging, are wastes. In terms of use and physical and moral wear and tear, products such in production (e. g. machinery and equipment), even in households (e. g. appliances, furniture) are degraded, lose their functionality. They become waste. In many cases, functional products that are superfluous for different reasons are waste, e.g. food after the expiration date.

Every term used in the scientific field should be precisely defined. In the field of waste management, we have different definitions of waste.

Basel Convention (1989) defines wastes as substances or objects which are disposed of or are intended to be disposed of or are required to be disposed of by the provisions of national law.

The United Nations (1997) define waste as "materials that are not prime products (that is, products produced for the market) for which the generator has no further use in terms of his/her own purposes of production, transformation or consumption, and of which he/she wants to dispose."

Under the Waste Framework Directive 2008/98/EC, Art. 3(1), the European Union defines waste as "an object the holder discards, intends to discard or is required to discard."

Waste can be defined as a by-product of manufacturing and service provision (Siváket al. 2011).

Definition of waste associated with its place of origin, in order to realize the control of the treatment and disposal. Table 1 defines four classes of waste (Pongrácz 2009).

"The quantities of solid waste have grown steadily along with Gross Domestic Products (GDPs) over the past decades. For example total quality of municipal waste per capita increased by 29% in North America, 35% in OECD, and 54% in the EU15 from 1980 to 2005" (Sjöström & Östblom 2010). Each waste affects the environment and human health.

The primary objective of waste management should be to minimize the negative effects of the generation and management of waste on human health and the environment Waste prevention should be the first priority of waste management. The European Union requires the application of the waste hierarchy as shown in Figure 1.

The Waste Framework Directive has brought a new philosophy into the waste management of the European Community. Emphasis is placed on the waste prevention and introducing an approach that takes into account the whole life cycle of products and materials, not just their waste phase.

Table 1.   The four classes of waste.

Class 1	Non-wanted objects, created not intended, or not avoided, with no purpose
Class 2	Objects that were given a finite purpose, thus destined to become useless after fulfilling it
Class 3	Objects with well defined purpose, but their performance ceased being acceptable
Class 4	Objects with well defined purpose, and acceptable performance, but their users failed to use them for the intended purpose

Source: Pongrácz, 2009.

Figure 1. Hierarchy of waste management.

Waste prevention is defined as measures taken before a substance, material or product has become waste, that reduce the quantity of waste, including through the re-use of products or the extension of the life span of products; reduce the adverse impacts of the generated waste on the environment and human health; or reduce the content of harmful substances in materials and products.

Preparing for re-use means checking, cleaning or repairing recovery operations, by which products or components of products that have become waste are prepared so that they can be re-used without any other pre-processing.

Recycling means any recovery operation by which waste materials are reprocessed into products, materials or substances whether for the original or other purposes. It includes the reprocessing of organic material but does not include energy recovery and the reprocessing into materials that are to be used as fuels or for backfilling operations.

The recycling rate in Japan was roughly the same in 2007 (20.5%) as it was in 2002 (19.9%). It appears recycling rates have reached some sort of steady state. The relevant policy question is whether the steady state is socially optimal (Kinnaman 2010).

The estimated socially optimal recycling rate is defined as the rate that minimizes the net social costs of managing reidnential solid waste (Kinnaman 2010).

The average share of waste for recycling in the member countries of the European Union, including Slovakia, should increase from the current 44% to 70% by 2030. European Union want to achieve 50% recycling by 2020.

2 MATERIAL AND METHODS

The object of our investigation are the analyze of the development of the total waste generated by households and businesses by economic activity (accoording to NACE Rev. 2 and year) and the analyze of the development of the development of the municipal waste generation and treatment, by type of treatment method in EU-28, in EU-15 and in ten countries that joined the EU in 2004 with an emphasis on Slovakia. The main objective of the contribution is to assess the development and current state of waste generate and waste management in the EU. Published data on waste generation in Europe varies considerably depending on selected data and assumptions. Given that there are different methodology for setting values in the field of waste management (Bräutigam et al 2014), we used only Eurostat data.

3 RESULTS AND DISCUSSION

In 2014, the total waste generated in the EU-28 by households and budinesses by economic activity (accoording to NACE Rev. 2 and year) amounted to 2502.89 million tonnes. In the EU-28, over the ten years analyzed, the total quantity of waste produced decreased. The volume of waste produced in the analyzed period grew, with the exception of 2008 when there was a decrease in waste generation of 5.46%.

The analysis of the development of waste generation in the EU-15 shows that during the analyzed period, the volume of total waste generate in 2014, as compared to 2004, increased in ten countries—in Belgium (24%), Germany (6%), France (9%), Italy (14%), Nederland (44%), Denmark (60%), Greece (109%), Finland (38%), Sweden (82%), Austria (5%) and decreased in five countries—Luxembourg (−15%), Ireland (−38%), United Kingdom (−16%), Spain (−31%), Portugal (−50%). The development of total waste production illustrated Figure 3.

Figure 4 shows the development of waste production in countries that joined the EU in 2004. The analysis of the development of waste generation in the EU-15 shows that during the analyzed period, the volume of total waste generate in 2014, as compared to 2004, increased in 3 countries—Estonia (9%), Latvia (109%), Poland (30%) and decreased in seven—Cyprus (−9%), Czech Republic

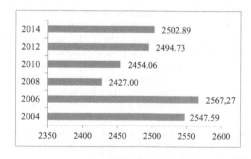

Figure 2. The development of waste production in EU-28 (in mil tones).
Source of data: Eurostat 2017.

18

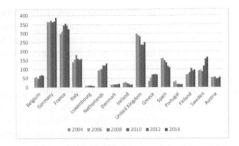

Figure 3. The development of waste production in EU 15 (in mil tones).
Source of data: Eurostat 2017.

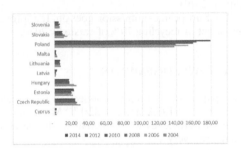

Figure 4. The development of waste production in countries that joined the EU in 2004 (in mil tones).
Source of data: Eurostat 2017.

(−20%), Lithuania (−12%), Malta (−47%), Slovakia (−17%), Slovenia (−19%).

Municipal waste consists to a large extent of waste generated by households, but may also include similar wastes generated by small businesses and public institutions and collected by the municipality; this part of municipal waste may vary from municipality to municipality and from country to country, depending on the local waste management system. For areas not covered by a municipal waste collection scheme the amount of waste generated is estimated by Eurostat.

In 2015, the municipal waste generated in the EU-28 by households and similar wastes generated by small businesses and public institutions and collected by the municipality amounted to 476 kg per capita.

Figure 5 shows the development of municipal waste generated in EU-15. The most municipal waste in the analyzed period 2004–2015 was generated in Denmark in 2008–830 kg per capita.

Figure 6 shows the development of municipal waste generated in in countries that joined the EU in 2004. The most municipal waste in the analyzed period 2004–2015 was generated in Cyprus in 2009–729 kg per capita, in 2008–728 kg per capita and in Malta in 2009–674 kg per capita.

Since 2005 in Slovakia the municipal waste generation has been increasing. A slight decrease

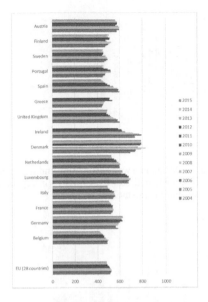

Figure 5. The development of municipal waste generated in EU 15 (in kg per capita).
Source of data: Eurostat 2017.

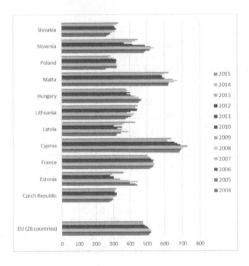

Figure 6. The development of municipal waste generated in countries that joined the EU in 2004 (in kg per capita).
Source of data: Eurostat.

was recorded in 2009 and 2011, which is mainly due to the economic situation of households. The same situation was also observed in the amount of municipal waste per capita. Municipal waste generation in Slovakia increased in 2015. It remained well below the EU average (329 kg per capita compared to around 476 kg per capita).

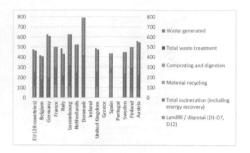

Figure 7. Municipal waste generation and treatment, by type of treatment method in EU 15 in 2015 (in kg per capita). Source of data: Eurostat 2017

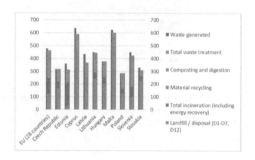

Figure 8. Municipal waste generation and treatment, by type of treatment method in countries that joined the EU in 2004 in 2015 (in kg per capita). Source of data: Eurostat 2017.

Figure 7 illustrates the municipal waste generation and treatment, by type of treatment method in EU 15 in 2015.

Figure 8 illustrates the municipal waste generation and treatment, by type of treatment method in countries that joined the EU in 2004 in 2015. The total amount of municipal waste generated in Slovakia in 2015 was 329 kg per capita. Statistics for 2015 show that in the EU on average recycled or composted 45% of municipal waste. Slovakia belongs among the Member Countries with the lowest proportion of recycled or composted waste, which is only 15%.

4 CONCLUSION

As shown in Figures 2 and 3, in the years 2004–2014 in the EU 28 was recorded different developments in total waste generated by households and budinesses by economic activity (according to NACE Rev. 2 and year). Total waste generation in the EU-28 in the year 2014 compared to 2004 decreased by 1.75% and in Slovakia degrased by 17%.

The Figures 4–7 show the situation in municipal waste generation and treatment, by type of treatment method.

In the EU-28 in 2015 compared with 2004, municipal waste generation decreased by 7% on average per person. There was another situation in Slovakia where in this period the generation of municipal waste grew by 26%.

The European Commission has approved the goal of 50% recycling of waste by 2020. Several countries of the EU-28 have already fulfilled this objective. It is questionable whether Slovakia manage to fulfill this goal. The question is also reporting waste statistics, as Member States can choose from up to four different methods.

Scotland has approved a national goal to become the wold's first zero-waste country by 2025. (Kinnaman 2010)

The European Commission goes on. It approved the target of recycling 70% of waste by 2030.

ACKNOWLEDGEMENTS

This paper was written in connection with scientific project VEGA no. 1/0582/17 "Modeling the economic efficiency of municipal waste material-energy recovery".

REFERENCES

Basel Convention, 1989. Basel Convention on the Control of Transboundary Movements of Hazardous Wastes and Their Disposal. http://www.basel.int/Portals/4/Basel%20Convention/docs/text/con-e-rev.pdf.

Bräutigam, K.R. et al. 2014 The extent of food waste generation across EU-27: Different calculation methods and the reliability of their results. In *Waste Management and Research*. 32(8), pp. 683–694. http://journals.sagepub.com/doi/10.1177/0734242X14545374.

Directive, 2008. DIRECTIVE 2008/98/EC OF THE EUROPEAN PARLIAMENT AND OF THE COUNCIL of 19 November 2008 on waste and repealing certain Directives Eurostat 2017.

Kinnaman, T.C. 2010 The Optimal Recycling Rate. *Other Faculty Research and Publications*. Paper 6. http://digitalcommons.bucknell.edu/fac_pubs/.

Pongrácz, E. 2009 Through Waste Prevention Towards Corporate Sustainability: Analysis of the Concept of Waste and a Review of Attitudes Towards Waste Prevention. In *Sustainable Development Sust. Dev.* 17, 92–101 (2009). (www.interscience.wiley.com) DOI: 10.1002/sd.402.

Sivák R. et al. 2011. *Slovník znalostnej ekonomiky*. Bratislava: SPRINT dva.

Sjöström, M. & Östblom, G. 2010. *Decoupling waste generation from economic growth - A CGE analysis of the Swedish case Ecological Economics* 69(7), pp. 1545–1552. http://www.hallbaravfallshantering.se/download/18.71afa2f11269da2a40580007855/1350483384244/Depcoupling+waste+generation+from+economic+growth++Sj%C3%B6 str%C3%B6 m+%26+%C3%96 stblom.pdf.

UN, 1997. Glossary of Environment Statistics, *Studies in Methods*, Series F, No. 67, United Nations, New York, 1997. Updated web version 2003.

New Trends in Process Control and Production Management – Štofová & Szaryszová (Eds)
© *2018 Taylor & Francis Group, London, ISBN 978-1-138-05885-9*

Management styles as a basic assumption of maximizing the company's value

P. Badura, K. Vavrová, M. Bikár & M. Kmeťko
Faculty of Business Management, University of Economics in Bratislava, Bratislava, Slovak Republic

ABSTRACT: Managers and their management styles affect critically the effectiveness, quality and the overall success of any organization. The goal of the article is to describe how to identify dominant management styles and how to find the most significant deviations from the desired state. Meeting the goal would make it possible to recommend such changes that would increase efficiency of any company and significantly improve the use of employees' full potential. Mainly analysis of current state and comparison of existing and desirable management styles have been used to achieve that goal. Multi-point scale and selected mathematical and statistical methods have been used to quantify the results. The used system of identification and correction of management styles has clearly proved its full functionality. Because the system is not limited to any geographical borders, it allows the human capital development in any company, whether within our country or abroad.

1 INTRODUCTION

If a business wants to survive, even under conditions of the strong competition in international markets, it needs to increase the efficiency of activities in several important areas. One of the most important ever is the process of managing employees. It is always connected to creating, using and strengthening of certain management styles. The concept of managers' work is a set of practical steps and actions to cope with core business problems faced by managers of each organization. It can be applied to questions of integration of the individual, as well as, the business needs and objectives, selection and application of particular model of power relations and control in the organization or the issue of coordination of separate groups' activities which have different goals. Each company from the wide range of instruments builds its own management system, which is constantly evolving. Some findings show that the organizational culture has also an impact on a manager's style, which is then forwarded to the followers (Shearer 2012). The management styles do not depend just on the organization culture, but the frame can be much wider. If we go to a different country, we could observe different management styles used in the management of employees. For example Nordic managers have been consistently reported as individualistic but also more 'feminine' and employee-oriented, than those further south (Smith et al. 2003). Talking about e.g Japanese management, they are "a highly disciplined people

with a relatively high standard of education, so management and leadership is more about objectives, strategies and performance, and less about crisis management, reskilling and trouble shooting" (Deveson 1993).

All these facts mean that while evaluating management styles within a company we must always take into account various specifics (of the country, of the company, of their field of business etc.).

If we want to measure the efficiency of human resources management and human capital development and express it by a coefficient we would also need to include the employees' frustration, mobbing, job dissatisfaction, high stuff turnover etc. These factors are much more difficult to quantify than the corporate expenses are. That means that although the management efficiency coefficient cannot be exactly quantified, we know how to increase the efficiency—by implementing the most suitable management styles, to an appropriate extent. As other research findings show: "stimulating innovation through the motivation and the appropriateness of management style is an effective way to increase the capacity to create and achieve new products with market demand" (Corăbieru 2010).

The managers are using many management styles while executing their activities. They are using the management styles so long that they are not even aware of using them—not to mention their suitability or unsuitability. Because of this, there are often used less appropriate management styles which lead to decreasing effectiveness of corporate management and consequently make harder to achieve business objectives. In such case, it is not possible to utilize the full potential of employees. The human capital development stagnates, which may be the same as the value of the company does. If the enterprise wants to fulfil the main goal of its existence, which is maximization of its market value, while taking into account the interests of all participants, it must necessarily focus on effective management of its employees.

The system of management styles is a combination of individual forms and management methods used by the manager in relationship to his subordinates, in order to guide their activities and fulfill the business goals. While there are relationships of superiority and subordination there must also exist certain management styles which the manager uses for leading and managing his subordinates.

The appropriate implementation of management styles is affected by many factors. These factors include e.g.: the age of employees, educational structure, competence of both managers, as well as, the employees, manager's personality, the ratio of men and women in the company, corporate life cycle (expansion, stagnation etc.), the environment in which the activities are carried out, corporate main business focus (or the main focus of the analyzed department in the company), the prevailing partial management activities, size of the enterprise etc. There may also be differences in the management styles used by men or women. As some studies show "the management style of the women managers as seen by majority of their subordinates is of the Autocratic style while the management style the women managers see themselves as possessing is the Laissez-Faire style of management" (Pioquinto 2016). The same conclusions were made by Kocher: "We also find that male managers employ a democratic style more often than women" (Kocher et al. 2013). Moreover, the authors' findings show that "managers who prefer efficiency are more likely to exercise an autocratic management style by ignoring preferences of their team members. Equality concerns have no significant impact on management styles. Elected managers have a higher propensity than exogenously assigned managers to use a democratic management style by reaching team consensus" (Kocher et al. 2013).

2 AIM AND METHODOLOGY

The aim of this paper is to describe how to identify dominant management styles in any organization, and how to find the most significant deviations from the desired state. This consequently makes it possible to offer some recommendations about changes that should be done. That would increase the efficiency of management and lead to better use of employees' potential. For this purpose also an analytical model has been created. This model is able to identify individual management styles, the degree of their use in any company, their strengths and weaknesses, as well as, their interconnection and overlapping. So we are able to uncover completely new areas and possibilities for increasing the management styles efficiency.

The objects of our research are primarily companies in the Slovak Republic. The research subjects are the used management styles, their combinations and the degree of their implementation in the organizations.

The results of our research can by used by managers in any organization to evaluate the actual status of applied management styles and consequently optimize the corporate operations in the field of managing their employees.

There are many different management systems in Slovak organizations that must be taken into account. To create an appropriate analytic model which covers all specifics of a particular company (e.g. the type of business, the specific activities done by managers, as well as the internal environment) is a very challenging task. To fulfil the task it was necessary to carry out some additional activities. Those partial activities were necessary for the successful achievement of the goal. The analytic model should reflect the particularities of each company. It should also reflect the fact that the used management styles could be quite appropriate might in some companies, but their implementation might be much less suitable in any others. As a good example can be used the management of a nuclear power station, security company, army, drugstores—in such enterprises must be used different management styles, that might be otherwise considered as too autocratic or directive. The same also applies in a case of emergency in any other company (e.g. when there is a fire or other imminent danger). This has also been proved e.g. by Alakbarov's study of hygiene and epidemiology centers' management styles: "management of hygiene and epidemiology centers by higher organs of sanitary-hygienic service has mainly a directive character. Such characters of management as coordinating (10.06%), consultative (12.8%), methodic (25.6%), normative or democratic are used relatively rare. Integrally—democratic style or couching in practical activity of those institutions is not applied" (Alakbarov 2004).

That is why it is much better to choose individual approach to individual enterprises, and not to impose management styles preferred by e.g. some world famous organizations to all other companies, although they could even carry out similar business activities.

The analytic model, described in this paper, is always focused on specifics of any analyzed company. Moreover, the participation of employees in the research takes into account the internal corporate factors which influence the use of appropriate management styles.

The research samples are either the employees of the analyzed company or the employees of selected department within the company (it depends on whether we analyze the whole company or just some selected departments—e.g. marketing department or human resources department).

If we want to analyze the dominant management styles of the general manager, the sample for our research would consist of employees from various departments but especially the managers, who are in direct contact with the general manager and report directly to him. These respondents have the best overview of the company's situation, the personality of the director, the core business activities and other factors that influence the management styles used by him.

If the object of our analysis would be a specific department of the company, the research sample would be employees of this particular department.

In both cases the more employees are involved in the research the more objective the results are and the more accurately they describe the reality.

3 ANALYTIC SYSTEM

Before developing the analytic model of management styles, it was necessary to fulfil two basic tasks. The first one was collecting all the existing management styles and their classification into various groups. The second task was to get basic values of the appropriate application of each style.

3.1 *Collecting all management styles*

The first step in management styles analysis was the review of all the styles that were currently used in business practice. In the initial stages of our research the book of František Lipták "Methods and styles of management" (1991) and many other books, mainly from foreign authors, have been used as the main literature.

After a thorough analysis of all available management styles, it has shown that the optimal number of management styles (which would still be possible to identify, while avoiding duplications) are 65 different styles. They were classified into four main groups according to their main features. Each group had its own name:

- Type of management (20 management styles);
- Orientation of management (15 styles);

- Dominant mean of management (15 styles);
- Used method of management (15 styles).

3.2 *Optimal values for management styles*

We presume that the results from the analysis of management styles in a particular company could be subjective to a certain level. To avoid an excessive subjectivity of the results (e.g. excessive emotions after disagreement between the manager and his subordinates), we will not take into account only the views of the employees, but also the view of selected external respondents.

It was therefore necessary to get some "basic values" for each of the management styles. These basic values would indicate the suitability or unsuitability of management styles used in the business practise, in general. In such way, the results of the management styles analysis of a particular company and our following recommendations how to increase the management styles effectiveness, could be objectified by outside respondents' view.

For this purpose we have created questionnaires which were then distributed to selected respondents. Those respondents had to determine what styles are appropriate or inappropriate for application in business practise, in general (not taking into account any business orientation).

The survey was conducted by assigning the basic point values on a scale from 1 to 10, where a minimum value of 1 meant the "least appropriate" and the value of 10 meant the "most appropriate" management style.

The number of respondents who participated in this phase of research was 80. After processing the questionnaires' results we gained the data which told us how the responders assess the suitability or unsuitability of different management styles application in the practice. By this way we have obtained the basic points for each management style.

In effectively managed companies the values for applied management styles should be very close to these basic point values. Of course, there will be an exception for some specific organizations which simply must to use different management styles (companies working with hazardous substances or offering waste services, transportation companies, security companies etc.).

But when a new research is realized, we will receive new additional data about appropriate use of management styles from all the research participants (employees of the analyzed company). These new data can be added to the previous basic points. So the basic points can be continuously specified and the number of respondents involved in the research can continuously rise. The main advantage of this approach is that we could create separate, independent categories of basic point values for specific types of businesses (for example—basic points of management styles just for chemical companies, financial organizations, educational institutions etc.). By such approach it would be possible to get basic point values e. g. only for companies with chemical production and solve the problem how to optimize the applied management styles in such specific organizations.

This approach to optimization of corporate management is not based on integration of management styles into any larger unit. On the contrary, its basic idea is to divide the management system into as such small parts as possible. This makes it possible to identify even the smallest deviations from the desirable state.

3.3 *System verification and results*

After we have collected all the existing management styles (65 styles) and obtained the basic point values for each of them (on the scale from 1 to 10) the analytic system could be finally developed. After then it was also tested in the practice of a certain organization to find out which management styles are used most and find the most important deviations from the optimum.

The analytic model for a selected company was developed in MS Excel software. It contained all the 65 management styles and two basic columns. The respondents (employees of the company) had to assign points which expressed their opinion on the real and also the appropriate application of management styles within their company into these two columns. The first column was for their opinion about the real application of a certain management style. The second column was for their opinion about the appropriate application of the same management style.

Of course, to make it as simple as possible, the employees had not to write the point values manually into the table. The file has also contained a Visual Basic script that allowed the respondents to evaluate the use of management styles by simple mouse click in the proper cell. When the respondent had clicked in any selected cell, the value "x" has shown in it. In the summary column has appeared a number value (just in line with the 'X' mark). If the respondent decided to change his selection, all he had to do was to click into another column. After this a new 'X' appeared and the value in the summary column had changed, as well. By this way the consumption of time of the employees had been reduced to the minimum.

There was also available a comment to each of the management styles. The comment explained the basic features of each of the styles and its nature. It had showed up when the respondent had moved the mouse pointer at any of the man-

agement style's title. In addition, with this Excel file there was also another Word file distributed. The file contained more detailed description of every management style. So, if a respondent had a problem to understand any of the style's nature, he could find it easily in the Word file. However, the management styles were named very suggestive right from the beginning. Their title should suggest immediately what the style is about and what its nature is. The evaluation of their application in a particular company was therefore possible just with "semantic spontaneity" (e. g. the democratic management, autocratic management, sanction management, management by objectives).

Several enterprises had been selected for verification of the analytic system functionality. The detailed results of the analysis go beyond the scope of this paper. We decided to present at least a brief conclusions and recommendations for one selected company. The company operates in a sector of engineering. Management styles analysis took place in the Department of Construction.

We present the most important results in the Table 1–4 (below), as well as the main deviations

Table 1. Type of management.

Management styles	Reality	Optimum	Deviation
Non-entrepreneurial	5.67	2.12	+3.55
Limited	4.89	2.43	+2.46
Strategic	5.89	8.79	−2.90

Table 2. Orientation of management.

Management styles	Reality	Optimum	Deviation
By alibism	5.11	1.65	+3.46
By promise	4.44	2.62	+1.83
By force	5.00	1.49	+3.51

Table 3. Dominant mean of management.

Management styles	Reality	Optimum	Deviation
By participation	5.22	8.07	−2.85
By stimulation	3.67	8.14	−4.48
By perambulation	8.78	4.33	+4.45

Table 4. Used method of management.

Management styles	Reality	Optimum	Deviation
Systematic	4.56	7.44	−2.88
Autocratic	4.78	1.92	+2.86
With foresight	5.78	8.82	−3.04

from the desired state. The management styles are divided into the four main groups. From each group we have selected three management styles with the most serious deviations.

As the most significant deviations from the appropriate application of management styles we have identified:

- High level of application the non-entrepreneurial management style;
- High level of using the limited management style;
- High level of using the management style by promise;
- High level of the autocratic management style;
- Low level of using the management by stimulation;
- Low level of the systematic management style;
- The company should focus first on improving the management styles mentioned above. So it can improve the efficiency of management within the department of construction. To remove the most serious deviations we would advise to focus;
- On higher degree of employees participation in the decision making process. This would largely avoid the feeling of excessive autocratic management and would have a positive effect on stimulation of employees;
- To increase the level of the strategic management and the long-term goals. That would increase the level of conception management style. The creation of conceptual framework, strategy, tactics and consequently identification with all these goals by employees are the basic assumption for the increasing of the management effectiveness and the company's success;
- To reduce the level of alibism of the manager. That would also have a positive effect on increasing the strategic management style;
- The level of decentralization should be also increased. And it is necessary to increase the level of decision-making power of the employees. Delegation of tasks without proper rights leads to dissatisfaction of employees and interfere with the smooth running of the business. The employees consider such system as a restrictive (limited management) and autocratic one (autocratic management);
- To reduce the use of coercive measures (threats and criticism in particular), as well as, the promises that will remain unfulfilled. This has a negative impact on the employees and also the systemic management;
- The company (and mainly the analyzed department) should focus more on predictive management style. The style will lead to an increased level of prevention, innovations and conversely

it will lower the rate of corrective management actions which increase the corporate expenses;
- The company should also reduce the rate of perambulation and inspections to control the employees in this department. The increased autonomy of employees should be accompanied by appropriate confidence in their ability and competence to solve relevant tasks;
- Reducing the coercive approach of management should reduce the application of management by push actions and increase the management of pull actions. It is possible to achieve by fulfilling all of the abovementioned recommendations: to increase the level of autonomy, increase the participation of employees and reduce the rate of alibism, unfulfilled promises and coercive measures, in particular.

The created analytic system allows us to analyze the results in many ways. For example, it is possible to compare the opinions of women versus men about the appropriateness of used management styles. It is also possible to compare the views of younger versus older employees or opinions of employees with secondary education with opinions of employees with university education etc.

4 CONCLUSIONS

The next analyzed research objects could be organizations not only in Slovakia but also abroad. The analytic system is able to analyse any management styles, regardless of the business orientation. We can analyze companies offering services as well as the manufactures. We can also analyze government organizations or private sector companies. The important thing is that the companies are different from various points of view. The most common differences are: average age of the employees (so they prefer partially different management styles), different women and men ratio in different companies, different business activities, conditions etc. Although the system of management styles, within similar departments respectively within similar companies, can be also very similar, it is impossible to generalize and give any universally applicable recommendations. The individual approach that takes into account the specifics of the company or the analyzed department is needed. The analytic model was designed with regard to these differences so it accesses each entity individually.

If the analysis is carried out in small businesses, it is possible to analyze the company as a whole. If it is done in medium-sized or big companies, then it is more appropriate to analyze management styles by particular departments. This model is flexible enough so it can take into account the specifics of each analyzed company. This just confirms our statement that the system is fully applicable, it should be used more in practice and thanks to it the enterprises may significantly improve their used system of management styles.

We admit that the created analytic model can be subjective to a certain extend, because it takes into account the views of the employees. These views can be influenced e.g. by recent disagreements and quarrels with a manager. However it is necessary to base the evaluation mainly on these views, because no external view of the company (or a particular department) can offer such a precise overview of the applied management styles as the employees can.

To eliminate partially the excessive degree of subjectivity the views of employees have been combined with the previously obtained views of external respondents (the obtained basic point values for each management style). The synthesis of subjective (internal) opinions of employees which reflected internal conditions in the company, with external opinions, is an appropriate solution which eliminates the adverse effects of subjectivity and also takes into account the specifics of the company. The specifics of each company would be otherwise omitted.

This model of analysis is also a kind of feedback, which can help the managers to highlight not only a higher rate of applying some inappropriate styles, but also point out to so far untold employees' requirements, respectively their opinions to the application of effective management styles. After caring out the analysis, managers can organize a meeting, where they can discuss all the adverse deviations which have escaped their attention on one hand and about the desired future changes in the management styles, on the other hand. At the same time the internal corporate factors are taking into account (the average age of employees, the ratio of men and women etc.).

System of management styles, as a whole, is too complicated and multi-dimensional for easy assessing its properties and proposing measures for its rationalization. It needs to be decomposed into smaller parts that are manageable. These parts are the individual management styles. All the individual management styles have certain autonomy, but each of them needs to be constantly coordinated with respect to the global objectives of the management system. The basic rule is that no subsystem (management style) should pursue action that would damage other subsystem or the management system as a whole.

ACKNOWLEDGEMENTS

The authors are grateful for the support of the Scientific Grant Agency: VEGA—Grant No. 1/0404/16.

REFERENCES

Alakbarov, M.M. 2004. Modern features of management style. In *Azerbaijan Medical Journal*. Issue 2, 2004, p. 48–53.

Badura, P. 2004. *Analýza využívania štýlov moderného manažmentu v praxi malých a stredných podnikov v SR (PhD. thesis)*. EF UMB, Banská Bystrica.

Corăbieru, A. 2010. Study on the need to adapt motivation and management style in the innovative process. In *Metalurgia International*. Volume 15, Issue 6, June 2010, p. 42–45.

Deveson I.A. 1993. Japanese Management Some Lessons for the Australian Management Style. In *Japanese Studies*. Volume 13, Issue 1, 1 May 1993, p. 89–92.

Kocher, M.G. et al. 2013. Other-regarding preferences and management styles. In *Journal of Economic Behavior & Organization*. Volume 88, April 2013, p. 109–132.

Lipták, F. 1991. *Metódy a štýly manažmentu (Príručka identifikovania a diagnostikovania)*. Bratislava: Bradlo.

Pioquinto, M.A.J.G. 2016. Management Styles of Women Managers: Its Effect on Job Performance as Prime Mover to the ASEAN Economic Community (AEC) Integration. In *Global Business & Finance Review* 2016 12; 21(2), p. 47–64.

Shearer, D.A. 2012. Management styles and motivation. In *Radiology management*. Volume 34, Issue 5, September 2012, p. 47–52.

Smith, P.B. et al. 2003. In search of Nordic management styles. In *Scandinavian Journal of Management*. Volume 19, Issue 4, December 2003, p. 491–507.

New Trends in Process Control and Production Management – Štofová & Szaryszová (Eds)
© *2018 Taylor & Francis Group, London, ISBN 978-1-138-05885-9*

Analysis of ergonomics risks using the RULA method for a selected profession

M. Balážiková & M. Dulebová
Faculty of Mechanical Engineering, Technical University of Košice, Košice, Slovak Republic

ABSTRACT: Ergonomics is the science of designing the workplace while keeping in mind the capabilities and limitations of the worker. Overall, poor ergonomic safety and health at work can cause serious damage to health or occupational illnesses. This article is focused on the issue of ergonomic risk analysis and occupational health at a workplace. The primary ergonomic conditions by using the ergonomic RULA (Rapid Upper Limb Assessment) method in a selected workplace; it subsequently proposes reasonable measures. In the first part of the article the basic concepts of ergonomics and the most frequently used methods that deal directly with ergonomic aspects are briefly characterized. The following section is characterized by an analysis of a job and the related evaluation of the ergonomic RULA method using worksheets. A description of the particular method and procedure are graphically displayed by illustrations. Acute measures were thenproposed that are actually applicable in practice.

1 INTRODUCTION

Analysis of a workplace from the viewpoint of occupational safety and health is essential in today's innovative industries. Increasing safety at a workplace or of work processes contributes to a significant increasing of the efficiency and quality of life of employees. With the continuous development of science and technology many innovate and reliable methods or software programs are available which can notably help improve safety and protection of health in a given workplace (Loriato 2016).

The application of these innovations has in the present become a rapidly advancing trend that has changed the angle of view on overall safety and protection of health at work. In contemporary industry the issue of ergonomics has come more and more to the forefront, proof of which lies in that the field of ergonomy has in recent years recorded its greatest progress. A complex evaluation of ergonomic conditions using verified methods and new software brings more and more efficient and exact results when assessing a workplace (Chundela 2001).

Lack of familiarity with basic ergonomic or safety aspects every day exposes many employees to unsuitable working conditions, and thusthey unknowingly damage their health gradually, and what is worse, through unsuitable work and ignorance they subsequently lower their own quality of life. Legislation points to the fact that an employee has the right to refuse work activities, if he/she is convinced that the given working conditions may endanger his/her health and safety. Harmony between ergonomy as such and legislation should be the responsibility of the employer, which should regularly inform and point out to workers the rights and duties provided to them by legislation. In the end resultit is logical that the health and safety of employees should be a priority of every good employer (Ahram 2012).

The aim of the contribution is assessment of ergonomic aspects at a selected work profession using the RULA method and subsequently the proposing of truly applicable measures which may actually improve the state of the assessed workplace.

2 METHODOLOGY

British ergonomists McAtammney and Corlett were the first to describe RULA as an integrated ergonomic method. The task of this specific method is to analyse in detail a selected workplace, primarily where there is a greater loading of the upper limbs on the human body, though this is not mandatory condition (Hovanec 2017). The RULA method is considered to be one of the most useful of the simple ergonomic methods. The main task of this method is the so-called screening of the requisite workplace and subsequently the assessment of the loading of the entire body of a person during the work process both from a biomechanical viewpoint as well as a postural viewpoint. This practical method is used mainly at those workplaces where the upper limbs, neck and torso are excessively overloaded. Implementation and evaluation of this ergonomic method is time-consuming. It is founded on the resulting four categories, where the 4th is actually the most serious. RULA for effective use should be applied as a component of a wider or deeper ergonomic study (Hatiar 2008).

3 APPLICATION OF THE SELECTED METHOD FOR A CHOSEN PROFESSION

3.1 *Description of the selected activity*

Three workers alternate at the job of swing drill operator, see Fig. 1. The operation of a swing drill is very demanding and important for the overall running of production. During the production of cable systems it is necessary to make and finish many accessory constructions or other support constructions. The worker in this job drills exactly given openings into a galvanized welded part of the support construction. The operator performs his activities according to drawings which are applied to each type of work piece (Maščenik 2016).

During the work shift the worked parts of the construction are continuously changed; therefore, it is exceedingly important to pay heed to precision and thoroughness in this profession (Majerník 2016). The size and type of these components depend on the given type of cabling system. All worked parts are checked at regular intervals by a manager. Therefore, employees at this job when drilling must constantly remeasure these drilled openings on the workpiece. Setting, remeasuring and drilling these workpieces takes up the greatest part of the work shift. When performing these work activities the employee is constantly bent over or even leaning through the workpiece. In the end this means that he spends most of the work time in unfavourable work positions that over-

load the support-locomotion system in particular. The loading during this work activity occurs on the torso and also on the upper limbs, which is the main reason why this job was assessed using the effective ergonomic RULA method, as it evaluates this very issue.

This job contains many work operations, however, which the worker does not do evenly; that is, he performs each work operation at a different time during the shift. This was the reason for creating a schedule of the day for the given job. This schedule was recorded for a common work day, where the shift lasted 8 hours, more specifically, from 8:00 to 16:30, with a lunch break during the course of the working day from 11:30 to 12:00. During monitoring of the work activity in this shift no malfunctions or other unforeseeable facts were recorded that could have interrupted the perfect course of the work shift and influence in some way the results of the time schedule for the work activity.

3.2 *Evaluation of the work position using the RULA method*

The position during work activities, which is the setting and drilling of holes into a metal component, is evaluated using the ergonomic method RULA. Even at a first glance at the workplace it is clear that the torso of the employee is dangerously bent for a long time during the work time and simultaneously even turned to the side, which further worsens this unfavourable work position. The RULA method thoroughly evaluates the position of the hands, torso and lower limbs (Skřehot 2009).

In the following Figure 2 and 3 the possible positions of the body are depicted, and appropriate values are assigned to them. After the compilation

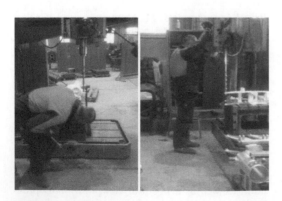

Figure 1.   The assessed job.

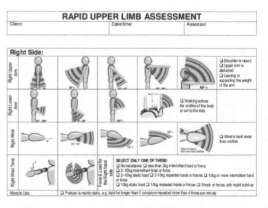

Figure 2.   Graphic depiction of the RULA method—right side.

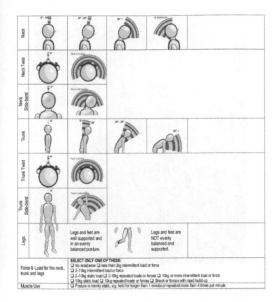

Figure 3. The RULA method graphically depicted—neck, torso, lower limbs.

Table 1. RULA—resulting categorization table.

1st category (1–2)	The assessed work is acceptable and not executed for the long-term
2nd category (3–4)	It is necessary to perform a different type of evaluation; certain requirements for changes are necessary
3rd category (5–6)	Necessity of making changes in the near future
4th category (7)	The making of changes is essential

head when performing the job, which in the end increases the point assessment for the given body part several fold. The entire RULA method is actually divided into worksheets, which evaluate the right and left sides of the body, neck and torso and the position of the lower limbs. Figure 2 shows the first sheet, according to which the right side of a person is assessed; it evaluates the position of the arm, forearm, wrist as well as the use of muscles, that is, their loading. With selection of the weight loading on the right hand, the average weight of the workpieces which workers machine at this workplace was used.

The position of the upper arm is estimated at between 30° to 40°, and the upper limb is found in abduction; we assigned the position of 100° + to the upper arm, and during this activity work is often done to the side or through the central axis of the body; the wrist is at 15° + and the wrist is often rotated outside the central axis of the body, thus we assigned a +1 to this value. The wrist is also rotated with a total repeating load of 2 kg to 10 kg, and in the end we assigned +1 to the use of the muscle, since holding is done more than 4 times per minute.

In Table 2 all the parts of the body evaluated in this worksheet are clearly depicted, and the relevant values are assigned to them.

Evaluation of the left side did not differ very much from the right side, since in this work position the position of the upper limbs, torso, legs and neck are the same for both sides. A worker at the given job loads both sides of the body symmetrically. It is possible that with some work operations this loading differs minimally, but these are negligible differences which are not important to differentiate or follow in detail with the given job. Therefore, the resulting score for the right and left

of these tables the values are compared with the resulting tables from the RULA method.

With the resulting value the assessed work position is assigned to a risk category, Table 1.

For evaluation of a workplace good knowledge on the work process and the activities at the workplace itself is required. (Nahavandi 2017) With detailed observation of the production process at the swing drill workplace a lot of details were acquired which were subsequently used with this ergonomic method. During evaluation of the workplace using the RULA method, it is necessary to focus on the aspects which are assessed by this method.

At first glance it may seem that certain body positions are unimportant, but in the end result these are very important aspects of the assessment. Examples include the rotating of the torso to the side, the rotating of the wrist, or the incline of the

Table 2. Resulting values of the RULA method—right hand.

Right hand		
Evaluated part of the body	Verbal evaluation	Point evaluation
Right hand	Upper limbs in abduction +1	4
Right forearm	Action though central axis of the body +1	3
Right wrist	Wrist rotated beyond central axis of the body +1	4
Right wrist rotated	Yes	2
Force and load for right hand	2–10 kg repeated load or force	1
Muscle used	Position mainly static	1

Table 3. Resulting values of the RULA method—left hand.

Left hand		
Evaluated part of the body	Verbal evaluation	Point evaluation
Left hand	Upper limbs in abduction +1	4
Left forearm	Action though central axis of the body +1	3
Left wrist	Wrist rotated beyond central axis of the body +1	4
Left wrist rotated	Yes	2
Force and load for left hand	2–10 kg repeated load or force	1
Muscle used	Position mainly static	1

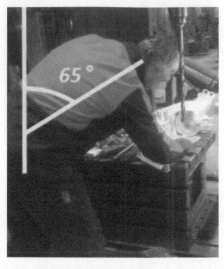

Figure 4. Depiction of the bending of the torso when working.

Table 4. Resulting values of the RULA method—neck, torso and lower limbs.

Neck, torso and lower limbs		
Evaluated part of the body	Verbal evaluation	Point evaluation
Neck	20+	3
Rotated neck	Yes	1
Neck bent to the side	Yes	1
Torso	65	4
Rotated torso	Yes	1
Torso bent to the side	Yes	1
Lower limbs	LL and feet are well supported and in evenly balanced positions	1
Force and load for neck, torso and LL	2–10 kg interrupted load or force	2
Muscle used	Position mainly static	1

side was the same in the final assessment. The final values are shown in Table 3. For both sides the final value of 9 was calculated, which is relatively high.

The neck, torso and lower limbs were evaluated as the last part of this control sheet. When performing the work operations the worker is exposed to unpleasant work positions, from which it follows that the RULA values are higher, especially with assessment of the torso and neck.

Therefore, these parts of the body were assigned the following increased values. The position of the neck is continuously turned at least 20° and more; in addition, the neck is turned to the side and tipped to one side. All the movements of the neck or torso are not captured in the photographs of the assessed work position, but in reality the worker is constantly rotating the torso and neck, and the leaning to the sides was captured there.

This work process is specific in that the worker must consequently follow and re-check all the drilled openings, and therefore these movements are essential. The position of the torso is 65° and the fact that it is rotated and bent to one side only increases the overall score. The lower limbs are supported by the floor and in a balanced position. For these parts we selected an interrupted load with 2 to 10 kg and muscle use is predominantly static.

Work with a swing drill is immensely demanding, and the position of the body is in the majority of cases in an unfavourable position. In Figure 4 the work position of an employee is depicted and the angle which is significant during this assessment by the RULA method is marked in colour. The angle of the bending torso during classic machining is shown.

This work activity is performed most often at this angle of bending the torso, but with some smaller or dimensionally longer workpieces this number is much greater.

It is important to note that the torso is not only dangerously bent during the work shift but also turned to the side, which multiplies many times the dangerous consequences on health when performing this job. The resulting evaluation of this part of the RULA method is depicted in Table 4.

Table 5. RULA—final evaluation and categorization.

Resulting evaluation	
Right side	9
Left side	9
Neck, torso, legs	11
Total evaluation	Category 4 – The making of changes is essential

The resulting numerical evaluation of this load is in the end very high.

Numerical evaluations were assigned according to the result tables of the RULA method. By categorization of the results it was found that the assessed work position belonged to category 4, which represents the application of corrective measures for minimizing risk. A brief summary of the evaluation is shown in Table 5. The result indicator points to the fact that it is more than essential to make changes at the workplace or in the work process when performing this activity. This result demonstrates that this job is a risk from an ergonomic point of view, and there is threat here of damaging the health of employees (Čanda 2016).

3.3 Proposal of measures

During assessment of ergonomic risks one must mainly pay heed to precision and the acquisition of the correct information or other contexts, the use of which will determine the most exact result possible. Ergonomic risks as such are very complex, and their reduction or complete removal is in some cases excessively difficult or impossible.

As the first measure for this job, the rotating of employees, or so-called "job rotation", was proposed. This is a very effective measure and in particular financially undemanding. The role of this rotation is that workers "rotate" during the shift, or in other words, they alternate at the given positions.

Upon assessment of ergonomic risks using the RULA method it was found that a worker at the assessed workplace of swing drill operator is exposed to unacceptable ergonomic risks during the entire eight-hour work day. By job rotation the ergonomic risk would be reduced to an acceptable level. The effects of this ergonomic loading would thus be divided among the three employees.

The total process of employee rotation must be performed such that in the end it does not become counterproductive. In the initial phase, criteria must be proposed, according to which the choice of employees and the workplace rotation will be made. See Figure 5.

The operator of the swing drill spends the predominate part of his working time on setting

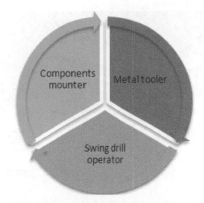

Figure 5. Graphic depiction of the functioning of the proposed rotation of employees.

Figure 6. Graphic depiction of attachments on the base of the swing drill.

the machined component and drilling, whereby his body is constantly in an unfavourable working position. For this reason it was necessary to focus on effective corrective measures which would change this position.

The primary reason for this unsuitable positioning of the body is the work table, which is static; that is, it is not possible to adjust it according to the needs of the worker.

Additional anchoring of an adjustable work surface to the load-bearing construction of the drill is impossible, because the construction is not inbuilt. Replacement of this swing drill by a new one would be very demanding financially, since the price of such a swing drill moves around ten thousand euro.

Therefore, replacement of the static work table with an adjustable hydraulic one, which can be ordered exactly according to the dimensions and other additional requirements, was proposed. The

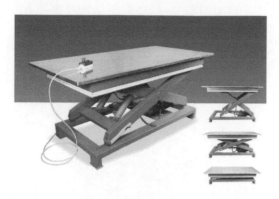

Figure 7. Adjustable work platform.

costs for this type of measure are significantly lower than the purchase of a new swing drill with an adjustable work table.

The whole construction is made from solid material intended for such use. The bottom iron base of the swing drill was designed so that it is possible to mount an additional work table and other items.

These attachments are highlighted in colour in Figure 6. The assembly of this work table (Fig. 7) is relatively simple and fast and guarantees solidness of the construction (Čanda 2016).

4 CONCLUSION

It is necessary to note that assessment of ergonomic risks is exceedingly important for each healthy workplace. Therefore, it is essential that the most effective instruments that are currently available be used for resolving this issue. And so this work was focused directly on the assessment of ergonomic aspects through the RULA method. A crucial aspect with this assessment was defining the correct input values of the workplace into the given ergonomic approach used.

The aim of the work was on the basis of the assessment of a selected job to describe and to propose relevant corrective measures which improve the current state of the ergonomic conditions in a selected metal-working or other similar operation.

ACKNOWLEDGEMENT

The paper was prepared within the project: The contribution was prepared in the scope of the project: VEGA no. 1/0150/15 Development of methods of implementation and verification of integrated safety systems for machines, machine systems and industrial technology and project: APVV-15-0351 Development and application of risk-management models in terms of technological systems in line with the Industry (Industry) 4.0.

REFERENCES

Ahram, T. & Karwowski, W. 2012. *Advances in Physical Ergonomics and Safety.* USA: CRC Press, 604 p.
Chundela, L. 2001. *Ergonomie.* Praha: ČVUT, 2001. 171 s.
Čanda, P. 2016. Posúdenie ergonomických rizík pri vybranej pracovnej profesii. In: *Technická Univerzita v Košiciach*, Strojnícka fakulta, KBaKP.
Hatiar, K. et al. 2008. Ergonómia, ergonomické programy a oblasť BOZP. In: Integrovaná bezpečnosť 2008: *Zborník z medzinárodnej vedeckej konferencie*, Trnava: Alumni Press, 2008. s. 27–36.
Hovanec, M. 2017. Digital factory as a prerequisite for successful application in the area of ergonomics and human factor. In: *Theoretical Issues in Ergonomics Science* 18(1), pp. 35–45.
Hovanec, M. et al. 2015. Tecnomatix for successful application in the area of simulation manufacturing and ergonomics. In: *International Multidisciplinary Scientific Geo Conference Surveying Geology and Mining Ecology Management*, SGEM 1(2), pp. 347–352.
Loriato, H. N. et al. 2016 Ergonomics and work: A study from the household informality In: *Espacios*.
Majerník, M. et al. 2016 Production management and engineering sciences. In: *Production Management and Engineering Sciences—Scientific Publication of the International Conference on Engineering Science and Production Management, ESPM*.
Marek, J. & Skřehot, P. 2009. Základy aplikované ergonomie. Praha: VÚBP, v. v. i., 2009. 118 p.
Maščenik, J. 2016. Experimental determination of cutting speed influence on cutting surface character in material laser cutting. In: *MM Science Journal*, 2016 (September), pp. 960–963.
Nahavandi, D. & Hossny, M. 2017. Skeleton-free task-specific rapid upper limb ergonomie assessment using depth imaging sensors. In: *Proceedings of IEEE Sensors*.
Polohovateľná pracovná plocha [online]. [24.04.2016] <http://www.imcslovakia.sk/produkty/polohovatelny-montazny-stol>.

New Trends in Process Control and Production Management – Štofová & Szaryszová (Eds)
© 2018 Taylor & Francis Group, London, ISBN 978-1-138-05885-9

Economic features of the V4 countries environment and impact on the business

A. Bartoš
Faculty of Business Management, University of Economics in Bratislava, Bratislava, Slovak Republic

ABSTRACT: The Visegrad group was founded in 1991 for the purpose of supporting the process of transforming economies, as well as the European integration process. In addition to the transition to market economy, gradual legislative and economic development also lead to changes in the business environment. Certain differentiations were also introduced when the states joined the European Union and by the adoption of the European currency in Slovakia. These factors had an impact on the subsequent development of the economic environment in these countries. The goal of this paper is to compare economic characteristics of individual states of the Visegrad group and determine the probable impact on the business environment.

1 THE V4 GROUP

The alliance of the Czech Republic, Hungary, Poland and Slovak Republic was established in 1991. Until 1993 it was the Visegrad three; the name was changed to the Visegrad four in 1993, when the Czech Republic and the Slovak Republic because two independent states. The first decade saw the fulfillment of basic goals in the form of implementation of the social-economic transformation, deeper regional cooperation reflected in the field of economy, energy, infrastructure, culture, education and enforcing joint interests against the European Union as well as third countries (MFAEA SR 2014).

Similar development of the social situation led to common goals of the three states. Specifically (Visegrad declaration 1991):

- Restoration of state sovereignty, democracy and freedom,
- Elimination of social, economic and spiritual remnants of the previous establishment,
- Building a parliamentary democracy, rule of law and respecting human rights,
- Creation of a marked economy,
- Full participation on the European political, security and economic system.

These tasks relate to the listed states. Therefore, it can be assumed that there will be identical issues in their implementation. Mutual cooperation thus facilitated the implementation of individual measures.

The highest form of government cooperation represents the meeting of Prime Ministers of the member countries. A lower form of cooperation is based on meetings of state secretaries, who meet twice a year to prepare the meeting of Prime Ministers and preparing proposals for further cooperation. The lowest for is based on meetings of the Visegrad coordinators. These meetings are focused especially on the assessment of the cooperation and preparation of the meetings of state secretaries.

The basic pillars of the states' cooperation consist of the fields of foreign policy, interior, youth, sport, science and technology, environment, infrastructure and cross-border cooperation (MC SR 2011).

The cooperation of the Visegrad four states helps further development of the region. It also allows for more efficient enforcing of joint proposals in European institutions.

2 BASIC ECONOMIC FEATURES OF THE V4 COUNTRIES

The autonomy acquired by these states by the end of the 20th century allows influencing basic economic features through the adoption of legislation and determining conditions. The European environment and its legislation influence them in a different way to a certain degree.

In this section, we will describe the development of selected economic indicators in the V4 countries.

2.1 Unemployment in the V4 countries

Žilová (2003) understands this phenomenon as a natural phenomenon and an attribute of a society based on the market principle and democracy.

Table1. Unemployment in the V4 countries and the EU average during 2006–2015 (%).

Period	2006	2007	2008	2009	2010	2011	2012	2013	2014	2015	Growth index
EU Average	8.2	7.2	7.0	9.0	9.6	9.7	10.5	10.9	10.2	9.4	1.15
Slovak Republic	13.5	11.2	9.6	12.1	14.5	13.7	14.0	14.2	13.2	11.5	0.85
Czech Republic	7.1	5.3	4.4	6.7	7.3	7.6	7.5	7.0	6.1	5.1	0.72
Poland	13.9	9.6	7.1	8.1	9.7	9.7	10.1	10.3	9.0	7.5	0.54
Hungary	7.5	7.4	7.8	10.0	11.2	11.0	11.0	10.2	7.7	6.8	0.91

Source: Eurostat: Available.http://ec.europa.eu/eurostat/tgm/table.do?tab=table&tableSelection=1&labeling=labels &footnotes=yes&layout=time,geo,cat&language=en&pcode=tsdec450&plugin=1>, own calculations https://www.ksh.hu/more_key_figures

Table 2. Development of GDP in the V4 countries compared with the EU average during 2005–2015.

Period	2006	2007	2008	2009	2010	2011	2012	2013	2014	2015	Dev. index
EU	3.3	3.0	0.4	−4.4	2.1	1.7	−0.5	0.2	1.5	2.2	0.66
Slovak Republic	8.5	10.8	5.7	−5.5	5.1	2.8	1.5	1.4	2.5	3.6	0.42
Czech Republic	6.9	5.5	2.7	−4.8	2.3	2.0	−0.8	−0.5	2.7	4.5	0.65
Hungary	3.8	0.4	0.8	−6.6	0.7	1.8	−1.7	1.9	3.7	2.9	0.76
Poland	6.2	7.0	4.2	2.8	3.6	5.0	1.6	1.3	3.3	3.6	0.58

Source: http://ec.europa.eu/eurostat/statisticsexplained/images/1/13/Real_GDP_growth%2C_2005%E2%80%932015_ %28%C2%B9%29_%28%25_change_compared_with_the_previous_year%3B_%25_per_annum%29_YB16.png

In addition to other, it is also based on the preference of the source of income of persons—i.e. it is a dependency on income from employment or, for example, social benefits. It is also based on the necessity of mobility of the work force (Table 1).

The economically active population is the sum of the employed and the unemployed. Students, housewives and pensioners are considered economically inactive (Brožová 2003).

Unemployment is also linked to the increase of social issues represented by decline of living standards of the family of the unemployed, possible health issues, increased crime, and alcoholism. It can be stated that the longer the unemployment, the more severe the consequences. (Martincová 2005).

Listed countries record a major increase of unemployment during 2009–2013 caused by the economic crisis. A gradual stabilization of this indicator can be usually observed after this period. During the last four years of the examined period, of the V4 countries, only the Slovak Republic has unemployment higher than the EU average. This situation is one of the long-term solved issues.

Unemployment in the Czech Republic during 2006–2015 did not exceed the EU average, not even once. On the contrary, our country is each year above the EU average in this area. The same situation is in Croatia, Greece, Spain, Portugal, and with the exception of 2006, in Lithuania as well.

The situation in Hungary is related to changes in the economy and government measures in the field of public works.

2.2 Gross domestic product

Gross domestic product is expressed in market prices and represents the result of the activity of domestic units, which was produced during the set period (SO SR 2017).

It is necessary to pay increased attention to make sure a product isn't counted twice. This is the so-called intermediate product, which is not counted individually, but as a part of the final product. The final production of goods includes products and services designed for final consumption (private consumption), purchase of tangible and intangible capital assets by businessmen (private investments), purchases by the state (governmental consumption) and net export (Roguľa 2011).

In the following table, we list GDP achieved in the Visegrad four states and the EU average (Table 2).

The gross domestic product of the Slovak republic reached higher levels during the monitored period in comparison with the EU average. 2009 was an exemption, when the impact of the economic crisis manifested not only in our country (Fig. 1). Hungary achieved the lowest level, but since 2013 the development begins to be positive. We assume the reasons for this are extensive reforms adopted by the government, as well as newly created jobs. In last two years GDP of the Czech Republic has been growing again. Compared to the EU average its levels were always higher with the exemption of 2009, 2012 and 2013

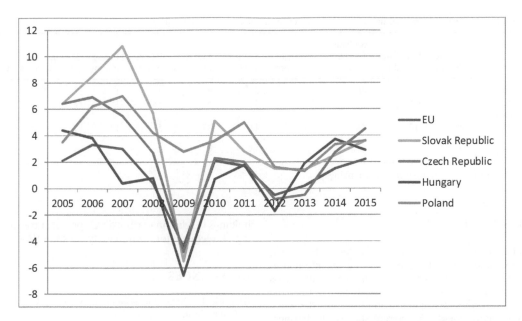

Figure 1. Development of GDP in the V4 countries compared with the EU average during 2005–2015. Source:http://ec.europa.eu/eurostat/statisticsexplained/images/1/13/Real_GDP_growth%2C_2005%E2%80%932015_%28%C2%B9%29_%28%25_change_compared_with_the_previous_year%3B_%25_per_annum%29_YB16.png.

during the monitored period. During the monitored period in Poland, positive economic growth was reflected in positive GDP. Years marked by the onset of economic crisis were not an exemption. A significant drop was recorded in this country in 2012, however it may be noted, that the country copied a trend typical for all observed countries.

3 BUSINESS ENVIRONMENT IN COUNTRIES BASED ON THE DOING BUSINESS INDICATOR

The Doing business indicator provides objective measurements of regulations affecting the business environment. It is based on stages of life cycle of small and medium enterprises. It focuses on the assessment of indicators, such as:

– Conditions for starting a business,
– Registration of property,
– Obtaining a loan,
– Availability of electrical power,
– Solving insolvency,
– Licensing procedures,
– Investor protection,
– Foreign trading,
– Law enforcement,
– Payment of taxes.

Doing business, a World Bank project, was established in 2002 and currently it is focused on 190 countries. By collecting and analyzing complex data it compares the business environment in selected economies and provides also the possibility of real-time monitoring of individual indicators over time.

It examines a total of 10 parameters, by which it is possible to assess the level of regulation and bureaucracy in the life cycle of a small or a medium enterprise. Each indicator has the same weight (Šúbertová 2016).

It serves as a source for academics, journalists, researchers and other interested in the business environment. It is also encouraging countries to constantly improve conditions for business development (Table 3).

In overall assessment of the business environment, the Slovak Republic ranked third. Poland is ranked first and the Czech Republic second. Hungary was ranked on the last place of the Visegrad four countries.

The Doing business indicator has the advantage that it is possible to compare the status of the examined issue between individual countries, because it is working based on set criteria. The disadvantages include that it is focusing only on limited liability companies. However this can be accepted due to the fact that this group of businesses is the most numerous in said countries. Another disadvantage is the absence of criteria such as quality of human capital or market size.

Table 3. Positions of countries in individual metrics of the doing business ranking in 2017.

Country	Starting a business	Licensing- procedures	Availability of electrical power	Registration of property	Obtaining a loan	Investor protection	Payment of taxes	Foreign trading	Closing contracts	Solving insolvency
Poland	107	46	46	38	20	42	47	1	55	27
Czech Republic	81	130	13	31	32	53	53	1	68	26
Slovak Republic	68	103	53	7	44	87	56	1	82	35
Hungary	75	69	121	28	20	81	77	1	8	63

Source: http://www.doingbusiness.org/rankings

4 CONCLUSION

The quality of business environment based on a specific indicator depends to a large extent on chosen criteria. It was proven that indicators like gross domestic product or unemployment are only the consequences of setting conditions of the business environment. The current state of the business environment expressed by a set of criteria making up Doing business is influence by a wide range of factors. Poland, the V4 country ranked the highest improved its position by one spot between 2016 and 2017. Greatest improvement was recorded in the field of licensing procedures and solving insolvency. The Czech Republic improved conditions to start a business, which improved its spot in this criterion by seven spots. However overall it dropped by one position. Measures in the field of taxes in Slovakia led to improvement by two spots in the field of paying taxes. However compared to 2016 it fell overall from the thirtieth spot to the thirty-third spot. Business environment in Hungary worsened by one spot, specifically from the fortieth spot to the forty-first. It improved by five spots in the field of closing contracts and by two spots in the field of paying taxes.

ACKNOWLEDGEMENT

Processed pursuant to the project VEGA 1/0709/15 "Assessment of efficiency of financing for projects to support the enterprise development of newly established and established active SMEs in Slovakia" in scale 50% and project VEGA 1/0128/15 "Factors on increasing of efficiency of agricultural holdings in relation to rural development and ensuring of adequate food sovereignty" in scale 50%.

REFERENCES

Brožová, D. 2003. *Společenské souvislosti trhu práce.* Praha: Slon.

Martincová, M. 2005. *Nezamestnanosť ako makroekonomický problém.* Bratislava: Iura Edition.

MC SR. 2017. Vyšehradská skupina. [online]. [cit. 2017-02-09]. Available at <http://www.culture.gov.sk/posobnost-ministerstva/medzinarodna-spolupraca/odbor-europskych-zalezitosti/vysehradska-skupina-v4–105.html>.

MFAEA SR. 2017. Vyšehradská skupina. [online]. [cit. 2017-02-09]. Available at <https://www.mzv.sk/zahranicna_politika/slovensko_a_v4-vysehradska_skupina>.

Roguľa, P. 2011. *Hrubý domáci produkt ako makroekonomický ukazovateľ výkonnosti ekonomiky Slovenska.* Prešov: Prešov University.

SO SR. 2017. Principles of calculating gross domestic product. [online]. [cit. 2017-02-11]. Available at <http://www.statistics.sk/pls/elisw/objekt.sendName?name=m_NUhdp>.

Šúbertová, E. 2014. *Podnikateľsképrostredie v Európskej únii.* Bratislava: KARTPRINT, 2014. 126 s. ISBN 978-80-89553-24-2.

Šúbertová, E. 2016. *Kvalitné podnikateľské prostredie ako základ pre dlhodobý pozitívny rozvoj podnikateľskej aktivity malých a stredných podnikov v SR.* Bratislava: Ekonóm.

Visegrad declaration. 1991. [online]. [cit. 2017–02–09]. Available at <http://www.visegradgroup.eu/documents/visegrad-declarations/visegrad-declaration-110412>.

Žilová, A. 2003. *Práca s komunitou s vysokou mierou nezamestnanosti.* Banská Bystrica: Pedagogická fakulta UMB.

New Trends in Process Control and Production Management – Štofová & Szaryszová (Eds)
© 2018 Taylor & Francis Group, London, ISBN 978-1-138-05885-9

Management of technological inovation introducing in mineral industry

V. Bauer, M. Herman & J. Zápach
Faculty of Mining, Ecology, Process Control and Geotechnologies, Technical University of Košice,
Košice, Slovak Republic

ABSTRACT: Industry of minerals extraction in Slovakia has passed in the last ten years many qualitative changes. In mining and quarrying of industry minerals, began to use more of modern technologies and implemented the new system management factors. The many technological mines operation have been innovated at the level of performed processes, as well as in existing technical systems of industry minerals extraction. Innovative changes are demonstrable also in the rock destruction process with using blasting energy of explosives. In the mining system is a primary process of rock destruction performed with blasting works, which is critical for the production of the both industry minerals quality and aggregates. The article described an innovative procedure for introducing the new concept of using modern and environmentally friendly explosives. Qualitative parameters of the process of rock disintegration, with the use of selected types of explosives are presented as outputs of practical and laboratory research.

1 TECHNOLOGICAL INNOVATIONS IN EXTRACTION OF INDUSTRIAL MINERALS

The shortage of some raw materials in the economic policy of the EU is starting to show still more pronounced, and re-evaluation process capacity and yield of domestic raw materials and mineral extraction are again very current. Currently, most discussion is about the need for extraction of raw materials of strategic importance respectively called critical raw materials, which need to increase consistently. Mining, processing and use of raw materials in the Slovak Republic based on the strategy of raw materials policy, which is oriented more in the area of non-metallic minerals. The primary task of the raw material policy mentioned materials is to ensure production volumes of extraction within the desired range of materials and products to ensure a sufficient amount of raw materials for various industries, especially for metallurgical and construction industries. From today's perspective remains a priority especially the production of building materials and products for building production. In contrast to previous periods when the primary importance of our resource base often overestimated, there is at present a more realistic perspective on technically and technologically available, i.e. available resources of minerals and raw materials. They are taken into account while actually existing economic conditions and potential opportunities, but also the requirements for exploitation deposits of these commodities (Bauer 2014, 2017).

Mining and quarrying of raw materials has a long tradition in Slovakia. Technical and technological development of the mining industry over the years have taken place in various and changing economic conditions, but also in different economic systems. Broadly conceived Slovak mining-ore, non-ore and coal, including the quarrying of raw materials, the stages of its development sought to capture all the developments, which occurred in the field of advanced mining technologies and modern approaches to the extraction of raw materials.

At the same time, it was in their economic possibilities always prepared, offered and available technical innovations, which gradually appeared in the developed world mining industry to adopt their technology and knowledge mining incubator and use them in real conditions of mining. In the case of mechanised drilling technologies and technologies to use the energy of explosives—explosives for disconnection of rock, it was not different.

Destruction of rocks by using energy explosion explosive is currently the most widespread and thus the most used technology of disintegration in the extraction of minerals and raw materials. In the technological system of drilling and breaking rocks, the technical result is blasting rocks in a particular volume is essential in this method optimal use of explosion energy. This is given by the explosive properties of chemical substances accumulated in explosives and explosive forces released into the rock, what is at present changing their original solid physical state which is subsequently destroyed. From a technical point of view is a technological system of drilling and breaking rock very

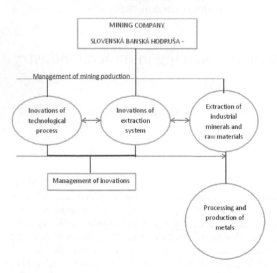

MINING COMPANY

SLOVENSKÁ BANSKÁ HODRUŠA -

Management of mining poduction

Inovations of technological process

Inovations of extraction system

Extraction of industrial minerals and raw materials

Management of inovations

Processing and production of metals

Figure 1. Levels of inovation management in Rozália mine.

specific interactive system, manifesting itself by the interaction of several factors characterising the natural rock and specific technical factors characterising the expolsives.

2 INNOVATIVE TRENDS IN THE USE OF INDUSTRIAL EXPLOSIVES

In Slovakia, the extraction of minerals and raw materials, both in underground mining for example. Raw magnesite, talc, polymetallic ores, etc., as well as surface mining quarry stone, for instance. Limestones, andesites and other natural stone, used in most cases of industrial explosives. The mechanical disintegration of rock with the occurrence of industrial minerals is hardly ever used. Justification of the use of industrial explosives is determined by the need for large volumes of extraction of minerals, which stands at several tens of millions of tonnes of raw materials annually. Required a large amount of mining technology corresponds to the relatively large amount of explosives. When choosing the explosives for use in a mine or on the surface, there is a choice of explosives by type, but also by manufacturers of industrial explosives. At present, our mining conditions used exclusively industrial explosives foreign producers. In the technical assessment of the advantages or disadvantages of the utilisation of the explosives which, in specific mining conditions, it is always important to perform necessary technological analysis of drilling and blasting at the workplace operated. Assessed primarily functional parameters respectively explosion characteristics

of various types of explosives, the primary consideration is handling security and technological efficiency of the proposed explosives.

Like other industries, the mining industry has undergone a period of the great technological innovation process. Similarly, it was the innovation of production and use of new types of explosives in mining minerals. The introduction of new kinds of industrial explosives, which also corresponds to a higher level of technical means of blasting techniques (detonators and firing systems), it is important, in addition to assessing functional explosion parameters of explosives, review and test the safety and handling effective technology enforcement of explosives. This applies to all types of explosives used today—plastic, emulsions and pumpable emulsion explosives, or for other unique types of explosives. The trend of innovation for industrial use of explosives in mines, quarries and tunnels, can be traced back two primary lines. When introducing a new type of explosive is always a critical parameter of safety and health risks explosives. And therefore, the research activities in managing the introduction of new types of explosives in mining operations, aimed at a comprehensive solution to the issue of destruction of rocks about the performance of blasting mining in a particular area. Modern technology of production of new explosives and introduction of new technical means for firing systems of industrial explosives require not only technological analysis of blasting, as well as extractive and technical analysis of the environment in which the blasting carried out (Bača 2015).

3 MANAGEMENT OF THE INTRODUCTION OF NEW TYPES OF EXPLOSIVES IN MINING OPERATION MINING ROSALIA

The deposit of polymetallic ores, nonferrous metals in Hodrusa-Hamre in mine Rozália using proven and production operations for years proven means of drilling and blasting techniques. Technological structure and layout selection mean used blasting techniques, such as detonators, industrial explosives, charging devices, or other devices over time the mine gradually changed to fit the needs of the mine. Due to the relatively high diversity and different levels of hardness of the rocks at the deposit are favoured more rock mining explosives/explosives, which are characterised by strong brizantnosťou and sufficiently large detonation capability. The technological process of blasting at mining and blasting operations rocks traditional way and the means in blasting techniques such as plastic rock mining explosives, electric detonators, and electric blasting machine. In exercising blasting are followed all the rules and principles of safety and operational safety. The actual rock blasting on work carried out by mining model drilling scheme launch. Before blasting rocks by assessing several factors, but mainly—geological environment factors, technical factors, but also predictive of Chemical Technology qualitative factors of the expected rock heap.

Due to developments in the use of industrial explosives, the management of mining company SB Hodrusa-Hamre decided to replace conventional plastic explosives, for which they are an essential component of liquid explosive nitrate esters, emulsion explosives, which contain the explosive component. An analysis of the technological process of disintegration of rock blasting work and the results of operational test performance parameters of plastic and emulsion explosives clearly confirmed the justification for the need of introducing new types of emulsion explosives. Pi testing new emulsion explosives have proved to be higher handling safety and reduced load on the human body flue gas explosion NOx and CO. They also compared the performance and functional parameters of new types of industrial emulsion explosives with classic plastic explosives. In individual cases they tested selected traditional plastic explosives and emulsion explosives in two groups (Jakubček 2015):

1. New plastic emulsion explosives that are more favourable values so "Parameter harmful"to health workers handling these explosives.
2. Modern classic plastic explosives, which are not inconsiderable. negative effects on the health of employees.

Tests have confirmed that unlike the standard, until now largely used conventional plastic explosives are plastic emulsion explosives/explosives characterised as progressive and modern mainly concerning:

– Structural (chemical composition of explosives),
– Technological (performance and functional parameters),

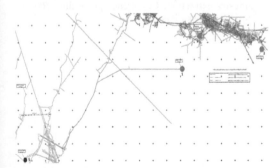

Figure 2. A selected workplaces for testing the emulsion explosives.

– Safety (more comfortable handling),
– Health and hygiene (cleaner work).

Therefore, selecting and securing technologically and operationally appropriate type of emulsion explosives for carrying out blasting at mine Rozália, directed mainly mining production following requirements:

– Increasing the detonation forces used explosives,
– Reduce toxicity for use explosives, to the exclusion of harmful nitroesters in explosives,
– Adjust the cost of blasting with a blasting rocks.

Management of mining company Slovak mining, Ltd. Hodruša-Hámre plans to technological innovation and the development of mining activities on the deposit, to incorporate the new concept of innovation blasting, including modernization of the means of blasting technique. The intention is to solve the problem of the disintegration of rocks on productive workplaces mine Rosalia in such a way that the mining blasting safe regarding performance blasting to be less harmful to health, i.e. to have favourable values so. "Parameter harmful" to health workers handling these explosives, and to achieve optimal technical parameters of the blast. Therefore, in the new concept of blasting also considering the introduction of advanced pumpable emulsion explosives, whose negative effects on the health of workers are almost negligible.

4 INNOVATION MANAGEMENT IN THE USE OF PUMPABLE EMULSION EXPLOSIVES

Due to the multiple benefits of using pumped respectively pumpable emulsion explosives in the underground mining of minerals were innovative plans assessed the possibility of using these explosives in conditions of SB Banská Hodruša-Hámre. Pumped emulsion explosives have a significant advantage in that it can be pumped continuously into wells, have stable performance parameters throughout the borehole, contain dangerous toxic substances and are easy to handle. Management of mining company prepared a draft of new explosives are to be assessed both production systems and the use of pumpable emulsion explosives:

1. The pumpable registration emulsion being prepared i.e. stirred at the place of consumption, referred to in the literature as the system SME (Side Mixed Emulsion).
2. The pumpable registration emulsion being prepared i.e. sensitising emulsions for local consumption referred to in the literature as a system of SSE (Side Sensitized Emulsion).

Both these systems of production of the explosives—SME and SSE system, use the principle of the preparation of emulsion explosive-based mixing of components, which are defined as non-explosive raw products. In our country, there is already mining experience with mixing and charging vehicle (e.g., From Austin Detonator, Orica, SSE emulsified), but only at the surface of quarries—in blasting, but also during the excavation of the tunnel tubes tunnelling. The new concept of blasting for mining of polymetallic ores counts with a similar system, which will be important to solve the technical design self-propelled charging unit.

The basis of SSE is the production and processing of non-explosive emulsion matrix in the factory respectively the stationary race said matrix is classified as non-combustible material in the form of oxidation agents. The system itself has the following parts:

– Charging unit,
– Storage tanks,
– Special pumps for pumping emulsion matrix and splyňujúcich ingredients that are mixed before charging and become explosive when pumped into the bore.

The basis of the system SME production and processing non-explosive emulsion matrix in the mobile charging device that is structurally composed:

– chassis of the truck,
– of special vehicles with the necessary supply of non-explosive materials in tanks,
– the mixing device for mixing the components
– a system of pumps and accessories,
– the charging of antistatic hose with the possibility to change the speed of pulling the hose out of the hole.

A possible design of a prototype of such a device assessed for entirely new technical and technological development, or continue a way of adapting the parameters of any of the above-presented mobile charging units for the pre-set service conditions of such deposits in Hodruša-Hámre.

5 CONCLUSION

The main aim in choosing strategies and to take a decision on how to proceed in this area further, it could be economic analysis and consideration of performance carried out of such deposits, especially during the drilling and blasting, but also questions of assessment increase of hygiene and culture of mining work underground mines. At friends of mine geology of the mine Rozália and taking into account the specific conditions of existing mining, the large abundance of mining workplaces in carrying out blasting, but also quite complicated transport respectively carrying explosives currently used, in this situation must be addressed by assessing the possibility of modernising the process of blasting in terms of usage: modern explosives, economic and cheap explosives, health safe explosives and explosives with minimal effort during handling. These conditions meet the pumpable respectively pumped emulsion explosives, for which you must select the appropriate technological means of blasting technique so that the economic result of blasting, measurable desired technological parameters TP i.e. fragmentation, efficiency and the total cost was in terms of the innovation plan Slovak mining company mining, Ltd. Hodruša—Hámre optimal.

REFERENCES

Bača, D. 2015. The blasting analysis and drilling pattern proposal for the various type of explosive at the mine Rozália, Diploma thesis. Košice: Technical university of Košice.
Bauer, V. 2014. *Drilling and blasting technics.* Košice: Equilibria.
Bauer, V. et al. 2017. *Rocks destruction by using of new pumpable explosive and technical design of mobile charging unit.* Košice: FBERG, Technical university of Košice.
Jakubček, E. et al. 2015. Quality and possbilities of plastic and pumpable emulsion explosive by mineral industrz extraction, Conference proceeding *Blasting Technic*, Banská Bystrica.

Consumer neuroscience as an effective tool of marketing management

J. Berčík

Faculty of Economics and Management, Slovak University of Agriculture in Nitra, Nitra, Slovak Republic

ABSTRACT: The paper explores the place and position of consumer neuroscience in marketing management while it points out its role in understanding consumer behavior. We clarify the main difficulties and obstacles of full use of neuroscience in marketing, comparing the traditional and modern approach to consumer persuasion models both with and without using consumer neuroscience methods. Finally, the paper points to the new possibilities of market research and underlines the growing need for their implementation in the form of innovative research solutions that can be used in different economic sectors (retail and services), too. These devices allow exploring the contentment (emotions, visual attention), but also the environmental factors acting not only with the awareness of the respondent (research vest), but even without his full knowledge (smart kiosks). The interpretation of the obtained data provides businesses new opportunities how to do more effective decisions in making management.

1 INTRODUCTION

1.1 *Marketing management*

Managers must carry out one of the very important tasks that relates to their customers, and that is to know their needs and wants, and to deliver their products and services to the market in a better and faster way than the competitors. Gallo (2010) defines marketing management as a process, which involves the planning and implementation of concepts, pricing, promotion and distribution of ideas or goods and services, the aim of which is the exchange that satisfies individuals or organizations. According to Kotler ct al. (2007) marketing management is defined as a science or art, in which target markets are selected and profitable relationships are built with them. Thus, managing demand and further managing customer relations includes marketing management. Bartošová & Krajníková (2011) state that we can speak about marketing management as a continuous process, which involves analysis, planning, decision-making, implementation and monitoring, which creates and maintains long-term relationships with target customers and allows achieving set goals Kubicová & Kádeková (2017).

1.2 *The importance of consumer neuroscience in research and marketing management*

The progress of time increases the demand for a deeper understanding of human knowledge and behavior Taylor et al. (2015). This led to the creation of synergy between the biological and social sciences. Joint research efforts of biological and social scientists supported major advances along the various fields of social, behavioral, biological, and management sciences Agarwal & Dutta (2015).

In recent years there has been a great progress of neuroscientists in their ability to directly study the cortical activity. Psychological and physiological sciences rapidly applied these techniques to get to the surprising discoveries in understanding the brain. However, most social sciences have yet to accept neuroimaging as a standard tool or procedure for research. Economy also began to use neurological techniques in their research, resulting in the creation of neuroeconomy. Marketing as a science began taking advantage of the benefits provided by consumer neuroscience later, despite the fact that both fields share many common challenges in terms of decision-making and exchange Lee et al. (2007).

Soon after the first findings related to neuroeconomy were published, marketing researchers have realized the potential of these new techniques in research, compared with the full range of classic and qualitative methods previously used. Academic studies in this area have highly interdisciplinary character. Knowledge of marketing management is linked to psychological knowledge and various medical disciplines Javor (2013).

Linking multiple scientific disciplines was essential because understanding consumer behavior in society has meant a complex study of many variables. Changes in social dynamics, new technologies, new lifestyles and increased consumer awareness have led to questioning the hegemony of mass communication tools, particularly advertising as a source of information Horská et al. (2015). It triggered a crisis of credibility of this form of communication. Today's recipient of messages is aware that advertisers are trying to persuade him to buy the product or service. Consumers are critical and skeptical, thereby diminishing the power of this communication tool Colaferro & Crescitelli (2014). A unique feature of consumer neuroscience is its direct relevance for practice. Industry has always followed developments with great interest. Neuroscience methods offer hope to solve one of the key problems for many marketing scholars that is how to reliably measure implicit response to marketing stimuli. Therefore, the area of marketing research quickly approved the adoption of neuroscience and the proof has been an increasing number of neuromarketing companies Plassmann & Ramsøy & Milosavljevic (2012).

Consumer neuroscience has a great benefit as it avoids relying entirely on the studied subject. The subject, in spite of his will, cannot accurately express his subconscious motives. Also, classic research is not likely to capture emotional reasons on which consumer preferences or decisions are based, unlike consumer neuroscience, which can affect the subconscious prejudices. Some experts believe that neuroimaging can even lead to "igniting" pleasurable feelings in consumers. Psychiatrists say that this information about the brain could help in practice to better customer segmentation. In other words, the potential of consumer neuroscience to provide valuable information for academia, as well as reduce marketing failures and increase marketing success looks quite promising, though debatable Fugate, (2007).

Research in the field of consumer neuroscience can be realized in laboratory but also in real conditions (Paluchová & Kleinová 2014). Based on what is the subject of research, the appropriate biometric method (e.g., monitoring heart rate variability (HRV), eye movements Eye tracker, facial expressions FA) or neuroimaging method (e.g., monitoring electrical brain activity EEG) is selected or a combination of these methods. When processing and interpreting some research data (e.g., about brain activity) an active cooperation with neurologists and psychologists is required, because working with this data requires involvement of experts also from this field. In addition, the quality of the research itself is influenced by many aspects that need to be considered (e.g., total fatigue of respondents or weather), which presents the need to repeat these types of interdisciplinary research in an effort to obtain more accurate data Horská, Berčík & Gálová (2015).

2 MATERIAL AND METHODS

The research paper is based on the study of existing knowledge in the following areas:

– Consumer neuroscience and neuromarketing
– Market research using biometric and neuroimaging methods
– Distribution and use of various research methods
– The need for the implementation of consumer neuroscience in market research and marketing management

The study results in the definition of the main reasons for the need to implement consumer neuroscience within market research to enhance the explanatory power of these studies, evaluation of their use in marketing management and the optimal size of the survey sample. The need to implement innovative research methods is also documented in an increasing percentage of biometric and neuroimaging methods within surveys conducted mostly abroad. In order to obtain information about the optimal size of the survey sample, 13 available researches from the field of consumer neuroscience carried out mainly abroad, but also in Slovakia were compared. The standard size of a representative sample of compared institutions was calculated based on the average of researches conducted by the institution. In addition to determining the optimal size, the paper also presents the most frequently used biometric and neuroimaging technologies of these institutions dealing with consumer studies. The paper highlights some examples of primary data collected from neuromarketing research useful in academic and commercial practice.

3 RESULTS AND DISCUSSION

In today's challenging competitive fight, companies that do not use the knowledge of these new innovative researches are hardly able to stand in the fight against companies which have this information, because traditional research tools including questionnaire surveys provide increasingly irrelevant information.

3.1 *The development of consumer neuroscience in the world*

At present there are neuromarketing agencies in the world offering their services to companies to determine how their products, brand, advertising and other elements of their marketing activities affect people. Interest in consumer neuroscience abroad is reflected also in the fact that it is a topic of lectures and conferences around the world, such as the 5th Neuromarketing World Forum 2016 held in Dubai or Neurobusiness ExpoForum held in 2014 in Brazil. In addition to neuromarketing agencies also associations are created that bring together agencies and organizations, such as Neuromarketing Science & Business Association, which conditions its membership on the acceptance of the Code of Ethics for the Application of Neuroscience in Business. NMSBA is a global trade association established by Carlou Nagel in 2012, who is also the founder of the mentioned Neuromarketing World Forum. The association is for everyone who have a professional interest in neuromarketing or consumer neuroscience.

The growing interest in this interdisciplinary field can be seen even in general in the form of an increasing number of searches (Fig. 1) of the term "neuromarketing". Looking at Google Scholar citations for neuromarketing and validation, the number of publications reporting "neuromarketing" and "validation" almost tripled compared with the 2011.

In general we can state that within the use of research tools there is a constantly increasing share of biometric and neuroimaging methods (see Figure 2), even despite the fact that these methods are presented separately not as part of neuromarketing or consumer neuroscience. The reason is that in many mainly developed countries these methods have been established enough to be considered a common standard.

In consumer studies with the use of biometric and neuroimaging methods there is no such term as "standard study". Each type of research in this field is unique in its own way and takes into consideration the needs of defined research presumptions. The implementation of the new technologies in research in the area of consumer behaviour and target segments represent an important time investment not only to cary out the testing itself but especially processing and then interpreting the results. As you can see in the Table 1, the size of the research sample depends not only on the utilized research tool but also on the institution which organizes the relevant research activities.

In some cases, the institutions active in research in the field of consumer neuroscience use only 6 respondents due to time and financial demands. In general it can be said that so far the most expensive research was carried out at Oxford University in 2004–2007, where 2,081 respondents were tested using functional magnetic resonance imaging. On the contrary, the smaller survey samples are used in companies Neurensics, Neuromarketing Labs, at the University of Bonn (Life & Brain GmbH), Copenhagen Business School (Neurons Inc.), as well as the Slovak Agricultural University in Nitra.

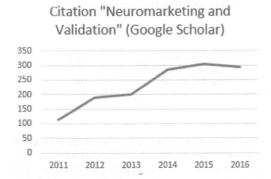

Figure 1. Google Trends: "Citation and validation". Source: Nagel (2017) Trends in Neuromerketing 2017.

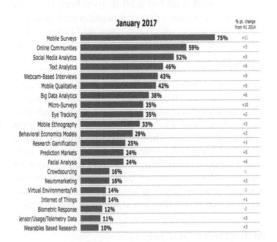

Figure 2. Adoption of emerging market research methods.
Source: GreenBook Industry Trends (GRIT) report 2017.

3.2 The development of consumer neuroscience in Slovakia

Establishing cooperation between marketers and neuroscientists in Slovakia is complicated due to lack of funds, interest or doubts about whether companies would really use it for strategy decision making. The research tools, as well as services of neuromarketing agencies are expensive and most Slovak firms cannot afford them. According to Šášiková (2013), although media and research agencies in Slovakia offer neuromarketing services, most are just additional service of the offer of traditional marketing research. One reason is that the equipment and facilities for such studies are expensive and also neuromarketing services for marketing research submitters are not cheap, thus for smaller companies infeasible. "If a company has three- or four thousand euros for conducting the research, they cannot count with neuroresearch" (Mlaková 2013). For such services a uniform price list cannot be set and the prices are calculated according to specifics of the client. GfK provides less expensive alternatives to these services, such as eye camera that monitors eye movements and their retention in certain parts of the display area

Table 1. Chosen researches and standard size of research. sample typical for consumer neuroscience.

Researcher (academic/business)	Institution	Research technics	Standard size of research sample
Elissa Moses	Ipsos Neuroscience	EEG, Eye tracker, IAT, Facial coding,	65
Bernd Weber	University of Bonn/Life & Brain GmbH	fMRI, Eye tracker	30–40
Thomas Ramsoy	Copenhagen Business School/ Neurons Inc.	Eye tracker, EEG, Facial coding, fMRI	40–50
Richard Silberstein	Swinburne University/Neuro-Insight	SST (EEG)	50–100
Jaime Romano	Neuromarketing S.A. de C.V. (NMKT)	EEG, Eye tracker, GSR, EKG, Blood volume pulse	18–96
Hirak Parikh	Neuromarketing Labs	EEG, Eye tracking, GSR, Respiration rates	25–40
Christopher Morin	Fielding Graduate University/ Salesbrain	Voice analysis, Biometric studies, Facial imaging	6–200
Martin de Munnik	Neurensics	fMRI, Eye tracker	20–30
Arnaud Petre	Universite Catholique de Lille/ Brain Impact	fMRI, EEG, Eye tracker, EKG, GSR	12–72
Gemma Calvert	Neurosence Limited	Implicit priming paradigms, fMRI	10–400
Martin Lindstrom	Oxfor university	fMRI, SST (EEG)	2081
Reada Montaguea	Human Neuroimaging Laboratory/ Virginia Tech Carilion School of Medicine	fMRI	67
Jakub Berčík	Slovak University of Agriculture in Nitra	EEG, Eye tracker, Facial coding	18–60
Graham Page	Millward Brown	EEG, Eye tracker	100

Source: Nagel 2014. Sample size in Neuromarketing.

(e.g., a poster or novelty Emo Scan, which can scan the face and responses to various stimuli through a webcam). The Slovak neuromarketing agency Dicio Marketing was established in 2009, while its first experiments with the conductivity of the skin started already in 2007. Currently the company Samo Europe Ltd. is operating in Slovakia, which was established in 2016 and in addition to common research in the field of consumer neuroscience provides also research with two technologies that have been developed at the SUA in Nitra and are the subject of international patent procedure.

4 CONCLUSION

In developed economies, investments of businesses in consumer neuroscience are generally higher than in Central and Eastern Europe. In Slovakia mostly syndicated types of surveys are sought, in which multiple parties participate. On the contrary, in Western countries in addition to syndicated studies tailored surveys are added which focus more on action steps, right on the implementation of changes e.g., in the marketing campaign, assortment, ways to set a new pricing strategy. Another difference can be seen in the sample of respondents. In addition to

nationwide studies targeted at the whole population (e.g., customer satisfaction), studies addressing specific customer groups are added with which the store chain plans to work and in which a potential for the development of a particular area is seen.

In general, we can say that research helps to understand the market and also all its members, from the position of the submitter, competition and monitoring of their strategies to customers and buyers. Market research using consumer neuroscience allows to reveal the gaps and the potential on the market that is very important for each stage of marketing management. Such studies provide more and more precise results than the traditional forms (e.g., questionnaire or group talks), therefore we can expect that they will become a common part of everyday lives in more areas, even though with a different name. A perfect knowledge when and how a certain individual reacts to something is of great importance for understanding consumer behaviour and decision-making. Future development of these disciplines depends mainly on the development of technologies.

ACKNOWLEDGEMENT

The paper is a part of the research project KEGA 038SPU-4/2016 "Implementation of new technologies and interdisciplinary relations in the practical learning of consumer studies", solved at the Department of Marketing and Trade (Faculty of Economics and Management, Slovak University of Agriculture in Nitra). As well as, paper was supported "AgroBioTech" Research Centre ITMS 26220220180.

REFERENCES

Agarwal, S. & Dutta, T. 2015 *Neuromarketing and consumer neuroscience: current understanding and the way forward*. DECISION 42:4, 457–462.

Bartošová, H. & Krajníková, P. 2011. *Základy marketingu*. Praha: Vysoká škola regionálního rozvoje.

Colaferro C.A. & Crescitelli E. 2014. *The Contribution of Neuromarketing to the Study of Consumer Behavior*. Brazilian Business Review, 2014. 11(3).

Fugate, D. 2007. *Neuromarketing*. In: *Journal of Consumer Marketing*. [online][cit. 2016-02-13] vol. 24, iss. 7, s. 385–394.

Gallo, P. 2010. 2011. *Marketing—Základy marketingu*. Košice: Dominanta.

GreenBook Industry Trends (GRIT). Adoption of Emerging Market Research Methods. 2017. [online]. [cit. 2016–03–22]. Retreived from: http://www.greenbookblog.org/2014/09/29/the-top-20-emerging-methods-in-market-research-a-grit-sneak-peek/.

Horská, E. et al. 2015. *Consumer neuroscience solutions: towards innovations, marketing effectiveness and customer driven strategies*. In 5th International Conference on Management 2015. Gödöllő: Szent István University.

Horská, E. et al. 2015. *Bioeconomics, neuroeconomics and neuromarketing: new approaches to customers and businesses*. In Globalization, economic development, and nation character building. Depok: Research Institute of Gunadarma University.

Javor A. et al. 2013. *Neuromarketing and consumer neuroscience: contributions to neurology* //BMC neurology. – 2013. – T. 13. – №. 1. – S. 13.

Kotler, P. et al. 2007. *Moderní Marketing*. Praha: Grada Publishing a.s.

Kubicová, Ľ. & Kádeková, Z. 2017. *Strategic marketing*. Nitra: SUA in Nitra.

Lee, N. et al. 2007. *What is _Neuromarketing? A Discussion and Agenda for Future Research*. In International Journal of Psychophysiology 63:2.

Mlaková, M. 2013. *Reklama získala ďalšiu zbraň. Skúma náš mozog*. 2013.

Nagel, C. 2014. *Sample Size in Neuromarketing*. In: Neuromarketing Theory & Practice, no. 9, July 2014, s. 6–9. Venlo: NMSBA (Neuromarketing Science & Business Association).

Nagel, C. 2017. *Trends in Neuromerketing 2017*. [online]. [cit.2016-03-22]. Retreived from: https://www.linkedin.com/pulse/trends-neuromarketing-2017-carla-nagel.

Paluchová, J. & Kleinová, K. 2014. *New trends in catering: How merchandising and other popular tools can attract customers*. 373 – 382 str. In: New trends in management in the 21st century—Cross-Atlantic perspective. 1st ed. Częstochowa: Sekcja Wydawnictw Wydziału Zarządzania Politechniki Częstochowskiej.

Plassmann, H. et al. 2013. *Branding the brain: a critical review and outlook*. Journal of Consumer Psychology, 22(1), 18–36.

Šášiková, M. 2013. *Neuromarketing na Slovensku a v zahraničí a jeho etické aspekty*. Bratislava: Ekonomická univerzita v Bratislave.

Taylor S.J. et al. 2015. Introduction to qualitative research methods: A guidebook and resource. John Wiley & Sons; 2015.

New Trends in Process Control and Production Management – Štofová & Szaryszová (Eds)
© 2018 Taylor & Francis Group, London, ISBN 978-1-138-05885-9

Changes to transportation systems and socio-economic impacts

H. Bínová, D. Heralová & R. Vokáč
Faculty of Transportation Sciences, Czech Technical University in Prague, Prague, Czech Republic

ABSTRACT: Transportation indicates the state of society in the areas of consumption of raw materials, resources or finished products or their parts. The existing imbalance of incomes of the population on a global scale leads to the transportation of products with low production value, and if travel costs are included, these products are cheaper than the products in the country of consumption. The result is a growing volume of freight transport, increasing environmental pollution and social and economic impacts. The level of congestion and how the transport system works thereby impacts the social and economic climate. Given that freight transport is operated on a scale that transcends the borders of individual countries, the specified impacts and effects have an international importance. In the future, it will be necessary to seek out transportation systems that will function optimally, efficiently use possible and acceptable sources, and will be environmentally friendly and safe.

1   INTRODUCTION

The basic guidelines of the "Freight Transport Logistics Action Plan" (2007) EU document also include a declaration about the fact that the concept of transport corridors is given by the concentration of freight transport between major hubs, and by relatively long transport distances. Cooperation between several types of transportation and modern technology will be supported in these corridors in order to cope with increasing traffic volumes, and to promote environmental sustainability and energy efficiency. Coastal voyages, railways, inland waterways and roads mutually complement each other. These corridors will be equipped with catchment terminals at strategic locations (e.g. seaports, inland ports, marshalling yards and other relevant logistic terminals and facilities), Čorejová, Al Kassiri & Valica (2015).

All of these factors have a subsequent impact on the formation of social ties, and thereby also changes in the economic level.

2   DEVELOPMENT OF TRANSPORT SYSTEMS

In line with societal challenges, Transport systems in Europe must efficiently utilize available resources and be environmentally friendly, safe and functional. Acceptance of all of these requirements can lead to intelligent, green and integrated transport.

This particularly concerns reduction of emissions, reduction of externalities caused by traffic congestion and a reduction in the number of fatal accidents, of Bínová (2016):

– Efficient use of resources and a reduction in their consumption, a reduction in greenhouse gas emissions, utilization of alternative fuels, optimal utilization of the existing infrastructure, etc.
– A reduction in traffic congestion, support of intermodality, implementation of intelligent solutions, creation of conditions for decreasing the accident rate,
– The development of a new generation of means of transport and potential decreasing of the development time, efficient production processes and cost optimization,
– Improving the creation of the policies that are necessary for promoting innovations and resolving of problems relating to transport.

This also deals with ensuring safety:

– Increasing public safety and the fight against crime and terrorism—preventing consequences and mitigation of incurred damages, preparation of new technologies for ensuring the proper functioning of society and the economy—health, food, water, the environment and other technologies for ensuring the necessary services, such as communications, transport, supplying, etc.,
– Protection and improving the resilience of critical infrastructure—new technologies for protecting critical infrastructures, systems and services, including roads, transport, health, food, water and the supply chain; analysis and securing of public and private networks and services against any threat.

Crisis situation is process by which an organization deals with a major event that threatens to harm the organization, its stakeholders, or the management. Three elements are common to most definitions of crisis: a threat to the organization, the element of surprise, a short decision time.

There are many approaches and methods (forecasting) for the quantification the creation of supply and demand functions in transport as a basis for decision-making. This particularly includes operational analysis methods, econometric models and mathematic-statistical methods.

2.1 Scenario method

Decision-making about future statuses can also be done using scenario methods. The creation of scenarios allows for taking into consideration a number of factors affecting transport, and to set priorities that may arise and affect further development.

The scenario method consists of sorting expected events from different prognoses and their immediate impact on the investigated environment. The aim is to show the possible development as a result of consecutive logical contexts of specific events. This is therefore a matter of connecting information in a logical and time sequence. This method can predict the development of situations and conditions that are currently being investigated. The following must be carried out:

– Analysis of the current situation with a focus on certain facts that it was possible to gather in relation to the analysed state, and determination of a goal/goals,
– Identification of critical points, places and situations,
– Identification of the relationships between different situations,
– A description of the consequences that might occur after the analysed state,
– Analysis of potential or real consequences which may affect the chosen mode of transport or related modes of transport,
– Evaluation of all of the consequences that may arise during the implementation of the proposed measures, including areas that may not be associated with the selected mode of transport,
– Selection of scenarios with a high probability of success,
– Assessment of the negative impacts on selected scenarios and analysis of the consequences. A reassessment is carried out with regard to the initial stage (analysis and determination of goals),
– Implementation of the selected scenario.

In order to forecast the transport of goods, and thus also transport systems, it is necessary to also determine critical events during which it will be necessary to decide on other solution alternatives. The results of such decision-making are therefore described as alternatives until the final goal.

2.2 Analysis of the time series trend

Using extrapolation of data arranged in a time series, it is possible to determine the likely future course of the investigated phenomenon, i.e. even the likely development of transport systems and other related events, of Bínová & Březina (2014).

This method is mainly used during forecasting of prospective intensities of transport streams. The following formula can be used to calculate prospective intensities:

$$M^v + M^s \cdot K \tag{1}$$

where M^V = prospective intensity, M^S = actual intensity, K = prospective coefficient, growth coefficient.

A disadvantage is the fact that individual coefficients do not take into account local conditions, and it is therefore necessary to use gross estimates.

2.3 Average growth coefficient method

It can be used in preparing transport intensity forecasts between selected areas that have a different growth coefficient, wherein the resulting growth coefficient will have the arithmetical average value, Bínová & Březina (2014).

The following formula can be used to calculate prospective intensity:

$$M^V = M^s . - \frac{Ki.Kj}{2} \tag{2}$$

where M^V = prospective intensity, M^S = actual intensity, Ki, Kj = prospective coefficients in both selected areas.

A disadvantage is the fact that the resulting coefficient may not be directly proportional to the growth in the volume of transport, and the volume of transport may increase due to factors other than increased population.

2.4 Gravitational model

According to authors Shepperd et all. (2013) the gravitational model predicts the movement of people, information and goods between cities and even continents. These models therefore measure the intensity of the relationships between two objects (a small relationship between small objects and a large relationship between large objects).

In terms of resolving transport networks, gravitational models are used for addressing the impact of the technical—economic parameters of individual edges of a transport network (length, price of transport, mode of transport, etc.). The Pater Formula can be used to make the calculation:

$$I_{AB} = k \cdot \frac{M_A \cdot M_B}{d_{AB}^{\alpha}} \qquad (3)$$

where M_A, M_B = sizes of sources or destinations in locations A and B, e.g. number of residents, number of economically-active residents, or the number of residents multiplied by the average income. Another impact may be the morphology of the landscape, climate, social aspects, features of the population, culture, level of education, etc.

d_{AB} = distance between sources; metric, time, i.e. distance or time

I = intensity of the relationship between points (localities) A and B

k = coefficient (it is not always used, only during calibration, i.e. comparison of empiricism (experience) with the model)

α = usually the second power

It is necessary to emphasize that values M_A, M_B and d_{AB} are variable in time.

3 TECHNICAL AND SOCIAL IMPACTS

In order to predict transport of goods, it is necessary to select a suitable method (see chapter 2) and evaluate the technical and social impacts that can affect a change in the transport system, and thereby also the economic situation in the relevant region.

Due to the fact that transport of goods or their volume is affected by supply and demand, it is necessary to use parameters other than for personal transport for making a prognosis.

Primarily the following are important factors:

– Number of residents in selected areas,
– Amount of GDP/resident,
– Employment rate,
– Originality of transported product,
– Product price,
– Transport time,
– Transport price,
– Social aspects.

Other factors:

– Political impacts,
– Changes in the banking sector,
– Changes in laws etc.

4 EVALUATION OF IMPACTS

It is possible and suitable to use the SWOT analysis to evaluate impacts or prepare an analysis, Bínová & Březina (2014).

The aim of the SWOT analysis is to evaluate the relevant intent from all viewpoints. After evalua-

tion, it is possible to determine strategic goals and the overall strategy for optimizing of the project, a specific process, a business plan or even individual companies.

This is because this is a very universal, and one of the most frequently used analytical techniques focused on the evaluation of internal and external factors affecting the functionality and success of the evaluated transport system, i.e. even of the strategic intent and its use or modification, Graf & Teichmann & Dorda (2016).

It is necessary to evaluate the following when preparing the SWOT analysis:

– Internal factors—Strengths and Weaknesses. Evaluation of the internal situation of the project and its causes, e.g. available industrial and commercial land plots, transport infrastructure and international connection, environmental damage, protection of landscape and nature, solid waste disposal, civic amenities, education and culture, tourism, attractive points of regional and national significance.
– External factors—Opportunities and Threats). Evaluation of threats and opportunities to the external environment, e.g. changes in laws, economic changes on the national level with relation to international bonds (EU) in relation to demographic, social changes and employment, infrastructure on the national level in relation to international bonds (e.g. transport corridors of international importance, highways and railways with regional importance).

Before beginning work on the SWOT analysis, external factors must be identified in very great detail.

The strategy should lead to:

– Maximizing opportunities, i.e. maximizing strengths and minimizing weaknesses,
– Minimizing threats, i.e. maximizing strengths and minimizing weaknesses.

It is necessary to evaluate the following aspects when preparing the SWOT analysis:

– Characteristics of the territory, internal and external relationships—e.g.: geographic area, coordination of the concept of the development of residential localities, dealing with property rights issues, conditions for business development, particularly for the creation of small and medium-sized companies, association of municipalities in order to create a stronger unit in the preparation and implementation of investments, and the existence of development areas for civic amenities,
– Infrastructure—e.g.: use of railway, watercourse, expressway or highway, airport, existing or planned logistics centres, transport links with a

large agglomerations, signal coverage for mobile phones, the existence of business development zones, the level of suburban integrated transport, the possibility of strategic planning in the region, and the merging of funds from different resources to build an infrastructure, of Teichmann & Dorda & Bínová & Ludvík (2015),

- Economy—e.g.: proximity to large agglomerations, conditions for business development—especially for small and medium-sized enterprises, conditions for the development of agricultural production, employment level, increasing the number of small and medium-sized enterprises, growth of the services sector, increase in employment opportunities, activities of various associations with support for entrepreneurship, comprehensive concept of the offer of buildings for entrepreneurship, and creating conditions for placement of foreign capital, of Čorejová & Rostášová (2014).
- Job market—e.g.: employment level, education levels of the population in the relevant region, possible jobs for a qualified workforce, retraining opportunities, support for small and medium business activities and creation of new jobs, and spending on active employment policy,
- Social infrastructure—e.g.: the social structure level of the population, the level of the economic situation, or the quantity of funds spent on education, culture, social and health areas,
- The environment—e.g.: the existence of protected landscape areas, the development stage of the implementation of territorial environmental stability systems, energy concept, amount of funds for the implementation of the concept and strategy for the protection of the environment, the level of cooperation between state administration bodies and private entities in the environmental sector, respecting of the principles of permanently sustainable development, the possibility of a conflict of development interests and protection of the environment, uncontrolled expansion of built-up areas into the landscape, the share of automobile transport with a negative impact on air pollution, etc.,
- Tourism—e.g.: attractiveness of nature with an offer of areas suitable for recreation, accessibility for tourists, tourism development concept, cooperation between municipalities, state institutions and entrepreneurs in order to develop tourism, support of the development of related tourism infrastructure activities.

5 RISK MODELLING

When proposing changes to the transport system, it is necessary to identify possible risks, analyse them and subsequently prepare a risk analysis, with Bínová (2016).

A risk can be defined as the likelihood of the occurrence of a loss, variability of results and uncertainty of whether they can be achieved, deviation of actual and expected results, risk of negative deviations from objectives, risk of bad decisions, the possibility of creation of losses or profits, investment risk, indeterminate result (at least one is undesirable).

The three approaches to a risk are: aversion, risk tendency, neutral stance.

5.1 *Risks analysis*

The risks analysis contains identification of assets and determination of their value, identification of threats and determination of their seriousness.

Types of risks:

- Technical,
- Technological,
- Project,
- Investment,
- Business,
- Financial.

Risk management is based on the rules of corporate governance.

Examples of risk factors—increase in investment costs, acceptability of investment costs, failure to adhere to project parameters, lack of funds, deadlines in administrative and permitting procedures, higher operating costs, failure to adhere to a construction schedule, and public opposition.

Table 1. Methods for country risk assessment.

Risk assessment method	Input measures	Output measures
Panel of experts	Perception of country risk	Agreed risk index
Discrete scoring model	Interval index for each risk attribute	Average risk index
Analytic hierarchy process	Judgmental assessment for each risk attribute	Relative weights of risk attributes
Simulation survey	Intention of early/late entry for different risk scenarios	Probability estimates for entry decision
Fuzzy scoring model	Categorical assessment of each risk variable	Fuzzy envelope for country risk

Source: Authors, with facts of Sviderske (2014).

Table 2. Likelihood of occurrence of risk factors and intensity of negative impact of risk factor.

Likelihood of occurrence of risk factor (I)		Intensity of negative impact of risk factors (L)	
Level	Occurrence of risk factor	Level	Impact of occurrence
1	Improbable	1	Negligible
2	Unlikely	2	Small
3	Significantly likely	8	Large
4	Almost certain	16	Critical

Source: Authors.

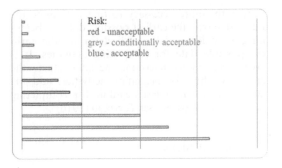

Figure 1. Model result of risk analysis.
Source: Authors.

Practical approaches to risk management are: not to risk more than is necessary, to think about probabilities, not to risk a lot for a little (a possible loss should not exceed a potential profit), of Fuchs & all (2015).

Advantages of Sviderske (2014):

– Panel of experts—combines experts' knowledge and practice, amenable to group decision process,
– Discrete scoring model—easy application of quantitative techniques, ease of comprehension, computation, and interpretation,
– Analytic hierarchy process—combines management judgment and intuition, amenable to group decision process,
– Simulation survey—flexible for scenario design, combines regression or discriminant analysis,
– Fuzzy scoring model—performs linguistic analysis, propagates complete information from stage to stage.

Disadvantages of Sviderske (2014):

– Panel of experts—time-consuming, nonobjective; difficulty identifying qualified experts,
– Discrete scoring model—arbitrariness in estimating weights of attributes for qualitative information,
– Analytic hierarchy process—possible inconsistency or bias in determining information categories,
– Simulation survey—time-consuming and costly for survey design, data collection, and analysis and evaluation,
– Fuzzy scoring model—user subjectively interprets fuzzy envelope, interpretation may vary among users.

If a transport alternative of connecting a significant transport node is evaluated, the issue is with what likelihood and with what impact risk factors may occur. A group of experts therefore makes assessments according to two perspectives.

The result of the risk analysis can subsequently be depicted graphically, wherein it is possible to define:

Risk = Likelihood × Impact (L × I).

An example is the evaluation of two alternatives of a road connection of a logistics centre to an external road network.

The L × I values for determining the acceptability of risks can be selected as follows:

– 1–8 risk is acceptable
– 16–24 risk is conditionally acceptable
– 32–64 risk is unacceptable

Risk factors can be defined as follows:

– Acceptability of investment costs,
– Exceeding of investment costs,
– Unforeseen investment costs,
– Change to project parameters,
– Statements and opinions of administrative bodies,
– Increased operating costs,
– Changes to construction time schedule,
– Failure to secure finance,
– Rejection by the public.

5.2 Critical path

The critical path method is used when proposing a change to a transport system so that it is possible to determine points which are critical in terms of preparation and implementation. These are critical points that may have a negative impact on the environment, conditions for the lives of residents, on business, etc.,—these are impacts that have economic and social consequences, with facts of Rosenau (2003).

This is the longest path from the beginning to the end, and it has no time tolerance, i.e. a path corresponding to the shortest period in which a change to the transport system can be completed.

It is therefore necessary to first determine the Earliest start—the earliest possible date for beginning activities, and the earliest finish—the earliest possible date for completing activities. It is also necessary to determine the latest start—the latest deadline for beginning activities, and the latest finish—the latest deadline for completing activities, of Plos, Soušek & Szabo (2016).

6 CONCLUSION

Changes to transportation and logistics freight transport systems, particularly in macro logistics, are influenced by supply and demand, and these factors vary depending on economic and social changes. However, changes to transport and logistics systems retroactively affect the economy and social aspects in society. Given that freight transport is operated at a scale that transcends the national borders of individual states, the specified effects and impacts have an international significance.

The text describes impacts that may cause changes in transport and logistics systems, as well as an evaluation and analysis thereof. The use of a risks analysis is part of this process, and the result is an evaluation of individual alternatives in terms of potential losses.

However, it is necessary to state that what is very risky is a fact whose occurrence may be unlikely, but which would have a critical impact. Even if the likelihood of occurrence is very low, it will still be necessary to evaluate such a case.

Examples of possible changes that affect the entire logistics system:

– Impact on commodity flows—social changes, changes in the way of life of the population and their consumption requirements, changes in conditions in the banking sector, policy changes, etc.,
– Impact on the technical—economic parameters of the logistics system—existing and projected external transport networks (roads, railways, waterways, airports), changes in the projected external transport networks (changes to costs, changes to the date of implementation), etc.

It can therefore be concluded that in view of their overlapping of the borders of regions and states, or with a global impact, transport systems have a significant impact on the economy and social aspects in society, and this is an important part of the functioning of society.

ACKNOWLEDGMENT

The authors acknowledge support from the project: "Increasing the Efficiency of Passenger Terminal Operation and Customer Satisfaction", SGS16/189/OHK2/2T/16.

REFERENCES

Bínová, H. 2016. *Model výběru dopravních módů v logistickém systému*, habilitation thesis, Žilina, Slovakia.
Bínová, H., Březina, E. 2014. *17DLOG—Dopravní logistika"*, Praha, Česká technika-nakladatelství ČVUT.
Čorejová, T. et al. 2015. Comparison Of Innovation Indexes Within Selected Eu Countries. *Proceedings: IX. International conference on applied business research (ICABR 2015)*, 14–18 September, Madrid, Spain.
Čorejová, T. & Rostášová, M. 2014. Regional Development, Innovation and Creativity, *Proceedings from IX. International conference on applied business research (ICABR 2014)*, 3–4 October, Talca, Chile.
EU. 2007. *Akční plán pro logistiku nákladní dopravy.*
Fuchs, P. et al. 2015. *The Assessment of Critical Infrastructure in the Czech Republic*, Transport Means, Kaunas: Kaunas University of Technology.
Graf, V. et al. 2016. Mathematical model of charter flight scheduling with predefined rime slots, *Proceedings: APLIMAT 2016–15th Conference on Applied Mathematics, 2–4 February 2016*, Bratislava, Slovakia.
Plos, V. et al. 2016. Risk-based indicators implementation and usage, *Proceedings: World Multi-Conference on Systemics, Cybernetics and Informatics*, Orlando, Florida, USA.
Rosenau, M.D. 2003. *Řízení projektů*, Brno, Computer Press.
Shepperd, B. 2013. *The Gravity Model of International Trade*. A User Quide, United Nations ESCAP.
Sviderske, T. 2014. *Country risk assessment in economic security and sustainability context*, doctoral dissertation. Vilnius Gediminas Technical University, Lithuania.
Teichmann, D. et al. 2015. Optimization of parking public transport vehicles, *PROMET—Traffic & Transportation*, 27(1): 69–75.

Financial reporting as a tool for enhancing the efficiency of SMEs

S. Blahútová
The School of Management, Trenčín, Slovak Republic

M. Sedliačiková
Faculty of Forestry, Technical University, Zvolen, Slovak Republic

ABSTRACT: Small and medium-sized enterprises are operating in an ever more complicated environment accompanied by turbulent changes, and they have to react promptly and flexibly to such changes. The aims of the present paper were to deal with a possibility of financial reporting employment as a tool for transferring the internal information within the enterprise, to study the effect of financial reporting on increasing enterprise efficiency and to present scientific and practical methods of measuring efficiency of SMEs' enterprises. The given issues were mapped by means of primary and secondary sources analysis. Theoretical basis was verified by means of empirical survey in a form of questionnaire. The aim of the questionnaire was to study financial reporting implementation in SMEs' management. The results of the study were evaluated by means of descriptive statistics in a form of graphical presentations. In conclusion, recommendations for successful financial reporting employment in practice were presented.

1 CURRENT STATUS OF THE ISSUE

1.1 Company's performance

SMEs currently operate in an environment characterized by a higher competitive pressure and by a number of changes which they need to respond to quickly and in time. In such environment they are not able to maintain their efficiency for a long time and prolong their life without a high-quality financial planning and management. According to Lesáková (*2004*), company's performance is such an ability which helps achieve the desired effects or outputs in measurable units. With regard to the above-stated, it is necessary to know:

– What are measurable outputs?
– How to measure them?
– How to transform data into information necessary for decision-making?

The aim of the present paper was to deal with the possibility to apply financial reporting as a tool for transmitting internal information in a company, reviewing the effect of financial reporting on enhancing the efficiency of companies and presenting scientific and practical methods of measuring the efficiency of SMEs.

The first part of the presented paper deals with scientific approaches to measuring company's performance. The second part of the paper deals with the issue of financial reporting as a means of transmitting information necessary for financial planning and management of a company.

The final part presents outcomes of a survey focused on implementation of financial reporting in SMEs as well as recommendations for its successful implementation in practice.

1.2 Scientific approaches to measuring company's performance

Original management systems focused on enhancing company's performance, such as *lean management, kaizen* or *flat organization,* did not meet basic conditions for corporate management since they were not targeted and could not comprehensively connect factors determining generation of value with the top goal of the company (Šatánová & Potkány 2004). This is the reason why systems based on exact financial indicators were getting to the foreground. In such systems, there were also revealed some shortcomings, and this is a proof that non-financial indicators which are not based on accounting standards need to be included appropriately in the measurement system. The process of heading towards value-based management—the basis of which consisted in generating value for company owners—was triggered by the so-called *value based management* which lays an emphasis on company's economic added value (Marinič 2008).

From the scientific, academic and business point of view, company's performance measurement indicators are divided into *common* or *traditional indicators* and *modern indicators* (Volčko 2011).

Traditional company assessment methods may be divided into

– Profit indicators;
– Profitability indicators;
– Cash flow indicators.

Profit indicators are the most frequently used methods of measuring company's performance. According to Sedláček (2007), there are various profit categories, such as:

Earnings After Taxes – EAT, i.e. after-taxation profit which is intended for allocation of profit among company owners or to replenish any retained earnings of the company (Volčko, 2011).

Earnings Before Taxes – EBT which is used to compare performance between individual periods or companies from countries with different tax rates (*Volčko 2011*).

Earnings Before Interests and Taxes—EBIT is focused on measuring economic performance of a company with no interests and taxes (*Sedliačiková, 2005*).

Earnings Before Interests, Taxes, Depreciation and Amortization – EBITA compares the economic efficiency of a company adjusted for interests, taxes and depreciation and amortization.

Profitability Ratios represent an outcome of business efforts. They are a means of assessing the overall company's efficiency. According to Vlachynský (2009), the best known profitability indicators include:

Return of Equity – ROE. It expresses the return of company's own resources using the relation (Hajdúchová 2011):

$$ROE = net\ profit/equity \qquad (1)$$

Return on Assets – ROA. The indicator is a ratio of net profit and total assets invested in business according to the following relation (Lesáková 2011):

$$ROA = net\ profit/total\ assets \qquad (2)$$

When comparing ROE and ROA we may find out if a company is able to use leverage for its own benefit.

– If ROE < ROA, it means that the leverage is too expensive, the invested funds will not return and the equity is forced to pay for it. The price of the leverage is higher than ROA.
– If ROE > ROA, the invested leverage returned and helped increase the value of the equity.
– The ROE = ROA relation means that the financial structure has no effect on earnings from equity.

Return on Investment – ROI expresses the intensity of reproduction of capital invested in a company. It is expressed by the following relation (Pavelková & Knapová 2010):

$$ROI = (net\ profit + interest)/total\ capital \qquad (3)$$

Return on Sales—ROS represents the ratio of an after-taxation economic result to company's revenues and it is given by the following relation (Kislingerová & Hlinica 2000):

$$ROS = net\ profit/revenues \qquad (4)$$

Cash flow indicators represent information about company's cash flow, i.e. its incomes and expenditures. *Cash flow* is an important element of company's financial management and financial analysis. The basis of cash flow monitoring is to watch any changes in the status of funds. According to Pavelková and Knapková (2010), cash flow reporting is divided into:

– The area of operational and economic activities;
– Investment area;
– Financial area.

According to Vavrová and Bikar (2016), traditional company assessment methods are supported only by accounting and financial indicators; they do not reflect the real need of the market and do not consider all items necessary for performing such assessment, e.g. the risk of inflation, time value of money, opportunity costs, etc. The main downside of the traditional approach is the fact that financial indicators show if a company performed well or not, i.e. they only assess its performance (Lueg & Schäffer, 2010).

Modern company performance assessment methods were created based on criticism of the traditional approach. According to Mařík & Maříková (2005) they should meet the following criteria:

– Use data available from accounting;
– Consider the risk and the amount of committed capital in their calculations;
– Enable identification of a connection to individual management levels;
– Enable assessment of company's performance and its evaluation.

Basic modern indicators of measuring company's performance include: Economic Value Added (EVA), Market Value Added (MVA), Cash Value Added (CVA), Cash Flow, Economic Value Added (CEVA), Shareholder Value Added (SVA), Return On Net Assets (RONA).

Most attention in the business world is paid to EVA. The *economic value added* is an indicator which arose from the need to find an economic

Figure 1. EVA variables scheme.

indicator which, according to Ismail (2011), would represent the closest connection to the value of stocks, enable to make the most of accounting data, include risk calculation and enable assessment of company's performance and evaluation. According to Gitman (2006), the EVA system is a rate used to determine if an investment contributes to generating wealth for owners. According to him, EVA is calculated by deducting costs of funds expended on financing the investment from its after-taxation operating profit.

The Figure 1 shows the understanding of variables which are used to calculate EVA.

While NOPAT stands for *Net Operating Profit After Taxes*. Net operating profit is its equivalent in traditional accounting. WACC means *Weighted Average Cost of Capital*. Simply said, NOPAT equals net income minus operating costs and expenditures including depreciation after deducting certain specific adjustments for every particular company and taxes.

1.3 *Financial reporting—intermediary of information*

Financial reports or financial statements providing an overview of financial indicators of a company represent an elementary tool of corporate management. According to Bendert (2013), it is necessary to concentrate financial reports on key factors which have an impact on company's performance and value. What is of importance for financial reports, is quality and relevance rather than extent of reported data. Otherwise they will not perform their function and they will not help the management identify problems and make decisions.

High-quality information related to finances requires a clearly determined framework and a common set of principles for financial purposes with the aim to ensure a common basis when presenting, measuring, assessing and monitoring financial results. The most important principles include:

– Adhering to the accounting framework;
– Preparing financial statements based on accounting accruals;
– Preparing statements with an assumption of a prosperous company;
– Providing understandable information;
– Relevance of information for users' decision-making needs;

– Importance of reported information;
– Reliability;
– Faithful representation;
– Essence over the form;
– Impartiality;
– Carefulness;
– Completeness;
– Comparability.

According to Fíberová (2001), reporting used to represent one of sub-systems of a company's information system coordinated by controlling. With regard to its importance, it is a separate corporate management discipline. It is one of modern corporate management techniques and with its essence it helps create a company's strategy and operational planning.

According to Harumová (2016), the main goal of financial reporting is to create a comprehensive system of information and indicators which characterize the company's economic activities in an understandable and user-friendly form. Information provided through financial reports is meaningful only if it has an impact on economic decisions of users by helping them assess the past, current and future events or by confirming or correcting their past assessments.

According to (Šlosárová 2015, Saxunová 2008, Harumová 2008), basic roles of information which is the subject of financial reporting may be summarized as follows:

– Predictive and confirming roles of information are related. E.g. information about the current level and structure of held assets is valuable for the user if it is trying to predict the company's ability to utilize the advantages of opportunities and its ability to respond to unfavourable situations. The same information plays a confirming role as far as forecasts are concerned. Information about a financial standing and past results are often used as a basis for forecasting a future financial standing, results or other issues such as the ability of the company to pay its liabilities when due.
– In order to achieve a predictive value, information needs to have a form of an explicit forecast. The ability to make forecasts from financial statements depends on how information about past transactions and events is presented. e.g. a predicative value of a profit-and-loss statement is higher if unusual, uncommon and rare items of a revenue or cost are presented separately.
– Relevance of information is influenced by its nature and significance. In some cases the nature of the information itself is sufficient to determine significance. In other cases both nature and significance are important. Information is significant if its omission or wrong presentation

could have an effect on economic decisions of users made based on financial statements.

– If any information is to be useful, it must also be reliable. Information is reliable only if it does not contain any major mistakes and it can be relied on. For any information to be reliable, it must faithfully display transactions and other events as of the reporting date. Most financial information is subject to a risk that its presentation is less faithful. It is not because of distortion but rather due to problems in identifying the transaction and other events which are measured or in designing and using measurement and presentation techniques which may intermediate messages corresponding to such transactions and events.

– If information is to represent transactions and other events faithfully, it must be accounted and presented in accordance with the economic reality, not only its legal form. The essence of transactions or of other events is not always consistent with what results from its legal or executed form.

– It is essential that information is unbiased and not distorted. Financial statements are unbiased if they—with their selection or presentation of information—do not have any influence on decision-making or assessment with the aim to achieve a pre-determined outcome.

– When preparing financial statements we must struggle with insecurities accompanying a number of events, such as recoverability of claims, probable life of buildings, equipment and the number of complaints. Such insecurities are accepted by disclosing their nature and extent when preparing financial statements.

– For information in financial statements to be reliable, it must be complete mainly in terms of significance and costs. Any omission may cause that information will be false or misleading and thus unreliable as far as its meaning is concerned.

– Financial statements should be comparable, i.e. users should be able to compare financial statements of a company in the course of time in order to identify trends and its financial standing and results. Users must also be able to compare financial statements of different companies with the aim to assess their relative financial standing, their results and any change in their financial standing.

– Measurement and presentation of an impact of similar transactions and other events must be made by a certain entity in a uniform manner and in the course of time for the particular entity, and in a uniform manner for various entities. Since users want to compare financial standings, results and changes in financial standings in the course of time, it is important that financial statements show corresponding information for previous periods.

– Efficiency of financial reports is related to the time aspect of reporting. According to Harumová (2016), a standard reporting cycle is primarily built on three elements:

– monthly current financial reporting, based on a monthly cycle of current reported data connected with filing more extensive quarterly and annual reports focused on comments on financial performance of the company;
– regular forecasts, while forecasts of revenues should be reported on a monthly basis while forecasts of financial results and other financial indicators are submitted on a quarterly basis;
– long-term financial plans for the upcoming period as well as preliminary financial plans for longer periods are submitted on an annual basis.

Financial reports are regularly a subject management meetings and top management meetings.

The form, type and schedule of financial reports are commonly laid down in internal regulations of companies. Most frequent forms of financial reports include:

– profit-and-loss statement such as analysis of revenues and costs of a company as per individual types, function and nature of the reported items. A profit-and-loss statement also contains calculation of a production or trading margin, calculation of EBIT, EBT or other indicators;
– balance sheet with a clear division into non-current assets, current assets, the so-called operating assets;
– cash flow statement;
– receivables and payables report;
– stocks report;
– revenues reports;
– other reports based on the status and the nature of the accounting unit;
– regular forecasts;
– regular financial planning.

2 METHODOLOGY

With the aim to meet the determined goal, the starting point of our survey included both primary and secondary sources dealing with the issue in question. The basis of processing secondary sources and making recommendations, conclusions and measures comprised methods of analysis, description, comparison, analogy, summarization, synthesis of knowledge, information and a systemic approach. The secondary source of the presented paper included both national and foreign magazines and publications. The survey was based on analysing primary sources through an interrogation method in the form of a questionnaire sent to SMEs. Judgements about basic groups were assessed using pivot tables and charts.

3 SURVEY OUTCOMES AND DISCUSSION

Primary survey was the source of data for the purposes of the presented paper. The object of the survey was to monitor implementation of financial reporting in SMEs. For the needs of the survey we focused on the SME category.

The basic source of primary information was a questionnaire as a means of an interrogative survey method. The questionnaire was sent in an electronic form to managers and company owners. It contained three questions focused on monitoring of the current status of the examined issue:

– Do you use financial reporting in your company as a tool for transferring information necessary for decision-making?
– Do you use financial analysis tools in your company?
– Do you use financial planning and forecasting tools in your company?

The survey outcomes were evaluated in a descriptive and graphic manner and summarized using pivot tables and charts focused on transformation of the discovered facts. The selected group was formed by 200 SMEs chosen based on available information. The return rate of the questionnaire was 16%, i.e. we received 32 filled-in questionnaires. Based on the above-stated we did not generalize the acquired information for the selected group but we described the reality which resulted from the questionnaire form of the survey. In the following part of the paper we will graphically present the survey outcomes.

Question 1 *Do you use financial reporting in your company as a tool for transferring information necessary for decision-making?*

Answers to the question if companies use financial reporting as a tool for transferring information necessary for decision-making are presented in the Figure 2.

The survey has shown that 50% companies use financial reporting as a corporate management tool, 34% companies stated that they did not use financial reporting as a decision-making tool and as many as 16% companies stated that they did not know such support tool.

The following chart presents the use of financial analysis tools.

Question 2 *Do you use financial analysis tools in your company?*

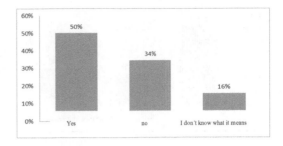

Figure 2. Use of financial reporting.

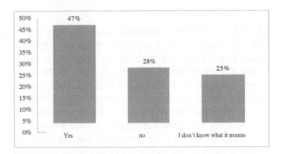

Figure 3. Use of financial analysis tools.

47% SMEs use financial analysis tools. As many as 28% companies do not use financial analysis tools and a comparable percentage, i.e. 25% companies do not have any knowledge about financial analysis tools. It has resulted from the survey that more than a half of respondents from the survey sample do not use or have no knowledge about financial analysis tools.

Question 3 je is focused on the use of financial planning and forecasting.

Questions 3 *Do you use financial planning and forecasting methods in your company?*

Survey outcomes are presented in the Figure 4.

Financial planning and forecasting is a basis of successful control and decision-making in management of SMEs. With our question we wanted to find out if companies used financial planning and decision-making tools. The Figure 4 presents outcomes comparable to previous cases. Almost a half of respondents from the survey sample do not use or have no knowledge about financial planning and forecasting tools. 53% respondents realize the need for forecasting and planning financial indicators in the company.

Theoretical knowledge of this issue confirms a clear need to implement financial reporting as a management tool for SMEs. According to Harumová (2016), financial reporting with its information value has an impact on competitiveness and performance of companies. According to Fíberová

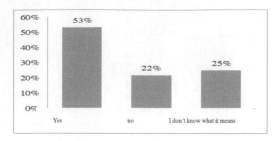

Figure 4. Use of financial planning and forecasting tools.

(2001), information intermediated through financial reporting forms a basis for efficient corporate management and for making reasonable decisions about economic and financial activities of the company. Lueg and Shäffer (2010) claim that assessment of performance of a company only based on accounting indicators does not offer true security in introducing key indicators necessary for assessment and management of the company. Such allegations help create new competitive management strategies based on generation of value which would help in corporate management. The main intermediary of information necessary for decision-making is exactly financial reporting. Through correctly set financial reporting management is provided with information at the right time and in the correct form, what enables companies to quickly and timely respond to all changes happening inside them and in their global external environment.

The survey outcomes show that in spite of benefits, positive effects and advantages of financial reporting for management of SMEs, it is not their common management tool. Despite all its benefits it is used by only a half of respondents from the survey sample. We suppose that companies realize how demanding the global competitive environment is but they do not realize the possibilities of acquiring competitive advantages resulting from correct financial analyses and financial planning.

4 CONCLUSION

With regard to the above-stated we may come to the conclusion that in spite of an increasingly demanding business environment, only a small number of companies use financial management tools to create business strategies which will enable them to survive and to successfully compete in the market. Financial reporting is a very efficient tool of financial management; it transmits information necessary for corporate management in time and thus helps company's managers identify problems and make decisions in real time.

ACKNOWLEDGEMENT

The present publication is a partial outcome of dealing with grant tasks: VEGA MŠ, and the Slovak Academy of Sciences, Grant No. 1/0010/17, No.1/0934/16, Grant APVV-14-0506 and APVV-16-05-90.

REFERENCES

Fibírová, J. 2001. *Reporting moderní metoda hodnocení výkonnosti uvnitř firmy.* Grada Publishing, Havlíčkův Brod 2003. ISBN 80-247-0482-X.
Hajdúchová, I. 2000. *Finančná analýza podniku.* Zvolen: Vydavateľstvo TU vo Zvolene, 2000. 54 s. ISBN 80-228-0961-6.
Harumová, A. a kol. 2016. *Účtovný a daňový reporting.* Bratislava: Ekonóm. 258 s., ISBN: 978-80-225-4195-4.
Ismail, I. 2011. The ability of EVA attributes in predicting company performance. In *African Journal of Business Management,* 2011, No. 5 (12). p. 4993–5000. ISSN 1993-8233.
Lesáková, Ľ. et al. 2011. *Finančno-ekonomická analýza podniku.*Banská Bystrica: Univerzita Mateja Bela, Ekonomická fakulta, 2007. 208 s. ISBN 978-80-8083-379-4.
Maříková, P., Mařík, M. 2005. Moderní metody hodnocení výkonnosti a oceňování podniku. Praha: Ekopress, 2005. 164 s. ISBN 80-86119-61-0.
Marinič, P. 2008. *Finanční analýza a finanční plánovaní ve firemní praxi.* Praha: Grada Publishing a.s., 2008. 240 s. ISBN 978-80-247-2432-4
Saxunová, D.2008. *Ako správne rozumieť informáciám z účtovnej závierky: rozdiely a podobnosti v USA a v SR* – 1. vyd. – Bratislava: Iura Edition, 2008 – 210 s. (Ekonómia) ISBN 978-80-8078-189-7, str. 14 – 19 52.
Sedláček, J. 2007.*Finanční analýza podniku.* Brno: Computer Press, 2007. 149 s. ISBN 978-80-251-1830-6.
Sedliačiková, M. & Drábek, J. 2005. The proposal of methodics of financial control. In *Intercathedra.* 2005, No 21, s. 122–125. ISSN 1640-3622.
Šlosárová, A. 2015. *Analýza účtovnej závierky. Aktuálne problémy v oblasti účtovnej závierky podnikateľov.* Bratislava: Vydavateľstvo EKONÓM, 2015. 94 s. ISBN 978-80-225-4068-1.
Vavrova, K. & Bikar, M. 2016. Effective tax administration as a factor affecting the competitiveness of Slovakia at the global level, Edited by: Kliestik, T.
Conference: 16th International Scientific Conference on Globalization and its Socio-Economic Consequences Location: Rajecke Teplice, SLOVAKIA Date: OCT 05-06, 2016.
Vlachynský, K. 2009. *Podnikové financie.* Bratislava: IURA EDITION, spol. s. r. o., 2009. 547 s. ISBN 978-80-8078-258-0.
Volčko I. 2011. Identifikácia faktorov rastu hodnoty vybraného podniku na základe pyramídovej analýzy EVA. *In Manažment podnikovo.* Roč. 1, č-2 (2011). ISSN 1338-4104
Volčko, I. 2011. Využitie pyramídového rozkladu EVA pre riadenie finančnej výkonnosti podniku. In *Ekonomika a manažment podnikov.* Zvolen: Technická univerzita vo Zvolene, 2011.

New Trends in Process Control and Production Management – Štofová & Szaryszová (Eds)
© 2018 Taylor & Francis Group, London, ISBN 978-1-138-05885-9

Management of production in the manufacturing enterprise

M. Bosák, A. Tarča & A. Belasová
Faculty of Business Economy with seat in Košice, University of Economics in Bratislava, Košice, Slovak Republic

V. Rudy
Faculty of Mechanical Engineering, Technical University of Košice, Košice, Slovak Republic

ABSTRACT: To improve the production process, each manufacturing enterprise should perform process mapping at least once per half-year. Mapping can reveal problems that cause process slowdown or downtime on lines. The Lean Production method used by manufacturing companies is Value Stream Mapping. Its use in Car-parts Slovakia Company has increased the effectiveness of the Painting line.

1 INTRODUCTION

The success of motor car company depends primarily on the efficient process of production, the quality of the offered products and the satisfaction of the customers. To achieve an efficient process of production, it needs to be constantly improved. Therefore, every manufacturing company should map at least once in a half year the current state when it identifies which processes work efficiently and which needs to be changed. Mapping can reveal for instance surplus stocks that are tied to funds that the business could use more efficiently. In addition, it is possible to detect problems that cause a slowdown in the process of production, or breakdowns on the lines. These negativities cause the extension of delivery of the product to the customer, which adversely affects the satisfaction of the customer.

The automobile industry is currently the largest employer in the Slovak Republic by comparison with other industries. In the last year 2015, the world leadership in the number of cars produced per thousand inhabitants was maintained, while the magical limit of millions of cars produced was exceeded. In the coming years, it is assumed that the number of cars produced would increase due to the entry of the Jaguar Land Rover into the Slovak market (Majerník et al. 2015).

2 USED METHOD—LEAN PRODUCTION

Toyota company, with its lean production, brought a new impulse to the automotive industry. The method introduced Taichi Ohno and Shigeo Shing in the middle of the last century as a set of different tools and methods, whereby their purpose from long-term side was not only the stability and productivity gains but also the efficiency of production. Toyota tried to prevent all kinds of wasting. Its effort was aimed at preventing waste of time to repair the resulting scrap, waste of resources and so on. (Bosák & Rudy 2016; Pružinský & Mihalčová 2016). Wastage, in Japanese MUDA, is an activity that does not add value. According to Lean Manufacturing Tools, we distinguish the following 7 mistakes:

– transportation
– surplus stocks,
– unnecessary movement,
– downtime,
– repair, faulty pieces
– overproduction,
– badly processed.

Largely used method by The Lean Production in manufacturing companies is the Value Stream Mapping (VSM), which is the fastest way to uncover the deficiencies that exist in the enterprise. Below reserves and deficiencies we can also understand the different kinds of wasting. (Erlach 2012)

Value Stream Mapping can be translated as a mapping flow of values, or we can talk about value chain analysis. The title suggests that this analytical technique serves to map the entire value flow, whether in production or administrative processes. The aim of this method is to quickly and easily describe both material and information flows in the enterprise, describing not only those processes that add value to a product or service, but also processes that do not add value to it. Added value is the process the customer requires and for which he pays. Below the value that is not added, it is necessary to understand the process that must be done to produce a quality product, but the customer does not care, so it does not pay for it. (Kyseľ 2011, Rovňák et al. 2013)

The mapping process includes the following steps:

a. selection of the mapping object, representative of a particular component or product,
b. drawing a current state map,
c. analysis based on the map of the current state and the identification of the insufficiencies of the map of the future state,
d. establishing a plan of proposals needed to achieve the future state. (King & King 2015)

3 MAPPING PROCESS

Before starting the mapping, a working group will be created to perform the mapping, with the purchasing, logistics, production and finance department being represented in the working group. Selecting the mapping object follows the selection of the mapping flow direction, Table 1. There are two lines of flow in the direction and counter-flow. While looking at the flow direction, it is possible to understand the process of turning the raw material into a product, and technical problems will emerge in that direction. When watching the counter-flow, everything is covered by the customer. At the beginning, the customer requests are made to see if the selected vendor can respond to these requirements, and there is a view of creating unnecessary stocks that slow down the process. (Nash & Poling 2011)

The map of the future state is a condition that the enterprise would want to achieve in the future. The goal is to streamline the process, to eliminate downtime, problems and "muddy" that disrupt the entire production process. The map is drawn based on outputs from the current state map and on the basis of found wasting.

4 MANUFACTURING ENTERPRISE

Car-parts Slovakia Company is one of the leading suppliers of the automotive industry in Slovakia and is an independent industrial concern, which is currently expanding its production almost on all continents. Its activity is mainly equipment of personal cars as well as trucks from design to the actual sale of individual components.

At Car-parts Slovakia Company several components are produced:

– keys (32%),
– locks (27%),
– handles (16%),
– electronic access (14%),
– electric steering lock (7%)
– powerful closing systems (4%).

For the production management analysis, the product of handle for the Renault Citroën company was selected as the third most important and most-produced product in the enterprise. The mapping process was performed upstream, that is, from customer to supplier. This direction is more complicated, but it is ideal for mapping because of the rapid detection of errors.

Key products, handles and locks make the car an important locking security system, which can be seen in the following Figure 1.

5 PRODUCTION LINE—PAINTING LINE

The manufacturing process begins by moving the components from the warehouse by operator before painting to the Painting line. The process in the line Painting is separate and operator intervention is only present if the line is not working. In the line, the components are dyed and painted in

Table 1. Flow direction for VSM.

In the direction of flow	In the opposite of flow
Physical flow of components.	Order flow—customer's view in relation to the supplier.
Product requirements.	Customer requirements.
Flow of raw material—it is necessary to contact the expert.	Product flow—it is possible to assess whether the individual stages of the process were necessary and useful.
Supplier restraint is also shifted to the customer.	Customer needs are monitored to see if the vendor meets customer needs.
Do not see the growing stock between the stages of the process. See a flow failure.	Stocks to see clearly.
Simpler flow tracking.	More streamlined flow tracking.

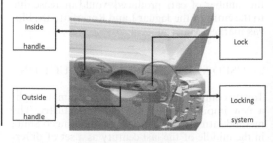

Figure 1. Locking security system.

colour that the customer asks for. Subsequently, the painted components are stored in the warehouse after painting next to the line and waiting for inspection.

When checked, faulty components that cannot be longer reprocessed become waste and those that can still be repaired are reprocessed (for example, the colour to pumice and components are repainted). The Table 2 below shows the individual line data.

When mapping the current state of the company, several drawbacks were found on the Painting line. Subsequently, a draft plan was drawn up, which identified solutions identified as deficiencies in the process, Table 3.

After implementing solution from the design proposals, the most common problems encountered in the process should be removed. These are mainly problems with excessive handling, time wasting, unnecessary movement, and inadequate output.

After implementing the suggestions on the Painting line, the condition on the line should be

improved. The Table 4 lists changes in results after the introduction of proposals, Figure 2.

At the time of mapping, it was found that the Painting line during a three-shift operation can

Table 4. Proposed state of the line after the introduction of proposals.

Painting line	Original state	Changed status
Overall efficiency of machines and workers	65%	68%
Non-productivity in production	9%	9%
Time needed to produce 1 piece of product	10.01 s/pc	9.57 s/pc
Down time the line	21%	18%

Table 2. Data on the link painting.

Painting line	Value
Overall efficiency of machines and workers	65%
Non-productivity in production	9%
Time needed to produce 1 piece of product	10.01 s/pc
Cancelling the line	21%

Table 3. Shortcomings and suggestions on the painting line.

No.	Shortcomings	Suggestions
1.	Insufficient output on the Painting line	Add another checkbox
2.	Insufficient output at the loading station	Adding a helping hand
3.	Large distance and opacity at the warehouse	Moving the storage of components closer to the warehouse after painting
4.	Excessive handling	Reorganization of stan-ding and sliding trucks
5.	Time span at the loading station	Add 2 sliding shelves to the loading station
6.	Time span due to material search for production	Adding 2 positions
7.	Unnecessary movements on unmatched components	Moving storage of unmatched components
8.	Problem with production planning and production itself	Booting the PC on the line

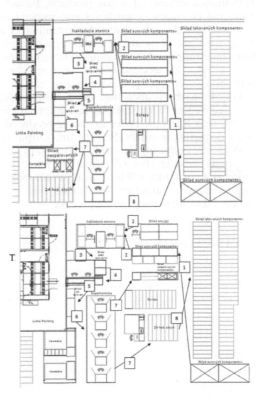

Figure 2. Painting line before and after the introduction of the proposals.
The legend:
1. Move to components store
2. Move to the loading station
3. Move to storage before painting
4. Move to the Painting line
5. Move to storage after painting
6. Move the check
7. Move to unpaired components store
8. Transfer to painted components

Table 5. Summary of the results before and after the introduction of the proposals.

Data		Painting line
Original state	Production in pieces/3 change	7822
	Scrap loss	704
	Output in units/3 changes	7118
New state	Production in pieces/3 change	8184
	Scrap loss	736
	Output in units/changes	7448
Increase status	Production	362
	Output	330

produce 7822 products a day, with an output of 7118 units after removal of non-products, Table 5. The total efficiency of machines and workers is now at 65% of the painting line, the remaining 35% being time out of production, the highest share being line stopping, 21% downtime. The remaining 14% includes, for example, product control, data entry into a computer, product shipment, etc.

Following the introduction of the proposals, the effectiveness of the Painting line will increase by 3%, with 8.184 pieces per day being produced. After deducting non-subsidies, the line output is 7448 pieces. Following the implementation of the proposals, the enterprise will re-map the manufacturing process, which will introduce further measures to increase output on the Painting line and ensure a smooth manufacturing process.

6 CONCLUSION

Every enterprise should take care of the satisfaction of their customers by producing quality products, thus ensuring a competitive advantage.

Making the required quality product can only be an efficient production process that needs to be constantly improved. A very often used method for improving the efficiency of a production process is the value flow analysis. The essence of the method is to understand the current state of the production process, identify the causes of slowing or non-functioning of the process, and suggest improvements.

REFERENCES

Bosák, M. & Rudy, V. 2016. Manažérstvo výroby. Ekonóm Bratislava, ISBN 978-80-225-4369-9.

Erlach, K. 2012. Value Stream Design: The Way Towards a Lean Factory. Stuttgart: Springer Science & Business Media, ISBN 978-36-421-2569-0.

King, P. & King, J. 2015. Value Stream Mapping for the Process Industries. New York, CRC Press, ISBN 978-14-822-4769-5.

Kyseľ, M. 2011. Mapovanie toku hodnôt vo výrobe. Žilina, IPA Slovakia, ISBN 978-80-89667-08-6.

Majernik, M et al. 2015. Integrated management of environmental-safety and technical risks of plants producing automobiles and automobile components. Published in: Communications - Scientific Letters of the University of Zilina info. Volume 17, Issue 1, p. 28–33.

Nash, M. & Poling, S. 2011. Mapping the Total Value Stream. New York, CRC Press, ISBN 978-14-200-9532-6.

Pružinský, M. & Mihalčová, B. 2016. Material flow in logistics management, Production Management and Engineering Sciences - Scientific Publication of the International Conference on Engineering Science and Production Management, ESPM 2015, pp. 523–528, ISBN 978-1-138-02856-2.

Rovňák, M., Chovancová, J., Bednárová, L., Adamišin, P. & Huttmanová, E. 2013. Managing environmental risks in production companies, International Multidisciplinary Scientific GeoConference Surveying Geology and Mining Ecology Management, SGEM-2013, pp. 651–658, ISSN 1314-2704.

New Trends in Process Control and Production Management – Štofová & Szaryszová (Eds)
© *2018 Taylor & Francis Group, London, ISBN 978-1-138-05885-9*

A new requirement of ISO 9001:2015 standard

K. Čekanová & M. Rusko
Faculty of Materials Science and Technology in Trnava, Slovak University of Technology in Bratislava, Trnava, Slovak Republic

ABSTRACT: ISO 9001 is the world's most commonly used standard for quality management systems. This standard is an agreement or best practice that an organization can apply voluntarily. A standard reflects a good level of professionalism. New ISO 9001:2015 standard was published on 23 September 2015. Organizations must implement the new ISO 9001:2015 standard before 23 September 2018. For this purpose, the aim of article is define of important differences between ISO 9001:2008 and ISO 9001:2015.

1   INTRODUCTION

The ISO 9000 standards originated in 1987 with a bulletin from the International Organization for standardization (ISO) (Ferguson 1996). The original ISO 9000 series consisted of five standards: ISO 9000, 9001, 9002, 9003 and 9004, plus ISO 8402 (which was published in 1986 and it focused on terminology).

All ISO standards are evaluated on a five-year schedule to determinate whether they remain suitable for their application or if they need to be revised or withdrawn. At the annual meeting of ISOs Technical Committee 176 in 1990 it was agreed that the series of ISO 9000:1987 should be revised and that the revision should be done in two phases (Tsiakals 2002). This approach was adopted because a great numerous of organizations were familiar with the 1987 standards and would likely be resistant to major structural changes. The results of these processes were a small revision of the ISO 9000 family in 1994 and a greater and more important revision with major changes to structure and content of the standards in the year 2000. After the 2000 revision ISO 9000 family consisted of the following three standards: ISO 9000:2000, ISO 9001:2000 and ISO 9004:2000. ISO 9000:2000 was the general standard that serves as an overall guide to the other standards. Its purpose was to provide definitions of terms and a basic explanation of the ISO 9000 standards. ISO 9001:2000 consolidates the former ISO 9001/9002/9003 standards into a single document and was the only standard to which certification was assessed. The 2000 version of ISO 9001 was written to be more user-friendly to small businesses and service organizations.

ISO 9004:2000 provided further guidance for continuous improvement of internal quality management systems. ISO 9001:2008 in essence re-narrates ISO 9001:2000. The 2008 version only introduced clarifications to the existing requirements of ISO 9001:2000. There were no new requirements. ISO 9001 was supplemented directly by two other standards of the family: ISO 9000:2005 and ISO 9004:2009. In 2012, ISO TC 176—responsible for ISO 9001 development concluded that it is necessary to create a new QMS model for the next 25 years. The revised standard ISO 9001:2015 was published by ISO on 23 September 2015. September 2015 is start of 3 years transition period to September 2018. Certifications to ISO 9001:2008 will no longer be valid after September 2018.

The standards are to be applied to any type of organizations; independent to the size of the organizations or the kind of products manufactured or services provided, in private and public organizations, including government services.

ISO does not decide when to develop a new standard, but responds to a request from industry or other stakeholders such as consumer groups. Typically, an industry sector or group communicates the need for a standard to its national member who then contacts ISO. ISO standards are developed by groups of experts from all over the world, that are part of larger groups called technical committees. These experts negotiate all aspects of the standard, including its scope, key definitions and content. The technical committees are made up of experts from the relevant industry, but also from consumer associations, academia, NGOs and government.

The aim since 1987 was to provide a series of international standards dealing with quality systems that could be used for external quality purposes. Another important consideration was the desire to provide information to organizations about how to design their own quality systems based on individual company marketplace needs (Aggelogiannopoulos 2007).

The paper conduct a theoretical framework in order to general characteristics of the QMS. A literature review is conducted to find information from various research articles which either justified or discouraged implementation quality management systems. The aim of the article is define of important differences between ISO 9001:2008 and ISO 9001:2015.

2 THEORETICAL FRAMEWORK

Quality management (QM) is business practice that may benefit companies. As several empirical studies have shown, implementing QM (Huarng and Chen 2002, Kaynak 2003, Parast 2011) may effectively have a positive influence on firm performance, improvement of productivity, operational efficiency, product quality, employee motivation or on external nature such as marketshare, customer satisfaction, delivery and organizational image-related factors (Sampaio, 2009, Kim, 2011). This positive effect may result from their impact on firm costs and differentiation levels. Firms that implement QM focus on providing superior value to the customer and on improving the efficiency of the processes. Continuous improvement of processes and product quality lead to increased revenues (through product reliability) and reduced costs (through process efficiency) (Bernardo 2015). Also, research has been done on how far this type of standard have a significant impact and positive influence on business performance (Chow-Chua 2003, Mokhtar 2012). As the literature shows, many scholars have analyzed the benefits derived from the ISO 9001 on several performance dimensions. In order to analyze the benefits arising from the ISO 9001 standard, some authors have used lists of benefits to examine its effects, whereas others have used or even proposed classifications of benefits, such as (1) internal benefits and external benefits (Casadesús 2001), (2) benefits related to operational performance and financial performance (Naveh 2004, Briscoe 2005), (3) benefits related to operational, customer, people, and financial results (Casadesús 2000, Karapetrovic 2010), (4) other classifications (e.g. Lee 1998, Nield 1999). Furthermore, the difficulties frequently associated with ISO 9001 standard adoption include the following: lack of top management involvement during the implementation process (Sampaio 2009), employee and middlemanagement resistance, lack of financial and human resources, insufficient knowledge about quality programs (Al-Najjar 2011, Magd 2012) and involvement of a long and bureaucratic documentation (Zeng 2007, Boiral 2011, Magd 2012).

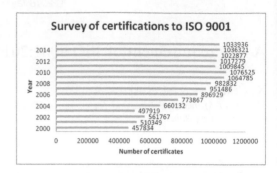

Figure 1. Survey of certifications to ISO 9001 (www. isi.org).

The ISO 9000 series was the fastest growing standards in history and was very popular from the start (Fig. 1). The implementation of this type of standards is voluntary, although in some sectors it has de facto become an obligatory measure, given the coercive influence of customers (Braun 2005, Mendel 2006).

A total of 1,519,952 certificates were issued worldwide in 2015, compared to 1,476,504 the previous year, an increase of 3%. Since 1987, were the ISO 9000 family was issued, it is unlikely that any other standards had more impact on international trade, on the relationship between suppliers and their customers and on the management of quality (Aggelogiannopoulos 2007).

3 IMPORTANT DIFFERENCES BETWEEN ISO 9001:2015 AND ISO 9001:2008

ISO standards are reviewed every five years and revised if needed. This helps ensure they remain useful tools for the marketplace. The challenges faced by business and organizations today are very different from a few decades ago and ISO 9001 has been updated to take this new environment into account. For example, increased globalization has changed the way we do business and organizations often operate more complex supply chains than they did in the past. In addition, there are increased expectations from customers and other interested parties and, with more access to information, today's wider society has a stronger voice than ever before. ISO 9001 needs to reflect these changes in order to remain relevant (BSI 2015a)

ISO 9001 needs to change, to:

– Adapt to a changing world,
– Reflect the increasingly complex environments in which organizations operate,
– Provide a consistent foundation for the future,

- Reflect the increasingly complex environments in which organizations operate,
- Ensure the new standard reflects the needs of all relevant interested parties,
- Ensure alignment with other management system standards.

The key changes in the standards are:

3.1 *New structure and new terms*

The most noticeable change to the standard is its new structure. ISO 9001:2015 now follows the same overall structure as other ISO management system standards (known as the High-Level Structure), making it easier for anyone using multiple management systems. All ISO management systems standard will be look the same since 2018, with the same structure (some deviations) and will be provide the option of integrating management systems. It means 10-clause structure and core text for all Management System Standards (MSS) and its more compatible with services and non-manufacturing users. At the same, the core definitions are standardized.

It has also changed terms: document and records—documented information, product—product and services, improvement—continual improvement, purchased product—externally provided products and services, outsourcing—external provision, supplier—external provider etc.

3.2 *Seven principles of quality management*

ISO 9001standard reduced of quality management principles from eight to seven (systematic to management).

3.3 *Relevant interested parties*

The organization shall identify the relevant interested parties and their relevant requirements.

3.4 *An explicit requirement for risk-based thinking*

Another major difference is the focus on risk-based thinking. The concept of risk has always been implicit in ISO 9001 such as "Prevention actions"– this edition makes it more explicit and builds it into the whole management system.

Risk-based thinking is inherent in all clause of a quality management system:

- Introduction – the concept of risk-based thinking is explained
- Clause 4 – organization is required to determine its QMS processes and address its risks and opportunities

- Clause 5 – top management is required to (promote awareness of risk-based thinking, determine and address risks and opportunities that can affect product/service conformity)
- Clause 6 – organization is required to identify risks and opportunities related to QMS performance and take appropriate actions to address them
- Clause 7 – organization is required to determine and provide necessary resources
- Clause 8 – organization is required to manage its operational processes
- Clause 9 – organization is required to monitor, measure, analyse and evaluate the effectiveness of actions taken to address risks and opportunities
- Clause 10 – organization is required to correct, prevent or reduce undesired effects and improve the QMS and update risks and opportunities
- Note, risk is implicit whenever suitable or appropriate is mentioned (clause 7 and 8) (ISO 9001 2015).

Organization shall use risk-based thinking to prioritize the way you manage of processes. It means organization shall identify of risk, plan actions to address these risks and evaluate the effectiveness of these actions. Risk-based thinking is part of the process approach. Not all the processes of a quality management system represent the same level of risk in terms of the organization's ability to meet its objectives. Some need more careful and formal planning and controls than others (BSI 2015b).

Principles and guidelines management risk are described in ISO 31000:2009 standard and tools used for risk assessment are described in ISO 31010:2010 standard. Risk management reduces the probability of negative results.

ISO 9001 standard combines words risk and opportunities. The risk can be either negative or positive. Opportunity is not the positive side of risk. An opportunity is a set of circumstances which makes it possible to do something.

3.5 *Fewer prescribed requirements and less emphasis on documents*

The 2015 version is also less prescriptive than its predecessors. It is stressed that ISO 9001 requires (and always has required) a "Documented quality management system", and not a "system of documents". The terms "document" and "records" have been replaced with the term "documented information". Documented procedure in ISO 9001:2008 have been replaced by maintained documented information and Documented record in ISO 9001:2008 have been replaced by retained documented information. The definition of documented information can be found in ISO 9000 clause 3.8. Documented information can be used to communicate a message, provide evidence of what was planned has actually been done, or knowledge sharing.

ISO 9001:2015 says that:

– Clause 4.4 Quality management systems and its processes requires an organization to "maintain documented information to the extent necessary to support the operation of processes and retain documented information to the extent necessary to have confident that the processes are being carried out as planned."
– Clause 7.5.1 General explains that the quality management system documentation shall include:
– documented information required by this International standard
– documented information determined by the organization as being necessary for the effectiveness of the quality management system (ISO 9001:2015)

The note after this Clause make it clear that the extent of the QMS documented information can differ from one organization to another due to the:

– Size of organization and its type of activities, processes, products and services,
– Complexity of processes and their interactions,
– Competence of persons.

All the documented information that forms part of the QMS has to be controlled in accordance with clause 7.5 Documented information. The type and extent of the documented information will depend on the nature of the organization's products and processes, the degree of formality of communication systems and the level of communication skills within the organization, and the organizational culture.

The documents may be in any form or type of medium such as paper, electronic or optical computer disc, photograph, master sample (BSI 2015c) In ISO 9001:2015 standard is no quality manual. Standard but does not prohibit it.

3.6 Planning and changes

When the organization determines the need for changes to the quality management system, the changes shall be carried out in a planned and systematic manner. The organization shall control planned changes and review the consequences of unintended changes, taking action to mitigate any adverse effects, as necessary.

There are many triggers that can cause a change to the Quality Management System such as cus-

tomer feedback, customer complaint, innovation, determined risk and opportunity, internal audit results, management review results, identified nonconformity etc. Some changes need to be carefully managed while others can be safely ignored.

Prior to making a change, the organization should consider: consequences of the change, likelihood of the consequence, impact on customers, impact on interested parties, impact on quality objectives, effectiveness of processes that are part of the QMS, etc.

Typical steps to implement changes:

– Define the specifics of what is to be changed,
– Have a plan (tasks, timeline, responsibilities, authorities, budget, resources, needed information, others),
– Engage other people as appropriate in the change process,
– Develop a communication plan (appropriate people within the organization, customers, suppliers, interested parties, etc. may need to be informed),
– Use a cross functional team review the plan to provide feedback related to the plan and associated risks,
– Train people,
– Measure the effectiveness (BSI 2015d).

Other generally requirements are:

– A requirement to define the boundaries of the QMS,
– A requirement that those at the top of an organization be involved and accountable, aligning quality with wider business strategy,
– Inclusion of Knowledge Management principles,
– In the field staffing there is no requirement of management representative.

4 CONCLUSIONS

The aim of the article was define of important differences between ISO 9001:2008 and ISO 9001:2015.

The most noticeable change to the standard was its new structure, requirement for risk-based thinking and requirement for documented information. The revised standard ISO 9001:2015 was published by ISO on 23 September 2015. September 2015 is start of 3 years transition period to September 2018.

As pointed out in the previous section, the 2015 version is less prescriptive than its predecessors and focuses on performance. This was achieved by combining the process approach with risk-based thinking, and employing the Plan-Do-Check-Act cycle at all levels in the organization.

For any organization the degree of change necessary will be dependent upon the maturity and effectiveness of the current management system, organizational structure and practices.

Organizations using ISO 9001:2008 are recommended to take the following actions: identify organizational gaps which need to be addressed to meet new requirements, develop an implementation plan, provide appropriate training and awareness for all parties that have an impact on the effectiveness of the organization, update the existing quality management system (QMS) to meet the revised requirements and provide verification of effectiveness, where applicable, liaise with their Certification Body for transition arrangements.

ACKNOWLEDGMENT

This paper was written with the financial support of the granting agency VEGA as a part of the project No. 1/0990/15 Readiness of industrial companies to implement the requirements of standards for quality management systems ISO 9001:2015 and environmental management systems ISO 14001: 2015.

REFERENCES

Aggelogiannopoulos, D. 2007. Implementation of a quality managemnt systems (QMS) according to the ISO 9000 family in a Greek small-sized winery: A case study. *Food control*. 18 (2007), 1077–1085.

Al-Najjar, S.M. 2011. ISO 9001 implementation barriers and misconceptions: an empirical study. Int. *J. Bus. Adm*. 2. 1923–4007.

Bernardo, M. et al. 2015. Benefit of managemnt systems integration: a literature review. Int. *Journal of Cleaner Prosuction*. 94. 260–267.

Boiral, O. 2011. Managing with ISO systems: lessons from practice. *Long. Range Plan*. 44 (3). 197–220.

Braun, B., 2005. Building global institutions: the diffusion of management standards in the world economy e an institutional perspective. In: *Alvstam, C.G., Schamp, E.W. (Eds.), Linking Industries across the World. Ashgate*, London. 3–27.

Briscoe, J.A. et al. 2005. The implementation and impact of ISO 9000 among small manufacturing enterprises. Int. *J. Small Bus. Manag*. 43 (3). 309–330.

BSI, 2015a. Moving from ISO 9001:2008 to ISO 9001:2015. *International Organization for Standardization*. Geneva, Switzerland. ISBN 978-92-67-10646-5.

BSI, 2015b. Risk-based thinking. *International Organization for Standardization*. Geneva, Switzerland. www.iso.org.

BSI, 2015c. Guidance on the requirements for Documented Information of ISO 9001:2015. *International Organization for Standardization*. Geneva, Switzerland. www.iso.org.

BSI, 2015d. How Change is addressed within ISO 9001:2015. *International Organization for Standardization*. Geneva, Switzerland. www.iso.org.

Casadesús, M. & Gimenez, G. 2000. The benefits of the implementation of the ISO 9000 standard: empirical research in 288 Spanish companies. *TQM Mag*. 12 (6). 432–441.

Casadesús, M. et al. 2001. Benefits of ISO 9000 implementation in Spanish industry. *Eur. Bus. Rev*. 13 (6). 327–335.

Chow-Chua, C. Et al. 2003. Does ISO 9000 certification improve business performance? Int. *J. Qual. Reliab. Manag*. 20 (8). 936–953.

Ferguson, W. 1996. Impact of the ISO 9000 series standards on industrial marketing. *Industrial Marketing Magazine*, 25,305–310.

Huarng, F. & Chen, Y.T. 2002. Relationships of TQM philosophy, methods and performance: a survey in Taiwan. *Ind. Manag. Data syst*. 102.226–234.

ISO 9001, 2015. Quality managemnt systems.

Karapetrovic, S. et al. 2010. What happened to the ISO 9000 lustre? an eight-year study. *Total Qual. Manag*. 21 (3). 245–267.

Kaynak, H. 2003. The relationship between total quality managemnt practices and their effects on firm performance. *J. Oper. Manag*. 21. 405–435.

Kim, D.Y. et al. 2011. A performance realization framework for implementing ISO 9000. Int. J. Qual. Reliab. Manag. 28 (4), 383–404.

Lee, T.Y. 1998. The development of ISO 9000 certification and the future of quality management: a survey of certification firms in Hong Kong. Int. *J. Qual. Reliab. Manag*. 15. 162–177.

Magd, D. & Nabulsi, F. 2012. The effectiveness of ISO 9000 in an emerging market as a business process management tool: the case of the UAE. *Procedia Econ. Finance*. 3. 158–165.

Mendel, P.J. 2006. The Making and Expansion of International Management Standards: the Global Diffusion of ISO 9000 Quality Management Certificates. Oxford University Press, New York.

Mokhtar, M.Z. & Muda, M.S. 2012. Comparative study on performance measure and attributes between ISO and non—ISO certification companies. Int. *J. Bus. Manag*. 7. 185–193.

Naveh, E. & Marcus, A.A. 2004. When does the ISO 9000 quality assurance standard lead to performance improvement? Assimilation and going beyond. *IEEE Trans. Eng. Manag*. 51. 352–363.

Nield, K. & Kozak, M. 1999. Quality certification in the hospitality industry: analyzing the benefits of ISO 9000. *Cornell Hotel Restaur. Adm. Q*. 40. 40–45.

Parast, M.M. et al. 2011. Improving operational and business performance in the petroleum industry through quality managemnt. Int. *J. Qual. Reliab. Manag*. 28 (4). 426–450.

Sampaio, P. et al. 2009. ISO 9001 certification research: questions, answers and approaches. Int. *J. Qual. Reliab. Manag*. 26 (1). 38–58.

Tsiakals, J.J. 2002. The need for change and the two-phase revision process. In. C.A. Cianfrani, J.J. Tsiakals & J.E. West (Eds.). *The ASQ ISO 9000:2000 handbook*. 63–72. Milwaukee: ASQ Quality Press.

Zeng, S.X. et al. 2007. A synergetic model for implementing an integrated management system: an empirical study in China. *J. Clean. Prod*. 15. 1760–1767.

Peculiarities of managing social enterprises

Ľ. Černá
Faculty of Education, Catholic University in Ružomberok, Ružomberok, Slovak Republic

J. Syrovátková
Faculty of Economy, Technical University in Liberec, Liberec, Czech Republic

ABSTRACT: Social entrepreneurship creates positive externality, it is market unvalued social benefit. The activity of social enterprise is primarily oriented towards social targets, to create workplace for disadvantaged groups. It shifts to the area of social needs displeased by the market of the state. Despite the fact, that social enterprise is an entrepreneurial body; its guidance has its own specifications. This article deals with the selected areas of social enterprise management. The aim is to identify specifications of social enterprise management in Slovakia and the Czech Republic, in terms of social economy subjects, in context with social responsibility as well as to explain how the following differences are perceived by the selected group of management and social work students

1 INTRODUCTION

Social enterprise is a competitive entrepreneurial body performing on the common market whose aim is to create work opportunities for disadvantaged groups on the market and to provide adequate work and psychosocial support. It develops entrepreneurial activity in order to fulfil its social aims. It is one of the models to employ disadvantaged people on the job market. It offers work place similar to regular employment, but also considers specifications of employees, offers to some extent, support and also aims to fully develop the potential of the employees, develops their skills and abilities.

That is why social entrepreneurship gets even stronger political support reflected in legal regulation. It is the key point of European social model and has key importance for the success of the strategy Europa 2020. European community supports social economy in the long term but since 2011 social entrepreneurship has become its priority and political aim. Social enterprises employ 14.5 million persons, which in European Union presents 6.5% of economically active population.

2 SOCIAL ENTERPRISE AND SOCIAL ENTREPRENEURSHIP

2.1 Historical development of social entrepreneurship

Social entrepreneurship has become a fashionable construct in recent years. The language of social entrepreneurship may be new, but the phenomenon is not. We have always had social entrepreneurs, even if we did not call them that. Bill Drayton, the founder of ASHOKA organization with the approach called: "social innovation school of thought", is considered to be a pioneer of social entrepreneurship in the USA. The concept of social enterprise became known in Europe in the 1990s, when the Italian government legally approved a new type of social cooperative organisation. In 2002, in Great Britain the Coalition for Social Enterprises was established in order to educate population about social entrepreneurship. The volume of social entrepreneurship understanding was unified by EMAS.

There is a difference between American and European perception of stakeholders. While in the USA there is prevalent orientation towards the individual social entrepreneur who with the help of innovative idea set up often hierarchically guided organization fulfilling his vision. In Europe this effort shows collective character, built on group initiative and common effort. That is why social economy in Europe becomes more known for its efforts to strengthen the ability to support social coherence, raise employment, support local development and keep social and economic structures. Second difference is the way they work with financial sources. American approach emphasises using market sources, while European social enterprises use non-market sources and public finances.

History of social entrepreneurship in the Czech Republic and Slovakia is based on strong civic society, development of alliance and church institutions. Social economy has long tradition mainly in the countryside, where SMEs, worker cooperatives, associations, mutual/municipal savings banks and cooperative agricultural banks developed under the Hapsburg Empire. These voluntary organisations helped to form Czech cultural and economic identity within the empire. Their activities intensified towards the end of the century and reached their peak after the First World War. In the 1920s and early 30 s, the number of associations and cooperatives was rising in all areas including student and interest organisations, sport clubs, associations of national minorities, etc. (Dohnalova 2009) The activity of alliance was interrupted by 2th world war with the following normalisation and has been re-established after 1989. After the Second World War, the communist regime was established in February 1948. While some of the associations and cooperatives survived this change, their activity lost certain important elements of the social economy. They lost their autonomy from the state and became politicised. Many were dissolved by the state because their social mission was not in accordance with the regime. Cooperatives had to produce goods according to the plans made by the state at the central level, rather than according to their own free will.

2.2 *Definition of social enterprise and social entrepreneurship*

Establishing an agreed definition of social entrepreneurship has not proved to be an easy task. Dees speaks for many when he declares: "Social entrepreneurs are one species in the genus entrepreneur" (Dees 1998).

The definition of social entrepreneurship should not extend to philanthropists, activists, and companies with foundations, or organizations that are simply socially responsible. While all these agents are needed and valued, they are not social entrepreneurs. Traditional entrepreneurship is mostly connected with economic result of activity. Social entrepreneurship is connected with double result—economic and social success and equilibrium creation in the society (Greblikaite 2012).

The literature on the subject uses three different terms which, at first sight, might seem linked in a very simple way: "social entrepreneurship" is the dynamic process through which specific types of individuals deserving the name of "social entrepreneurs" create and develop organizations that may be defined as "social enterprises" (Defourny & Nyssens 2008). However, the use of one term or the other is often linked to a different focus and/ or understanding of the phenomenon depending on context and perspective (Huybrechts & Nicholls 2012).

Korosec and Berman (2006) define social entrepreneurs as "individuals or private organizations that take the initiative to identify and address important social problem in their communities". Bacq and Janssen (2011) noted 17 different definitions of "social entrepreneurs", 12 definitions of "social entrepreneurship" and 18 definitions of "social enterprise", "social entrepreneurial venture" or "social entrepreneurship organization".

European Union defines social entrepreneurship as an operator in the social economy whose main objective is to have a social impact rather than make a profit for their owners or shareholders. The Commission uses the term "social enterprise" to cover the following types of business: a) Those for who the social or societal objective of the common good is the reason for the commercial activity, often in the form of a high level of social innovation. b) Those where profits are mainly reinvested with a view to achieving this social objective. c) Those where the method of organisation or ownership system reflects the enterprise's mission, using democratic or participatory principles or focusing on social justice.

The Social Business Initiative (SBI) definition incorporates the three key dimensions of a social enterprise that have been developed and refined over the last decade or so through a body of European academic and policy literature: a) an entrepreneurial dimension, i.e. engagement in continuous economic activity, which distinguishes social enterprises from traditional non-profit organisations/ social economy entities (pursuing a social aim and generating some form of self-financing, but not necessarily engaged in regular trading activity); b) a social dimension, i.e. a primary and explicit social purpose, which distinguishes social enterprises from mainstream (for-profit) enterprises; c) a governance dimension, i.e. the existence of mechanisms to 'lock in' the social goals of the organisation. The governance dimension, thus, distinguishes social enterprises even more sharply from mainstream enterprises and traditional non-profit organisations/social economy entities. (European Commission 2015).

Social entrepreneurship is gaining stronger political support which is projected into its legal regulation. Social entrepreneurship is the key element of European social model and is of key importance for success of the Europa 2020strategy. European committee supports social economy in the long term, but since 2011social entrepreneurship has become its priority and political aim.

2.3 *Social entrepreneurship in the Czech Republic and Slovakia*

The term social entrepreneurship is still not legislatively bound in the Czech Republic. Generally it is a legal body based on private law, or its part, or a person who matches principles of social entrepreneurship, fulfils social aims, stated in establishment documents. Social entrepreneurship does not have its own legal form or legal register. At the moment a law about social entrepreneurship is being created. Standards of social entrepreneurship have been established in order to define and acknowledge social entrepreneurship in the Czech Republic and in this way to differentiate it from other models of employing disadvantaged groups. It should help to explain main characteristics of social entrepreneurship and understand basic values established by social entrepreneurship. To support social economy and social entrepreneurship in the Czech Republic TESSEA—Thematic Net for the development of social entrepreneurship has been created. It has established principles and indicators of social entrepreneurship. For general social enterprise altogether 19 indicators have been started out of which 8 mandatory. For integrative social enterprise there have been stated 20 indicators, out of which 10 mandatory. At this moment the directory of TESSEA comprises around 200 enterprises, 24% in the field of gardening, environmental upgrading and cleaning services. These concentrate on the equality of opportunities by disabled persons.

In the Slovak republic social enterprises support integration of disabled applicants in the job market. The aim of its existence is explicitly concluded in the law, as well as providing support and help with finding occupation at the job market. The law Act No. 5/2004 legislatively amends support and maintaining work positions of disadvantaged applicants. Social enterprise is entitled to ask for a benefit to support creating and maintain work places. The category of social entrepreneurship includes also protected workshop or sheltered working place having at least 50% of its employees the people with disabilities (Pongráczova 2013). In reality, as argued by Strečansky and Stolárikova (2012), even before the amendment of the law that defined social enterprises in Slovakia, there were entities acting like social enterprises, for instance some associations and foundations that were engaged in economic activity and pursued social goals at the same time. Importantly, once the law came into force, only few of them registered as social enterprise.

At the moment registry of social enterprise up to 20th of January 2017 comprises 10 legal bodies established between 2009–2012, out of which 95 have been cancelled and 5 have stopped. With an effect since 1st of May 2013 has been providing of support benefit cancelled. Up to December 2015 the term social entrepreneurship has been used which since 15th of December 2015 has changed into "social entrepreneurship for work integration". The support of social entrepreneurship is the key question in all countries of European Union. After annulment of these benefits we can generally state that the Slovak republic does not defend social enterprise, but it does not support it, either. Explicit legislative and financial support is missing, except from utilisation of socially responsible public procurement. Specific Slovak problem in this field is previous negative experience with the project of pilot social enterprise which creates negative impact on the term social enterprise itself. The concept of social economy and social enterprising suffered in Slovakia a serious blow in reputation due to a scandal with government support to eight "pilot social enterprises" in 2008. The scandal had a profound influence on how the concept is perceived in Slovakia today; therefore we stay with it in next paragraphs to provide basic facts. (Strečanský & Stoláriková 2014).

Problem in many Slovak communities and in the establishment and management of social enterprise is their lack of capacity to identify real business opportunity and then develop a sustainable business plan, which in addition to standard business environment also fulfilling the conditions of social entrepreneurship. The problem is the common perception of social enterprise as an institution whose establishment and operation shall be one hundred percent publicly funded. Obstacles of social entrepreneurship development in the Czech Republic and Slovakia are similar:

– social entrepreneurship is not visible, practically without public relation;
– social entrepreneurship is missing acknowledgement and understanding;
– social entrepreneurship has low and ineffective support, there is no systematic policy support for social entrepreneurship at national and regional policy levels.;
– social entrepreneurship is not legislatively bound and is missing control. The existing support offer focuses more on support during the first steps and overlooks support during subsequent stages of the enterprise's lifecycle, support in crisis situations etc.;
– support is offered mostly to NGOs and its scope is usually limited to only few organisations;
– there is still a lack of support organisations for social enterprises, especially those free of charge;
– there are no common criteria that would enable to compare the quality of the existing support services;
– systemic support structure with regional branches is needed because it would help to stabilise the existing social enterprises and help with the development of the new ones.

The list of potential opportunities for the sector included: a) Emergence of the new economically viable social enterprises: Although slowly and in a small scale, there has been number of emerging grass-root initiatives, examples of socially oriented and yet sustainable business models. And the development of dynamic structures which combine the strong entrepreneurial dimension and the social objectives may have stimulating and refreshing effect on the whole sector. b) Growing demand for social services: Growing demand for social services has been seen as one of the opportunities for the sector.

Corporate social responsibility (CSR) is such activity of the enterprise where company deals reliably within legal norms, when this behaviour is not rare, but permanent and becomes natural part if it's strategic operating. Social policy of the enterprise is one of the parts of CSR and present specific entrepreneurial activity as a part of internal activities of CSR. However, this type of company does not have to fulfil demands imposed on such social enterprise.

The rise of social enterprise, corporate social responsibility, social investing, and sustainable development are all examples of how various actors are pursuing a blend of financial, social, and environmental value. (Fig. 1) Corporate social responsibility is not classified as social enterprise, although philanthropic activities may support social enterprises, make a positive social impact, or contribute significantly to a public good (Alter 2007).

3 PECULIARITIES OF SOCIAL ENTERPRISE MANAGEMENT

Managers who bear responsibility and risk have strategical positions in social enterprises. The first year is critical because 87% of small and middle companies finish their activity within one year of functioning.

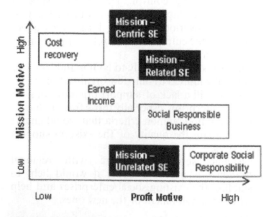

Figure 1. The relationship between the type of organization and its motives. Note: SE – social enterprises. Source: Alter, 2007.

Management of social entrepreneurship, economy, financing of knowledge of social enterprise, social marketing, utilization of social capital, basics of personal work with disadvantaged person on the job market, specifications of social enterprise supervising. Specifically, leaders of social enterprises face manifold challenges: many leaders do not have a formal business education and were driven by the passion to solve a social cause. Their intrinsic motivation comes first and economic reasons rank low. But business procedures need to be implemented, teams built and money earned to run a social enterprise successfully. Necessary skills of a manager are: leadership, team working, communication, problem solving and financial skills. Managers follow so called "Diversity Management". Diversity constitutes theoretical and practical extension to equal opportunities. According to EMES is social entrepreneurship defined by the following basic economic and social criteria.

A detailed description of the criteria can be found in Defourny (2001): a) a continuous activity producing goods and/or selling services; b) a high degree of autonomy; c) a significant level of economic risk; d) a minimum number of employees; e) an explicit aim to benefit the community; f) an initiative launched by a group of citizens; g) a decision-making power not based on capital ownership; h) a participatory nature, which involves the persons affected by the activity; i) limited profit distribution. These criteria differentiate them from commercial enterprises.

It is important to note that financial objectives differ among organizations. Social enterprises don't need to be profitable to be worthwhile. They can improve efficiency and effectiveness of the organization by: a) reducing the need for donated funds; b) providing a more reliable, diversified funding base; or c) enhancing the quality of programs by increasing market discipline (Dees 1998).

Economic activity of social enterprise of work integration is the key to accomplish or fulfils its social aim. The target group is disadvantaged population. Disadvantaged persons are considered disabled persons (bodily and mental disadvantage, person with combined disability and mentally ill), socially disabled (long term unemployed, persons returning from prison homeless, former drug addicts, persons without cultural and family background, ethnic and national minorities and illiterate persons). Employment in social enterprise is suitable for persons who are expected to need long term or repeatable work and psychosocial support. Work conditions are equal for all employees.

Social enterprise has stated ethical principles and rules of internal communication and cares about awareness of its employees. Ethic principles comprise respecting of dignity, privacy, declaration and keeping the principle of equality and solidarity. These principles are stated in the regulations of internal communication. Employees when entering the company meet and understand ethical principles and regulations of internal communication. Finances are obtained from various sources. Money provided in the form of state grants, allocations or European funds submit to regular control.

Social enterprises are also required to inform the public about its activities mostly in the form of annual report. Within this type of management, health problem or other type of disability must be considered, as well as social status of employees. Workplace and working time must be adjusted to the type of disability of employees. Social enterprise manager has to consider differences in terms of work production, norms of work expenditure, time management. He or she has to focus on creation of motivational factors and rewarding, to introduce innovation into practice. He or she has to keep in mind creation of rehabilitative and health improving programmes and utilizes environmental approaches in favour of social entrepreneurship.

4 PREPARATION OF SOCIAL ENTERPRISE MANAGERS

In order to be able to transform social entrepreneurship into practice in the form of "classical business" we need to create flexible attractive and acceptable system of professional education where the end product becomes a social enterprise manager (Pčolinská 2014). Management operating in social enterprises often does not have professional university education level which would educate social enterprise managers. There is very little research on social enterprises/social economy and no specific degree in the higher education curricula. In the academic sphere recently two institutions in Slovakia produce research outputs on the development of the concept: Department of Sociology at Comenius University in Bratislava or Centre for Research and Development of Social Economy and Social Entrepreneurship at the Economic Faculty of University of Matej Bel in Banská Bystrica. Technical university in Liberec Economical faculty, Department of entrepreneurial economy and management with the Association of non-state, non-profit organizations of Liberec region (ANNOLK) in the years 2011–2012 realized the project and altogether created conditions for raising employment in Liberec region by offering other forms of education of target group in the area of social entrepreneurship and gaining skills for establishing social enterprises. This subject is taught in Slovakia at the Economical University, Faculty of economics, Comenius University and University of P. J. Šafárik both at Faculty of law. Non—formal education is realized abroad, for example job and skills training program that provides clients with the skills and tools that they need to successfully seek employment in their efforts to labour market.

5 RESEARCH AND DIAGNOSIS OF THE LEVEL OF COMPETENCE

Objectives of the conducted research and diagnosis of the level of competence of youth and the employers' (social enterprises) needs include: a) investigation and determination of the level of personal and social competences young people; b) collecting information about employers' assessment of the level of personal and social competences of young people who want to work in social enterprises; c) collecting information about skills and competences of young people expected and required by employers (social enterprises); d) study whether the personal and social competencies of students are the same competencies that the employers search for.

Tests of competences in the Technical universities in Liberec (Czech republic) and Catholic University in Ružomberok (Slovak Republic) were produced as an online and paper questionnaire (n = 63). We tested 3 industry companies, 2 agriculture and food companies, 5 service and trade companies, 5 marketing companies and 1 public company (n = 16). Date of the research: March 2016 – June 2016. Competence Model was identified which includes the following 14 competencies and for university students and employers 24 competencies.

Based on the research we see that the disproportion between the real needs of employers and training and education of students. Social enterprises have the following priorities: Customer orientation, be able to think and act economically/economic competence, the ability to identify and solve problems, ability to communicate and negotiate with people, presentation skills, ability to resolve conflicts, the level of writing skills. Students, who want to work in social enterprises, think that they will receive the following competences at the university: presentation skills, ability to take responsibility, the ability to engage in teamwork, the ability of creative and flexible thinking and acting, the ability of independent decision-making. Almost all Slovak students expressed negative reception of social enterprises.

6 CONCLUSION

However different social enterprises may seem in their activities they follow these targets—mutual interconnection of economic and social activity with environmental targets, offering social and environmental products and services, using such methods and principles which provide important corporate contribution. The growth and the development of social enterprises in the Czech Republic and Slovak republic is an issue that is underdeveloped. There is a rising need in the society to put social entrepreneurship on the agenda but there is a lack of political support and the whole social enterprise sector is still weak. Many social enterprises started their functioning with the structural funds funding that has already finished or is coming to its end. These social enterprises are endangered and their sustainability us under threat. Ministry of Labour and Social Affairs originally planned an accompanying support project that would help these social enterprises to survive but it somehow did not happen. To sum up the situation in the Czech Republic and Slovak Republic, social entrepreneurship is on the rise with growing interest of public authorities and civil society. Its promotion is associated on both sides mainly with devoted individuals who are able to cooperate in partnership.

Employee becomes in a special way a client who cannot be implemented, at the beginning of this process of socialization, revitalization or inclusion, standard principle stated by a work law. Employee—client of a social enterprise is first of all a person, citizen in need. The fact creates a significant discrepancy between an employer, in this case a helper in need and an employee who is in this relationship very vulnerable. Customers will buy from a business because it offers high, consistent quality and excellent value. A social enterprise is no different and should not expect customers to flock to it because of its social purpose or its links to the parent body. We cannot forget disadvantages of social enterprises: a) as with all businesses, there is a degree of risk. They can fail in difficult market conditions, if not properly run or just because a business idea doesn't work out; b) it can be difficult for everyone (employees, clients, the parent board, the public) to see a social enterprise as both a business and a community organisation; c) keeping the balance between business (income generation) and social purposes and activities can be difficult in both the short and long terms; d) social and financial benefits can take a long time (years rather than months) to be fully delivered so there needs to be an up-front understanding of this.

ACKNOWLEDGMENTS

This contribution is a component presenting the meeting results of project VEGA 2/0038/14 Adaptation strategies on natural and social disturbances in forest country.

REFERENCES

Alter, K. 2007. Social Enterprise Typology. *Virtue Ventures* LLC. 124p.

Bacq, S. & Janssen, F. 2011. The multiple faces of social entrepreneurship: A review of definitional issues based on geographical and thematic criteria. *Entrepreneurship & Regional Development: International Journal,* 23(5): pp. 373–403. ISSN 0898-5626.

Dees, J. G.1998. Enterprising Nonprofits. *Harvard Business Review*, January-February 1998, pp. 54–67. ISSN 0017-8012.

Dees, J. G. 1998. The Meaning of "Social Entrepreneurship". Stanford University: *Draft Report for the Kauffman Center for Entrepreneurial Leadership*: 6 pp. [online], https://entrepreneurship.duke.edu/news-item/the-meaning-of-social-entrepreneurship/ [cit. 2017–03–22].

Defourny, J. & Nyssens, M. 2008. Social Enterprise in Europe: Recent Trends and Developments, *EMES Working Paper*, 08/0. [online], http://www.emes.net/site/wp-content/uploads/WP_08_01_SE_WEB.pdf. [cit. 2017–05–27].

Defourny, J. 2001. From Third Sector to Social Enterprise, In Borzaga, C. and Defourny, J. (Eds.), *The Emergence of Social Enterprise,* Routledge, London and New York, pp. 1–28.

Dohnalová, M. 2009. Sociální ekonomika—vybrané otázky. Výzkumný ústav práce a sociálních věcí, v.v.i. Praha. ISBN 978-80-7416-052-3.

European Commission. 2015. A map of social enterprises and their eco-systems in Europe. Luxembourg: Publications Office of the European Union, 2015. p. 169. ISBN 978-92-79-48814-6.

Greblikaite, J. 2012. Development of social entrepreneurship: challenge for Lithuanian researchers. *European Integration Studies* 6: pp. 210–215. ISSN 1822-8402.

Huybrechts, B. & Nicholls, A. 2012. Social entrepreneurship: definitions, drivers and challenges. In: Christine K. Volkmann, Kim Oliver Tokarski, Kati Ernst (eds.). *Social entrepreneurship and social business: an introduction and discussion with case studies Wiesbaden:* Gabler; London: Springer 2012. pp 31–48. ISBN 978-3-8349-7093-0.

Korosec, R. & Berman, E. 2006. Municipal support for social entrepreneurship. *Public Administration Review* 66, pp. 446–460, ISSN: 1540-6210.

Pčolinská, L. 2014. Návrh implementácie predmetu "sociálne podnikanie" do vyučovacieho procesu In: *Sociálna ekonomika a vzdelávanie* Zborník vedeckých štúdií. Banská Bystrica. ISBN 978-80-557-0623-8.

Phillips, W. et al. 2015. Social innovation and social entrepreneurship: A systematic review. *Group & Organization Management*, 40 (3). pp. 428–461. ISSN 1059-6011.

Pongráczová, E. 2013. Social entrepreneurship and corporate social responsibility in Slovakia. In: *2nd International Scientific Conference Economic and Social Development. Paris:* Varazdin Development and Entrepreneurship Agency, Faculty of Commercial and Business Sciences, 2013. pp. 711–717, ISBN 978-961-6825-73-3.

Strečansky, B. & Stolárikova, K. 2012. Social Economy and Social Enterprises in Slovakia. *Civil Szemle (Civil Review)* 9(4), pp. 87–100 [online], http://www.cpf.sk/files/ files/Pages%20from%20CivSzle_2012_4_web.pdf.

Vaňová, J. et al. 2009. Company success dependency on management. *Annals of DAAAM and Proceedings of DAAAM Symposium* 2009: Vol. 20, No. 1. pp. 741–742. ISSN 1726-9679.

Walk, M. et al. 2015. Social Return on Investment Analysis: A Case Study of a Job and Skills Training Program Offered by a Social Enterprise. *Nonprofit Management and Leadership,* 26(2), Winter 2015, pp. 129–144. ISSN 1542-7854.

New Trends in Process Control and Production Management – Štofová & Szaryszová (Eds)
© 2018 Taylor & Francis Group, London, ISBN 978-1-138-05885-9

Internal and external audit: Relationship and effects

M. Daňová
Faculty of Management, The University of Presov, Prešov, Slovak Republic

ABSTRACT: The development of auditing in companies confirms the increase in its importance and use. Evolution is evident in the subject of the audit, the objectives, the form, the auditor's requirements, the obligation to audit. The reason for the disapproving attitude to its use can be considered: improper perception of the functions of its forms, resulting in a change of attitude of management towards the use of its forms, failure to comply with standards for audit processes and its source provision, problematic identification of the effects of its application and their quantification. The findings are based on some published opinions and findings of their own research. There are few studies that classify audit effects. We analyze similarities and differences in the functionalities of both forms and identify factors that limit the effectiveness of the audit.

1 INTRODUCTION

The developing an internal and external form of auditing, focusing on both the financial and non-financial nature and its role as a tool for verifying truthfulness of information, a tool for verifying compliance with reality, a tool for streamlining business activities over time is the result of its development.

From the written sources, it is possible to take into account several stages of development. In each of them, the role and form of audit as a management tool is given by the need of individuals or groups in a company looking for information or assurance about the behaviour or performance of entities in which they have a legitimate interest (Flint 1988):

Historically, the earliest form can be considered as the use of audit as a tool for verifying the transmission of true information to the population as well as a system of protection of state treasury interests. The newer form of development is the implementation of audit in enterprise systems, the audit is used as a tool for verifying the accuracy of financial transactions and the truthfulness of the accounting documentation. An increase in the scale and complexity of business activities is considered to be the cause of the formal internal audit function. An increase in market environment competition results in the expansion of financial, as well as system and process audits carried out in the corporate environment by a third party.

Developments in auditor requirements mean shifting the emphasis from purely personal assumptions to verifying identifiable facts (indicative of the oldest developmental stage) to a set of personal and professional assumptions. The requirement for expertise comes in the context of the development of financial audits, followed by the use of different sampling, monitoring and evaluation methods. The requirement of impartiality and independence of audit activities results in a gradual separation of the audit function from the audited entity, in the case of an external audit, clearly dependent on the organizational independence of the auditor from the audited entity.

The result of this development is the development of internal and external audit. There are two different management tools used with very similar goals. Their content and purpose to verify compliance with the requirement and to testify to it makes it possible to increase their effectiveness by providing mutually beneficial information and accepting findings. If we explain the effectiveness as a measure of evaluating the resources spent to achieve the objectives, then the internal and external audit functions imply a requirement to optimize the consumption of the resources spent to implement it. This presupposes the combination of the specific advantages of both forms of audit to achieve the stated goal. By limiting the interconnection of both forms, it is a requirement to perform an internal audit to the extent and in a manner that minimizes all the risks associated with the existence of the enterprise, also the requirement to perform an external audit to the extent and in a way that enables independent, objective and professional certification of fact to the extent of the requirements of the contracting entity.

2 RESEARCH PROBLEM

The article is based on the study of literary sources and the comparison of published opinions. In his writing, both academic and non-academic resources were used. Based on our search, we strongly believe that:

- A large part of research on the relationship between external and internal forms of audit is linked to financial audit;
- Factors that significantly affect the relationship between internal and external audit are referred to as objective and subjective factors;
- As objective factors, the external auditor's limitations are identified, relying on the findings of the internal audit and the amount of resources available for auditing;
- The nature of the relationship also depends on the subjective perception of the importance of the work of internal auditors by company management.

An important finding of the analysis is that there is no academic determination of the acceptable scope of internal and external audit coordination. Since identification with academic opinion is real only under the (rational) acceptance of the proposed solution, the subject of this material is to identify a set of indicators suitable for quantifying the effects of external and internal audit coordination.

3 THE DEVELOPING OF GOALS AND AUDITING IN COMPANIES

The beginnings of the audit are not well documented. However, it was found that the audit had developed in several phases. There are differences between these phases in a distinct role, staffing and the use of audit findings (Lee 2008):

Historically, the oldest period of it is dated back to the period of Mesopotamia, Old Egypt and Rome, when the auditor's role was to carry out control activities and testify the truth and truth of the information. Lee (2008) considers the main role of audit in these season detection of fraudulent fraud actions. He links the ways of implementation with the existence of auditors subject to owners, who are independent of the audited entity.

The tasks of audit are oriented on controlling and verification the truth of accounting records, documents and transactions. In the age of The Industrial Revolution the role of protection ownership can be considered as secondary cause the aim was the verify probity of persons entrusted with fiscal responsibility. Given the size of the audited entities and the volume of their accounting transactions, as well as a small commercial use doesn't exists specific requirements for auditor expertise and the relationship to the audited entity (Porter et al. 2005). During the Industrial Revolution we can see an increase in the scope and complexity of business activities and the requirement for the creation of a stand-alone internal security function raised. This was raised to verify the veracity of the information used for decision-making. A need for a proprietary rights protection tool is increasing. The solution is to establish a formal internal auditing function, which is delegated to provide reliable and trustworthy owners with both financial and non-financial data. Auditors are shareholders chosen by their colleagues. Their duty is to perform full transaction control and prepare the correct accounts and financial statements. Developments in the audit approach there is replacement of the full control of transactions, auditors use sampling techniques, rely on internal control of the company. Davies (2001) talks about changing the "book transaction verification" approach to "relying on the system". Audit is used as a tool to increase the efficiency of business operations, processes, and systems. The functionality and efficiency of the enterprise system, its subsystems, processes, and business operations become subject to auditors. The employment relationship of the internal auditor with the audited company can be considered as a risk factor. A tool for eliminating the risk of maladministration and the risk of misrepresentation of facts, related to possible incompetence of internal auditors and their dependence on company management, is the third party's performance of audits. Their implementation extends the set of management tools does not mean a retreat from internal audits. It is assumed that the maximum effectiveness of the audit can be achieved by assuming the interconnection of the internal and external forms of audit. To achieve that objective, the audit must be a multilateral and comprehensive examination and assessment of the enterprise as an object with respect to the purpose for which it is carried out. The task of the audit is to evaluate and verify the truthfulness and accuracy of information, the regularity, adequacy and effectiveness of risk management processes, controls, and management processes. The goal is continuous improvement. Information on the effectiveness of management processes is gained through their systematic implementation. Verification is carried out in terms of the processes leading to the realization of the desired reality as well as the results.

4 RELATIONSHIPS BETWEEN EXTERNAL AND INTERNAL AUDITS

To clarify aspects of the relationship between external and internal audits has been the focus of several studies. The most frequent subject of analyses is the willingness and restrictions of external auditors to use the information obtained during the internal audit processes. Despite the conditions under which the external auditor may use the Internal Auditor's findings are According to ISA 610 international auditing standards, limited by the requirement of objectivity and competence, also by knowledge of the audited environment. First, the findings of interactions between internal and external audit are published by Schneider (1984) and Mautz (1984). Schneider (1984) considers the decisive factors of the interrelationship of internal and external audit: objectivity, competence and volume of work done. The results of Mautz (1984) show only a seemingly good relationship of internal and external audit. However, the cause of this condition was not found in his study. His findings are identical to the findings of Peacock & Pelfrey (1989). The most significant of them state that:

- Only 50% of internal auditors perceived the relationship between internal and external audits as excellent. Another 31% of internal auditors consider this relationship to be good;
- The relationship between internal and external auditors is good if corporate policy allows the internal auditor to respond to the criticism of external auditors;
- External auditors generally do not acknowledge the added value of internal auditors. At the same time, 92% of internal auditors appreciate that external auditors use their experiences. Internal auditors consider the most important operational audits and to be of less importance to their external audit support function. Same, the internal auditors believe that they carry out more than one third of the external audit work;
- The relationship of internal auditors to the management of the company is also different: The external audit firm has a Board of Directors and, where it is set up, an Audit Committee. Internal audit reporting at 80% is realized through the Audit Committee, with the Internal Audit Department having a defined relationship with the Committee.

Both of publications from Kaplan et al. (2008), Wang et al. (2015) analyze the possibility of coordinating both forms of audit later. As a reason for coordinating both forms, they point to the need to optimize the consumption of resources to improve business processes. They consider the objective characteristics of the audit as objective factors that influence the ability to coordinate internal and external audit activities.

According to ISO 19011, the differences can be identified in the subject matter and scope of the audit, the environmental and risk assessment of the audited entity by internal and external auditors and the degree of independence of the auditor (Table 1).

Table 1. Significant differences between internal and external audit.

Internal audit	External audit
Scope and focus	
Affected by the interests of the organization	Conformity of fact with the requirement according to the requirements of the interested parties
Term/execution interval	
continuously	at regular intervals
Role	
Compliance with fact, demand, reality and internal documentation	The conformity of fact with the requirement, fact and internal documentation, Assessing the severity of the impact of the identified disagreement on the functionality of the enterprise system (enterprise units, processes)
The auditor	
Professionally competent employees of the company. A unique external company	An independent external entity
Focus on	
The future of the company	Historical information that triggered the current state
Methods	
Risk analysis. Evaluation of the effectiveness of the control system	Monitoring the presence of steering mechanisms. Evaluation of the effectiveness of management tools
Subject of assessment	
The subject of the financial and non-financial aspects of the company's activities	The company as a whole, its organizational components, processes or activities
Knowledge of the functioning of the company and its processes	
Good	Obtained by observing, analyzing business documentation, communicating with the management of the organization, process owners and responsible employees

Source: processed according to ISO 19011.

5 AUDIT EFFECTS AND EFFECTIVENESS

Audit implementation into a management tools is enforced by a legal requirement in some types of companies. Many companies are deciding on their use on a voluntary basis. The reasons for such a decision are:

- Need to be assured of compliance with reality. The external auditor, as an impartial, expert, can identify areas where internal documentation is not in line with the new regulations. It can also identify the areas in which it finds disagreements of fact with the requirement caused by subjective factors;
- The interest in having a management tool to identify the occurrence of business or other business risks. In the case of financial audit, particular emphasis is placed on the risk of fraud. The internal audit controls activities at relatively short intervals, and improvements in control systems are based on their findings;
- The interest in having a management tool that forces business processes to improve and streamline its activities;
- Involvement to publicly present its socially responsible behavior in order to maintain your market position or gain a competitive advantage.

According to Survey IIA (2009), senior management and internal auditors consider it the most important contribution of internal audit to provide objective assurance that:

- Major business risks are managed in an appropriate manner and
- Risk management and the internal control framework function effectively.

They believe that internal audit is more likely to bring value by concentrating on its reassurance role. They do not require consulting activities. This is consistent with the definition of an internal audit: The internal audit consists in ascertaining whether the information is accurate and reliable, whether the risks identified and minimized, whether the external rules and generally accepted internal policies and procedures are complied with, fulfil the relevant criteria, resources are used economically, and the organization's goals are fulfilled effectively.

Using audit in practice confirms, that practice considers audit to be a useful management tool. Positive attitude to its use will only be maintained by the entity until the application of the audit in business management yields no benefit, or if the benefits received are not effective from the point of view of corporate management (or eligible external stakeholders). Based on the concept of effectiveness, the effectiveness of the audit is influenced by the merits of the activities performed in the audit process and the amount of benefits that audit implementation will entail in the management process. It is related to the possibility to coordinate internal and external audit activities. The risk of non-compliance with reality makes

auditing efficiency understandable. According to Wiliamson & Hobbs (2015), an effective audit process must provide the auditor's correct opinion, supported by appropriate audit evidence and professional judgment.

In order to determine the effectiveness of the audit, the internal audit activities and their justification should be specifically monitored, as well as specifically monitoring and analyzing the justification of the scope of external audit activities. Laker (2006) as the core elements of internal audit efficiency refers to resource security and competencies, independence, planning, monitoring and reporting. Similarly, Woolf (1992), Philip (1996), Dandago (2005) are of the opinion, that internal audit functions include, in particular, an assessment of the internal control system, whether the required principles were applied throughout the period in a prescribed manner, and a conformity assessment with the requirement.

The published opinions have led to the conduct of an impact assessment:

1. Internal audit of the effectiveness of internal control systems.
2. External audit of the effectiveness of internal audits in a set of public administration institutions.

Primary data was collected through a survey conducted by external audit units in public administration institutions. The main findings of the study are that:

- The Internal Audit Service in the selected sample of institutions is not effective because it is not sufficiently independent, has insufficient staff and is not prepared to submit internal audit reports on time. The findings of the survey point to an incorrect set-up of the internal control system, particularly in the area of approval, supervision and personnel controls. This contribution contributes to the inefficiency of the institution's internal control system.
- The implementation of the external audit positively influences the effectiveness of internal audit in the evaluated institutions. Respondents' reactions may be judged to be the reason for the critical external auditor's findings.

From this it can be concluded that effectiveness affects the quality of the auditor's work and the optimization of the activities and the forms of audit. Co-ordination of the work of the internal and external auditors improving efficiency enables. Coordination options follow from the auditor's work standards and the relevant ISO guidelines (ISO 19011). Under the provisions of the Standards, the external auditor must sufficiently understand the internal audit activities to help with planning the audit and developing an effective auditing approach. During the audit planning, the external auditor must perform a preliminary assessment of the internal audit function if the internal audit activity appears to be relevant for external audit in specific areas of audit. If it intends to rely on a specific internal audit work, the external auditor must assess and test the internal audit work to confirm its adequacy for the purposes of the external auditor.

The effectiveness assessment under the ISO provisions is to be carried out regularly as part of the audit process. The person who manages the audit program is responsible for evaluating the effectiveness of the audit. Its role is to identify and evaluate stakeholders' attitudes towards auditing activities and to respond to audit findings. The most commonly used methods of internal audit performance include the assessment of the percentage of the completed audit plan, the adoption and implementation of recommendations, feedback from the board (audit committee, senior management), feedback from the departments audited, ensuring proper risk management, reliance on external auditors for internal audit work.

6 CONCLUSION

Internal and external auditors have common interests. They have different choices and ways to achieve them.

The role of internal audit is determined by management. The usual requirement of management is to ensure that business systems (with an emphasis on the control system) are working. This will improve business processes. The fact that the internal audit is part of the entity causes the risk of non-evaluative evaluation.

The primary concern of the external auditor is to find out whether the internal control and management system in an entity is designed and operated to allow improvement of business processes over time. Unlike the internal audit, the entire business system may not be the subject of an external auditor's interest. The outsourcing method of conducting an external audit in the company means some limitations and advantages for the external auditor: The greatest constraint is the ignorance of the business environment, its conditions. The added value of the external auditor's evaluation, which allows the auditor's independence from the auditee, is beneficial. However, the external relationship between the auditor and the audited company entails a requirement to cover the extent of the monitoring and evaluation performed.

Some of the means to achieve external and internal audit objectives are similar. This allows coordi-

nation of activities. Internal audit's activities may affect audit risk and, as a result, the nature, timing and extent of external audit procedures. Regardless of the degree of independence and objectivity, the internal audit cannot achieve the same independence as an external auditor when expressing its opinion on its findings. Responsibility for expressing audit findings has an external auditor. Its responsibility is not diminished when using the internal audit. This fact indicates the willingness and limitations of external auditors to use internal audit findings in the audit process. At the same time, the implementation of an external audit in an entity positively influences the effectiveness of internal audit in the rated institutions. Respondents' reactions may be judged to be the reason for the critical external auditor's findings. These findings correlate with previously published opinions on the mutually beneficial partnership relationship between internal and external audits in entities.

REFERENCES

Accounting Research Foundation. 1995. Considering the Work of Internal Auditing. Australian Accounting Research Foundation. p. 13. Available at: http://www.auasb.gov.au/admin/file/content102/c3/AUS604_10-95.pdf.

Dandago, K. I. & Suleiman, D. M. 2005. The Role of Internal Auditors in Establishing Honesty and Integrity: Are the Watchdogs Asleep. Proceedings of the Third National Conference on Ethical Issues in Accounting, BUK, pp. 56–67.

Flint, D. 1988. Philosophy and principles of auditing. Hampshire: Macmillan Education Ltd.

ISO 19011. Guidelines for auditing management systems.

Kaplan, S. E., O'Donnell, E. F., Arel, B. M.. 2008. The influence of auditor experience on the persuasiveness of information provided by management. *Auditing*, 27(1), 67–83. DOI: 10.2308/aud.2008.27.1.67.

Laker, J. F. 2006. The Role of Internal Audit a Prudential Perspective. The Institute of Internal auditors.

Lee, T. 2008. The evolution of auditing: An analysis of the historical development. *Journal of Modern Accounting and Auditing*, 4(12). ISSN1548-6583. Available at: http://www.davidpublishing.com/Upfile/12/2/2012/2012120283233169.pdf.

Mautz, R. 1984. Internal and external auditors: how do they relate?, *Corporate Accounting*, 3(4), pp.56–58.

Peacock, E. & Pelfrey, S. 1989. How internal auditors view the external audit?, The Internal Auditor, Vol.46, pp.48–54.

Philip, S. 1996. The new face of Auditing. *Management Accounting and Auditing Journal*. 74(5), pp 26.

Porter, B., Simon, J., Hatherly, D. 2005. Principles of external auditing. John Wiley & Sons, Ltd.

Schneider, A. 1984. Modelling external auditors' evaluations of internal auditing, *Journal of Accounting Research*, 22(2), pp. 657–678.

The Institute of Internal Auditors Research Foundation, 2009. Supplemental Guidance: The Role of Auditing in Public Sector Governance. pp 26. https://na.theiia.org/standards-guidance/Public%20Documents/Public_Sector_Governance

Wang, X., Wang., Y., Yu, L., Zhao, Y., Zhang, Z. 2015. Engagement audit partner experience and audit quality, China *Journal of Accounting Studies*, 3(3), pp. 230–253, Available at: http://dx.doi.org/10.1080/21697213.2015.1055776.

Wiliamson, K. & Hobbs, A. 2015. Assessing the effectiveness of the external audit process. Ernst & Young LLP. http://www.ey.com/Publication/vwLUAssets/EY_Assessing_the_effectiveness_of_the_external_audit_process.

Woolf, E. 1992. Auditing to Day Second Edition. Prentice Hall international London (U.K) Department of Accounting.

New Trends in Process Control and Production Management – Štofová & Szaryszová (Eds)
© *2018 Taylor & Francis Group, London, ISBN 978-1-138-05885-9*

Key factors of mining sustainability of chosen mineral raw material

L. Domaracká, M. Taušová, M. Shejbalová Muchová & B. Benčőová
Faculty of Mining, Ecology, Process Control and Geotechnologies, Technical University of Košice,
Košice, Slovak Republic

ABSTRACT: In view of strategy we have to see raw material importance for state's progress. The exploitation of own raw material is efficient and the ignorance of this possibility would have been denial of property, which is evaluated by The Constitution of Slovak Republic and very meaningfully. It is important to define factors of sustainability of mining well, which will lead to right exploitation of mineral raw materials also in changed conditions. Within the scope of factor's formation it is necessary to take account of macroeconomic view, microeconomic view, peculiarities and to respect natural characteristics of the country and with regard to specifications of chosen raw material. The goal of this article is to define key factors influencing the mining undertaking regard to specifications given by character of industry. Significant factors of this age are the influence of science a technology.

1 INTRODUCTION

In Slovakia, a history of mining extends to 13rd century (Zámora 2003). From this time, mining went through important turn from alluvial mining and extraction mainly of surface mineral raw materials to underground mining by the using of modern techniques and technologies. The whole development was influenced expressively by changes in individual features of macro-environment and industry environment, which affected and still affect mining undertakings (Šimková et al. 2016). By the observation of sources and consequences we were able to set key factors manipulating the area of mining industry in Slovakia and the world, too. The article is about key factors of sustainability of petroleum raw material's mining.

2 THE FACTORS OF MACRO-ENVIRONMENT AND INDUSTRY

The macro-environment is a part of surrounding, which influences all business subjects equal not focusing on their purposes. It is the important work in a process of choosing a country for opening business activity and foundation of settlement. The individual features of macro-environment and assigned factors evaluate Table 1.

In recent years, the factor of scientific-technical environment is very important because modernization of this area allow to decrease investments costs and it helps to improve process of raw material's extraction. It leads to necessity to re-evaluate

Table 1. Key factors resulting from macro-environment.

Macroenvironment	Factors
Economical	- Inflation - Unemployment in regions - Market rates - Level of taxation
Scientific-technical	- New Technologies of extraction and processing of raw materials
Social-cultural	- Minimum wage - Knowledge of population
Demographic	- Age structure - Knowledge structure
Political and legal	- Support of mining - Restrictions of mining
Ecological	- Requests of recultivation and revitalization

Source: Own processing.

also viable deposits and to review their level of efficiency.

According to Porter (1994), industry environment influencing all undertakings in this industry can be defined through a model of 5 competitive forces. It is an operation of suppliers, users, substitutes, established and new undertakings (see Table 2).

The activity of mining undertakings is influenced by many factors resulting from specification of this industry. High investment requirements and a long preparatory phase of project's realization of raw material's mining could be considered as the most important.

Table 2. Key factors of industry environment.

Factors of industry		
Porter model	Suppliers	- Structure and amount of suppliers
	Users	- Structure and amount of users
	Substitutes	- Scarcity of raw material
	Established undertakings	- Amount and power of competitiveness
	Entry of new undertakings	- The existing barriers of entry at market

Source: Own elaboration.

3 FACTORS OF MICRO-ENVIRONMENT

3.1 A technological process of mining and processing of brash and gravel sound

The technological process of mining and processing of brash and gravel sound we can realize by two methods. There is a dry extraction and a water extraction. A choice of method is conditioned by natural conditions and setting of focus in environment (Beer et al. 2016).

In a case, that natural conditions allow to use more methods of mining and processing of raw materials, there is a serious problem for management of undertakings to consider all factors and to make right decision for security of effective choice of variant. A projecting of suitable technology has impact at the whole economy of service (Laciak & Šofranko 2013). Nowadays, in these cases, we use a methodology of multi-criteria evaluation of variant. By the example of choice of method of gravel sand's extraction, according to key criteria of evaluation in two undertakings LB Minerals, s.r.o., we will show the application of variant methodology evaluation with minimisation of subjective influences of evaluator.

3.2 Dry extraction

A mining of gravel sand by dry extraction is using in a case that a part of focus is above a level of groundwater. In this case there is made a chamber, from which we decrease the level of water by usage of pump. This chamber is bounded by focus from one side and by protective barrier by the other side, and it isolates it from body of water. In this chamber we can mine by using a caterpillar excavator, bucket wheel excavator or digging wheel excavator in a combination with a freight or also conveyor.

A withdrawing of water from chamber is continual during the mining to stop the increase of the level of groundwater. The material given by mining is taken to dump and then it is taken to primary processing.

3.3 Extraction from water

In the mining from water we can distinguish two ways of mining:

– A mining from water, when an excavator is set up at ground part,
– A mining from water by swimming excavator.

The mining from water when the excavator is set up at a ground part:
The mining by this way provides the option to mine into bigger depth—until 100 m. From the point of view of investment costs a significant advantage is a possibility to use classical extracting machines (caterpillar excavator, dragline, bucket wheel excavator). By comparison to swimming excavators they are cheaper. A disadvantage of mining, when the excavator is at the ground is an ineffective capacity utilization of focus. It is caused by pollution of water by the activity of excavator.

The extraction from water by swimming excavators:
In this case, there are more types of swimming excavators:

a. Swimming shovel excavator—a disadvantage of them is a limited scope of excavator, it is a level about 6–8 m. In the mining of bigger depth, there is necessity of bigger machines and it also means higher costs.
b. Swimming clamshell excavator—their advantage is a illimitable depth, but if depth is bigger, the mining is slower.
c. Swimming bucket wheel excavator—by comparison to bucket wheel excavators using in quarries the swimming bucket wheel excavator is specific by adverse movement of buckets.
d. Suction excavators—this type of excavator produces powerful suction. It is consisted of wide pipe, air intake tube and pump. A depth of mining is limited by capacity of engine. An advantage of this type is a low mechanic waste of excavator. The factuality of mining is about 5 cm and because of that there is possible correct regulation of pollution.
e. Hydro-pneumatic excavators (airlift)—they are quite similar to suction excavators but they don't use pipe for disintegration of material. They use water, which is lead to blast tube by high pressure. The advantage of them is that the mining by them is not limited by depth. On the other hand the disadvantage is a big waste of these types of excavators.

3.4 Multi-criteria evaluation of variant through a method of equal comparison

Multi-criteria evaluation belongs to methods of complex evaluation, through which we minimise the level of subjectivity by a choosing a right alternative. The objective of multi-criteria evaluation of variant is to define objective reality in the choice of standards policies and to formalize a problem. It means to assign it at mathematical model of more criteria situation. Methods of multi-criteria have the same goal—to evaluate number of solution's variants of problem according to chosen criteria and establishment of their policy (Straka et al. 2014).

Multi-criteria evaluation was done in this way: at first we defined key criteria for evaluation of variants. By the chosen method of equal comparison we designed weights for individual variants. After it we counted the weights of value for individual variants. We designed the most adequate variant for these conditions. Within the scope of multi-criteria methods we evaluated two concrete variants. The result should be the choice of optimal variant, by which we insure a right usage of mineral raw material by applicable technology. In technology of extraction we consider two options—V1 and V2:

– V1. Shovel excavator.
– V2. Bucket wheel excavator.

For evaluation and comparison we have defined four most important criteria. The evaluation was done by maximize approach. If criteria get more points from cardinal scale, the variant will fulfil defined criteria. The criteria are as follows:

A. mobility of machines,
B. mining from water,
C. costs for repair,
D. cost for mining.

A creation of table of criteria, where is defined criteria written down in a first line and column. Into patch of top triangle part we will write the criteria, which is in comparison between criteria in a patch with criteria in column more important (see Table 3).

Into the column k_i "number of criteria's presence" we will write a total number of criteria's presence in the whole top triangle part. If there is a

situation that there is a similar number of criteria's presence in number of criteria (Criteria A, B, C, D), there is necessity to use editing $k_i = n + 1 - p_i$, in this case, n is a number of criteria and p_i is order of i—criteria. It is from the most important. See the Table 4.

Calculation of weight values of criteria to standard design according to:

$$\alpha_i = \frac{k_i}{\sum_{i=1}^{n} k_i} \tag{1}$$

By equal comparison we will count the values of usefulness and then we change it to standard values of every criteria u_{ij}. Into lines of top triangle part we will write down the variant, which fulfil criteria better. The comparison of variant for individual criteria is in a Table 5. The usefulness of individual variant U_j was counted by:

$$U_j = \sum_{i=1}^{n} \alpha_i \cdot u_{ij} \tag{2}$$

where j = 1, 2, ..., m; m = the amount of evaluated variant; n = the amount of defined criteria; α_i = standard weight of i criteria; u_{ij} – the usefulness

Table 4. The equal comparison for individual criteria.

A	V1	V2	Number of occurrences of variants	u_{ij}
V1	–	V1	2	1
V2	V1	–	0	0
Σ			2	1

B	V1	V2	Number of occurrences of variants	u_{ij}
V1	–	V2	0	0
V2	V2	–	2	1
Σ			2	1

C	V1	V2	Number of occurrences of variants	u_{ij}
V1	–	V1	2	1
V2	V1	–	0	0
Σ			2	1

D	V1	V2	Number of occurrences of variants	u_{ij}
V1	–	V1	2	1
V2	V1	–	0	0
Σ			2	1

Source: Own processing.

Table 3. The creation of table's criteria.

	A	B	C	D	k_i	$k_i = n + 1 - p_i$	Standard scales
A	–	A	A	D	2	4 + 1 – 1 = 4	0.4
B	–	–	C	D	0	4 + 1 – 4 = 1	0.1
C	–	–	–	C	2	4 + 1 – 2 = 3	0.3
D	–	–	–	–	2	4 + 1 – 3 = 2	0.2
			Σ			10	1

Source: Own processing.

Table 5. A comparison at the base of the whole usefulness of variant.

Criteria	Weights α_i	V1 Shovel excavator		V2 Bucket wheel excavator	
		U_{ij}	$\alpha_i * U_{ij}$	U_{ij}	$\alpha_i * U_{ij}$
A	0.4	1	0.4	0	0
B	0.1	0	0	1	0.1556
C	0.3	1	0.3	0	0
D	0.2	1	0.2	0	0
Σ	1		0.9		0.1556

Source: Own processing.

of j variant according to i criteria; U_j = the whole usefulness of variant.

The problem was solved as maximalized, the variant with the most higher value of the whole usefulness is the most suitable solution. In this case, it showed that the better option is the variant with the usage of shovel excavator. The value 0.9 given for shovel excavator is higher than value 0.16 counted for bucket wheel excavator.

By this way, we can also evaluate the impact of other factors influencing a concrete mining undertaking.

4 CONCLUSION

The goal of this article was to define key factors influencing the mining undertaking regard to specifications given by character of this industry. We summarized the factors of macro-environment and industry. One of the most significant factors

of this age is the influence of science a technology—it highly decreases high investment seriousness typical for this industry (Kudelas et al. 2013). The changes in technologies help us to use new methods of extraction also at focuses marked as viable. The head of undertaking is often asked which of these methods of extraction is optimal and they have to consider more criteria. By the examples of gravel sand mining by the usage of multi-criteria methods we judged two variants of mining. We considered 4 key criteria of evaluation and we wanted to check possibilities of decreasing quantitative and qualitative variables, too.

REFERENCES

Beer, M. et al. 2016. Development of heat storage unit based on the phase change materials for mining machinery with combustion engines. *Acta Montanistica Slovaca,* 21(4): 280–286.

Kudelas, D. et al. 2013. Possible complications at transport of polymetallic nodules from seabed by means of self-propelled mining machine with flexible riser. In *Proceeding SGEM 2013,* Albena.

Laciak, M. & Šofranko, M. 2013. Designing of the technological line in the SCADA system PROMOTIC. In *Proceeding International Carpathian Control Conference, ICCC 2013, Rytro, Poland.*

Porter, M. E. 1994. *Konkurenční strategie: metody pro analýzu odvětví a konkurentů.* Praha: Victoria Publishing.

Straka, M. et al. 2014. Utilization of the multicriteria decision-making methods for the needs of mining industry. *Acta Montanistica Slovaca,* 19(4): 199–206.

Šimková, Z. et al. 2016. Strategy of point out relevance of responsible exploitation of mineral resources. *Acta Montanistica Slovaca,* 21(3): 208–216.

Zámora, P. 2003. *Dejiny baníctva na Slovensku.* Košice: Zväz hutníctva, ťažobného priemyslu a geológie Slovenskej republiky.

New Trends in Process Control and Production Management – Štofová & Szaryszová (Eds)
© *2018 Taylor & Francis Group, London, ISBN 978-1-138-05885-9*

High-growth firms: Multivariate estimation of growth potential evidence from Russia

E.V. Dragunova
Faculty of Business, Novosibirsk State Technical University, Novosibirsk, Russian Federation

ABSTRACT: This article presets an attempt to identify the features of fast-growing Russian companies and estimate their growth potential. In this study the companies operating in market for a long time (more than 5 years) are examined. The results of analyzing several National Rankings of the most successful High-Growth Firms (HGFs) are presented. There are highlighted the main restrictions and determinants of HGFs' growth. In this article is shown that main possibilities of the growth have to be relevantly described using the algorithm of a multidimensional assessment including formation of external environment profile (the analysis of deviating GEI-index values from the expected level according to the GDP of the country) and company's profile (the analysis of determinants). The study of growth potential provides an in-depth analysis of successful Russian ICT HGFs. The main types, intensity and trajectories of Russian ICT companies' growth are revealed.

1 INTRODUCTION

In conditions of globalisation any State is interested in the increase in the number of companies which demonstrate rapid growth rate of revenues and jobs. Nevertheless rapid growth rate of the company is hard to achieve and even harder to sustain within a long period of time. By identifying patterns and characteristics of high-growing company development, researchers give "ordinary" firms the opportunity to improve their performance.

The theoretical foundation is laid in writings of Greiner (1972), Penrose (1959), Storey (1994), Adizes (1988), Birch (1994), Delmar & Wiklund (2008), McMahon (2001), Moran (1998), Moreno & Casillas (2007), Phillips & Kirchhoff (1989), Gilbert et al. (2006). However in Russia interest in the considered subject arose quite recently and didn't reach desirable scales. In recent years, the economic nature of fast-growing firms has been readjusted and firms with high achievements in the target markets demonstrating a sustained positive trend of development over the past few years are considered to be "potential champions". Moreover national economic policies are gradually reoriented towards active and stable generation of such companies—potential leaders. In studies devoted to the analysis of problems and growth mechanisms is usually considered a category of firms showing fast and super-fast growth— gorillas, gazelles and high-growth enterprises. Our study focuses on enterprises that have been working in the market for relatively long time (more than five years), that is, "firms with history". The growth of these enterprises is the result of changes in policies, actions, behavior, etc., these changes which are drastic (qualitative, evolutionary leap) can significantly accelerate the growth of the firm. This is a clear example of a strategic renovation which enables a company to reconfigure its product as well as market position and to alter the structure of the resources and opportunities it used to have in the past to have better prospects for growth.

2 PORTRAIT OF A SUCCESSFUL COMPANY: BARRIERS AND DETERMINANTS

The notion of "development" is viewed by the representatives of technological companies as multiple changes in such performances as the creation of innovative and technologically complex products, start of new business directions; increased revenues, profitability; dominance in "their" market segment; entry into the world market; increase in the number of highly skilled jobs. The reasons that encourage firms to grow are the following: "long-term" welfare and protection against unfavourable business conditions, opportunity to familiarise themselves with the latest technologies, prestige and market power, expansion of the market and improvement of competitive position in the market through diversification of the product line, increase in profits and creation of resources for further reinvestment in business, access to subsidies, tax relieves and other incentives provided by support bodies at different levels. The analysis of the results of RosBusinessConsulting's (RBC) rating "50 fast-growing companies in Russia" for 2013–2016 enabled us to point out the characteristics of the most successful Russian companies development. Thus, it may be noted that industry structure of companies changes annually. For example, in 2013 about 70% of the RBC's ratings belonged to retail trade (chain stores) and construction, while in 2016 their share made up about 28%, but there were also power engineering, agriculture, information technology, and oil-gas sector represented on the list. In 2013 the average annual revenue growth rate (for three years) of all the companies in the rating was 65.74% and in 2016 it was 71%. The analysts of the RBC company in cooperation with the Association of Russian innovative regions and their partners identified key competencies of the fast growing companies such as availability of technologically innovative ideas, a strong research and development team, a high quality product and its full compliance with consumer demands, a successful business model of the company, modern production facilities and a strong leadership team (Rating "TexUp", Medovnikov et al. 2015). Average quantitative characteristics of technological companies with high growth rates are presented in Figure 1.

It is noteworthy that Russian entrepreneurs put the opportunity to receive "governmental support" and "additional financing" in the last place, thereby demonstrating their willingness to move to a new quality level at the expense of their own capabilities, i.e. to implement organic growth strategies. At the same time the leaders do not plan any takeovers or mergers with other companies in the coming years, that is, they do not demonstrate any desire to "grow at the expense of acquisition".

Among the factors encouraging growth in the medium term, the entrepreneurs point out: innovations available, contracts with large companies, reception of governmental order, introduction of new production capacity, optimisation of costs and readjustment of the company management, growth of consumer purchasing power, increase in sales in the world market.

Many of researchers pay attention to barriers to growth. They reveal institutional, internal, social barriers and barriers for globalization. According to the survey, Russian entrepreneurs often called the following barriers to growth: macroeconomic instability, bad infrastructure, legislation and regulations within taxation, labor market, strengthening of administrative pressure (bureaucratic processes), negative attitude of society to business, unwillingness to expend, strong competitors, limited access to long-term financial resources and operation space, limited access to market and productive resources (business services), and small loans for opening and development of the business, "criminal" component, lack of qualified employees, lack of standard procedures and methods of effective management of firms, competence of owner-managers, outdated equipment and technologies (RCSME 2010). Foreign researchers (OECD 2015) identify key obstacles for companies' growth, focusing on "improving the weaknesses of the firm": low availability and cost of external financing, limited internal sources of revenue, general management skills, financial management skills, local regulation, the relationship between the owners, shortage of qualified personnel, marketing skills, the state of information systems, search of innovation, accounting standards, employee health and safety regulations, customs duties and excise duties, institutional environment, depreciation of machinery and equipment, the low share of exports, competitive pressures, the complexity of implementation and compliance with quality standards.

3 THEORETICAL ASPECTS OF A MULTIDIMENSIONAL ASSESSMENT OF COMPANY GROWTH POTENTIAL

The growth prospects of the companies which are already operating are proposed to be determined by applying algorithm of multidimensional assessment of the company's growth potential, which includes the following aspects:

1. Evaluating the quality of the business environment (environment profile) and analysing the company's development dynamics at regional, national, global levels (Acs & Laszlo

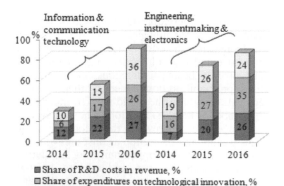

Figure 1. Average quantitative characteristics of technological companies with high growth rates for 2012–2016.

2009, Khayrullina 2014, Pushkar & Dragunova 2014).

2. Setting (adjusting) the goal and objectives of the company's growth, localising and clarifying the desirable areas of growth with respect to the company.
3. Identifying patterns and relationships between growth, business process management (Brush et al. 2009, Khayrullina et al. 2015).
4. Determining the enterprise growth potential (profile), comparing "benchmarking potential" with real performances and identification of deviations; providing recommendations for improving growth potential (Dragunova 2015).

In shaping the enterprise profile, we provide for considering both static (assessment of organic skills, indicators of growth, demographic variables) and dynamic components (type of growth, growth intensity and trajectory). The following alternative types of growth are highlighted: active and reactive types; short-term growth and long-term growth; increasing, steady, stagnating, faltering, declining and super growth (Bullok et al. 2004, Kemp & Verhoeven 2002). We investigate four growth trajectories developed by Brush et al. (2009): rapid, incremental, plateau, episodic pathways. We also calculate the EIM growth rate identified by Bangma & Verhoeven (2000) and divide firms into four groups (growth intensity): fast-growing companies with an EIM growth rate $>=1.5$, normal growers ($0.05<=$ EIM <1.5), stable companies ($-0.05<$EIM <0.05), shrinking companies (EIM $<=-0.05$). EIM growth rate is (1):

$$EIM = \left|empl_t - empl_{t-4}\right|^{0.25} \times \frac{\left(empl_t - empl_{t-4}\right)}{empl_{t-4}} \quad (1)$$

where $empl_t$, $=$ employment in year t; $empl_{t-4} =$ employment in year t–4.

Depending on the goal and objectives of growth, a great number of indicators which are applied to assess the determinants of a company's growth potential are defined as a set of growth indicators (changes in sales, profit, revenues, market share, export structure and volume, employment, image), a great deal of demographic variables (the age of the company, the stage of the life cycle, industry, regional affiliation), and numerous variables that characterise the company's organic capacity: entrepreneurial profile (ambition index and the owner's personal characteristics), marketing component, organisational change, information power and control level, financial conditions, innovation, material base (BIS Research Paper 2002, Dragunova 2006, Gupta et al. 2013, Mateev & Anastasov 2010, McMahon 2001, Morone & Testa 2008, Zhou &

de Wit 2009). This approach includes both financial and non-financial indicators, which is more in line with the real nature of economic growth. According to defined value of the growth potential, a company belongs to one of the four groups: regressive growth, stagnation, minor growth, significant growth (growing company).

When defining the quality of the business environment (building an environment profile), PEST characteristics of the environment can be studied, for example, as well as the overall level of business activity maintenance and other factors that the company's managers are not able to influence directly. Measurement of the entrepreneurship environment quality (entrepreneurship environment profile) involves studying the scale of the entrepreneurship development support in a country. The emphasis is on measuring the impact of innovation, technology quality, education and the availability of venture capital. One of the complex indicators displaying quality of the entrepreneurship in a country is the Global Entrepreneurship Index (GEI) (Acs & Laszlo 2009, Acs et al. 2015, Global Entrepreneurship Index 2013–2017). GEI is the tool which offers an in-depth view of difficult interactions of individual and institutional variables. These variables reflect the level of entrepreneurship development. The GEI is aggregated index made up of three sub-indexes for a country, such as Entrepreneurial Attitudes, Entrepreneurial Abilities and Entrepreneurial aspirations. These three sub-indexes stand on 14 pillars: opportunity perception, startup skills, risk acceptance, networking, cultural support, opportunity startup, technology absorption, human capital, competition, product innovation, process innovation, high growth, internationalization and risk capital.

4 TEST OF THE MULTIDIMENSIONAL ASSESSMENT OF THE COMPANY'S GROWTH POTENTIAL

4.1 Creation of an environment auspiciousness profile

We have conducted analysis of the index and its sub-indexes for the period of 2013 and 2017 to evaluate the opportunities of dynamic enterprise in Russia. According to the study at the beginning of 2017 the Russian Federation took the 72nd place from 138 countries to be evaluated. Meanwhile the United States, Canada, Switzerland and Sweden demonstrate the highest level of enterprise environment development. In comparison with BRICS countries, Russia fell behind China, South Africa, India, and outpaced only Brazil (for comparison in 2016 Russia was ahead of India and took the

68th place). Comparison of pillars for BRICS and European countries is shown in Figure 2. As the picture shows, Russia's greatest lag is observed in such pillars as "product innovations", "process innovations", "internationalization" and "technology absorption", which indirectly reveals relatively more complex conditions for the development of technological firms. In the year 2017 the "high growth" performance, which in 2016 favored Russia from other BRICS countries, showed deterioration, but it is still at a relatively high level.

Figure 3 shows the GEI of the countries in relation to the trend line describing the dependency between GEI and GDP per capita.

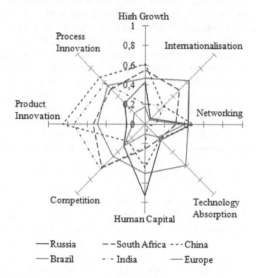

Figure 2. The quality of entrepreneurship (GEI) in BRICS & Europe, 2017.

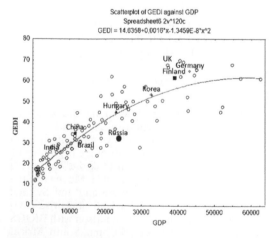

Figure 3. Distribution of countries GEI values in 2016 (according to GDP per capita, $R^2 = 0.81$).

On the basis of the graph, it can be assumed that the expected GEI value for Russia, with its GDP per capita ($22,452) taken into account, will make up 44.86, but the real value turns out to be substantially lower. The analysis of global enterprise indices from 2013 to 2017 leads to a conclusion that the assessment of the conditions for enterprise development in Russia (compared to other countries) used to be at the "below average" level up to 2014, but by 2016 it had almost reached the average level.

By building the same trends for the sub-indexes, we defined their expected value depending on the GDP level as well as the deviations from the real results (Table 1).

On the basis of Russia's GDP in comparison with other countries which participate in the rating, it can be concluded that the enterprise development potential in Russia has not been fully realised so far. It should be noted that, Russia surpasses even the developed countries that lead the rating, such as the US (1st), Australia (7th place), Sweden (4th place) in "human capital" performance. The strengths of the Russian business environment are presented by the following performances: "networking", "high growth", and "startups skills" (is not presented in Figure 2), which shows that there are potentially favorable conditions for the creation and existence of new companies. Nevertheless it is necessary to take measures to create and introduce innovations, to encourage marketing agglomeration and internationalisation by supporting the export of technological enterprises which will enable these companies to achieve high growth rate and absolutely new level.

4.2 Analysis of the development dynamics of successful enterprises (based on the example of IT industry)

In this paper, we have studied the type and growth trajectories of the firms that make up the "list of the largest Russian companies in the field of ICT", which is compiled annually by the agency "ExpertRA". It was decided to analyse company growth distinguishing between two time periods (separated by the crisis point). We have previously identified

Table 1. Estimation of sub-indexes GEI (Russia, 2016).

Denotation	Real value	Expected value	Relative deviation %
GEDI	32.2	44.86	28.22
Entrepreneurial ability	36.67	43.06	14.84
Entrepreneurial aspirations	26.08	44.89	41.90

growth trajectories that are characterised by a certain type of growth in each period. The description of the selected trajectories is presented in Table 2.

By forming the source database, we introduced some restrictions: businesses were supposed to have been rated for at least three consecutive years in the period from 2001 to 2008 (excluded) and from 2008 (included) to 2013, i. e. in two time intervals separated by the crisis. The firm was required to provide information on both the "revenues" and "employment", that is, there should be an opportunity to compare the growth rate of "owners interest" and increase in the number of jobs. These restrictions allowed to make calculations using "absolute" and "relative" growth criteria, i. e. not only to determine the direction of growth but also to measure its rate.

After verification of the restriction, out of 195 firms that had ever been involved in the rating, we selected only 31 enterprises according to the performance "revenues" and 28 enterprises on the performance "employment". The results of the companies distribution according to the type of growth in different time periods are presented in Table 3.

In conclusion, we calculated the EIM growth rate. Table 4 presents the growth intensity types of the Russian ICT firms.

The graphical interpretation of the trajectories (rapid, episodic, plateau) was made only for those firms whose continuous series of data from 2010 to 2013 were available (Fig. 4). Having analysed the development dynamics of ICT firms, we have drawn conclusions about the existence of firms which have been demonstrating growth in the Russian market for a long period of time, but their growth rates are different and are mostly related to slow growth with graphical visualisation "Plateau". The revenue growth rate is ahead of the employment growth rate in most cases. It should be noted that firms managed to survive during the economic crisis (2008), but it had a different impact on the further path of development, and only a few were affected positively.

In addition, we have marked a tendency to downsize activities (to focus on 1–2 directions), which are typical of almost all the companies firms that have been under scrutiny. The firms that had at least one strategic foreign partner and a project supported at the federal level took the best positions.

Table 2. The description of the growth trajectories.

Growth trajectories	Description	
	Period I	Period II
Super growth	Positive fast growth (PF)	PF
Increasing growth	Positive slow growth(PS)	PF
Faltering growth	PF or PS	Negative
Declining growth	PF	PS
Stagnating growth	N	N
Steady growth	PS	PS

Table 4. The growth intensity types of the Russian ICT firms.

Growth patterns (due to EIM estimation)	Number of firms
Fast-growing company	16
Normal growers	6
Shrinking companies	4
Steady companies	2

Table 3. The companies' distribution according to the type of growth.

Type of growth	Number of firms:			
	Revenue		Employment	
	Period I*	Period II**	Period I	Period II
Negative growth	0	4	0	5
Positive growth:	31	27	27	22
– PF	16	5	8	2
– PS	15	22	19	20
Zero	0	0	1	1
Total	31	31	28	28

*2001–2008 (excluded),
**2008 (included)–2013.

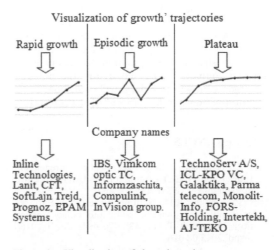

Figure 4. Visualization of the trajectories.

4.3 Creating a company profile

As it has previously been stated, a lot of indicators applied to assess the determinants of the company growth potential (company profile) are defined by a number of growth indicators, demographic variables and indicators that characterise the company's organic skills. Most of their values are defined for the long period and present a multitude of qualitative and quantitative indicators. To conduct the research and process the results the following methods were used: economic, benchmarking, logic-structural analysis, interviews, questionnaires (Dragunova 2016). The achieved values are compared with "benchmarking potential" and the most critical deviations are analysed.

5 CONCLUSIONS

The study focuses on the explanation of growth, its prerequisites and prospects for the "companies with history". It is proposed to analyse the growth prospects of such companies through a multidimensional assessment of the company's growth potential, which includes considering both static and dynamic performances and the creation of environment and company auspiciousness profiles. The environment auspiciousness profile is proposed to be evaluated by analysing the deviations of the GEI index values for a particular country from the expected level of its GDP. Dynamic characteristics for the company profile are formed by defining its growth trajectory, the growth type, the EIM growth rate. Static characteristics are provided by the calculation and analysis of growth indicators, demographic variables, and the indicators that characterise the firm's organic skills. The assessment of determinants should include information on changes in resource base, motivational activity, strategic adaptation, company's architecture. Considering them provides for multidimensional analysis of the potential for organic growth. The accumulation of structured information on the change in the determinants in particular and the potential of organic growth in general will contribute to the creation of "corporate memory" of the enterprise.

REFFERENCES

Acs, Z.J. & Laszlo, S. 2009. The global entrepreneurship Index (GEINDEX). Foundations and Trends in entrepreneurship 5(5): 341–435.

Acs, Z.J. et. al. 2014. National systems of Entrepreneurship: Measurement issues and policy implications. Research Policy 43: 476–494. doi:10.1016/j.respol.2013.08.016

Adizes, I. 1988. Corporate Lifecycles: How and Why Corporations Grow and Die and What to Do About It. Englewood Cliffs, N.J.: Prentice Hall.

Bangma, K.L. & Verhoeven, W.H.J. 2000. Growth patterns of Dutch firms ('Groeipatronen van bedrijven'). Zoetermeer: EIM.

Birch, D. L. & Medoff, J. 1994. "Gazelles". In L. C. Solmon & A. R. Levenson (eds), Labor Markets, Employment Policy and Job Creation: 159–167. Westview Press: Boulder, CO.

Brush, C.G., Ceru, D.J. & Blackburn, R. 2009. Pathways to entrepreneurial growth: The influence of management, marketing, and money. Business Horizons 52(5): 481–491.

Bullock, A. et. al. 2004. SME Growth Trajectories. A pilot study of UK SME growth and survival using the CBR panel data Report prepared for the Small Business Service, DTI.

Business growth ambitions amongst SMEs. 2012. Final report. BIS Research Paper 215.

Delmar, F. & Wiklund, J. 2008. The Effect of Small Business Managers' Growth Motivation on Firm Growth: A Longitudinal Study. Entrepreneurship: Theory and Practice 32(3): 437–453.

Dragunova, E.V. 2006. National features of Russian entrepreneurs: International Journal of Entrepreneurship and Small Business 3(5): 607–620.

Dragunova, E.V. 2015. Types and trajectories of enterprise growth. Problems of modern economics 3 (55): 179–182.

Dragunova, E.V. 2016. Multidimensional analysis of high-tech companies' growth potential. In Proc. of 13th intern. scientific-technical conf. on Actual Problems of Electronics Instrument Engineering (APEIE). Novosibirsk, October 2016. Novosibirsk: NSTU. V.3: 232–237.

Gilbert, B.A., McDougall, P.P. & Audretsch, D.B. 2006. New venture growth: A review and extension. Journal of Management 32(6): 926–950.

Global Entrepreneurship Development Index. Available from http://www.thegedi.org/.

Greiner, L.E. 1972. Evolution and Revolution as Organizations' Grow. Harvard Business Review 50(4): 37–46.

Growing small and medium businesses in Russia and abroad: the role and place in the economy. 2010. M.: RCSME.

Gupta, P.D. et. al. 2013. Firm growth and its determinants. Journal of Innovation and Entrepreneurship. Available from http://www.innovation-entrepreneurship.com/content/2/1/15

Kemp, R.G.M. & Verhoeven, W.H.J. 2002. Growth patterns of medium-sized, fast-growing firms. The optimal resource bundles for organizational growth and performance. Research Report H200111. Zoetermeer.

Khayrullina, M. 2014. Innovative Territorial Clusters as Instruments of Russian Regions Development in Global Economy. In Proc. of 21st Intern. Economic conf. of Sibiu—Prospects of Economic Recovery in a Volatile International Context—Major Obstacles, Initiatives and Projects (IECS): Sibiu, ROMANIA; May 2014. Procedia Economics and Finance. V.16: 88–94.

Khayrullina, M. et. al. 2015. Production systems continuous improvement modelling. Quality Innovation Prosperity 19(2): 73–86.

Mateev, M. & Anastasov, Y. 2010. Determinants of small and medium sized fast growing enterprises in central and eastern Europe: a panel data analysis. Financial Theory and Practice 34(3): 269–295.

McMahon, R.G.P. 2001. Growth and performance of manufacturing SMEs: The influence of financial management characteristics. International Small Business Journal 19(3): 10–28.

Medovnikov, D. et. al. 2015. The candidates for champions: the peculiarities of rapidly growing Russian technological companies, their development strategies and the potential of the State for supporting the implementation of these strategies. M.: NRU HSE.

Moran, P. 1998. Personality characteristics and growth-orientation of the Small Business Owner-Manager. International Small Business Journal 16(3): 17–38.

Moreno, A.M. & Casillas, J.C. 2007. High-growth SMEs versus non-high-growth SMEs: a discriminate analysis. Entrepreneurship & Regional Development. An International Journal 19(1): 69–88.

Morone P. & Testa, G. 2008. Firms' growth, size and innovation—an investigation into the Italian manufacturing sector. Economics of Innovation and New Technology 17(4): 311–329.

Penrose, E. 1959. The theory of the growth of the firm. New York: Oxford University Press.

Phillips, D.B. & Kirchhoff, B.A. 1989. Formation, growth and survival; small firm dynamics in the US economy. Small Business Economics 1(1): 65–74.

Pushkar, D.I. & Dragunova, E.V. 2014. Development of information system of innovation evaluation at the meso level. In Proceedings of the 12th intern. scientific-technical conf. on Actual Problems of Electronic Instrument Engineering (APEIE). Novosibirsk, October 2014. Novosibirsk: NSTU. V.1: 651–655.

Russia`s fast-growing high-tech companies National Rating ("TECHUP") Available from http://www.ratingtechup.ru/.

Russian Federation: Key issues and policies, OECD studies on SMEs and entrepreneurship, OECD Publishing, Paris. 2015 Available from http://dx.doi.org/10.1787/9789264232907-en.

Russian Information and Communication Technologies Ranking. Available from www.raexpert.ru

Storey, D.J. 1994. Understanding the Small Business Sector, London: Routledge.

Zhou, H. & de Wit, G. Determinants and dimensions of firm growth (February 12, 2009). SCALES EIM Research Reports (H200903). Available from: https://ssrn.com/abstract=1443897.

Implementation of business intelligence tools in companies

J. Dugas
Faculty of Business Economics with seat in Košice, University of Economics in Bratislava,
Košice, Slovak Republic

A. Seňová & B. Kršák
Faculty of Mining, Ecology, Process Control and Geotechnology, Technical University of Košice,
Košice, Slovak Republic

V. Ferencz
Ministry of Economy of the Slovak Republic, Bratislava, Slovak Republic

ABSTRACT: This paper presents the results of our own poll-based research on the implementation and usage of Business Intelligence (BI) tools in business management. The purpose of this paper is to show the most significant and frequent issues arising from the implementation of BI tools in Slovak businesses and, at the same time, identify the factors that are key to the most effective deployment of these tools in businesses. The paper also gives a brief run-down of key factors that determine a successful implementation of BI tools, organized into an implementation management flow-chart to be used by corporate management.

1 PRESENTATION OF RESULTS OF THE RESEARCH ON IMPLEMENTATION AND USAGE OF BI TOOLS IN BUSINESSES

1.1 *Poll-based research*

We carried out an online poll research with 110 participating businesses in Slovakia that have BI tools deployed and rely on them for their decision-making processes, regardless of the effectiveness of their usage. Table 1 shows the types of businesses that took part in our online poll research.

1.2 *Most frequent issues with the implementation of BI tools*

The outcome of the research carried out on a sample of Slovak businesses confirms that many

Table 1.   Types of businesses in research.

Business size:	(%)
24 Small business (49 employees)	21.9%
38 Medium-sized business (50–249 employees)	34.5%
48 Large enterprise (> 250 employees)	43.6%
Industry:	
49 Manufacturing	44.5%
22 ICT	20.0%
18 Wholesale and retail	16.4%
11 Hospitality services	10.0%
10 Finance and insurance	9.1%

businesses have encountered several problems and hurdles before and also during the actual implementation of BI tools, which prevented the optimal use of these tools in real life. The most frequent and serious issues that businesses have encountered include the following:

- The value, purpose, requirements and objectives of such implementation are not clearly and intelligibly defined,
- The end effect and expectations regarding BI tools are not realistically defined and/or the method to assess and measure the latter is not defined,
- There is no apparent connection between the BI tool usage strategy and the overall corporate strategy,
- Issues with relevance of input data,
- Lack of integration of already existing corporate systems,
- Doubts on the part of the users arising from the choice of BI tools,
- Use of wrong BI tools,
- Overly complicated BI tools,
- Preference for different tools,
- Weak or no support and/or lack of involvement of managers into the implementation of the BI project,
- Top-tier managers don't recognize the need to implement and rely of BI tools,
- Lack of skilled employees during BI tool implementation,

– Distrust of users towards a newly implemented tool or poor acquisition of skills required for proper BI use,
– Absence of an overall innovation-driven corporate culture.

These issues during implementation of BI tools have a direct impact on the success of each implementation that the management of each business must be aware of in order to achieve the best possible results with the deployment and day-to-day use of BI tools (Majerník et al. 2014).

2 KEY FACTORS FOR THE MANAGEMENT OF BI TOOL IMPLEMENTATION

Based on the most serious issues arising with the implementation of BI tools, we determined the factors that are key to the successful management of the implementation and the subsequent use of BI tools in businesses, based on which it is possible to identify the relevant aspects and formulate simple and logical recommendations for the implementation of these tools in businesses, as shown in Figure 1.

In order to achieve the most effective implementation of Bi tools and to ensure their best possible use in the managerial decision-making process in the day-to-day running of businesses, the following factors are key:

1. Vision, strategy and clearly defined targets for all areas relevant to the BI tools. This means defining—as clearly and precisely as possible— the value, the purpose and the requirements, as well as the main objectives pursued by the implementation of BI tools, including the metrics and methods of assessment that will allow for a successful implementation and allow management to meet business targets.
2. Availability and quality of relevant data. Business managers must strive towards a sustainable quality of all data stored in corporate systems, which can only be achieved by way of adherence to the tried and tested principled of corporate data quality management and the implementation of an appropriate database, document and knowledge-base management system.
3. Full-scale deployment following BI tool implementation. This means the involvement of all business units in an overall company-wide implementation of BI tools. This also ensures the most effective data collection from all available sources within the company.
4. Segmentation of all BI tool users. This involves identifying all relevant groups of users within the company and selecting the most appropriate BI tools for them to use. As businesses rely on a multitude of different employees and skills,

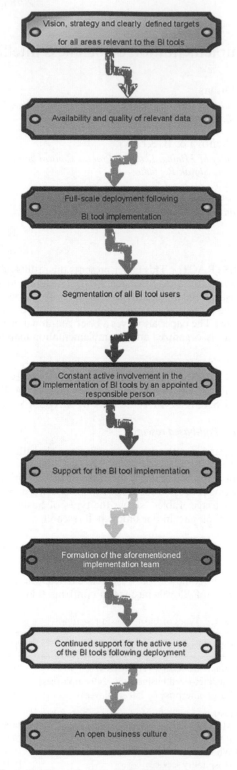

Figure 1. Flow-chart showing effective management of the implementation of BI tools in businesses.

the BI tools need to be selected and structured accordingly (i.e. technology, manufacturing, finance, personnel, etc.).

5. Constant active involvement in the implementation of BI tools by an appointed responsible person—so-called BI Implementation Team Leader. The main role of this person is to drive the process and ensure that proper support and funding for the implementation. In addition to having decision-making authority, the BI Implementation Team Leader should have be natural leader with the requisite respect, influence and dedication to drive a successful implementation. Business owners also need to be committed to a successful implementation and they need to be ready to back their commitment with appropriate financial support.

6. Support for the BI tool implementation not only on the part of business owners but also all managers in the company.

7. Formation of the aforementioned implementation team and appointment of qualified employees to lead to the process. The implementation team should consist of IT specialists as well as representatives of all business units.

8. Continued support for the active use of the BI tools following deployment. The idea is to gradually increase the active and independent use of available BI tools that have been deployed during the day-to-day work of all business units.

9. An open business culture that is open to change. It also involves the possibility for all users to participate in the implementation BI tools within the company. On the part of the business owners, it is important to ensure a timely preparation, communication of the planned changes and mostly the encouragement to facilitate information sharing within the company.

3 CONCLUSIONS

As indicated in various significant international papers and as indicated by our own research, the most frequent and most serious issued with the implementation and usage of BI tools in businesses can be avoided by way of identifying and managing the key factors that have an influence on such an implementation because of their fundamental effect on the success of such an implementation. The flowchart presented above shows the key factors that determine a successful implementation of BI tools in businesses and as such, it can be used to facilitate the implementation process and, following implementation, allow businesses to make the best use of these tools in their decision-making processes in order to better achieve their business targets.

REFERENCES

Čarnický Š. et al. 2016. *Faktory úspešnosti zavádzania a využívania Business Intelligence v riadení podnikov*, 1.vyd.—Košice: VÚSI, 2016.—108 s. ISBN 978-80-89383-35-1.

Čarnický Š. et al. 2015. Benefits and barriers to the introduction of Business Intelligence in Slovak enterprises—Empirical study. In *SGEM 2015*.- ISBN 978-619-7105-47-6. – ISSN 2367-5659.

Habiňaková, M. 2016. Factors Critical to a Successful Management of Business Intelligence Projects in Slovak Business Practice. In *Trends of education and research in biomedical technologies*. Univerzita Pavla Jozefa Šafárika v Košiciach, 2016, ISBN 9788081524707.

Majernik, M. et al. 2014. Innovative model of integrated energy management in companies. *Quality Innovation Prosperity*, 19(1), 22–32.

Pour, J. 2005. Faktoryúspěšnosti Business Intelligence—užití a řešení. In *Systémová Integrace*. ISSN 1210-9479.

Škanta, J. 2010. Kritické factory úspěšnosti Business Intelligence. In *IT Systems*. ISSN 1212-4567.

New Trends in Process Control and Production Management – Štofová & Szaryszová (Eds)
© 2018 Taylor & Francis Group, London, ISBN 978-1-138-05885-9

The use of the geographic information system for instrument approach design

S. Ďurčo, J. Sabo, D. Čekanová & T. Puškáš
Faculty of Aeronautics, Technical University of Košice, Košice, Slovak Republic

ABSTRACT: The idea of the article is to assess the use of satellite navigation systems in cases when aircraft approach aerodromes and heliports in the Slovak Republic. It is advantageous to use systems that do not need ground equipment located within the airport, for the activities with regard to the potential use of existing navigation systems for the non-precision approach procedures, approach with vertical guidance and precision approach of airplanes and helicopters at a small airport, or precision approach of helicopters to an appropriate landing area of the required dimensions—Final Approach and Take-Off area (FATO). The article further describes the general principles of construction of approach procedures and options for visualization and qualified estimation of instrument approach using geographical information systems.

1   INTRODUCTION

The Slovak Republic currently registers 27 public and private airports with regular or irregular operations, 11 heliports and 8 airports for aircraft work in agriculture, forestry and water management applicable for aircraft operations after obtaining airport operation permission. Instrument landing approach can be performed only at six of these airports.

In future years we can expect the growth of registered carriers and the number of registered aircraft, especially in the segment of small private helicopters and airplanes of light category weight and airplanes of A manoeuvring capabilities which will be used for sports, tourism, commercial flights, ambulance services and flights related to the implementation of aerial work in agriculture, forestry and water management and other sectors of the national economy. The development of the aviation activities will require expansion of the number of existing airports useful for operations in low visibility and building of new heliports and contact surfaces for Helicopter Air Ambulance Service, which enable operation even under difficult weather conditions (Antosko 2015).

It is advantageous to use systems that do not need ground equipment located within the airport, for the activities with regard to the potential use of existing avigation systems for the non-precision approach procedures, approach with vertical guidance and precision approach of airplanes and helicopters at a small airport, or precision approach of helicopters to an appropriate landing area of the required dimensions—Final approach and take-off area (FATO).

Designed approach procedures using satellite navigation systems are appropriate solution for operations at small airports with a limited number of movements in case of decreasing visibility (DOC 8168 2009).

2   THE USE OF SATELLITE NAVIGATION SYSTEMS

The requirement to use satellite navigation systems to secure navigation in airport terminals and for the use of exact approach methods in low visibility, increases demands for their navigation performance in terms of accuracy of the position information and in terms of the integrity of the values currently meet only by systems for precision approach ILS MLS.

Satellite systems GPS, GLONASS are nowadays used in the air traffic en route flights and for the navigation in the terminal areas, and at the end of last year the system GALILEO launched a limited operation. Especially, the system GALILEO, fully operated and with the functions for all planned services, estimated in 2020, will also bring a higher level of navigation performance in air navigation.

The current global satellite navigation systems in order to increase their performance on the level of existing terrestrial systems utilize advanced on-board equipment, ground equipment and other satellite systems. These extensions are called Airborne Based Augmentation Systems—ABAS (system expanded to include on-board circuits of integrity monitoring and using other on-board navigation sensors), Ground Based Augmentation System—GBAS (systems with the extension of terrestrial resources, including a network of terrestrial radio transmitters) and Satellite-based Augmentation System—SBAS (systems with the extension of land resources and utilizing communication satellite systems).

Exactly the SBAS systems are advantageous for use in navigation the aircraft on approaching small airfields, because it does not require placing of any ground airport equipment in the premises of an airport and related energy facilities, management, protection, and other activities that are economically and organisationally challenging and in many cases the planned use is infeasible.

SBAS is essential GNSS (Global Navigation Satellite System) extended by terrestrial means (designed to calculate and distribute corrections) and transmitters placed on the satellite of a communication system (used for transmitting the corrections and other navigation and control data). The principle of operation is based on monitoring the stations carrying out measurements on all visible satellites and transferring the found data of the station. The main station processes the data, and calculated values of corrections are transferred on aircraft board by means of geostationary satellites, where the corrections are used to calculate more accurate positions. For information transmission a signal (SIS Signal-in-Space) is used for frequencies in the L-band (L1/L2 and L5/L2C—planned frequencies).

In the Slovak Republic it is now possible to use the navigation information system called EGNOS (European Geostationary Navigation Overlay Service), which is the European version of SBAS. This system provides coverage of the entire territory with a signal of sufficient power for the execution of non-precision approach and the first category precision approach. We can assume that the system performance will significantly increase after introduction the fully operational satellite system GALILEO (Sabo 2014).

3 DESIGNS OF APPROACH PROCEDURES

The construction of aircraft approach procedures is a process to create a safe route, descent and approach to airport for landing. Each procedure depends on the navigation performance of aircraft systems that interact with ground navigation means of conventional landing systems, or with the space and ground segments of satellite navigation systems. In the procedure of designing, the engineer has to evaluate the possibilities of technical means to navigate through the areas of the procedures and safety area parameters for planned aircraft categories. This activity is carried out using theoretical calculations and suitable methods

of constructing, with the help of technical means, software tools and data sets characterizing the terrain and obstacle environment situation. The quality of these factors, that underpin the process of construction, is crucial to the final product that will meet all safety criteria.

The most common design of software products include professional software applications that allow computers assisted design—CAD and contain all the necessary tools for creating instrument procedures. The tool for proposing the common assessment of the surface planes barriers PANS-OPS Obstacle Assessment Surface Software and a design tool PANS-OPS Software (CD-101) belong to professional tools recommended by ICAO. The standards of ICAO PANS-OPS are a part of these applications.

Currently one of the most elaborate and most widely used software in the world for creating a comprehensive design of environment based on the standard documents ICAO PANS-OPS and the FAA TERPS is the WX1 Series™.

The procedure calculation, accomplished by a certified authority using these instruments, is very expensive and therefore problematic for the operator of a small airport. The initial evaluation of the most suitable approach procedure using satellite navigation systems due to the particular configuration of the terrain, obstacle and operational situation around the airport can be done by a qualified estimation using a geographic information system (Vagner 2014).

The Geographic Information System is a set of hardware and software for storing, processing and the use of geographic information in two forms—graphics and data interconnected and topologically arranged. It uses the information obtained from satellite and aerial photos and the GIS information system for imaging the earth's surface. It also enables measurement of the geographical parameters, inserting the overlapping maps, defining and displaying of track lines, flat and three-dimensional objects.

The companies ESRI, Intergraph, MapInfo and others belong to the most famous producers of GIS software. Software packages of listed companies provide a professional environment for surveying and for a wide range of other activities. But there is also free software such as GRASS GIS, or Internet Google Earth used to view and assess the possibility of underlying the track structure.

Google Earth Pro is a virtual global geographic information system developed by Google; it uses the information obtained from satellite and aerial photos and the GIS information system for imaging the earth's surface. It also enables measurement of the geographical parameters, inserting the overlapping maps, defining and displaying of track lines, flat and three-dimensional objects (DOC 9906 2009).

4 POSSIBILITIES FOR THE USE OF THE GEOGRAPHIC INFORMATION SYSTEM WITH REGARD TO ACCURACY

The possibilities of precision display and transformation of geographic positions are necessary to establish for visualization of evaluating obstacles planes OAS and qualified estimation of operating minima in GIS environment.

To determine the accuracy of displaying and transformation of geographic positions in Google Earth Pro, we chose the method of direct comparison of coordinate sets of identical points and calculated distances and directions from these coordinates with values (with the required accuracy) officially published in AIP SR. The tests were performed to determine the positional deviation dimensions between points, directions of positional deviation and the difference between measured amounts of points (Socha, 2016).

Each test variable has a calculated average position error, standard deviation (selective) and the confidence interval (with probability of 95%) for the mean value (Vittek, 2015).

The average positional deviation expresses the mean value, which is the position in the middle of groups of numbers in a statistical distribution. In this case, it is expressed by the arithmetic mean of the absolute values:

$$\bar{x} = \frac{\sum_{i=1}^{n}|x|_i}{n} \tag{1}$$

Standard deviation (mean square deviation) measures the dispersion of the values from the average (a mean value). It is defined as the square root of the sample deviation. We use it when we cannot determine the standard deviation because of a small number of measurements. For the function the following formula is used:

$$s = \sqrt{\frac{1}{1-n}\sum_{i=1}^{n}(x_i - \bar{x})^2} \tag{2}$$

The confidence interval for the mean value is the range of values in which the mean value of the sample x is in the centre of the screen, the range is $\pm$ x value of the confidence interval. The interval indicates the area where the actual parameter is likely to be. In our case we consider that reliability coefficient is equal to 95%. The critical value of the standardized distribution for reliability coefficient $P = 0.95$ is $z = 1.96$ and then the confidence interval can be expressed as:

$$\bar{x} \pm z\left(\frac{s}{\sqrt{n}}\right) \tag{3}$$

Coordinates of points, direction and distances of the Aeronautical Information Publication of the Slovak Republic (AIP), edition 19/11/2009, a part of Airports—the physical characteristics of the runway, air traffic barriers were used as an etalon for comparing the accuracy display of points, measuring the directions and distances in the orthophotomap of Google Earth Pro (observed shifts of an orthophotomap to the system of geographical coordinates) (see Tables 2, 3).

The points and directions were chosen to meet the conditions of the real situation and were identified in the two applications, so that the coordinates THR RWY, ARP and point objects published as barriers to air traffic. These points and directions are published with sufficient accuracy—the geographical orientation of runways is given with accuracy to the hundredth of a degree, the position of a displaced threshold to the nearest meter or foot, threshold elevation of the runway and geoid undulation to the nearest half meter or foot, geographical coordinates threshold runway in degrees, minutes, seconds and hundredths of seconds, the geographical coordinates of obstacles in degrees, minutes, seconds and tenths of seconds. The length of runways were calculated from the coordinates of THR RWY and compared with the appropriate measured distance in Google. Earth Pro. In the calculation of positional deviation the Vincenty algorithm was used for the distance calculation (see Table 1).

For comparison, the law of cosines and substitution radius of the sphere of values derived

Table 1. The calculation of deviation values of runway lengths calculated from published values of runway lengths at AIP SR using the Vincenty algorithm.

Mean linear deviation [m]		0.6560
Standard deviation [m]		0.2777
Confidence interval (95%) for the mean value [m]		0.3143
Interval of values with 95% confidence	FROM [m]	0.3417
	TO [m]	0.9703

Table 2. The determination of the positional deviation of points in the orthophotomap of Google Earth from coordinates of points in the AIP SR.

Type of deviation		dD [m]	d SMzRWY [°]
Mean deviation		2.4	0.05
Standard deviation		1.2	0.03
Confidence interval (95%) for the mean value		0.8	0.02
Interval of values with 95% confidence	FROM	1.6	0.03
	TO	3.2	0.08

99

Table 3. Establishment of distance and direction measured in the orthophotomap of Google Earth from a distance calculated from coordinates of points AIP and direction published in AIP.

Type of deviation		Distance [m]	Direction [°]	Height [m]
Mean deviation		6.1	184	3.9
Standard deviation		4.4	78	1.5
Confidence interval (95%) for the mean value [m]		2.0	36	0.8
Interval of values with 95% confidence	FROM	4.0	148	3.1
	TO	8.1	220	4.7

from the WGS-84 ellipsoid were used to calculate the lengths of runway deviations calculated from values published in the AIP runway lengths. The following calculated values were used sequentially: R = 6378.137 km; R = 6356.752 km; R = 6371.00 km; R = 6366.234 km; R = 6380.7105 km. After comparing calculations using Wincenty algorithm the assumption that the use of secondary radius is less accurate has been confirmed (Vincenty 1975).

5 CONCLUSIONS

The mean positional deviation determines a shift of an orthophotomap toward a system of geographical coordinates. The average value of the shift is within the range (four meters, 8 meters) at the prevailing southerly direction. This shift value calculated in the tested areas is satisfactory for simulation purposes of envelopes position for minimum safety height and other elements of approach manoeuvres in the area of Google Earth, but to preserve the objectivity of measurements, due to the expected uneven characteristics of the orthophotomap shifting, it is necessary to compare the results of interactions of envelopes with terrain map data in digital and classical maps.

The measurement of angles is possible to consider accurate due to the expected utilization of distances. Measuring of heights, according to the results, is sufficiently accurate, but this information is the most important as to the safety and therefore should be compared to the height data in the digital map and the classic map.

Simulation of positioning the envelopes of protective planes carried out in the Google Earth environment demonstrates the usefulness of the qualified estimation of minima of a small airport or helipad in the conditions of low visibility.

REFERENCES

Antoško, M. et al. 2015. One runway airport separations In *SGEM. Sofia: STEF92 Technology Ltd.*, 2015.
DOC 8168 2009. *Procedures for Air Navigation Services—Flight Procedures.* Vol. 1., 5th edition: ICAO.
DOC 9906 2009. *Quality Assurance Manual for Flight Procedure Design.* Vol. 1., *Flight Procedure Design Quality Assurance System.* 1st edition: ICAO
Sabo, J. et al. 2014. New trends of using GNSS in the area navigation. *Our Sea, 61(1-2)*: 1–4.
Socha, V. et al. 2016. Training of pilots using flight simulator and its impact on piloting precision. In *Transport Means 2016.* Juodkrante: Kansas University of Technology.
Vagner, J. & Jenčová, E. 2014. Comparison of Radar Simulator for Air Traffic Control. *Our Sea,* 61(1–2): 31–35.
Vincenty T. 1975. Direct and inverse solutions of geodesics on the ellipsoid with application of nested equations. *Survey Review,* 176: 88–93.
Vittek, P. & Lališ, A. 2015. Safety Key Performance Indicators system for Air Navigation Services of the Czech Republic, In *Transport Means—Proceedings of the International Conference.*

New Trends in Process Control and Production Management – Štofová & Szaryszová (Eds)
© 2018 Taylor & Francis Group, London, ISBN 978-1-138-05885-9

The issue of sustainable industrial growth concerning mineral resources

J. Dvořáček, R. Sousedíková & S. Matušková
Faculty of Mining and Geology, VŠB-Technical University of Ostrava, Ostrava, Czech Republic

Z. Kudelová
Ministry of Finance of the Czech Republic, Prague, Czech Republic

ABSTRACT: This paper has focused on relevance of mineral resources as regards the issue of sustainable industrial growth. Theoretical assumptions are made that underlie reliable delivery of minerals. History is given of the mineral mining industries in the Czech Republic. The authors analysed availability of selected raw materials and identified possible risks. The importance of minerals is relative to time, which fact has been highlighted. As such, possible future utilisations of minerals are put in context with the mine abandonment methods, and recommendations are made and action proposed for the choice of the best mine closure and abandonment practice. The main result of contribution is new approach to the mine closure taking into account possible future utilization.

1 INTRODUCTION

Mineral resources constitute the very foundations of human society, its function and development. Names of principal periods of human history derive from utilisations of dominant materials—stone, bronze, iron ages.

Published in 1972, The Club of Rome raised worldwide attention to mineral resources by its report, Limit of Growth. Raising the issue of limited availability of mineral resources paralleled by oil and raw material crises initiated asking questions about economic growth related to consumption of minerals. The concept of sustainable growth, which was a kind of reaction to the trend, requires growth that can satisfy not only the current but also future needs of mankind maintaining natural environmental balance (Law No. 17/1992 Coll., On the Environment). Exploitation of mineral resources goes hand in hand with mining industry. Let us focus our special attention to ore mining in the Czech Republic.

2 MINING INDUSTRY IN THE CZECH REPUBLIC

Archaeological finds of the historical Czech lands demonstrate that mineral deposits were exploited for production of various objects from the very beginning of human history. The ores played principal roles as humans learned about smelting of metals. Tin panning dates as far back as the seventh century B. C. Metal ores were excavated from the seventh to third century B.C. In the second century B. C., minting of gold coins started. In the ninth and tenth century A. D., mining and processing of silver ores began. The 15th and 16th centuries represent culmination of the Czech mining industries. Silver and gold mining dominated the scene. As economic development progressed, the importance of non-ferrous metals and iron were steadily growing. The Thirty Year War had a detrimental effect on mining industries. The ore mining balance was redressed in the 18th century in relation to reorganisations of the state government, armed forces, education, and justice. At the beginning of the 19th century, uranium was extracted to provide colour for glass and ceramics. During the World War I, many deposits of mineral resources of strategic metals were exploited. These were abandoned after the war. Also the post-World War I recession witnessed a decline of mining, which, during the World War II, began to reverse.

The post-WW II reconstruction asked for maximum exploitation of domestic resources, inclusive uranium. Disregarding profitability, intensive mining of ores took place until the sixties of the twentieth century. Later, efforts were made to increase profitability of mining industries. Economic and political changes after 1989 were of fundamental importance, as regards ore and uranium mining. In 1990, the Government introduced restrictive programmes for all mining industries. The extraction of uranium was stopped altogether, which was also the case of ore mining in 1994. The mining organisations must provide for controlled abandonment of mines, inclusive mine rehabilitation measures to mitigate environmental effects of mining activities.

Currently, there is no ore mining existent in the Czech Republic. Will it always be the case?

3 CONCEPT OF SUSTAINABLE GROWTH

Arguing about mineral resources after the Club of Rome report, Limits of Growth, led to polarisation of two camps—optimists and pessimists.

The optimists maintain: (i) Mineral resources are vast as the Earth is very large; (ii) Even extraction of poor ores is practicable thanks to innovative mining and processing technologies; (iii) Nuclear fusion will provide energy for exploitation of energy demanding mineral resources.

The pessimists on the other hand say: (i) Resources are vast but limited; (ii) Exploitation of poor ores asks for lot of energy; (iii) Mining of many poor ores inevitably implies nonrecurring detrimental effects in the environment.

The fact is that mineral resources are limited. In essence, a steady mineral supply for sustainable growth can be ensured by:

– Significant reduction of exploitation of resources in question, for example by employing innovative technologies (i.e. the exhaustion time limit is prolonged),
– Replacement of one mineral by another (the problem is transferred on another mineral resource),
– Continued recycling of products of specific mineral commodity, and recovery of the commodity from scrapped or decommissioned products (this option is not always practicable, and technical/financial feasibility must be taken into account).

Contemplation of mineral resources in relevance to country economies was on topic also in the past. In 1988, the Institute for Forecasting of the Czechoslovak Academy of Sciences issued Draft on Comprehensive Forecast for the Scientific, Economic, and Social Development in the Czechoslovak Socialist Republic until 2010. The Draft was based on the assumption: "... there is natural resource abundance worldwide, and it is only a financial issue to what extent and which structure domestic or foreign resources are exploited."

Let us focus our attention on the situation of some mineral commodities nowadays. Some mineral commodities were designated as "critical material with regard to security of supply" by the U. S. Department of Foreign Affairs. The British Geological Survey and the European Commission ranked some mineral commodities among "critical raw materials". The latter are especially those ores on which are based the following products:

– Originate mostly or exclusively abroad,
– Substitution is difficult,
– Important for national economy, and especially for defence of the country.

The EU listed these critical materials: Antimony, beryllium, borates, chromium, cobalt, coking coal, fluorite, gallium, germanium, indium, magnesite, magnesium, natural graphite, niobium, phosphates, platinum group metals, rare earth elements, silicon, tungsten (European Commission 2014).

Substantiation for such listing of individual mineral commodities can be given by the following facts:

Tungsten: China dominates both production and consumption of tungsten worldwide and such situation has exercised influence over the world economy for many years. The nineties of the 20th century were characterized by Chine's high and relatively lowly priced exports. This implied reduction or termination of activities of many mining companies operating outside China. China's subsidized production of tungsten led to limitation of mining projects for the commodity in other parts of the world. Since 2004, the global demand for tungsten has been rising. The Chinese government identified tungsten as strategic raw material and started exercising influence over world markets by application of extraction and export limits, reversed authorisation of export rebate, and introduction of export duties. China limited export of tungsten and increased import of it (Wolf Minerals 2013). Since 2012, China has been importing 50% of tungsten available worldwide, and their demand is fastest-growing.

Lithium: This mineral commodity is extracted from igneous rocks (i.e. ores) or continental brines. Regarding supply, lithium market concentration is high—four companies produce about 90% of lithium worldwide (Ebensperger et al. 2005). Lithium production from ores is much more expensive than that from brines but the latter takes much longer time to complete and as such, it cannot keep pace with fast change of demand. That is why the increasing demand will be rather satisfied by ore production, which may imply increasing prices.

Niobium: Many technologies need niobium, which cannot be substituted by any other element. The world biggest producer of niobium is Brazil, where it is extracted by surface mining. Canada follows with its underground mining. In the period 2009–2012, two Brazilian and a single Canadian mine produced 99% of niobium traded worldwide. In spite of the vast deposits of this raw material, it is listed among top risk natural resources by the U.S. Government vis-a-vis of high supply concentration and the importance of the element for the U.S. industry (NioCorp Development Ltd. 2016).

Graphite: Global trade in graphite is characterized by:

– Strong demand given by ever growing importance of "green technologies",
– Restricted supply caused by diminished production of western countries and Chinese control of the 70% of the market implying production quality decline, increasing costs of extraction, introduction of an export duty (20%) and VAT (17%).

Apart from economic measures, in the period 2011–2013, China applied restrictions on production of graphite for environmental reasons, and overtook control over some small private mines. Such policies and measures imposed a limitation on exports and imply serious difficulties in balancing supply and demand of graphite worldwide.

4 MINE ABANDONMENT AND ITS FUTURE OUTLOOK

Ore extraction in the Czech Republic was terminated in 1994. Underground workings were abandoned in 1995. The mine surface openings were closed by:

– Backfill of the mine shaft accompanied by land reclamation measures or backfilling of all underground workings,
– Placing a steel-reinforced concrete construction a few meters under the surface opening and filling the gap by waste material up to the surface level,
– Construction of a steel-reinforced concrete cap parallel to surface.

In each case, underground cavities are flooded. The first case excludes any mine reopening, which in general is also valid for the second case. The third case leaves an option open for removing of water from underground workings by pumping. It may seem that any reopening of abandoned mines is highly unlikely or just purely theoretical.

After several decades, apprehensions about safety of future supplies of raw materials and rising prices of mineral commodities led to renewal of ore exploration. In the Czech Republic, localities of Krusne hory and Slavkovsky les were or have been explored. Geological survey of the locality, Cinovec, is the most popular. The latter is performed by the Czech company, Geomet, which is supported by the Australian mining company, European Metals Holding. At Cinovec, extraction of tin and later tungsten goes back to Middle Ages. The current survey has concentrated on lithium, tin, and other elements. The Cinovec deposit of lithium has been identified as the biggest in Europe. Lithium carbonate of the locality is best suited for the production of lithium batteries, and as such its market price might be the highest.

Another case: utilisation of water from the flooded mines is in preliminary evaluation stages. This water might help solving problems of rainfall deficiency.

5 CONCLUSION

There exists no ore mining in the Czech Republic whatsoever. Nevertheless, in the framework of abating safety and environmental risks related to processes of mine abandonment, some changes of abandonment policies and techniques have been made recently. The paragenetic sequence in mineral formation implies that exploitability and value of minerals change with time and the abandonment process should consider the following factors:

– If a mineral deposit is depleted and there is no other mineral of interest in the rock massif, the abandonment implementation should be oriented by elimination or at least substantial abatement of safety and environmental risks,
– If minerals of interest have been left in a mine, and the mining activities were terminated for economic or environmental reasons, the choice of the abandonment practice should take into account possible renewal of activities in future, as economic conditions change, and extraction and environmental technologies develop,
– If other minerals have been left that have no utilisation at the moment, the abandonment practice should be the same as in the preceding case. Future developments of extraction and processing technologies may change the situation making a demanded commodity from an uninteresting one.

It is obvious that an outlook scale is that of decades. It can be assumed that current practices of mine abandonment, which often imply safety and environmental risks, should take into account future possibilities of mineral utilisation in the framework of sustainable industrial growth.

ACKNOWLEDGEMENTS

We thank organizers of the Project, MERIDA, Research Fund for Coal and Steel, Grant Agreement, No. RFCR-CT-2015-00004 that made investigation of this paper possible.

REFERENCES

About Niobium 2016. NioCorp Developments Ltd. 2016. [on-line] http://niocorp.com/index.php/about-niobium.

Cinovec lithium project: production of battery grade lithium carbonate from sodium sulphate roast 2016. European Metals 13 December 2016. [on-line] http://europeanmet.com/assets/13_Dec_2016_-MH_ASX_Lithium_Carbonate.pdf.

Demand and Pricing 2013. Wolf Minerals, 30. 1. 2013. [on-line]. http://www.wolfminerals.com.au/tungsten/demand-and-pricing.

Ebensperger, A. et al. 2005. The lithium industry: Its recent evolution and future prospects. [on-line] http://repositorio.uchile.cl/bitstream/handle/2250/124612/Ebersperger_A.pdf?sequence=1.

Iványi, K. et al. 2003. Rudné a uranové hornictví České republiky (Ore and Uranium Mining in the Czech Republic). Ostrava: Anagram.

Law No. 17/1992 Coll., On Environment.

Report on critical raw materials for the EU 2014. Report of the Ad hoc Working Group on defining critical raw materials. European Commission, May 2014.

Souhrnná prognóza vědeckotechnického, ekonomického a sociálního rozvoje ČSSR do roku 2010 (Draft on Comprehensive Forecast for Scientific, Economic, and Social Development in the Czechoslovak Socialist Republic until 2010) 2008. Prognostický ústav ČSAV (Institute for Forecasting of the Czechoslovak Academy of Sciences), August 1988.

New Trends in Process Control and Production Management – Štofová & Szaryszová (Eds)
© 2018 Taylor & Francis Group, London, ISBN 978-1-138-05885-9

Influence of new regulations of the law on costs of SMEs in eastern Poland

T. Dziechciarz

State Higher Vocational School Memorial of Prof. Stanislaw Tarnowski, Tarnobrzeg, Poland

ABSTRACT: The year 2016 was a period of considerable changes in the law in Poland. They concerned also tax regulations and modifications of conducting business activities by SMEs. The author of the present article carried out the questionnaire survey amongst owners of companies from Eastern Poland with the use of electronic mail and the Internet. It concerned the influence of mentioned changes on costs of functioning SMEs. The results of surveys presenting problems of the stability of the conducting business activities and conclusions following them were included in the article. They suggest that in spite of increasing fixed costs in companies entrepreneurs are prone to accept some new regulations of law because they minimize direct inspections in companies. However, other regulations—being a novelty—cause anxiety that they will contribute to the uncertainty in investing and the rise in bureaucracy.

1 INTRODUCTION

The beginning of the 90 s of the twentieth century, which was the socio-political and economic breakthrough for Poland began the rapid development of small and medium-sized enterprises (SMEs). They have become in the coming years the lifeblood of the country new market economy. It is these small, often family businesses that form the basis of gross domestic product (GDP). A very important element in the development of companies, often critical for their profitability, is the tax system occurring in their country of their operation. Stable and friendly tax system is characterized not only by low rates of income tax, financial or local but most of all their clarity and simplicity of calculation. Sometimes too complicated process of tax calculation in the company is not compensated by their low rate, contributing to the increase in operating costs of enterprises.

In Poland, over the last twenty years, the rate of tax on goods and services (VAT) has changed rarely and slightly (an increase in the base rate since 2011 of 22% to 23% and the reduced rate of 7% to 8%). Corporate tax CIT since the Polish entry into the European Union was at a constant level 19%. Excise duty rates and the most important local taxes changed only slightly. The most significant revenues to the state budget provides VAT, as with every year, on average, it represents nearly 50% of the volume of budget revenue (Wolański 2008).

The high degree of complexity of the Polish tax law, and in particular the interpretation of VAT and CIT, contributes to the problem of its collection by the tax authorities. Examples of this are the return of the input VAT from transactions of steel prod-ucts, scrap metal, petroleum products, and hard drive. According to estimates by audit company PriceWaterhouseCoopers International Limited, the so-called tax gap for VAT in Poland in 2016 was around 10.5 billion EUR, which accounted for about 2.8% of national GDP (<www.strefabiznesu.pl>).

To reduce and even in the future eliminate such high losses of the state budget in this respect, the Ministry of Finance introduced in 2016 a new law containing regulations involving the obligation of companies to transfer via the Internet the so-called Standard Audit File-Tax (SAF-T) to the tax authorities. It is within a few years to reduce the VAT gap. The obligation to transmit SAF-T applies to all private entities (excluding persons who are not VAT payers). At this point, a question arises about the costs that entrepreneurs must bear in connection with these regulations (Dobija 2001).

The introduction of SAF-T in 2016 coincided with the implementation of two innovative ideas of the government—the introduction of the new lower rate of 15% CIT and reforms involving a combination of tax administration, fiscal control and the Customs Service into a single body, called the National Tax Administration (KAS).

The new rate of 15% CIT from 2017 can only be used by taxpayers starting a business (only in the year it was launched) and the so-called "small taxpayers" whose income for the previous year did not exceed 1.2 million EUR. Implementation of the new regulations on CIT is undoubtedly beneficial for entrepreneurs. In contrast, the introduction of KAS on 1 March 2017 caused a number of negative organizational and social events such as the resistance of customs officers who lost some earlier privileges and confusion among the employees of the tax administration, fearing job losses.

In this paper, the author presents the results of research, whose leitmotif was primarily Standard Audit File-Tax. The examined data consisted of the opinions of entrepreneurs and SME owners on the financial impact and the cost of introducing SAF-T in their companies. In addition, entrepreneurs were asked for their opinion on the introduction of KAS and 15% CIT rate, as well as on improving of the transparency of their company to the Revenue and financial control authority.

2 THE RESULTS OF THE SURVEY

2.1 *Research methodology*

The research was conducted between November 2016 and March 2017. It was a questionnaire interview conducted directly by the author and through some companies' websites and e-mail. Entrepreneurs were asked some questions about the overall financial situation of business entities and effects of recent changes in regulations. Questions were sent to 67 randomly selected small and medium-sized companies concerned with production, commercial and service located in three voivodeships of Eastern Poland. The interviewees were chief executives, chief financial officers and chief accountants of these companies. Research has guaranteed anonymity and discretion.

46 questions were answered in complete answers. These were companies of all sizes and different ownership types (micro and small businesses):

- up to 9 employees – 5 companies,
- from 10 to 50 employees – 28 companies,
- more than 50 employees – 13 companies.

Among 46 entities—there were 39 share-holding companies, 5 natural persons conducting a business activity, 2 foreign companies. All business entities were VAT payers. Full accountancy was maintained by 41 companies, while 5 led a tax account of revenues and expenditures.

2.2 *The significance of tax changes resulting from the introduction of Standard Audit File-Tax*

Standard Audit File-Tax (SAF-T) is a new, revolutionary—according to the Ministry of Finance—IT solution for all Polish business entities. It is a pattern (standard) of the electronic file in which accounting and financial data documenting company's activities are stored. The data are divided into the following 7 groups (<www.sage.com.pl>):

- Records of purchase and sale of VAT,
- VAT invoices,
- Account books,
- Tax revenue and expense ledger,
- Revenue account,
- Warehouse,
- Bank statement.

Each entity is required to submit SAF-T—only via the Internet—to the tax authorities for tax inspection. Thus, the tax controls will become more efficient and rapid, often they will be done remotely—via the Internet.

Each of the documents sent by the company must be pre-compressed, encrypted and converted into a separate XML file. The size of the file cannot exceed 60 MB. The files are sent electronically to the database of the Ministry of Finance in the so-called "cloud". From there, they are downloaded and decrypted only by officials of the tax authorities.

The deadlines for introducing Standard Audit File-Tax are as follows:

- For large enterprises: from 1 July 2016—obligatory reporting in the SAF-T standard in relation to VAT records—every month together with VAT-7 declaration; other SAF-T structures must be reported on the tax authority's request
- For small and medium-sized enterprises (SMEs): from 1 January 2017—obligatory reporting in the SAF-T standard in relation to VAT registry—every month together with VAT-7 declaration; reporting of other SAFT-T structures (on the summoning of taxing authorities) will become obligatory from 1 July 2018.
- For microenterprises: from 1 January 2018—obligatory reporting in the SAF-T standard in relation to VAT registry—every month together with VAT-7 declaration; reporting of other SAFT-T structures (on the summoning of taxing authorities) will become obligatory 1 July 2018.

2.3 *The importance of the changes resulting from the introduction of KAS and a 15% CIT rate*

From 1 March 2017, the structure of the Polish tax administration has changed. The Act on National Tax Administration introduces a new—next to the tax authorities—control entity, which is the customs-tax office (Pokojska 2016). All old and new tax authorities are required to control companies using SAF-T. The introduction of KAS is, in author's opinion, an unusual change of organization linking the so-called uniformed services (customs officers) and tax offices—both until now subjected to other ministries. This can cause organizational and legal chaos resulting from the diverse privileges of these two different institutions. From the point of view of the entrepreneur in whom KAS is interested—the organizational chaos may be the reason for different interpretations of tax law, and consequently the instability of legal regulations. This is a very unfavorable change for companies and people doing business.

Lowering the CIT rate from 19% to 15% in 2016 only applies to so-called "small taxpayers" and newly established companies, and functions only for the first year after the start of business. Hence the scale of this new change is small. However, according to the author, this may be the first step towards preventing the transfer of profits of international corporations from Poland through tax optimization. This can be done provided the rate is extended to all companies operating in Poland.

Table 1. The results of the survey in SMEs—answers of the board members of companies.

	Affirmative responses	
	Number	%
1. Did the company had financial problems over the last 3–5 years due to the adaptation of the accounting system to the changing legal regulations?	23	50.0
2. Was/will the introduction of SAF-T be an additional burden on the company organization (e.g. training, changing the structure of the company, etc.)?	38	82.6
3. Was/will the effect of the introduction of SAF-T be an increase in fixed costs (e.g. hiring new employees with financial and accounting education and IT education, purchasing new software, etc.)?	38	82.6
4. Could SAF-T become for the company a tool for facilitating the tax and financial settlement?	28	60.9
5. Is it possible for SAF-T to improve or restore the transparency of the company in relation to national financial and treasury control institutions?	43	93.5
6. Will the introduction of the new 15% CIT rate contribute to a better functioning and growth in the number of SMEs?	40	86.9
7. Will the creation of the National Fiscal Administration in 2017 contribute to the efficient and objective service to entrepreneurs by officials?	6	13.0
The number of companies participating in the survey, which answered to the above questions	46	100.0

2.4 Research results

The results of the studies (Table 1) show that the introduction of new legislation for the vast majority of companies (82.6% of respondents)—some of which have been experiencing financial problems for some years—have increased business costs and deteriorated earnings.

Positive feedback from entrepreneurs was aroused by the introduction of a 15% CIT rate (86.9%). On the contrary, the negative effect in their opinion had KAS, which contributed to the chaos in their service at the tax offices.

Most important, however, are the conclusions of replies 4 and 5 in the questionnaire regarding the introduction of SAF-T. Despite the higher costs, entrepreneurs are willing to accept them considering that the functioning of SAF-T will contribute to greater transparency of their companies (93.5%) and improve the settlement of CIT, PIT and other local taxes (60.9%). The amount of these costs—the employment of an additional employee (computer scientist or accountant) or the purchase of an additional module in the company's financial and accounting system—may fluctuate around PLN 4,000–5,000 per month. For large companies, these are not significant operating costs, unlike for SMEs as they exceed 50 thousand zlotys a year.

3 CONCLUSIONS

Summarizing the questionnaire surveys conducted in SMEs in Eastern Poland, the author of this paper divides the three legislative changes described above into three subjects—SAF-T, 15% CIT and KAS (Table 1).

The most important change was the introduction of SAF-T in 2016. However, in the opinion of the author, it was late at least a dozen years. This should have happened earlier—when the world, as well as Poland, was in a period of economic prosperity (2002–2006). It would be possible to prevent many tax crimes then, and above all, to bring out billions of zlotys of VAT and CIT outside of Poland.

The latest financial data from the Ministry of Finance for the first two months of 2017show that the effects of the introduction of the SAF-T become visible fast. State budget revenues for January and February 2017 increased by PLN 4.7 billion and amounted to PLN 60.9 billion (comparing year to year). Such big sums The Ministry of Finance owes to a very high VAT receipts. Since the beginning of the year, they amounted to PLN 33.5 billion and were as much as 40% higher than in the same period of 2016 (<www.bankier.pl>).

The cost of introducing SAF-T has increased in most companies, resulting in a deterioration of their gross financial result. Despite this, entrepreneurs are willing to accept this situation, hoping that in the future a way of control in the form of SAF-T will eliminate direct control in the company. Until now, such inspections—carried out even with the earlier announcement, but at the headquarters—often disorganized the work of many people from the financial and accounting department, the warehouse staff, and above all from the board and management of the company. This contributed to the excessive workload of many employees, freezing of funds, deterioration of financial liquidity and, in extreme cases, even bankruptcy of the company, despite the fact that the defendants' allegations were untrue and the company was conducted in a fair manner. Until now, the rights of entrepreneurs have been less important than of the officials controlling them.

On the other hand, when considering whether SAF-T will stimulate or inhibit the development of SMEs, it can be stated that if it becomes a tool for a reliable control exercised discretely by tax and fiscal authorities—in the future it may be a stimulus for the development of all businesses. This is the expectation of the vast majority of people doing business in Poland.

The second change introduced—15% CIT rate for SMEs and startup companies—is significantly less important. This includes a very small number of companies. This is due to the fact that the company bears the highest operating costs in its initial period of operation. Hence, in the first year of operation, the gross financial result and the so-called CIT tax base are most often negative. So there is no tax, and the revenue of the Treasury in this respect is negligible. Despite this, in the opinion of both the author of the article and the entrepreneurs who responded to the questionnaire—this is a good, promising action of the Ministry of Finance in tax policy (Wala 2013).

However, the CIT problem concerns not SMEs, but international corporations that annually transfer from Poland profits in the form of dividends in billions of zlotys. According to the data of the Ministry of Finance, annual transfer of profits of foreign corporations operating in Poland to their subsidiaries abroad—consists of about 5% of the invested capital (Stańczuk 2016). In order to keep this money in Poland, it would be necessary to change the tax rules for foreign corporations based on the ones existing in Estonia. Foreign companies are totally exempt from CIT provided they reinvest profits in their country of business.

The third change—the introduction of KAS on March 1, 2017—is in the opinion of the author

and entrepreneurs most debatable issue. Theoretically, this change should simplify tax controls and tax services of companies. However, the way of its introduction results in organizational chaos and the lack of competence of treasury officials towards entrepreneurs. Such a situation of chaos usually occurs in the early stages of revolutionary change. Officials take the most wrong decisions at that time, interpretation of the rules is varied. All of this contributes to the feeling of uncertainty and instability among the owners and managers of companies.

It should be added here that together with the introduction of new rules on KAS an important rule on conducting inspections in companies was eliminated. So far, all the checks had to be preceded by a few days' notice from the controlling institution about the date and scope of the check. The new laws on KAS abolish this rule. In the opinion of entrepreneurs, it is contrary to the principles of economic freedom. This will increase the company's own costs and will reinforce a sense of instability.

In conclusion, it should be stated that entrepreneurs are most concerned about the chaos and variability of regulations, their ambiguity and different interpretations. Only the stability and immutability of legislation guaranteed over a long period of time will contribute to the economic growth of the country and the well-being of its citizens. These legislative changes do not fully guarantee this stability.

REFERENCES

Bankier. 2017. <www.bankier.pl/wiadomosc/drugi-miesiac-nadwyzki-w-budzecie-panstwa->. [retr. 2017–03–03].

Dobija, M. 2001. *Rachunkowość zarządcza i controlling*. Warszawa: Wydawnictwo Naukowe PWN.

Pokojska, A. 2016. Reforma KAS już ruszyła. *Dziennik Gazeta Prawna* (4386): B3.

Sage. 2016. <www.sage.com.pl/male-i-srednie/system-symfonia-2013>. [retr. 10–11–2016].

Stańczuk, M. 2016. Program ślepej wiary w państwo. *Rzeczpospolita* (10419): A7.

Strefabiznesu. 2016.<www.strefabiznesu.pl/wiadomosci/a/luka-podatkowa-vat-w-polsce-w-2016-r->. [retr. 2016–12–27].

Wala, M. 2013. *Zbiór przepisów podatkowych na 2013 rok*. Gorzów Wielkopolski: Wydawnictwo Podatkowe Gofin.

Wolański, R. 2008. *System podatkowy w Polsce*. Warszawa: Wolters Kluwer Polska.

New Trends in Process Control and Production Management – Štofová & Szaryszová (Eds)
© 2018 Taylor & Francis Group, London, ISBN 978-1-138-05885-9

Differences in the financial situation of municipalities in the Świętokrzyskie region

P. Dziekański

Institute of Law, Economics and Administration, Jan Kochanowski University in Kielce, Kielce, Poland

ABSTRACT: The management of financial resources, which affects the economic situation of the local government and the efficient achievement of goals takes on special meaning in the functioning of a unit territorial government. The article presents the financial condition and the level of socio-economic development of rural municipalities of świętokrzyskie voivodeship on the basis of synthetic index. Synthetic measure of financial condition and development indicate various level of studied units. Their value depends from the economic character of the region, as well as from the financial independence, the level of own income, local taxes or ongoing investment expenditure. The value of financial situation index fluctuated between 0.06 (Słupia Konecka) and 0.51 (Sitkówka Nowiny) in 2005 and 0.11 (Łubnice) and 0.49 (Sitkówka Nowiny) in 2014. A measure describing the development concluded in the range of 0.14 (Moskorzew) and 0.52 (Morawica) in 2005 and 0.16 (Imielno) and 0.65 (Morawica) in 2014.

1 INTRODUCTION

Finances are a synthetic expression of the potential economic development of the region. Prism of finance enables to make a comprehensive assessment of the functioning of local government units and its development abilities (Wojciechowski 2012, p. 234). Finances form the basis of public tasks implementation and determine the economic conditions of the local development (Glova & Gavurová 2012, Szabo 2015). They affect the economic situation of self-government and the achievement of targets or the ability to timely fulfill the obligations or to finance the development (Łukomska-Szarek 2012), understood as a process of qualitative and quantitative transformations regarding a given area. Rational management of financial resources requires independence in action and instruments which increase the efficiency and effectiveness of budget policy (Filipiak & Flejterski 2008, p. 22).

Each spatial unit has defined conditions of its development (inter alia, natural resources, human resources, fixed assets, financial resources). They are the resource base of the region, which contains natural, cultural, human and economic resources. Part of them is of external (egzogenous) nature, independent or almost independent from actions and activities of the local community, while other part are the internal (endogenous) condtions connected with the cumulating features of a given area.

2 AIM AND RESEARCH METHOD

The aim of the article is the description of financial situation and the level of socio-economic development of rural municipalities of świętokrzyskie voivodeship (71 units) with the synthetic measure. It indicates the problem of internal differentiation of the region and the spatial distribution according to the studied areas of action of local government units. The analysis is of statistic and dynamic nature. The basic source of data for the assessment of financial situation was Local Data Bank of Central Statistical Office for the years 2005 and 2014.

The financial situation or the development are processes the assessment of which may be done in multidimensional approach. Their analysis was preceded by the process of selecting diagnostic features (stimulants, destimulants). Their selected set was analysed in terms of variability (threshold value 0.15) and correlation (value 0.75), in order to eliminate those which are poorly differentiated and those which contain repeating information. If the feature is over-correlated with other features, the diagonal elements of the inverse matrix R-1 significantly exceed the value of 10, which means bad numerical conditioning of matrix R. The variables described in the Table 1 were used to assess the financial situation and the development of municipalities.

Diagnostic variables tend to have different titers, which prevents their direct comparison, let alone adding. Therefore, the normalization of the values of features was done, obtaining values contained

Table 1. Variables describing the financial condition and the development of municipalities.

Variables describing finance	Variables describing development
(1) share of own income in total income, S,	(1) population using water supply system, S,
(2) share of local taxes in total income, S,	(2) population using the sewage system, S,
(3) share of stamp duty in total income, S,	(3) population using the gas network, S,
(4) share of income from assets in total income, S,	(4) population using waste water treatment, S,
(5) share of income from PIT and CIT in total income, S,	(5) water supply network in km2, S,
(6) share of subvention in total income, D,	(6) sewage system in km2, S,
(7) share of donation in total income, D,	(7) gas network in km2, S,
(8) share of investment expenditure in total expenditure, S,	(8) balance of internal migration, S,
(9) share of current expenses in total expenditure, D,	(9) balance of external migration, S,
(10) share of expenses on agriculture and hunting in total expenditure, S,	(10) birth rate, S,
(11) share of expenses on transport and communication in total expenditure, S,	(11) population per km2, S,
(12) share of expenses on public utilities and environmental protection in total expenditure, S,	(12) demographic dependency ratio (in non-working age per 1,000 people of working age), D,
(13) share of expenses on housing in total expenditure, S,	(13) schooling ratio—primary schools, D,
(14) share of administrative expenditure in total expenditure, D,	(14) registered unemployed, D,
	(15) economic operators registered in Register of National Economy, S,
	(16) natural persons conducting economic activity, S,
	(17) area of forest land, S,
	(18) protected areas, S,

s-stimulant; d-destimulant.
Source: Own elaboration.

within the range [0,1] (Kukuła 2000). For that purpose, the zero unitarization method was used. Stimulants were unitarized according to the equation:

$$z_{ij} = \frac{x_{ij} - \min_i x_i}{\max_i x_i - \min_i x_i} \quad (1)$$

for destimulants

$$z_{ij} = \frac{\max_i x_i - x_{ij}}{\max_i x_i - \min_i x_i} \quad (2)$$

where: i = 1,2,...N; j = 1,2,...,p (N is the number of objects (municipalities), and p—number of features); z_{ij}—refers to the unitarized value of a feature for a studied unit, x_{ij}—refers to the value of j feature for a studied unit, max—maximum value of j feature, min—minimum value of j feature (Wysocki, Lira 2005).

Synthetic measure is the arithmetic mean of the transformed data. It assumes that achieved aggregate variable contains all the information provided by specific measures of structure. Synthetic measure is based on the non-model method by using the formula:

$$s_i = \frac{1}{p}\sum_{j=1}^{p} z_{ij} (i = 1,2,...,p) \quad (3)$$

where: si—synthetic measure in the studied period, zij—features of the structure of the synthetic index, p—number of features. The index takes values from the range [0,1]. Value closer to unity means that the object is characterized by a high level of the analysed phenomenon, whereas, the more the values are closer to 0, the less developed the object in the studied terms (Malina 2004, Młodek 2006, Pawlik 2011, Olak & Pawlik 2013).

At the end, the ordering and classification of objects according to the level of phenomenon was done on the basis of the values of the meter. Studied objects were divided into 4 quartile groups. At this stage, the assessment of disproportion in the level of financial condition index was also done (Bury, Dziekański 2012, pp. 7–29, Satoła 2015, pp. 115–123, Dziekański 2016).

3 THE FINANCIAL SITUATION OF A LOCAL DEVELOPMENT. SYNTHETIC DESCRIPTION

The analysis of budgetary income and expenditure provides information which enable to take decisions safe for the functioning of the self-government unit, regarding both current and future activity (Dylewski et. al. 2010, p. 74). The revenue base, which the self-government dispose of, and its stability is important for the process of socio-economic development of self-government territorial unit. The development capabilities of local governments depend on their own resources budget (Sokołowski & Żabiński 2011).

Financial situation shapes the process of development of self-government. It is a synthetic picture of current level of economic development of a given unit. Own income of municipalities indicate the foresight of the boards and the economic activity of the residents and their holdings. Investment expenditure indicates the endeavour of the municipalities to expand their possessions, which contribute to the improvement of living conditions of the residents and to the general socio-economic development.

Synthetic measure of financial condition indicates various level of studied unit, which is affected by the economic nature of the unit and function of the area, as well as the financial independence, the level of own income, local taxes or ongoing expenditure. The value of measure of financial situation fluctuated between 0.06 (Słupia Konecka) and 0.51 (Sitkówka Nowiny) in 2005 and 0.11 (Łubnice) and 0.49 (Sitkówka Nowiny) in 2014. The measure describing the development was contained in the range from 0.14 (Moskorzew) to 0.52 (Morawica) in 2005 and from 0.16 (Imielno) to 0.65 (Morawica) in 2014. In the analyzed period of time, in relation

of 2014 to 2005, in case of development measure 15 units decreased the value of measure (i.e. Sitkówka-Nowiny, Pacanów, Wilczyce), 5 units had the same measure (i.e. Wodzisław, Czarnocin), and 51 units improved the value (i.e. Kije, Bałtów, Morawica). In case of financial situation measure, 56 units presents higher measure (i.e. Nowy Korczyn, Fałków, Tuczępy), 15 units decreased the value (i.e. Iwaniska, Waśniów, Pacanów), while 4 did not change (i.e. Klimontów, Bieliny).

The analysis enabled the division of rural municipalities into 4 groups on the basis of the value of quartiles according to the synthetic measure of financial condition. In 2005 19 units (average value of measure 0.28) were attached to group A (the best), 17 units (0.19) to group B, 18 (0.15) to group C and 17 (0.11) to group D; in 2014 accordingly 21 (0.30) to group A, 15 (0.23) to B, 21 (0.18) to C and 14 (0.14) to D. Time displacement may be noticed between the groups. Information about the affiliation of a unit to a group is important for an investor who is looking for location for a company and it may have practical meaning both for the authorities of the unit, and for the central authorities (Table 2; also see Fig. 1).

In 2014 in comparison to 2010, the differentiation according to the financial condition as well as the development did not change (standard deviation

Table 2. The level of synthetic index of financial condition of rural municipalities of Świętokrzyskie Voivodeship (2005, 2014).

	Financial situation measure		Development measure	
	2005	2014	2005	2014
A very good	Sitkówka-N. 0.51	Sitkówka N. 0.49	Morawica 0.52	Morawica 0.65
	Masłów 0.37	Tuczępy 0.40	Sitkówka N. 0.49	Zagnańsk 0.49
	Morawica 0.35	Morawica 0.39	Brody 0.48	Strawczyn 0.47
	19/0.28	21/0.30	19/0.39	18/0.43
B Good	Szydłów 0.22	Piekoszów 0.24	Raków 0.31	Mniów 0.33
	Samborzec 0.21	Bejsce 0.24	Solec Z. 0.31	Mirzec 0.33
	Pacanów 0.20	Michałów 0.24	Bieliny 0.30	Tuczępy 0.33
	17/0.19	15/0.23	17/0.29	22/0.30
C Weak	Bieliny 0.16	Solec Z. 0.21	Radoszyce 0.26	Łopuszna 0.26
	Ruda M. 0.16	Moskorzew 0.21	Lipnik 0.26	Skarżysko K. 0.26
	Wiślica 0.16	Radków 0.21	Wilczyce 0.26	Sadowie 0.26
	18/0.15	21/0.18	17/0.24	13/0.25
D Bad	Radoszyce 0.13	Waśniów 0.15	Słupia K. 0.21	Smyków 0.23
	Pawłów 0.13	Pacanów 0.15	Skarżysko K. 0.21	Złota 0.23
	Solec Z. 0.13	Klimontów 0.15	Czarnocin 0.21	Ruda M. 0.22
	17/0.11	14/0.14	18/0.19	18/0.20
Minimum	Słupia K. 0.06	Łubnice 0.11	Moskorzew 0.14	Imielno 0.16
Maximum	Sitkówka N. 0.51	Sitkówka N. 0.49	Morawica 0.52	Morawica 0.65
Standard deviation	0.07	0.07	0.09	0.09
Interval	0.45	0.38	0.38	0.49
Variability	0.40	0.32	0.31	0.31

Source: own elaboration based on data from Local Data Bank of Central Statistical Office (in Table 3 the best units in a group; number of units and average value of measure in a group).

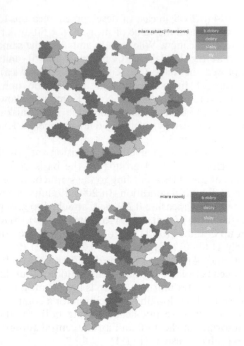

Figure 1. Cartogram of spatial distribution of synthetic measure of the financial situation and development in 2014.

0.07 and 0.09), while at the same time the interval decreased from 0.45 in 2010 to 0.8 in 2014 (financial situation measure; decrease of differentiation of the municipalities). At the same time, we observe the increase of growth of the interval of development measure from 0.38 to 0.49 (which may be interpreted as the increase of differentiation of the units; Table 2).

In the group of rural municipalities of świętokrzyskie voivodeship, there is a positive and negative correlative dependency between the synthetic index of financial condition and the development measure (Fig. 2). Leading units of the studied area are suburban municipalities with well-developed area of small and medium-sized enterprises.

The value of correlation between the measure of financial situation and development amounted to 0.586 (in 2005)–0.476 (in 2014). This may suggest that each of analysed measures indicates that in this period of time we had to deal with divergence and the spatial differentiation according to the studied measures was rather stable (Fig. 2, Table 3).

In Figure 3, the correlograms describing relations between the taxonomic transformations of development measure and its level were presented. From them we can conclude that the measures were subjected to divergence in the years 2005–2014 (Pearson correlation coefficients in the studied period of time and the level decreased; and in case of development measure they had negative value, Table 3).

The analysis of correlation indicates positive and negative correlation between the index of financial condition and of development and the selected financial measures. However, it may be indicated that the economic development is a result of the level of own income, income from PIT and CIT, subvention and investment expenditure.

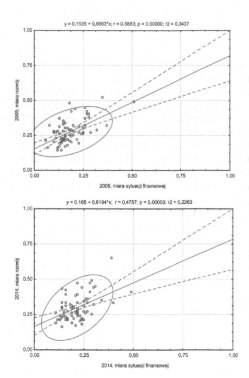

Figure 2. Relation financial-development measure in the context of rural municipalities of Świętokrzyskie Voivodeship.
Source: Own elaboration.

Table 3. The value of the correlation of taxonomic measure and its changes.

		R Pearson	Tau spearman	kendalla	Gamma
Relation	2005	0.586	0.518	0.379	0.392
F—R	2014	0.476	0.391	0.272	0.283
Relation	2005	0.519	0.522	0.381	0.389
F—dF	2014	0.347	0.339	0.232	0.237
Relation	2005	0.252	0.256	0.205	0.210
R—dR	2014	−0.016	−0.026	−0.039	−0.040

F—financial situation measure; R—development measure; dF—change of financial situation measure; dR—change of development measure
Source: Own elaboration based on data from Local Data Bank of Central Statistical Office.

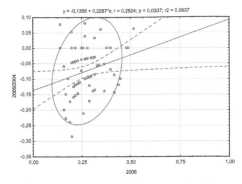

y = -0,1368 + 0,2287*x; r = 0,2524; p = 0,0337; r2 = 0,0637

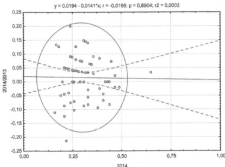

y = 0,0194 - 0,0141*x; r = -0,0166; p = 0,8904; r2 = 0,0003

Figure 3. Relation of development measure in the context of the dynamics of its transformations. Source: own elaboration.

4 CONCLUSIONS

Financing the developmental activity becomes incredibly difficult because of the limited financial resources in the sector of public finance and the possibility of indebtedness of municipalities. Today, the recognition of the range of social, economic and spatial development at the local level becomes particularly difficult, because the municipality operates and develops as an integral part of the whole.

The distribution of the assessment of the financial situation and the development of rural municipalities of świętokrzyskie voivodeship is spatially polarized. We are observing a positive correlation between the measure of financial condition and the level of development of municipalities. The economic nature of the region affects the financial situation and the level of development (agriculture—worse situation, rural area with majority of industry—better situation) and the location in relation to strong urban units.

The use of the method of multidimensional comparative analysis to the recognition of the intraregional diversity enables the evaluation and comparison of the units. This knowledge can be used by local authorities of the region to assess the effectiveness of past development instruments or financial management.

Collected measures depend on the number and type of variables taken for study. They give a comparative picture between municipalities which are subjects of analysis and enable to identify weaker and better areas of functioning of the unit.

REFERENCES

Bury, P. & Dziekański, P. 2012. *Porównanie wybranych elementów budżetów gmin województwa świętokrzyskiego*, pp. 7–29. In: P. Dziekański (ed.), *Gospodarka lokalna drogą rozwoju regionu*, Wyd. Stowarzyszenie Nauka, Edukacja, Rozwój, Ostrowiec Świętokrzyski.

Dylewski, M. et al. 2010. *Metody analityczne w działalności jednostek podsektora samorządowego*. Difin. Warszawa. 74.

Dziekański, P. 2016. *Spatial Differentiation of the Financial Condition of the Świętokrzyskie Voivodship Counties*. Barometr Regionalny. Tom 14. nr 3. pp. 79–91.

Filipiak, B. & Flejterski, S. 2008. *Bankowo-finansowa obsługa jednostek samorządu terytorialnego*. Wydawnictwo CeDeWu.pl. Warszawa. 22.

Glova, J. & Gavurová, B. 2012. *Perspektívy hodnotenia intelektuálneho kapitálu*. In: Hradecké ekonomické dny 2012. *Ekonomický rozvoj a management regionů: sborník recenzovaných příspěvků*: 31.1.–1.2.2012, Hradec Králové. Gaudeamus, pp. 73–78. http://fim.uhk.cz/hed/data/sbornik/HED%202012%20I%20cz.pdf.

Kukuła, K. 2000. *Metoda unitaryzacji zerowanej*. PWN, Warszawa.

Łukomska-Szarek, J. 2012. *Analiza wskaźnikowa w procesie zarządzania finansami samorządów lokalnych*. Studia i Materiały. Miscellanea Oeconomicae Rok 16, Nr 2.

Malina, A. 2004. *Wielowymiarowa analiza przestrzennego zróżnicowania struktury gospodarki Polski według województw*. Wyd. AE w Krakowie. Kraków. pp. 96–97.

Młodak, A. 2006. *Analiza taksonomiczna w statystyce regionalnej*. Difin. Warszawa. pp. 28–32.

Olak, A. & Pawlik, A. 2013. *Wrażliwość regionu na zmiany*. Ostrowiec Św.: Wyd. WSBiP

Pawlik, A. 2011. *Zróżnicowanie rozwoju społeczno-gospodarczego w województwie świętokrzyskim*. In. Wiadomości Statystyczne. nr 11. PTS. GUS. Warszawa. pp. 60–70.

Satoła, Ł. 2015. *Kondycja finansowa gmin w warunkach zmiennej koniunktury gospodarczej*. In. Journal of Agribusiness and Rural Development, 1(35), 115–123.

Sokołowski, J. & Żabiński, A. (ed.) 2011. *Finanse publiczne*. Wyd. UE we Wrocławiu. Wrocław 2011.

Szabo, S. et al. 2013. *Trust, innovation and prosperity In: Quality Innovation Prosperity*. Vol. 17, no. 2, p. 1–8.

Szabo, S. 2015. *Determinants of Supplier Selection in E-procurement Tenders* In: Journal of Applied Economic Sciences. Vol. 10, no. 7(37), pp. 1153–1159.

Wojciechowski, E. 2012. *Zarządzanie w samorządzie terytorialnym*. Warszawa.

Wysocki, F. & Lira, J. 2005. *Statystyka opisowa*. Wyd. AR im. A. Cieszkowskiego w Poznaniu, Poznań.

New Trends in Process Control and Production Management – Štofová & Szaryszová (Eds)
© 2018 Taylor & Francis Group, London, ISBN 978-1-138-05885-9

Management for operation of the aviation electronic security systems

M. Džunda, D. Čekanová, Ž. Miženková & Z. Šusterová
Faculty of Aeronautics, Technical University of Košice, Košice, Slovak Republic

ABSTRACT: In the presented paper we are introducing the principles of modelling and simulation for operation of an aviation electronic security system. We have created the specific model for an air operation of an electronic security system in the program MATLAB-Simulink, which we use for the simulation of economic efficiency in the process of air traffic control. Further, based on the method of Monte Carlo, we model the mean time trouble-free operation of an air electronic security system and by simulation of mean time to failure in operation, we find out the time in which the system is available during the observed period, as well as the time in which the system, due to failure or periodic inspection, is out of service. Based on this data, we can control aviation electronic security systems, determine their reliability and financial losses on the disposal of these systems out of service.

1 INTRODUCTION

A mathematical and computer simulation in today's modern world is an essential part of scientific—research activities (Vagner 2014). All computer simulations and simulation languages have their basic structure and rules to be followed. The aim of this paper is to describe the basic principles of mode-profiling and simulation of mean time to failure of an aviation electronic security system (AESS). We have created a particular model of such a system and performed a simulation of the mean time to in operation in the programming language and MATLAB-Simulink.

When creating a model of mean time to failure in operation of AESS we use some types of probability distribution of random variables, which are suitable for its modelling, while we are considering the mean time to failure in operation of AESS a random process. Modelling of AESS operation requires the use of statistical method synthesis of electronic warfare systems and adherence to the basic principles of the mathematical—computer simulation. The creation of algorithms, by which we can model the mean time to failure in operation of AESS, requires perfect knowledge of such systems and the management of their operations at the airport (Melnikova 2016). Based on the evaluation of operation management processes AESS at the airport and knowledge of their operating parameters it is possible to create algorithms by which we can model the mean time to in operation AESS (Sebescakova 2013). Consequently, it is necessary to make an entry model created in the selected language of simulation and perform the simulation of the investigated process. To simulate

the mean time to in operation it is appropriate to use Matlab Simulink because of its clarity and general availability. MATLAB is a high performance language for technical computing. It integrates computation, visualization and programming in an easy to use environment (Socha 2016). It can be used mainly for engineering calculations, development of algorithms, modelling, and simulation etc. The connection of MS Excel with MATLAB is the progressive enlargement for easy data entry and presentation of results (Smith 2014).

Using performed simulations of mean time to failure in operation we can detect the time at which the system is available for the observed period of time in which the system is out of service due to failure or periodic inspection. Based on these data, we can find out its reliability and hence financial losses in the inability to use the system (Tobisova 2014).

2 THE THEORY OF RELIABILITY

Although AESS has a reliable design, its reliability may not be satisfactory in the operation. The reason for this low reliability may be its poor production. Despite the fact that AESS has a reliable design, the operation may be unreliable. The reason for this may be a non-standard manufacturing process (Cekan 2014). As an example we can mention cold solder joints that are a part of AESS. These connections could pass the initial testing by the manufacturer, but the failure takes its effect in operations as a result of cyclic thermal stress or vibration. This type of fault has not appeared as a result of improper design, but rather is the result of substandard manufacturing processes. Therefore, regarding a particular type of AESS, the design may be reliable, but its quality is unacceptable because of the production process.

2.1 Introduction to the theory of reliability

The language of technical sciences is mathematics. Learning about each specialized technology is so defined by a set of specific mathematical procedures. For special control reliability, availability and maintainability (RAM) the theory is built on mathematical probability and statistics.

The rationale for the use of this concept is an inherent uncertainty in predicting failures. Possible fault models are based on physical or chemical reactions, results therefore do not exist as a failure of a certain part, but as the time given by a ratio of non-performing components in percentage or probability that a given part will fail in the given time (Gajdos 2014). The individual components will fail according to their individual characteristics which will vary from parts to parts and are virtually undetectable. Similarly, the time to repairing the faults will vary depending on a number of factors, and its values in individual cases are also virtually undetectable.

2.2 Theory of reliability

Reliability is defined in terms of probability. Parameters of probability used in the theory of reliability are random variables, distribution and density distribution functions. Studies on the reliability deal with discrete and continuous random variables. An example of a discrete variable is the number of failures in a given time interval. Examples of a continuous random variable are the time of component installation to failure or the time between successive failures.

The distinction between a discrete and continuous variable (or function) generally depends on what the problem is solved, and not necessarily the basic physical processes that cause it. For example, when analysing the "one-off" systems such as missile, usually a discrete function is used which represents the number of successful attempts at "n" launches. Whether yes or not the shot successfully launched may be a function of the age, including the time of storage. This knowledge is then processed as a continuous function.

2.3 Mean time to failure

Mean time to failure (MTTF—Mean-Time-To-Failure) is nothing more than the estimated value of time to failure and is based on the basic statistical theory as follows:

$$MTTF = \int_0^\infty t.f(t).dt = \int_0^\infty t.\left[\frac{-dR(t)}{dt}\right]dt = \int_0^\infty R(t)dt,$$

$$(1)$$

after using Lopital rules for integration.

In many cases, this relationship allows simplified calculation of MTTF. If you know, or can determine the value of probability from the data, the set of fail-free operations R(t), MTTF can be calculated by direct integration of the R(t). If R(t) is mathematically described, we determine MTTF by the graphic approximation or by the Monte Carlo simulation. For repairable equipment MTTF is defined as the mean time to the first failure.

2.4 Lifetime

Lifetime (Mean life) - θ takes into account the overall cast of the elements considered in the device. For example, the initial cast of n elements, if they are all able to operate until their failure, mean lifetime (θ) is the average mean time to failure of total occupancy, given by the equation:

$$\varnothing = \frac{\sum_{i=1}^n t_i}{n}$$

$$(2)$$

where t_i is the time to failure of i-the element of the observed file and n is the total number of elements of the file.

2.5 Mean time between failures

The notion of MTBF (MTBF—Mean-Time-Between-Failure) is the most frequent in the literature about the reliability of AESS. It belongs to the elements that are renewable after a failure.

Relationship for MTBF is as follows:

$$MTBF = \frac{T(t)}{r}$$

$$(3)$$

where T(t) are total working hours and r is the number of failures.

It is important to recall that MTBF is relevant only for the serviceable components and for such cases MTBF represents exactly the same parameters as the mean lifetime (θ). More important is the fact that the constant failure λ of AESS is assumed. For these two assumptions—replacement of the damaged component and a constant failure—the probability to failure is given by: (Spall 2003)

$$R(t) = e^{-\lambda t} = e^{-\frac{t}{\varnothing}} = e^{-\frac{t}{MTBF}}$$

$$(4)$$

and for this case is true:

$$\lambda = \frac{1}{MTBF} \qquad (5)$$

3 SIMULATION OF THE MEAN TIME TO FAILURE-FREE ACTIVITY BY METHODS OF MONTE CARLO

Generally speaking, the Monte Carlo method has a wide range of uses. It can be used everywhere we look for a solution several times by repeated random experiments.

The Monte Carlo method is a class of algorithms for simulation systems. These are stochastic methods using random or pseudo-random numbers. Typically they are used to calculate integrals, especially multidimensional, where conventional methods are not effective.

The Monte Carlo method is widely used from simulations of experiments through counting of definite integrals, up to differential equations. The basic idea of the method is very simple: we want to determine the mean quality of values that is the result of on-worthy event. A computer model of the event is created, and after a sufficient amount of racing simulations the data can be processed by conventional statistical methods, for example, to determine the average and standard deviations.

The method is named after Monte Carlo known for its casinos and especially roulette. The term was first used in 1940 by physicists working to construct a US atomic bomb. The two variants of this Monte Carlo method are distinguished: an analogue and non-analogue model.

3.1 Analogue model

We must be able to model the whole situation on the computer, meaning, for example, to know all the probable distributions of investigated phenomena and physical laws they are governing. Performing this simulation we get a result, implementation of a sort of random variables.

3.2 Non analogue model

This is a name for a case where we do not use a model of a real action in calculating, for example, the calculation of the certain integral or content of a limited unit (Del Moral 2013).

3.3 Description of simulation

Using the method of Monte Carlo we have created a simulation entry which is required to fill in the basic parameters: the number of experiments which subsequently expresses a period of observation the simulation. For the needs of our simulation of mean time to failure in operation the researcher must manually enter the following input parameters. The first input parameter to the simulation of N—number of attempts and this number determines the simulation time. When you enter N = 1000, our simulation time has changed to 16 years. For the needs of our simulation we have entered 600 attempts, the simulation time is 10 years.

Other input parameters for our simulation data are MTTF (Mean Time to Failure) and MTTR (Mean Time to Repair). It depends on our decision to give the expected approximate time (in hours) to failure or repairs. The MTTR parameter must always be shorter because this notion rather means repairs and maintenance of the system in order to prevent disturbances, and therefore the time to failure of MTTF is greater.

Parameters Tup and Tdn are vectors with components of artificial history. Further, we enter the three-zero-matrix, and the cycle is repeated by N (number of times), where the numbers are generated by exponential random distribution based on the data of MTTR and MTTF. The simulation generates a random variable, and then places it in a random variable (i). In the next step, the generated random variables are divided by each other according to the formula and continue to the next cycle in which another number is generates and divides once again, and stores it in the other cell, etc. Then calculates the amount of variables and converts to the amount of years (simulation time). In the final phase the listing of percentage expressing availability and non-availability of the system

The data of MTTF and MTTR express time in hours to repairs or changing of system components.

3.4 Simulation printout

After running the simulation we get the statement in the language Matlab, which can be further used for calculation of efficiency, economic profitability and other parameters of AESS.

Printout of simulation results:
Simulation time (years): 10.095605307556992
Availability = 0.6562
Availability = 65.616020%
Unavailability = 0.3438
Unavailability = 34.383980%

We have introduced the facts which result in expression of availability time AESS which is equal to 65.6% of the total time of observation period. Non availability AESS time is equal to 34.4% of the total time period of observation. Simulation results have showed that the disposal AESS hours of operation is large, which requires from the operation management of AESS to seek compensation for the system and hence additional costs.

4 CONCLUSIONS

Using a computer we can create simulations of processes in management operation of AESS. Before we proceed to the actual creation we need to have enough knowledge about the investigated processes and modelling and simulation. As the basis we will consider knowledge and control of the programming language we want to use to create the simulation.

When creating simulations we start with creation of the algorithm that describes the model of an examined process. It is important to establish conditions under which a simulation will be run.

The advantage the use of computer simulation in management operations of AESS are some benefits that simulations offer. The price of a computer simulation is lower compared to the costs, for example, of the practical experiment. So we can also work with compression and expansion of time, while in a real time system we cannot control the time. Therefore, mathematical and computer simulations should be used when introducing new AESS in aviation.

We conducted a simulation operation of AESS provided so that the total time of the reviewed period was equal to 10,095 years. We set parameters of time to failure (MTTF) every 100 hours and the time to repair (MTTR) every 50 hours and at 600 times (N) - 10 years of the simulation time. We came out with the following data on the availability and non-availability of a system.

Simulation results of operation AESS show that AESS was able to operate for 65.6% of the total time of observation period. As a result of failures and repairs AESS was out of service 34.4% of the total observed time. From the above simulation results, it is clear that the time of AESS elimination from the operation due to failures and repairs is large, which requires from the management of AESS in operation to seek compensation for the system and hence additional costs.

REFERENCES

Cekan, P. et al. 2014. Human Factor in Aviation—Models Eliminating Errors. *Kaunas University of Technology Press Conference*. Lithuania: 464–467.

Del Moral, P. 2013. Mean field simulation for Monte Carlo integration. *Chapman & Hall/CRC Press*: 626.

Gajdos, J. et al. 2014. The Use of Penalty Functions in Logistics In: *Our Sea*, International Journal of Maritime Science and Technology. Vol. 61, no. 1–2 (2014), p. 7–10.

Melnikova, L. et al. 2016. Building a training airport for pilots. In: *SGEM 2016*. Sofia: STEF92 Technology Ltd. Vol. 1–2–3. p. 109–116.

Sebescakova, I. et al. 2013. Maintaining Quality Management System at the Faculty of Aeronautics, Technical University in Košice In: *Exclusive e-journal, Economy and Society and Environment*. Vol. 1. p. 1–9.

Smith, C. 2014. Simulating confidence for the Ellison-Glaeser Index In: *Journal of Urban Economics*: Vol. 81, p. 85–103.

Socha, V. 2016. Training of pilots using flight simulator and its impact on piloting precision In: Transport Means. *Kansas University of Technology*: 374–379.

Spall, J.C. 2003. Estimation, Simulation, and Control. In: *Introduction to Stochastic Search and Optimization*. John Wiley & Sons. p. 580–583.

Tobisova, A. & Pappova, E. 2014. Základy ekonomiky leteckej dopravy. *Košice: TU v Košiciach*.

Vagner, J. & Jencova, E. 2014. Comparison of Radar Simulator for Air Traffic Control In: *Our Sea*, 61(1–2): 351–35.

New Trends in Process Control and Production Management – Štofová & Szaryszová (Eds)
© 2018 Taylor & Francis Group, London, ISBN 978-1-138-05885-9

Development and trends in airlines business models

E. Endrizalová, M. Novák & I. Kameníková
Faculty of Transportation Sciences, Czech Technical University in Prague, Prague, Czech Republic

ABSTRACT: Article describes innovations in business models driven by high competitive markets. It unfolds the reasons behind Hybrid Business Model uprising and describes periods of air transportation evolution which triggered changes in business models of air transport providers. It is focused on segment structure and air transport demand changes. Article introduces transformation of business models on case studies and outlines possible future predictions.

1 INTRODUCTION

Business activities in air transport were from the beginning influenced by economic and political situation. Natural development of this segment started after deregulation of air transport, first in United States and later in Europe. Airlines began with adjusting their business models to market conditions, where competition started to play significant role. Endeavor to survive on market brought low-cost business model to life. Next milestone in business model development was set by economic crisis which exhausted performance of air transportation until 2009. In order to stay on the market, especially in high competitive markets, companies had to innovate their business models. This article aims to describe the background and details behind those changes.

2 WHAT IS A BUSINESS MODEL

Business model is often mentioned in relationship with airlines. Services in this area must operate in highly competitive neighborhood. In order to survive in such word, it is needed to have mindful and well tuned business model. *„A business model is a mechanism for turning ideas into revenue at reasonable cost."* Baden-Fuller & Morgan (2010) According to authors Demil & Lecoq (2010) concept of business model is rather connected with revenue and costs. Formation of airline's business model development is continuously driven by competition and endeavor to remain on the market. In business model definition we cannot miss out the relation to competition strategy: *„A business model is a system designed for competing in a specific marketplace."* Jones & Robinson (2012) Business model of airlines is possible to defined as an essential part of business strategy of company, basic template of

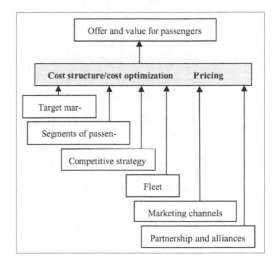

Figure 1. Basics elements of airlines business model.

entrepreneur's strategy oriented on economic value generation. According to monitored development in this sector and researched resources we consider following pillars (Fig. 1) as core of business model of an operating airlines:

3 DEVELOPMENT

In this section milestones in airline's business model development are described. In general following main branch change induction factors can be defined:

- Political and economical background: deregulation of air transport sector, alliances creation (cooperation of airlines in providing customer service).

– Effort to remain on the market: pricing revenue management implementation to airlines, low-cost model systems development, merging and fusing of airlines and hybrid business model creation.
– Evolution of demand after air transport.

Common for each of above factors it that need for fast travel and transport of cargo across the world always existed. Except two specific periods—terrorist attacks in 2001 and consequences of crisis in 2009—air transport demand curve is constantly ascending.

3.1 *Deregulation*

Before deregulation of air transport there was no business models in place—air transport has been fully regulated and controlled by politics of each country. Open economic contest and competition across the market practically did not existed. After release of restricting rules, regular market competition in air travel services providing started to evolve. Economic benefits of free market are available to everyone involved from customers to companies.

Main milestone of air transport deregulation was Airline Deregulation Act in USA (ADA), in 1978. Deregulation process was not coincidental. It was preceded by strong voices about malfunctioning air market, which was unpleasant for travelers as well as for airlines. The Travel Insider (2010). Load Factor of flights was only 50%, but airline's offers were limited to either full tariff or only one "slightly cheaper" tariff, such as "student fare" for instance. On the other side, public perceived air tariffs as unavailable and it was not clear, why fares are so high, even when aircraft is half empty. Airline's systems did not correctly work, did not react to customer's requests nor on demand (The Travel Insider 2010).

Whole situation has been escalated by Arab Oil Embargo in 1973, followed by high growth of fuel prices, resulting in fare prices growth and decreasing of travelers count and related revenue lost. That was one of many reasons which contributed to idea that aviation will better prosper on deregulated market on its own.

Main purpose of ADA was to allow mutual market competition between individual airlines. The idea was based on experience from other market segments—competition between companies brings high efficiency, innovations and low prices while keeping high service standards. Complete free market entry was accomplished on 31st of December 1981 Morris (2013).

System in Europe before deregulation was impenetrable and fragmented. Ticket prices were subordinated to bilateral agreements between states and airline market entry was strictly controlled. Tariffs must have been agreed by both states with IATA oversight. European deregulation was adopted consecutively by accepting three deregulation packages (1987, 1990 and 1992). Third package replaced existing airlines with "European community airlines" and as a core rule it was set that European community airlines are allowed to set prices for passengers as well as cargo and may use any air way in EU without additional permissions or approvals. Truly restriction-free access to all air ways inside union for all scheduled and charter flights began on 1th of April, 1997 (Eighth freedom of the air).

Enhancements resulted from deregulation on air transport markets allowed creation of airline business model itself. Business model as was defined in the beginning of this paper could not exist before deregulation. Air transport deregulation resulted into era of strong competition. Main national airlines started to compete with smaller regional airlines and many companies, operating in one country had to spread abroad. Due to restructuration, price battles and discount tariffs have been popularized, which directly led to air transport demand increase.

Air market release brought many companies to bankruptcies, which were until than protected and donated by individual governments. Some of the companies has been acquired by stronger airlines. Impacts on passengers were very positive—competitive markets pushed prices down and for air transport it was huge improvement and significant enhancement.

3.2 *Revolutionary idea of low-cost*

Profit for the customer brought the low-cost business model. This concept was developed in USA in seventies, later in nineties it evolved in Europe and the rest of the world. Deregulation of air transport allowed airlines to choose its own pricing strategy. Competitive environment allowed creation of low-cost operation offering cheaper and thus more available tariff which requires certain prerequisity: maximal cost descrease while following defined operation concept. Very first successfully operated low-cost air transport model belongs to Southwest Airlines. Company began to offer cheap tickets on frequent short haul flights (1971). In first two years the company struggled with finance issues and negative numbers. In 1973 it's popularity increased hand in hand with its performance. Today it belongs to one of the biggest airlines.

Case study 1: Business model of Southwest Airlines (Texas, USA) was changed in November 2007. Thanks to high competition, Southwest

Airlines management decided to left behind the motto "*One size fits all*" which was in place for many decades and created new product: "*Business select*". Business customers obtain benefits, such as more credit with loyalty program or welcome drink on board. Those steps differentiate Southwest airlines from typical low-cost company, because loyalty program as well as wide business products portfolio is characteristic sign of network airlines. Other sign of Southwest model change is operation on international flights (since July 2014). Southwest claims: "*After five years of intense work on five strategic initiatives, 2015 was the first full year to demonstrate results: record traffic, record revenues, record profits. Southwest boosted its available seat miles (capacity) 43 percent since 2010, driven by the acquisition of AirTran in 2011*" Southwest (2016).

3.3 *Alliances*

One of the characteristic of business model of full service airlines is cooperation. At the end of seventies, during liberalization of air transport, head of CAB (now nonexistent agency) introduced new politics so called Encirclement Strategy. Levine (1979) Through this politics international strategy of United States was able to set more liberal conditions with smaller European countries. This improvement in worldwide market liberalization led to creation of global alliances between airlines. The reason behind is to share capacities on flights of cooperative companies. Reservation and booking systems are shared as well. Frequent Flyer program is common. Such cooperation brings wider offering to passengers. Cooperation of network airlines causes performance increase of cooperative members.

Alliance cooperation is not typical for low-cost airlines, but new information from February 2017 indicates new trend in shaping of business model of low-cost airlines: The Telegraph reports that a group of Europe's largest budget airlines— EasyJet, Norwegian and Ryanair have been toying with the idea of forming their own low-cost alliance, which could pose a major threat to the business of full-service carriers like British Airways and Emirates. The Telegraph (2017) "*The collaboration would be a momentous U-turn for the rival airlines, usually fighting for business in an increasingly competitive marketplace.*" (The Telegraph 2017).

3.4 *Bankruptcies and mergers*

Personal air transport went through various bankruptcies and mergers. USA: US Airways announced bankruptcy in 2002 and came back to market in 2005 after merge with America West.

Delta Airlines and Northwest Airlines declared bankruptcy at the same day and in 2009 they merged to avoid liquidation and now continue as Delta Airlines. Another big juncture occurred in 2010 between United Airlines and Continental Airlines now operating under common name United Airlines. To important fusions belongs also successful merge of Southwest Airlines with four smaller airlines between years 1985 and 2011 which shifted Southwest Airlines to top four major carriers in United States. At the turn of the century around 10 another American carrier got bankrupted or merged.

In Europe British Caledonian, Laker Skytrain, Sabena, Swissair, Olympic, Malev and Spanair went bankrupt. British airline British Airways and Spanish airline Iberia signed fusion in 2010. Air France and KLM agreed on 784 million euro fusion: Air France took over smaller KLM in 2003. British Airways and Spanish Iberia fused in 2010. Fusion is step to avoid bankruptcy and obtain stronger market position, which can be marked as competitive strategy of airlines connected with new company business model raised from two single operating companies.

3.5 *Price war and revenue management*

Pricing is one of main parts of airline business model. Price war first started on short haul flights which are strategic targets of low-cost airlines. On short haul flights is not easy for network provider to find optimal price versus cost coverage not only because of low-cost competition but also other types of transportation. Network carrier must keep short haul flights in order to keep network character and to keep share of transfer passengers. Lufthansa solved short haul flights problem with its subsidiary company Germanwings. Transfer of short haul flights to Germanwings came into being during spring 2013 and autumn 2014.

Airline's finance management, similar to other enterprise businesses, is based on its sales strategy which is specifically important in high competitive markets. Economic performance of airline is directly influenced by demand change factors knowledge, market segmentation, segment's demand elasticity and presumptions of expectations of passengers which already bought the ticket. Airline which in detail monitors those elements converts data into strategic decisions: expands tariff groups to effectively react on demand elasticity change for various income groups or to reflect price politics of its competition and market condition changes. Carrier's system revenue management has been created to set optimal tariff structures. Revenue management has been set up in USA as a reaction of network airlines to their low-cost competition. Efforts to fill in the free capacities with price sensitive passengers was first time applied by the American Airlines. Today revenue management is used by every airline in the world. It is system which ensures to offer optimal ticket price structure based on the historical data from the reservation system to achieve highest possible revenue. Of course not only the price but also service quality, line connections, arrival times and other factors plays role in travel choice. Wide proposition is amplified with multiple distribution channels. The tickets can be purchased online directly from the airline using its own Computer Reservation System (CRS) or trough Online Travel Agency (OTA). In the distribution chain Global Distribution Systems (GDS) and online Meta Searches can be found. The simple schema based on diploma research Eiselt (2015) is presented on Figure 2. In tourism each distributor has its own price. The distribution chain cost is dependent on its length and number of levels. Multiple subjects and broad proposition may cause confusing and difficult selection for customer.

For the low-cost airline is typical direct sale through its own website, which naturally eliminates middle man provisions. This fact opens question, if creation of own OTA and CRS for the low-cost airline alliance will become new factor of upcoming hybridization in air segment. The low-cost alliance can have its united alliance CRS or common OTA which simplify choice and shopping for customers.

Pricing and revenue management and nowadays widely supported by various sophisticated software customized for sales in the air transport sector. For example *airRM* and *airRMexpress* of *Revenue Management Systems Inc.* (customers: Blue Air, Germania, Ryanair, Air Asia, Meridiana and others) was developed for small and middle size businesses. Except standard outputs such as forecasting and optimization it may be expanded adding extra modules for managing and directing revenues and capacities.

Another company providing revenue management software systems for airlines is *PROS Holdings* based in Houston (Texas). Its flagship product *Origin and Destination Revenue Management* oversight specific situation on single line for 180 days before take-off and evaluates and suggests optimal ticket price. Final decision about price lays on human operator which also takes actual ticket price development into account. *PROS O&D Network Hybrid Optimizer* is used for example by Lufthansa, Austrian Airlines, British Airways and Air France-KLM. In 2009 company ČSA (the Czech republic) started to use product *Origin & Destination*. *"System is setting ticket prices for each segment individually but at the same time it works more plastic according to demand on the whole network. Moving free capacity is done very effectively—the system increases the likelihood of sales of free seat. It can separately conduct point-to-point air ticket price creation for local market as well as on other markets using network with transfer in Prague"*. Tomáš Holan (2009). To implement this system Czech Airlines invested about 1,9 million EUR and expected about 250–300 000 new customers (2009).

3.6 *Hybrid*

From the beginning of 2008 after financial crisis which significantly impacted transport performance, airlines has been forced to adjust their business models to market situation. Many carriers struggled with their finance situation which forced them to change their business models. Standard network carriers started to offer new products, reorganizing and rationalizing operations and cut down costs. Low-cost airlines started to operate on market segments, which was before privileged to network carriers only. Business model which before allowed low-cost airlines to reach on 50% of costs of network carrier was not available anymore. This way so called hybrid airlines have emerged. Hybrid carrier business model is combination of cost saving strategy, product proposition, flexibility and wider airway structure Sabre (2010).

Case study 2: Air Berlin on the beginning of its exstence operated charter flights. In nineties it began to offer point-to-point based on low-cost model and operate flights from many secondary airports near big cities. Based on this concept it attracted business travelers as well. Period between 2006 and 2010 was characterized by fusions and acquisitions (for example Niki, LTU, Belair, TUI-fly). Thanks to mentioned operations, carrier was able to increase traffic by 45% and in 2010 transported 35 million passengers. Final phase of Air Berlin change was its entrance to global alliance OneWorld in 2012. Air Berlin also changed its tariff structure. *"Since the beginning of its hybridization process in 2006 where the airline had identified itself as a "low-cost with frills", Air Berlin has registered almost continuous operating losses and a net loss in six out of the ten years. Air Berlin has faced*

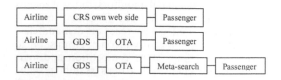

Figure 2. Possibilities of online distribution chain used in air transport.

is the continuous necessity to attract new demand to cover the extra costs of the added frills while being unable to generate significant savings" Corbo (2012).

4 FUTURE

What will be the next business model development direction? Most probably we will witness further hybridization and efforts to cost cutting way searching. High density of competition on the market will not allow stable adaptation to everybody. Low-cost alliances will be the key players and biggest competitors for network carriers. Customers will experience financial benefits at the expense of fading differences of low-cost and full service airlines and reduction of distribution chain subjects, which eliminates choice ambiguity. The main future hypotheses:

- New low-cost alliance,
- Common CRS or OTA,
- Marketing strategies—clear and simple identification of carriers,
- The transition from the lowest price to the fair price,
- Effort of network airlines to shorten the distribution chain,
- Clear ticket price break down for passengers.

5 CONCLUSION

In this article are named the main parts witch create the airlines business models. A further section describes the development of this area—main branch change induction factors: deregulation of air transport industry, low-cost business model, alliances of airlines, their pricing and distribution chain, mergers as one of changes and formation of hybrid business model. As shows the Case study 2, transition to the hybrid business model is not guarantee of yield growth.

We think, that the continuing hybridization will have resulted in unclear identification of airlines business model. In the article are some options that could in the near future in development occur. Competitive struggle will bring the new used term—fair price.

REFERENCES

Baden-Fuller, CH. & Morgan, M.S. 2010. Business models as models. *Long Range Planning.* 43(2–3), pp. 156–171.

British Airways. Stronger together—British Airways and Iberia. [online] Available at: <https://www.britishairways.com/en-us/information/about-ba/iag>.

Corbo, L. 2016 In search of business model configurations that work: Lessons from the hybridization of Air Berlin and JetBlue. *Journal of Air Transport Management.* <https://www.researchgate.net/publication/309181234_In_search_of_business_model_configurations_that_work_Lessons_from_the_hybridization_of_Air_Berlin_and_JetBlue>.

Eiselt, D. 2015. The analysis of online distribution of airlines tickets. Master thesis. Department of air transport, CTU in Prague, Faculty of transportation Sciences.

Holan, T. [online] 2009. Available at: <http://www.csa.cz/cs/portal/quicklinks/news/news_tz/news_tzarchiv_data/tz_21092009.htm>.

Jones & Robinson. 2012. Operations Management: Chapter 12 Operations strategy. 2012 [online] Available at: <http://slideplayer.com/slide/8809553/>.

Lecocq, CH. & X. Demil, B. 2010. Business model evolution: in search of dynamic consistency. *Long Range Planning.* 43(2–3), pp. 214–225.

Levine, M.E. 1979. Civil aeronautics memo. Aviation daily.

Moriss, D. 2013. Airline deregulation: A triumph of ideology over evidence. 2013 [online] Available at: <http://www.huffingtonpost.com/david-morris/airline-deregulation-ideology-over-evidence_b_4399150.html>.

Sabre. The Evolution of the airline business model. 2010 [online] Available at: <https://www.sabreairlinesolutions.com/images/uploads/Hybrid_Model_Brochure.pdf>.

Southwest. 2017. Report on the triple bottom line. 2017 [online] Available at: <https://www.southwest.com/html/southwest-difference/southwest-citizenship/one-report.html>.

The Telegraph. How Europe's three biggest budget airlines plan to take over the world. 2017 [online] Available at: <http://www.telegraph.co.uk/travel/news/norwegian-plans-to-join-forces-with-ryanair-easyJet-low-cost-alliance/>.

The Travel Insider. 2010. A history of US Airlines deregulation: part 3. 2010 [online]. Available at: <http://thetravelinsider.info/airlinemismanagement/airlinederegulation1.htm>.

Verbal expressions of mental imbalance among air traffic controllers

L. Fábry, R. Rozenberg & L. Socha
Faculty of Aeronautics, Technical University of Košice, Košice, Slovak Republic

V. Socha
Faculty of Transportation Sciences, Czech Technical University in Prague, Prague, Czech Republic

ABSTRACT: The article discusses an environment and lifestyle impact on activities of air traffic controllers (ATC), the possibilities to observe their performance and focuses on their verbal expressions during work activities. The results of air traffic controllers' verbal communication gained and evaluated, recorded in air traffic procedural and radar control training, point to a possible relationship between stress, fatigue, limited ability to perform a controlling activity on one side and verbal expression characterized by variations in the rate, volume, colour or delay responses to impulses on the other. The discovery of this dependence opens up the possibilities to look at a hypothetical solution of non-invasive, non-contact observing their skills and limitations in direct activities by monitoring their main working tool—own voice.

1 INTRODUCTION

It is generally known that air traffic control is one of the most stressful professions. The actual control is not a fundamental problem. In fact, several circumstances which grow in power and affect the controllers so that the stress has an acceptable form and more positive influence. This period is referred to as the first period of appropriate load, but there are already the basic signs of stress, especially in external manifestations. The second load phase is already regular or irregular accumulation of problems, with the lack of time for their solutions, and the possible relaxation of the body will be shortened (Antoško et al. 2014, 2015).

Most often it is a combination of problems with ensuring of separation between the operation and communications and, of course, in this period of stress already receiving wrong decisions, especially in priority of link connectinuity and order of flow and safety solutions. This period is not yet known to be critical, because, on average, occurs only a few times per working cycle, followed by enough time for regeneration. There may however, occur partly incorrect operations, which result in their accumulation and delays may increase and disrupt the order of flow, and in extreme cases, there could also be potential security violation. In this case, the controller could cause a higher risk. Despite regular medical examinations and various tests for resistance to stress each individuality deals with their own lives and problems. Moreover, all the tests (Čekan et al. 2014) are based on the average

values, but non-controlled home environment has an unforeseen development, and thus the impact. Then, controllers very easy get into an internal conflict or is not able to perform the activities and come to work with the concerns and often have such high stress, that controlling dangerous for them becomes and the surrounding. This is already the maximum stress limit that is dangerous for the human health, even life-threatening with heart attack threatening or unexpected stroke. It is necessary to talk about what stress can actually cause to a person and to point to possible, e.g. for a head of a shift, perceptible changes in attitude and behaviour of the controller, and thus prevent the final phase. (Socha et al. 2014).

Stress causes the body alert reaction, which activates the brain, endocrine glands, nervous and immune systems. The result is the increased blood supply to skeletal muscles needed for an attack or escape reaction. The circulation of blood volume increases, there is an increase in heart rate and energy releases. Stress is an awkward situation for the body because it gets it from equilibrium to disequilibrium condition, so it involves all the body mechanisms and reacts somehow to "survive" in stressful situations (Szabo & Sidor 2014).

For example, if a person is frequently exposed to pressure hypertension stress over time the persistent hypertension occurs. If, due to stress, bile flows poorly for a long time, they will form gallstones and so on. The very chronic stress weakens the immune system and people are sick more often. Stress can completely paralyze a person and limit their performance comparing to activities in a peaceful state. Observable signs of stress can be divided into three basic groups.

2 PHYSICAL SYMPTOMS OF STRESS GETTING STARTED

2.1 *Situation hypertension*

Stress and excitement in most people cause a sudden increase in blood pressure when the pressure is "shot" to high values. Frequent acute increasing of pressure during stress is felt by people who are already taking antihypertensive drugs which keep the pressure under control most of the time, but nevertheless it suddenly increases under stress.

2.2 *A weakened immune system*

Stressful situations can trigger spasms in the gall duct, resulting in gall bladder pain and stinging below the right costal margin. The gall can not freely leave the gall bladder, which causes overpressure in it. This symptom of stress particularly troubles people with choleric nature and ones susceptible to the formation of gallstones.

2.3 *Muscle spasms and pain in the lower limbs*

During stress the exchange of magnesia and calcium ions occurs in the muscles causing a spasm of skeletal muscle, as in the legs. In the aftermath of the stress the spasm releases and pain appears.

2.4 *Frequent urination*

The spasm in the bladder causes frequent urge to urinate several hours prior to the situation which is the reason for stress (e.g. test, public appearance ...).

3 PSYCHOLOGICAL SYMPTOMS OF STRESS

During stress most people manifest irritability, nervousness, restlessness, inability to delegate tasks. Later it gets sometimes to utter depression and apathy, a form of stress when the body is not able to respond and defend to negative stimulus.

3.1 *Anxiety*

Anxiety is a negative emotional state that results from the various vulnerable situations, but also in uncertainty or failure. It may also be one of the responses to stressful situations. The psyche is expressed in various concerns and fears that

the body maintained for relatively long time in the aftermath of stressful situations. It can cause insomnia, as people are anxious and stressed out at night when it's quiet; all the duties urge them and cannot sleep.

3.2 *Panic*

This is a serious condition that often occurs in stressful situations and manifests a disordered human behaviour as a result of disturbance of mental equilibrium. People in panic do not realistically assess the situation and are a subject to external influences. The stress is a condition that induces emotional lability, man is more sensitive and therefore can response to the bad news or stress irritably, with apathy, and panic—"What to do now, how to handle it."

3.3 *Stage fright*

Stage fright is uncomfortable tension that occurs before an expected event as a public performance, exam or presentation. This is reflected especially by palpitations, pallor, tight throat feelings, fainting.

4 OTHER SYMPTOMS

- Clutching diarrheal and abdominal pain—caused by a spasm in the abdomen.
- Palpitations and/or arrhythmia.
- Pain and tightness of the sternum.
- Bloating, loss of appetite.
- Muscle tension in the neck and low back associated with the pain.
- Persistent headaches that often start at the nape and spread on top of the head and face.
- Discomfort in the neck (bump).
- Difficulty concentrating on one point vision, double vision.
- Rash on the face.
- Decreasing sexual desire, impotence, frigidity.
- Changes in the menstrual cycle.
- Sharp and pronounced mood swings.
- Excessive tiredness.
- Inattention, difficulty concentrating.
- Irritability, anxiety.
- Feeling tired in the morning after waking up.
- Increased consumption of alcohol, cigarettes or drugs.

So far we have mentioned only stressful environment and its basic levels. Other factors have been less discussed as their overall effect is similar or even identical. It is not necessary to speak about the substances causing dependence, it is a special case discussed in a number of areas where the impact is always the same—complete degradation of the individual. We must not forget the common condition which can be generally described as psychologically uncomfortable where we can include completely normal states of a body such as a fatigue, home, family and financial situation, starting disease, a woman's period.... All of these conditions have a direct or indirect impact on performance and stress resistance.

Normal, healthy fatigue that occurs after a mental or physical activity and through which the body makes it clear that we need to regenerate, will disappear (if appropriate rest) within hours. Unless fatigue persists (days, weeks, and months) may be the cause of poor lifestyle (stress, stress, boredom, poor diet ...) or a previous illness. Another alternative is one of the diseases of modern times—chronic fatigue syndrome (Chronic Fatigue Syndrome—CFS).

Frequent excesses that lead to fatigue, discomfort and loss of overall fitness also include, inter alia, the required working conditions for controllers and the resulting limitations (Szabo & Sidor 2014):

- *Lack of liquid*. Most people drink until the feeling of thirst, so with some degree of dehydration which causes fatigue.
- *Poor nutrition*. Fatigue, often the result of overeating, is falling on the entire body and complicates the administration of oxygen. The problem is, of course, a diet and lack of nutrients. There is a warning for lovers that coffee and regular coffee consumption can lead to a feeling of permanent fatigue.
- *Unpleasant environment*. Poorly ventilated, noisy, uncomfortable and unsightly areas contribute to fatigue significantly. There is evidence that the dark interiors of most people literally suffocate and unkept environment literally robs the energy and encourages depressed mood. A good alternative, however, are neither ultra-modern rooms with lots of shiny glass and metal whose reflections constantly "emphasis" on human senses and evoke the feeling of exhaustion.
- *Lack of physical activity*. The lack of movement complicates the blood circulation and excretion of waste products from the body, resulting in "unidentified" fatigue.
- *Medications*. Frequently and arbitrary use of medicines can make you feel very tired. This is one reason why doctors are reluctant to use drugs—have can see too often what such a "menu" can cause.
- *Monitoring of radar information*. It is actually a source of information and can indicate almost by 90% that it is the same as TV. To watch television, a source of information and relaxation, where just 10% of those are dedicated to relax,

is literally "a vacuum cleaner of energy". When people with persistent symptoms of fatigue were given strict prohibition of "television culture" they visibly regenerated within three weeks, but it is not possible to do with controllers.
- *Stereotype and boredom*. The boring daily program, the amount of duties, lack of free time and hobbies, rush and constant waste of energy leads to a feeling of emptiness, "burnout" and constant fatigue which is,, according to experts, "spreading like a plague" ....

If, after a rigorous lifestyle modification, the feeling of permanent fatigue has not been lost, it may be a chronic fatigue syndrome which belongs to the so-called diseases of the 20th century. This group of diseases includes more syndrome sensitivity to chemicals (particularly polluting substances); syndrome of sick buildings (i.e., response of the body to a dry, overheating air conditioning, fungi ...), and fibromyalgia (muscle aches and stiffness, headache, chest pain, sleeping troubles).

The Chronic Fatigue Syndrome is quite difficult to diagnose and treat: it is sometimes more or less bearable and some patients are completely cured and sometimes patients move between relative health of the period of the disease, and at other times there is a gradual deterioration, up in permanent disability.

The stress is known factor which weakens the immune system and increases the susceptibility to various diseases—physical and emotional stress often precedes or appears right at the beginning of morbid fatigue and causes a vicious circle between fatigue and stress.

The most common symptoms of chronic fatigue syndrome (CFS) are (Vagner & Jenčová 2014):

- Permanent fatigue that lasts more than six months;
- Increasing body temperature after physical or mental stress;
- Muscle pain;
- Long-term pain in the neck node enlargement and tenderness;
- Headache, especially in forehead;
- Swelling and joint pain;
- Photophobia, and visual disturbances, distractibility, depression, sleep disturbances;
- Weakened ability to remember;
- Tension, restlessness, irritability, anxiety.

5 VERBAL COMMUNICATION IN OCCUPATIONAL ACTIVITY FOR AIR TRAFFIC CONTROLLERS

The previous descriptions of symptoms and the resultant danger for a body make clear that none of the above in air traffic controlling, can be underestimated or downplayed even partly. Therefore the fundamental question how to ensure prevention has also been raised, as treatment solutions have been already mentioned. How to monitor the performance of the controllers during their work without causing any physical or other restrictions or partial reduction in their already limited comfort as affixed to the instrument, regular testing of body fluids, or other unpopular measures? In this case it is necessary to have a look into the working environment of a controller and focus on what exists within the natural environment and thus the controller's routine work. This element, in which signs of stress, fatigue, or other psychological discomfort or imbalance are shown, is the voice which is actually one of the main working tools of the controllers who have to work with it willingly or not. It is therefore probably the most advantageous solution to focus on the controller's speech in the normal communication and comparison of the voice sample in the same situation. There is, in fact, one of the advantages and it is overwritten phraseology and some standard phrases for each specific workplace which may constitute just such a sample. Where to find options for comparison? Already in everyday life the change of voice of a particular person in different life situations can be noticed and thus the whole verbal expression. It is possible to observe anger in voice—increasing volume of speech; lack of communication with the other side—bored curt response in the lower level voice; fatigue—extending call to voice tone changes; stress, lack of time or disagreement with the other side—the acceleration of expression … and so on. Of course, there is also the particular combination of these symptoms. Similar symptoms in specifically and regularly recurring situations can be investigated with controllers, but what is essential, the predetermined expressions can be expected and to measure them is much easier than in a general conversation, so possible solutions directly arise (Vittek et al. 2016).

As a part of students' reactions during practical training in procedural and radar air traffic control in the period from 2010 with a sample of 1595 records it was possible to create a simple table, which confirms the hypothesis of a possible measurement of speech. Of course, student's readiness for an exercise also plays a role for, but because of the continuity of training, the imbalance is negligible. Monitored signs considering the absence of measuring devices were partly of instructor's subjective character and, therefore, the record of only "yes or no" questions and any ambiguity were taken as a natural state i.e. without recording. The current student's mental state was assessed in an interview before the exercise. The instructor during

Table 1. Legend.

Monitored signs abbreviation	
Exercise done with results marked as	OK
Increasing of voice intensity marked as	UP
Got lost marked as	0
By using any words (crutch) marked as	HLP
Yawning marked as	OU
Talking quieter than usual marked as	LOW
Accelerating the speech marked as	SPD
Started to get out breath during the communication marked as	AIR
Varied voice tune marked as	WAV
Slow response to the stimulus marked as	TM
Confirmed fatigue before the exercise marked as	FTG
Evident stress marked as	ST

	OK	UP	0	HLP	OU	LO	SPD	AIR	WA	TM	FTG	ST
OK	1389	46	0	56	0	0	421	89	2	0	1	24
UP		210	52	23	2	-	198	100	153	140	0	208
0			54	52	15	11	54	50	50	51	12	11
HLP				180	25	2	5	174	123	180	157	51
OU					61	13	0	55	61	53	61	0
LO						14	2	6	8	8	9	0
SPD							992	442	900	80	121	410
AIR								563	520	65	210	423
WA									1200	2	233	401
TM										156	142	2
FTG											210	2
ST												423

Exercise done with results marked as	OK
Increasing of voice intensity marked as	UP
Got lost marked as	0
By using any words (crutch) marked as	HLP
Yawning marked as	OU
Talking quieter than usual marked as	LOW
Accelerating the speech marked as	SPD
Started to get out breath during the communication marked as	AIR
Varied voice tune marked as	WAV
Slow response to the stimulus marked as	TM
Confirmed fatigue before the exercise marked as	FTG
Evident stress marked as	ST

Figure 1. Observed student's responses to the load during training.
Source: Own processing.

the exercise enrolled, in addition to evaluating, the assessment of individual areas and a verbal language with the state of student's assignment.

6 CONCLUSION

Based on the real facts it is expected that the research will expend up to confirming the hypothesis using more precise measurement devices. The possible recordings and measurements by invasive procedures such as a subsequent check of the heartbeat in artificially induced stress conditions and the fatigue situation regime within several independent groups are expected. The confirmation and follow up testing will help find the optimal use of either the software application validating verbal speech based on a comparison of samples and subsequent way of alerting the head of the work shift to a possible increase in uncomfortable controller's environment and to ensure the follow-up measures to such advance so that the controller would not get in a longer-term exposure to stress and failure to cope with situations or to provide other adequate situations with the same result.

REFERENCES

Antoško, M. et al. 2014. Psychological readiness of air traffic controllers for their job. *Our Sea,* 61(1–2): 5–8.

Antoško, M. et al. 2015. Ergonomy of ATCO training workplace. In *Proceedings SGEM 2015,* Sofia: STEF92 Technology Ltd.

Čekan, P. et al. 2014. Human Factor in Aviation—Models Eliminating Errors. In *Transport Means 2014, Proceedings of 18th International Conference, October 23–24, Kaunas.* Lithuania: Kaunas Univeristy of Technology.

Socha, V. et al. 2014. Evaluation of the variability of respiratory rate as a marker of stress changes In *Transport Means 2014, Proceedings of 18th International Conference, October 23–24, Kaunas.* Lithuania: Kaunas University of Technology.

Szabo, S. & Sidor, J. 2014. The Performance Measurement System – Potentials and Barriers for its Implementation in Healthcare Facilities. *Journal of Applied Economic Sciences,* 9(4): 728–735.

Vagner, J. & Jenčová, E. 2014. Comparison of Radar Simulator for Air Traffic Control. *Our Sea,* 61(1–2): 31–35.

Vittek, P. et al. 2016. Challenges of implementation and practical deployment of aviation safety knowledge management software. *Communications in Computer and Information Science,* 649: 316–327.

Logistics management of airports

J. Ferencová
University College of Business in Prague, Prague, Czech Republic

P. Koščák, J. Ferenc & P. Puliš
Department of Air Transport Management, Faculty of Aeronautics, Technical University of Košice, Košice, Slovak Republic

ABSTRACT: The article describes some methods, procedures and styles of airport logistics management appropriate for the international airports of Slovak republic. They are described in several categories that can be broken down to simple questions. Who controls? Who are decision makers in airport logistics management (single person makes the decision, team decision, collective decision)? By what means the control is carried out? Which are predo minantmeans ofcontrol? Means that are, or should be, applied more extensively than others (material means, conceptual means, organizational means, informational means and ideological means). According to what is management carried out? Which are various types (by objectives, by models, by occasion, by decision, by function, by project, by rules), forms and styles of management? What practices do we distinguish by the nature, priorities and orientation (preventive management, corrective management, intensive management, process-based management, expansive management, conceptual management, strategic and tactical control, operational management)?

1 INTRODUCTION

Man is the most important and indispensable component of the management process. During this process the ongoing sub-set of management activities take place. in our case, management of airport logistics is just an empty notion by which the action cannot be named without people. it explicitly follows that the process of management is the one of socio-technical and informatics nature. Thorough in-depth research and analysis of airport logistics management processes, we can say that the individual management varies according to: who manages? – analysis is concerned with the management; by what ? – focus is on the predominant means of control; according to what? – examined is the form of management; nature of management? – interest lies on priorities of management orientation; way of management? – discussed is the method of management (Ferenc et al. 2013).

2 MANAGEMENT MEANS OF CONTROL

Apart from a wide range of material means, this group also involves a set of conceptual, organizational, informational and ideological means. We selected only those, the application of which should have a higher frequency, or those which have been extensively applied (Šebo 2007).

There are many ways "by which to control", but we only deal with those considered to be suitable for airports, such as: command, plan, act, penalty, balance, motivation, participation, innovation, decentralization, delegation, communication and democratization. Managing by:

a. Orders—is usually associated with the prescriptive ways of governance. Initiative is bound even in those cases where it is demanded. Issuing orders is justified for risk management processes in situations of threats, but even in cases, when by participatory way, it was agreed on a desirable conduct of individuals or groups or organizational units. The large number of issued commands reduces their importance and effectiveness. On the positive side, an order, if subject to conditions and clearly formulated, compact and of robust guidance enables rapid control of subjects. Negative aspect is that the staff of the managed unit is not involved in the management and decision-making because management by commands is mostly directive. Control by command necessitates well-defined channels of communication through which orders are sent from the commanding to the performing part.

b. Plan—is based on a plan. This method of control is equally concerned with national economic and corporate level management team. The more severe is the administration, the more directive is the plan. Plans for the user are meant to highlight what, when and how to do. The style of management by plan as a support tool has many advantages with regard to the systematic building of management guidance, as well as the management entity deductible tasks.

c. Regulations—a lot of management entities act by issuing regulations and their control activity is concentrated only on whether the regulation complies or fails to comply, even without regard to the favorable or adverse effects based on its compliance.

d. Sanctions—managing an object is based on different sanctions if it does not perform certain functions and tasks or behave according to the vision of management entity. Nowadays, when it comes to individuals, dominant sanctions are the financial penalty, loss of freedom for a certain period, loss of property or functional status.

e. Balance—the company needs for its proper operation and management by balance, which is bound to the maximum possibilities, abilities, while the rules are based on the economics of shortage. The management entity is required to promote and perfect balancing knowing the importance of the outcomes to be expected. The application of this method should not be used for power-based directive management actions. The elaboration of the balance by the computer enables the manager almost continuous view of the financial balance or imbalance of the company managed.

f. Motivation—the essence of motivation lies in the fact that management uses all the available ideas that drives to such behavior, as a necessary aspect of proper operation and behavior of the object. Thus, company management is specifically focused on the human element of management, using primarily knowledge of psychology and social psychology.

g. Participation—interprets the participation of those who are managed in the management process. It indicates that the subordinate employee participates in the management process performed by the manager for region management.

h. Innovation—is a way of managing by the constant change of management, which becomes the most stable feature of the business. The changes are becoming more violent in nature, that is, they are bound to risky opportunities, while others temporarily stagnate. The identical time phase leads to a situation where management process takes place in a highly turbulent environment.

i. Decentralization of decision-making—is about breaking down authority and responsibility from central to lower organizational levels. Decentralization raises the level of self-management of its organizational units as well.

j. Delegation—is widely applied in the management and decentralization when the managing subject authorizes the object of management, or any component in the management activities, decision making, but also in areas that it previously performed itself. The advantage consists in speeding up the authorization process, which is directly associated with management participation.

k. Democratization—takes full advantage of participation, motivation, delegation, decentralization and management innovation. Democratization is a way of management, which is the basis of modeling the atmosphere of mutual trust, cooperation and understanding.

l. Communication—is occasionally characterized as a process of communication of information, reports, data, or as a process of balancing the level of information. They are all about giving emphasis to the following characteristics of management by communication:

- Freedom of information with respect to the subject and object of management; subject, object of management and environment; object of management and environment;
- Information rate, timeliness of information, reporting true information content, maintaining a high substantiated information dynamics.

3 TYPES OF MANAGEMENT

Classification of this group brings us to some ways of management, which naturally are bound to the appropriate style of management. From among a large number, we selected only those which follow the direct line and provide answer to the question—management by what? The methods for airport logistics management are based on: objective, design, opportunities, decisions, functions, and project rules. If there are correctly and appropriately defined goals of the firm as a whole, it is possible to derive the goal of lower organizational structures and thus gradually continue downwards as low as to the level of the individuals (Lipták 1991).

Types of Management, by:

- Objectives enables building a basis for live and constructive development of individuals and groups, or development of motivation, participation in management, improving the ranking of individuals, groups, departments and increasing fairness of pay. The disadvantage is in the high intensity, long duration of implementation, administrative burden. There are computers, labor intensive, rigorous continual care about the formation, improvement and application of this method. Management by objectives must stand for long periods and may not be disposable.

- Models are one of the oldest and most commonly used forms of governance. Manager observes, examines, and evaluates the management of others, taking them from those parts of their lifestyle that are important to him, or admires and emulates practices in the management process observed from the others.
- Occasion—this management seeks market opportunities and its importance is growing directly in the fluctuating markets. It is referred to in many cases as a method of management with flexible targets. This way they are managed for objectives to be achieved by the object of direction.
- Decision—this includes the administration of those cases where, after a decision it is necessary to follow with direct management activities. At this point, managers need to systematically or accidentally, intentionally or unintentionally choose means and the method for the purpose of implementing a decision. Management by decision making process is related to other activities, making it a line between the partial control activities and the application of these methods.
- Function—emphasizes the nature of the position held and is based on the choice of management methods, their group, association or combination. Many management methods can be used in any case and require only some modifications that are related to assigned functions.
- Project—as a matter of fact, it is the modified form of management by objectives. It is the method based on network planning, continued extension of its sphere of influence to other reliable method, such as present-day expert systems, as well as human behavior, the way of harmony of people in complex social systems with multidisciplinary and multi-professional foci. This way of management can be rated among large and extensive ones.

Rules—there is an impressive amount of rules to become effective, efficient, successful, entrepreneurial, innovative, but there are also lots of other management methods. Of course, the universal rules of administrative procedures should be supplemented by specific rules, as the management of human activity.

4 APPLICATION OF THE METHOD OF MANAGEMENT

The classification group (Table 1.) gives emphasis on the methodological essence of how to manage, whereby emphasizing the subject's cooperation with its immediate surroundings in the management process. Airport logistic can be controlled:

Table 1. Application of the method of airport logistic management.

Exactly	This recognizes only exact methods of management, and is based on mathematics, use of computer technology, modeling and design methods for solving management problems. It emphasizes the use and application of precise methods of management wherever possible, reasonably, helping to successfully manage the process of management
Heuristic	Such management is based on seeking new ways and methods of solving problems of management. It combines empirical, intuitive and exact knowledge and methodologies
Dynamically	Characterized by thoughtful, variable management, it is bound to situations and opportunities not to stereotypes or routine. The term to manage dynamically is understood as orientation onto the future, striving for continuous maintenance of the dynamics of events
Synergy	Synergy means to compose a new unusual ways of cooperation and interaction of a partnership. Joint venture is just one of the branches of the methodology set. This control makes it necessary to combine various dynamics and conceptually oriented styles
Consultation	One of the promising methods of management. It is also linked with managements based on team, participatory and other forms. The management entity, when performing management, consults with advisors who are they colleagues, or defined advisory boards of various external organizations

exactly, heuristically, dynamically, synergistically and by consultancy (Petruf et al. 2015).

Proposal for airport logistics management—it is impossible to find or choose only one way of management and expect that it will comply with all requirements and will bring us one hundred percent satisfaction (Gajdoš et al. 2014).

As there is no such method, no airport can earn enormous profit, satisfy employees and have no problems. The methods and styles of management must be continuously innovated, improved, developed, and made more flexible, because only in this way one can achieve success at airports, but also other businesses in the face of competition. Awareness of the specificity and uniqueness of each case is leading to its identification and diagnosis of "customizing to one's own requirements" (Husáková & Marasová 2012).

5 CONCLUSION

When analyzing and identifying a certain method and style of management for airport logistics in real life, we realize that they are not found in pure forms, forming complex connections with features typical to concrete cases. Such connection must be assessed separately with regard to the particularity of each case, in which it is applied. Management is and must remain multifaceted; otherwise it becomes valid only for a tiny category of objects. Of the relatively large number of methods, styles, types, means and forms of managements, this paper describes only those that would be appropriate and promising to manage logistics at the airport. (Ferenc et al. 2013).

ACKNOWLEDGEMENT

This work has been supported by KEGA, the Grant Agency of Ministry of Education Slovak Republic under the grant "Design of the specialized training concept oriented to the development of experimental skills within the frame of education in the study branch logistics" No. 009. TUKE-4/2016.

REFERENCES

Ambriško, L. & et al. 2015. Application of logistics principles when designing the process of transportation of raw materials. In *Acta Montanistica Slovaca*. Vol. 20, no. 2 (2015), s. 141–147.—ISSN 1335-1788.

Drahotský, I. & Řezníček, B. 2003. *Logistika—procesy a jejich řízení*. Copyright © Computer® Press 2003, s. 1, 6–11, 87–130. ISBN 80-7226-521-0.

Ferenc, J. et al. 2013. *Prevádzka letísk*. Letecká fakulta Technickej univerzity Košice 2013. ISBN 978-80-553-1377-1.

Gajdoš, J. et al. 2014.The Use of Penalty Functions in Logistics. In *Our Sea, International Journal of Maritime Science and Technology*. Vol. 61, no. 1–2 (2014), p. 7–10. ISSN 1848-6320.

Husáková, N. & Marasová, D. 2012. Application of reverse logistics in the field of car wrecks. In *Carpathian Logistics Congress*. 7th-9th November 2012, SpaPriessnitz, Jeseník, Czech Republic. TANGER, 2012 p. 1–5. ISBN 978-80-87294-33-8,

Lipták, F. 1991. *Metódy a štýly riadenia – Príručka identifikovania a diagnostikovania*. Bradlo. Bratislava. ISBN 80-7127-006-7.

Petruf, M. & Kolesár, J. 2011. Computer Aided Acquisition and Logistic Support. In *LINDI 2011: 3rd IEEE International Symposium on Logistics and Industrial Informatics*: August 25–27, 2011, Budapest, Hungary. Budapest: Óbuda University, 2011 P. 241–244. ISBN 978-1-4577-1840-3.

Petruf, M. et al. 2015. Roles of Logistics in Air Transportation. In *Our Sea, International Journal of Maritime Science and Technology*. Vol. 62, no. 3 (2015), p. 215–218. ISSN 0469-6255.

Pudło, P. & Szabo, S. 2014. Logistic costs of Quality and Their Impact on Degree of Operationg Leverage. *In Journal of Applied Economic Sciences*. Vol. 9, no. 3 (29) (2014), p. 470–476. ISSN 1843-6110.

Šebo, D. 2007. *Logistika ako nástroj manažérskej činnosti*. Technická univerzita v Košiciach, Strojnícka fakulta, Edícia vedeckej a odbornej literatúry, 2007, s. 7–9. ISBN 978-80-8073-776-4.

New Trends in Process Control and Production Management – Štofová & Szaryszová (Eds)
© 2018 Taylor & Francis Group, London, ISBN 978-1-138-05885-9

Use of selected indicators of the control of agricultural farms in Slovakia

J. Fiľarská
Faculty of Management, University of Prešov, Prešov, Slovak Republic

ABSTRACT: Agriculture in Slovakia has a long-lasting tradition. Although the subjects usually prosper with high sales, business in agriculture is significantly influenced by external factors, in particular the foreign competition. In order to fulfil its objectives, the subject not only needs to collect and evaluate the information, but also to put the controlling in praxis. The aim of the paper was to investigate the sales of agricultural subjects in Slovakia and to analyze the chosen indicators of cost controlling. In the paper, we used method of analysis of chosen indicators during the years 2007 to 2015. Based on the evaluation of the analysis results, it was stated that while increasing the number of employees, the subjects could contribute to the increase of employment in Slovakia; they would not be entirely dependent on the government subsidies and at the same time they could contribute to the increase of food sovereignty in Slovakia.

1 INTRODUCTION

While in the past more than 17% of the population of Slovakia worked in agriculture, nowadays it is only 4% of the total employment rate in Slovakia. The situation is similar also with farming businesses, which have been ceasing to exist after 1990. Their revenues from sales are gradually decreasing as a result of the import of cheap and poor quality food. Therefore, the managers are being forced to constantly monitor meeting the basic objective of each subject, which represents the management of sales and costs influencing the final prices of the subject. Agricultural subjects are interested in growth of their sale networks and strategies in the field of economic competition. However, their effective business in the field of ecological farming in Slovakia is mostly beneficial with the use of bank loans or with the help of the state. In December 2002 at the Copenhagen summit, The European Union approved historically the biggest enlargement to ten more countries, including Slovakia (Sedlák 2003).

That enabled Slovak farming businesses to claim for subsidies and other subventions. The financial assistance to business firms may be inter alia directed to the completion of national structural policy plans, the stabilization and the development of agriculture, the development of economic activities in economically underdeveloped regions, and financing the activities related to the entering of the subjects into EU (Vlachynský 2006).

2 MATERIAL AND METHOD

The aim of the paper was to highlight the importance of the use of controlling in agribusiness. In the paper we examined the chosen indicators of sales controlling and cost controlling. The first criterion to achieve the objective was to compare the influence of subsidies on achieved sales. Another criterion was to compare the effect of costs on the sales; we specifically compared the interannual indicator of cost-effectiveness. The research was conducted on the chosen samples during the years 2007 to 2015. The selected data required for the calculations were obtained from The Research Institute of Agricultural and Food Economics. In the paper, we used the scientific methods of investigation, comparison, analysis, and synthesis.

3 CONTROLLING

One of the tools of financial and management controlling is data obtained from balancing books that can be used mainly in vertical analysis of financial statements. The vertical analysis consists in seeing the individual items in relation to some variable (Kislingerová & Hnilica 2008).

In thorough management of sales, managers can't just rely on indicators derived from the financial statements, but they also carefully need to monitor the business plans, price calculations, sales strategy etc. Knowledge and information have been becoming the basis for many new products and services (Čarnický & Mesároš 2009). A need for constant searching for new ways how to deal with prosperity of agricultural businesses arises for the managers, as well as the need for efficient use of all tools of improving the position on market.

One of the tools to achieve the goals is represented by business controlling. The situation in company changes rapidly under the influence of many factors, therefore the controller's task is to constantly solve problems, or to prevent their occurrence in time. The controller is involved in the management of costs and determining pricing.

He also monitors sales and budgets plans, follows new trends and market situation.

He is helpful in whole process running in the business, from acceptance of orders to their implementation on market.

3.1 Sales controlling

The aim of any business is to track and especially increase the sales. One of controller's tasks in operative operative business management is to interact and cooperate with the executive managers, who are responsible for sales management (Pružinský & Lajoš 2011).

The business needs to check its sales amounts systematically, and for ensuring the sales it needs to efficiently use a special control, particularly focused on one of the business areas (Mihalčová & Pružinský, 2006).

One of the tasks of controller is also the cooperation in the development of sales plans and the forecasts of their implementation on market. In the process of management, business managers efficiently use the controlling of revenues from the sales consisting of sales controlling and revenues controlling, their structure and their occurrence in time. The effectiveness of subsidies on sales is expressed by the ratio of sales to the subsidies.

$$SbE = S/Sb \qquad (1)$$

where $SbE = Subsidy\text{-}effectiveness$; $S = Sale$; $Sb = Subsidy$.

For the purpose of this paper, the indicators of 69 Slovak agricultural subjects were used, expressed in euro per 1 hectare of agricultural land.

In 2007, the ratio of sales to subsidies had the value of 9.11 €. With the decline of 5.83 €, the value of the indicator was calculated at 3.27 €. In 2009, the indicator was, again with a fall, calculated at 2.40 €. In 2010, the indicator value was evaluated with growth of 0.20 € at 2.60 €. Next year, with an increase of 0.75 €, the value of the indicator was evaluated at 3.34 €. In 2012, the ratio of sales to subsidies increased slightly by 0.23 to 3.57 €. In 2013, the indicator showed a tendency of moderate growth of 0.06 €. Its value was calculated

at 3.63 €. In 2014, the efficacy of the subsidy was recorded with a decline of 0.48 € to 3.15 €. In the last year, the indicator had a value of 4.18 €, which represented an increase of 1.03 €.

While comparing the data, it is clearly visible that the indicator of the subsidies efficacy on sales has in average had the tendency to increase. Its values show a high share of the subsidies on the sales of agricultural businesses. Therefore, the subsidy policy should focus more on the objective allocation of grant funding and thorough supervision of its use, because the subsidies are also a competitive advantage compared to other companies operating in non-agricultural areas.

3.2 Cost controlling

Cost management represents an integral part of cost controlling. It is the main indicator of a decrease or an increase of sales and their close link to the quality of production. In order to carry the basic business activity, the company has to be ensured also by the performance of other supportive and management activities, on whose securing the costs have their significant share (Zalai 2010).

Kotulič points to the necessity of the relationship of comparing the changes of inputs to outputs and vice versa. The above changes are necessary for the purpose of quantification of business effectiveness or, if you like, efficiency of a reproduction process in the company and they should reach an appropriate level for its successful development (Kotulič 2010).

Cost analysis represents an important indicator of the controlling. It points out the cost-effectiveness and its use in the input area and the realisation itself. For the purpose of the paper we used the indicators of cost efficacy and cost ratio.

3.3 Cost ratio

One of the important indicators of cost controlling that the controllers deal with in cost management is cost ratio. This indicator expresses the ratio of total costs to sales. Its value should not be higher than 1.0. Otherwise, the value of the company predicts a loss. The indicator of cost ratio is defined as follows

$$CR = TC/S \qquad (2)$$

where $CR = Cost\ ratio$; $TC = The\ total\ costs$; $S = Subsidy$.

This indicator can also be expressed for each particular individual type of business costs—labor costs, energies, deprecations etc.

According to Table 1, it is possible to state that the cost ratio values have the value, which is higher than 1, representing a loss for the businesses, in

Table 1. Chosen indicators of agricultural business firms in Slovakia in 2007–2015. The values are presented in euros (EUR).

Indicator	2007	2008	2009	2010	2011	2012	2013	2014	2015
Costs (C)	1248.16	1336.06	1177.14	1113.28	1193.27	1224.02	1302.3	1375.72	1183.07
Sales (Sa)	860.85	844.79	650.8	681.79	829.74	854.92	858.38	900.87	781.57
Subsidies (Su)	94.54	258.22	271.73	262.65	248.16	239.56	236.7	286.01	187.12
Cost-effectiveness CE (Sa/C)	0.69	0.63	0.55	0.61	0.70	0.70	0.66	0.65	0.66
Cost Ratio CR (C/Sa)	1.45	1.58	1.81	1.63	1.44	1.43	1.52	1.53	1.51
Efficiency of Subsidies ES (Sa/Su)	9.11	3.27	2.40	2.60	3.34	3.57	3.63	3.15	4.18

other words, the increased amount of costs. However, it is not the purpose of the paper to follow the values of cost ratio.

3.4 Cost efficacy

Another indicator analyzed in the paper is the cost efficacy. It is a reciprocal indicator to cost ratio. Cost efficacy is a criterion of costs invested to reaching the health of a business and it is defined as the ratio of sales to total costs. Cost efficacy indicates how much money from revenues one Euro of costs produces (Kotulič, 2010).

$$SbE = (S)/Tc \qquad (3)$$

where $SbE = Subsidy\text{-}effectinevess; S = Subsidy; TC = The\ total\ costs$

Cost efficacy expresses the volume of sales that one Euro of costs invested to sales produces. In 2007, the ratio of sales to costs of monitored businesses had the value of 0.69 €. In 2008, the indicator showed a slight decrease of 0.66 € and reached the value of 0.63 €. In 2009, cost efficacy showed a tendency to decline by 0.08 € and was quantified at 0.55 €. In 2010, the indicator had a value of 0.61, which represented an increase of 0.06 €.

In 2011, the ratio of sales to subsidies showed a tendency to increase again by 0.08 €. It was quantified by the value of 0.70 €. In 2012, the value of the indicator didn't change. In 2013, the cost efficacy was quantified by the value of 0.66 €, which meant that the ratio of sales and costs decreased by 0.04 €. In 2014, the change of the indicator value was almost identical. The change was enumerated at 0.01 € with the value of 0.65 €. The value of the indicator remained almost unchanged in 2015. With a very small increase of 0.01 € it was enumerated at 0.66 €, which is the same value as in 2014.

During the monitored period, the indicators of cost efficacy were always enumerated by value less than 1; therefore we can assume that the agricultural businesses in Slovakia are not in good financial situation. Again, it was pointed out that the businesses prosper also thanks to the state subsidies or contributions from The European Union funds.

Cost controlling should be aimed more to the increase of costs in human resources area and sales controlling should be aimed more to a better utilization of subsidies. The increase in employment in agriculture, especially in expert and technical work, represents an increase in business performance, leads to the increase of sales and possible return to food self-sufficiency in Slovakia.

4 CONCLUSION

Agricultural businesses in Slovakia have only a small representation comparing to the businesses operating in other than agricultural activity. The indicators of costs and subsidies represent significant business management tools, because they point out the level of company management in a crucial margin. For Slovak agricultural businesses, it is important to draw subsidies and other subventions from the national or European funds. By means of chosen indicators of subsidies efficacy and cost efficacy we showed that the businesses prosper in a long-term thanks to the subsidy policy. Apart from subsidies, the costs, as an important input, are not effectively utilized as well. After 2009, the subsidy effectiveness has manifested itself rather by the increase, thus the businesses have used the possibility to get subsidies more.

In average, the cost efficacy manifested itself on the same level throughout the whole monitored period. For Slovak businesses this means that the utilization of costs is always on the same level.

The businesses should address the cost controlling more and they should use its tools effectively. For Slovak agricultural businesses, it would be appropriate to utilize the subsidies more effectively and to increase the costs in the field of personnel costs. This effect relating to the employment can manifest itself in gross domestic product and in employment in Slovakia.

REFERENCES

Cellini, S.R. et al. 2015 Cost-Effectiveness and Cost-Benefit Analysis Handbook of Practical Program Evaluation: Fourth Edition, pp. 636–672. [online] Available on: https://www.scopus.com/inward/record.uri?eid=2-s2.0–84977080490&doi=10.1002%2f9781119171386.ch24&partnerID=40&md5=61550607d7a4695b727cf082363bfc1a.

Čarnický, Š. & Mesároš. P. (2009). *Informačné systémy podnikov.* Bratislava: Ekonóm. ISBN 979-80-225-2676-01.

Kislingerová, A. & Hnilica, J. 2008. *Finanční analýza— krok za krokem.* 2. vydanie. Praha:C.H. Beck. ISBN 978-80-7179-713-5.

Korneeva, T.A. & Potasheva, O.N. 2016. Developing the methods for the analysis of costs associated with quality (2016). Actual Problems of Economics, 178 (4), pp. 294–303. [online] Available on: https://www.scopus.com/inward/record.uri?eid=2-s2.0–84964892716&partnerID=40&md5=a6be19c5cfb5d7ab90056a5fff6777d9.

Kotulič, R. 2010. The chosen aspects applied by considering the prosperity and the effectiveness of the business activities in the management of the agricultural enterprises [Vybrané aspekty pri posudzovaní prosperity a efektívnosti podnikateľských činností v manažmente poľnohospodárskych podnikov] [online] Available on: http://pulib.sk/elpub2/FM/Kotulic13/pdf_doc/06.pdf.

Kotulič, R. et al. 2010. *Finančná analýza podniku.* 2nd ed. Bratislava: Iura Edition. ISBN 978-80-8078-342-6.

Krings, U. et al. 2005. Controlling and performance measurement within networks IEEE International Technology Management Conference, ICE 2005, art. no. 7461255, [online] Available on: https://www.scopus.com/inward/record.uri?eid=2-s2.0–84971326386&doi=10.1109%2fITMC.2005.7461255&partnerID=40&md5=155cd16eecb4924ab3d7f10dab200a82.

Majerník, M. 2009. *Environmentálne manažérske systémy.* Skalica: Západoslovenské tlačiarne Skalica, s.r.o. ISBN 978-80-89391-05-9.

Mihalčová, B. & Pružinský, M. 2006. *O manažmente a manažovaní.* Ružomberok: Katolícka univerzita. ISBN 80-8084-122-5.

Pružinský, M. & Lajoš, B. 2011. *Podnikový controlling.* Technická univerzita v Košiciach: Typopres 1. Vydanie. ISBN 978-80-553-0648-3.

Sedlák, J. 2003. *Slovenské poľnohospodárstvo v Európskej únii.* Bratislava: Delegácia Európskej komisie v Slovenskej republike. ISBN 80-89102-04-2.

Szabo, L. & Grznár, M. 2002. Subsidies and efficiency in the agricultural sector [Dotácie a efektívnosť' v agrárnom sektore] Ekonomicky casopis, 50 (6), pp. 971–988. Cited 3 times. [online] Available on: https://www.scopus.com/inward/record.uri?eid=2-s2.0–0036987573&partnerID=40&md5=60f57754ba1be12dcf1fcbd0be16c8d3.

Šipikal, M. Et Al. 2013. Are subsidies really needed? The case of EU regional policy in the Czech and Slovak Republics [Sú dotácie naozaj potrebné? Prípadová štúdia regionálnej politiky EÚ v Českej republike a na Slovensku][online] E a M: Ekonomie a Management, 16 (4), pp. 30–41. Cited 2 times. Available on: https://www.scopus.com/inward/record.uri?eid=2-s2.0–84890335151&partnerID=40&md5=818309bcfebe0bdf04d92bef 2b43bf98.

Vanková, V. & Baláž, I. 2005. *Ekológia environmentálnych poľnohospodárskych systémov.* FPV UKF v Nitre. ISBN 80-8050-908-5.

Vlachynský, K. 2006. *Podnikové financie.* Bratislava: Iura Edition, ISBN 80-8078-029-3.

Zalai, K. 2010. *Finančno-ekonomická analýza podniku.* Bratislava: Sprint dva. ISBN 978-80-89393-15-2.

New Trends in Process Control and Production Management – Štofová & Szaryszová (Eds)
© *2018 Taylor & Francis Group, London, ISBN 978-1-138-05885-9*

Investigation of innovation outputs of SMEs financed by venture capital

M. Freňáková
Faculty of Business Economics with seat in Košice, University of Economics in Bratislava,
Košice, Slovak Republic

ABSTRACT: This article contains the analysis of innovation outputs of Small and Medium sized Enterprises (SMEs) financed by venture capital in the Slovak Republic. On the results of our analysis, conducted on the sample of 60 SMEs, we can conclude that Slovak enterprises financed by venture capital did not apply considerable innovation activity in form of innovation outputs as patents, designs, utility models, trademarks in their performance. The share of enterprises with innovation activity (which created some of innovation outputs) of all 60 SMEs financed by venture capital in our research sample was only 21.67% (13 enterprises). On the basis of analysis of our research sample we can conclude that the rate of innovation outputs is related to affiliation of enterprise to industry branch following its technological or knowledge demandingness. The affiliation to high-tech or medium high-tech sector was related with higher rate of innovation outputs in our research sample.

1 INTRODUCTION

Private equity (including venture capital) in all its forms represents an external capital invested by investor or fund (venture capitalist) in company, when invested financial sources became a part of equity capital of the company, into which the investor (fund) came. The investment of private equity capital means for the company not only financial sources but as well non-financial benefits in form of time, effort, experience and contacts of private equity investor (Freňáková 2011).

According to Invest Europe's 2015 European Private Equity Activity (Invest Europe 2016) private equity is equity capital provided to enterprises not quoted on a stock market. Private equity includes the following investment stages: venture capital, growth capital, replacement capital, rescue/turnaround and buyouts. Venture capital is a subset of private equity and refers to equity investments made for launch (seed), early development (start-up), or expansion (later stage venture) of business (Invest Europe 2016).

Venture capital is primarily a source of funding for start-ups, often technology-oriented small and medium sized enterprises (SMEs), with innovative ideas (Gompers & Lerner 2001), allowing them faster progress.

Authors (Lyasnikov et al. 2017) derived the following major inferences: "Venture capital financing is a modern institution whose activity is aimed at accumulating and redistributing temporarily available investment resources that are sought after in the sphere of innovation entrepreneurship; Countries whose economy may currently be recognized as transitive are characterized by a set of uniform issues: underdeveloped infrastructure in the national innovation system; lack of sources of venture capital financing; businesses reporting decreased innovation activity levels due to lack of economic incentives; lack of personnel resources; The evidence from the experience of more economically developed countries suggests that to enable the proper making of the institution of venture capital financing in countries with a transitive economy a set of interrelated objectives may need to be undertaken, namely: Ensuring legal optimization; boosting investment attractiveness; altering the nature of partnership between the state, business, and science-and-education sector; reducing state participation in economic and research activity."

As economy of the Slovak Republic can be recognized as transitive, the objective of the article is investigation of innovation outputs of SMEs financed by venture capital in the Slovak Republic.

2 METHODOLOGY AND MATERIAL

Based on the assumption that venture capital is a priority source of funding for innovations, we were interested in whether we can find the relationship between innovation outputs of venture-backed SMEs and the particular investor of venture capital as well as relationship between innovation outputs and technological or knowledge demandingness of the industry branch in which venture-backed SME operates.

For our investigation we selected four innovation outputs: patents, designs, utility models and trademarks. According to authors (Ját et al. 2005) trademarks can be considered as basic types of intellectual and industrial property.

Analysis of innovation outputs of SMEs financed by venture capital in the Slovak Republic was conducted on the sample of 60 SMEs. SMEs in our research sample were financed (venture-backed) by two investors: Fond fondov and SAEF (Slovak American Enterprise Fund).

Technological or knowledge demandingness of the industry branch in which venture-backed SME operates was divided into four groups: high-tech, medium high-tech, medium low-tech and low-tech.

In this article we understand "venture capital" as venture capital investments, which as an external financial source become a part of equity capital of company with innovative or extraordinary idea and these investments are put into the company in seed, start-up or early development stages. We investigated venture capital investments realized in time period 2001–2006.

3 RESULTS

3.1 Relationship between innovation outputs of venture-backed SMEs and the particular investor of venture capital

In the surveyed enterprises (60 SMEs), we detected through the online registers of the Industrial Property Office of the Slovak Republic (IPO SR 2008) whether the enterprises are owners or applicants for innovation outputs, only valid innovation outputs and applications for these innovation outputs were taken into account. Outdated innovation outputs were not taken into account. In this analysis, we did not distinguish between actual and historical investments. The findings are contained in Table 1.

Table 1. Innovation outputs of SMEs in a research sample by the particular investor.

Investor	Number of innovation outputs	Number of SMEs with innovation outputs
Fond fondov	6	6
SAEF	20	7
Total	26	13

The proportion of SMEs with innovation activity (in the form of patents, designs, utility models and trademarks) out of the total number of SMEs in the research sample is only 21.67%.

Table 2 contains the more detailed information about type of innovation outputs (patents, designs, utility models and trademarks) according to the type of investor.

The sum of SMEs with innovation outputs supported by SAEF, specifically 7 SMEs (in the Table 1) is not equal to the sum of SMEs for each type of innovation output (in the Table 2), because in one case the same enterprise was both the applicant for the patent and the proprietor of the trademark and in the second case, other enterprise was at the same time the owner of utility model and trademark. Both enterprises were included only once in the final number of innovative SMEs.

Due to the low innovation activity of the SMEs in the research sample and the resulting low numbers in the individual groups of innovation outputs, a special contingent table (Table 3) containing the binary variable (SME with innovation output = success or no innovation output = failure) in the two groups by type of investor (Fond fondov respectively SAEF).

Row profiles (Table 3) represent the share of successes and failures in groups by investor type (calculated row profiles are included in Table 4).

Table 2. Type of innovation output of SMEs in a research sample by the particular investor.

	Type of innovation output						
	Patents		Designs		Utility models		Trademarks
Investor	No.	No. of SMEs	No.	No. of SMEs	No.	No. of SMEs	No.	No. of SMEs
Fond fondov	0	0	0	0	1	1	5	5
SAEF	1	1	0	0	3	1	16	7
Sum	1	1	0	0	4	2	21	12

Table 3. SMEs with an innovation output, respectively without an innovation output in a research sample by the particular investor.

Investor	SMEs with innovation outputs (success)	SMEs without innovation outputs (failure)	Total
Fond fondov	6	32	38
SAEF	7	15	22
Total	13	47	60

Table 4. The difference in shares, the ratio of shares and the odds ratio of SMEs in research sample by investor type.

Investor	Share of successes within group	Share of failures within group	Difference in shares of successes	Ratio of shares of successes	Odd within the group	Odds ratio between groups
Fond fondov	0.1579	0.8421	−0.1603	0.4962	0.1875	0.4018
SAEF	0.3182	0.6818			0.4667	

To compare the binary variable shares by investor type, we used three characteristics, namely the difference in shares, the ratio of shares and the odds ratio (Table 4).

On the basis of the calculated characteristics (Table 4), it can be said that the share of SMEs with innovation output in a group of SMEs supported by the Fond fondov in our research sample is 15.79% and the share of SMEs without innovation output in this group is up to 84.21%. The situation in the SME group supported by SAEF is significantly different. The share of successful SMEs (with innovation output) is up to 31.82% and the share of unsuccessful SMEs (without innovation output) is 68.18%.

Based on the difference in shares of successes (−0.1603), we can conclude that the share of SMEs with innovation output in the SME group supported by Fond fondov in our research sample is 16.03% lower than in the group of SMEs supported by SAEF, respectively the share of SMEs without an innovation output supported by Fond fondov is 16.03% higher than in the SME group supported by SAEF.

Ratio of shares of successes (0.4962) indicates that the share of SMEs with innovation output in the group funded by Fond fondov is only 49.62% of the share of SMEs with innovation output financed by SAEF. Ratio of shares of successes (0.4962) can also be written as 15.79%: 31.82%, meaning that 15.79% of successful SMEs in the first group (SMEs funded by Fond fondov) to 31.82% of successful SMEs in the second group (SMEs funded by SAEF). The odd within a group is calculated as the ratio of the share of successes (SMEs with innovation output) and the share of failures (SMEs without innovation output) within the group. The odd (ratio of success rate to failure rate) in the group SMEs funded by Fond fondov in our research sample is only 0.1875, while the odd in the group SMEs funded by SAEF in our research sample is significantly higher up to 0.4667. The odds ratio (between Fond fondov and SAEF) calculated as a share of the odds in both groups is 0.4018.

On the basis of this, we can state in our research sample that SMEs funded by SAEF have a higher level of innovation output in the form of patents, utility models, trademarks compared to SMEs funded by Fond fondov. No design was recorded in either of the groups. We believe that the SMEs funded by SAEF are more innovative (up to 76.92% of all innovation outputs in the research sample) due to fact that the SAEF is more careful in project selection and more focuses on the innovation potential of supported companies.

3.2 Relationship between innovation outputs of venture-backed SMEs and technological or knowledge demandingness of the industry branch

The reason for the different innovation activity can also be the affiliation of enterprise to industry branch following its technological or knowledge demandingness. The classification of SMEs in the research sample by type of investor and by the industry branch according to technological or knowledge demandingness is contained in Table 5.

The row and column percentages (profiles), for easier realization of conclusions about the relationship between the investor type and technological or knowledge demandingness of the industry branch, to which the investment was directed, are contained in Table 6.

In the sector high-tech (Table 6), there are 2.63% investments by Fond fondov, while in this sector there are up to 4.55% of the investments by SAEF. In the medium high-tech sector, there are 18.42% investments by Fond fondov and up to 27.27% of investments by SAEF. Fond fondov has made the most investments in the medium low-tech sector (up to 52.63%), while SAEF's share in this sector accounts for 31.82%. Different investment is also in the low-tech group, where Fond fondov share of investments is 26.32%, SAEF's investments represent the largest share of 36.36% in this sector. In the set of analyzed investments, which were directed to 60 SMEs, there is a relationship between the type of investor and the technological or knowledge demandingness of SMEs.

We try to establish the strength of the relationship between technological or knowledge demandingness of the industry branch in which the SME operates and the type of investor, which made the investment. We calculated Cramer's contingency coefficient V, which takes values from 0 (no relation) to 1 (perfect relationship). For its interpretation we

Table 5. Technological or knowledge demandingness of SMEs in the research sample by the particular investor.

| | Technological or knowledge demandingness of the industry branch | | | | |
Investor	High-tech	Medium high-tech	Medium low-tech	Low-tech	Total
Fond fondov	1	7	20	10	38
SAEF	1	6	7	8	22
Total	2	13	27	18	60

Table 6. Profile of technological or knowledge demandingness of SMEs by the particular investor (%).

| | Technological or knowledge demandingness of the industry branch | | | | |
Investor	High-tech	Medium high-tech	Medium low-tech	Low-tech	Total
Fond fondov	2,63	18,42	52,63	26,32	100,00
SAEF	4,55	27,27	31,82	36,36	100,00

used the Cohen scale. Based on the Cramer coefficient value calculated for all 60 SMEs, respectively investments (Cramer's $V = 0.2027799$) we can say that there is only a small relationship between the two categorical variables (investor type and technological or knowledge demandingness of the industry branch).

After the analysis of the technological or knowledge demandingness of the industry branch of all SMEs in the research sample, we investigate only SMEs with innovation outputs, from the technological or knowledge demandingness point of view (the sector in which they operate) and by the type of investor (Table 7).

In the next step, we calculated the share of SMEs with an innovation output in the number of SMEs in the sector in terms of technological or knowledge demandingness and by the type of investor. The results are shown in Table 8.

The share of SMEs with innovation outputs in the high-tech sector in the total number of high-tech enterprises is 50%, the share of SMEs with innovation outputs in the medium high-tech sector is 15.38%, in the medium low-tech sector there is 18.52% of innovative SMEs and the share of SMEs with innovative outputs in the low-tech sector is 27.78%.

Based on the results of our research sample, contained in Table 9, we can conclude that in case of the surveyed SMEs is confirmed the assumption that affiliation to the high-tech or medium

Table 7. SMEs with innovation output in the research sample according to the technological or knowledge demandingness and according to the investor.

| | Number of SMEs with innovation outputs according to technological or knowledge demandingness of the industry branch | | | |
Investor	High-tech	Medium high-tech	Medium low-tech	Low-tech
Fond fondov	0	0	2	4
SAEF	1	2	3	1
Total	1	2	5	5

Table 8. Share of SMEs with innovation output on the total number of SMEs in individual groups (%).

| | Share of SMEs with innovation output on the total number of SMEs in the particular sectors | | | | |
Investor	High-tech	Medium high-tech	Medium low-tech	Low-tech	For the investor
Fond fondov	0.00	0.00	10.00	40.00	15.79
SAEF	100.00	33.33	42.86	12.50	31.82
For the industry	50.00	15.38	18.52	27.78	21.67

Table 9. Rate of innovation outputs of SMEs with innovation output in the research sample according to the technological or knowledge demandingness of the industry branch.

| | Technological or knowledge demandingness of the industry branch | | | |
Indicator	High-tech	Medium high-tech	Medium low-tech	Low-tech
Number of innovation outputs	3	7	10	6
Number of innovative SMEs	1	2	5	5
Rate of innovation outputs	3	3.5	2	1.2

high-tech industry branch would also mean higher production of innovation outputs. Because if we take into account number of innovation outputs created by total number of innovative SMEs in our research sample (total 26 innovation outputs created by 13 SMEs), the situation is as follows: 3 innovation high-tech outputs (created by 1 SME),

7 innovation medium high-tech outputs (created by 2 SMEs), 10 innovation outputs in the medium low-tech sector (created by 5 SMEs) and 6 innovation outputs in the low-tech sector (created by 5 SMEs). If the number of innovation outputs created by one SME in the relevant industry branch would be considered to be a measure of innovation activity as the rate of innovation outputs, then this rate decreases (with the exception of medium high-tech industry branch) with a declining technological or knowledge demandingness of the industry branch.

4 CONCLUSIONS

Venture capital as a form of business finance is suitable especially for young innovative and fast growing SMEs, which struggle with insufficient equity capital, managerial experience, possibly they are not able to attract debt financing and thus are not able to finance their growth and launch their innovative thoughts to the market.

Despite the argument that firms in technology-intensive or knowledge-intensive sectors receiving venture capital much more than firms operating in less technology and knowledge-intensive sectors, the situation in the Slovak Republic is reversed in this regard. In the Slovak Republic is in fact a lack of investment opportunities in high-tech sectors and in addition priority financial resources of venture capital in the Slovak Republic are public resources, which in most cases prefer the promotion of employment prior to the innovation.

On the results of our analysis, conducted on the sample of 60 SMEs, we can conclude that Slovak enterprises financed by venture capital did not apply considerable innovation activity in form of innovation outputs as patents, designs, utility models, trademarks in their performance. The share of enterprises which created some of innovation outputs in our research sample was only 21.67% (13 enterprises).

The share of SMEs with innovation outputs in our research sample in the group of SMEs supported by Fond fondov was 15.79% and in the group of SMEs supported by SAEF was 31.82%. Based on the difference in shares of successes (−0.1603), we can conclude that the share of SMEs with innovation output in the group of SMEs supported by Fond fondov in our research sample was 16.03% lower than in the group of SMEs supported by SAEF.

On the basis of this, we can state in our research sample that SMEs funded by SAEF had a higher level of innovation output in the form of patents, utility models and trademarks compared to SMEs funded by Fond fondov. No design was recorded in either of the groups.

We try to establish the strength of the relationship between technological or knowledge demandingness of the industry branch in which the SME operates and the type of investor, which made the investment. Based on the Cramer's contingency coefficient V, calculated for all 60 SMEs, respectively investments (Cramer's V = 0.2027799), we can say that there is only a small relationship between the two categorical variables (investor type and technological or knowledge demandingness of the industry branch).

On the basis of analysis of our research sample we can conclude that the rate of innovation outputs is related to affiliation of enterprise to industry branch following its technological or knowledge demandingness. The affiliation to high-tech or medium high-tech sector was related with higher rate of innovation outputs in our research sample.

REFERENCES

Freňáková, M. 2011. *Venture kapitál a rozvojový kapitál pre váš biznis.* Bratislava: Vydavateľstvo TREND.

Gompers, P.A. & Lerner, J. 2001. *The money of invention: How Venture Capital Creates New Wealth.* Boston: Harvard Business School Press.

Invest Europe. 2016. *Invest Europe's 2015 European Private Equity Activity.* Brussels: Invest Europe.

IPO SR. 2008. *The online registers of the Industrial Property Office of the Slovak Republic* 2008. Retrieved from https://www.indprop.gov.sk.

Jáč, I., Rydvalová, P. & Žižka, M. 2005. *Inovace v malém a středním podnikání.* Brno: Computer Press.

Lyasnikov, N.V., Frolova, E.E., Mamedov, A.A., Zinkovskii, S.B. & Voikova, N.A. 2017. Venture Capital Financing as a Mechanism for Impelling Innovation Activity. In *European Research Studies Journal* 20(2): 111–122.

New Trends in Process Control and Production Management – Štofová & Szaryszová (Eds)
© 2018 Taylor & Francis Group, London, ISBN 978-1-138-05885-9

Modern business models and prediction in tourism

P. Gallo, M. Karahuta, D. Matušíková & A. Šenková
Faculty of Management, University of Prešov, Prešov, Slovak Republic

ABSTRACT: The paper addresses the issue of management decision-making using artificial neural networks and their application in hotel management. Balanced ranking and prediction model using financial and non-financial indicators with the application of artificial intelligence, allows us to reach high level of effectivity and accuracy in evaluation of financial and non-financial health of companies operating in this segment.

1 INTRODUCTION

Market instability (in the form of numerous and frequent changes in the economy and business) affects the economy and industry all over the world. These changes require from the managers to use increasingly sophisticated decision-making and prediction tools in everyday' decisions. Even partial knowledge of the consequences of complex decisions can create a competitive advantage. For tourism sphere is specific that although companies are frequently and strongly linked to local markets, they compete globally and with similar service offer around the world. Therefore, one of the important information that manager should know is the overall financial health of the company—and not only the absolute one, but especially relative when comparing to the competition. Prediction of business health problems identification is an extremely complex area requiring an in-depth analysis of the amount of information, data and context. In tourism as well as in the hotel industry the situation is complicated by significant element in the form of the human factor, whose behaviour and decisions are often based on feelings and it's strongly subjective.

For housing facilities, management needs in order to increase the efficiency of their work, competitiveness and success in terms of management. There is the possibility of a comprehensive assessment of a company's health, using artificial intelligence in the form of artificial neural networks. Neural networks, as a tool of universal approximation, is a comprehensive and sufficiently robust tool for managing subjectivity, complexity of data and situations in tourism.

2 DESIGN OF MODEL FOR THE EVALUATION OF FINANCIAL HEALTH

Proposition of the model for the evaluation of the financial health of businesses operating in the tourism sector is based on the basic knowledge evaluation of enterprises based on financial and prediction models. Proposed innovative methodology of research based on benchmarking has two following parameters:

– Financial performance represented by selected financial indicators
– Successfulness represented by selected prediction models.

The created model of evaluating the performance of enterprises uses the following parameters:

– Financial performance:
– the period of accounts receivable turnover
– the period of accounts payable turnover
– the period of inventory turnover
– the degree of recapitalization
– total liabilities
– current liabilities
– profitability of one's own capital
– profitability of sales
– total liquidity
– current liquidity.

These indicators make selected indicators of a financial analysis and this in the area of activity, profitability, capital structure and liquidity (Horváthová & Mokrišová 2014).

Successfulness:

– Quick test (the share of one's own capital, the period of paying debts from cash flow, the share of cash flow from revenues, the profitability of total assets)
– Solvency Index
– Z-score
– Tafler's Index.

The result of the model can be construction of business success portfolio and status of its financial health. As an example, the position of Slovak spas in this portfolio can be seen in Figure 1.

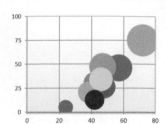

Figure 1. Financial health of Slovak spas. Source: own processing based on www.indexpodnikatela.sk.

3 FINANCIAL HEALTH OF THE ESTABLISHMENT AND ARTIFICIAL NEURON NETWORKS

Artificial neural network (ANN) is used to model and copy operation of the human brain. It operates on a very similar principle as a network of neurons in the brain. ANN view of the concept is not entirely uniform because it is used for biological and non-biological networks. ANN belongs among statistically learning algorithms and are used to estimate, respectively approximation of functions that can depend on a large number of inputs that are generally unknown. Possibilities of using ANN in management and tourism are quite broad. They have a number of advantages and versatility. They are therefore also applicable to the management of tourism. Task areas often solved by ANN are (Kelemenová et al. 2010):

– Approximations of functions,
– prediction problems,
– classification,
– process management,
– association problems, simulation of memory, transformation of signals.

In addition to the areas mentioned in Table 1 ANN is also used in agriculture, astronomy, meteorology and many other areas. They have wide application also in the entrepreneurships. Hakimpoor et al. (2011) analysed and summarized wide list of activities in the business that use ANN. In the practical application of ANN in business is highlighted the importance of the impact of management in proposing the solutions. They have practical applicability in any field of management in tourism. Requirements must be well defined and context should make sense. For managers, it is then important to understand the basic principles and management process, not software development. ANN can then serve as an effective tool for decision-making.

Managerial decision-making and development prediction from the aspect of financial health in hotel management can therefore play an important

Table 1. Application of ANN in management of hotel.

Establishments Area/Activity	Source
Prediction of consumers reactions	Fiesle a Beale 1997 Bruce Curry a Luiz Mountinho 1993
Analysis of financial health	R.C. Lacher 1995 Rafiei a Manzari 2011 Caron H. St. John 2000
Prediction of bankruptcy	Tang a Chi 2005 Wilson a Sharda 1994 du Jardin a Séverin 2011
On-line hotel booking	Corazza, Fasano a Mason 2014
Prediction in tourism	Palmer, José Montaño a Sesé 2006
Analysis of demand in tourism	Berenguer a kol. 2015 Uysal a Roubi 1999 Cankurt a Subasi 2015
Analysis of hotel on-line rating	Phillips a kol. 2015
Analysis of air condition consumption on hotel room	Moon a kol. 2015
Evaluation and location of hotel	Yang a kol. 2015

role. Franklin (2003) defines tourism as a comprehensive set of social and cultural phenomena. In 2014, tourism created 10% of global GDP with the amount of 7.6 trillion dollars and it offered 277 million of working positions (World Travel & Tourism Council 2015b). The trend of tourism development is growing.

An integral part of the tourism is a hotel industry. This does not include only hotels, but also other types of accommodation establishments such as guesthouses, but also so-called para-hotel devices such as camping or hiking hostel etc. In Slovakia, the hotel industry is an important part of tourism. Annual revenues from the provision of accommodation services in 2014 in Slovakia were 267,550,285 euro, and from it 81.5% (SACR 2015) accounted sales of hotel style accommodation (SACR 2015). 2015 and especially 2016 were years where revenues increased again and the growth trend has not changed.

Tourism and the accommodation establishments have global impact on almost every area of human activity. Economically, accommodation establishments play crucial role. By the provision of accommodation to the consumer, allow longer stay that in the area and thus create conditions for the efficient operation of the other sub-areas (catering, retail, souvenirs, sports, etc.). Another important effect of accommodation establishments is the creation of jobs and economic development in specific areas where other activities such as tourism are not feasible. In terms of social significance, hotels and hotel industry can be defined as an area of a wide range of activities with social importance. Place capacity and catering services allow implementing social events such as weddings, celebrations, meetings, meetings, training, conferences and many others.

4 DECISION-MAKING AND PREDICTION IN HOTEL MANAGEMENT—ANN APPROACH

Each area of human activity uses some form of decision-making. Hotel management requires complex decision-making processes, since the whole essence of service is dependent on two elements which behaviour is often very difficult to predict—the people and the weather. In the hotel management, there are defined classification markings for accommodations establishments for their classification into categories and classes. Not every accommodation establishment is affected by the weather, but most of them do. Of course, people and weather are not the only input into decision-making processes in hotel management. O'Halloran (2015) argues that the manager of the accommodation establishment daily makes more decisions than the manager in the other one (area). Decision-making situations often require immediate decision and use semantic rather than factual language. This complicates the use of standard statistical methods. The other fact is that often all the information for decision-making are not available.

Another major problem is the complexity of the situations and a number of unknown variables and relationships. Complexity and multi-criteria of decision and prediction situations in hotel management creates space for the application of the most modern techniques and methods. Although the bankruptcy of accommodation establishments is a relatively common phenomenon, surprisingly few authors are devoted to the prediction of this fact (Adams 1991, Gu and Gao 2000, Fernandez et al. 2013, Kim and Gu 2010).

In an ideal world would such problems be spread to small parts in which we could be able to decide on the basis of the main criteria. In reality, the problems are complicated and multi-criterial. Neural networks are suitable to solve problems, where we do not know the complete context. This phenomenon is common when working with people and in prediction of their behaviour (Kwortnik 2003). The hotel business is significantly based on emotions, subjective assessments and feelings, and it complicates even the standard management tools, such as positioning. Modern techniques such as fuzzy logic and neural networks are applicable in hotel management and also in predicting the success of the hotel as an organization.

5 MODERN APPROACH TO HOTEL MANAGEMENT DECISION-MAKING

Currently there are mainly classical conventional bankruptcy and creditworthy predictive models, modern financial indicators and in-depth financial analysis. The current development of information technology allows us to create algorithms already capable to handle the management issues in the hotel management at least in part. One of them are the aforementioned artificial neural networks and interpretation of results via dashboards.

The main and most important contribution of the model is a dashboard of accommodation establishment. Outcome of the model is meaningful only if it can be properly and easily interpreted. Although the model output is an index (real number) in the range from 0 to 1, the value of which defines the level of health of the company, for the application into the management processes it is needed to be properly visualized and complemented by context. Figure 2 shows the screenshot of online web application that allows managers to have an interactive overview of the state of company's' health. Report is in a form of customizable dashboard. It allows the manager to customize it to be the most suitable for their working process. It includes several basic components:

– Index ANN. Dominant information as a result of the proposed numerical model and its word interpretation (for example, 0.72 – very good). Manager can immediately identify the percentage change compared to the previous period. Based on the analysis of previous years together with the application predicts the future development (positive, negative or no significant change).
– Visual benchmarking. Client has the opportunity to compare the development of their own business compared to the industries in which it operates. It can monitor various indices, models and indicators. It is also possible to set different filters, dates and scales.
– Financial indicators and their comparison. Report effectively uses input data of the model and clearly displays them using various forms of graphs and visual elements. One of the possibilities is a radar chart that visually very effectively displays position of the company within the sector through selected indicators. The bar and line graphs show the development of the main indicators in the last period.

– Overall condition. Overall condition, respectively business health represents a percentage of the outcome of our model and benchmarking with companies with similar characteristics. The key element is the index ANN converted to a percentage of adjusted overall condition of the sector. Colours provide immediate information about the company's condition.

– Recommendations. Recommendations are an important part of the report. These are partially automated and warn of thresholds values indicators. Application of them allows to make recommendations to measure. There is a space for cooperation of commercial and academic world. Together with the index ANN they generate the most added value to the report.

6 SCALABILITY AND FURTHER DEVELOPMENT

In addition to the practical benefits for the practice, model based on ANN also offers benefits to the expansion of the theory and science. Hotel management is significantly practical field and its scientific development is connected with the practice.

Proposed model allows a strong link between the two dimensions. Scientists and academics can apply their extensive knowledge in real situations, analyse real data, and assist managers with the development of their business. Managers can also confront scientific theories with practice and provide results and data. Visually the model is shown in Figure 2. The advantage of the core model is reflected based on artificial neural networks—learning ability is reflected here. Within the principle of GIGO (garbage in, garbage out) it is important that the input data are true and reliable. How much more accurate and detailed data will add the practice, better advice, analysis and inputs delivers science, and so much better reporting model is able to generate and discover new connections and improve results over itself. From this point of view, there is a space for symbiotic cooperation of practice and science.

While maintaining the principles and methodology applied in the design of the model, the model can be extended to new input data, thus exponentially expand the number of combinations of elements and potential new context. By summarizing of the benefits we can argued that the model offers benefits on multiple levels and dimensions. The fact that the proposed model is based on real data and

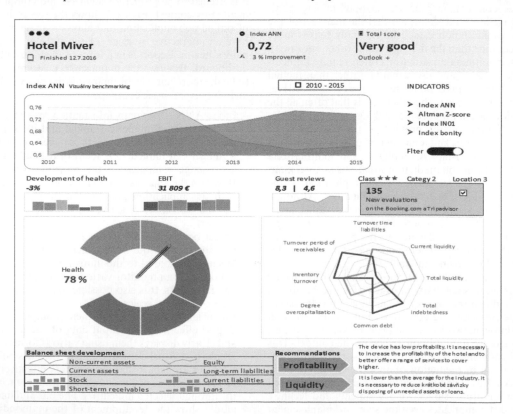

Figure 2. Dashboard of hotel establishment.

real analysis, allows its easy application into practice. Created database of accommodation establishments and implemented analysis of this model of the variables creates a solid foundation for deeper scientific knowledge and research in the future.

6 CONCLUSION

Tourism is from a scientific and managerial point of view fascinating area connecting many unions. Examination of tourism and all its components requires high-quality interdisciplinary cooperation. Analysis and evaluation of the company in the hospitality sector requires a comprehensive view of the surrounding body and the consideration of several factors. Universality of neural networks offers a wide range of applicability, but emphasizes the principle of GIGO—garbage in, garbage out. Mechanic functioning of used ANN is based on learning with a teacher. The analysis and definition of the input variables and data quality became a key part of the design of the model. Input data significantly affect the ability of the model, management approach and view on the issue under consideration here plays the most important role. The model presented in Figure 2 is a balanced model that equally uses both financial and non-financial data.

The paper was written within a project KEGA n. 020PU-4/2015 "Tvorba multimediálnych web dokumentov pre e-learningové vzdelávanie pre zvyšovanie kvality vedomostí manažérov a študentov" and project VEGA n.1/0791/16 "Moderné prístupy zvyšovania podnikateľskej výkonnosti a konkurencieschopnosti s využitím inovatívneho modelu—Enterprise Performance. Model pre zefektívnenie manažérskeho rozhodovania (Modern approaches to improving enterprise performance and competitiveness using the innovative model—Enterprise Performance Model to streamline Management Decision-Making Processes) solved at the Faculty of Management, the University of Prešov in Prešov.

REFERENCES

Adams, D. J. 1991. Do Corporate Failure Prediction Models Work? In: *International Journal of Contemporary Hospitality Management* [online]. Roc. 3, c. 4 [cit. 08.11.2015]. Dostupné z: http://www.emeraldinsight.com/doi/ abs/10.1108/EUM0000000001681. ISSN 0959-6119.

Fernandez, F., G. et al, 2013. *Use of Artificial Neural Networks to Predict The Business Success or Failure of Start-Up Firms* [online]. [cit. 08.11.2015]. Dostupné z:http://cdn.intechopen.com/pdfs/38255/InTechuse of artificial neural networks to predict the business success or failure of start up firms.pdf. ISBN 978-953-51-0935-8.

Franklin, A. 2003. *Tourism: An Introduction.* London: SAGE. 306 str. ISBN 978-1-84860-525-1.

Gu, Z. & Luyuan G. 2000. A Multivariate Model for Predicting Business Failures of Hospitality Firms. In: *Tourism and Hospitality Research* [online]. Roc. 2, c. 1 [cit. 08.11.2015], str. 37–49. Dostupné z: http://thr.sagepub. com/content/2/1/37. ISSN 1467-3584.

Hakimpoor, H. et al. 2011. Artificial neural networks' applications in management. In: *World Applied Sciences Journal.* Roc. 14, c. 7, str. 1008-1019. ISSN 1818-4952.

Horváthová, J. & Mokrišová, M. 2014. Diagnostika výkonnosti podnikov s aplikáciou moderných metód hodnotenia financnej výkonnosti. In: *Economics Management Innovation.* Roc. 6, c. 3.Dostupné z: http://emi.mvso.cz/EMI/2014–03/05horvathova mokrisovadiagnostikavykonnosti_podnikov.pdf. ISSN 1805-353X.

Kelemenová, A. et al. 2010. *Umelý život, vybrané modely, metódy a prostriedky.* Ružomberok: Verbum. ISBN 978-80-8084-642-8.

KIM, H. & GU, Z. 2010. A Logistic Regression Analysis for Predicting Bankruptcy in the Hospitality Industry. In: *Journal of Hospitality Financial Management* [online]. Roc. 14, c. 1 [cit. 02.11.2015]. Dostupné z: http://scholarworks.umass.edu/jhfm/vol14/iss1/24. ISSN 2152-2790.

Kwortnik JR. & Robert, J. 2003. Clarifying "fuzzy" hospitality-management problems with Depth interviews and qualitative analysis. In: *The Cornell Hotel and Restaurant Administration Quarterly* [online]. Roc. 44, c. 2 [cit. 08.11.2015], str. 117–129. Dostupné z: http://www.sciencedirect.com/science/article/pii/S0010880403900255. ISSN 0010-8804.

O'halloran, R. 2015. *Strategies for Decision Making* [online]. [cit. 08.11.2015]. Dostupné z: http://hotel-executive.com/business_review/2146/strategies-for-decisionmaking.

World Travel & Tourism Council, 2015b. *Travel & Tourism—economic impact 2015 world.* [online]. [cit. 03.11.2105]. Dostupné z: http://www.wttc.org/-/media/files/re-ports/economic%20impact%20 research/regional%202015/world2015.pdf.

Using the balanced scorecard concept with the mining industry as example

P. Gallo Jr.
Faculty of Management, University of Prešov, Prešov, Slovak Republic

B. Mihalčová
Faculty of Business Economics with seat in Košice, University of Economics in Bratislava, Košice,
Slovak Republic

ABSTRACT: The article is aimed at the Balanced Scorecard (further only BSC) concept as a means of strategic management applied in the Slovak mining industry. The main objective of paper in mining companies is to determine the current state of the use of BSC there and identify the reasons that prevent its use. Provided in it is a statistically documented application of BSC with the mining companies in Slovakia, as well as recommendations of implementation the concept for improving the quality of strategic management in this branch of national economy. To evaluate the questionnaire were used statistical methods and line graph that compares the development trends of individual indicators of the Balanced Scorecard. Substantial part of this contribution is devoted to a questionnaire-based survey and the data obtained in this branch, namely in the mining industry.

1   INTRODUCTION

A businesses operating on the market today is characterized by unstable and constantly changing business environment. It is these constant changes must be able to anticipate business and they know the necessary modifications. The importance of quality improvement is nowadays also undeniable. The mining branch of national economy is adaptively challenging to changes. It is even more important for the reason of steadily changing legislation and limitations. In this regard, company management is primarily affected by two basic aspects of company operation. The first one is representing the financial side of company performance, which consists in the owner's view as a financial investment, expected to provide an added value. The second view understands a company as a socio-economical system, which is made up of a complex network of internal and external relations that have to be managed in balance. The two views form the basis of the modern systems of company management, with the concept of Balanced Scorecard among them. This paper focuses on the using of the BSC method in mining companies and it trying to uncover the reasons why businesses do not use it. Generally, the concept is not so widespread in the Slovak business community, and it is its implementation that can be of great help in the mining industry, bringing improvements in

performance indicators thereby meeting the requirements of both the state-owner and shareholders. The industry of mining along with the subsystem of extracting minerals forms part of the international business of every country. In 2015, mining of mineral ores contributed to the gross domestic product (GDP) by € 330.30 million in current prices, making up 0.46% share of the GDP. Substantial portion of products is non-metallic raw materials, energy resources and those for civil engineering. Mining and processing of most non-metallic and construction materials (i.e. magnesite, limestone, gypsum, stones and others) essentially cover domestic consumption. Minerals and mineral-based products are an important item within foreign trade. Foreign trade in mineral resources is permanently passive owing to the large volume of imported mineral fuels (i.e. oil, gas, coal) and metallic raw materials (iron ore, raw materials for metallurgy aluminum, iron and ferro-alloys). The geological reserves found in 629 sole deposits as of 1st of January 2015 have been priced to 16,461 thousand millions.

According to (Grabara 2013) efficient business management under contemporary conditions becomes a complex task. Within this task, managers must not only timely identify capabilities of the external environment, but also look for internal growth reserves. When applying these reserves, manager should take into account the evaluability of the company and possible threats to the company development. There are many ways, methods and approaches to improve the strategic business management of contemporary companies. At that, the technique, which is based on the Balanced Scorecard, remains the most appropriate approach to build and structure company business management, as well as elaborate the long-term development strategy. The Balanced Scorecard is a special set of indicators that are of equal importance for analyzing and evaluating the operation and business development of companies. The Balanced Scorecard is focused on meeting the information needs of strategic planning and management. The Balanced Scorecard has become a widely known management tool and recent surveys have indicated that many organizations use, or intend to use, the Balanced Scorecard (Kaplan & Norton 2005).

The Balanced Scorecard concept was first published in 1992 in the Harvard Business people Review its main actors Robert S. Kaplan and David P. Norton. In the definition of Balanced Scorecard based Norton Kaplan (1996) of three basic concepts that constitute its essence: Balanced—a balance, Score—indicator Card—letter (seen as reviewed in conjunction with the indicators). Balanced Scorecard is a conception of how to transfer the vision and strategy into objectives and metrics so comprehensively provide not only the financial results of the company, but also non-financial area (Meyer 2002). The BSC, therefore, explicitly adopts a multi-dimensional framework by combining financial and non-financial performance measures (Štefko & Gallo 2015). Hence, the BSC allows a more structured approach to performance management while also avoiding some of the concerns associated with the more traditional control methods. Many organizations, both in the private and public sectors, have embraced the concept and implemented it in an attempt to improve performance (Dudin & Frolová 2015; Štofová & Szaryszová 2016). The scorecard has floundered as a device for measuring and re-warding performance (Bostan & Grosu 2011). Pursuant to Kaplan and Norton (Balogh & Golea 2015) the scorecard lets managers introduce four new processes that help companies make important link. The first process—translating the vision—helps managers build a consensus concerning a company's strategy and express it in terms that can guide action at the local level. The second—communicating and linking—calls for communicating a strategy at all levels of the organization and linking it with unit and individual goals. The third—business planning—enables companies to integrate their business plans with their financial plans. The fourth—feedback and learning—gives companies the capacity for strategic learning, which consists of gathering feedback, testing the hypotheses on which a strategy is based, and making necessary adjustments.

Wagner recommends a corporate performance parameters divided into four groups designated by Kaplan and Norton as the perspectives that through answers to four basic questions will enable a coherent and balanced way to evaluate corporate performance (Kosábková 2008).

Financial Perspectives—The current emphasis on financials leads to the "unbalanced" situation with regard to other perspectives. Let us look, therefore the answer to the question, "as it appears to our share-holders in order to be financially successful?" (Sales growth, increase return on equity, earnings, revenue, cash flow, EVA, cost reduction).

Customer perspective—Recent management philosophy has shown an increasing realization of the importance of customer focus and customer satisfaction in any business. These are leading indicators: if customers are not satisfied, they will eventually find other suppliers that will meet their needs (Itnner & Larker 2003). Therefore, we should answer the question: "How do you create a relationship with customers to achieve our vision?" (Satisfaction, price, quality, market position, repeated demand).

The perspective of internal processes—This perspective refers to internal business processes to answer the question: "In which business processes we must excel to satisfy our shareholders and customers?" (I.e. measurement of processes affecting customer satisfaction, the percentage of sales from new products in total sales, the number of our new products to a number of new competitive products, ROAS, shorter processes) (Ittner & Larker 2003).

Learning and growth perspective—This perspective includes employee training and corporate cultural attitudes related to both individual and corporate self-improvement. Let's ask: "How will we sustain our ability to change and improve in order to achieve our vision?" (i.e. knowledge, skills, education, motivation, information systems, fluctuations). At the beginning of year 2010, a research was conducted aimed at verifying the functionality, or non-functionality of the BSC in selected companies in Slovakia. In view of the nature of the methodology, it was a quality-based/oriented research focused on opinions, standpoints and experiences with the practical implementation of the BSC. Currently it is used by as much as 53% of European and American companies and organizations, whereas in Slovakia, BSC has been implemented by less than 20% of companies. But the fact that either the company is owned by a Slovak or foreign proprietor has no real effect on the utilization of the BSC concept (Gallo & Mihalčová 2016).

2 METHODOLOGY OF RESEARCH FOCUSED ON THE BALANCED SCORECARD CONCEPT REALIZED IN THE MINING INDUSTRY OF SLOVAKIA

The questionnaire was distributed to companies via internet for the references. The questions in it were formulated as those requiring multiple-choice answers and establishing priorities of a 5-point Likert's scale (Marcheová et al. 2011). The companies involved have been selected from the annual review of the industry: Nerastné suroviny SR 2015 published by the State Geological Institute. The survey taking the form of a questionnaire was distributed to 87 mining companies in Slovakia. The sample was collected by simple random sampling. It was generated 16 enterprises, representing a 18,40% rate of return. In the survey questionnaire prevailed as core business of extraction of building stone, gravel and limestone mining. These companies in terms of staff belonging to large and medium-sized enterprises and were mostly owned by Slovak owners.

Evaluation of the data obtained involved application of research methods such as descriptive statistics, contingency tables etc. while making use of analyses, comparisons, selections, inductions and deductions. When developing the questionnaire we started with the formulation of the hypotheses and professional questions, which were to be verified by statistical operations. Verification of the hypotheses and the questions involved the method of proportion of the given phenomena in the population, arithmetic mean, standard deviation and the rate of occurrence of the most frequent answers. The follow-up hypothesis and research question were focused on making use of the BSC concept and causes of not using it. The formulas and characteristics are shown in Table 2 and Table 3.

Hypothesis No. 1: Let us assume that more than 20% of mining companies in Slovakia makes use of the Balanced Scorecard method to measure company performance. To verify the hypothesis we have stated question focused on the knowledge and application of the BSC concept. With the help of the question we were trying to find out the real status of applying the Balanced Scorecard in Slovak mining companies. The data obtained by via a questionnaire-based survey have been subsequently processed with the aim to verify or reject the hypothesis formulated by us. We displayed the questions in Table 4.

Table 1. Basic information of the questionnaire-based survey.

Questionnaire-based survey on Balanced Scorecard

Questionnaires sent	87
Questionnaires filled in	16
Rate of return %	18.40%

Table 2. Statistical methods.

Method proportions of the phenomenon in the population

$$p = \hat{p} + z_\alpha * \sqrt{\frac{\hat{p} * \hat{q}}{n}}$$

$\hat{p}$ – proportion of the phenomenon
   in the selected sample

q – proportion of the opposite phenomenon,
   in the selected sample

n – Scale selection

z_α – confidence level

Table 3. Standard deviation.

Standard deviation

$$S = \sqrt{\frac{1}{n}\sum_{i=1}^{n}\left(x_i - \overline{x}\right)^2}.$$

S – standard deviation
n – The total number of samples
x_i – median
$n - 1$ – accuracy rating

Table 4. Question 2 focuses on understanding the concept of Balanced Scorecard.

Do you know the system performance measurement Balanced Scorecard?	a) Yes, our company uses it b) Yes, our company is considering the introduction of a Balanced Scorecard c) Yes, but our company does not use it d) No, but we are interested in the Balanced Scorecard System

Table 5. Reasons against the use of Balanced Scorecard.

The reasons why companies use the Balanced Scorecard	Likert scale (1–5)
1. The failure to clarify the current strategy (O18A)	1 – Completely disagree
2. The high financial cost of the operation (O18B)	2 – Disagree
3. Lack of personal resources (O18C)	3 – I do not know
4. An enterprise considers the Balanced Scorecard as a current trend (O18D)	4 – I agree
5. The discrepancy between the ideas of managers in the company (O18E)	5 – Strongly agree

Data collected survey, we then statistically processed and determined that the hypotheses. Identifying the reasons which prevent the use of the Balanced Scorecard in mining companies in Slovakia. When creating questionnaires were based on surveys that were conducted in the past and based on the data obtained, we expect that in this survey will be represented by companies which do not use the Balanced Scorecard concept. The questionnaire was therefore included a question aimed at finding reasons that a major proportion of the cutback of Balanced Scorecard and was posed as follows: What are the reasons that might hinder application of the Balanced Scorecard concepts? The question was stated in a form of a five-point Likert's scale, whereon the companies might assign a score to causes preventing use of the Balanced Scorecard concept, by setting the grades along the scale. When evaluating the answers we paid attention to the arithmetical mean, standard deviation and the rate of occurrence of the most frequent answers shown in Table 5.

On the basis of these the questions we get data that we processed and evaluated by using statistical methods as follows.

3 THE RESULTS OF THE QUESTIONNAIRE-BASED SURVEY FOCUSED ON THE BALANCED SCORECARD CONCEPT AS APPLIED IN SLOVAK MINING COMPANIES

The aim of the research was to determine the current review of the Balanced Scorecard using in Slovakia mining companies. From the data that was available we noted that the BSC is not in Slovak companies widespread, therefore we were interested in the reasons which prevent the use of the concept by mining companies.

3.1 Results and verification of the hypothesis on utilizing the Balance Scorecard method in Slovak mining companies

On the basis of the data obtained we went on to verify the hypothesis by method of proportion of the phenomena in the population with a determined co-efficient of reliability. In our case, reliability coefficient was 1.96 in value. Table 6 shows the figures.

By verifying the hypothesis applying the statistical method, the value from 18.73% to 37.87% was obtained. Based on the calculations it follows that the value of 20% falls within the interval found, which means that the Balance Scorecard concept is not used by more than 20% of mining companies. By the results obtained, the hypothesis formulated must be rejected.

Table 6. Results of the proportion method applied to the share of the phenomenon on the population.

Method of proportion of the phenomenon on the population	
Calculation	$\hat{p} = 0,1875$
	$\hat{q} = 0,8125$
	$p = 0,1875 \pm 1,96 * \sqrt{\dfrac{0,1875 * 0,8125}{16}}$
	$p = 0,1875 \mp 0,1912$
	$0,1873 \leq p \leq 0,3787$

Table 7. Statistical evaluation of the causes of non-using the Balanced Scorecard concept in mining companies.

Abbreviation of the cause	Average	Standard deviation	The frequency of the most common answers
(O18 A)	2.83	0.98	8
(O18B)	3.17	0.97	8
(O18C)	3.17	1.33	6
(O18D)	2.67	1.03	6
(O18E)	2.67	0.82	10

3.2 Results of the questionnaire-based survey focused on the causes that hinder use of the Balanced Scorecard in Slovak mining companies

Evaluation of the reason that prevent the BSC method from being applied by mining companies involved arithmetical average, standard deviation and the rate of occurrence of the most frequent answers. Answers from surveyed businesses are treated in Table 7.

Based on the questionnaires that have been returned and the subsequent analysis of the data it follows that the most frequent causes why companies make no use of the Balanced Scorecard concept is in high financial costs of its operation and lack of human resources. When evaluating the level of importance we can make good use of the standard deviation, which is the lowest at answers referring to high financial costs of operation and the disharmony among the ideas of the managers of the company. With the help of the standard variation, we can state that among the significant causes of non-using the Balanced Scorecard is the disharmony already mentioned. The standard deviation is lower, however, due to lack of resources. Thus, we can conclude that the most significant reason for non-use BSC mining companies consider the lack of financial resources for the implementation of the method.

4 CONCLUSION

Balanced Scorecard method in Slovakia compared to Western European countries relatively little used. Using a questionnaire survey we investigated what is its actual use in the environment of mining companies in Slovakia. For use of the method, we hypothesized that we are using statistical calculations verified. By realizing the questionnaire-based survey we have found out the actual status in using the Balance Scorecard concept as well as the main causes that hinder the application of it. After having processed the data obtained we can state that utilization of the BSC in the mining industry does not achieve even the rate of 20%. Based on this finding, we have tried to focus on the causes, which are instrumental in the rather low level of utilization of this concept. Among the causes resulting in the low preference to BSC is in high financial costs of operating the system and lack of human resources. Another important factor is the disharmony among the managerial ideas in the companies involved. According to the obtained results we arrived at the following recommendations: Procedures of strategic management of mining companies are to be based mostly on reports and detailed accounts of the company business. The documents have to be developed in compliance with the conceptual documentation approved for mining companies. Successful operation of these companies require establishment of a post for a coordinator, who will be responsible for the implementation of BSC in mining companies. The concept of the BSC represents a tool with which a company can achieve improvements in all areas of company operation. The essence of the method consists in utilizing both financial and non-financial indicators. By adopting the recommendations as part of the process of implementing the BSC in mining companies, this concept can be used to their benefit so that they could meet the expectations of both the owners and other interest groups, which both affect companies and are affected by them as well.

REFERENCES

Balogh, P. & Golea, P. 2015. Balanced scorecard in human resource management: Quality—access to success, Vol. 16, No. 137.

Bostan, I. & Grosu, V. 2011. Contribution of Balance Scorecard Model in Efficiency of Managerial Control. In *Romanian Journal of Economic Forecasting*, Vol. 14, p. 178–199.

Dudin, M. & Frolová, E. 2015. The Balanced Scorecard as a Basis for Strategic Company Management in the Context of the World Economy Transformation. In *Asian Social Science*; Vol. 11, No. 3.

Gallo, P. & Mihalčová, B. 2016. Knowledge and use of the Balanced Scorecard concept in Slovakia related to company proprietorship: Quality—access to success, Vol. 17, No. 151.

Grabara, J. 2013. Employer's expectations towards the employees from the marketing and management department. In *Polish Journal of Management Studies*, 7, p. 58–70.

Ittner, C.D. & Larker D.F. 2003. Coming Up Short on Nonfinancial Performance Measurement. In *Harvard Business Review*: 1–9.

Kaplan, R.S. & Norton, D.P. 2007. Using the Balanced Scorecard as a Strategic Management System. [online] https://hbr.org/2007/07/using-the-balanced-scorecard-as-a-strategic-management-system.

Kosábková, L. 2008. Balanced Scorecard—ako nástroj pre efektívne strategické riadenie podniku. Medzinárodná vedecká konferencia Globalizácia a jej sociálno ekonomické dôsledky.

Marcheová, D et al. 2011. Základy štatistiky pre pedagógov, UKF Nitra.

Meyer, M.W. 2002. Rethinking Performance Measurement, Beyond the Balanced Scorecard, Cambridge University Press, p. 81–113.

Nerastné suroviny. 2015. Štátny geologický ústav Dionýza Štúra, ISBN 978-80-89343-26-3.

Štefko, R. & Gallo, P. 2015. "Using management tools to manage network organizations and network models", Management of Network Organizations: Theoretical problems and the dilemmas in practice, p. 249–263.

Štofová, L. & Szaryszová, P. 2016. Environmental Criteria of Public Procurement as a Tool of Development Sustainability. Calitatea, 17(152), 67.

New Trends in Process Control and Production Management – Štofová & Szaryszová (Eds)
© 2018 Taylor & Francis Group, London, ISBN 978-1-138-05885-9

Research and development of a new ergonomic tool in the meaning of the Industry 4.0 concept

M. Gašová
CEIT, n.o., Žilina, Slovak Republic

M. Gašo & A. Štefánik
Faculty of Mechanical Engineering, University of Žilina, Žilina, Slovak Republic

ABSTRACT: Along with the development and the application of innovative solutions in production and logistics, interest in modern tools in the ergonomic area does not abate. Thanks to them, in an effort to fit between companies that set out to the way Industry 4.0, it is possible to improve working conditions and the quality level of workplaces. Ceit Ergonomics Analysis Application, which is described in article, is output of our own research and development. It is a mobile application developed in CEIT Company in collaboration with the University of Žilina and Slovak ergonomic association. It is a screening evaluation of space conditions and work positions of workers at potentially risky workplaces. It is developed at the base of legislation and technical norms, at our own platform, with the support of virtual and augmented reality. It is an innovative way of applied augmented reality tools during the ergonomic evaluation of chosen workplaces.

1 INTRODUCTION

Over the years' approach focusing in ergonomics has changed. We still talk about identification—analysis—elimination of the risks on the workplaces. But differences are at the possibilities of modern ergonomics, movement of science and technical possibilities. The options of using a mobile application, Internet of Things, data gathering and their real-time evaluation and their sharing. We present those solutions that combine traditional knowledge and modern technologies. The results are innovative and advanced ergonomic tools based on Industry 4.0 concept. Electronic tools are a new direction in ergonomics. With the support of mobile applications, we see a way to create healthy conditions at work for production and also non-production workers, assembly and logistics.

With the growing development of society, we got to stadium when, luckily, majority of companies—employers even know the meaning of ergonomics or work risks, about risks at their workplaces and establish their evaluation and try to eliminate them.

We have many methods and tools of modern ergonomics which enable us to realize analysis and optimization of employee's work to their benefit. Considering experience, we can surely claim that we know the main problem of these days. It is req-uisite to realize ergonomic evaluation perfectly, extensively and mainly quickly. Slowness of some solutions discourages managers and directors and makes effective improving of work conditions impossible.

The idea of mobile application developing which works as a screening tool came with demands from big companies that have dozens of workplaces and cannot identify work risks by themselves.

2 OWN RESEARCH AND DEVELOPMENT

The field of ergonomics has had a long-lasting tradition in Žilina. It focuses mainly on detailed workplace project engineering with the use of tools of virtual reality. In last year's we have been using our own methodical approach (algorithm) named CEIT ErgoDesign (published in 5th International Ergonomics Conference—ERGONOMICS 2013, Croatian Ergonomics Society, Zagreb, Croatia. ISSN 1848-9699) by which we identify and analyze risks of physical load in praxis. It has been updated and supplemented by new systems of data collection and ways of evaluation. With the usage of digital factory's tools, we implement evaluation of working positions, manipulation with loads, repetitive operations and other fields at virtual models of workplaces and chosen work operations. These virtual models are obtained and made by 3D laser scanning and following digitalization. If we speak about our own research, we regard design of our own systems, connection of tools or their development. The area of connection the human motion scanning system during work (Motion Capture, Kinect) with the Tecnomatix software and consecutive linking with the immersive technology CAVE is worth mentioning. The research in the domain focused on sensory system of data collection regarding psychophysiological response of load of workers is also running, exactly to monitor physical load of logistic workers. Substantial lack is absence of possibility to collect and process pressure forces of fingers. The newest direction of the research is focusing right on the field in question. The tools of digital factory, virtual reality, augmented reality and sensory systems form the whole concept of the modern ergonomics while we interconnect and evolve them with the use of specific mobile apps. All solutions have been made with the only one aim. This is the effective use in praxis to improve ergonomics at workplaces and during working operations. Currently, the whole research and development in ergonomics turns toward the new system of data collection and applications for their processing. (Ďurica et al. 2015, Gregor et al. 2016, Furmann & Krajčovič 2011)

3 QUICK RISK ASSESSMENT TOOLS

The modern ergonomics direction leads ahead in the meaning of fulfilling the actual needs which arise in industry. One of those needs is also quick evaluation of workplace from the view of ergonomics. Fulfilment of this need is becoming state of art scientific problem of modern ergonomics. Solution to this problem is the use of mobile applications for a quick evaluation of risk factors at workspaces. Speaking about so-called electronical tools—QRA (quick risk assessment) tools. The support of mobile applications, is vision the way to help both, manufacturing and non-manufacturing organizations, or specialists who utilize this type of evaluation. Looking for solutions with the use of some new and effective approaches is of the great importance. Making use of ergonomics in organization is not only about showing activities at the Human Resources department, but, first of all, about looking for and use of new effective tools and methods. In modern ergonomics, in our midst are used progressive tools from the area of digital factory, virtual or augmented reality for purpose of quick, effective and efficient implementation of technical norms ISO, European and our legislation and relevant ergonomic implements and methods.

4 DEVELOPMENT OF A NEW SCREENING TOOL

Ceit Ergonomics Analysis Application is mobile application developed in CEIT a.s. company in collaboration with the Slovak ergonomic association and the University of Žilina. CERAA is the user-friendly and quick mobile application developed on the base of legislation. It is dedicated to screening evaluation of spatial conditions and working positions of workers at potentially risky workspaces. (Gašová & Dulina 2016)

One of the greatest advantages is that user does not need to be expert in ergonomics, basic knowledge from this field and from detailed workplace designing which customer acquires during training is sufficient. For the effective usage of the application customer needs a tablet, installed application, marker and completed provided training. In case of need, the extensive user guide is available. Except Slovak language, language variations in Czech and English are finished. The Help in application and also at the website edu4industry is ready for users, where they can address authors with their remarks and suggestions either technical or expert.

The main goal of evaluation with the use of CERAA is to determine if the workspace is from the view of ergonomics risky and if detailed evaluation is necessary—by the second level tools and design of correctional arrangements, alternatively what health risk threatens workers. It can finally lead to worsen quality of production and work effectivity, productivity decrease, and so on. Evaluation arises from spatial conditions and working position. The application is based on the use of virtual and augmented reality items. The core of evaluation results mainly from legislation and European technical norms.

The first area that application offers is evaluation of working conditions for workers. This evaluation (Fig. 1) contents rating of working area during sitting or standing on the base of knowledge from anthropometry. We can find basic rules stated by legislation there, for example minimal vacant area for legs, defined handling space (space division for frequent and occasional movements by upper limbs).

All this is re-counted and visualized for men as for women and in both cases for different heights of workers too. The second area is the evaluation of chosen working positions (Fig. 2). It includes characteristics and criterions intended to determine about admissibility of individual positions with regards to torso position and positon of head, neck and upper limbs. In order to develop application determined for effective evaluation, the series of virtual models of humans and elements of working environment has been created. From those, various alternates are accurately modelled

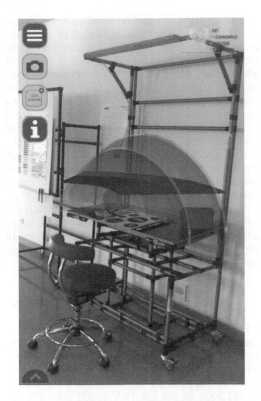

Figure 1. Evaluation by augmented reality (CERAA ver.1).

Figure 2. Evaluation of chosen working positions (CERAA ver.1).

for different workers in regard to working position (sitting, standing), sex, height and principles of workspace projecting, as to verification of angles of chosen parts of real workers with the help of augmented reality and for quick comparison with legal regulations and norms. For application controlling the series of buttons was made and they lead user through the whole evaluation.

Dimensions [mm] of particular body parts, from TNI CEN ISO/TR 7250-2 were as input data inserted to formulas for working and manipulating level dimensions' calculation. Information about needed height of space for lower limbs, recommended height of seat, height of footstool, minimal space for lower limbs, needful width of space for lower limbs, optimal height of working area, maximal height of working level, maximal width of working area, optimal depth of working area, maximal depth of working area and other information about body posture during working action were computed.

5 PRACTICAL VERIFICATION

The first version is finalized from the second half of 2016 and used by companies as Lear Corporation Seating Slovakia s.r.o., Lear Corporation Seating Slovakia s.r.o., Syráreň Bel Slovensko a.s., SMC Vyškov CZ, Nemak Slovakia, s.r.o., Prcmcdis s.r.o CZ, Magna Slovteca s.r.o., Eltek s.r.o., Continental Matador Rubber, s.r.o., etc. Application has gone through the phase of professional preparations, practical testing and pilot deployment into praxis directly at some customers. Users are industrial firms as yet, however, interest is observed also from the side of Health Services, Security Technicians and technical universities. By the means of this tool we evaluate regular working operations and workspaces at which worker performs his activities for more than half of working shift. This tool can be used during projecting of new workplaces and also during proactive ergonomics (Fig. 3).

Figure 3. Practical verification in industry (CERAA ver.1).

User choses within main menu—ergonomic workspace, height of working level, reaching zones or working positions. Choice is more specified on the basis of the sex, height and working position of worker. After input parameters enter, defined subject displays in the middle of the tablet screen. That object can be manipulated and in case of need it provides various additional information. They contain description of basic principles about detailed workplace projecting with regards to legislation.

After opening and launching the application, launching the camera and marker scanning, with the support of tools for augmented reality in working environment, virtual model of worker will be displayed. This model is intended to compare virtual and real workspace and consecutive identification of individual risks. It may be considered as an innovative way of application tools of augmented reality in the course of single workplaces evaluations from the ergonomic point.

This solution presents an innovative approach to screening evaluation of workplaces and working operations. In quite short time organizations confirmed ergonomics of their workspaces and subsequently decided about following steps which need to be realized.

6 SUB-MODULE: ADMINISTRATIVE WORKSPACE

After choosing this sub-module, user will be offered by options to evaluate an administrative workspace entirely (whole test) or to choose

Figure 4.   Sub-module for evaluation of administrative workplaces (CERAA ver.2).

one from eight groups. Altogether test consists of 77 questions that are dedicated to problematics of functional and dimensional workspace characteristics, working chair, table, keyboard and mouse, monitor, workplace hygiene, mode of sleep and rest. The audit is processed to an unequivocal form in which answers to questions are only "yes" or "no". Questions are visualized by virtual models (Fig. 4) for greater clarity and understanding. Answers cumulates and at the end of every test is a partial evaluation and, of course, at the end of complex test is an complex evaluation from which user gets knowledge if there is a risk and to what extend (semaphore colors), how many percent of fundamentals for administrative workspace set workspace keeps and what is the most essential, user can see in report which fundamentals are not kept and recommendations how to correct them.

7 SUB-MODULE: LOAD MANIPULATION

CERAA ver. 3 will contain two sub-modules of the evaluation of the manipulation with load. The first of them is the subject of ongoing research and development. The evaluation will be prepared on the basis of the Government Regulation n. 281/2006 Statue law (NV č. 281/2006 Zb. z.), which evaluates lifting and lowering of load. The augmented reality will not be used in the application. It will use some virtual models, calculator and camera. Regarding the button choice, user chooses from the options sex, working position, age and thereafter conditions of work. There will be 11 inappropriate working conditions (Fig. 5) e.g. wrong grip, unacceptable working postures, instability of the load, manipulation above the heart level, sporadic load in working shift, insufficient physical condition of workers and so on. Afterwards user chooses from three ways of evaluations. The first will be screening of maximal weight of the load. On the base of previous choice, system assigns allowed weight limit. Subsequently user writes real weight of the load, writes the name of the load, takes a photo of it and assigns visualization. By this way report will be made and user will be able to save it into an internal storage or send it via mail. Either user screens some loads and evaluation or can continue by the special button to the second way of evaluation. In this evaluation, user will assign number of manipulations in rhythm and number of rhythms in working shift. By recalculations maximal entire-shift manipulated weight will be determined. It will be checked again by the limit. One of reports will be also information about workers who must not manipulate with given loads under such conditions.

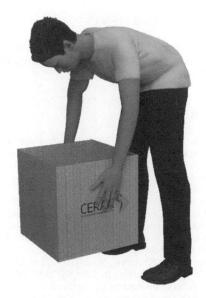

Figure 5. Sub-module for evaluation of load manipulation (CERAA ver.3).

Third option of evaluation is screening of weight limits for outreach. This evaluation is according to our legislation possible for women only. On the basis of the virtual model of woman—worker and of four buttons we can realize this evaluation. User will know from—to evaluated worker moves the load and according to this, will click combination from two to four points. After click, he will obtain information written in a line about how heavy load is in kilograms and how often evaluated worker can manipulate the load, for example: "Limit of load for given height of lowering and lifting is 5 kg. The load can be manipulated with 6x/min frequency. After 10–15 min. of manipulation 10–15 min. lasting break is needed."

The vision is to add sub-module for evaluation of intervertebral disc L4/L5 according to NIOSH study. These evaluations will be developed with the aim to simplify evaluation of loads of logistic workers. Because it is not possible to use classic systems of physical load evaluation for logistic activities.

8 SUB-MODULE: ACTION FORCES AND REPETITIVE ACTIONS

Evaluation of local muscle load is realized on the base of electromyography, what is difficult to interconnect with application. Therefore, we decided to evaluate cyclically repetitive operations by the OCRA method which is included in EN 1005-5. It is the most difficult way of evaluating of cyclically repetitive operations by upper limbs and that is why it will be necessary to simplify this method for purposes of screening. Generated forces and ways of grip enter into this process of evaluation. So, we decided to interconnect the screening application CERAA with sensory system of data collection (glove for pressure forces measurement) at the turn of 2017 and 2018, which is being developed in partnership with the Department of electromagnetic and biomedical engineering of the Faculty of electrical engineering, the University of Žilina.

Modul will be prepared in three steps:

1. Testing of action forces regarding grip,
2. Testing of repetition depending on outcomes from 1. step,
3. Testing of load by OCRA methodic.

Development runs along with development of ergonomic glove for pressure hand forces sensing. The glove will sense forces from more than 20 sensors and evaluation will run real-time with synchronized video analysis. Methodic of evaluation is still matter of own research and development. Assumed planned system of evaluation is following. User will get through the option of sex, working posture and the type of load. Chosen technical assembly grips will be available to choose. The maximal and limit forces on the base of research results will be assigned to the grips. After the synchronization of video analysis with measured data—by developed force, user choose risky time period, assign the technical grip and check if the limited force is overstepped or not. By this solution unique system of data collection and current evaluation of risks at workplaces with complicated action forces of upper limbs will come into existence.

9 ERGONOMIC IN THE MEANING OF THE INDUSTRY 4.0

The new scientific problem accrues even from the requirement for processing the large amounts of data being the result of individual assessments and analysis. Making inputs and outputs of databases analyses and very risks has a great meaning for further evaluations. The same as sharing those data within departments and interconnection of systems of the first and the second level evaluation. (Bubeník & Horák 2014, Plinta & Krajčovič 2016)

In the future, a complex logical system for identification and evaluation of risks related to ergonomics must be made (Fig. 6). Tools of the first and the second, maybe also the third level, must be interconnected and the whole system adjusted by the way that during every change at workplace or working operation, the whole long-lasting evaluation system will not have to run from all over again,

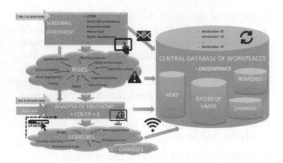

Figure 6.   Connection of ergonomic tools in the meaning of the Industry 4.0 (authors).

but calculations will be automatically renewed and results will be generated depending on this change. (Mičieta et al. 2005, Matuszek et al. 2009)

Accordingly, the whole system should be on-line connected to application visual interface. On this base, involved workers will be immediately informed about new conditions that appeared and about their influence on operators in accordance with realized analyses and about pointing on requirements for organizationally-technical changes. (Krajčovič et al. 2013).

And it is well-founded. Operators in production/montage should not be only ones discussed, but also support actions of production. Those are logistics and other indirect working activities (maintenance or administration). Only by this way it will be possible, to the future, within constantly rising demands on volume and diversity of products and services, to ensure adequate working environment and realization of activities in accordance with fundamentals of ergonomics and productivity. Exactly opportunities of digitalization of all activities and processes, modern sensory systems and growth of computing power of ICT form all needed assumption of that.

10   CONCLUSION

Long-term vision is to create the complex application that will contain detailed and exact visualization of all relevant methods of evaluation and legislative regulations used in ergonomics. The application represents an innovative solution to the scientific problem of requirement of rapid assessment of workplaces in terms of ergonomics. The main benefit of such solution is possibility to provide fast evaluation at real workplaces based on actual legislative.

ACKNOWLEDGEMENTS

"This work was supported by the Slovak Research and Development Agency under the contract No. APVV-0755-12".

REFERENCES

Bubeník, P. & Horák, F. 2014. Proactive approach to manufacturing planning. In *Quality Innovation Prosperity*. Vol. 18, No. 1 (2014), ISSN 1335-1745. ISSN 1338-984X—online. Košice: Technical University of Kosice. p. 23–32.

Ďurica, L. & Mičieta, B. & Bubeník, P. & Biňasová, V. 2015. Manufacturing multi-agent system with bio-inspired techniques: CODESA-Prime. In *MM science journal*. ISSN 1803-1269. ISSN 1805-0476—online. Prague: MM Science Journal. p. 829–837.

Furmann, R. & Krajčovič, M. 2011. Modern Approach of 3D Layout Design. In *TRANSCOM 2011* – 9-th European Conference of young research and scientific workers, Section 2 (Economics and Management)—Part 1, ISBN 978-80-554-0370-0, Žilina: EDIS—Printing House of the University of Žilina: p. 43–46.

Gašová, M. & Dulina, Ľ. 2016. Mobile applications for screening assessment of assembly. In *Journal of safety research and applications* (JOSRA). ISSN 1803-3687 - online. Vol. 9, No. special (21.11.2016), Prague: JOSRA.

Gregor, M. & Haluška, M. & Fusko, M. & Grznár, P. 2016. Model of intelligent maintenance systems. In *Annals of DAAAM* Proceedings of the 26-th DAAAM International symposium on intelligent manufacturing and automation. ISSN 1726-9679. ISBN 978-3-902734-07-5 – online, Vienna: DAAAM International Vienna: p. 1097–1101.

Krajčovič, M. et al. 2013. Intelligent manufacturing systems in concept of digital factory In: *Communications—Scientific letters of the University of Žilina*. Vol. 15, no. 2, ISSN 1335-4205. Žilina: EDIS—Printing House of the University of Žilina: p. 77–87.

Matuszek J. & Więcek D. & Więcek D. 2009. Estimating prime costs of producing machine elements at the stage of production processes design. In *Modelling ang designing in production engineering* / ed. A. Świć, J. Lipski. Vol.5, No2, ISBN: 978-80-89333-15-8. Lublin: Lubelskie Towarzystwo Naukowe. p. 80–92.

Mičieta, B. & Dulina, Ľ. & Malcho, M. 2005. Main factors of the selection jobs for the work study. In *Annals of DAAAM for 2005 & Proceedings of the 16th International DAAAM Symposium*: Manufacturing & automation: Focus on young researches and scientists. Opatija. Croatia. ISBN: 978-3-901509-46-9. Vienna: Danube Adria Association for Automation and Manufacturing, DAAAM. p. 249–250.

Plinta, D. & Krajčovič, M. 2016. Production system designing with the use of digital factory and augmented reality technologies. In *Advances in Intelligent Systems and Computing*. Vol. 350 (2016). ISSN 2194-5357. Berlin: Springer Verlag. p. 187–196.

New Trends in Process Control and Production Management – Štofová & Szaryszová (Eds)
© 2018 Taylor & Francis Group, London, ISBN 978-1-138-05885-9

Dependence quantification of comparative advantage on V4 export performance

M. Hambálková & Z. Kádeková
Faculty of Economics and Management, Slovak University of Agriculture in Nitra, Nitra, Slovak Republic

ABSTRACT: Foreign trade represents the import of those commodities which in the country fail to grow because of the climate or cannot be produced for various different reasons. On the other hand, imported foreign commodities are competition in domestic market. The effort of each country is to achieve a positive trade balance. In this respect, it is very essential to follow the export and import of foreign trade. The aim of the submitted paper is dependence quantification of comparative advantage on the export performance of the V4 countries. Reaching this aim is supported by following partial aims: Slovak Republic as part of the V4, Global index of competitiveness of V4, Agrarian Trade of V4, Agrarian Trade of the Slovak Republic and the Czech Republic, Agrarian Trade of the Slovak Republic and Hungary, Agrarian Trade of the Slovak Republic and Poland.

1 INTRODUCTION

Integrating factor of the world economy is globalization, which also creates an increasing specifics of local business environment (Lipková et al. 2011). Nowadays almost every region, every company in any part of the world has access to the Internet, mobile network, or the ability to raise funds under the same conditions in the global financial markets and the companies from the so-called peripheral regions of the world can become global players and also can be systematically involved in international division of labor (Nagyová et al. 2016).

Globalization of today's world brings different tools by which it is possible to mobilize international trade and Slovakia as a member state of European Union use different gains deriving from the Common commercial policy, in particular to strengthen the position of a foreign trade relations, which focus mainly on the EU Member States especially to neighbouring countries which are Poland, Czech Republic and Hungary that create the Visegrad Group, also called the Visegrad Four (V4).

2 THEORETICAL AND METHODOLOGICAL APPROACHES, ACHIEVED RESULTS

2.1 Aim, data and methods

Paper points at dependence quantification of comparative advantage on the export performance of the V4 countries, the main aim of submitted paper is to evaluate the position of the Slovak Republic as a part of the V4 as well as to estimate V4 countries based on the global index of competitiveness and quantify trends in economy and trade of the most important trade partners of Slovakia.

Estimated status of foreign trade in 2020 is based on the use of statistical—mathematical methods using the database of UNCTAD, OECD, World Bank, WTO. The data are based on estimate of growth of global GDP and GDP of the SR World Bank (2013–2015) and OECD (2016 to 2020). Estimate of global growth of GDP of the Slovak Republic in 2013—based on data by Ministry of Finance SR. Trade elasticity is based on the methodology of WTO.

2.2 Assessing the position of the Slovak Republic as part of V4

The Slovak Republic is a landlocked country in Central Europe. It is bordered by the Czech Republic and Austria to the west, Poland to the north, Ukraine to the east and Hungary to the south. Slovakia's territory spans about 49,036 square kilometres and is mostly mountainous. The population is 5.4 million and comprises mostly ethnic Slovaks. Population density (people per sq. km) in Slovakia was last measured at 112.79 in 2015. The capital and largest city is Bratislava. Slovakia is a parliamentary democracy with president as the head of the government. The official language is Slovak (Ružeková et al. 2013).

Slovakia became an independent state on 1 January 1993 after the peaceful dissolution of Czechoslovakia. Slovakia is a high-income advanced economy (Baláž et al. 2010). The country maintains a combination of market economy with universal health care and a comprehensive social security system (Šedík 2016). The country joined the European Union on 1st May 2004, the Schengen Area on 21st December 2007 and the Eurozone on 1st January 2009 by adopting Euro and so became officially the 16th country of eurozone. This is the economic part of the EU, which includes a free trade area, a customs union between the EU member states, free movement of persons, goods, services and capital within the EU member states, economic policy coordination between EU member states and the common euro currency. The eurozone is a monetary union of those European Union member states which have adopted the Euro (€) as their common currency and sole legal tender. The monetary authority of the eurozone is the Eurosystem (Kalínska et al. 2010).

These countries have representation in the European Central Bank (ECB) or in the Eurogroup. The basic macroeconomic indicators include GDP, annual change in GDP, unemployment rate, trade balance and inflation. Table 1 shows the most important trade partners of the Slovak Republic in 2015.

2.3 Evaluation of individual V4 countries based on the global competitiveness index

Global Competitiveness Index (GCI) has been regularly publishes by Global economic forum since 2004 (Kašťáková & Ružeková 2012). This index ranks countries according to their level of competitiveness and takes into account the macroeconomic and microeconomic bases of national competitiveness. GCI captures the openness of criteria that are in the cross-correlation while not contradictory. Evaluation is taking a weighted average of many different components, each reflects the aspect of a complex reality and it is called competitiveness (Kubicová & Kádeková 2011). All the components are grouped into 12 pillars of competitiveness. These criteria are in the cross-correlation and each one influences another one. The pillars divided into the following categories: quality of public institutions, infrastructure, macroeconomic environment, health and primary education, higher education and training, the effectiveness of the type of goods, labor market efficiency, financial market, technological readiness, market size, business sophistication and innovation.

The Slovak Republic has moved again in competitiveness rankings, which is publishes by World Economic Forum in Switzerland (WEF) based on a survey among entrepreneurs (Jeníček & Krepl 2009). This year is Slovakia placed at 65.th position (Table 3). Compared to the year 2015–2016 it is a shift up of two positions—as it was announced in the Global Competitiveness Report for 2016–2017. In this aspect, the most noticeable improvement of Slovakia started in 2014, when shifted by 13 points from the historical worst 78.th position. Anyway, it is still far away from 37.th position where it was in times of economic reforms in 2006.

WEF ranks 138 countries from all continents, list of Top ten countries (Table 2) has not changed that much. Still prevail the European economies, the first place took Switzerland—already eight years in a row. Singapore maintains second place and the United States also defended third place. The largest improvement in the top ten noticed the United Kingdom, Holland that reached the fourth place continues the climb up. Japan (8th) drop by two places, Hong Kong took 9th place and Finland continued the descent already on 10th place. As for other countries should be noted pronounced jump

Table 1. Top 10—the most important trade partners of the Slovak Republic in 2015 (in billion EUR).

	Country	Import	Export	Balance	Turnover
	SR Total	64,562.3	67,865.2	3,302.9	132,427.6
1.	Germany	9,800.5	15,392.1	5,591.7	25,192.6
2.	Czech Republic	7,004.8	8,436.4	1,431.7	15,441.2
3.	Poland	3,172.8	5,620.7	2,447.9	8,793.5
4.	Hungary	3,165.5	3,786.5	621.0	6,952.0
5.	China	5,429.6	1,019.6	−4,410.0	6,449.1
6.	France	2,089.2	3,832.2	1,743.0	5,921.4
7.	Austria	1,619.7	3,862.8	2,243.1	5,482.5
8.	Italy	2,099.2	3,067.0	967.8	5,166.2
9.	Russian Fed.	3,486.1	1,461.0	−2,025.0	4,947.1
10.	Great Britain	1,018.2	3,756.2	2,738.1	4,774.4

Source: database of UNCTAD, OECD, World Bank, WTO, own proceeding.

Table 2. Global competitiveness index of top 10 countries (positions).

Country	GCI 2016	GCI 2015	Change
Switzerland	1	1	0
Singapore	2	2	0
USA	3	3	0
Netherlands	4	5	+
Germany	5	4	−
Sweden	6	6	0
Great Britain	7	10	+
Japan	8	6	−
Hong Kong	9	7	−
Finland	10	8	−

Table 3. Global competitiveness index of the V4 countries (positions).

Country	GCI 2016	GCI 2015	Change
Czech Republic	31	31	0
Poland	36	41	+
Slovakia	65	67	+
Hungary	69	63	–

of New Zealand (13th) and Austria (19th). The Czech Republic is on the 31st place after last year's jump. Rapid progress upwards continues in Poland (36th) India, Malta, Mexico. Conversely, Portugal, Turkey, Romania and Hungary noticed a change for the worse. Competitiveness Index terminate the perspective of achieving sustainable economic growth in the medium term, annually assessing the quality of public institutions, government policies and other factors which determine the level of productivity and prosperity in 138 countries.

In the Slovak Republic, the issue of competitiveness had been opened during the transformation of the Slovak economy and during its preparation for EU accession. For Slovakia as well as for another new EU member states, the entry into the EU started to adopt the benefits associated with the liberalization of mutual trade exchange and establishment of systems for the support provided by the Common Agricultural Policy. Slovak Republic as well as the newly adopted countries try to stabilize their position in the agri-food market of the EU, which is reflected increased competitive pressure, leading to diversification of farming activities within the EU (Hambalková et al. 2011).

Since 2004, the agri-food market of the Slovak Republic took over all rights and obligations of the Common Agricultural Policy (Nagyová et al.). The agri-food commodities are governed by the rules that streamline and organize the market. The main objectives of the Common Agricultural Policy are to improve agricultural productivity, to stabilize markets, to ensure adequate living standards of farmers, to guarantee the regular supply to the consumers while keeping the adequate market equilibrium of the country (Holota et al. 2016). In the long term it is more appropriate to focus on the export of products with higher added value, i.e. food products and direct the agricultural production to ensure the basic raw materials for the food industry.

2.4 Development trends of economies and agreement of main trade partners of Slovakia

Based on the long-term development of the territorial structure of the foreign trade, the major trade partners of the Slovak Republic are the EU member states including V4, the Russian Federation, China and USA (Table 4).

Based on long-term projections of trade of the main global players can be concluded that the most important involved countries are USA, EU, and China. There will be no significant change in export specialization of these countries and trade groups. China will become a net exporter of machines and will remain a major exporter of electrical devices. As for import. There will dominate the primary commodities, gradually should even increase the share of agri-food products. In the USA and in the EU countries will increase services export, whereby the USA should become an exporter of natural gas.

Estimated status of foreign trade in 2020 is based on the use of statistical—mathematical methods using the database of UNCTAD, OECD, World Bank, WTO. There were developed five scenarios of the foreign trade development of the Slovak Republic in 2020, from which the most appropriate is middle one (Tables 5 and 6).

The data for the middle scenario are based on estimate of growth of global GDP and GDP of the SR World Bank (2013–2015) and OECD (2016 to 2020). Estimate of global growth of GDP of the Slovak Republic in 2013—Ministry of Finance SR. Trade elasticity is based on the methodology of WTO.

According the middle scenario, the estimated volume of foreign trade in 2020 will reach 184% value in comparison with year 2012, in real terms, 225 billion EUR in prices of year 2012—from which exports will take 124 billion EUR and imports 101 billion EUR. This corresponds to an average annual growth of export by 8.8% and import by 6.9%. The positive balance of foreign trade will increase from 3.6 billion EUR in 2012 to 23 billion EUR in 2020. When expected level

Table 4. Expected development of real GDP growth of the major trade partners of the Slovak Republic (annual average in %).

Country	2014–2017	2018–2020
1. Germany	1.6	1.1
2. Czech Republic	2.4	2.9
3. Poland	3.3	2.2
4. Hungary	1.8	2.9
5. Austria	1.7	1.4
6. France	1.8	2.1
7. Italy	0.6	1.6
8. Russian Fed.	3.6	2.7
9. Great Britain	1.5	2.1
10. Netherlands	1.6	1.9
11. China	8.9	5.5
12. USA	2.1	2.4
13. Spain	1.5	2.3

Table 5. Five scenarios of the foreign trade development of the Slovak Republic in 2020.

Scenario	Avg. Annual Increase in GDP	Avg. Annual Growth of World Trade	Exp. Volume of Foreign Trade of SR in 2020
Scenario 1 Global Crisis	1%	0%	123 bln. EUR
Scenario 2 Local crisis of SR	1%	9%	194 bln. EUR
Scenario 3 Middle scenario	2.9%	4.8%	225 bln. EUR
Scenario 4 Local boom of SR	6%	0%	310 bln. EUR
Scenario 5 Global boom	6%	9%	463 bln.EUR

Note: the volume of foreign trade—data in billion EUR in the prices of 2012.

Table 6. Expected development of GDP and foreign trade of the Slovak Republic—middle scenario.

	2013	2014	2015	2016	2017	2018	2019	2020
HDP	71.82	74.07	76.58	79.10	81.71	84.41	87.20	90.07
Exp.	64.31	69.53	76.49	84.21	92.72	102.10	112.41	123.78
Imp.	59.03	62.80	67.97	73.60	79.69	86.29	93.42	101.15

Note: data in billion EUR in prices of 2012.

of GDP is 90 billion EUR, the openness of the Slovak economy will grow at 250%.

Above scenario seems to be slightly optimistic. View of past development between 2004 and 2012 shows that in observed period the volume of foreign trade in real terms increased by 47%, corresponding to annual growth of 4.9%. Development in the given period was highly volatile and was strongly influenced by the economic crisis, in particular the slump of world trade in the year 2009. If it were not for this slump, the growth of the volume of foreign trade of Slovakia would be approximately doubled.

Expectations in 2020 (Table 7) are that within V4 Slovak export will be 11.27% to the Czech Republic import from the Czech Republic to Slovakia will be 6.09%. Export to Poland will be represented by 8.2% and import from Poland 4.40%. To Hungary will route the exports of 4.35% and from Hungary to Slovakia will be import 3.27%. Dominant trading partner remains Germany, who will participate in the Slovak export 24.78% and 21.93% in Slovak import.

Further diversification of the structure of Slovak export in favor of products with the highest

Table 7. Top 10—the largest import and export markets of the Slovak Republic in 2020.

Export of Slovakia		Import of Slovakia	
No. Country	Share	No. Country	Share
1. Germany	24.78%	1. Germany	21.93%
2. Czech Rep.	11.27%	2. Republic of Korea	11.16%
3. Polan	8.02%	3. Russian Federation	11.14%
4. Austria	6.84%	4. China	10.33%
5. France	6.34%	5. Czech Republic	9.16%
6. Russian Fed.	4.35%	6. Poland	4.40%
7. Hungary	4.35%	7. France	3.59%
8. UK	4.09%	8. Hungary	3.27%
9. China	3.37%	9. Austria	2.73%
10. USA	2.65%	10. Taiwan	1.99%

added value is considered essential also in accordance with the structure of Slovak industry. On the other hand, should increase the share of Section XVIII: Optical, photographic, cinematographic, measuring, checking, precision, medical or surgical instruments and apparatus; clocks and watches; musical instruments; parts and accessories, mainly Chapter 90 instruments and apparatus. Another section which should be Slovak export focused on is Section VI. Products of the chemical or allied industries, mainly due to developments in the demographic structure of the world's population and development of individual business partners, suggesting growth in demand for products of this class. Due to the expected growth in demand for basic foodstuffs can be considered the increase of the share of Sections I. Live animals and animal products and II. Plant products, also given to the expected development in the territorial structure of Slovak export, which foresees a decline in the share of EU countries and increase the share of Asian economies. The main imported items should only be products necessary for manufacturing of finishing products in Slovakia.

3 CONCLUSIONS

Slovak Republic as a country in the heart of Europe belongs to the smaller countries but nevertheless seeks to achieve a level of economies of neighboring countries. The Slovak Republic has moved again in competitiveness rankings, which is publishes by World Economic Forum in Switzerland (WEF) based on a survey among entrepreneurs. This year is Slovakia placed at 65.th position. In the Slovak Republic, the issue of competitiveness had been opened during the transformation of the Slovak economy and during its preparation for EU accession. For Slovakia as well as for another new EU member states, the entry into the EU started to adopt the benefits associated with the liberalization of mutual trade exchange and establishment of systems for the support provided by the Common Agricultural Policy. Slovak Republic as well as the newly adopted countries try to stabilize their position in the agri-food market of the EU, which is reflected increased competitive pressure, leading to diversification of farming activities within the EU.

Since 2004, the agri-food market of the Slovak Republic took over all rights and obligations of the Common Agricultural Policy. The agri-food commodities are governed by the rules that streamline and organize the market. The main objectives of the Common Agricultural Policy are to improve agricultural productivity, to stabilize markets, to ensure adequate living standards of farmers, to guarantee the regular supply to the consumers while keeping the adequate market equilibrium of the country.

As for the foreign trade, based on the long-term development, the major trade partners of the Slovak Republic are the EU member states including V4, the Russian Federation, China and USA.

Slovakia should seek all possible financial resources and use them to benefit domestic producers and exporters to achieve better competitiveness in foreign markets. The Czech Republic has the largest share in Slovak export and import of among all V4 countries.

The estimated volume of foreign trade in 2020 will reach 184% value in comparison with year 2012, in real terms, 225 billion EUR in prices of year 2012- from which exports will take 124 billion EUR and imports 101 billion EUR. This corresponds to an average annual growth of export by 8.8% and import by 6.9%. The positive balance of foreign trade will increase from 3.6 billion EUR in 2012 to 23 billion EUR in 2020. When expected level of GDP is 90 billion EUR, the openness of the Slovak economy will grow at 250%.

Expectations in 2020 are that within V4 Slovak export will be 11.27% to the Czech Republic import from the Czech Republic to Slovakia will be 6.09%. Export to Poland will be represented by 8.2% and import from Poland 4.40%. To Hungary will route the exports of 4.35% and from Hungary to Slovakia will be import 3.27%. Dominant trading partner remains Germany, who will participate in the Slovak export 24.78% and 21.93% in Slovak import. Further diversification of the structure of Slovak export in favor of products with the highest added value is considered essential also in accordance with the structure of Slovak industry.

Overall it is apparent that the Slovak Republic is an independent and competitive country.

REFERENCES

Baláž, P. et al. 2010. *Medzinárodné podnikanie. Na vlne globalizujúcej sa svetovej ekonomiky.* 5. vydanie. Bratislava: Sprit dva.

Hambalková, M. et al. 2011. *Systémy podpory zahraničného agrárneho obchodu v SR.* Nitra: SPU.

Holota, T. et al. 2016. The management of quality costs analysis model. In *Serbian journal of management.* Vol. 11, no. 1.

Jeníček, V. & Krepl, V. 2009. The role of foreign trade and its effects. In *Zemědělská ekonomika.* p. 211–220 Praha: Česká akademie změdělských věd.

Kalínska, E. et al. 2010. *Medzinárodní obchod v 21. století.* Praha: Grada Publishing, 2010. 228 s. ISBN 978-80-247-3396-8.

Kašťáková, E. & Ružeková, V. 2012. *Operácie v zahraničnom obchode, teória a prax.* Bratislava: EKONÓM.

Kubicová, Ľ. & Kádeková, Z. 2011. Comparison of the income development and the food demand elasticities of the private households in Slovakia. In *Agricultural economics.* Vol. 57, no. 8, pp. 404–411.

Lipková, Ľ. et al. 2011. *Medzinárodné hospodárske vzťahy.* Bratislava: Sprint dva.

Nagyová, Ľ. et al. 2016. Corporate social responsibility in food manufacturing companies—environmental dimensions. In *Acta Universitatis Agriculturae et Silviculturae Mendelianae Brunensis.* Vol. 64, no. 3.

Nagyová, Ľ. et al. 2016. Food security drivers: economic sustainability of primary agricultural production in the Slovak Republic. In *Journal of security and sustainability issues.* Vol. 6, no. 2.

Oesd. 2017. [retr. 2017-03-30]. Available online at: <http://www.oecd.org/>.

Ružeková, V. et al. 2013. *Analýza zahraničného obchodu Slovenskej republiky.* Bratislava: EKONÓM.

Šedík, P. 2016. Upravlenije povedenijem pokupatelej i potrebitelej na rynke mjoda v Rossii i Slovakii. In *XI nedelja nauki molodjoži SVAO.* Moskva: Nauka.

UNCTAD. 2017. [retr. 2017-03-30]. Available online at: <http://unctad.org/en/Pages/statistics.aspx>.

WEF. 2017. [retr. 2017-03-30]. Available online at: <https://www.weforum.org/>.

WORLD BANK. 2017. [retr. 2017-03-30]. Available online at: <http://data.worldbank.org>.

WTO. 2017. [retr. 2017-03-30]. Available online at: <https://www.wto.org/>.

New Trends in Process Control and Production Management – Štofová & Szaryszová (Eds)
© *2018 Taylor & Francis Group, London, ISBN 978-1-138-05885-9*

Ergonomic programs based on the HCS model 3E as integral part of sustainable CSR strategy

K. Hatiar, H. Fidlerová & P. Sakál
Faculty of Materials Science and Technology in Trnava, Slovak University of Technology in Bratislava, Trnava, Slovak Republic

ABSTRACT: The paper presents author's contribution to tackling the underlying paradigm of strategic management in the 21st century that seeks and finds the answer to the question of sustainability of human civilization on the planet Earth. The only alternative is change in creation and distribution of wealth/GDP on the planet/on state level/ business level. However, as shown by critical analysis of empirical data from 90's in Slovakia; privatization of strategic enterprises has caused irreversible strategic decisions. Enterprises are in hands of multinational corporations, with support of state subsidies, grants and tax breaks, while often produced added—value/profit are directed into tax havens or their national economies. For these multinational corporations, our citizens are just cheap labor, "raw material" that after use becomes "waste". Ergonomics programs based on the HCS model 3E, as an integral part sustainable corporate social responsibility, suggest solution.

1 INTRODUCTION

For corporate social responsibility (further CSR) in companies it is necessary to provide a direct impact on the sustainable production quality without negative impacts on employees. As far as a man will be not replaced in the production by the automated systems, it is need systematically create conditions for a sustainable level of human labor effectiveness. It should be based on the principle of positive Health Effect and Cost Benefit in enterprises. In our opinion, this is possible to achieve in practice by sustainable way only through a systematic application of ergonomic programs based on participatory principle.

2 ERGONOMIC PROGRAMS BASED ON THE CONCEPT OF HCS MODEL 3E

2.1 HCS model 3E of local ergonomics program

The HCS model 3E of local ergonomics program, specific for countries of the Central and Eastern Europe, was created by authors K. Hatiar, T. M. Cook and P. Sakál within 4-years USA—Slovak cooperation project APVV No. 019/2001:" Transforming Industry in Slovakia Through Participatory Ergonomic. "(Hatiar 2008) The proposed HCS model 3E focused *on man as object and subject of all efforts and the concept where working environment should contribute to building the quality of working life for everybody with sustainable quality of environment and adequate economic conditions for the overall quality of human life.*

This objective is interconnection of three E (HCS model 3E) is not understood or admitted by many stakeholders and lobbyists in Slovak industrial companies for mostly their own economic reasons.

The HCS model 3E, which can be defined as a micro solution of macro—problem (in terms of well known theorem „think globally—act locally").

The definition of the *HCS model 3E* concept means that (Hrdinová 2013, Sakál et al 2016)

$$HCS\ model\ 3E \approx (SWQ \wedge SPQ \wedge SLQ), \qquad (1)$$

where:
$\approx$ – symbol of equivalence,
$\wedge$ – (and), symbol of conjunction,
SWQ – sustainable work quality,
SPQ – sustainable production quality (of goods and services),
SLQ – sustainable life quality.

3 APPLYING ERGONOMIC PROGRAMS IN SLOVAKIA

In general, one can expect a full and sustainable quality of man—employee work performance only if he is healthy, rested and satisfied. One of basic requirements in developed countries is that work and working conditions should not cause human pain and injury. The work should not be a punishment for man, but he should enjoy it, meet and participate in its development. In addition, the work must have such quality that ensures the benefits of a cost (Cost Benefit). To meet these requirements the companies systematically focuses ergonomic program.

Slovak legislation does not deal with the systematic applying of ergonomic programs in companies through targeted preventive ergonomic programs based on participatory principle (Kyzek & Hatiar 2011).

In this field focus in developed countries institutions known as the "Health and Safety Institutes" where this whole issue is integrated into practical shape through the area of ergonomics and ergonomic programs based on Participatory Ergonomics.

Authors Driessen et al. (2010) considers possible barriers and facilitators to implementation of a Participatory Ergonomics programme.

Under the label 'participatory ergonomics' has been a central issue within ergonomics the idea of establishing changes in working conditions through participatory approaches. (Jensens 1997) This integration role of ergonomics through ergonomics programs is shown in Figure 1.

Fundamental difference between work safety (WS) and the field of preventive occupational medicine (POM) is also in the process of risk assessment. Nonetheless both above mentioned areas together create conditions to deal with the efficiency of human labor through integration

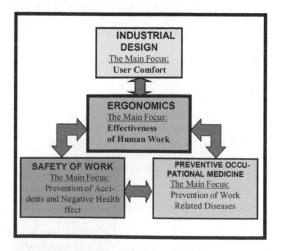

Figure 1. Integrating role of ergonomics in the concept of full-featured solutions of work-organizational system in the company (microergonomics) taken as an integrated occupational health and safety.

requirements for safety and health at work and the environmental requirements and requirements of necessary working comfort through ergonomics and ergonomic programs focused on efficiency of human labor through positive health effect and cost benefit.

The problem in Slovakia is, that Degree of Risk in area of Safety of Work (R_{sw}) is considered as probability of risk (R) multiplicities by its consequence (C)

$$R_{SW} = P \times C \qquad (2)$$

In the medical field where there is valid a Hippocratic oath, and in ergonomics, the risk level ($R_{med+erg}$) directly reflects the probability of risk (P).

$$R_{med+erg} = P \qquad (3)$$

For assuming the direct probability of risk ought to be used epidemiological methods and indicators. From a methodological point of view, it is a big issue in Slovakia that is not distinguished the scope of ergonomics.

Area of ergonomics, which focuses on solving everyday problems in companies using preventively oriented ergonomics programs based on participatory principle in foreign countries, is called microergonomics.

Area of ergonomics, which focuses on the application of ergonomics in the design of new items of tools and systems are equipped abroad, is called macroergonomics.

In macroergonomics we know the target population only at a certain degree of probability based on available databases.

The advantage of microergonomics in comparison with macroergonomics is, that during solution we can address our attention directly to persons having MSS problems as indicators of workplace shortcomings from point of view ergonomics. In cooperation with these workers we can propose appropriate solution.

From experience in enterprises in Slovakia is mostly unaware of this ergonomic classification, so it happens that in microergonomics are inadequately applied single methodology related to the design of new facilities and systems, which is more the scope of macroergonomics.

In addressing the diploma thesis was an opportunity to compare the sensitivity of methodology Ergonomics Assembly Worksheet (further EAWS) and ergonomic analysis using epidemiological methods that are an integral part of the HCS model 3E in enterprise (Hatiar et al. 2017).

The comparison of results (Fig. 2) obtained using the methodology EAWS and results obtained

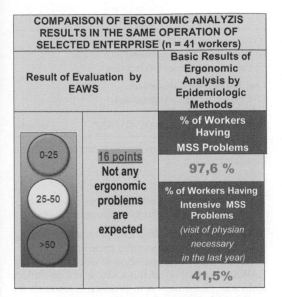

COMPARISON OF ERGONOMIC ANALYZIS RESULTS IN THE SAME OPERATION OF SELECTED ENTERPRISE (n = 41 workers)		
Result of Evaluation by EAWS		Basic Results of Ergonomic Analysis by Epidemiologic Methods
0-25	16 points Not any ergonomic problems are expected	% of Workers Having MSS Problems
		97,6 %
25-50		% of Workers Having Intensive MSS Problems
>50		*(visit of physian necessary in the last year)*
		41,5%

Figure 2. Comparison of results ergonomic evaluation in operation of selected using methodology EAWS and using ergonomic analysis based on epidemiological methods.

by ergonomic analysis applying epidemiological methods. The result of analysis based on methodology EAWS represents 16 points, according to the traffic light assessment falls within the group of 0–25 points indicated by green, which means that they do not foresee any substantial shortcomings in terms of ergonomics, which could have a negative impact on employees.

Nonetheless, the preliminary results of ergonomic analysis using epidemiological methods, showed the 97.6% of employees having difficulties located in musculoskeletal system (further MSS) which indicates the shortcomings of work places from point of view of ergonomics. In addition, it was found that 41.4% of employees had such a degree of MSS system problems that was necessary visit of physician and therapy during the last year.

The results of comparison both methods indicate a serious problem, because there is a necessary to perform ergonomic analysis to look for the cause of this high incidence of health problems among employees and ensure the elimination of these causes by technical and organizational solutions.

This also defines the potential in applying of above mentioned methodologies and techniques. EAWS methodology has a typical application in the field of macroergonomics, for designing and planning new plants with new technologies. It based on model assumptions based on practical experience. Its advantage is that it is linked to MTM method and allows for a new facility and operation to create time study.

The ergonomic analysis using epidemiological methods, as part of microergonomics, asseses the real situation and its impact on employees.

Methodology EAWS is by linking to the MTM also an important tool for interconnection the solutions macroergonomics in microergonomics, where it is an application of solutions in business practice and its adaptation to the specific conditions of enterprise that is trying to succeed in the competition and respecting the state of offer and demand in the global market.

The serious methodological problems in Slovakia practice is, that these methodologies are applied incorrectly.

3.1 Methods

Methodology EAWS (Ergonomics Assembly Worksheet)

Represents a first line ergonomic tool designed for quick overall assessment of biomechanical risks to which workers may be exposed during their professional activity (Bruder, R., et al, 2013).

This is a method of "pencil—paper" which is based on four evaluation sheets (1 sheet of the range of the general description, the second sheet with scales for evaluating job positions, 3. sheet with scales for assessing expended forces, 4. sheet—with scales for evaluating the load of upper extremities.

Total score is based on the conjunction of the size of the workload (intensity I) and its duration (D):

$$R = I \times D \qquad (4)$$

The result of the analysis performed by methodology EAWS is a score which is divided into three categories through semaphore evaluation (Fig. 2).

The 1st category is presented by gaining from 0 to 25 points. This category is marked by green colour and indicates recommended status of operation, when not any measures are required.

The 2nd category is presented by gaining from 25 to 50 points is marked by yellow colour and indicates not recommended status of operation, when preventive measure s are required.

The 3rd category is presented by gaining more point than 50, is marked by red colour and indicates unsuitable working conditions in evaluated operation, where appropriate measures implementation is necessary.

Ergonomic analysis by using epidemiological methods

Data collection for ergonomic analysis based on epidemiological methods is performed by modified "Nordic Questionnaire" (Kuorinka, 1987, Hatiar et al. 2004).

The basis for further progress of solution and future results interpretation is to find out how homogenous are data obtained in working groups from point of view selected unmodifiable factors.

Occurrence, location and intensity of work related MSS problems in workers are indicators of their work and workplaces shortcomings.

Epidemiological studies in the field of occupational medicine are focused on quantifying occurrence of work related diseases. In field of ergonomics are epidemiological studies focused on occurrence, localization and intensity of MSS problems as symptoms that precede occurrence of painful syndromes, and they are indicators of work, workplace and working environment shortcomings.

In analysis of impact potential risk factors both, modifiable and unmodiafiable are used epidemiological methods represented by cohort studies.

In the process of ergonomic program initiation is used retrospective cohort study design. This calculates Odds Ratio for all suspected risk factors related to MSS problems of workers. This indicator informs us how many times are in higher risk person exposed by assessed risk factor in comparison with persons unexposed. This analysis represents base line for further cohort studies, performed after e.g.one year time intervals.

This later cohort studies calculates Risk Ratio, which similar like Ods Ratio in form in how many times is in in higher risk worker exposed by risk factor to unexposed worker.

This process is described in article Hatiar (2004).

Cost-benefit analysis
Dealing with a project, an overall assessment of cost-benefits can be realized through the cost benefit analysis methodology developed at STU Bratislava, MTF in Trnava, and successfully tested in practice (Mrvová & Sakál 2005).

Regarding an immediate appreciation of the benefits of cost-return on investment for the implementation of ergonomic solutions in practice, it can be used e.g. following method recommended by the Cornell University (Hedge 2001) Table 1.

The condition for use is that available data are the actual costs. If are available information about losses due to occupational accidents and diseases related to work and working conditions and shortcomings of work, and work organization in terms of ergonomics, before the implementation of ergonomic rationalization and cost of these reasons for implementing solutions to ergonomic rationalization can be calculated return on investment.

For comparison, it is necessary to use the same time period before and after the implementation of solutions (eg. Annual cost, 6-month costs, monthly costs).

Table 1. Easy calculation of Cost Benefit (Hedge 2001).

A) Injury costs before ergonomic intervention
B) Injury costs after ergonomic intervention
C) Cost of ergonomics intervention

Cost Justification Calculations
D) Cost savings from Ergonomics
   Intervention (A—B)
E) Return on Investment ((C/D) x 12)*   months

[* You may need to convert your cost data to annual data for this calculation

3.2 Overview of proceeding process of solution enterprise problems by ergonomics programs

Next part of papers describes proposal of finding solution applying ergonomics programs.

It is desirable in preparation of documents using EAWS methodology, if it is a whole new operation and its implementation in practice, forasmuch its results are helpful in the initial period of the study. The operation should then start up and a detailed ergonomic analysis using epidemiological methods can be carried out at least one year after the production initialization, to obtain information and to analyze real-acting factors in production process and their impact on production efficiency.

After creation of conditions for ergonomics program initiation the can proceed the data collection (Marková & Hatiar 2010).

The "Nordic Questionnaire" is consideed as a basic method for data collection. This questionnaire, as has been demonstrated by authors Rosencrance et al. (2002), provides stable results also in condition of multiple data acquisition repeated at weekly intervals.

Following step is storing the data from questionnaire to the database and their verifying. of their accuracy.

Next step is the descriptive analyses to find out how homogenous are group of employees conducting the workload of employees.

Tables characterizing the occurrence of MSS problems are the basis for visit and interviewing employees, focusing on the causes of the detected levels of MSS problems. At the same time is there a chance to make videos of work cycles on the representative types of work places which allows precisely analyze body segment positions, expended forces and the pace of work (ergonomic triangle.) by applying of legislation requirements (Announcement of the Slovak Ministry of Health No. 542/2007, registered in collection of laws, about mental and physical workload).

Characteristics of modifiable factors based on data obtained by "Nordic questionnaire", supported by data obtained by video analysis and interviews with employees provides us results for identifying impact of workers by MSS problems both, by unmodifiable and modifiable risk factors of work and working conditions.

The next step is applying of epidemiological cohort study, which enables to asses potential risk level of identified factors in terms of ergonomics.

If above mentioned ergonomic analysis is performed for the first time and when we intend initiation of ergonomic program in already working production, it allows aply study with design of a retrospective cohort to assess the impact of prior work and working conditions to employees and this way prepare a baseline for next comparative studies based on cohort design.

Then is implemented solutions aimed either to eliminate or minimize the impact of factors, which are proved through ergonomic analysis as risky.

About a year after implementation of ergonomic rationalization measures, it is in ergonomics program time to perform cohort study, in which can be followed the dynamics of the impact of rationalization ergonomic solutions.

The result of our primary retrospective cohort study for all assessed suspect risk factors is to calculate Odds Ratio, which allows us sort evaluated factors by the intensity of their impact on the incidence of MMS problems.

Unlike a retrospective cohort study in a cohort study, after one year the focus is on monitoring the dynamics of impact identified factors for the implementation of ergonomic rationalization solutions.

There may appear following expected alternatives of ergonomic solution on effect of observed risk factors to worker's health and cost benefit (Fig. 4):

Continuation of an ergonomics program should focus on preventative solution regarding ergonomic

Figure 4. Matrix of biomechanic risk factors (ergonomics triangle—posture, force, frequency) combined evaluation, which focuses ergonomic "FMEA", This results are then reflected in planning the tasks and deadlines for their solution.

rationalization for group of employees, which have negative impacts on the occurrence of MSS problems and proved applied solutions as ineffective.

As stated by Fonesca et al (2016 and 2016 a) after implementation of ergonomic program based on the principles of Participatory Ergonomics, the workers' satisfaction has increased considerably after the implementation of the suggested measures.

At the same time, it is also necessary to evaluate the benefits of the cost of implemented solutions.

In order to achieve a full program of ergonomic efficiency, it is necessary to be interconnected with the Quality Management System (QMS) or should be integrated through integrated management system (IMS). In the QMS is used Failure Mode and Effects Analysis (FMEA).

For the area of ergonomics, the company may create a custom form (Potential Ergonomics Issues List), based on ergonomic analysis based on epidemiological methods assessing individual risk moments on individual positions and allocate tasks and deadlines to address the situation (Hatiar & Eisenberg 2015).

The focus here is the assessment of factors related to the basic biomechanical risk factors (posture, force, frequency of movements), also referred to as the ergonomic triangle.

4 CONCLUSIONS

If the humans will be in factories not replaced by automated systems and robots, it will be necessary in practice to create the appropriate conditions for people—employees who work in these enterprises. At the current level of science and technology is an anachronism when one has to experience pain at work.

Human work you should continue to hold meaning, which had so far in the evolution and in the

Alternatives of Ergonomic Solution Results

(0 = not any occurrence of MSS problems;
+ = MSS problems occurrence;
++ = intensive MSS problems occurrence
; – = decreasing of MSS problems occurrence ;)

Results of 1st study (retrospective)	Results of 2nd study after one year (cohort)	FINAL EVALUATION
0	+	Damaging Solution
+	++	
+	+	Inefficient Solution
0	0	
++	+	Efficient Solution
+	–	

Figure 3. Expected alternative results of ergonomic solutions in framework of HCS model 3E ergonomics program. (Source: Own Processing).

future. Certainly, it should not harm human, but it should be further developed by work and work ought to be the possibility of self-realization. This can be sustainable way achieved only when in the process of corporate social responsibility will be paid attention to the man and his work to the ergonomics through sustainable programs based on participatory basis. This is involving also a work in the industry.

If there is a replacement of man at work in enterprises by automatic machines, it can be assumed that it will be necessary to create suitable working conditions without undue stressing and endangerment in science, culture and services. The subject of ergonomics should move towards more homework, sports and recreational activities, and necessary comfort.

ACKNOWLEDGEMENT

This paper is an part of the project KEGA No. 006STU-4/2017 *"Innovation of the structure, content and teaching methods for the subject "Strategic Management" in the context of new business model for the 21th century based on "win-win" strategy."* and VEGA No.1/0235/17 *"System identification of complex assumptions to support industrial innovation and employment in the less developed regions of Slovakia"* and builds on results of successfully finished project KEGA No. 037STU-4/2012 *Implementation of the subject of "Sustainable corporate social responsibility" into the study programme of Industrial Management in the second degree of study at STU MTF Trnava"* and APVV No. LPP-0384-09 *"Concept of HCS model 3E vs. concept Corporate Social Responsibility."*

REFERENCES

Bruder, R. et al., 2013. EAWS, Course Materials, MTM Institut/IAD Mladá Boleslav, Sdružení MTM pro ČR and SR, 213 p.

Driessen, M. T. et al. 2010. What are possible barriers and facilitators to implementation of a Participatory Ergonomics programme? In *Implementation Science*, Volume 5, Issue 1, Article number 64.

Fonseca, H. M. et al. 2016. *Integrating human factors and ergonomics in a participatory program for improvements of work systems: An effectiveness study*. IEEE International Conference on Industrial Engineering and Engineering Management. Volume 2016-December, 27 December 2016, pp. 1579–1583.

Fonseca, H. (a). 2016. Participatory ergonomic approach for workplace improvements: A case study in an industrial plant In *Advances in Intelligent Systems and Computing*, 491, pp. 407–419.

Hatiar, K.. et al. 2004. Ergonómia a preventívne ergonomické programy: Ergonomická analýza pomocou modifikovaného dotazníka "NORDIC QUESTIONNAIRE".

Ergonomics and preventive ergonomics programs: Ergonomics analysis by modified NORDIC QUESTIONNAIRE. In *Bezpečná práca*, 35, 4, pp. 20–28.

Hatiar, K. 2004. Ergonómia a preventívne ergonomické programy: Hodnotenie rizík v pracovnom procese z hľadiska ergonómie (*Ergonomics and preventive ergonomics programs, Risk Assessment of Working Process from Ergonomics point of vie)*, In Bezpečná práca, 35, 3, pp. 3–10.

Hatiar, K. 2008. *Ergonomics and Technology effectivness.* Köthen: Hochschule Anhalt,. 83 p.

Hatiar, K. et al. 2007. Ergonómia ako súčasť podnikových procesov, In Journal *Procesný manažér,* SAPRIA, Poprad, II, 02, pp. 23–30.

Hatiar, K. & Eisenberg, G. 2015. Ergonómia pri prevencii chorôb súvisiacich s prácou v rámci systému bezpečnosti a ochrany zdravia pri práci v podniku Johnson Controls International, spol. s r.o.—OZ Lozorno, In: Jurkovičová, J,. Štefániková, Z.: *Životné podmienky a zdravie, proceeding*, Úrad verejného zdravotníctva SR, Bratislava, pp. 257–264.

Hatiar, K. et al. 2017. Preventívne zameraný ergonomický program v podniku, In: Jurkovičová, J,. Štefániková, Z.: *Životné podmienky a zdravie, proceeding* 2017, Úrad verejného zdravotníctva SR, Bratislava, 2017.

Hedge, A. 2001. *Ergonomics Cost-Justification Worksheet,* Cornell University, July, 2001, (online, 28-3-2017) http://ergo.human.cornell.edu/ahECW.html

Hrdinová, G. 2013. *Koncept HCS modelu 3E vs. Koncept Corporate Social Responsibility (CSR).* [Dissertation thesis] – STU MTF in Trnava;—Supervisor: Prof. Ing. Peter Sakál, CSc.—Trnava: MtF STU, 2013. 228 p.

Jensen, P. 1997. Can participatory ergonomics become 'the way we do things in this firm'—The Scandinavian approach to participatory ergonomics (Conference Paper) Ergonomics Volume 40, Issue 10, pp. 1078–1087.

Kuorinka, B., et al. 1987, Standardized Nordic Questionnaires for the Analysis of Musculoskeletal Symptoms. In *Applied Ergonomics*, 18, pp. 233–237.

Kyzek, J. & Hatiar, K. 2011. Ergonomic program as a tool for enhancing efficiency of human work. In *Annals of DAAAM and Proceedings of DAAAM Symposium.* pp.1573–1574.

Marková, P. & Hatiar, K. 2010, Návrh modifikácie všeobecného modelu ergonomického programu "HCS 3E" pre priemyselné podniky na Slovensku. In: *Ergonómia 2010: Progresívne metódy v ergonómii.* Žilina,: Slovenská ergonomická spoločnosť, 2010, pp. 121–127.

Mrvová, Ľ. & Sakál, P. 2005. Využitie metodiky CBA v environmentálnych a ergonomických projektoch a programoch. Utilization of using CBA in environmental and ergonomics projects and programs. In *Research papers* Faculty of Materials Science and Technology Slovak University of Technology in Trnava. No. 19, pp. 71–77.

Rosecrance, J.C. et al. 2002. Test-retest Reliability of Self-Administrated Musculoskeletal Symptoms and Job Factors Questionnaire Used in Ergonomics Research. In *Applied Occupational and Environmental Hygiene*. 17, 2002, 9, pp. 1–9.

Sakál, P. et al. 2016 Transformation of HCS model 3E in IMS in context of sustainable CSR. In *Production Management and Engineering Sciences* London, UK: CRC Press. Taylor & Francis Group, 2016, pp. 259–263.

New Trends in Process Control and Production Management – Štofová & Szaryszová (Eds)
© 2018 Taylor & Francis Group, London, ISBN 978-1-138-05885-9

Cross-border M&A activity within the European area

J. Hečková, A. Chapčáková, E. Litavcová & S. Marková
Faculty of Management, The University of Presov, Prešov, Slovak Republic

ABSTRACT: Cross-border Mergers and Acquisitions (M&As) have become the dominant mode of growth for enterprises seeking competitive advantage in an increasingly complex and global business economy. The questions cross-border M&As raise and the reactions they elicit are not new, but they are clearly growing in importance with the rise in foreign takeovers. The emergence of multinational enterprises from developing countries has also added a new dimension. It is often pointed out that cross-border capital reallocation is partly the result of financial liberalization policies and regional agreements. The aim of this paper is to contribute knowledge on the issues of cross-border mergers and acquisitions more objective perspective on selected aspects determined the development of cross-border mergers and acquisitions in selected countries of the European area and Turkey.

1 INTRODUCTION

Over the last two decades, the volume of foreign direct investment has increased significantly and is considered one of the key drivers of global economic growth (UNCTAD 2017). The largest share of foreign direct investment is covered by cross-border mergers and acquisitions. Accordng to UNCTAD (2015), the cross-border transactions currently have around 66% share of foreign direct investment, whereas before the crisis (2005–2007), their share was up to 80%.

Cross-border mergers and acquisitions thus belong to one of the most significant phenomena of the last two decades in the European as well as the global scale. Advancing globalization of the world economy, turbulent and unpredictable changes in the global environment, liberalization of trade, finance and investment, deregulation of the service sector, technological change, changes in the cost structure, regional trade agreements and integration groupings created conditions that provide businesses with more and more opportunities for international expansion.

Cross-border mergers and acquisitions can be seen as one of the main forms of integration that enables businesses to create synergies, gain economies of scale, reduce costs, increase market power, expand and diversify business, and globally create competitive superiority (Frankovský et al. 2016).

Processes of the deepening financial and trade liberalization in the European Union and the European Monetary Union (EMU) support the implementation of cross-border mergers and acquisitions. The current outlook for cross-border mergers and acquisitions in the European area is optimistic (more

Deloitte, 2016). The global financial and economic crisis of 2007–2008 also helped to significant changes in the European area and cross-border transactions have become one of the ways accelerating the corporate and financial restructuring. Enterprises in this environment, with cheaper funding opportunities, greater cash reserves and better market conditions, thus focus more and more attention on achieving their strategic objectives and growth strategies through cross-border mergers and acquisitions.

In the twentieth century, an eminent position within the cross-border mergers and acquisitions was taken by the developed European economies. However, the situation in the last decade has changed and in the field of international mergers and acquisitions, the emerging economies of Central and Eastern Europe are starting to play a key role. The volume of cross-border transactions there is currently growing twice as fast compared to the Western European countries. Thanks to the functional legal and political systems based on democratic principles, transparency of the business environment, as well as being the part of a vast single market of the European Union, they have the dynamic growth potential and attractiveness, especially for foreign investors and the future periods (Kotulič et al. 2016; Kiseľáková & Šofranková 2015).

The aim of this paper is to identify the impact of the selected predictors for the development of the volume of cross-border mergers and acquisitions implemented in the manufacturing sector on the selected sample of countries of the European area broken down geographically into two groups, namely the countries of Western Europe, and Central and Eastern Europe in the period 1998–2012.

The paper was prepared in the scope of implementing the project VEGA No. 1/0031/17 "Cross-border mergers and acquisitions in the context of economic and social determinants in the European area".

2 DATA AND METHODOLOGY

The data set for the purpose of this research included 85,510 data about the implemented cross-border mergers and acquisitions in the European space and in Turkey in the period from 1998 to 2012 in 16 source countries (Belgium, Cyprus, Denmark, Finland, France, Greece, Netherlands, Luxembourg, Malta, Germany, Poland, Portugal, Austria, Spain, Italy, United Kingdom) and in 25 target countries (Belgium, Bulgaria, Cyprus, Czech Republic, Denmark, Estonia, Finland, France, Greece, Netherlands, Lithuania, Latvia, Luxembourg, Hungary, Malta, Germany, Portugal, Austria, Romania, Slovakia, Slovenia, Spain, Italy, Turkey, United Kingdom). The source data were drawn from databases: Zephyr (Bureau van Dijk 2014), Eurostat (European Commission), and Freedom House (Freedom House, 1998–2012). The research emphasis was placed on the sample of 25 target countries into which, within the European area, reallocation of capital was carried out through cross-border mergers and acquisitions, and this sample was geographically divided into two groups consisting of 15 Western European countries (Belgium, Cyprus, Denmark, Finland, France, Greece, Netherlands, Luxembourg, Malta, Germany, Portugal, Austria, Spain, Italy, and United Kingdom) and 10 countries of Central and Eastern Europe (Bulgaria, Czech Republic, Estonia, Latvia, Lithuania, Hungary, Romania, Slovakia, Slovenia, and Turkey).

Of the total number of entries of the data set, 11,583 records relate to cross-border mergers and acquisitions, 4395 entries have the indicated value of the volume of cross-border activities of which 4285 also have values of other selected predictors. After excluding ambiguous, error data and the highest extreme values (204.7M Euro of the volume of cross-border assets, which is about four times higher than the next highest value of the volume of cross-border assets of 51.3M Euro), 4260 entries remained.

To quantify the impact of the considered predictors on the volume of cross-border mergers and acquisitions, a generalized regression model has been used (Hušek 2007; Agresti 2002; Anderson et al. 1993). The research results processing was carried out by the software MS Excel, Statistica 12 and IBM SPSS Statistics 20.

$M\&A_{ij,s,t}$ refers to the total value of cross-border assets acquired through cross-border mergers and acquisitions by the source country i in the target country j in the sector s at the time t. An important predictor affecting the volume of cross-border mergers and acquisitions may be considered to be the gross domestic product of the source (i) and the target (j) country in the sector s at the time t ($GDP_{j,s,t}$, $GDP_{i,s,t}$). Using the logarithm of their product eliminates their disparity of elasticity and does not affect the overall result. *Market capitalization* is the annual average market capitalization of the considered sector, which was obtained from the database Zephyr (Bureau van Dijk 2014). The ratio of market capitalization to GDP has been used as an indicator of the development of the stock market. The proximity of the countries, the specificity of their culture and language affinity should not be omitted. The proximity of the source and the target country was quantified by geographical distance between their capitals, labeled as $Distance_{ij}$, sharing a common border of the countries was quantified by the binary variable $Border_{ij}$, taking the value of 1 in the positive and 0 in the negative case. Binary variable $CommonLanguage_{ij}$ acquiring the value of 1 in the case of identical language, and otherwise 0, was considered to quantify the impact of the relative language on the volume of cross-border assets.

Another considered predictor in the above equation is the variable $CivilLiberties_{i,t}$, which was quantified through the index of civil liberties— The Civil Liberties Index (Freedom House, 1998–2012) and assesses the quality of institutions in the source countries (i) for individual years t. This Civil Liberties Index reflects the state of political democracy, individual freedoms and good government practices. The scale of the index is from 1 (best country) to 7 (worst country), wherein the analyzed data set reaches the maximum value of 3. For the low frequency of the 3, occurring in seven records only, the values 2 and 3 were merged in the database. Similarly, the variable $CivilLiberties_{j,t}$ of the target country (j) at the time t reached in the database the maximum value of 5. However, due to the infrequent presence of the acquired values of 4 and 5, the values of 3, 4 and 5 were merged. It is assumed that improvements in the civil rights reduces the costs of the capital and supports investment in these countries because of the reliability of institutions, transparency of the business environment especially for foreign investors, and a functional legal and political system based on democratic principles. Other predictors of the considered model are dummy variables related to membership of the source and the target countries in the European Union and the European Monetary Union, namely $EU_{i,t}EU_{j,t}$ takes the value of 1 if the source country i as well as the target country j were members of the European Union at the time *t*, otherwise the value of 0; $NonEU_{i,t}EU_{j,t}$ takes the value of 1 if the source country (*i*) was not and the target country (*j*) was a member of the European Union at the time *t*, otherwise the value of 0. For the other two additional options $NonEU_{i,t}NonEU_{j,t}$ and $EU_{i,t}NonEU_{j,t}$, dummy variables were not used, and in the analyses they are considered as the reference category. The $EMU_{i,t}EMU_{j,t}$ variable takes the value of 1 if the source country *i* as well as the target country *j* were members of the European Monetary Union at the time *t*, otherwise the value of 0; $NonEMU_{i,t}EMU_{j,t}$ takes the value of 1 if the source country (*i*) was not and the target country (*j*) was a member of the European Monetary Union at the time *t*, otherwise the value of 0. As in the previous case, for the other two additional options $NonEMU_{i,t}NonEMU_{j,t}$ and $EMU_{i,t}NonEMU_{j,t}$, dummy variables were not used, and in the analyses they are considered as the reference category. The last term of the equation $e_{ij,s,t}$ denotes a random component.

Applied was a generalized linear model, gamma distribution, link function log. According to Nelder (2000), generalized linear models (GLM) include two extensions of the classical linear model for the explained variable *y*, which assumes a normal distribution of random errors and for the mean of the systematic component, consisting of a set of explanatory variables, is true that $y = \Sigma \beta_j x_j$, whereas β_j arc the estimated parameters. First, the group of distributed random errors is extended to an exponential family with one parameter, which includes a Poisson distribution, binomial, gamma, and inverse Gaussian. Second, the systematic component assumes additivity of the predictors on the transformed scale, given by $\eta = g(\mu)$, where $\eta = \Sigma \beta_j x_j$, whereas η, is the linear predictor, and $g(\mu)$ is the linking function. GLM uses the algorithm of a maximum likelihood for the model fitting, namely the weighted least square method acting on the dependent variable *z* defined by $z = \eta + (y - \mu)(d\,\eta\,/d\,\mu)$ and the weight *W* is given by $1/W = var(z) = var(y)(d\,\eta/d\,\mu)^2$. A fitting statistics derived from the logarithm of the credibility ratio test is called *Deviance* (more Nelder & Wedderburn, 1972; McCullagh & Nelder, 1989). Generalized regression models, linear and nonlinear, cover a wide range of statistical methods with different types of variables which are widely used in the economic and management disciplines. For example, the probit model with the dichotomous explained variable used by Hindls & Hronová (2002), regression with the nominal explained variable used by Jenčová & Maťovčíková (2013), or a special case where the nominal explained variable is not clearly established and the log-linear analysis is used by Litavcová & Butoracová-Šindleryová (2009) (more Agresti, 2002). Several regression

models were tested and their results in terms of significance of the coefficients were comparable. The GLM model with a normal distribution and the linking function of identity was, in terms of significance of the predictors and their effect in the model, identical with the selected model. A similar result was provided by the classical linear model with appropriately modified categorical predictors found by the stepwise regression, with the result $p = 0.000$ for the *F* test, adjusted R^2 with the value of 0.880. The quality of each regression has been verified by the regression diagnostics (more Fox & Weisberg, 2010). Experimentation with more complex patterns of interactions did not clearly lead to a better outcome in terms of quality; ultimately, the satisfactory "parsimony" model was selected. To identify the relative importance of significant predictors, a Wald chi-square test of the effect in the model was observed. Using robust estimates to eliminate the impact of deviating values changed the order of the predictors in terms of the observed effect. Among the eleven variables, ten showed significant contribution to the considered model.

3 RESULTS

The aim of the research was to estimate the weights of the considered predictors on the total value of assets purchased through cross-border mergers and acquisitions $M\&A_{ij,s,t}$ by the source country *i*, in the target country *j*, in the sector *s*, at the time *t*.

The estimated regression coefficients are in the form of the following equation:

$$
\begin{aligned}
log\,(M\&A_{ij,s,t}) = &\; 1.265 + 0.413log(GDP_{i,s,t}GDP_{j,s,t}) \\
&+ 0.862log(MarketCapitalisation/GDP_{j,s,t}) \\
&- 0.074log(Distance_{ij}) + 0.089(Border_{ij}) \\
&- 0.122(CommonLanguage_{ij}) \\
&+ 1.085(EU_{i,t}EU_{j,t}) + 0.239(EMU_{i,t}EMU_{j,t}) \\
&- 0.471(CivilLiberties_{i,t}\ is\ middle) \\
&+ 1.301(CivilLiberties_{j,t}\ is\ low) \\
&+ 0.352(CivilLiberties_{j,t}\ is\ middle)
\end{aligned}
\tag{1}
$$

where

- log(M&Aij,s,t) denotes log of the total value of assets purchased through cross-border mergers and acquisitions in the target country j by firms in the sector s, resident of the country i, in year t.
- log(GDPi,s,tGDPj,s,t) denotes log of the product of the two GDPs at the time t, which restricts the elasticity to be the same for the country i and the country j, but none of the results depend on this restriction.
- log(MarketCapitalisation/GDPj,s,t) denotes log of market capitalization to GDP ratio of the acquirer and the target sectors. It is used there as

an indicator of stock market development and can help controlling the equity bubbles. Data on market capitalization are the yearly average market values of the sector from the Zephyr database (Bureau van Dijk 2014).

- log(Distanceij) denotes log of bilateral geographical distance between the capitals of the source country i and the target country j, which could be considered as negligible too, as well as the proximity of the countries and the relationship of their languages. Borderij is the dummy variable which equals one when the two countries shared the common border and the dummy CommonLanguageij equals one if the two countries share a common language.
- EUi,tEUj,t is the dummy variable equal to one if both countries belong to the EU at the time t, and zero if otherwise. Similarly, the dummy EMUi,tEMUj,t is equal to one if both countries belong to the EMU at the time t, and zero if otherwise. For the complementary possibilities, the dummy variables were not introduced. They are handled in the analyses as reference categories.
- CivilLibertiesi,t (or CivilLibertiesj,t) control the quality of institutions in the source (host) country by means of an indicator of civil liberties at the time t, which measures over time and across countries the freedom of expression and belief, the association and organization rights, the rule of law and human rights, personal autonomy and economic rights. The civil liberty index is taken from Freedom House (1998–2012) and ranges between one (the best country) and seven (the worst country). In our data set, CivilLibertiesi,t ranges only between 1 and 3 (with values 2 and 3 merged due to the low frequency of values 3 with

only 7 occurrences), and CivilLibertiesj,t ranges between 1 and 5 (with values 3, 4 and 5 merged due to the low frequencies of higher values).

In the paper, the focus is on quantification of relationship between M&As and its above presented predictors in the manufacturing sector only and for two disjoint groups of the target countries. The first group of the target countries consists of Western European countries and the second one consists of the target countries of Central and Eastern Europe.

Regression results presented by the values of regression coefficients are illustrated in Table 1. The same generalized linear model was used in the previous work by Hečková et al. (2016) in both analyzed groups of 16 source and 25 target countries (Table 1).

Further presented are the models with significant coefficients in the explanatory variables. Not significant predictors were excluded from the model. Among 4260 entries, which are all considered as valid variables, 1272 records were in the manufacturing sector in the group of the target countries of Western Europe, and 268 entries were in the group of the target countries in Eastern Europe (+ Turkey).

Following are the results of the regression equations presented by the values of regression coefficients for 1272 entries in the group of the targeted Western European countries:

$$
\begin{aligned}
log\,(M\&A_{ij,s,t}) = {} & 2.463 + 0.391log(GDP_{i,s,t}GDP_{j,s,t}) \\
& + 0.902log(MarketCapitalisation/GDP_{j,s,t}) \\
& + 0.188(Border_{ij}\,exists) \\
& - 0.555(CivilLiberties_{i,t}\,is\,middle) \\
& + 0.386(CivilLiberties_{j,t}\,is\,middle)
\end{aligned} \tag{2}
$$

with standard errors of 0.3639 for the intercept and 0.0182, 0.0139, 0.0592, 0.0513, 0.0498 for the predictors in the same order as in the equation. Impact of all the variables listed in the equation is significant at the significance level of 0.01. Significant was also the impact of the variable $EU_{i,t}EU_{j,t}$, but because of the extremely low frequency, only in the case of 4 entries, in which neither the source nor the target country were members of the EU, the variable has been omitted from the model.

In the case of the target countries of Central and Eastern Europe including Turkey, the explanatory variables have been modified. There were 100 entries where both the source and the target country belong to the EU, and 168 entries where this was not the case. Membership in the European Union ($EU_{i,t}EU_{j,t}$) has no significant impact here, and the variable related to the common border ($Border_{ij}$) has almost no significant impact here. The $CommonLanguage_{ij}$ variable is not here because a common language did not occur in either case and the variables related to the membership in EMU ($EMU_{i,t}EMU_{j,t}$) were, except for one record, in this file filled with only a constant value of 0.

Table 1. Regression model with the dependent variable of the volume of cross-border mergers and acquisitions for the manufacturing sector.

Predictor	Coefficient β	Standard error, robust estimation
(Constant)	.793**	.3264
log(GDP_{ist}GDP_{jst})	.414***	.0145
log(MarketCapitalisation /GDP_{j,s,t})	.886***	.0133
Border_{ij} (yes)	.131**	.0627
EU_{it} EU_{jt} (yes)	1.151***	.1564
EMU_{it} EMU_{jt} (yes)	.136***	.0553
CivilLiberties_{it} (middle)	−.359***	.0582
CivilLiberties_{jt} (low)	1.131***	.2260
CivilLiberties_{jt} (middle)	.190***	.0683

Source: Own calculation

Note: The GLM model was used. Statistical significance at the 10% (or 5% and 1%) significance level is marked with * (or ** and ***). Number of observations is 1540.

178

Result of the regression equation for 268 entries in the group of the target countries in Central, Eastern Europe and Turkey is following:

$$log\,(M\&A_{ij,s,t}) = 0.502 + 0.371log(GDP_{i,s,t}GDP_{j,s,t})$$
$$+\,0.915log(MarketCapitalisation/GDP_{j,s,t})$$
$$+\,0.186log(Distance_{ij})$$
$$+\,0.262(Border_{ij}\,exists) \hspace{2cm} (3)$$
$$-\,0.392(CivilLiberties_{i,t}\,is\,middle)$$
$$+\,0.660(CivilLiberties_{j,t}\,is\,low)$$
$$-\,0.408(CivilLiberties_{j,t}\,is\,middle)$$

with standard errors of 0.6500 for the intercept and 0.0307, 0.0224, 0.0558, 0.1408, 0.1195, 0.1925, 0.1749 for the predictors in the same order as in the equation. The intercept is not significant and the coefficient of $Border_{ij}\,exists$ is significant only at the significance level of 0.1. The coefficients of all the other predictors in this equation are significant, $CivilLiberties_{j,t}\,is\,middle$ at the level of 0.05, and others at the significance level of 0.01.

4 CONCLUSION

In a narrow view on the selected predictors, results of the research analyses have shown that the volume of cross-border activities in the manufacturing sector have the strongest impact on the countries of Western Europe (variables shown in the equation 1) and on the countries of Central and Eastern Europe (variables in the equation 2). In the case of the countries of Central and Eastern Europe including Turkey, as opposed to the group of Western European countries, it was shown that the European Union membership has no significant impact on the volume of cross-border transactions. To conclude, these results may be helpful for managers, board members, community, and other stakeholder groups for a deeper understanding of the relationship between the volume of mergers and acquisitions and their continuous individual predictors.

ACKNOWLEDGEMENT

The paper was prepared in the scope of implementing the project VEGA No. 1/0031/17 "Cross-border mergers and acquisitions in the context of economic and social determinants in the European area".

REFERENCES

Agresti, A. 2002. *Categorical Data Analysis.* Wiley, 2003. Second Edition. University of Florida. Wiley Interscience, John Wiley & Sons, Inc., Hoboken, New Jersey, 2002.

Anderson, D. R. et al. 1993. *Statistics for Business and Economics.* St. Paul: West Publishing Company.

Bureau van Dijk. 2014. *Zephyr.* (zakúpené údaje z databázy Zephyr za obdobie rokov 1998–2012). http://www.bvdinfo.com/en-gb/our-products/economic-and-m-a/m-a-data/zephyr

Deloitte. 2016. M&A Trends Report, Mid-year 2016. Our annual comprehensive look at the M&A market. Available at: https://www2.deloitte.com/content/dam/Deloitte/us/Documents/mergers-acqisitions/us-deloitte-mergers-acquisitions-report-trends-2016.pdf

European Commission: http://epp.eurostat.ec.europa.eu/portal/page/portal/eurostat/home/

Fox, J. & Weisberg, S. 2010. *An R Companion to Applied Regression.* Sage Publications, Inc., 2011.

Frankovský, M. et al. 2016. Assessment of occurrence predictors of cognitive distortions in managerial decisions. *Polish Journal of Management Studies.* 14 (2): 61–70.

Freedom House. 1998–2012. Reports. Available at: https://freedomhouse.org/reports#.VS0 WBPAp5 sI

Hečková, J. et al. 2016. Ekonomická integrácia a jej vplyv na cezhraničné fúzie a akvizície v európskom priestore. *Politická ekonomie* 64 (1): 19–33.

Hindls, R. & Hronová, S. 2002. Detekce a prognóza bodů obratu v ekonomickém vývoji. *Politická ekonomie* 50 (2): 217–227.

Hušek, R. 2007. *Ekonometrická analýza.* Praha: Oeconomica, 2007.

Jenčová, S. & Maťovčíková, D. 2013. *Medzinárodný obchod.* Prešov: Bookman, 2013.

Kiseľáková D. & Šofranková B., 2015. Effects and risks of mergers and acquisitions on entrepreneurship in banking and finance: Empirical study from Slovakia. *Review of European studies* 7: 23–35.

Kotulič R. et al. 2016. Status of foreign direct investments and their relationship to selected economic indicator of the sustainable development in the Slovak Republic—localization factors of foreign direct investment allocation and their spatial differentiation. *Journal of applied economic sciences* 11 (1): 86–96. Available at: http://cesmaa.eu/journals/jaes/files/JAES_1(39)%20Spring%202016_online.pdf

Litavcová, E. et al. 2009. Aplikácia log-lineárnej analýzy v marketingovom výskume. *Ekonómia a proces poznávania: Vedecká konferencia doktorandov a mladých vedeckých pracovníkov.* Prešov: FM PU v Prešove: 95–103.

McCullagh, P. & Nelder, J. A. 1989. *Generalized Linear Models.* 2nd ed. London: Chapman & Hall.

Nelder, J. A. & Wedderburn, R. W. M. 1972. Generalized Linear Models. *Journal of the Royal Statistical Society.* Series A (General) 135 (3): 370–384. http://www.jstor.org/stable/2344614.

Nelder, J. A. 2000. Quasi-likelihood and pseudo-likelihood are not the same thing. *Journal of Applied Statistics* 27 (8): 1007–1011.

Štefko R. et al. 2010. Marketingové inštrumentárium v procese akceptácie projektov pri akcelerácii rozvoja zaostávajúcich regiónov. *Ekonomický časopis* 58 (5): 512–526.

Unctad. 2015. World investment report 2015. Available at: http://unctad.org/en/PublicationsLibrary/wir2015_en.pdf

Unctad. 2017. www.unctad.org/fdistatistics.

New Trends in Process Control and Production Management – Štofová & Szaryszová (Eds)
© 2018 Taylor & Francis Group, London, ISBN 978-1-138-05885-9

Business performance improvement applying controlling tools

J. Horváthová & M. Mokrišová
Faculty of Management, University of Prešov, Prešov, Slovak Republic

ABSTRACT: Nowadays there are new fields in concept of controlling such as risk controlling; performance controlling or value based controlling. Application of controlling in these areas is given by dynamic environment and by the effort of management of businesses to increase company value. In terms of historical development business performance measurement has passed through several time periods, from reporting of profit margin through profit maximization and various types of profitability indicators to the criteria for achieving value for owners. The aim of this paper is to calculate business performance and to identify performance indicators applying selected methods. Subject for objective fulfilment were secondary data from financial statements of businesses operating in Slovak heat industry. Based on data analysis we determined key performance indicators of selected sample of businesses. The benefit of this paper is the selection of appropriate model for performance evaluation of Slovak businesses.

1 THEORETICAL BACKGROUND

Controlling, which is according to Freiberg (1996) defined as a complex function of economic management, coordination of planning, control and information security, begins to apply nowadays in new functional areas of business management and with new content specification. These are controlling of business performance, controlling of business value as well as controlling of internal and external risks. Establishment of new controlling orientation is given by an effort of businesses to survive in constantly changing competitive environment. It is also given by the development of theory and practice of business management, by promoting the theory of business value management, business performance criteria, methods of risks elimination and methods of predicting future development.

In accordance with the above-mentioned the application of dynamic ScoreCard within controlling is preferred. Dynamic ScoreCard is flexible open system of tools, methods, reports of financial controlling, risk controlling, performance controlling, business value controlling and strategic controlling. The aim of dynamic ScoreCard is to ensure dynamic view of business in order to increase its value (Horváthová et al. 2016, Gallo 2013). The role of financial controlling within this system is to ensure business financial balance (Mussing & Faschlang 1999). The most common method of assessing business performance within financial controlling is method of fundamental and technical analysis, which evaluates the enterprise in economic terms based on a detailed study and analysis of financial statements (Fisher 1992). In the opinion of many Slovak and foreign authors (Ittner et al. 2003, Pavelková & Knápková 2009, Synek et al. 2007, Petřík 2009) as the most common controlling tools to measure the performance of companies are used financial indicators.

An important component of integrated system of performance and risk management is controlling of business value. Currently the best known and most utilized modern indicator of performance measurement is Economic Value Added (EVA). This model is known from 1980s. Authors of the EVA model are representatives of Stern Stewart & Co., American researchers Joel M. Stern and G. Bennett Stewart III. The main task of EVA model is the measurement of business economic profit. Extensive use of the EVA model dates back to 1989. We can summarize that EVA indicator represents a shift towards indicators focused on maximizing value for owners (Šofranková & Čabinová 2016). Benefit of the EVA indicators is that it takes into account risks as well as evaluation of Equity. Fundamental difference of EVA indicator compared to conventional indicators, which are based on the information of financial statements, consists of two basic facts: EVA indicator introduces to performance measurement the idea of so-called opportunity cost, which are a price, resp. cost of capital (WACC—Weighted Average Cost of Capital) and works with operating profit or loss (NOPAT—Net Profit After Tax). Therefore the introduction of EVA indicator was significant and anticipated step in measuring performance and market valuation of business and this indicator has become an important indicator applied in business controlling. Another positive shift in controlling and especially in the strategic controlling is Balanced Scorecard (Kaplan & Norton 2007). The strategic map of Balanced Scorecard (BSC) is the most important output of integrated system of business performance and risk analysis (Gallo & Mihalčová, 2016). Nowadays, especially in Slovakia, there is third generation of development of BSC system as a strategic management system. Numerous findings about the issue of this generation of BSC summarize Gavurová in her study (Gavurová 2011). The third generation BSC systems are typical for the interlinking strategy and the management of competitive advantages as well as the management of transformational changes (Antošová et al. 2014).

In addition to these methods in the area of business performance measurement, attention should be paid to new innovative approaches to measuring business performance. There are many studies dealing with the problems of calculation and analysis of business performance with the use of different methods, especially mathematical and statistical methods (correlation analysis, regression analysis, principal component analysis, factor analysis, sensitivity analysis, etc.). These studies are based on the assumption that there is a strong correlation between business performance and quantitative variables and their aim is to define performance as a function of these variables. According to Kocmanová et al. (2013), business performance can be measured by various methods, which can be expressed by different types of models. Performance can be measured also by a set of indicators used to calculate effectiveness and efficiency of various activities. Significant benefit in this area is performance measurement with the use of matrix models processed in contributions of Grell & Hyránek (2012, 2014). According to these authors, conventional indicators are very good basis for more precise performance research applying mathematical methods. Based on mentioned facts business performance can be analysed using matrix system of indicators, in which are applied indicators of inputs and outputs and their various combinations. One of the significant results of matrix model is the definition of new indicators, which may be beneficial for controlling of business performance. Another positive aspect of this approach is that indicators measuring business efficiency, effectiveness and performance create a network with strong relations between them (Štefko & Gallo 2015).

2 RESEARCH SAMPLE AND METHODS OF PROCESSING

Research was carried out on a sample of 30 businesses operating in Slovak heat industry. This industry is important in social terms and plays significant role in daily life of consumers. Typical for these businesses is that they are local systems of central heat supply system. As a starting point, input controlling analysis of selected businesses was processed. Within it these indicators were analysed: Return on Assets (ROA), Return on Equity (ROE), Return on Sales (ROS), Current liquidity (CL), Total liquidity (TL), Assets turnover (AT), Average collection period (ACP), Creditors payment period (CPP), Equity ratio (ER), Debt to equity ratio (DER), Overcapitalization (OC), Times interest earned ratio (TIER). Results of the analysis are shown in abbreviated form in Table 1. Based on these data it is obvious, that these businesses achieve low profitability—it may be related to particularities of given industry.

Performance of analysed businesses was measured and analysed applying a number of methods and procedures (EVA equity, EVA entity, EVA_{ROS}, matrix model, linear programming model, correlation matrix, principal component analysis—PCA). Due to the limited extent of the contribution, there are presented only results of selected above-mentioned methods.

To calculate performance of research sample, we used indicators EVA Entity and EVA Equity. Cost of Equity, which enters into performance value, was calculated applying CAPM with the acceptance of external systematic risks (ERP—Equity Risk Premium, CRP—Country Risk Premium, β—coefficient of systematic risk). This model should be complemented by selected internal business risk in order to ensure specificity of Equity valuation in given business or industry. Analysis of the relationship between indicators was realised applying correlation matrix. Because of its compatibility with other financial indicators, we selected EVA_{ROS} as performance indicator. As the multicollinearity among indicators was confirmed, we applied multivariate analysis PCA (Principal Component

Table 1. Input analysis of selected financial controlling indicators.

Variable	Average	Min	Max
Return on Assets	8%	−6%	30%
Return on Equity	18%	−15%	80%
Return on Sales	5%	−5%	18%
Current liquidity	1.25	0.01	5.91
Total liquidity	1.32	0.01	5.91
Assets turnover	1.26	0.41	6.01
Average collection period	79 days	18 days	331 days
Creditors payment period	115 days	25 days	259 days
Equity ratio	44%	2%	92%
Debt to equity ratio	3.84	0.09	56.61
Overcapitalization	1.02	0.02	10.44
Times interest earned ratio	1340.44	−5.74	39,827.32

Analysis). With the use of this method, we verified the results obtained applying correlation matrix and identified principal components for business performance analysis.

Research problem was focused on the quantification and analysis of EVA indicator as performance measure and on the identification of key business performance indicators:

Is EVA indicators appropriate measure of performance? Is it possible to identify key business performance indicators? Which indicators can be considered as key performance indicators? Can we reduce identified key performance indicators to principal components of business performance?

In accordance with research problem, these hypotheses were determined:

H1: We suppose that there are statistically significant correlation relationships between

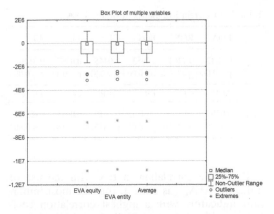

Figure 1.   Comparison of EVA equity, EVA entity and average EVA.

Table 2.   Performance order according to the average value of EVA indicator in €.

Performance order	EVA	Business
1	1,030 459,04	TP27*
2	680 302,84	TP18
3	578,664.36	TP23
4	405,066.84	TP26
5	218,376.07	TP1
6	207,150.04	TP12
7	162,527.71	TP21
8	156,363.52	TP19
9	139,277.75	TP10
10	106,125.07	TP9
11	92,659.24	TP11
12	85,311.65	TP29
13	20,151.42	TP24
14	15,386.69	TP14
15	−53,951.18	TP17
16	−62,354.13	TP30
17	−65,433.47	TP25
18	−171,040.64	TP13
19	−381,744.28	TP28
20	−444,771.66	TP16
21	−543,312.24	TP20
22	−601,006.17	TP22
23	−868,713.57	TP8
24	−961,824.52	TP15
25	−1,646,568.88	TP7
26	−2,525,477.51	TP2
27	−2,639,398.38	TP6
28	−3,121,312.82	TP5
29	−6,659,719.95	TP4
30	−10,767,355.59	TP3

* TP – Identification of analysed businesses.

the selected financial indicators and business performance.

H2: We suppose that it is possible to identify principle performance components and to apply them to analyse business performance.

3   QUANTIFICATION OF PERFORMANCE OF ANALYSED BUSINESSES

As a starting point in addressing given issue, it was necessary to quantify the value of EVA indicator. Table 2 shows the average value of performance of businesses and their ranking according to the achieved performance. From 30 analysed businesses, 14 businesses showed positive value of EVA indicator and 16 businesses had a negative value of this indicator. The highest achieved value of EVA indicator was 1 million € and the lowest value was—10 million €. Based on these results we can say that analysed sample of businesses has difficulties in achieving optimum values of performance. The main cause of these problems is the low profitability of companies, which does not cover risks of external environment.

When analysing the values of EVA equity and EVA entity, we found out minimum differences (Fig. 1). It means that analysed businesses do not have high debt. This is confirmed also by high values of Times interest earned ratio. Average value of this indicator is at the level of 1340.

4   ANALYSIS OF THE RELATIONSHIPS BETWEEN FINANCIAL INDICATORS

We selected a group of financial indicators, in which were represented financial controlling indicators, indicators of profitability or liquidity, solvency, activity, as well as indicators of capital structure.

Table 3. Correlation matrix applying EVA$_{ROS}$.

	ROA	ROE	ROS	CL	TL	AT
EVA$_{ROS}$	0.6947	0.7264	0.7083	0.0445	0.0367	0.3126
	p = 0.000	p = 0.000	p = 0.000	p = 0.816	p = 0.847	p = 0.093

	ACP	CPP	ER	DER	OC	TIER
EVA$_{ROS}$	0.0261	−0.1936	−0.4145	0.1710	0.0441	0.0972
	p = 0.891	p = 0.305	p = 0.023	p = 0.366	p = 0.817	p = 0.609

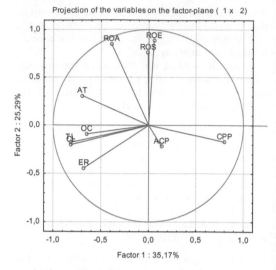

Figure 2. Projection of the variables on the factor—plane.

From the correlation matrix with the use of indicator EVA$_{ROS}$ is evident, that the most important indicators with a highest correlation coefficient with respect to EVA$_{ROS}$ were indicators Return on Equity, Return on Sales and Return on Assets. Relationships between these indicators and EVA$_{ROS}$ are expected, because EVA$_{ROS}$ is indicator of business profitability. We can say that it is a modification of profitability indicators. Indicator with medium correlation coefficient with respect to EVA$_{ROS}$, is Equity ratio. There is inverse relationship between these two indicators. With an increase in the value of Equity ratio, synthetic indicator EVA$_{ROS}$ decreases as indicated by the Du Pont equations. Mentioned relationships resulting from correlation matrix confirmed hypothesis H1, which assumed that there are statistically significant relationships between selected financial indicators and business performance.

To reduce the number of performance indicators, we applied the method of principal component analysis, with the use of which the original variables were replaced by new independent variables called principle components. With the use of this method, we confirmed the hypothesis H2: We suppose that it is possible to identify principle performance components and to apply them to analyse business performance.

The result of this analysis was the substitution of financial controlling indicators by 3 principal components. In the case of component 1 there is directly proportional relationship to Creditors payment period. In the case of this component there is also indirect proportional relationship to the indicators Current liquidity, Total liquidity, Assets turnover, Equity ratio and Overcapitalization. Component 1 evaluates liquidity, activity and capital structure. Between component 2 and the indicators Return on Equity, Return on Sales and Return on Assets is strong direct relationship. It means that component 2 evaluates business profitability. The higher is its value, the higher is business profitability. Strong direct relationship is also between component 3 and the indicator Average collection period. This indicator is important performance indicator of analysed sample of businesses. This

conclusion was expected in view of the specifics of analysed industry.

We constructed the chart of component scales for the two principal components (Fig. 2), which shows original variables in new coordinate system of the components 1 and 2. We assess the relationship between indicators and components by comparing the vectors of indicators. The longer is the vector, the stronger is their relationship. The smaller is the angle inclination of the vector, the stronger is the relationship between indicator and component.

Figure 2 shows that In the case of component 1 there is strong indirect proportional relationship to indicators of liquidity and activity. In the case of this component there is also strong directly proportional relationship to the indicator Creditors payment period. This implies that reducing liquidity and increasing the volume of liabilities lead to increase in profitability and business performance. Between component 2 and profitability indicators is strong direct relationship, which can in relation to business performance be considered as obvious result of this analysis. For the comparison of performance of businesses, we constructed also plot of scores (see Figure 3). It represents the coordinates of businesses in the new area defined by the principal components. With the use of plot of scores we can find out the relationship among chosen businesses, whether they are similar or whether they differ from each other. If businesses are very far from the beginning of the coordinate system and away from the clusters, they are the extremes.

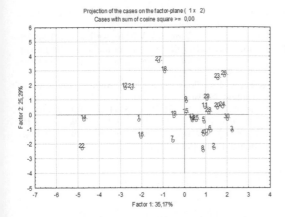

Figure 3. Plot of scores.

Enterprises located immediately in the beginning of the coordinate system are considered to be the most typical for the group of businesses.

In quadrant A of plot of scores (top left quadrant in Fig. 3) there are businesses, which achieve very good results in terms of liquidity and profitability. These companies achieve the best results in efficiency and performance of all analysed businesses, which was confirmed by matrix system of indicators. They also achieve positive values of EVA indicator.

In quadrant B of plot of scores (top right quadrant in Fig. 3) are located businesses that achieve very good results in terms of liquidity and worse results in terms of profitability. These businesses have problems with second principal component—profitability. Very good results in terms of profitability and problems with liquidity show businesses located in quadrant C of plot of scores (bottom left quadrant in Fig. 3). Businesses located in quadrant D of plot of scores (bottom right quadrant in Fig. 3) have problems with the first and the second principal component.

5 CONCLUSION

Based on detailed analyses and calculations, we can conclude that businesses operating in Slovak heat industry have difficulties in improving performance. The average value of performance measured by EVA indicator was—920,000 €. Negative value of performance was in the case of several businesses caused by low liquidity. Some businesses reached a negative value of profitability. In general we can say that profitability of analysed businesses is low in relation to the Cost of Equity. To improve performance, we propose to focus on key performance indicators of businesses operating in Slovak heat industry. We can find these indicators applying the selected controlling tools. The paper pointed out new methods and techniques in controlling of performance, especially with a focus on business performance improvement. EVA indicator is appropriate measure for business performance evaluation and instrument of performance controlling, which connects all functional areas of business.

Similarly, this indicator expresses the impact of internal and external business risks. It's flexible, market-oriented indicator, which is strongly correlated with capital market.

ACKNOWLEDGEMENTS

This paper was prepared within the grant scheme VEGA no. 1/0887/17—Increasing the competitiveness of Slovakia within the EU by improving efficiency and performance of production systems and grant scheme VEGA no. 1/0791/16—Modern approaches to improving enterprise performance and competitiveness using the innovative model— Enterprise Performance Model to streamline Management Decision-Making Processes.

REFERENCES

Antošová, M. et al. 2014. Assessment of the Balanced Scorecard system functionality in Slovak companies. *Journal of Applied Economics and Science,* 9(1): 15–25.
Economic Value Added 2012. [on-line] www.dominanta. sk.
Fisher, J. 1992. Use of Non-Financial Performance Measures. *Journal of Cost Management,* 6(1): 1–8.
Freiberg, F. 1996. *Finančný controlling.* Bratislava: Elita.
Gallo, P. 2013. *Management and controlling analyses.* Prešov: Dominanta.
Gallo, P. & Mihalčová, B. 2016. Knowledge and use of the Balanced Scorecard concept in Slovakia related to company proprietorship. *Quality – Access to success,* 17(151): 64–68.
Gavurová, B. 2011. Balanced Scorecard in corporate governance. *Economic journal,* 2011(2): 163–167.
Grell, M. & Hyránek, E. 2012. Maticové modely na meranie výkonnosti produkčných systémov. *Ekonomika a management* 5(1): 73–88. [on-line] www.ekonomie-management.cz/download/1379590470_6eb0/2012_01+Maticove+modely+na+meranie+vykonnosti+produkcnych+systemov.pdf.
Horváthová, J. et al. 2016. *Meranie a hodnotenie výkonnosti podniku.* Prešov: Bookman.
Hyránek, E. et al. 2014. *Nové trendy merania výkonnosti podniku pre potreby finančných rozhodnutí.* Bratislava: Ekonóm.
Ittner, C. et al. 2003. Performance implications of strategic performance measurement in financial services firms. *Accounting, Organizations & Society,* 28(7/8): 715–741.

Kaplan, R.S. & Norton, D.P. 2007. *Balanced Scorecard. Strategický systém měření výkonnosti podniku.* Praha: Management Press.

Kocmanová, A. et al. 2013. *Měření podnikové výkonnosti.* Brno: Litera.

Mussing, W. & Faschlang, A. 1999. Operatives controlling II – Finanzcontrolling – Liquiditätsanalyse/ Liquiditätsmanagement. *Controlling:* 110–144.

Pavelková, D. & Knápková, A. 2009. *Výkonnost podniku z pohledu finančního manažera.* Praha: LINDE.

Petřík, T. 2009. *Ekonomické a finanční řízení firmy.* Praha: Grada Publishing.

Synek, M. et al. 2007. Manažerská ekonomika. Praha: Grada Publishing.

Šofranková, B. & Čabinová, V. 2016. Aplikácia vybraných moderných nástrojov merania podnikovej výkonnosti. *Exclusive journal: economy and society and environment,* 4(3): 60–68.

Štefko, R. & Gallo, P. 2015. Using Management Tools to Manage Network Organizations and Network Models. In Sroka, W., Hittmár, Š. et al. *Management of Network Organizations Theoretical Problems and the Dilemmas in Practice:* 249–264. Switzerland: Springer International Publishing.

Manager's ability to make the right decisions in industrial enterprises

H. Hrablik Chovanová & D. Babčanová
Faculty of Materials Science and Technology in Trnava, Slovak University of Technology in Bratislava, Trnava, Slovak Republic

S.A. Firsova
Engineering and Economics Faculty, Izhevsk State Technical University of the Name M.T. Kalashnikov, Izhevsk, Russian Federation

J. Samáková
Faculty of Materials Science and Technology in Trnava, Slovak University of Technology in Bratislava, Slovak Republic

ABSTRACT: The purpose of this paper is to emphasize managers' function to decide correctly. It describes managers' personal features which influence the course of right decision making; most frequent barriers of effective problem solving and most often appearing mistakes in decision making. When decision-making, managers need to have the relevant information and they need to have them in the "right quantity and quality." The paper stresses the importance of mathematical programming in developing of management thinking. According to current knowledge, mathematical programming (one field of the operational analysis) has got a direct relevance as a toolbox for solving optimization problems (methods are being used in economics, medicine, insurance sector, banking and architecture). In close relation to the system approach it has helped to optimize managerial thinking—in theory and in practice, too. This logic modelling provides a very organized approach for solving decision problems.

1 INTRODUCTION

At present, managers should not only rely on their own experience and intuition in managerial decision-making (one of the main important competences in 21st—century (Suciu 2014)). In today's rapidly changing globalized environment where managers face uncertain (turbulent) market conditions; demanding customer requirements and more aggressive market competition, the right decision is no longer own judgment of many years experiences, but it has to be based on correct current information on one hand (which is available in due time, in right quality and location) and science-based on the other hand.

Corporations look for employees—managers, who are resistant to stress and make decisions quickly—only those right of course. (Smolag 2015).

2 MANAGERS AND THEIR DECISION-MAKING

"Managers are people who are responsible for decision-making and implementation of decisions in the organization with the purpose to co-ordinate activities of subordinates to meet organizational objectives." This means that the manager is the person who directly supervises one or more people in the organization and guides them in order to achieve organizational objectives. (Sedlák 1997).

The decision-making process is influenced by certain personality traits as abilities, temperament, interpersonal skills, experience and age, gender and others. Looking through current researches, the most widely featured personality traits or tendencies of the managers' function are as following (Míka 2006):

1. The way of thinking. Some managers excel in analytical thinking, i.e. in a convergent way of thinking, which aims to target definition, when the judgment derives from one another; when looking for relationships between variables; when verifying thesis and looking for serious arguments and drawing logical conclusions. Another way of thinking represents a creative thinking based on divergent thinking, which departs from the logical connections by bringing new ideas, looking for unusual new solutions, new targets. When comparing these two types of thinking, it is appropriate to use the first way of thinking in a stable situation. In the

situation of unexpected changes it is better to apply the second type of thinking.

2. Risk-taking tendency. In taking decisions, there is rarely a situation where all the variables are known influencing the processes of decision-making. Thus, the manager in decision-making process always takes into account a risk that the result will be more or less different than expected. Although individual decisions may be based on more clear-known variables, and less reliance on estimation or prediction of the variables that impact is unclear. Optimistic decision-makers without major concerns are at risk—are more likely to risk.

3. Susceptibility to doubt. To some extent, it should be a part of managerial thinking (Mrvová 2013), because doubting can bring us to a new argument or new evidence. Yet impact has got the importance, psychological or existential meaning of the decision. Doubting may be of positive impact when there are several seemingly close variants in decision-making process.

4. Ability to lead decision-making team is an important assumption for effective decision-making in the case of poorly structured decision-making problems. In such a situation is important to involve more experts when to reap the benefits of group decision-making, such as in particular:

– opportunity to consider more views on the issue of decision-making;
– application of broader set of knowledge;
– taking into account several aspects of the issue of decision-making;
– opportunity to develop a variety of creative approaches in the team;
– collective assessment of possible options for solutions.

Head of the decision-making team must be able to exclude the impact of a sense of higher status or dominant people, and at the same time to eliminate unwanted conformism and the idea of the necessary unity.

5. Barriers to effective decision making. An ideal situation can be referred to a state when a manager or a decision-making team:

– have all of relevant information at the right time;
– conditions are relevant for an objective assessment;
– evaluation of the information is consistent;
– clear objectives are set;
– nothing prevents to reach set objectives.

However, the reality is based on other better facts. Managers have different levels of knowledge and experience, the ability of intuitive decision-making, ability to lead the team, ability to support creative thinking, etc. Based on today's reality there is mostly less relevant information at the right time available and the resources are also limited. Valuation of the information is not clear—inconsistency evaluation is influenced by differences in preferences between decision-makers, and between representatives of different departments or other bodies (shareholders, managers, employees, unions, customers, the public, environmental challenges, etc.)

The most common barriers to effective problem solving include in particular (Nôllke 2003):

– deficiencies in the quantity, content, relevance and accessibility of information (usually the number is small, incomplete, late acquired, rapidly obsolescent);
– errors in perception (in perception of reality; content of information; in perception of the problem), negative impact of personality traits, lack of experience, personal preferences and attitudes, prejudices, stereotypes, emotional relationship with the originator of the information to the object information, inadequate presentation of information, etc.;
– shortcomings in thinking—limited memory, slow processing of information, inability to assess information needs (including the effects of fatigue), relationships with their own decisions (or, conversely revaluation of their doubting, fear of risk taking;
– negative effects of a group—group thinking, conformity;
– organizational impacts—interference from superiors; adaptation to the interest of some workplaces; impact of different working operations; impact of the remuneration system (support of creativity and respect of rules and standards;
– lack of time (Saniuk 2010);
– stress—caused by: an extreme urgency; number of tasks; lack of information; the importance of the problem; contradictory goals, etc.

Therefore managers' errors of judgment have their objective and subjective reasons. In order to avoid subjective reasons of wrong decisions, the managers have to overcome some of their personality tendencies. The most common mistakes in decision-making are as following (Míka 2006, Nôllke 2003):

1. Dealing with simple problems. On one hand problems are relatively easy to solve, the result is easy to follow, and on the other hand, significant problems remain unsolved, which produces the following error.
2. Postponing decisions. An expression of responsibility for the results, but also the fear to

embark on solving a complex problem often associated with the belief that the problem will somehow solve itself.

3. Hasty decisions. The opposite of the previous errors, resulting mostly from the overvaluation of managers' abilities, judgments and conclusions; sometimes the inability to assess the problem with the participation of more competent people, or for fear of weakening their own position ("I'm good, because I know how to quickly decide").

4. Overvaluation of intuition, feelings, and emotions. Feelings and emotions can sometimes be over valued due to the rationality of the decision.

5. Inability to determine the complex nature of the problem. Inability to separate irrelevant information from relevant. This can lead to a loss of an overview, to the selection of inadequate criteria and thus inappropriate variants of problem solving.

6. Putting effort to less important problem. Inadequate use of time and other resources. As a consequence, for solving of important problems remains less time; the solution of small problem is then disproportionately expensive.

7. Reliance on the recommendations of experts indicates own uncertainty. The manager decides on the recommendation of various commissions and experts, however, they may have other preferences, other experiences, and other criteria. They may largely refer only to certain aspects of the solution.

3 MANAGERS AND OPERATIONAL RESEARCH

Operational research isn't just another management buzz-word—it's been around since the 1940s. But in our cost-conscious, productivity-driven age, operational research professionals continue to find new ways to use operational research to increase revenues and profits, streamline processes and save organisations big money. (The guide to Operational research)

According to current knowledge, mathematical programming (one field of the operational analysis) has got a direct relevance as a toolbox for solving optimization problems (methods are being used in economics, medicine, insurance sector, banking and architecture (Łegowik 2015)). In close relation to the system approach it has helped to optimize managerial thinking—in theory and in practice, too. This logic modelling provides a very organized approach for solving decision problems (Witkowski 2011).

For example, it is often positively evaluated approach, which helps managers to define (www.fsi.uniza.sk):

– understanding of the acceptable solutions in the context of decision-making variants;

– understanding of optimization, i.e. selection of the best option according to the predetermined decision criterion (purpose function, objective function);

– the need to know (if it is permissible content) how to transform and reduce aims or purpose solutions to simplify the understanding of the optimization of objective function;

– understanding the pitfalls of management solutions that have multiple targets (set of objectives, a hierarchy of objectives, incorrectly defined or arranged targets, etc.) and where the solution of decision has to be selected (pitfalls of the simplifications, problems of vector optimization, etc.);

– understanding the problems of sensitivity of optimal or suboptimal solutions to change the default assumptions (e.g. sensitivity analysis of the structure of an optimal solution to the restrictive conditions or objective function, valuation of changes of restrictive conditions by a system of "shadow prices");

– understanding of how often it is difficult, even impossible for some economic considerations or plans to quantify and fit into the abstraction model of classical decision problem (e.g. the impact of psychology and actions of people in general, the effects of the uncertainty of surrounding world).

The prosperity of enterprises in developed economies depends on:

– timely and correct decision-making of the top management in obtaining financial (Bestvinová 2011), raw material and material flows,

– work efficiency of individual departments and employees in production (Fidlerová 2014),

– inventory management and distribution,

– efficient use of fixed capital (machinery and equipment),

– location of customers and transport,

– environment in which is enterprise located, as well as other factors (http://ep.tuke.sk/pdata).

Managers may in certain situations choose an option according to their specific feeling that their decision is exactly the best one. (Bednár 2013). Based on various surveys/studies it is needed that especially leaders/managers of several types of organizations (large, small, private, public, non-profit, etc.) cannot decide by such a feeling. The recommendation is to structure their "problems", to find the significance (value) of acquired/existing data, so they are able to model complex systems so they can make better decisions with less risk.

When decision-making, managers need to have the relevant information and they need to have

189

them in the "right quantity and quality." Decision making does not depend on available information only, it also depends on knowledge, skills and experience of "decision maker", which enable need of less information in problem solving in cases they met similar problem already and their decision that time was correct, or in case of an incorrect decision feedback was afterwards successfully used.

During the survey most important features of information (Fig. 1) were being ascertained, which are (Hrablik Chovanová 2014, Szabo 2008):

– timeliness (16.67%),
– completeness (16.12%),
– availability (12.93%),
– reliability (11.16%),
– clarity (10.06%),
– controllability (8.68%),
– truthfulness (6.20%),
– relevance (5.65%),
– flexibility (4.27%),
– accuracy (4.27%),
– exactness (2.34%),
– low costs (1.65%).

From Figure 2 we can see that more than three fourths of managers consider available amount of information for decision is sufficient for them. Some of surveyed managers feel that part of given information is not relevant for them and it unnecessarily loads them. Results of another survey are interesting, because they report that additional

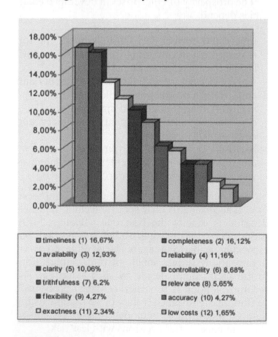

Figure 1.   Most important features of information in% (Szabo 2008).

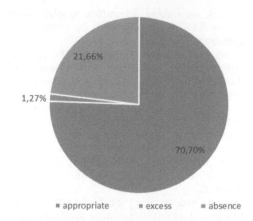

Figure 2.   Amount of information for making a decision (Szabo 2008).

information must not bring a better decision. (Hrablik Chovanová 2014, Szabo 2008).

The survey (Szabo 2008) has been detecting also the way how managers use information systems (Table 1).

According to A. Nowicki, the basic functions of the information system (IS) include (Kuzdowicz 2014):

– collecting data, which is achieved through recording of the data concerning economic events which occur in the enterprise and its environment and affect its operation,

– data processing, which consists in performance of a variety of operations (arithmetic, logic, indexing or sorting) on initial data in order to achieve output information,

– data warehousing, which is aimed at storing data that allows for using them in the future and storage of a copy in case of damage or loss of the original; additionally, the data are stored in read-only data storage devices,

– data transfer, which includes information and communication processes which occur between a sender and recipient of the data, among which there are the processes of ordering, transformation and adaptation of data to proper information channel,

– searching for and presentation of data: this function consists in reading users' queries concerning performance of a specific task, collecting necessary data for the formulation of answers to the question and presentation of this answer.

Based on survey answers (Szabo 2008), 91.07% of companies have got an IS, but they are using it in a small extent for support of decision making (some managers are not using it at all).

Table 1. Availability and use of IS to support decision-making processes. (Szabo, 2008).

Type of information system	IS are available (no. of managers)	IS are used by (no. of managers)	Relative use of IS (use/availability in%)
Enterprise information system (PIS)	217	213	98.04
Managerial information system (MIS)	91	75	82.81
Decision supporting system (DSS)	45	39	87.50
Group decision supporting system (GDSS)	13	12	88.89
Expert system (ES)	18	11	61.54
Executive information system (EIS)	10	7	71.43

Based on further similar surveys, possible reasons of incorrect decisions made by managers are as following (Hrablik Chovanová 2014, Szabo 2008):

- low possibility of finding whether industrial enterprise uses all its opportunities in management and deciding,
- insufficient availability or abilities of suitable "decision makers",
- communication problems,
- in some cases still remain insufficient support of industrial enterprise management,
- lack of compelling vision/strategy,
- IT/IS is not used for supporting of decision making,
- no use of "simulation environment" in decision making; simulation is still being done on real models however it would significantly decrease costs in incorrect decisions,
- "decision makers" are not trained enough for methods for supporting decision making, vast majority of them decides "only" on their own experience and knowledge.

4 CONCLUSION

By the process of managing and making decisions (Bajdor 2012) are always present people. "The quality of the decision" depends on managers who participate directly on decision making process.

Therefore, the preparation of their staff for decision making process is the most important step (especially top managers and middle management level) in their practice. The staff should be able to make such decisions, which will not "damage" the enterprise but will increase business success in today's turbulent competitive global markets.

In order to make the right decision, it is necessary to have right information available, the right amount of information and all of them at the right time. This "type" of information provides support system for decision making. Managers should use the right information in order to make right decisions. Appropriate selection of operational analysis method (most often applied in practice are linear programming, network analysis and structural models) will help them to choose optimal solution under given conditions.

ACKNOWLEDGEMENT

The contribution is sponsored by VEGA MŠ SR No 1/0367/15 prepared project "Research and development of a new autonomous system for checking a trajectory of a robot" and project KEGA No 006STU-4/2015 prepared project University textbook "The means of automated production" by interactive multimedia format for STU Bratislava and Košice. Special thanks to the leader of the project VEGA 1/2601/05 "Development of company management and decision making processes in conditions of integration" doc. Ing. Ľuboslav Szabo, CSc. for access to research results of his survey (results arc interpreted with no direct attendance of authors of the article within given research).

REFERENCES

Bajdor, P. 2012. Comparison between sustainable development concept and green logistics—the literature review. In *Polish journal of management studies*. Volume 5, pp. 236–244.

Bednár, R. et al. 2013. Implementation procedure of Lean methods in logistics processes. *In METAL 2013: 22nd International Conference on Metallurgy and Materials. May 15th–17th 2013, Brno, Czech Republic.* Ostrava: TANGER.

Bestvinová, V. et al. 2011. Financial management of small and medium sized enterprises in Slovakia during financial crisis. In *Annals of DAAAM and Proceedings of DAAAM Symposium.* pp. 1029–1030.

Fidlerová, H. et al. 2014. Application of material requirements planning as method for enhancement of production logistics in industrial company. In *Applied Mechanics and Materials: Novel Trends in Production Devices and Systems.* Vol. 474, pp. 49–54. ISSN 1660–9336.

Hrablik Chovanová, H. & Babčanová, D. 2014. Meaning of quantitative methods in managerial decision making. In *Časopis výzkumu a aplikací v profesionální bezpečnosti*, Information on: http://www.bozpinfo.cz/josra/josra-02-03-2014/kvant-metody-rozhodovani.html>. [online 03.03.2017].

Kuzdowicz, P. et al. 2014. The role of reverse logistics in creating added value in metallurgy. In *METAL 2014: 23rd International Conference on Metallurgy and Materials. May 21st - 23rd 2014, Brno, Czech Republic. Proceedings*. Ostrava: TANGER. pp. 1953–1958.

Łęgowik-Świącik, S., 2015. Evaluation of decision-making processes with reference to cost information management. In *Polish journal of management studies*. Information on: http://pjms.zim.pcz.pl/PDF/PJMS112/Evaluation%20of%20Decision%20Making%20Processes%20with%20Reference%20to%20Cost%20Information%20Management.pdf. [online 03.03.2017].

Metódy ekonomickej analýzy—skriptá. Information on: http://ep.tuke.sk/pdata/11195/documents/metody_ekonomickej_analyzy__pomocne_materialy_/metody_ekonomickej_analyzy-skripta.doc. [online 03.03.2017].

Metódy operačného výskumu. Information on: http://www.fsi.uniza.sk/ktvi/leitner/2_predmety/OA/00_Metody_operacneho_vyskumu.pdf. [online 03.03.2017].

Míka, V.T., 2006. Základy manažmentu. Virtuálna učebnica. Vybrané kapitoly pre študentov externého štúdia, Žilina: FŠI ŽU. Information on: http://fsi.uniza.sk/kkm/files/publikacie/mika_ma.html. [online 03.03.2016].

Mrvová, Ľ. & Púčiková, L. 2012. Using methods of CBA in the context of CSR, focusing on the social projects. In *Applied Mechanics and Materials: 3rd Central European Conference on Logistics (CECOL 2012), November 28–30, 2012, Trnava, Slovak Republic*. Vol. 309. pp. 177–184.

Nôllke, M. 2003. *Rozhodování. Jak činit správná a rychlá rozhodnutí*. Praha: Grada.

Sedlák, M., 1997, Manažment, Bratislava: Elita.

Smoląg K. et al. 2015. Contemporary conditions of engineers education process management. In *Polish journal of management studies*. Information on: http://pjms.zim.pcz.pl/PDF/PJMS112/Contemporary%20Conditions%20of%20Engineers%20Education%20Process%20Management.pdf. [online 23.03.2017].

Suciu M.C. & Lacatus M.L. 2014. Soft skills and ecomonic education. In *Polish journal of management studies*. Information on: http://www.pjms.zim.pcz.pl/files/Soft-Skills-and-Economic-Education.pdf. [online 02.03.2017].

Szabo, Ľ. et al. 2008. Informačné zabezpečenie rozhodovania v podnikovom manažmente, in: *Acta oeconomica et informatica*. Information on: http://spu.fem.uniag.sk/acta/download.php?id=466. [online 23.03.2017].

The guide to operational research, http://www.scienceofbetter.co.uk/or_executive_guide.pdf. [online 03.03.2017].

Witkowski, K. 2011. The innovativeness in logistics infrastructure management of the city for sustainable development. In *Skuteczność w biznesie. Współpraca Terytorialna w Euroregionie Pro Europa Viadrina*: Tom III. Rozwój regionów transgranicznych, Gorzów Wielkopolski. pp. 141–159.

New Trends in Process Control and Production Management – Štofová & Szaryszová (Eds)
© 2018 Taylor & Francis Group, London, ISBN 978-1-138-05885-9

Building competitive advantage

D. Hrušovská & M. Matušovič

Faculty of Business Management, University of Economics in Bratislava, Bratislava, Slovak Republic

ABSTRACT: This article focuses on the analysis of the sustainability of the competitive advantages associated with understanding the changes in customer behavior and expectations. The competitiveness of enterprises is determined by a combination of factors. We define and analyze the four factors affecting the intensity of competition in a given market. The paper presents competitiveness research of enterprises in Slovakia, which is based directly on the Barometer24. Data collection, currently conducted by authors of this paper and cooperating universities from Poland, Slovakia, Czech Republic, Finland and Spain. We analyzed relations on the sample of 660 companies operating in Slovakia. Based on an analysis we suggest draft of model of Factors Competitive Advantages (FCA).

1 INTRODUCTION

Strategy to build competitive advantage must be a priority for each company. During the information age, the impact of the new opportunities of the information communication technology, it is important to clearly identify your market the uniqueness of its competitive advantages. Basing on the previous research and their results described in the literature it is possible to reckon that competitiveness is often associated with product price, quality, resource productivity, costs of production, as well as competitive advantage (Flak & Glód 2014). However, competitiveness can be also understood as an ability to compete. Companies compete in order to survive market competition (Flak & Glód 2012). It may as well mean the ability and a way of coping with competition. One can define competition mechanisms and tools for a long-term and a short-term perspective (Lombana 2006). The strategic approach, as stated by Porcu (2012) is explained by determining opportunities for planning marketing activities and development of businesses that will create long-term competitive advantages. The company competitiveness is a set of features of the company which are multidimensional and which let describe a strength of the company to compete in a market sector (Olszewska & Piwoni-Krzeszowska 2004). If an enterprise fails to define its uniqueness is not entitled to long-term presence in the market. Term strategy to build competitive advantage is closely linked to long-term goals that can be characterized as a future state firms that are trying to reach business respectively meet, through its existence in its activities. The basic condition of building competitive advantage is that it is under active strategic management. The tactical approach is concerned with the short-term activities that will help to fulfil the strategic marketing objectives of the company. Kotler (2001) holds the view that the strength of integrated marketing is also based on the cooperation of particular company's departments while aiming at meeting the customers' needs (product management, sales, advertising, marketing research), but other departments also have to plan their concepts in order to meet the customer needs.

The success of competitive strategy depends largely on the knowledge of strategic competitors is what resources they have available to rivals, and are able and willing to invest in their competitive advantage. The strategy aims to achieve a market share of the company includes offensive actions that lead to obtaining certain benefits and a better position in the market, also includes defensive moves existing venture aimed at defending its current position. The competitiveness of enterprises is determined by various factors with the successful enterprise it can be assumed that competition always has a certain combination of factors. Find the universal combination of factors that would ensure the success of businesses in the market, which would be valid for each business entity is complex. A combination of factors that make the company successful for a variety of industries for a variety of large enterprises, as well as any undertaking specific. The most frequently occurring factors of competitiveness are presented in the following table:

Internal factors:

- Innovation activity,
- The ability of flexible adaptation to customer requirements,
- Production quality,

- Access to finance,
- Lower labor cost and other costs,
- Qualification of staff,
- Name, trademark.

External factors:

- Bargaining power of buyer and supplier,
- Competitive battle,
- Intends to enter into employment in the enterprise,
- Corruption environment,
- Support from state bodies and public authorities.

Companies must be able to create a sufficiently high value with low costs compared with competitors, allowing them to gain a competitive advantage. Competitive advantage is a unique position in the company, thanks to which it differs from the competition. Competitive advantage is not only the strength of the company, but is the result of the strengths of the functional areas of business such as production, marketing, human resources, finance and the like. Competitive advantage is manifested in the ability to product design, develop, manufacture, but also to sell at favorable prices or in greater amounts than competitors. Most competitive advantage is temporary, that is only for a certain time until given a competitive advantage does not imitate competitors. According to Porter (1990), there are two basic types of competitive advantage. It is a focus on low cost or differentiation. The enterprise to gain competitive advantage, the individual value chain activities realized through lower cost than competitors and differentiate themselves from the competition, so differentiation. An important feature of industry are competitive forces that are operating in the sector. When analyzing competitors should focus on business analysis:

- Current position in the market over competitors,
- The profitability of the industry compared to other comparable industries,
- Critical life stages of products in the industry,
- The availability, quality and prices of substitutes,
- Changes and evolution of competition, entry of new competitors into the industry, the departure of current competitors, the strategy and positioning of competitors.

Drivers of competition in a particular industry describes Porter (1990). This model shows that the strategic position of an undertaking operating in a particular market affects five fundamental factors. Competitive rivalry can take on varying degrees of intensity, which is influenced by several factors: the number and strength of competitors. Rivalry in the environment where it operates a large number of equally strong businesses is greater than oligopolistic rivalry in a monopoly environment. Competitive strength can be measured using *Her-findal-Hirschman Index,* which measures the evolution of the absolute concentration of the sector or an index concentration that is easier.

$$HHI = \sum_{i=1}^{n} xi^2 \qquad (1)$$

where n = number of all enterprises in the sector; x_i = share of $_i$th on the overall performance of the sector in %.

It can take values from 0–10 000 points:

0–1 000—concentrated sector
1 000–1 800—moderately concentrated sector
1 800—highly concentrated sector

The ratio of the concentration—CR = sum n strongest market share in the sector of enterprises (recommended n = 2, 4, 6).

2 METHODOLOGY

Crucial to building a competitive advantage is to understand the changes in customer behavior and expectations. Porter (1994) emphasises that a competitive advantage cannot be understood unless we link disciplines that play a role in its establishment (marketing, production, management and control, finance…). Therefore competitiveness is related to evaluation of competitors´ business outcomes, as well as their abilities to gain future profits in a changeable micro-environment (Bossak & Bienkowski 2004). We deal with the company competitiveness as the set of company´s multidimensional features, determining its ability to compete (Flak & Glód 2012).

Competitivenes has some constitutional elements, which are also included in the following Cimpetitiveness Integrated Model (Flak & Glód 2014), shown in Figure 1.

We apply this model through Barometer 24 research tool. It´s based on the questionnaire method, reflecting directly on the mentioned five integrated competitiveness model areas: Competitive potential (12 questions), competitive strategy (10 questions), competitive advantage (8 questions), positioning (6 questions), and the competition platform (9 questions). The answers given by respondents were expressed on a 5-step scale from very low (1pt.) to very high (5pt.).

The research tool is available on http://www.barometer24.org. Methodology of the Barometer24.org has two independent algorithms to calculate the results. The theoretical aims of the project are as follow:

a. To verify relations between the company competitiveness elements in the context of different countries and market sector (the examples of verification see: Flak & Głód 2014).

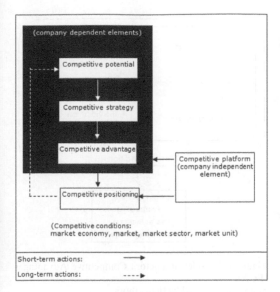

Figure 1. The integrated model of company competitiveness.
Source: Elaborated by the authors.

b. To assess the most important factors which influence on each element of the company competitiveness in different countries and market sectors.

Methodology of the Barometer24.org has two independent algorithms to calculate the results. The first is used for statistical calculations using the collected data from the respondents. The second algorithm is used to indicate the respondent, after completing the questionnaire, the degree of competitiveness of the company he or she represents is ready to be checked out automatically.

The second algorithm deserves attention because its operation can be checked immediately after completing the questionnaire on Barometer24.org. Additionally, it concerned the method of calculating the measure of the competitiveness of the audited company (from 0 to 300 points, the more points the company is more competitive).

The participants of the project will be some Universities from the previously selected European countries that will be coordinating research among companies in their countries. In this way, there would be the possibility of analyzing the results of research not only on the national level, but also on the international level concerning many European countries.

3 RESULTS

We define and analyze the four factors affecting the intensity of competition in a given market technology, pricing, creativity, individual offer for customer and flexible working hours. On the basis of the analysis we can conclude:

Technology we review depending on the extent to which the last five years has changed the technology used in your company. Technology remained unchanged at all, has not changed too much, has introduced major changes, it has changed a lot, or has undergone a radical change. We can conclude that the level of profitability increased with changes in technology. For a factor if technology not changed too much, or there were minimal changes, the profitability Increases. In the case when technology has undergone a radical change profitability even decreasing. The highest profitability is in the case that technology changed too much. Detailed results are displayed in the Table 1.

The answers given by respondents were expressed on a 5-step scale from very low (1pt.) to very high (5pt.).

Price: When analyzing how often a customer can negotiate the price, we can conclude that even if the price does not change the profitability is there. The highest profitability we can see if you can negotiate a price rarely, then it gradually decreasing. High profitability is always where customers can negotiate the price. Profitability is also suitable when the price never changes. If the price changes rarely or occasionally or often profitability is high. Profitability declines in the case that the price can vary each time. Detailed results are displayed in the Table 1.

Table 1. Frequency of profitability and rentability of equity by competitive advantages.

Competitive advantages		Profitability					Rentability of equity					SUM
		1	2	3	4	5	1	2	3	4	5	
Technology	1	-	2	3	2	1	-	-	5	2	1	8
	2	-	-	8	12	6	-	2	9	13	2	26
	3	-	-	13	11	8	-	3	15	14	-	32
	4	-	2	9	23	6	-	-	14	24	2	40
	5	-	1	4	3	4	-	2	3	4	3	12
Pricing	1	-	-	6	4	5	-	-	10	3	2	15
	2	-	-	8	23	4	-	4	12	18	1	35
	3	-	2	14	11	9	-	2	14	17	3	36
	4	-	-	6	11	5	-	1	4	16	1	22
	5	-	3	3	2	2	-	-	6	3	1	10
Kreativity	1	-	-	2	2	2	-	-	4	1	1	6
	2	-	-	3	3	1	-	4	2	1	-	7
	3	-	2	15	22	8	-	2	22	22	1	47
	4	-	-	9	16	7	-	1	11	18	2	32
	5	-	3	8	8	7	-	-	7	15	4	26
Individ. offer for client	1	-	10	20	7	-	-	2	17	16	2	37
	2	-	5	27	31	18	-	5	29	41	6	81
Flexible working hours	1	-	-	4	2	-	-	1	3	2	-	6
	2	-	-	12	21	4	-	3	14	20	-	37
	3	-	2	6	10	7	-	1	9	13	2	25
	4	-	2	14	16	9	-	2	17	18	4	41
	5	-	1	1	2	5	-	3	4	2	2	9

Source: Research results by authors.

Creativity: At very low creativity the gain is minimal. It starts to rise when creativity is suitable or is high, the profit is high as well. When the creativity at very high level profit drops by half. As for the dependence rentability of equity and creativity, if creativity is high and suitable, rentability of equity is appropriate. At the high and very high creativity are gradually decreasing profits. Detailed results are displayed in the Table 1.

Individual offer for the client: If we examine the relationship of profitability and return on capital from the point of view as to whether a firm makes the bid for client individually, we can conclude that higher profitability is when the offer to create individual client. Even if we do not create individual offer profit is high. The rentability of equity is higher when we create individual offer. Detailed results are displayed in the Table 1.

Flexible working hours: We examine what extent has your business a benefit while employees having flexible working hours. Even at low possibilities to use flexible working hours profitability is high, then falls and the high possibility of profits is high again with possibility having flexible working hours. With maximum use of flexible working hours is the minimum profitability. Profitability is high already at low possibilities to use flexible working hours, than slightly declining and afterwords is very high again if there is a high possibility of using flexible working time. Detailed results are displayed in the Table 1.

The comparable results are in percentage. The results are displayed in table 3. The technology in

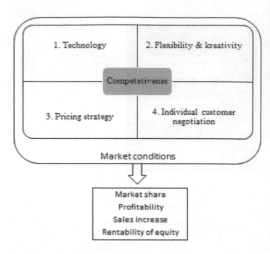

Figure 2. Model of Factors Competitive Advantages (FCA).
Source: Elaborated by the authors.

last five years has changed a lot (33.9% in scale 4). Pricing has changed rarely (29.7% in scale 2), the creativity (39.83% in scale 3), individual offer for client (68.64% yes) and flexible working hours for employers (34.74%) there is a high possibility of using flexible working time.

In previous parts were presented links between some competitive advantages mostly focused on technology, pricing, creativity, individual offer for the customer and flexible working hours and profitability and rentability of equity. For better understanding of these patterns we created the Model of Factors Competitive Advantages (FCA), shown in Figure 2. This model links all these significant competitive advantages in one formula.

4 DISCUSSION

Porter (1994) emphasizes that a competitive advantage cannot be understood unless we link disciplines that play a role in its establishment (marketing, production, management and control, finance...).

Any organization that wants to succeed and survive in the market, it must have customers. Customers demand value products, based on its needs for this value to decide. The value of goods or services that are provided to customers is not only the result of the efforts of producer, respectively service provider, but also all those who are with the product or service connection. The new market conditions force managers to develop new strategies, build sustainable competitive advantages and expand their product portfolio and market segments in order to survive, to ensure sufficient company performance and become competitive (Dědková & Blažková

Table 2. The average and order of competitive advantages.

Competitive advantages	Average	Order
Creativity	3,55	1.
Technology	3,18	2.
Flexible working hours	3,08	3.
Pricing	2,8	4.
Individual offer	1,68	5.

Source: Research results by authors.

Table 3. The percentage of competitive advantages.

Competitive advantages	Scale from 1 to 5(%)				
	1	2	3	4	5
Technology		6,78	22,03	27,12	33,9	10,17
Pricing	12,7	29,7	30,5	18,63	8,47
Creativity		5,08	5,93	39,83	27,12	22,04
Individual offer	31,36	68,64	...	...	...
Flexible working hours		5,08	31,36	21,19	34,74	7,63

Source: Research results by authors.

2014). When the analysis of internal and external environment, you can choose one of three basic approaches to build competitive strategy:

– Find the company for such a position in the industry to be given their resources can most effectively defend against competitive forces existing files.
– Affect the balance of competing forces through strategic measures to improve the position of the opponent.
– Anticipate changes in factors affecting the competitive forces and respond to the new strategy, rather than competitors.

Competitive advantage is formed by integrating exceptional abilities in several functional areas of business. Competitive advantage arises with integration capabilities in several functional areas of business that are linked and-activated capabilities. Outwardly, it may exhibit as a higher rate of profit growth in market share, growth of average wages of employees, or employee benefits or high-status managers etc. For the creation of competitive advantage must be favorable internal and external conditions. External conditions are price fluctuations in raw materials, changes in exchange rates, tightening environmental standards and so forth. Their changes will result in a competitive advantage only in case they have different effects on businesses as a result of their different resources and capabilities in a relatively stable industry with almost as powerful businesses are generally found small competitive advantage, profitable differences are slight.

Internal conditions that are important for the development of competitive advantage are represented by innovation. In the context of business innovations are in addition to new products and production processes and new approaches to the implementation of business and new ways of competing. Innovative business strategies have become a means to achieve excellent results, sometimes more than technical innovation. Size competitive advantage depends on whether the resources and capabilities are valuable and relevant. If resources are freely available to any company in the industry, then they are not a source of competitive advantage. Sustainable competitive advantage depends on the sustainability of resources, the capabilities on which it is based, as well as the ability of competitors to imitate successful strategy through the acquisition of resources and capabilities that form the basis of competitive advantage. Creating a competitive advantage is generally the object of competitive attacks. Forms of competition or weakening the competitive advantage associated with supply and demand in the market. Possible forms of competition:

Competition across the market (supply and demand). For this form of competition is charac-

terized by the fact that manufacturers are interested to sell the produced with the greatest profit, consumers want to turn to the greatest extent possible to meet their needs by buying products or services at the lowest price. Steady state occurs when the offer is equal to demand.

Competition on the demand side. It is a competition among consumers. This form of competition is reflected in periods when demand exceeds supply, that is a condition that occurs at the time of shortage of goods in the market. Competition among consumers leads to price increases.

Competition on the supply side. It is a competition between manufacturers who are trying to maximize profits and strengthen its market position. In the period when supply exceeds demand there is one competitor effort to oust the other from the market.

A competitive company is therefore one which could in the long term, in carrying out various activities, income and expand while maintaining its position in the domestic and foreign markets.

The objectives of focusing on long-term sustainable competitiveness:

– *High quality products.* One of the most important goals of the company is to produce high quality products that are made by special technology, distinct from the competition. To provide not only quality but also the services associated with the product by understanding the client's needs. Focus on customer-oriented culture and a long duration of customer relationships.
– *Conviction customers about product quality, increase product awareness.* Another major objective of the company is to convince customers of the quality of products. For the basis of this objective, the company is planning various promotions, which will be in direct contact with your potential customers. On promotions company will present its product lines, through ensuring adequate information about production program products, in order to give potential customers with information about their quality. Potential customers have the opportunity to persuade about the quality of product demonstrations and tasting buffet, where live and direct a particular product can be tested and submitted feedback that is used to analyze customer satisfaction.
– *Pricing products for lower income groups.* Another objective of the company is fixing prices of products suitable for lower income groups. The price is also adapted for customers with lower incomes, in order to cover a larger market share and more customer segments.
– *Strengthening competitiveness.* Another of marketing objectives is to strengthen the company's competitiveness. Thus, expansion of product lines similar competitors and promote them on the market so that what most impressed poten-

tial customers and distract them from the competition, expand product lines and sales in the international market, expand sales of gift collections through the website.

- *Efficient use of resources, implementation of cutting-edge production technologies.* A competitive advantage is also affects by the dimension range. We are talking about segmented scale, vertical scale, geographical scope and sectorial scope. Value chains within the segment range may be different for different segments for highly demanding customers and customer service-oriented. The same value chain for customers to purchase luxury products and conventional products is different. This is an adaptation to a particular segment of the value chain and its use among segments. When the vertical range of activities are divided between the company and its suppliers, distributors and buyers to what extent an activity held firm itself, or enter an independent company. Company can contract to ensure endurance own service organization. Geographical scope allows you to place, research and development centers and manufacturing in one country and serve several countries. Competitive advantage can be identified in any activity, service, technology development joint procurement of common inputs etc.

5 CONCLUSIONS

The main goal of this paper was to identify patterns in competitiveness research—Barometer 24, currently conducted by authors of this paper and cooperating universities from Poland, Slovakia, Czech Republic, Finland and Spain.

To achieve our goal we analyzed relations between Barometer24 questionnaire questions, focused on company competitiveness, on the sample of 660 companies operating in Slovakia. For better understanding of these patterns we created the Model of Factors Competitive Advantages (FCA), shown in Figure 2. This model links all these significant competitive advantages in one formula. Competitive advantage arises from the value that the company is able to create for its customers. It may take the form of lower prices for identical products than competitors, or providing special benefits that make up the higher price. For successful operation of the enterprise in market environment, it seeks to identify high competitiveness in strategic plans already in research, development and marketing, continued relationships with suppliers and supply strategy.

Competition is also essential for the functioning of an efficient market mechanism. The competition works on the market. The competition is a priority part of the market balance and the healthy functioning of the economic system. In terms of this system it acts as an incentive element of competition which effectively encourages businesses.

It should reflect the fact that the environment for business is changing, as is the method of achieving a competitive advantage. In the conditions of globalisation the world is more and more connected and is still more predisposed to unbalanced states.

ACKNOWLEDGEMENT

The contribution is part of the project outputs VEGA No. 1/0546/15 Evaluation of the performance of modern marketing approaches and their impact on business performance.

REFERENCES

Barney, J. 1991. "Firm resources and sustained competitive advantage". *Journal of Management*, 17: 99–120.

Carney, M. 2005. Corporate Governance and Competitive Advantage in Family controlled Firms. *Entrepreneurship: Theory and Practice*, 29(3), 249–265. DOI: 10.1111/j.1540-6520.2005.00081.x

Dědková, J. & Blažková, K. 2014. The competitive environment among companies in the Czech art of Euroregion Neisse-Nisa-Nysa. *E+M. Economics and Management*, 17(3), 86–99. ISSN 1212-3609. DOI: dx.doi.org/10.15240/tul/001/2014-3-008

Flak, O. & Głód, G. 2014. Barometr Konkurencyjności Przedsiębiorstw. Wyniki badań empirycznych. "Przegląd Organizacji", nr 1.

Flak, O. & Głód, G. 2012. Konkurencyjni przetrwają, Difin

Lombana, J.E. 2006. Competitiveness and Trade Policy Problems in Agricultural Export, University of Gotingen

Navarro, A. et al. 2009. Implications of perceived competitive advantages, adaption of marketing tactics and export commitment on export performance. *J. World Bus*, 45 (1): 49–58.

Olszewska, B. & Piwoni-Krezeszowska, E. 2004. Partnerstwo z klintami szansa zwiekszenia konkurencyjności przedsiebiorstw. In: A. Szplit (Ed.), *Przedsiebiorstwo I region w zjednoczonej Europie*. Kielce: Wydawnictwo Akademii Swietokrzyskiej.

Porcu, L. et al. 2012. How Integrated Marketing Communications. *Journal of Advertising Research*, 47 (3), 222–236.

Porter, M.E. 1990. The competitive advantage of nations. New York: Freee Press.

Porter, M.E. 1994. Konkurenční výhoda. Výhoda vyšší schopnosti konkurence. Praha: Victoria Publishing.

Power, D. 2010. The difference principle? Shaping competitive advantage in the cultural product industries. *Geografiska Annaler. Series B, Human Geography*, 92(2), 145–158. doi:10.1111/j.1468-0467.2010.00339.x.

Moral decision-making of business managers

J. Hvastová

Theological Faculty of Košice, Catholic University of Ružomberok, Ružomberok, Slovak Republic

ABSTRACT: This paper focuses on the problem of interaction between business motives and the ethical perspective of an individual approach. The research reflects the attitudes towards special ethical problems regarding individual value systems and decision making. The sample consists of 167 business school students who were questioned and then divided into 6 types of managers. Three scenarios were administered to them. They regard topics such as manipulation of prices as compensation for loss, empathy versus efficiency, and bribery as charity. Then four moral scales were measured (justice, relativism, utilitarianism, and egoism). Consequently, some individual factors (such as age, political orientation, and religiousness) were tested. The results confirm that some of the individual factors are related to moral decision making not excluding the types of managers. The most of the significant correlations are found among future managers labeled as judges and teachers.

1 THEORETICAL BACKGROUND

1.1 Ethical behavior in business

The business world is basically built in gaining maximum financial profit. Profit maximization is as the core drive of a lot of business organizations. Still many firms understand it as the principal motivation in their management and the ultimate measure of their success. This is why it is necessary to incorporate ethical and moral values into the business world.

Managers have a direct responsibility basically consisting in making as much money as it is possible because they are mere employees of the stockholders. This assumption states that any use of resources for social purposes implies the imposition of an "illegal type of tax on the company" (Friedman 1970).

It is possible to identify two kinds of behaviour of companies regarding the profit maximization (Khandwalla 2004): 1. "greedy kind"—is the one where ethics, corporate social responsibility, and the concern for stakeholders other than the promoters do not matter; 2. "responsible kind"—is subject to ethical restraints, appropriate company's social responsibility, and a balanced concern for the needs of all the stakeholders, not just the promoters. The responsible kind is considered the future of civilization.

1.2 Socially responsible behavior

Modern theoretical and empirical research indicates that business firms can strategically engage in socially responsible behavior to increase private profits (Hernández-Murillo & Martinek 2009). Corporate social responsibility is a commitment to behave ethically and to increase the quality of life of employers and people of the local community. At the same time, it means that the company contributes to the economic growth and protects the environment (Mihalčová & Jeleňová 2013). Social responsibility of business encompasses has these four components: economic, legal, ethical, and discretionary expectations that society has of organizations at a given point in time. These four basic expectations reflect some of the definitions offered earlier but they categorize the social responsibilities of businesses in a more exhaustive manner (Carrroll 1991).

It is clear that there is much to do in the area of ethics and morals in business. It is of a great importance to consider not only the economic background of what is happening in today's world but also the ethical and moral principles which places the problem in a more interpersonal and altruistic position considering motives and actions leading to the general welfare of mankind. We have to mention that the concept of corporate social responsibility is still not well known in Slovakia. One survey which was done among companies of the Eastern Slovakia says that some companies had already been doing corporate social responsibility but they did not call it like that (Ubrežiová et al. 2013).

1.3 Ethical dilemmas

Managers are sometimes faced with business choices that create internal tensions between ethics and profits, or between their private gain and the public good. Robinson & Yeh (2007) distinguish two perspectives when ethical issues appear in business. The first is the social perspective when it is related to virtue ethics and the second is the legalistic perspective when the ethic is rule—based. Any decision where moral considerations are relevant can potentially give rise to an ethical dilemma. An ethical dilemma is a kind of problem, situation, or opportunity that requires an individual, group or business organization to choose among several possible actions that must be evaluated as right or wrong, ethical or unethical. According to Geva (2006), an ethical dilemma is one of the four moral problems which are: genuine dilemmas, compliance problems, moral laxity, and no-problem problems.

This paper focuses on how future managers are capable to solve certain ethical ambiguities and what factors are related to their decision making.

2 METHODOLOGY

2.1 Quantitative approaches in business ethics

Researchers in business ethics are faced with two main challenges. First, it is necessary to develop adequate language and theories on which to base the discussion and research. Second, quantitative approaches to the study of ethics in business must be developed to gain the respect of other researchers in areas that are more numbers oriented (Bay 2002). The following research is a quantitative approach to studying decision making in certain ethical dilemmas. The measurement of the ethics of managers is based on the Multidimensional Ethics Scale and managers in the research sample are labeled according to the MoralDNA approach into six categories.

2.2 Multidimensional ethics scale

In the 1990s, an empirical approach emerged in business sciences that relied on the Multidimensional Ethics Scale (MES). This was elaborated by Reidenbach & Robin (1990) and was based on a survey of moral philosophy literature. The authors of this psychometric instrument identify five normative modes of moral reasoning: justice, relativism, utilitarianism, deontology, and egoism.

The problem of justice was already discussed in the writings of Aristotle who promoted that equals should be treated equally. Relativism, on the other side, presents an idea that ethical rules are not universal. It means that each society or an individual has its own ethics, values, and rules. The deontological attitude determines only what is correct while the teleological attitude attempts to determine what is good for humans. Finally, the

rightness of an action in terms of consequences is viewed in egoism and utilitarianism. The egoism defines "right" behavior in terms of its consequences for the individual. The utilitarianism seeks to achieve the greatest good for the greatest number of people. In the presented research, only four of the scales are taken into account. Namely, justice, relativism, egoism, and utilitarianism. They are measured on the 8-point scale.

2.3 MoralDNA

MoralDNA is an online psychometric instrument which measures two dimensions of human morality: ethical perspectives (or how we prefer to make moral decisions) and moral values (what moral values we prefer to consider when making moral decisions). The first version of MoralDNA was elaborated in 2008 and it was based on Kohlberg's idea of moral development regarding moral decision making. An interesting thing about the instrument is that it does not make a positive or negative value judgment on people and their decision making. It is rather descriptive which means that it merely shows differences and similarities between measures.

According to the authors (Steare et al. 2014) of the instrument, there are three ethical perspectives when making moral decisions: obedience (which is labelled as "law"), care (which is labelled as "love"), and reason (which is labelled as "logic"). The extent, to which each of the perspectives influences our moral decision making, varies among people and depends also on the particular situation. Moral decisions, based on obedience or disobedience (ethic of obedience), exist early in childhood and are closely related to consequences (reward and punishment). In the workplace, the ethics based on obedience are driven by laws, rules, and regulations. The ethic of care (labelled as "love") results from making decisions based on empathy and is related to the well-being of the individual and the group. The ethic of care is crucial to any human community. It is also crucial to real engagement with all stakeholders in any enterprise and in economic terms, it is crucial to the fair distribution of scarce resources. However, the ethic of care is often suppressed and ignored. Finally, the ethic of reason (labelled as "logic") is based on critical reasoning of the individual to make the right choice although it would be irrespective of the rules or the consequences. There is certain relationship between the ethic of obedience and the ethic of reason: the more people blindly comply with rules, the less they think logically about their actions.

The three ethical preferences explored by MoralDNA provide a good model for the process of ethical decision-making. However, it has been

Table 1. Types of managers and their preferences for ethics.

Type of manager	Preferences for ethics		
	First	Second	Third
Philosophers	Logic	Care	Obedience
Judges	Logic	Obedience	Care
Angels	Care	Logic	Obedience
Teachers	Care	Obedience	Logic
Enforcers	Obedience	Logic	Care
Guardians	Obedience	Care	Logic

Source: Elaborated according to STEARE, R. et al. 2014. *Managers and their MoralDNA. Better Values, Better Business*, pp. 9–10.

argued by some scientists, especially philosophers and psychologists, that moral values are the drivers of these ethical preferences. Therefore, MoralDNA examines the influence of a set of ten moral values, which are wisdom, fairness, courage, self-control, trust, hope, humility, love, honesty, and excellence. These values provide a more detailed insight into the moral composition of a specific individual. The choice of these values is based on both academic research and practical experience working on ethical decision-making with organizations.

Managers who complete the MoralDNA survey are identified as one of six character types based on their preferences for the ethics of obedience, care, and reason. Thus, the character types are defined according to the priority of these three ethical perspectives.

The authors of the instrument then studied some relationships and correlations between measured variables and other factors. For example, they identified the effect of age on the three ethics at work. They also compared the intensity of measured moral values between genders. In the proposed questionnaire they asked managers about their religious beliefs and political preferences (left or right) and they identified some relationships regarding both ethical perspectives and moral values. There were also other factors taken into consideration such as leadership styles or membership in organizations. Generally speaking, the instrument is easy to use and practical in deriving outcomes.

2.4 Scenarios

In the research, three scenarios are used as different situation to measure the ethics of managers. To reflect the reality in business in Slovakia, three topics were chosen. The first of them focuses on the problem of minority gipsy communities and doing business with them. The second one touches

the problem of employment—unemployment. The third one involves the question of bribery. All three scenarios were developed independently and do not copy any of scenarios used in other test methods.

Scenario 1: A food industry company has one of its butchery shops located very close to a gipsy settlement. That is why the shop is not doing well and is confronted with different troubles and unpredictable expenses. Due to this the shop manager always increases the prices of the product during the periods of time when the habitants of the settlement are given unemployment pay.

Scenario 2: A firm is heavily affected by recession and has to reduce the number of employees. A productivity analysis of employees says that they should dismiss a long-time worker who was often absent because of health problems with his family member. If the employer lets him be, a young and very competent employee will have to be sent away. Therefore, the manager decides to dismiss a long-time worker.

Scenario 3: A director of a private company thinks of withdrawing from an agreement with one of his suppliers because he sees a better and financially preferable one. However, the supplier asks to meet the director and gives him a financial gift. The director changes his mind and does not withdraw from the agreement. He accepts the financial gift but he decides to use it for charitable activities.

After each of the scenarios there are scales confronting the respondent while asking whether the decision of the actor was just or unjust, acceptable or unacceptable, harmful or useful, empathetic or apathetic. The answers represent four scales: justice, relativism, utilitarianism, and egoism. The scales are measured on 8-point interval and are computed as the mean of all three scenarios.

2.5 Research sample

A number of university students of a business school were chosen as respondents. The students were between 19 and 35 years of the age. There were 47 men (28.1%) and 120 women (71.9%). The average age was 22.65. Before the students were asked the questions on scenarios, they were asked about their political orientation and religious practices. Both of the characteristics were evaluated on a seven-point scale. The lower the point the more the political orientation of the student inclines to the left. Similarly, the lower the point the less the student practices his/her religious faith. The overall characteristic of the sample shows that in the selected group there are students mostly neutral regarding the political orientation or slightly to the right (the mean is 4.22 on the seven-point scale). As to the religious practices, the majority of the students are practicing their religious faith (with the mean at 5.94 on the seven-point scale).

3 RESEARCH RESULTS AND DISCUSSION

3.1 *Research results*

The tables show how certain factors such as age, political orientation, and practicing religion are related to scales of justice (JUS), relativism (REL), utilitarianism (UTI), and egoism (EGO). The

Table 2. Type of manager: Angel. *Correlation matrix.*

N = 25	JUS	REL	UTI	EGO
Age	0.233	0.002	−0.197	−0,103
Political orient.	−0.133	−0.134	0.169	−0,298
Practicing religion	0.357	−0.386	−0.328	−0,102

Pearson (Spearman)[1] coefficient (correlation is significant at the 0.05* or 0.01** level).

Table 3. Type of manager: Guardian. *Correlation matrix.*

N = 24	JUS	REL	UTI	EGO
Age	0.274	−0.207	−0.056	−0.208
Political orient.	0.086	−0.166	−0.239	−0.183
Practicing religion	0.160	−0.149	−0.039	−0.216

Pearson (Spearman)[1] coefficient (correlation is significant at the 0.05* or 0.01** level).

Table 4. Type of manager: Philosopher. *Correlation matrix.*

N = 13	JUS	REL	UTI	EGO
Age	0.260	−0.462	−0.370	−0.264
Political orient.	−0.072	0.427	0.167	−0.298
Practicing religion	0.466	−0.587*	−0.379	−0.274

Pearson (Spearman)[1] coefficient (correlation is significant at the 0.05* or 0.01** level).

Table 5. Type of manager: Judge. *Correlation matrix.*

N = 41	JUS	REL	UTI	EGO
Age	0.217	−0.359*	−0.334*	0.152
Political orient.	−0.184	0.226	0.250	0.413**
Practicing religion	0.335*	0.265	−0.137	−0.291

Pearson (Spearman)[1] coefficient (correlation is significant at the 0.05* or 0.01** level).

---

1. For the variables Political orientation and Practicing religion the Spearman coefficient of correlation was used because these two variables do not have normal distribution. For other variables the Pearson coefficient of correlation was used because they have normal distribution.

Table 6. Type of manager: Teacher. *Correlation matrix.*

N = 20	JUS	REL	UTI	EGO
Age	0.271	−0.564*	−0.227	−0.469*
Political orient.	−0.104	−0.069	0.071	0.055
Practicing religion	0.408	−0.156	−0.040	−0.453*

Pearson (Spearman)[1] coefficient (correlation is significant at the 0.05* or 0.01** level).

Table 7. Type of manager: Enforcer. *Correlation matrix.*

N = 44	JUS	REL	UTI	EGO
Age	0.353*	−0.197	−0.228	−0.128
Political orient.	0.029	−0.198	−0.220	0.041
Practicing religion	−0.064	0.256	0.309*	−0.121

Pearson (Spearman)[1] coefficient (correlation is significant at the 0.05* or 0.01** level).

tables are designed according to the types of manager: philosopher, judge, angel, teacher, enforcer, and guardian.

There are no significant correlations in groups of angels and guardians. However, among philosophers there is a strong and statistically significant negative relationship between practicing religion and the scale of relativism. This means that those philosophers who are more practicing religion are also less relativistic while solving ethical dilemmas.

In the group of judges all the independent variables show some significant relationships. Namely, it is the age of judges which is negatively related to the scales of relativism and utilitarianism. Those judges—managers who are older are also less relativistic and less utilitaristic. In addition, those judges who are right—oriented in their political orientation are also more egoistic in their decision making while solving ethical dilemmas in business. Practicing religion among judges also means that they are more just.

Among managers—teachers the scale of egoism is negatively and statistically significantly related to the age and practicing religion. This means that older and more religious teachers show less egoism. There is also negative significant correlation between the age and relativism. Thus, older teachers less relative in their ethical decisions in business.

Finally, among enforcers, the results suggest that older managers are more influenced by justice while deciding in ethical dilemmas. An interesting finding is that the more the managers—enforcers practice their religion the more are utilitaristic in their decision making.

Generally speaking, we may conclude that there are certain differences among types of manager regarding their decision making related to some specific factor such as the age, political orientation, and practicing religion. This strengthens the idea of better preparation of those who will in the future be subjects of decision in situations which are to some extent ambiguous in their ethical sense. The variability of their personality and their personal attitudes is also an important factor that has to be considered.

3.2 *Discussion*

One of the reasons why the ethics research in business world is interesting is that managers face extensive competitive pressure which might predispose them to participate in unethical behaviour. There are, of course, other reasons but the effect of competitive environment on ethical sensitivity is evident (Phatshwane 2013). While it is natural for business environment to be competitive it is impossible to change it. However, it is possible to explore the personality of managers and to examine how their personality is related to their decision making. Their way of thinking and deciding becomes apparent when they are confronted with problematic situations such as ethical dilemmas in business. As it is obvious from the presented research, different types of managers react differently to the situations. It would be even more interesting to explore how they speak about morally problematic situations or how they would argue for their decisions. The more profound reasons of the managers are related to their kind of morality they produce (Kujala 2003).

Ethical principles are as important in managerial work as in any other area of life. However, ethical principles of managers act together with other factors. As Weber & Wasieleski (2001) found in their research, significant differences were identified when considering the context of the ethical dilemmas, managers' type of work, and industry membership. According to them age was not a significant factor. In our study, not only age but also political orientation and practicing religion emerge as significant factor among some types of managers.

Unethical behavior of managers may have various negative consequences ranging from a kind of damage to the organization's reputation or lack of public trust to the resignation of some staff members (Hiekkataipale & Lamsa 2016). We have to admit that the research done among students has its limitations because they are not in a real state of employment and they do not bear responsibilities. In real situations, moral reasoning of managers might be influenced by avoiding responsibilities, relationship problems, hidden agendas and others (Hiekkataipale & Lamsa 2016). However, it is necessary to educate business students since their time

at university to increase permanently their moral intelligence as the moral intelligence has a significant and positive effect on business performance (Hosseini et al. 2013).

4 CONCLUSION

The paper investigated some factors which are related to moral reasoning of students of business schools. The preliminary assumption was that every student—future manager has different kind of personality and therefore also unique moral principles. Thus, the research sample was stratified per certain types of managers. Some personal factors were identified as statistically significant while decision making in ethical dilemmas. It is obvious that it is necessary to explore the problem of moral reasoning in business in larger proportions of research samples. Future research must consider various factors related to situational decision making by presenting managers—participants with multiple moral problems and ethical dilemmas.

REFERENCES

Bay, D. 2002. A critical evaluation of the use of the DIT in accounting ethics research. *Critical Perspectives on Accounting* 13(2): 159–177.

Carroll, A.B. 1991. The Pyramid of Corporate Social Responsibility: Toward the Moral Management of Organizational Stakeholders. *Business Horizons* 34(4): 39–48.

Friedman, M. 1970. The Social Responsibility of Business is to Increase its Profits. *The New York Times Magazine*, 13. September 1970: 122–126.

Geva, A. 2006. A Typology of Moral Problems in Business: A Framework for Ethical Management. *Journal of Business Ethics* 69(2): 133–147.

Hernández-Murillo, R. & Martinek, CH.J. 2009. Corporate Social Responsibility. Can Be Profitable. *The Regional Economist* (April): 4–5.

Hiekkataipale, M.-M. & Lamsa, A.-M. 2016. The Ethical Problems of Middle Managers and Their Perceived Organizasational Consequences. *Transformations in Business & Economics* 15(3): 36–52.

Hosseini, S.A. et al. 2013. The Effect of Managers' Moral Intelligence on Business Performance. *International Journal of Organizational Leadership* 2(2): 62–71.

Khandwalla, P.N. 2004. Management Paradigms Beyond Profit Maximization. *Vikalpa – The Journal for Decision Makers* 29(3): 97–117.

Kujala, J. 2003. Understanding Managers' Moral Decision-Making. *International Journal of Value-Based Management* 16(1): 37–52.

Mihalčová, B. & Jeleňová, I. 2013. Východická spoločenskej zodpovednosti podnikania. *Marketing manažment, obchod a sociálne aspekty podnikania; Zborník recenzovaných príspevkov z 1. medzinárodnej konferencie, 2013:* 363–373. Košice: Podnikovohospodárska fakulta so sídlom v Košiciach, Ekonomická univerzita v Bratislave.

Phatshwane, P.M.D. 2013. Ethical Perceptions of Managers: A Preliminary Study of Small and Medium Enterprises in Botswana. *American International Journal of Contemporary Research* 3(2): 41–49.

Reidenbach, R.E. & Robin, D.P. 1990. Toward the Development of a Multidimensional Scale for Improving Evaluations of Business Ethics. *Journal of Business Ethics* 9(8): 639–653.

Robinson, D.A. & Yeh, K.S. 2007. *Managing Ethical Dilemmas in Non-Profit Organizations.* http://

epublications.bond.edu.au/business_pubs/11. Accessed 25. February, 2017.

Steare, R. et al. 2014. *Managers and their MoralDNA. Better Values, Better Business.* London: Chartered Management Institute.

Ubrežiová, I. et al. 2013. Perception of Corporate Social Responsibility in companies of Eastern Slovakia region in 2009 and 2010. *Acta Universitatis Agriculturae et Silviculturae Mendelianae Brunensis* LXI(7): 2903–2910.

Weber, J. & Wasieleski, D. 2001. Investigating Influences on Managers' Moral Reasoning. *Business & Society* 40(1): 79–110.

New Trends in Process Control and Production Management – Štofová & Szaryszová (Eds)
© 2018 Taylor & Francis Group, London, ISBN 978-1-138-05885-9

RFID as innovation for improvement of manufacturing plant performance

E. Hyránek
Faculty of Business Management, University of Economics in Bratislava, Bratislava, Slovak Republic

A. Sorokač & B. Mišota
Institute of Management, Slovak University of Technology in Bratislava, Bratislava, Slovak Republic

ABSTRACT: Radio-Frequency Identification (RFID) has been successful verified by different application of different industry in the long term. Therefore, if we need to solve improvement of traceability of material flow in manufacture process or within hole logistic chain, one solution is implementation of technology RFID. Main advantage given of RFID technology is currently product portfolio, which is offered by leading companies of industry automation means. Key factor of concept Industry 4.0 is enough quality information. Material flow in manufacture process or specify technology operation is describe by moving elements, so that RFID. Article deal with case study of alternative for barcode technology by RFID technology. Implementation of given solution wasn't realized, but with simulation model was verified by specialist from manufacture plant. Output of simulation was assessed by economics evaluation metrics effect on manufacturing plant performance possible.

1 INTRODUCTION

Purpose for increasing efficiency of traceability of material flow is automatic collecting data. If we need to eliminate human factor for identification of elements of moving materials in manufacture process, then we need to use contactless identification (Dai et al. 2015). Technology of contactless identification is solving via RFID (radio-frequency identification). RFID offers wide portfolio of products for industry application with direct implementation to automation means with possibility database process and data analyze (Evangelista et al. 2013). Therefore, this technology is suitable for concept Industry 4.0. Next advantage of RFID is identification of entities in manufacture logistic without limitation speed of moving material (Herrmann et al. 2015). With this advantage we can prevent of bottleneck (Casalino et al. 2012).

Permanently implementation of RFID stimulate demand for this technology and affects to research in this sector, this way expands facilities of RFID. Minimalization dimension of electronic elements allow applications at high temperature, high dust or waterproof with many alternatives of fix to track's element. RFID tag can be extended of temperature sensor, sensor of humidity or other physical value.

From attribute describe above, RFID is using not only in industry, but in logistics, health care, aerospace, apparel, energy sector, defense and retail.

Increasing of overall productivity of the company is main factor enabling the achieving long-term growth of the company sectors (Tekulová et al. 2015). The main contribution of RFID implementation is increasing efficiency of economic processes. Modelling and simulation of economic processes and phenomena takes place gradually in the all sectors (Janáková & Zatrochová 2015).

Despite the lowering trend of price RFID technology can be enforced investment strenuous, for that is one way verification rentability via simulation model (Zhou et al. 2011). In this article have been created simulation model in software Witness from Lanner Group. Simulation model represents material flow within manufacture of brakes for car. During creation, this simulation model was using barcode technology print on paper for provide traceability. Steps for ensuring information of position or content of tracked entity were join with operator's job of scanned barcode or printed new paper card with barcode. These jobs were specifying as a bottleneck of material flow. From this reason, have been created simulation model for interpretation advantage of used RFID.

2 RESEARCH GOALS

Base on description in introduction was specified goals:

A. Identification types of RFID with limit parameters for implementation within logistic information system
B. Creation of simulation model of currently logistic performance and simulation model with RFID. Define benefits with RFID

Implementation of RFID within LIS presuppose:

– Reduce time demands on operations of positioning of material in production
– Reduce time demands on inventory check or complaint

3 METHODOLOGY OF IMPLEMENTATION OF RFID TECHNOLOGY

RFID use radio waves for transfer information. Transfer information run between RFID tag and RFID reader (Curty, et al. 2005). On RFID tag are written needed information, which reader can read and write. Tag have unique identifier—ID. We recognize two base group of tags:

a. *Passive tag*—tag gains energy via inductive coupling from reader. Main advantage is working without own power supply. On other side, disadvantage is smaller read radius and lower speed of transfer information.
b. *Active tag*—tag has own power supply. Tag has bigger read radius than passive tag. From power supply, we can supply sensors of physical value. Of course, disadvantage is battery, but life of battery may be up to 10 years.

On start of choice type of tag is important define limit parameters:

– Speed of read information. Material flow is performing by elements of logistic, where every element has given speed for performance of logistic. Next, we define speed of transfer (read and write) information between reader and tag.
– Distance between tag and reader. In ideal case, distances are in centimeters. Therefore, is necessary delimit clearance and imprecision of placing between moving material with tag and reader.
– Quantity of information. If we want to use RFID to maximum rate, we need to specify not only information for describe material flow, but also, we should specify information for data analysis of problem's prediction. Technologist

of given manufacture determines set of parameters, which affect to quality of product. On the other side, data's analyst with intelligent and "learner" mathematic model define state for servicing.

After careful consideration of limit parameters with enough reserve we identify type of tag.

4 IMPLEMENTATION OF RFID TECHNOLOGY IN PLANT CONTINENTAL ZVOLEN

Continental design and product brakes for personal cars. System of logistics, warehousing and its transfer is in the plant Continental high optimized. The structure base on request of client and preparatory zones is reliable for several years. Changing the structure of logistics is not necessary, but it is necessary to improve information flow leading to smoother operations in production. Close look at the actual performance of logistics, we can define a few deficiencies:

– Long shifts coordinators and warehouse operators. Manual recording of logistics movement in the SAP system.
– Relatively large discrepancies in inventory.
– The need for a larger workforce for finding material and billing of individual operations.
– The interface between logistics and production is not sufficient.

4.1 *Describe the process of gaining information*

Basic element in the material flow is a container with part of brakes. Each container has paper card with the data about it. Operator after performing the activity must to bill a change, because this operation increases added value of the product. He scans a barcode for write information to logistic system and following system prints a new paper card. With labels (RFID) is not need scan a barcode, because after the operation worker confirms last activity and remaining process of reading and writing of information about product will be automatic.

Automation of loading data about material flow is essential of RFID technology. Implementation of this technology we can be removed from the production paper cards with printed barcodes that traveled with containers during production.

Currently, there are on market a many producer who deliver many structural modifications of RFID tags. In our case is the most appropriate variant active tags with magnetic strip intended for the metal container. So, RFID tag with a magnetic strip will be properly anchored to container.

Table 1. Number of part per month.

Name of parts	Brakes	Holder	Casts	Case
Quantity	9 148	4 602	20 340	9 183
Quantity with RFID	9 233	4 637	21 180	9 295

Table 2. Selected value of machine per month.

Machine	% Idle	% Busy	% Waiting for operator
Holders 1	0.01	98.57	1.42
Assembly 4	0.28	98.49	1.23
Case 1	0.01	97.95	2.3
Galvanization	37.20	31.84	1.26
Model with RFID			
Holders 1	0.01	99.27	0.71
Assembly 4	0.28	99.11	0.62
Case 1	0.01	98.96	1.3
Galvanization	52.50	32.17	0.32

It cannot be fixedly attached on container due to a change container at the end of production. In production, they will be placed RFID reader of tags.

4.2 Consider implementation of RFID technology

In this paper, it doesn't describe detail procedure for define solution with RFID. In short, they were created two simulation models—current and proposed with RFID technology. Then, results were approved by experts from Continental. The simulation model was created by using a software Witness.

4.3 Results of simulation models

Comparison of results of both simulation models are in tables above.

Benefit follow from the tables:

- Average 50% reduction in waiting times for machine operators during production
- Reduce the average time of operator's work (Table 2)
- Increasing the number of input casts (Table 1)

All these benefits are reflected in increase of production of brakes, one container (Table 1) to shift (8 hour).

Simulation directly quantified only one of the major benefit of RFID technology. Other important are:

- Simplify search data during inventory
- Preventing loss of paper cards
- Improving traceability of material flow

Table 3. Innovation costs.

Product	Qtr.	Price [€]
RFID tags	100	1 600
Readers	36	44 000
Operation panel	28	24 000
Readers with PC	2	2 200
Wi-Fi access point	6	300
Database with SW	1	14 000
Installation HW with tests system	1	3 000
Integration to manufacture SAP	1	3 000
Total		92 100

- Improving transparency stores
- Relieving operators in production
  - Saving on paper cards. For one package of brakes consume 10 A4 paper (on one day then it's over 3300 A4)
  - Implementation of RFID technology would not need to stop machines, if so then only minimally

Estimated costs see Table 3.

We assume increase performance of hole logistic process even performance of hole company via substitution currently technology by RFID technology within LIS with time saving of manufacture operations of material positioning, with time saving of inventory check or products complain and with lowering costs of providing material traceability.

We achieved goals and presuppose base on results assessment of simulation and next quantify benefits.

5 CONCLUSION

Concept of Industry 4.0 consider about information in real time. Starting point is quality and quantity of information in real time given by decentral system. In case of logistic, we are talking about logistic information system—LIS. LIS is in Industry 4.0 decentral system, where will be ensure interoperability for exchange information between every control system of manufacture.

Article analyze opportunities of used simulation model as support for assess benefits of implementation RFID to manufacture of brakes. Simulation model results was evaluating by expert of production. Results of simulation model after implementation of RFID showed other benefits, like improve performance of production. Risk of rentability investment was minimalize by demonstrate of quantitative benefits.

Industry 4.0 give us new possibilities in testing by simulation models. Idea is join simulation model and control system, which control and monitor as a decentral periphery technological process. From this control system, we can to have real data which describe process in real time and after transferred real data to simulation model with innovation, we have exact results of LIS in process with new technology with direct impact on improve performance of company.

ACKNOWLEDGEMENT

The civil association EDUCATION-SCIENCE-RESEARCH has supported this paper. Conference paper was prepared within grant task VEGA number 1/0067/15 *"Verification and implementation of modeling company performance in finance decision tools"*.

REFERENCES

Casalino, N. et al. 2012. *ICT adoption and organizational change an innovative training system on industrial automation systems for enhancing competitiveness of SMEs.* Wroclaw, Inst. Syst. Technol. Inf., Control Commun, pp. 283–288.

Curty, J.P. et al. 2005. Remotely powered addressable UHF RFID integrated system. In *IEEE Journal of Solid-State Circuits,* November, 40(11), pp. 2193–2202.

Dai, H. et al. 2015. Design of traceability systems for product recall. In *International Journal of Production Research*, January, 53(2), pp. 511–531.

Evangelista, P. et al. 2013. Technology adoption in small and medium-sized logistics providers. In *Industrial Management and Data Systems*, 113(7), pp. 967–989.

Herrmann, S. et al. 2015. Co-creating value in the automotive supply chain: An RFID application for processing finished vehicles. In *Production Planning and Control*, 26(12), pp. 981–993.

Janáková, H., & Zatrochová, M., 2016. The numerical methods used in the simulation of economic processec. *In Aplimat-15th Conference on Applied Mathematics 2016 Proceedings*, pp. 583–590.

Liu, M. et al. 2015. RFID-enabled real-time manufacturing operation management system for the assembly process of mechanical products. In *International Journal of RF Technologies: Research and Applications*, June, 6(4), pp. 185–205.

Tekulová, Z. et al. 2015. Analysis of productivity in enterprises automotive production. In *Production Management and Engineering Sciences*: Proceedings of the International Conference on Engineering Science and Production Management (ESPM 2015), Tatranská Štrba, High Tatras Mountains, Slovak Republic, 16th-17th April 2015, pp. 545–550. CRC Press.

Zhou, R. et al. 2011. Production logistics system simulation about automobile parts shop based on witness and e-factory. In *Sanya, Trans Tech Publications*, pp. 3175–3179.

New Trends in Process Control and Production Management – Štofová & Szaryszová (Eds)
© 2018 Taylor & Francis Group, London, ISBN 978-1-138-05885-9

Suppression of unfair competition in Poland

K. Chochowski
State Higher Vocational School Memorial of Prof. Stanislaw Tarnowski, Tarnobrzeg, Poland

ABSTRACT: This article presents reflections on the legal aspects of suppressing unfair competition in Poland. The main normative act in this matter is The Suppression of Unfair Competition Act. It defines the concept of the act of unfair competition, identifies the individual types and forms the legal responsibility for commitment of act of unfair competition. In addition, it defines a catalog of claims for entities that have been affected by unfair competition.

1 INTRODUCTION

Competition between business is one of the foundations of a free-market economy. It contributes to the quality of the goods and services offered on the market. Unfortunately, competition on economic level is often unfair and brings harm to both costumer and business. Therefore it is the role of the state to create such a legal framework that would guarantee a fair competition between businesses. Fair competition is crucial for the proper development of both the free-market economy and the entrepreneurial spirit. It is beyond doubt that the state should interfere in the activities of market players, however, the scope of this interference can not be excessive. Legal regulations created by the state have to, at the same time, enable the development of entrepreneurship and eliminate pathological practices form the market. Honesty is universally accepted and desirable value in every sphere of social life, including in the economy.

Here comes the question, how to understand the concept of fair competition? There is no written definition in the Polish legal system. Nevertheless, it is assumed that competition in the legal sense is "(...) the pursuit of independent businessmen (business entity) on the same (common) market to reach the same economic goal, in particular to achieve profits by doing business with suppliers, customers and employees" (Szwaja 2016). Further notion, that it should be fair competition, refers us to ethical standards, which we should understand intuitively. It is clear that spheres of law and ethics penetrate each other in concept of fair competition.

2 PRINCIPLE OF FAIR COMPETITION IN POLISH LAW

The legal system of the Republic of Poland has recognized and appreciated the importance of fair competition, in the Constitution itself; we find rules that are devoted to this issue. It is mentioned in article 76 of the Constitution, according to which public authorities should protect consumers, users and tenants from unfair market practices and activities that threaten their health, privacy or security. The scope of this protection is determined by the normative act. At the level of the legislation in Poland, we have three leading normative acts: The Protection of Competition and Consumers Act, The Counteracting of Counterfeit Market Practices Act, The Suppression of Unfair Competition Act. Due to the wide scope of the discussed issue, I shall limit my deliberations to the last of the aforementioned legal regulations.

3 APPLICATION OF THE SUPPRESSION OF UNFAIR COMPETITION ACT

The Suppression of Unfair Competition Act (hereinafter abbreviated as TSoUC) covers prevention and means of suppressing of unfair competition in business activity, in particular industrial and agricultural production, construction, trade and services—in the interest of society, business and customers (article 1, TSoUC). As we can see, it is directed not only to repression but also to prevention, therefore, to build ethical attitudes among entrepreneurs. Often the preventive function is underestimated, but in the author's opinion, it has an important role in the overall activity of the state and its organs in the context of suppressing unfair competition. Building a proper pattern of behavior in a market-based game is extremely important, as it will help to avoid many pathological events in the future, including those known as unfair competition.

The aim of TSoUC is therefore, as stated in the verdict of the Białystok Appeal Court of 13 November 2014, (I ACa 479/14, OSA 2015, No 5, sec. 12) to eliminate unfair business practices that constitute misuse of competition and lead to deformation and misrepresentation.

The statements of Suppression of Unfair Competition Act refer to business activities. We naturally associate these activities with entrepreneurs, but also other entities can run business i. e.: non-profit organizations, foundations and associations. This means that the act in question applies not only to entrepreneurs but also to other legal entities, if they pursue an economic activity. In the light of Polish legislation an economic activity is, in accordance with Article 4 of The Freedom of Business Activity Act (hereinafter abbreviated as TFoBA), commercial manufacturing, construction, trade, service and search, exploration and extraction of minerals from the fields, as well as professional activity, carried out in an organized and continuous manner.

The leading feature of this business is its focus on profit. Of course, this profit does not have to appear, economic activity can bring a loss, but it is important that the aim of the entity performing this activity was to achieve it. The definition of entrepreneur in article 2 of TSoUC differs from that contained in article 4 of TFoBA. In the first case, entrepreneurs are considered as natural persons, legal entities and organizational units without legal personality, who engage in economic activity by professional or any other gainful activity. In the other case, the entrepreneur is a natural person, a legal entity and an organizational unit without legal personality, whose separate law grants legal capacity—acting on its own behalf and engaging in economic activity. Moreover, as entrepreneurs are also considered partners of a civil partnership in the scope of their economic activity.

Despite some differences in the definition of entrepreneur in the aforementioned laws, it is not problematic to define when we deal with an entrepreneur within the meaning of TSoUC.

Prevention and suppressing of unfair competition is to be done in the interest of society, entrepreneurs and customers. The fairness of trade is in the interest both of the state and of all entities offering and acquiring goods and services. Eradication of pathological phenomena, taking the form of acts of unfair competition is desirable, because it gives competition to entrepreneurs proper form.

4 THE ACT OF UNFAIR COMPETITION

Legislator gives a legal definition of the act of unfair competition in art. 3 sec. 1 TSoUC. It is a behavior that is against the law or good practices if it threatens or violates the interests of another entrepreneur or customer. It is noteworthy that this is also the case when there has been no infringement of the interests of another entrepreneur or customer yet, but it has been threatened. Therefore, it is sufficient, that the infringement is possible, to regarded behavior as unfair competition.

In the Polish doctrine of law, there is no doubt that the thesis contained in art. 3 sec. 1 TSoUC the legal norm is a general clause. It refers to the system of assessments and extra-legal norms, that is "good practices" leaving the assessment of the facts to the authority exercising the law (Sieradzka & Zdyb 2011, Szwaja 2016).

As the act of unfair competition, in accordance with art. 3 sec. 2 TSoUC, is recognized in particular: misleading usage of trademark, false or fraudulent designation of geographical origin of goods or services, misleading designation of goods or services, violation of company secrets, inducing to terminate or non-performance of the contract, imitation of products, mocking or unfairly praising, hindering access to the market, bribery of a person performing a public function, unfair or prohibited advertising, organizing the avalanche sales system and running or organizing activities in the consortium.

The above-mentioned catalog of unfair competition acts is only an example, and therefore the act not explicitly mentioned in Art. 3 sec. 2 TSoUC, and meeting the conditions set out in art. 3 sec. 1 TSoUC, should also be considered as an act of unfair competition. According to the verdict of the Court of Appeal in Katowice of 27 February 2009, (V ACa 308/08, LEX no. 519276), an act of unfair competition may also take place where the perpetrator is unaware that his conduct is contrary to good practices and violates the sphere of another entrepreneur.

Legislator in art. 5–17 d typifies some acts of unfair competition. This includes, for example, designation of a company in a way that is misleading to customers as to its identity, by using the company, name, emblem, abbreviation, or other distinctive symbol previously legally used to designate another company (Article 5 TSoUC); labeling the goods or services with a false or fraudulent geographical indication indicating, directly or indirectly, the country, region or locality of origin or use of such sign in commercial activities, advertising, trade letters, accounts or other documents (Article 8 TSoUC); labeling of the goods or services, or the absence of labeling, which may mislead customers as to the origin, quantity, quality, ingredients, performance, suitability, applicability, repair, maintenance or other relevant features of the goods or services, as well as concealing the risks associated with their exploitation (Article 10 sec. 1 TSoUC); dissemination of false or misleading news of your or any other entrepreneur or company, to gain or cause harm (Article 14 sec. 1 TSoUC); hindering other entrepreneurs from gaining access to the market (Article 15 sec. 1 TSoUC); Organizing an avalanche sale system by proposing the acquisition of goods or services by offering purchasers of those goods or services the promise of obtaining material benefits in return for persuading others to make the same transactions persuading other persons to participate in the system (Article 17c sec. 1 TSoUC).

Due to the immense impact of contemporary advertising on consumers, the legislator paid much attention to this issue. The regulations contained in Art. 16 sec. 1 TSoUC, which indicate examples of acts of unfair competition in the field of advertising. These include: advertisement that is against the law, good practices, or that diverges from human dignity; advertisement that may mislead customers and may affect their decision to purchase a product or service; advertisement referring to the feelings of customers by creating fear, exploiting superstitions or credulity of children; a statement that encourages the acquisition of goods or services, giving the impression of neutral information; advertisement, which is a significant interference in the sphere of privacy, in particular by inducement in public places, sending unsolicited goods at customer's expense or excessive using of technical means of communication. Also, advertisement that directly or indirectly identifies a competitor or goods and services offered by a competitor, known as "comparative advertising", is an act of unfair competition if it is in conflict with good practices (Article 16 sec 3 TSoUC).

Advertising, as the instrument of competition for the customer, as seen above, has been subject to numerous legal requirements, which entrepreneurs have to meet. If they are not met, then such advertisement will most likely be an act of unfair competition. State activity in this matter is, in the author's opinion, fully reasonable; the free market does not mean unrestricted market. Absolutization of freedom leads to anarchization of economic relations and a number of pathologies. Obviously, the scope of this interference can not be excessive and lead to the actual negation of the free market and the freedom of economic activity. A wide variety of examples for unfair competition are designed to help entrepreneurs and clients identify and counteract them.

5   CIVIL LIABILITY FOR ACTS OF UNFAIR COMPETITION

The liability for the act of unfair competition has, in principle, a nature of a civil law and therefore of compensation. Based on Article. 18 sec. 1 TSoUC, an entrepreneur whose business has been threatened or affected by an act of unfair competition may demand: cessation of prohibited activities; removal of effects of prohibited activities; filing one or more repeated statements of appropriate content and in appropriate form; damage repaid on general terms; Return unjustifiably obtained benefits, on a general terms; awarding the appropriate amount of money for a specific social purpose related to the promotion of Polish culture or the protection of national heritage—if the act of unfair competition was committed deliberately. Moreover, at the request of the entitled person, the court may also judge the products, their packaging, advertising materials and other objects directly related to the act of unfair competition. In particular, the court may order their destruction or credit for compensation (Article 18 sec 2 TSoUC). It is significant that in the case of matters concerning the truthfulness of the indications, information on the goods and their packaging or the expressions contained in the advertisement, the burden of proof is transferred to the defendant (Article 18a TSoUC). He must show that the act of unfair competition has not been committed. Adoption of this solution raised some concerns as to privileges of the plaintiff's position. Therefore, in order to slightly balance the position of both parties, it is considered that, if a manifestly unfounded unfair competition proceeding is filed, the court, at the request of the defendant, may require the plaintiff to file a single or multiple statements of appropriate content in the appropriate form. In addition, the defendant, who suffered damage as a result of this action, may claim damages to be repaired under general terms (Article 22 TSoUC). As far as the margin of discussion is concerned, mention should be made that claims for unfair competition are time-barred by the time of three years (Article 20 TSoUC).

The entity entitled to claims referred to in Article 18 TSoUC, is primarily an entrepreneur that suffered damage. However, on the base of Article 19 sec. 1 TSoUC, national and regional organizations whose statutory purpose is to protect the interests of entrepreneurs, also possesses the said legitimization, except that it is of limited character. These organizations can claim; cessation of prohibited activities; removal of effects of prohibited activities; filing one or more repeated statements of appropriate content and in appropriate form awarding the appropriate amount of money for a specific social purpose related to the promotion of Polish culture or the protection of national heritage—if the act of unfair competition was committed deliberately.

Another limitation of the active legitimacy of national and regional organizations, whose statutory purpose is to protect the interests of entrepreneurs in the sphere of suppression of unfair competition, is that they can not claim claims for unfair competition defined in Article 5–7, Article 11, Article 14 and Article 15a TSoUC. It is worth mentioning, that in the light of TSoUC, the organizations referred to above do not have to show that there has been a violation or threat to their interest. The only requirement is to show that their statutory purpose is to protect the interests of entrepreneurs. Unfortunately, TSoUC does not grant individual consumers the power to pursue their claims, even if the act directly affects their interests (Nowińska & du Vall 2008).

It has to be reminded, that according to Article 55 of The Code of Civil Procedure, also a prosecutor can file a claim in cases of unfair competition.

6 CRIMINAL LIABILITY FOR ACTS OF UNFAIR COMPETITION

In addition to civil liability, in TSoUC we can also find criminal liability in some cases marked as criminal offenses (Articles 23–26 TSoUC). Depending on the act of unfair competition, it can take the form of: fines, restriction of liberty, detention or imprisonment for up to 2 years, in case of organizer or manager of avalanche sale system up to 8 years of imprisonment.

By introducing penal liability of entrepreneurs or the unfairness of their actions, legislator postulated in favor of the fair competition on the market and defined its shape in which honesty has a significant place.

It is worth mentioning that foreign natural and legal persons exercise their rights under the TSoUC, based on international agreements binding on the Republic of Poland on the basis of reciprocity.

7 CONCLUSION

Fair competition is one of the key principles of the Polish legal system. There is no doubt that competition is a desirable phenomenon as entrepreneurs competing for a specific market and customer's have to constantly increase the quality of the offered services and goods. One can say that one of the conditions for the proper functioning and development of a free market economy is the fair competition of entrepreneurs. As aptly stated by M. Zdyb "Fair competition should be and attribute of every free economic activity" (Sieradzka & Zdyb 2011). It is not surprising that this issue was in the sphere of state interest. This is because the state determines the rules and conditions of conducting business activity, giving the activity of entrepreneurs desired shape.

It should be emphasized that the law should not be a barrier to entrepreneurship, on the contrary it should create entrepreneurial attitudes and stimulate economic development of the country. It is therefore pleasing that the Polish legislator recognized the importance of the principle of fair competition and put it firmly into the legal order of our country. Ensuring fair competition contributes to the creation of a broadly defined public order in the state, and therefore the role of the state and its organs is to create the conditions for the development of fair competition and its protection

REFERENCES

Nowińska, E. & du Vall, M. 2008. Komentarz do ustawy o zwalczaniu nieuczciwej konkurencji.
Sieradzka, M. & Zdyb, M. 2011. *Ustawa o zwalczaniu nieuczciwej konkurencji. Komentarz.*
Szwaja, J. 2016. Ustawa o zwalczaniu nieuczciwej konkurencji. Komentarz.

New Trends in Process Control and Production Management – Štofová & Szaryszová (Eds)
© 2018 Taylor & Francis Group, London, ISBN 978-1-138-05885-9

Econometric modeling as a support for the process of creating production technology

M. Iskra
Faculty of Business Economics with seat in Košice, University of Economics in Bratislava, Košice, Slovak Republic

O. Krasa
Faculty of Production Engineering and Materials Technology, Czestochowa University of Technology, Czestochowa, Poland

ABSTRACT: Customers often demand a price offer for a product in a very short period of time, so it is necessary to prepare, as soon as possible, the amount of technological hours which a component of technology and the price. The paper presents the possibility of using econometric tools to shorten the time of the process of creating production technology hours for steel constructions. In particular authors presented econometric methods of forecasting the technological time of machining, welding, painting and assembly processes.

1 INTRODUCTION

Globalization of the world economy, international competition in the world market are some of the reasons for improving managing systems by enterprises and making the enterprises maximum effective. Continuous development of science and technology, especially information technology, makes possible to achieve the progress in all the processes in the enterprise e.g. production, distribution of products and services, human resource management, financial resources management (Wrycza 2010).

There is a lot of data and information related to the processes in enterprises. Many enterprises have business management software implemented and therefore it is quite easy to get, save, analyze and report large amounts of data. A high-class well implemented information system is very helpful in the effective management of the enterprise but often there are some parts of business activity which demand additional support (Konishi & Nomura 2015).

The aim of this paper is to present the tool for supporting the process pricing the product. There is often a need to make a reply to the price inquiry in a very short time. The purpose of the tool is to calculate the number of technological hours needed to produce the product which is one of the main factors of product pricing.

2 STEEL CONSTRUCTION MANUFACTURING

2.1 *Characterization of manufacturing processes*

The process of manufacturing steel structures includes many stages of production such as: preparation of components for welding, welding, machining, painting, assembly, interpretational controls and final inspection and packaging. In many cases, there are also processes of thermal treatment and application of protective coatings. These stages, depending on the complexity of the structure are configured in different ways into components or monolithic constructions. The stages are also divided into operations, in the case of welding operations, they are: profile cutting, sheet metal firing and welding phases. The welding process involves operations such as routing, component coupling, component heating, welding, and relaxation. Machining is also divided into: assembly and disassembly operations on machines, alignment, routing, roughing, finishing and control operations. The painting stage includes surface preparation for painting, washing, drying, painting, drying and disassembly of the security. The next step is the assembly and packaging of the construction for shipping (Majerník et al. 2015b).

The standardization of these processes is difficult because of the unit production profile (Orea &

Kumbhakar 2004). Large dimensions of steel construction have higher tolerance, which results in less manufacturing repetition. This dependence also applies to the steel profiles from which the structures are manufactured. Incorporation of tolerance fields of components makes it difficult to maintain consistent geometry for the whole structure.

2.2 Process of pricing

The current market expects a quick response to the inquiry and requires that the cost of construction is not overestimated. This need makes the process of calculating labor costs for companies very important for obtaining orders (Dilworth 1993). It is also important that the calculated labor costs are as precise as possible because of the cost-effectiveness of production (Majerník et al. 2015a). Due to such demand, the need arises for the development of a new calculation method that will speed up and minimize the time consuming workload for individual operations. This paper attempts to use statistical and econometric methods based on multi-year calculations to estimate models or tools that will calculate the amount of technological hours and thus labor cost and price of steel construction based on several characteristics. As indicators for the calculation can take the weight of construction and degree of complexity (Battese et al. 2004).

On the basis of data from the enterprise which is a steel construction manufacturer with seat in Poland, the authors proposed econometric tools for calculating technological hours. The modeling is made for one type of steel construction. For the purposes of calculations, a variable called the degree of difficulty was developed. Based on the customer's technical drawings on welded constructions, patterns were developed on a scale from 1 to 8.

3 ECONOMETRIC MODELING OF TECHNOLOGY FOR A-TYPE PRODUCTS

3.1 Modeling of machining technological hours for product type A

There are many tools for forecasting and simulation in this paper econometric modeling is used (Maciag 2013). Data on the number of machining, welding, painting and assembly technological hours for 28 A-type products were collected. The above variables are the dependent variables in econometric model. The independent variables are weight of the finished product and degree of complexity. Data for independent variables was also obtained. Calculated Pearson correlation coefficients for the variables are presented in Table 1.

Table 1. Pearson correlation coefficients for the variables.

Pearson's r	Weight	Degree of complexity	Machining	Welding
Weight	1.000			
Degree of complexity	0.681	1.000		
Machining	0.894	0.791	1.000	
Welding	0.958	0.711	0.934	1.000
Painting and assembly	0.899	0.811	0.941	0.918

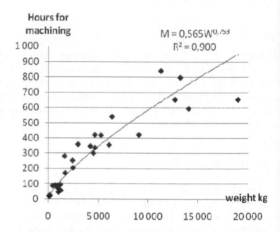

Figure 1. Econometric model for technological hours of machining and weight.

As the first analysis, technological hours of machining (M) were modeled in relation to product weight (W). Distribution of empirical points suggested that power function is the most suitable for the relation between weight and machining hours. See Figure 1.

Below proposed econometric model of power function for the variables (1),

$$M_i = \alpha_0 W_i^{\alpha_1} + \varepsilon_i \qquad (1)$$

where M_i = number of technological hours of machining, W_i = weight of product in kg, α_0 = regression coefficient (constant), α_1 = regression coefficient, ε_i = error.

In order to estimate the model using ordinary least squares method, the model was linearized by multiplying both sides of the equation by ln. See Equation 2 below:

$$\ln M_i = \ln \alpha_0 + \alpha_1 \cdot \ln W_i + \varepsilon_i \qquad (2)$$

Estimated model $M_i = 0.565 W_i^{0.753}$ with high coefficient of determination $R^2 = 0.90$ has got statistically significant coefficient α_1 with p-value = 1.56×10^{-14}, but α_0 is not significant with p-value = 0.157. Therefore another model was build without a constant but with degree of complexity variable—Equation 3.

$$M_i = W_i^{\alpha_1} \cdot C_i^{\alpha_2} + \varepsilon_i \quad (3)$$

where C_i = degree of complexity. After linearization to the form Equation (4) it was estimated.

$$\ln M_i = \alpha_1 \ln W_i + \alpha_2 \cdot \ln C_i + \varepsilon_i \quad (4)$$

Estimated model $M_i = W_i^{0.604} + C_i^{0.470}$ with high coefficient of determination $R^2 = 0.997$ has got statistically significant coefficient α_1 with p-value = 3.07×10^{-23}, and for α_0 p-value = 5.07×10^{-5}.

3.2 Modeling of welding technological hours for product type A

At first, there was distribution of empirical points analysis made, the conclusion was that power function is the most suitable for estimating number of technological hours for welding (K). Distribution of empirical points and the model is presented on Figure 2.

Estimated model $K_i = 0.555 W_i^{0.749}$ with high coefficient of determination $R^2 = 0.94$ has got statistically significant coefficient α_1 with p-value = 8.02×10^{-18}, and α_0 with p-value = 0.048. However another model was build without a constant but with degree of complexity variable—Equation 5.

$$K_i = W_i^{\alpha_1} \cdot C_i^{\alpha_2} + \varepsilon_i \quad (5)$$

After linearization—see Equation 6—the model was estimated.

$$\ln K_i = \alpha_1 \cdot \ln W_i + \alpha_2 \cdot \ln C_i + \varepsilon_i \quad (6)$$

Estimated model $M_i = W_i^{0.604} + C_i^{0.470}$ with high coefficient of determination $R^2 = 0.998$ has got statistically significant coefficient α_1 with p-value = 0.001, and for α_0 p-value = 1.56×10^{-25}.

3.3 Modeling of painting and assembly technological hours for product type A

Distribution of empirical points analysis made for technological hours for painting and assembly was presented on Figure 3, the conclusion was that power function is also the most suitable for estimating number of technological hours for painting and assembly (A).

Estimated model $A_i = 0.91 W_i^{0.595}$ with high coefficient of determination $R^2 = 0.735$ has got statistically significant coefficient α_1 with p-value = 5.64×10^{-9}, but α_0 is not significant with p-value = 0.867. Therefore another model was build without a constant but with degree of complexity variable—Equation 7.

$$A_i = W_i^{\alpha_1} \cdot C_i^{\alpha_2} + \varepsilon_i \quad (7)$$

After linearization—see Equation 8—the model was estimated.

$$\ln A_i = \alpha_1 \cdot \ln W_i + \alpha_2 \cdot \ln C_i + \varepsilon_i \quad (8)$$

Estimated model $A_i = W_i^{0.537} + C_i^{0.276}$ with high coefficient of determination $R^2 = 0.990$ has got statistically significant coefficient α_1 with

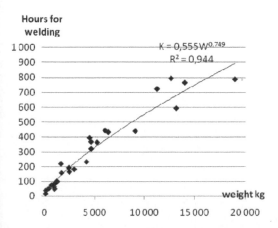

Figure 2.   Econometric model for technological hours of welding and weight.

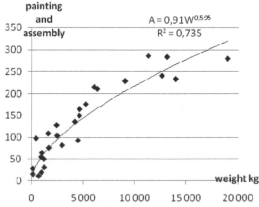

Figure 3.   Econometric model for technological hours of painting and assembly in relation to weight.

p-value $= 1.37 \times 10^{-15}$, and not significant α_2 p-value $= 0.126$.

Although in this model degree of complexity variable is nit significant the authors concluded that other function for this variable should be proposed. Another reason is that there is a need to give a degree of complexity variable for technological hours for painting and assembly.

4 CONCLUSIONS

The main conclusion is that econometric modeling is a great tool for improving or at least for reducing the time of work needed for preparation amount of technological hours. And since technological hours are an important part of pricing the product, the entire pricing process is much shorter. Therefore the company can react quicker for inquiries.

Presented models can be implemented as a part of the process of pricing or as a support for verification the amount of technological hours. In case of a large difference between the hours given by model and by technology department can be the basis for checking the technology more detailed.

Another conclusion is that the authors noticed that degree of complexity variable should be made separately for at least machining, welding, painting and assembly. It can be made for more specific technological operations. There is need to make this variable more objective by calculating on the basis of some specific data from the inquiry. This could be the number of small items of the steel construction in relation to the weight of the construction.

The presented models and conclusions are the basis for further work on this subject.

REFERENCES

Battese, G. et al. 2004. A metafrontier production function for estimation of technical efficiencies and technology gaps for firms operating under different technologies, Journal of Productivity Analysis 21, 91–103.

Dilworth, J 1993. Production and operations management. Manufacturing and Services, James B. Dilworth, ISBN 0-07-016987-X. McGraw-Hill, Inc. 1993

Konishi, Y. & Nomura, K. 2015. "Energy Efficiency Improvement and Technical Changes in Japanese Industries, 1955–2012," Discussion papers 15058, Research Institute of Economy, Trade and Industry (RIETI).

Maciąg, A. et al. 2013. Prognozowanie i symulacja w przedsiębiorstwie. ISBN: 978-83-208-2061-4. PWE. Warszawa 2013.

Majerník, M. et al 2015a. *Process innovations and quality measurement in automotive manufacturing. In Production Management and Engineering Sciences: Proceedings of the International Conference on Engineering Science and Production Management (ESPM 2015), Tatranská Štrba, High Tatras Mountains, Slovak Republic, 16th-17th April 2015* (p. 179). CRC Press.

Majernik, M. et al. 2015b. Integrated management of environmental-safety and technical risks of plants producing automobiles and automobile components. Published in: Communications—Scientific Letters of the University of Zilina info. Volume 17, Issue 1, p. 28–33.

Orea, L. & Kumbhakar, S.C. 2004. Efficiency measurement using a latent class stochastic frontier model, Empirical Economics 29, 169–183.

Wrycza, S. 2010. *Technologie Informatyka ekonomiczna. Podręcznik akademicki.* Warszawa: PWE. ISBN 978-83-208-1863-5.

New Trends in Process Control and Production Management – Štofová & Szaryszová (Eds)
© 2018 Taylor & Francis Group, London, ISBN 978-1-138-05885-9

Analysis of reported information about liabilities in financial statements in Slovakia

Z. Juhászová & M. Mateášová
Faculty of Economic Informatics, University of Economics in Bratislava, Bratislava, Slovak Republic

M. Kicová
Faculty of Business Economics with seat in Košice, University of Economics in Bratislava, Košice, Slovak Republic

ABSTRACT: Accounting provides information on identifying economic facts to their users through the financial statements. Financial statements should be prepared considering the precautionary principle, and should give a true and fair view of the assets, liabilities of an entity, its financial position and results of operations. The primary purpose of financial statements is that the information in the financial statements is useful to users. For users to properly assess the information presented in the financial statements, they are based on quantitative information in the financial statements that are associated with other information presented in the notes to the financial statements (Pakšiová & Kubaščíková 2015). In this article we will focus on the definition of responsibilities among businesses accounting in double entry bookkeeping, further on the reporting structure in the accounts of business entities, misstatements in the financial statements and their impact on the decisions of users of financial statements.

1 INTRODUCTION

Accounting entity during the accounting period is required to collect all facts involving a change in the asset, the source of assets under which income or expense may occur. All these facts are recorded in the accounting, namely account books. Subsequently, the last day of the financial year, the entity prepares financial statement, which presents summarized information on the economic entity with the desired content and in the prescribed format. Financial statement represents one unit and its basic aim is to provide information on the financial position, performance and cash flows that are useful for a wide range of users to decide. Information on financial position of accounting entity is primarily on the balance sheet that is a source of information on the amount and structure of assets, liabilities and equity as of the date on which the financial statement was compiled and the date on which the financial statement was compiled for the previous financial year. Balance sheet of its vertical alignment data for two consecutive immediately successive periods makes already at first sight basic and simple comparison of the development, either in absolute or relative terms, in terms of gain or loss of various forms of asset and their resources. These differences may indicate some consideration and analysis, without regard to other phenomena and processes. Some debt ratios, respectively financial independence are quantified from the structure of liabilities. One of the basic tasks of an entity is determining the correct ratio of own and external sources of financing activities of the entity, i.e. its financial structure concurrently with the determination of the total amount of resources needed. Medium and large entities are common not to finance their assets only by their own equity. However, it is not real for entities to finance their asset only by foreign sources. When exceeding the own resources, the entity is stable, independent and has a greater ability to survive a possible crisis. Indebted entity is indeed stable and less dependent on their creditors, but could do better viability and profitability as foreign capital is cheaper than their own. In this article, we will discuss the recognition of liabilities in the financial statements of businesses with an emphasis on the obligations inaccuracy of their recognition and their impact on decisions of users of financial statements (Tumpach & Baštincová 2014).

1.1 Definition of obligations under the rules of the SR

Pursuant to the Accounting Act, the liability means the existing obligation for an entity (§ 2 par. 4 letter b) Act No. 431/2002 Coll. on accounting, as amended by later regulations), arising from past events, it is possible that future economic benefits will reduce the entity and can be measured reliably. Definitions of various concepts are based on the definition in the International Financial Reporting Standards (IFRS). Obligations meeting such characteristic are recognized in the financial statements on the balance sheet as current liabilities, with equity. According to the nature we can divide the commitments into these basic groups: liabilities from business relationships, liabilities to employees and institutions of social security and health insurance, liabilities resulting from taxes and subsidies, liabilities from partners, members and participants in associations, financial liabilities (bank loans, borrowings) and other types of liabilities. Each group is further divided into more detailed individual types of liabilities. A special group of liabilities are reserves, which are also defined in the Accounting Act as follows *"Reserves are liabilities of uncertain timing or amount."* (§ 26 par. 5. Act No. 431/2002 Coll. on accounting, as amended by later regulations). A reserve is a liability representing the existing obligation of an entity arising from past events, it is possible to decrease in future economic benefits of an entity, while if the exact amount of this liability is not known, it will be evaluated by estimation in sufficient amount to meet the existing obligations of the date on which the balance sheet was compiled taking into account the risks and uncertainties. Reserves are created on the date of preparation of the financial statements and constitute the basis of the precautionary principle, and always for a purpose. They must have a passive balance. The term liability has a direct relevance to the accounting term, which is also defined in the Law on Accounting. Liabilities are sources of assets, which represent the total amount of liabilities of an entity, including other liabilities and the difference between assets and liabilities. The above definitions show that the liabilities are part of the entity's liabilities. Liabilities and other liabilities are obligations of the entity. The definition of other liabilities is mentioned in the accounting law. Other liabilities refer to an obligation of the entity that does not meet the conditions for its accounting ledger accounts and which is recognized in the notes of the financial statements. In other liabilities the most common reason to consider an item other than liability, is the lack of ability to award a particular item.

Interesting here is that conditions of posting are not mentioned in the law at all, only states that the general ledger shall demonstrate posting of all transactions in the accounts of assets, liabilities, the difference between assets and liabilities, income and expenses in accounting period. However, it explicitly does not follow not to find also

another liability in the general ledger because § 12 par. 1 letter b) refers only what must appear in the general ledger.

1.2 *Recognizing liabilities on the balance sheet*

The information in financial statements has to be useful to users; they are assessed in terms of their significance, which should be coherent, comparable and reliable. Information is considered significant if its omission in the financial statements or the misstatement in the financial statements could influence the judgment or decision of a user. Significant symptom is such information that it might be reasonable to assume that the omission or misstatement of such information would affect the decision that users adopt based on the financial statements of the enterprise. When preparing the financial statements, the entity must take into account not only all recorded facts, but also those, which have an impact on its future existence, taking into account whether in the future will be able to carry out their activities and achieve positive economic results (Simonidesová & Manová & Stašková 2015). The structure of assets, liabilities, equity, income and expenses depends upon the legislative framework in accordance with which the financial statements are prepared. In the case of the Slovak Republic it is mainly the highest legislation and the Act No. 431/2002 Coll. on accounting, as amended by later regulations.

Liabilities of an entity are over time divided into long-term liabilities and current liabilities. Long-term liability is a commitment, which the agreed maturity or settlement in any other way of accounting case is longer than one year. Conversely, short-term liability is a commitment, which the agreed maturity or settlement in any other way of accounting case is no more than one year. According to the agreed maturity liabilities are recognized when an accounting case occurs, but on the balance sheet are reported by the remaining period of their maturity as at the date of preparation of the financial statements. The remaining maturity period of liabilities is equal to the difference between the agreed maturity date and the date on which the financial statements is compiled.

2 ANALYSIS OF RECOGNIZING LIABILITIES ON THE BALANCE SHEET

We can often hear the sentence: "The information has the price of gold," because good information is the biggest source of wealth; information as essential resources as well as, for example, tangible fixed assets or employees. Technology diffusion and acceptance of information as goods changes the role of information to feed its significance equivalent to traditional economic resources such as labor, land and capital. Does information of accounting and financial statements also have such great importance? If so, for whom? Is it always the 'right' information? Can users obtain relevant information from this referred in the financial statements, can they read it, compare, do they know what principles have been used, know what individual items represent and what are possible hidden pitfalls for financial analysis? We have asked several questions, apparently simple, but in fact complicated. In the next part of the article we will try to evaluate information based on the collected data that the financial statement in the part of liability provides us. Published financial statements were used for the structure analysis of liabilities for the period from 1.1.2014 to 31.12.2014, which are posted on the register of financial statements and on the website www.finstat.sk.

The breakdown of liabilities in the period of 2014 was a change in the structure of financial statements. Entities were legislative, based on size criteria, divided into micro entities, small entities and large entities. In each of these size types of companies the breakdown of liabilities is the same, they are just shown on other lines:

Notwithstanding the breakdown of commitments, based on theoretical data, it is possible to suggest two hypotheses related to the reporting of liabilities in the financial statements:

1. Liabilities in accounting are monitored by increasing manner on the part of Credit side of individual liabilities accounts and their payment is monitored on the part of Debit side. It follows:

H1: Height of liabilities recorded in company accounting is positive or zero.

2. Liabilities on the balance sheet are shown not only by category but also in total. It follows:

Table 1. Micro entity.

mark	line of the balance sheet
B	1.34: Liabilities 1.35 + 1.36 + 1.37 + 1.38 + 1.43 + 1.44 + 1.45
B.I	1.35: Long-term liabilities
B.II	1.36: Long-term reserves
B.III	1.37: Long-term bank loans
B.IV	1.38: Current liabilities, excluding reserves, loans and borrowings
B.V	1.43: Short-term reserves
B.VI	1.44: Current bank loans
B.VII	1.45: Short-term borrowings

Table 2. Small entity.

mark	line of the balance sheet
B	1.101: Liabilities 1.102 + 1.118 + 1.121 + 1.122 + 1.136 + 1.139 + 1.140
B.I	1.102: Long-term liabilities 1.103 + 1.107 − 1.117
B.II	1.118: Long-term reserves 1.119 + 1.120
B.III	1.121: Long-term bank loans
B.IV	1.122: Short-term liabilities 1.123 + 1.127 − 1.135
B.V	1.136: Short-term reserves 1.137 + 1.138
B.VI	1.139: Current bank loans
B.VII	1.140: Short-term borrowings

Table 3. Large entity.

mark	line of the balance sheet
B	1.101: Liabilities 1.102 + 1.118 + 1.121 + 1.122 + 1.136 + 1.139 + 1.140
B.I	1.102: Long-term liabilities 1.103 + 1.107 − 1.117
B.II	1.118: Long-term reserves 1.119 + 1.120
B.III	1.121: Long-term bank loans
B.IV	1.122: Short-term liabilities 1.123 + 1.127 − 1.135
B.V	1.136: Short-term reserves 1.137 + 1.138
B.VI	1.139: Current bank loans
B.VII	1.140: Short-term borrowings

H2: The percentage of each liabilities category gives total 100%.

From 180,230 financial statements returned in 2015 and published on www.finstat.sk for the 2014 accounting period were selected 13,378 that were prepared for the reporting period from 1.1.2014 to 31.12.2014. The value of assets in selected entities in 2014 and in the previous period of 2013 was positive and the difference of assets and liabilities is zero (not every entity in the financial statements respected basic accounting formula). One of the possible solutions to make data available to users would be to use XBRL for corporate data communication (Hoffman & Strand 2001) and especially data in the financial statements of the enterprise, which impacts not only on increasing the efficiency of the exchange of such information, but also can have an impact of the conceptual nature on the overall concept of reporting and verification of data included by auditors in the financial statements (Ištvánfyová & Mejzlík 2006). With such methods of data transfer (Andrejkovič et al. 2011) can be assumed that entities shall not recognize the liabilities in the financial statement with a negative sign, respectively, in the data transformation there would be incorporated also the requirement to respect basic accounting formula—assets are at the same level as liabilities.

Given that this article is focused on the presentation of liabilities in the financial statements, listed entities are divided according to the type of ownership and at the same time there was mentioned the existence of debts, which the entities have (debt and arrears to health insurances—Dôvera, Union, Všeobecná Health Insurance Company, debt and arrears to the Social insurance Agency, tax debts to the Financial Administration):

The structure of liabilities is based on the division of liabilities, which was before 2014, for this analysis was such a breakdown sufficient. Before

2014, the reserves were also observed in aggregate together and then they were divided into long-term and short-term reserves, as well as bank loans.

In these entities, the structure of liabilities is from the value −157.177 € to 915,152,000 €, average value of liabilities in these entities is 1,042,512.66 €, which is higher compared to the value in 2013 (1,208,948.72 €).

Already in the selection of samples for analysis it was necessary to clean the entire set of those entities that showed in their financial statements a negative value of the total amount payable. As there have been identified a number of such companies, the hypothesis H1 must be rejected and it must adopt an alternative that entities recognize liabilities in the financial statements, namely in the balance sheet also in negative values, possibly due to incorrectly reported data (e.g. file a claim and the need to return cash to the customer) or due to an incorrect initial recognition.

Of the total liabilities we have undergone analysis as reserves, long-term and short-term liabilities. Reserves in the non-zero showed from 13,378 entities 5716 entities, while the average amount of reserves in 2014 was worth 115,806.27 €, in the previous year it showed 6 518 entities in the average value of 98,275.35 €. The average share of reserves in the total amount of liabilities is 5.68% however, the structure is as follows:

Long-term liabilities to non-zero showed from 13,378 companies 8259 companies, while the average amount of long-term liabilities in 2014 was valued at 231,210.21 €, in the previous year showed them 8314 companies in the average value of 237,650.98 €. The average share of long-term

Table 4.   Breakdown of entities.

Breakdown of entities according to ownership	Count	Liabilities
Private domestic	10,427	800
Coop	700	31
National	32	0
ownership of local self-government	79	5
ownership of associations, political parties and churches	47	3
foreign	1207	62
International prevailing private sector	879	46
N/A	7	0
TOTAL	13,378	947

Table 5.   The structure of entities depending on the legal form.

	Jsc	Coop	Lp	Ltd	State enterprise	Gp	Foreign LP	BO	LP	Non-profit Org.	Non-inv. Fund
A	522	3	11	9,761		129				1	
B	3	690		7							
C	11				19			1	1		
D	6			63	2			7	1		
E	2			41						3	1
F	56		3	1146			1	1			
G	78		5	791		5					
N/A	2			5							
TOTAL	680	693	19	11,814	21	135	1	8	2	4	1

Note:
A Private domestic
B Coop
C National
D ownership of local self-government
E ownership of associations, political parties and churches
F foreign
G International prevailing private sector
N/A

220

liabilities to the amount of liabilities is 13.26% however, the structure is as follows:

Current liabilities at non-zero amount showed from 13,378 companies 12,302 companies, while the average amount of current liabilities in 2014 was valued at 595,439.72 € in the previous year showed them 10,752 companies in the average value of 720,444.48 €. The average share of current liabilities at the amount of the liabilities is 79.37% however, the structure is as follows:

The sum of reserves of long-term and short-term liabilities has been conducted in all 13,378 companies. Apart from short-term borrowings (2458 companies keep record of the existence of short-term borrowings) and bank loans (3823 companies keep record of long-term bank loans), the sum of liabilities category is as follows:

Table 6. The business entity in 2014.

B	1.88: Liabilities 1.89 + 1.94 + 1.106 + 1.117 + 1.118
B.I	1.89: Reserves 1.90–1.93
B.II	1.94: Long-term liabilities 1.95–1.105
B.III	1.106: Short-term liabilities 1. 107–1.116
B.IV	1.117: Short-term borrowings
B.V	1.118: Bank loans 1.119 + 1.120

Table 7. The share of reserves in the total amount of liabilities.

	Jsc	Coop	Lp	Ltd	state ent.	Gp	foreign LP	BO
negative share more than 10%				3				
negative share 0–10%	1	2		26				
0% share	2	2		11				
0–10% share	420	467	5	3884	1	27	1	4
10.01–20% share	55	49	1	371	2	2		1
20.01–30% share	16	15	1	142	3			
30.01–40% share	9	8		55	1			
40.01–50% share	3	1		40	2	1		
50.01–60% share	3	1		26		1		
60.01–70% share	2			12				
70.01–80% share	1			14				
80.01–90% share		1		8				
90.01–99.99% share	2			6				
100% share				3				
more than 100% share				2				
TOTAL	514	546	7	4603	9	31	1	5

Table 8. The share of long-term debt to total liabilities.

	Jsc	Coop	Lp	Ltd	state ent.	Gp	foreign LP	BO	LP	Non-profit Org.
negative share more than 10%		2		13		1				
negative share 0–10%	1	3		54		3				
0% share	10	7	1	307		6				
0–10% share	350	212	5	4688	8	40	1		1	
10.01–20% share	66	131	1	539	1					
20.01–30% share	47	89		277	2					
30.01–40% share	21	66		185					1	
40.01–50% share	19	28		148	1	1				
50.01–60% share	22	29	1	135	1	1				
60.01–70% share	12	19		99				1		
70.01–80% share	5	10		92		2				
80.01–90% share	12	7		90						
90.01–99.99% share	21	12		142		2				
100% share	5			27		3				
more than 100% share				8						
N/A	4	3		150		7			1	
TOTAL	595	618	8	6954	13	66	1	1	2	1

Table 9. The share of current liabilities to total liabilities.

	Jsc	Coop	Lp	Ltd	state ent.	Gp	foreign LP	BO	LP	Non-profit Org.
negative share more than 10%		1		15						
negative share 0–10%		2		7						
0% share		2		5						
0–10% share	40	26		294	1	3				
10.01–20% share	39	45		267		2				
20.01–30% share	39	72		265	2	2				
30.01–40% share	44	93		291	2			1		
40.01–50% share	64	77	1	399	3	4				
50.01–60% share	53	81		448	2	5				
60.01–70% share	48	70		535	1				1	
70.01–80% share	58	56	2	680	3	4				
80.01–90% share	84	55	1	951		9		1		
90.01–99.99% share	144	54	6	3209	2	37	1		1	
100% share	47	37	7	3171	2	42		4		1
more than 100% share	1	3		70		4				
N/A	4	5		181		10		1		2
TOTAL	665	679	17	10788	18	122	1	7	2	3

Table 10. The sum of the reserves, long-term and short-term liabilities.

	Jsc	Coop	Lp	Ltd	state ent.	Gp	foreign LP	BO	LP	Non-profit Org.	Non-invest. f.
negative share more than10%		2		9							
negative share 0–10%		1		1							
0% share	5	4		197	1	10		1		3	1
0–10% share	5	3		83		1					
10.01–20% share	14	12		106		1					
20.01–30% share	19	16		109		2					
30.01–40% share	17	28		128	1						
40.01–50% share	34	43		193		1					
50.01–60% share	30	41		241	1	4					
60.01–70% share	25	73		341	3						
70.01–80% share	42	67		416	1	3					
80.01–90% share	49	58	1	462	1	7					
90.01–99.99% share	72	83		914	8						
100% share	352	249	16	7594	11	87	1	6	2	1	
more than 100% share	2	2		41		1					
N/A	14	11	2	979	2	10		1			
TOTAL	680	693	19	11,814	29	127	1	8	2	4	1

The entity does not need to have the sum of reserves, long-term and short-term liabilities more than 100% share of liabilities, but it is noteworthy that there are companies that have a higher percentage than 100% – 1063 companies, which is almost 8% of the analyzed companies. This fact also rejects the second hypothesis.

3 CONCLUSION

After the rejection of both hypotheses, it is necessary to look at another important principle, the principle of continuation of the company as a going concern. The entity in preparing the financial statements is required to apply accounting principles and methods in a way that assumes that it will continue as a going concern and that there are no facts that could restrict or prevent it from being a going concern in the foreseeable future, at least 12 months from the date on which the preparation of the ordinary financial statements was compiled. If the entity has information that this fact occurs, is obliged to apply the corresponding accounting methodology and provide information on the method used in the financial statements in notes

(§ 7 par. 4 Act No. 431/2002 Coll. on accounting, as amended by later regulations).

The given examples document the fact which the entity must consider in preparing the financial statements, as they may significantly affect the financial position of each entity (Markovič et al. 2015).

The requirement to evaluate the entity's ability to continue as a going concern of entity results not only from national accounting adjustments, but also from supranational legislation, if the entity prepares financial statements in accordance with the International Financial Reporting Standards (IFRS).

The importance of this requirement in preparing the financial statements and independent audit demonstrates ISA 570 *Going Concern in the activity* to verify the financial statements by the statutory auditor. Examples of events or conditions that may be individual or going concern can be divided into several groups: financial, operational and others. Violating accounting commitment (Hypothesis 1) and failure to comply with reporting obligations in administrative structure and height (Hypothesis 2) is not possible for the entity to comply with the principles of compliance with the going concern basis.

Implementing Decision (EU) 2016/120 was issued at the beginning of 2016 (28 January 2016), in which the European multilateral platform for ICT standardization evaluated extensible business reporting language to version 2.1 (XBRL 2.1) based on the requirements set out in Annex II to Regulation (EU) No. 1025/2012, and issued favorable recommendations for them intended for marketing in public procurement. XBRL 2.1 valuating was subsequently submitted for consultation with industry experts, who have confirmed positive recommendation to determine this. XBRL 2.1 is a technical specification for digital business reporting that manages global consortium of non-profit and private sectors around the world. The aim of this consortium is to improve reporting in the public interest. XBRL International. The consortium consists of approximately 600 institutional members from the public and private sectors around the world. The aim of this consortium is to improve reporting in the public interest. XBRL 2.1 can be used in the event of a wide range of business and financial data. It simplifies the preparation of business and financial reports for internal and external decision making. By using XBRL 2.1, companies and other entities generating financial data and company reports can automate the process of collecting data (Official Journal of 29.01.2016, L23/77).

ACKNOWLEDGEMENT

This article is the output of grant project VEGA 1/0935/16 XBRL based on implementation of the electronic financial reporting.

REFERENCES

Act No. 431/2002 Coll. on accounting, as amended by later regulations.

Andrejkovič, M. et al. 2011. Adjustment of the pension system in Slovakia. Registered: Web of Science. In *International days of statistics and economics at VŠE, Prague*: conference proceedings, September 22 and 23, 2011. - Prague: University of Economics, Prague, 2011. ISBN 978-80-86175-77-5, pp. 1–7.

Commission Implementing Decision (EU) 2016/120 of 28 January 2016 on the identification of the extensible business reporting language 2.1 for referencing in public procurement.

Hoffman, C. & Strand, C. 2001. *XBRL Essentials.* New York, American Institute of Certified Public Accountants, 2001.

Ištvánfyová, J. & Mejzlík, L. 2006. Reporting of business data in electronic form – XBRL. In *Czech financial and accounting journal*, 2006, vol. 1, no. 2, pp. 124–129.

Markovič, P. et al. 2015. Disadvantages of the traditional profitability ratios. Registered: Web of Science. In *Financial management of firms and financial institutions.* International scientific conference. Financial management of firms and financial institutions: proceedings: 10th international scientific conference: 7th–8th september 2015, Ostrava, Czech Republic [elektronický zdroj]. Ostrava: VŠB - Technical university of Ostrava, 2015. ISBN 978-80-248-3865-6. ISSN 2336-162X, pp. 756–762 CD-ROM.

Pakšiová, R. & Kubaščíková, Z. 2015. Business property of company and investments. In *Annual conference on finance and accounting* Procedia economics and finance: 16th Annual conference on finance and Accounting, ACFA, Prague 2015, 29th May 2015. Netherlands: Elsevier B.V., 2015. ISSN 2212–5671, 2015, vol. 25, pp. 70–78 online.

Simonidesová, J. et al. 2015. Societas Europaea as a new form of enhancement the quality of business from the perspective of the application of tax optimization. In *Investment management and financial innovations: international research journal.* Ukraine: Limited liability company "Consulting publishing company" "Business perspectives", vol. 12, no. 4, pp. 171–175.

Tumpach, M. & Baštincová, A. 2014. Cost and Benefit of Accounting Information in Slovakia: Do We Need to Redefine Relevance? Registered: Web of Science. In *European Financial Systems 2014: proceedings of the 11th International Scientific Conference*: June 12–13, 2014 Lednice, Czech Republic. Brno: Masaryk University, 2014. pp. 655–676. <http://is.muni.cz/do/econ/sborniky/2014/proceedings-EFS-2014.pdf>.

The creation of a competency model of employees of SMEs in the context of SD and SCSR

L. Jurík & P. Sakál
Faculty of Materials Science and Technology in Trnava, Slovak University of Technology in Bratislava, Trnava, Slovak Republic

ABSTRACT: The desired quality of human capital is expressed by the competency models. The current state of economic and social situation requires that the criteria and principles of sustainable development were include into the creation of competency models of employees of industrial enterprises, because employee is the one, who is the foundation and pillar in the dissemination of the ideas of sustainable development. The contribution follows on the contribution Jurík & Sakál (2015), in which design of the methodology of the creation of a competency model of employees of industrial enterprises. The ambition of this paper is to create of the methodology of the creation of competency model of employees of industrial enterprises in Slovakia with the utilization of the AHP method in the context of sustainable development, while mainly authors are focusing on conceptual bases formation of competency model in the context of sustainable development and sustainable corporate social responsibility.

1 INTRODUCTION

According to Drabek (2012), political, economic, social, ecological, moral and other manifestations of the global crisis development has highlighted the question of the relevance of the current neoliberal model of capitalism, which for 350 years has constituted a basic paradigm of development of the prevalent part of global society. There is to the formation of several concepts that would eliminate the shortcomings and problems in the current socio-economic system unfair distribution of wealth, usurpation of power in the hands of a few "chosen ones", arrogant plundering of nature, natural resources and the man himself.

For the conditions of industrial enterprises is available potentially several concepts which supported sustainable development (SD) a sustainable corporate social responsibility (SCSR).

These concepts have been created by independent of each other protagonists UR and CSR. In the design of the methodology of competency model we are based on the concept Sustainable Corporate Social Responsibility (SCSR) (Hrdinová 2013, Sakál et. al. 2016), concept Humanistic Economy (CHE) (Haluška, 2015), Bhutanese development concept (BDC) (Gross National Happiness 2016, Centre for Bhutan Studies & GNH Research 2016) a creating sustainable shared values (CSSV) (Šmida et al. 2011, Porter & Kramer 2011, Šmida 2015, Šmida et. al. 2011).

Competency model provides the tool for the strategic management of human resources, because only with thorough management it is possible to provide a quality human capital, who will be able to fulfil the tasks and simultaneously disseminate and implement the ideas of sustainable development and corporate social responsibility concept. The human factor is becoming a differentiating factor. enterprises which at their disposal its high quality, it must be protected because the human factor forms the basis of competitive advantage based on human creativity, thought, and knowledge, i.e. on competencies.

The importance of the human factor is increasing in connection with the transformation to sustainable development and sustainable corporate social responsibility.

Based on the theory of social learning in the organizational environment, managers and leaders influence the behaviour of their subordinates through their example and pattern. And therefore managers are key people on the way to sustainability in enterprises.

The basis for a successful transformation from the strategy of unlimited economic growth to strategies of sustainable development sustainable corporate social responsibility in industrial enterprises is necessary to change the employee's competency models. In the creation of competency models it will be used social, environmental and economic criteria.

2 SUSTAINABLE HUMAN RESOURCE MANAGEMENT

An important feature of human resources management in general, and particularly in industrial companies is that they should be strategic. This objective requirement is expressed by an integrated approach to creating HR strategies, enabling businesses to meet their strategic objectives. According to Koubek (2009) strategic human resource management is practical culmination of personnel strategy of enterprise. It is a specific activity, specific efforts towards achieving the objectives set out in the personnel strategy.

Koubek (2009), however, does not explicitly state whether they are sustainable goals or strategy of sustainable development.

Armstrong (2009) adds that the strategic human resource management access to decision-making about the intentions and plans of the organization in the form of policies, programs and practices related to employment of people, obtaining, selection and stabilization of employees, employee training and development, performance management and remuneration.

We would argue that neither Armstrong is not focused to sustainable development and sustainable corporate social responsibility.

Thus defined strategic human resource management can then generally be seen as a long-term approach to addressing issues in HR, which is part of the strategic management of the company, without the accent on the aforementioned attributes sustainable development and sustainable corporate social responsibility. Hrdinová (2013) has established a new business model through a "draft methodology for the creation of a sustainable CSR strategy for SMEs", which emphasizes the historical continuity of the concept of CSR in relation to the concept of HCS model 3E, and whose working conditions are to contribute to improving the quality of life of every employee in sustainable and acceptable conditions of environmental quality and efficient economic conditions. The mentioned concept we aim to support in human resources management.

Strategic human resource management is situated in the creation of sustainable strategy CSR system's on the level of sustainable functional strategies in the form of sustainable human resources strategy.

Because only with thorough management, we can ensure a quality of the human capital that will be able to perform the tasks and to disseminate and implement the concept of sustainable development and sustainable corporate social responsibility.

As reported Minarik (2012) human factor becomes a differentiating factor and companies that are available to its high quality, it must be protected because it is the basis of competitive advantage based on human creativity, thought and knowledge, i.e. on competencies.

Also in this work, however, lacks emphasis on sustainable development and sustainable corporate social responsibility.

In contrast with him, according to Jurík & Sakál (2015) the fundamental tool for sustainable human

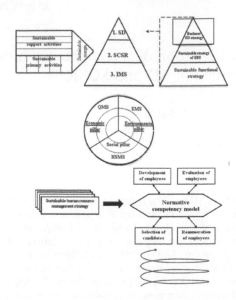

Figure 1. Proposal of the methodics for sustainable CSR strategy system creation for SMEs having regard to sustainable human resource management and normative competency model (own work on based Jurík & Sakál 2015).

resources strategy should be normative competency model created based on sustainable human resources strategy which based on the criteria of sustainable development and sustainable corporate social responsibility (Fig. 1).

Thus created normative competency model will be, along with the individual areas (selection, development, evaluation and renumeration of employees), open system evolving on the basis of feedback and also on spiral based on Deming's PDCA cycle.

3 CONCEPTS FOR THE CREATION OF SUSTAINABLE COMPETENCY MODELS OF EMPLOYEES IN THE CONTEXT OF SD AND SCSR

Term employee in the Slovak Republic defines Law No. 311/2001 Collection of Laws Labour Code as follow: *"An employee shall be a natural person who in labour-law relations and, if stipulated by special regulation also in similar labour relations, performs dependent work for the employer."*

In Labour Code is defined dependent work: *"Dependent work is work carried out in a relation where the employer is superior and the employee is subordinate, and in which the employee carries out work personally for the employer, according to the employer's instructions, in the employer's name, during working time set by the employer."*

Thus, the law defines the relationship of employer and employee in the form of superiority and subordination.

In 1949 Albert Einstein in his essay *"Why social-ism"* highlights the shortcomings of the capitalist establishment in relation businessman—employee, which are the exploitation of workers, payment of wages to the minimum amount, etc.

In recent years, in relation employee—employer there are several negative symptoms (Steingart 2008, Švihlíková 2015, Staněk & Ivanová 2016, Haluška 2011): trade union have lost any power, they do not have any means to an improvement in working and living conditions of employees, strengthens the rights of employers (especially transnational corporations), decreases real wages (decreases the size of the middle layer), labour productivity growth is not reflected in wage growth, it is big difference between the wages of ordinary employees and the wages of top managers.

According to Stead J. G. & Stead W. E. (2012) a manager is the one, who creates and disseminates the awareness of sustainability towards other employees, who acts as a role model and who contributes to sustainability by own decisions. Besides strategic decisions of a manager, which contribute to the sustainable development of the enterprises, it is necessary that the manager disposes of certain competencies. This requires that during strategic decisions, s/he takes into account the benefit of all stakeholders, and also that s/he includes into the decision-making processes the factors such as nature, society and ultimately the economic part.

Based on the theory of social learning in the organizational environment, managers and leaders influence the behaviour of their subordinates through their example (Remišová et al., 2015). And therefore managers are key people on the way to sustainability in enterprises.

Haluška (2011) states that from the three factors of production (human factor, natural resources, production resources) arose by the fourth factor, and the top management, whose weight and role increased.

And on the basis of these concepts we will be in the creation of employee's competency model specifically focus on the position of manager.

4 THE DESIGN OF THE METHODOLOGY OF THE CREATION OF A COMPETENCY MODEL OF EMPLOYEES OF INDUSTRIAL ENTERPRISES IN SLOVAKIA WITH THE UTILIZATION OF THE AHP METHOD IN THE CONTEXT OF SD AND SCSR

Nowadays, management by competences is considered the most progressive system (instrument and method) human resource management in enterprises. All members of the enterprise— employees and managers, often also external consultants, which specialize in creating competency models and their implementation in enterprises— are involved in management by competencies (Veteška & Tureckiová 2008).

Competency can be simplistically understood as the ability to behave in a certain way. Competency exhibits certain human behaviour. The behaviour of a particular person in a specific situation is a result of the dynamics of his personality, which consists of several elements. Some parts form a relatively stable characteristic of a person, such as attitudes, values and motives. Other parts contain competence, knowledge and skills. The observed behaviour is the result of collaboration of these and other factors (Vodák & Kucharčíková 2007).

Kubeš, Spillerová & Kurnický (2004, p. 27) state: *"...competency model describes a specific combination of knowledge, skills and other personality characteristics that are necessary for the effective performance of tasks in the organization".*

According to Hroník (2007), competency model is also bridge between the values shared of enterprise and job description. The enterprise usually has one set of values which govern functioning of enterprise. Competency model is a practical tool which links all activities of human resource management.

For the current time it is typical for competency models orientation of managers to achieve economic results, thus preferring those competencies whose disposition at managers contribute to maximizing profit.

Managers at any level of management will play an irreplaceable role in the application of the sustainable development in industrial enterprises. Managers and their characteristics (competencies) represent an assumption of the fulfilment of the activities directed towards the sustainability of the enterprises. Therefore, it is necessary to create a "new" competency model, which will promote the development of the sustainability and its principles into the activities of industrial enterprises. We suggest for systemic inclusion of sustainable development criteria's to competency models enterprises have to defined a strategy SD enterprise, vision SD enterprise and mission SD enterprise. Important in the creation of competency models is then their compliance with the strategy of sustainable development of enterprises, vision and mission of sustainable development of enterprise. Several authors agree that the proper setting up of model must be based on corporate culture. If the enterprise has included in the corporate culture values and principles of sustainable development, they are subsequently reflected in the competency model, thereby enhancing the quality of human resources in the direction of sustainability.

This contribution is a continuation of contribution Jurík & Sakál (2015), where we design the methodology of creation of competency model employees of industrial enterprises. The proposal was based on criteria Model ZET. Model ZET is the assessment methodology of the CSR in the competition *"Social Responsibility Award of the Slovak Republic"*.

In this design methodology of creation of employees' competency model of industrial enterprises in Slovakia, we propose to use the analytic hierarchy process (AHP) method for the creation of competency model as in various fields of its use (selection, development, evaluation, compensation of employees).

The creator of analytic hierarchy process method is an American professor of mathematics acting at the University of Pittsburgh. AHP allows preparing effective decisions in complex situations, to simplify and quicken the natural process of decision-making. AHP provides a comprehensive and logical concept for structuring a problem, for quantifying its components, which are related to the final objectives, and for the evaluation of alternative solutions. AHP method can be used in several different areas. It is used worldwide in various decision situations, in the fields such as state administration, business, industry, health, education. It is a suitable method for the evaluation of enterprises, where many criteria lead to the objectification of their evaluation. AHP was used in a number of decisions in the area of economics, energetic, management, environmental science, transport, agriculture, industry and army. By the usage of AHP, a decision maker knowingly and intentionally directs to increasing the quality and efficiency of all his decisions (Saaty 2008).

The decision-making according to AHP method is based on the three principles of analytic thinking (Saaty 2010):

1. *The principle of structuring a hierarchy.*
2. *The principle of determining priorities.*
3. *The principle of logical consistency.*

In the competency approach has the AHP method the following options of using:

1. *Selection of candidates.*
2. *Development of employees.*
3. *Evaluating and reward of employees.*

Within our previous scientific research activities, teaching activities and publishing activities we applied AHP method in the field of human resources management in industrial companies through the following final thesis:

1. *Bachelor thesis: Rauchová (2014), Vlková (2015), Smoláriková (2016), Špaček (2016).*
2. *Master thesis: Jurík (2013), Schiffel (2014).*

The methodology of the creation of a competency model of employees of industrial enterprises in Slovakia with the utilization of the AHP method in the context of SD and SCSR comprising the steps of (Fig. 2):

1. *Defining the objective of the project.*
2. *Building of an expert team.*
3. *Assess of compliance competency model with the strategy of SD enterprise, vision of SD enterprise and mission of SD enterprise.*

4. *Identifying and describing the general, specific and key management competencies.*
5. *Defining of the intent successful application of competency model.*
6. *Drafting the job description.*
7. *Drafting of competency profile.*
8. *Application of competency model in individual areas of HRM.*
9. *Evaluation of application of competency model.*

At the modified method were added philosophical and conceptual determinants (Fig. 2).

Penetration of these concepts develops criteria for creating competency models of managers in industrial enterprises, which are:

1. *Sustainable quality of life.*
2. *Sustainable quality of work.*
3. *Sustainable environmental quality.*
4. *Interest of enterprise for decision-making and the business activities about stakeholders.*
5. *Transferring principles SD (SCSR, CHE, BDC) to primary business activities.*
6. *Environmental Protection.*
7. *Participative management of enterprise.*
8. *A fair remuneration of employees based on merit, participating in creation of added value and labour productivity growth.*

Based on the above criteria, managers should have the following competencies:

1. *The ability of creative system thinking and decision-making in regard to the society, environment and economy.*

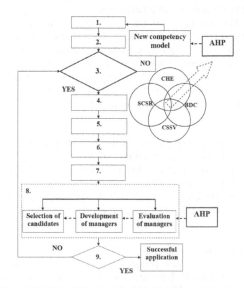

Figure 2. The design of the methodology of the creation of a competency model of employees of industrial enterprises in Slovakia with the utilization of the AHP method in the context of SD and SCSR.

2. *The ability to lead and motivate people towards SD.*
3. *The ability of empathy, compassion and solidarity.*
4. *The acceptance of the interests and needs of all stakeholders.*
5. *Knowledge about SD/CSR.*
6. *The support of the incorporation of the ecological aspect into all activities and products.*
7. *The ability of the value orientation for SD.*
8. *The ability of divergence of thought processes and creation of alternatives and visions towards SD.*

5 CONCLUSION

The ability to transform the current strategy of unlimited economic growth to strategy of SD and SCSR is the only basis for the continued existence of life on planet Earth. The basis for this transformation is a paradigm shift of thinking in the field. Owners and employees of industrial enterprises, as well as all stakeholders must be adapted to this reality. By the transition to a new business model which is based on the UR and USZP and strategies "win-win" will created equal relationship between people, nature and the economy. It will create a fair relationship between employees and employers. Managers will have a specific position in this relationship. They will have to dispose to the "new" competencies that support the SD and SCSR in industrial enterprises. AHP method is suitable tool for creating of competency model and then for their application for selection, development, evaluation and renumeration of employees.

During our research activities and in processing the above mentioned grants we have accumulated theoretical and practical experience with the application of AHP method in other areas of management and decision making, e.g.:

1. *In environmental management (Moravčík et al. 2012).*
2. *In risk strategic management (Naňo & Sakál 2011).*
3. *In the strategy of corporate social responsibility (Drieniková et al. 2011).*
4. *In process evaluation competency profile of managers in industrial enterprises and university employees (Jurík & Sakál 2014).*
5. *In ergonomy (Beňo et al. 2012).*
6. *In sustainable logistics (Fidlerová et al. 2016).*

ACKNOWLEDGMENT

The contribution follows on the results of APVV project No. LPP-0384-09: "Concept HCS model 3E vs. concept Corporate Social Responsibility (CSR)." and KEGA project No. 037STU-4/2012:
Implemen-tation of the subject of "Corporate Social Responsi-bility Entrepreneurship" into the study programme of Industrial Management in the second degree of study at STU MTF Trnava.".

The contribution is part of the upcoming KEGA project No. 006STU-4/2017 "Innovation of the structure, content and teaching methods for the subject "Strategic Management" in the context of new business model for the 21th century based on "win-win" strategy.".

REFERENCES

Armstrong, M., 2009. Armstrong's handbook of human resource management practice. London: Kogan Page. ISBN 978-0-7494-5242-1.

Beňo, R. et al. 2012. Multicriteria assessment of the ergonomic risk probability creation by chosen groups of stakeholders with using AHP method within the context of CSR. In Quantitative methods in economics. Multiple Criteria Decision Making XVI: Proceedings of the International Scientific Conference. Bratislava, Slovakia. Bratislava: Ekonóm, pp. 7–11.

Centre for Bhutan Studies & GNH Research, 2016. 2015 GNH Survey Report: A Compass Towards a Just and Harmonious Society. Thimpu. ISBN 978-99936-14-86-9.

Drábek, J., 2012. Nový model podnikání a vývoj paradigmatu managementu a Marketingu. (The new model of business and development of paradigma of management and marketing). Available at http://www.cutn.sk/Library/proceedings/mch_2012/editovane_prispevky/Dr%C3%A1bek%20%281%29.pdf.

Drieniková, K. et al. 2011. Case studies of using the analytic hierarchy process method in corporate social responsibility and environmental risk management. In Materials Science and Technology, Vol. 1, 2011, pp. 1–10.

Einstein, A. 1949. Why Socialism?. Available at: http://www.monthlyreview.org/598einst.htm

European Parlament. 2013. Report on corporate social responsibility: accountable, transparent and responsible business behaviour and sustainable growth. Štrasburg: EP. 2012/2098(INI).

Fidlerová, H. Application of AHP in the process of sustainable packaging in company. In Production Management and Engineering Sciences. London, UK: CRC Press. Taylor & Francis Group, 2016. ISBN 978-1-138-02856-2.

Gross National Happiness, 2016. Available at: http://www.grossnationalhappiness.com

Haluška, I. 2015. Manifest Humanistickej (ľuďom slúžiacej) ekonomiky. Available at: http://www.noveslovo.sk/c/Manifest_Humanistickej_ludom_sluziacej_ekonomiky

Hrdinová, G. 2013. Koncept HCS modelu 3E vs. koncept Corporate Social Responsibility (CSR). Dizertačná práca. Trnava: MTF STU. Evidenčné číslo: MTF-10904-31238

Hroník, F., 2007. Rozvoj a vzdělávání pracovníků. Praha: Grada Publishing. 240 s. ISBN 978-80-247-1457-8.

Jurík, L. 2013. Návrh využitia metódy AHP pre určenie kompetenčného profilu personálneho pracovníka podniku Delta Electronics (Slovakia), s.r.o. (The proposal of using AHP method for determine of recruiter's competency profile of company Delta Electronics

(Slovakia), s.r.o.") Diplomová práca. (Master thesis). Trnava: Trnava STU, 2013.

Jurík, L. & Sakál, P. 2014. Competencies of Managers, as part of the Intellectual Capital in Industrial Enterprises. In ECIC 2014: proceedings of the 6th European Conference on Intellectual Capital. Academic Conferences and Publishing International Limited, 2014, pp. 368–376.

Jurík, L. & Sakál, P. 2015. Náčrt metodiky tvorby kompetenčného modelu zamestnancov priemyselných podnikov na Slovensku s využitím metódy AHP v kontexte s udržateľným rozvojom. In Progresívne prístupy a metódy zvyšovania efektívnosti a výkonnosti organizácií. SAPRIA, 2015, pp. 1–12. ISBN 978-80-969519-8-7.

Koubek, J., 2007. Řízení lidských zdrojů: Základy moderní personalistiky. Praha: Management Press. ISBN 978-80-7261-168-3.

Kubeš, M. et al. 2004. Manažerské kompetence: Způsobilosti výjimečných manažerů. Praha: Grada Publishing. ISBN 80-247-0698-9.

Minárik, M. 2012. Trvalo udržateľný ekonomický rast a konkurencieschopnosť ako výsledok pôsobenia nerivalitných produkčných faktorov. Available at: http://www.cutn.sk/Library/proceedings/mch_2012/editovane_prispevky/Min%C3%A1rik.pdf.

Moravčík, O. et al. 2012. Perspectives for Utilization of Multicriteria Decision Methods AHP/ANP to Create a National Energy Strategy in Terms of Sustainable Development. Advanced Materials Research, Volume 616–618: Sustainable Development of Natural Resources. pp. 1585–1590.

Naňo, T. & Sakál, P. 2011. Suggestion of the AHP (Analytic Hierarchy Process) method utilization in risk strategic management of industrial companies. In Radioelektronika, elektrotechnika i energetika: 17. meždunarodnaja naučno-techničeskaja konferencija studentov i aspirantov, Tom 2. Moskva: Moskovskij energetičeskij institut, pp. 302–303.

Porter, M. & Kramer, M. 2011. How to Fix Capitalism? Creating Shared Value. In: Harvard Business Review. Január – Február 2011, pp. 63–67. ISSN 0017-8012.

Rauchová, I. 2014. Návrhy využitia metódy AHP pre určenie kompetenčného profilu manažéra podniku MIKROMAT spol. s r.o. (The proposal of using AHP method for determine of recruiter's competency profile of company Mikromat (Slovakia), spol. s.r.o.). Bakalárska práca. (Bachelor thesis) Trnava: Trnava STU, 2014.

Remišová, A. et al. 2015. Etické vedenie ľudí v slovenskom podnikateľskom prostredí. Bratislava: Wolters Kluwer. ISBN 978-80-8168-199.8.

Saaty, T.L. 2008. Decision Making for Leaders: The Analytic Hierarchy Process for Decisions in a Complex World. Pittsburgh, Pennsylvania: RWS Publications. ISBN 0-9620317-8-X.

Saaty, T.L. 2010. Principia Mathematica Decernendi: Mathematical Principles of Decision Making. Pittsburgh, Pennsylvania: RWS Publications. ISBN 978-1-888603-10-1.

Sakál, P. et al. 2016. Transformation of HCS model 3E in IMS in context of sustainable CSR. In Production Management and Engineering Sciences. London, UK: CRC Press. Taylor & Francis Group, 2016, pp. 259–263. ISBN 978-1-138-02856-2.

Schiffel, M. 2014. Návrh využitia metódy AHP pre určenie kompetenčného profilu manažéra podniku PCA Slovakia, s.r.o. v kontexte UR a USZP. (The proposal of using AHP method to determine the competency profile manager company PCA Slovakia, s.r.o in the context of SD and SCSR). Diplomová práca. (Master thesis) Trnava: Trnava STU, 2014.

Smoláriková, N. 2016. Návrh opatrení na zdokonalenie hodnotenia kompetencií zamestnancov prostredníctvom metódy AHP v podniku HYCA s.r.o. (The proposal of measures to improve the evaluation of workers skills through the AHP method in the company HYCA s.r.o.). Bakalárska práca. (Bachelor thesis). Trnava: Trnava STU. 2016.

Šmida, Ľ. 2015. Návrh tvorby systému udržateľných zdieľaných hodnôt priemyselného podniku v kontexte s udržateľným spoločensky zodpovedným podnikaním. Dizertačná práca. Trnava: Trnava STU, 2015.

Šmida, Ľ. et al. 2011. A possibility of social inclusion by responsible entrepreneurship in Slovakia. In Annals of DAAAM and Proceedings of DAAAM Symposium. ISSN 1726-9679.

Šmida, Ľ. et al. 2011. Predpoklady budovania spoločensky zodpovedného podnikania ako súčasti udržateľnej spoločnosti v zmysle konceptu udržateľného rozvoja. In: Transfer inovácií. XIII (21), pp. 198–203. ISSN 1337-7094.

Špaček, M. 2016. Návrh využitia metódy AHP pri tvorbe kompetenčného modelu v podniku PROTHERM PRODUCTION s.r.o. (The proposal of using AHP method to develop the competency model in the company PROTHERM PRODUCTION s.r.o.). Bakalárska práca. (Bachelor thesis). Trnava: Trnava STU. 2016.

Švihlíková, I. 2015. Jak jsme se stali kolonií. Praha: Rybka Publishers. ISBN 978-80-87950-17-3.

Staněk, P. & Ivanová, P. 2016. Štvrtá priemyselná revolúcia a piaty civilizačný zlom. Bratislava: Elita. ISBN 978-80-970135-8-5.

Stead, J.G. & Stead, W.E. 2012. Manažment pre malú planétu. Bratislava: Eastone Group. ISBN 978-80-8109-216-9.

Steingart, G. 2008. Globální válka o blahobyt. Praha: Knižní klub. ISBN 978-80-242-2301-8.

Veteška, J. & Tureckiová, M., 2008. Kompetence ve vzdělávání. Praha: Grada Publishing. ISBN 978-80-247-1770-8.

Vlková, A. 2015. Návrh využitia metódy AHP pre hodnotenie prvolíniových manažérov v kontexte UR v podniku ŽOS Trnava, a.s. (The proposal of using AHP method for the first-line managers in the context of sustainable development in the company ŽOS Trnava, a.s.). Bakalárska práca. (Bachelor thesis). Trnava: Trnava STU. 2015.

Vodák, J. & Kucharčíková, A., 2007. Efektivní vzdělávání zaměstnanců. Praha: Grada Publishing. 2007. ISBN 978-80-247-1904-7.

Zákon č. 311/2001 Z. z. ZÁKONNÍK PRÁCE (Law No. 311/2001 Collection of Laws Labour Code Slovak Republic).

Modeling of automobile components and spare parts demand by probability theory methods

K. Kabdi, T. Suleimenov & A. Abay
The L.N. Gumilyov Eurasian National University, Astana, Kazakhstan

M. Bosák
Faculty of Business Economy with seat in Košice, University of Economics in Bratislava, Košice, Slovak Republic

ABSTRACT: One of the trends in the automotive industry is moving to carmakers and suppliers by increasing the efficiency of the supply chain, each characterized by the introduction of new information and communication technologies, leading to the transfer of responsibility for different tasks. Key tasks which have the car manufacturers, currently largely take over suppliers, thereby increasing their share of participation in the tasks of production and the development of the car. The authors of the article are as the output of its research activity modeling method of optimizing the availability of individual components suppliers for automotive manufacturing companies.

1 INTRODUCTION

In the development of the global automotive industry, important factors include contractors, whereas the production car is a sophisticated system of interconnected enterprises cooperating within the supply chain. In many cases, the suppliers indicate the direction and trends due to the rapidly developing innovations that offer the final car manufacturer. Many of the parts and components of a car are developed, designed and tested in collaboration with supplier companies, which take over full responsibility for their product and also for all subcontractors at the various stages up to the producers of the raw materials, materials and semi-finished products (Boghani & Brown 2000).

Economic liberalization as the result of rigid planning abolition and the arrival of market regulation in post-communist countries at the end of the last century brought a strong competition to the automotive industry market raising the rate of progress of readiness and average efficiency of the motor vehicle operating at these companies.

In some of these countries, in 2014 the volume of automobile freight transportation increased in comparison with 2010 by 16–18% and the transportation of passengers by 21–22%. During the realization of these figures, it was necessary to, first of all, improve conditions of vehicular transport readiness which are characterized by the output coefficient of vehicles. Currently, this characteristic in these countries is about 19% lower than in the East European countries (Yousefi & Hadi-Vencheh 2010).

One of the main reasons of this backlog of vehicle operational readiness and keeping vehicles on the road is the problem of identifying and providing theoretically grounded expenses of spare parts for running the repair and maintenance service for vehicles in use. The correct solution to this important task can really improve the operational efficiency of automobile operating company, thereby reducing economic losses in all countries (Bosák 2010).

The overall requirements of automobile components and vehicular transport spare parts demand in a given country should be primarily defined in terms of the natural and operational features, and climatic conditions of the certain region. In addition, in new economic conditions not only the practice but also questions will be of a great importance, as well as the theory of definition of automobile components and spare parts demand, as they will directly influence the commercial profit of these private firms (Safaei M.H et al. 2009). For this reason automotive companies keep their spare parts data accurately a secret and protect their methods of prediction of requirements and try not to publish them at all (Majerník et al. 2014).

Modern-day complex transportation problems put the correct determination of automobile components and spare parts requirements and their effective production technologies in these countries to the forefront using the probability theory approximation methods. The optimal solution of this problem will improve the quality of transport services, improving its capacity and environmental situation in the cities (Valenčík 2011).

2 MODELLING OF AUTOMOBILE COMPONENTS AND SPARE PARTS DEMAND

Modelling of automobile component and spare parts expenses for vehicles in use must be based on selected operational characteristics of overall reliability, which depend on the fail-safe features, actual operating conditions and running repair technologies of vehicles (Agarski et al. 2012). Adequate formulation of modelling allows evaluating the average reliability of each of the missing range of items, such as the requirements of spare parts for vehicles in use in a certain region (Landmann et al. 2001). Territorial division of a country on the actual operating conditions at each area in the country is carried out in accordance with existing principles, based on a special screening method.

For non-restored elements of vehicles, the average durability of the original detail in the i area of vehicle operating in a certain region is determined by the following equation:

$$\bar{\ell}_i = \frac{1}{d} \sum_{j=1}^{d} \ell_{ij}, \tag{1}$$

The necessary formulations are during the preparation of model selected only for a specific area of vehicle operating in a certain region. Therefore, in order to simplify the notation, index i, which indicates the serial number of the region, will be omitted, and the durability for this area will be calculated in average value as ℓ instead of ℓ_i, and etc.

In the case of restored elements of vehicles, the average durability is determined according to the results of experimental observation of restored original elements by the equation (1). From the structure of the average durability of the restorable elements j it is necessary to separate duration up to the first restoration ℓ_{j1}. Then durability ℓ_j can be expressed as:

$$\ell_j = \ell_{j1} + \sum_{m=2}^{M_j} \ell_{jm} \tag{2}$$

2.1 Determination of the expected average requirements of spare parts as a function of the durability of automobile components

In the case when $\bar{\ell}_i > 0.2 L_{am_i}$ it is necessary to choose the version of the law of the original elements durability distribution and determine the value of $\bar{Z}$ as the average between the upper and lower limits, that is:

$$\bar{Z} \approx 0.5\left(\bar{Z}^h + \bar{Z}^d\right), \tag{3}$$

Using the formula (3) we may find out that:

$$\bar{Z} \approx 0.5\left[2\bar{Z}^h - q(L_{am})\right] = \bar{Z}^h - 0.5q(L_{am}). \tag{4}$$

This formula (3) can also be written as:

$$\bar{Z} \approx \frac{L_{am} * q(L_{am}) - A(L_{am})}{\bar{\ell}_n} + 0.5q(L_{am}). \tag{5}$$

In the same formula (3) we might define:

$$\bar{Z} \approx \frac{L_{am} - B(L_{am})}{\bar{\ell}_n} + 0.5q(L_{am}). \tag{6}$$

Formula expressing the values of the $A(L_{am})$, $B(L_{am})$ and $q(L_{am})$ determines the original elements of durability of the automobile. Studies have shown that when the coefficient of variation $V_i = 0.5$ the law of the vehicle original elements durability distribution in the i area of vehicle operating in a certain region is conform to the normal distribution, and when $0.5 < V < 1$ the distribution corresponds to the Weibull distribution of distribution and when $V = 1$ it results in an exponential distribution.

2.2 Requirements of vehicle spare parts for Gaussian distribution law

The equation of probability density function of the normal distribution is valid only when the coefficient of variation $V \leq 0.5$. And when the coefficient of variation $V > 0.5$ the result is the truncated normal distribution, which can't be used in this case for the determination of vehicle spare parts requirements. In equations define:

$$A(L_{am}) = \int_0^{L_{am}} \frac{t}{\sigma\sqrt{2\pi}} \exp\left[-\frac{(t-\bar{\ell})^2}{2\sigma^2}\right] dt.$$

Accepting an auxiliary variable of integration:

$$x = \frac{t - \bar{\ell}}{\sigma},$$

and quantile of normalized and centered Gaussian distribution with $t = L_{am}$ is as:

$$U_q = \frac{L_{am} - \bar{\ell}}{\sigma}. \tag{7}$$

then:

$$A(L_{am}) = \int_{-\frac{1}{v}}^{U_q} \frac{\bar{\ell} + \sigma x}{\sqrt{2\pi}} \exp\left[-\frac{x^2}{2}\right] dx =$$

$$\bar{\ell} \int_{-\frac{1}{v}}^{U_q} \varphi_0(x) dx + \sigma \int_{-\frac{1}{v}}^{U_q} \frac{x}{2\pi} \exp\left[-\frac{x^2}{2}\right] dx. \tag{8}$$

where $\varphi_0(x)$ is the density of normalized and centered Gaussian distribution.

It may be considered separately the first and the second terms of the first part of the equation (8) as:

$$\bar{\ell} \int_{-\frac{1}{V}}^{U_q} \varphi_0(x) dx = \bar{\ell}\left[F_0(U_q) - F_0\left(-\frac{1}{V}\right)\right] =$$

$$\bar{\ell}\left[q(L_{am}) + F_0\left(\frac{1}{V}\right) - 1\right].$$

where $F_0(x)$ is the function of normalized and centered Gaussian distribution:

$$\sigma \int_{-\frac{1}{V}}^{U_q} \frac{x}{2\pi} \exp\left[-\frac{x^2}{2}\right] dx = -\sigma \int_{-\frac{1}{V}}^{U_q} d\left[\frac{1}{\sqrt{2\pi}} \exp\left(-\frac{x^2}{x}\right)\right]$$

$$= \sigma\left[\varphi_0\left(\frac{1}{V}\right) - \varphi_0(U_q)\right].$$

Then:

$$A(L_{am}) = \bar{\ell}\left[q(L_{am}) + F_0\left(\frac{1}{V}\right) - 1\right] + \sigma\left[\varphi_0\left(\frac{1}{V}\right) - \varphi_0(U_q)\right]. \tag{9}$$

Substituting the equation (9) we define:

$$\bar{Z}^h = \frac{L_{am}}{\ell_n} q(L_{am}) - \frac{\bar{\ell}}{\ell_n}\left[q(L_{am}) + F_0\left(\frac{1}{V}\right) - 1\right] - \frac{\sigma}{\ell_n}$$

$$\left[\varphi_0\left(\frac{1}{V}\right) - \varphi_0(U_q)\right] + q(L_{am}) = \left(\frac{L_{am} - \bar{\ell}}{\ell_n} + 1\right) q(L_{am})$$

$$+ \frac{1}{K}\left[1 - F_0\left(\frac{1}{V}\right)\right] + \frac{\sigma}{\ell} * \frac{\bar{\ell}}{\ell_n}\left[\varphi_0(U_q) - \varphi_0\left(\frac{1}{V}\right)\right] =$$

$$\left(\frac{L_{am} - \bar{\ell}}{\ell_n} + 1\right) q(L_{am}) + \frac{1}{K}\left[1 - F_0\left(\frac{1}{V}\right)\right] + \frac{V}{K}$$

$$\left[\varphi_0(U_q) - \varphi_0\left(\frac{1}{V}\right)\right].$$

When the coefficient of variation $V \leq 0,5$ it is $F_0\left(\frac{1}{V}\right) \approx 1$, and $\varphi_0\left(\frac{1}{V}\right) \approx 0$, then:

$$\bar{Z}^h \approx \left(\frac{L_{am} - \bar{\ell}}{\ell_n} + 1\right) q(L_{am}) + \frac{V}{K} \varphi_0(U_q). \tag{10}$$

It can be noted that:

$$\frac{L_{am} - \bar{\ell}}{\ell_n} = \frac{L_{am} - \bar{\ell}}{\sigma} * \frac{\sigma}{\bar{\ell}} * \frac{\bar{\ell}}{\ell_n} = U_q * \frac{V}{K},$$

then the equation (10) is as follows:

$$\bar{Z}^h = \left(\frac{V}{K} U_q + 1\right) q(L_{am}) + \frac{V}{K} \varphi_0(U_q) = \frac{V}{K}$$

$$\left[U_q * q(L_{am}) + \varphi_0(U_q)\right] + q(L_{am}). \tag{11}$$

Here we can indicate that:

$$Z_q = q(L_{am}) * U_q + \varphi_0(U_q). \tag{12}$$

and then:

$$\bar{Z}^h = \frac{V}{K} Z_q + q(L_{am}). \tag{13}$$

Assigning a value of quantile for normalized and cantered Gaussian distribution Uq in tables, we may obtain the corresponding values of $\varphi_0(U_q)$ and $q(L_{am}) = F_o(U_q)$. Next, it can be obtained Zq from the equation (12).

Substituting the equation (13) into the equation (4) we can obtain:

$$\bar{Z} \approx \frac{V}{K} Z_q + 0,5 q(L_{am}). \tag{14}$$

From the equation (14) we can obtain the standard expenses of spare parts hi for conditions of the i area for a year operation of 100 vehicles, containing with n spare pieces in each:

$$h_i = \frac{100 * n}{t_{am_i}}\left[\frac{V_i}{K_i} Z_q + 0,5 q(L_{am_i})\right]. \tag{15}$$

2.3 Requirements of vehicle spare parts for Weibull distribution law

In a case when the vehicle original part durability is a Weibull distribution, the probability of failure-free car operation in the range of (0,t) is as followed:

$$P(t) = \exp\left[-\left(\frac{t}{a}\right)^b\right], \tag{16}$$

where the parameter of distribution is:

$$a = \frac{\bar{\ell}}{\Gamma\left(1+\frac{1}{b}\right)} = \frac{\bar{\ell}}{\Gamma(x)}. \tag{17}$$

Values of the distribution parameter b, as well as the gamma-functions $\Gamma(x) = \Gamma\left(1+\frac{1}{b}\right)$ are given in (6).

Using the equation (17) into the equation (16) leads to:

$$P(t) = \exp\left[-\frac{t}{\ell}\Gamma(x)\right]^b. \tag{18}$$

Denote that $y = \frac{t}{\ell}$ and $y_o = \frac{L_{am}}{\ell}$, then:

$$q(L_{am}) = q(y_0, b) = 1 - \exp\left\{-\left[y_0, \Gamma(x)\right]^b\right\}. \tag{19}$$

Values $q(y_0, b)$ are also given in [6]. Using the equation (18), can be defined:

$$B(L_{am}) = \int_0^{L_{am}} \exp\left[-\frac{t}{\ell}\Gamma(x)\right]^b dt.$$

If here we introduce a new variable of integration y, then:

$$B(L_{am}) = \bar{\ell}\int_0^{y_0} \exp\left\{-\left[y, \Gamma(x)\right]^b\right\} dt.$$

Using a relative value of $B_o(L_{am}) = B_o(y_o, b)$, the values which determine the numerical integration method are carried out in (6):

$$B_0(L_{am}) = \frac{B(L_{am})}{\bar{\ell}} = \int_0^{y_0} \exp\left\{-\left[y, \Gamma(x)\right]^b\right\} dt. \tag{20}$$

As $B(L_{am}) = \bar{\ell}B_0(y_0, b)$, the equation (6) can be written as:

$$\bar{Z} \approx \frac{L_{am} - \bar{\ell}B_0(y_0, b)}{\bar{\ell}_n} + 0.5q(y_0, b)$$

$$= \frac{\frac{L_{am}}{\bar{\ell}} - B_0(y_0, b)}{\frac{\bar{\ell}_n}{\bar{\ell}}} + 0.5q(y_0, b) \tag{21}$$

$$= \frac{y_0 - B_0(y_0, b)}{K} + 0.5q(y_0, b).$$

On the basis of the equation (21) we can define the standard expenses of spare parts h_i for condi-

tions of the i area for a year operation of 100 vehicles, containing with n spare pieces in each:

$$h_i = \frac{100 * n}{t_{am_i}}\left[\frac{y_0 - B_0(y_0, b)}{K_i} + 0.5q(y_0, b)\right]. \tag{22}$$

2.4 Requirements of vehicle spare parts for exponential distribution law

In a case when the coefficient of variation V = 1 there is an exponential distribution of the original elements durability and it can be assumed that the spare elements durability of the same range will also distribute exponentially. And it differs from the original elements durability in only an average value ℓ_n.

Therefore, it can be written:

$$\lambda_n = \frac{1}{\ell_n},$$

and accordingly:

$$\bar{m}(\ell) = \frac{L_{am} - \ell}{\ell_n},$$

$$\bar{Z}(\ell) = \bar{m}(\ell) + 1 = \frac{L_{am} - \ell}{\ell_n} + 1,$$

$$\bar{Z} = \int_0^{L_{am}}\left(\frac{L_{am} - t}{\ell_n} + 1\right) f(t) dt = \bar{Z}^h.$$

Next, using the equation in the case of exponential distribution will be as:

$$q(L_{am}) = 1 - \exp\left(-\frac{L_{am}}{\ell}\right) = 1 - \exp(-y_0). \tag{23}$$

$$B(L_{am}) = \int_0^{L_{am}} \exp\left(-\frac{t}{\ell}\right) dt$$
$$= \bar{\ell}\left[1 - \exp\left(-\frac{L_{am}}{\ell}\right)\right] = \bar{\ell}q(L_{am}). \tag{24}$$

$$\bar{Z} = \frac{L_{am} - \bar{\ell}q(L_{am})}{\bar{\ell}_n} + q(L_{am})$$

$$= \frac{\frac{L_{am}}{\bar{\ell}} - q(L_{am})}{\frac{\bar{\ell}_n}{\bar{\ell}}} + q(L_{am}) \tag{25}$$

$$= \frac{y_0 - q(L_{am})}{K} + q(L_{am}).$$

Using the equation (23) into the equation (25), it can be defined:

$$\overline{Z} = \frac{y_0}{K} - \frac{1}{K} + \frac{1}{K}\exp(-y_0) + 1 - \exp(-y_0)$$
$$= \frac{1}{K}(y_0 - 1) + 1 + \left(\frac{1}{K} - 1\right)\exp(-y_0). \qquad (26)$$

The obtained values $\overline{Z}$ from the equation (26) where given in (6).

On the basis of these data it is possible to determine the standard expenses of spare parts hi for conditions of the i area for a year operation of 100 vehicles, containing with n spare pieces in each:

$$h_i = \frac{100 * n}{t_{am_i}} * \overline{Z}_i. \qquad (27)$$

3 CONCLUSION

Economic liberalisation in post-communist countries engendered a strong competition in their vehicle spare parts markets and brought along a steep reduce of the level of state participation as an administrative structure in this area. In these new economic conditions the automobile operating companies have begun to realize an attempt to accurately determine spare parts expenses and to optimize inventory management of automobile components in their stocks. Thus, the subject has became even more urgent as dealer firms have an economic interest in the spare parts business and have a direct impact on the commercial profits of these private companies (Kabdi 2008).

The authors, as a result of their research activities, present a method for improving the operational readiness of automobile operating companies by increasing the working efficiency of vehicles.

1. Development of the probability theory and studies of vehicle durability allowed creating preconditions for further improvement of the fixed normative systems used in prediction of spare parts expenses, both in operation and in service and repair, as well as in planning the production capacity of automobile components and spare parts. However, the existing fixed normative systems of vehicle spare parts expenses do not adapt to the current market realities of vehicle operation, service and repair. Laws of automobile parts durability prefer using the probability theory methods of calculation to determine the supposed expenses of spare parts. Therefore, it's necessary to revalue the methodological approaches of development of these fixed expenses of spare parts.

2. Due to a large labour-intensiveness of the test duration study process in determining the longevity of every item of vehicle in detail, there is a need to use the fixed normative expenses offered by automobile producers only in the first years of operation of the new models with the relative correction, taking into account specific local conditions.

3. There is recommended the differentiation of the spare parts normative expenses for the practice of the particular country, firstly, for expenses under actual operating conditions, and secondly, for expenses to vehicle repair and maintenance services. Spare parts average expenses should be determined taking into account the renovation and repair of the worn-down parts along with the used parts. And it should be adjusted to the achieved level of reliability for automobile every 2–4 years of operation.

4. Following the studies of automobile production inherent reliability by the automobile exporters' main parameters that characterize the level of functional reliability of automobile components and their requirements as spare parts were selected.

5. The relationship between selected automobile reliability parameters and forecast of spare parts expenses was formed. This dependence has been expressed in the form of probability models for determining automobile spare parts expenses in each item separately. The proposed probability models allow not only to determine the operation, repair and service spare parts expenses, but also to establish the optimal production capacity of automobile components and spare parts. The correspondence between the calculated and actual expenses of spare parts is possible when the probability models reflect all real factors of existing physical process. Development of probability models determining the automobile spare parts expenses is not completed.

6. The proposed probability method for determining the automobile spare parts expenses allows composing the nomenclature of vehicle parts that limit the reliability of whole automobile, and which in turn makes it possible to concentrate in expenses production. The method can be used to determine spare parts expenses of other novel types of machines and mechanisms.

7. Solving the automobile spare parts problem in any country will need a complex approach, including the probability models and computer controlled systems of production and stock resources management through the modern logistic centres. The production of automobile spare parts will be, first of all, oriented to details with a low durability, when $\ell_i \leq 0,2L_{ami}$, or towards details about expenses which correspond with the normal distribution law and Weibull laws of distribution.

REFERENCES

Agarski, B. et al. 2012. Application of multi-criteria assessment in evaluation of motor vehicles' environmental performances, *Technical gazette*, vol. 19, no. 2 pp. 221–226, ISSN 1330-3651.

Boghani, B.A. & Brown, A. 2000. Meeting the Technology Management Challenges in the Automotive Industry. Warrendale: *Society of Automotive Engineers*, 132 p.

Bosák, M. 2010. Materiálové zhodnocovanie starých vozidiel, TU Košice, 28 p., ISBN 978-80-553-0418-2.

Kabdi, K. 2008 Metóda určovania spotreby a technológií výroby náhradných dielov strojov ako príspevok k teórií obnovy, In *Acta Mechanica Slovaca*, Košice, 4/ 2008, pp. 57–68.

Landmann, R. et al. 2001. The Future of the Automotive Industry, Warrendale: *Society of Automotive Engineers*, 262 p. ISBN 0768006880.

Majernik, M. et al. 2014. Innovative model of integrated energy management in companies. Quality Innovation Prosperity, 19(1), 22–32.

Safaei, M.H. et al. 2009. Development of a multi-criteria assessment model for ranking of renewable and non-renewable transportation fuel vehicles, *Energy*, 34 (1), pp. 112–125.

Valenčík, Š. 2011. Methodology for recovery machines. EVaOL TU Kosice, 330 p. ISBN 978-80-533-0679-7.

Yousefi, A. & Hadi-Vencheh, A. 2010. An integrated group decision making model and its evaluation by DEA for automobile industry, *Expert Systems with Applications*, 37 (12), pp. 8543–8556.

Development of a pricing system providing efficient price management

A.M. Kashirina
Business Faculty, Novosibirsk State Technical University, Novosibirsk, Russia

ABSTRACT: It is usual to think of price as the amount of money we must sacrifice to acquire something we desire. But from the producers point of view price is the sophisticated instrument of the commercial firm's policy. There are two approaches to pricing: the cost-based pricing taking into account the actual production and sale costs and the value-based pricing that take into consideration the value of the goods for the buyer. The majority of Russian enterprises has been using the cost-based pricing till now. Obviously the second way is more efficient because knowing buyer's needs you will sell more. But it is not right practice to move from cost-based practice to value-based practice ignoring old traditions. That is why author proposes using economic-and-mathematical models to combine these two approaches. The parametric price model is such a method that set price depending on the basic consumer parameters of the goods.

1 INTRODUCTION

It is necessary to pay attention to increase the competitiveness of national economy of Russia and to implement their activity for getting own place and integration the world economy considering market requirements.

The issues of pricing are particularly important for any industry sector in the modern conditions of managing. It is necessary to research the market and its features in order to implement an efficient pricing policy. The pricing practice established within the centralized management system has become practically insignificant. A wide range of pricing strategies and techniques developed abroad cannot be fully transferred to Russian enterprises.

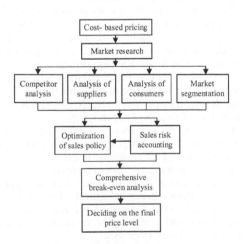

Figure 1. Algorithm for pricing.

Based on the pricing history in Russia and abroad and taking into account the current pricing situation I propose the following algorithm for pricing (see Fig. 1) (Dragunova 2006, Revetria & Maxamadaminovich 2014).

2 PROBLEM FORMULATION

There are two approaches to pricing (Utkin 1997):

- The cost-based pricing taking into account the actual production and sale costs;
- The value-based pricing that take into consideration the value of the goods for the buyer.

The majority of Russian enterprises has been using the cost-based pricing till now. Obviously the second way is more efficient because knowing buyer's needs you will sell more.

But it is not right practice to move from cost-based practice to value-based practice ignoring old traditions. That is why I propose using economic-and-mathematical models to combine these two approaches.

3 PROBLEM SOLUTION

3.1 *Parametric price model*

To build a mathematical pricing model I are going to use the parametric pricing approach. This approach makes it possible to calculate the price for a new modification of an existing item, that is associated with a change in a parameter, and explains why the price of the item is higher than the price for its previous version.

Parametric pricing is a method for setting prices for a parametric series of products according to a formalized model of price dependence on the values of the main consumer parameters of these products.

The first stage is selecting the main consumer parameters that are of some significance to the consumer.

Author considers the use of such models to the garment industry. Any garment is characterized by a number of parameters that have both direct and indirect effects on the price. The following characteristics affect consumer buying decisions: sales terms (discounts, delivery, storage, type of payment, store location), season, color, size, quality of fabric and accessories, garment quality, fashion trends. The parameters that make it possible to build a parameter tree are as follows:

- Parameters that affect the price from the buyer's point of view: sales terms, marketing circuit, mode of payment, delivery, storage, discounts, store location, model characteristics, size, height, accessories, price, consumption, quality, company, country, sex, fashion, season, class;
- Parameters that affect the price range from the manufacturer's point of view: prime cost, business expenses, actual cost of raw material and materials, consumption, processing price, salary, surcharge, accruals, shop costs, operating expenses, equipment maintenance costs, profitability, VAT, volume of production, paying capacity, store location, materials consumption, labour input.

The second step is to collect the source data for all the parameters. You must get all the values that the parameters can take.

The third stage is that of selecting the model and evaluating the model parameters by using regression analysis and statistics packages (see equations 1 and 2).

$$P = a \times X_1^{b1} \times X_2^{b2} \times ... \times X_i^{bi} \qquad (1)$$

or

$$LnP = a_o + b_1 \times lnX_1 + b_2 \times lnX_2 + ... + b_i \times lnX_i \qquad (2)$$

where P = price of goods; X_i = basic consumer parameters of goods; b_i = regression factors showing the influence of i-factor on price; a = a factor showing the influence of other parameters on price.

Author analyzes the relationships between each pair of factor features and between each of the factor features and the resultant one. We have collected and processed data on the 107 articles of Novosibirsk sewing company "Sinar". The Table 1 contains a description of the coded variables (because of the large amount of data in the table, only a portion of the information is provided).

Author analyzes the paired correlation coefficients presented in the Table 2.

Table 1. Correlation matrix.

Variable	Index	Variable	Index
X1	cost of materials	X9	profit
X2	consumption	X10	profitability
X3	accessories	X11	release price
X4	fare	X12	processing complexity
X5	total processing	X13	fashion trends
X6	salary	X14	exclusivity
X7	business expenses	X15	return rate
X8	prime cost	Y	product price

To understand how strong the linear relationship is between the features we compare the calculated coefficients with the information.

The value of correlation coefficients (Table 3):

Author is now going to give an economic interpretation of the correlation coefficients (Table 4).

As it can be seen from the Table 2, the resultant feature Y is noticeably related to the factor features of $X_1, X_2, X_4, X_7, X_8, X_9, X_{11}, X_{12}, X_{13}$ and X_{14}, it is very weakly related to the rest of the features. With regard to the strength of the relationship between the factor features, it should be noted that there is a very close relationship between such features as $X_1, X_2, X_4, X_7, X_8, X_9$ and X_{11}. As the relationship between the factor features stated is very strong, some of them should be excluded from the further study. The factor feature that meets the content of the task and that is more related to the resultant feature remains. The factor of this kind is factor X_{11} (the release price). Similarly, factor X_{12} (processing complexity) remains for analysis, but factors X_{13} and X_{14} are excluded.

So, on the basis of the statistical evaluation of the initial information we found out that it is roughly homogeneous, normally distributed grouping. The correlation analysis showed that the relationships between Y and X_{11} and X_{12} are substantial.

Thus, as a result of the correlation analysis the task stated above was reduced from finding multiple dependency of $Y = f (X_1, X_2, X_3, ..., X_{15})$ to finding $Y = f (X_{11}, X_{12})$ dependency.

In deciding on a regression equation, it is necessary to consider the purpose of its developing and the possibility of its economic interpretation. The most common way to develop a regression equation is to solve several types of that. Then, the equation that shows better statistical characteristics, corresponds to the content of the economic task and its aim, and lends itself to an economic interpretation is chosen from them.

The model of linearly exponential equation shows the best characteristics:

Table 2.   Correlation matrix.

	X1	X2	X3	X4	X5	X6	X7	X8	X9	X10	X11	X12	X13	X14	X15	Y
X1	1,0															
X2	–0,1	1,0														
X3	0,1	–0,6	1,0													
X4	**1,0**	0,0	0,1	1,0												
X5	0,0	0,3	–0,1	0,0	1,0											
X6	0,0	**0,5**	–0,2	0,1	**0,8**	1,0										
X7	**1,0**	0,0	0,1	1,0	0,0	0,1	1,0									
X8	**1,0**	0,1	–0,1	1,0	0,0	0,1	1,0	1,0								
X9	**0,5**	0,2	–0,3	**0,5**	–0,4	–0,3	**0,5**	0,6	1,0							
X10	–0,1	0,2	–0,3	–0,1	–0,5	–0,5	–0,1	–0,1	**0,8**	1,0						
X11	**0,9**	0,1	–0,1	0,9	–0,1	0,0	**0,9**	1,0	**0,7**	0,2	1,0					
X12	0,1	**0,6**	–0,9	0,1	0,2	0,2	0,2	0,3	**0,5**	0,3	0,4	1,0				
X13	0,3	**0,5**	–0,8	0,3	0,2	0,3	0,3	0,4	0,4	0,2	0,4	**0,9**	1,0			
X14	0,1	0,4	–0,1	0,2	0,1	0,4	0,2	0,2	0,3	0,2	0,3	**0,6**	0,6	1,0		
X15	–0,1	–0,1	0,1	–0,1	0,0	0,3	–0,1	–0,1	0,0	0,0	–0,1	0,0	–0,1	0,4	1,0	
Y	**0,5**	**0,5**	**–0,7**	**0,5**	0,1	0,2	**0,5**	**0,6**	**0,7**	0,3	**0,7**	**0,9**	**0,9**	**0,6**	0,0	1,0

Table 3.   Correlation ratio.

Correlation ratio	Communications
$0 \leq r_{ji} < 0.2$	almost no relationship
$0.2 \leq r_{ij} < 0.5$	a weak relationship
$0.5 \leq r_{ij} < 0.75$	a perceptible relationship
$0.75 \leq r_{ij} < 0.95$	a close relationship
$0.95 \leq r_{ij} \leq 1$	a relationship close to the functional one

Table 4.   Economic interpretation of the correlation coefficients.

Correlation ratio	The effect of the parameter on the price
$0 \leq r_{ji} < 0$	no effect
$0.2 \leq r_{ij} < 0.5$	a very weak effect, this parameter is not recommended for later use
$0.5 \leq r_{ij} < 0.75$	the parameter has a perceptible effect on the price, and the necessity to use it is determined just within the pricing process depending on its aims
$0.75 \leq r_{ij} < 0.95$	a relationship close to the functional one
$0.95 \leq r_{ij} \leq 1$	the parameter directly affects the price, and must be taken into account when the price is calculated.

– Multiple correlation coefficient R = 0.8975,
– Determination coefficient D = 0.8056,
– Error ε = 0.1564 (15.64%).

The fourth stage: the model analysis and the analysis of whether there is a possibility of its practical application for decision-making.

After the parametric pricing model has been built, that is that we know the parameters of the product, their values, to what extent they affect the price, and, therefore, the buyer decision, the whole bundle of goods can be divided into several segments. Here, the price for the products that have the worst characteristics will be a minimum one as the parameter values are low. This market segment is attractive to the poorest segment of the population. The products that have the best characteristics constitute the most expensive segment that is intended for wealthy customers. Here, the price will be a maximum one. Such characteristics as well as the prestige factor, in our view, will determine the buyer decision and make for spending money on this very product.

The reasons for using the parametric pricing method are not only to build the model and calculate the price using it. The ability to take advantage of the information obtained will make it possible to have positive results from the use of such models. Who should have knowledge of the parameters, their values, how relevant they are to the consumer and, consequently, the product worth? Should it be a specialist (or a department) responsible for calculating the final price?—Yes, definitely. Should it be a buyer?—It is hardly probable. It is hard, firstly, to distribute this information on each product to the ultimate consumer, and, secondly, to make customers willing to perceive and understand the figures, formulas and models. Therefore, there should be someone whose job is to convert the information that is in the form of figures into the one that is comprehensible to the customer. The best one who can do this job is a person who deals directly with the customer, that is a shop assistant. Consequently, it would be reasonable to talk about creating a network of own shops with specially trained shop assistants who are aware of their product, its strengths and weaknesses in contrast with other products and competitor's products.

Thus, the use of the parametric pricing method provides a means to calculate the price not only from the manufacturer's point of view, but also adjusted for the consumer demands and helps consumers choose from different products and manufacturers relying on how significant for them this product is.

The parametric pricing model will bring the manufacturer closer to the ultimate consumer. Shop assistants (not customers) should know where the prices come from to answer all the customer's questions. After the parametric model has been built, this is the easiest way to do this because it directly shows how the product parameters affect the price.

After the price has been calculated, it is necessary to optimize the distribution of output among the sales customers with the greatest benefit to the manufacturer.

3.2 Sale policy

It is advisable to optimize the price policy of an enterprise together with the sales policy because they greatly influence each other. The optimization model we propose does not aim at optimal prices, but, first of all, makes it possible to balance the output distribution taking into account the price policy of the enterprise.

Price policy directly depends on the choice of marketing circuit. Any well-designed pricing policy may fail if the distribution policy is not optimally selected. Marketing circuit should be selected based on market segmentation, product type, buyers, and etc. If necessary, intermediaries (dealers) should be included in the distribution pattern, despite the fact that expenses might increase.

Most enterprises are known to have the following marketing circuit: the distribution network, trade organizations, private entrepreneurs.

Price control for enterprises that have their own retail trade network is much easier. The introduction of the bar code system will make it possible to monitor price levels of final sales even hourly. Obviously, because of the current economic situation it is impossible even for companies that have their own distribution network to create the ideal product distribution system, but it is one of the steps in the right direction that should be taken, perfected and steps should be taken further. The fact that there are difficulties in controlling end prices gives cause for optimizing the sales policy of an enterprise at the stage of product distribution and price setting by the manufacturer.

So I suppose to optimize the distribution of goods through different sale cannels with the taking in consideration the price policy (see equation 3) (Cochovich 1994, Madatova 2002).

$$
\begin{cases}
F = \sum_{i=1}^{n} \\
\sum_{j=1}^{m} \left(\dfrac{M_{ij}}{M_{max}} \times \alpha + \dfrac{R_{ij}}{R_{max}} \times \beta + \dfrac{1 - RI_i}{(1 - RI)\,max} \times \gamma \right) \times V_{ij} \to max \\
\sum_{i=1}^{n} \sum_{j=1}^{m} V_{ij} \leq V \\
V_{ij} \geq 0, i = \overline{1, n}, j = \overline{1, m} \\
\sum_{j=1}^{m} P_{ij} \times V_{i,j} \leq O_i\, i = \overline{1, n}
\end{cases}
$$

$$(3)$$

where F = target function; M_{ij} = maneuverability of j-good price through i-channel; α = weight factor of maneuverability; R_{ij} = profitable of j-good price through i-channel; β – weight factor of profitable; RI_i = sale risk of i-channel; γ = weight factor of sale risk; $M_{max}, R_{max}, (1 - RI)_{max}$ = maximum value (from all channels) for corresponding factor; n = number of sale channels; m—number of goods; V_{ij} = sale volume of j-good through i-channel; V = whole sale of firm-producer; P_{ij} = price of j-good through i-channel; O_i = maximum income of i-channel.

Flexibility and profitability indexes, price, sales risks, sales volume of a manufacturing company, maximum turnover rates, and weighting factors are the specified values. Sales volumes for each channel are the target values that should be optimized.

We understand price flexibility as the width of a price change interval, the lower bound of which is a release price of a manufacturing company and the upper bound is a break-even price of a corresponding channel.

Alpha, beta, and gamma coefficients are weighting factors of the corresponding indexes demonstrating their significance for a manufacturer and adding up to one. They are evaluated by means of an experiment.

Risk evaluation involved a four-step process

1. Evaluation of each of the three risk components by means of the mathematical theory of probability.
2. Evaluation of risk probability. Each risk is reduced to a unified scale. The maximum probability is 100% (for the largest risk in absolute magnitude), the minimum probability is zero.
3. Identification of risk priorities. The priorities were obtained with the method of expert evaluation by the sales department personnel. They were asked to fill in a questionnaire.
4. Evaluation of the total risk as a weighted average adjusted for the priorities.

At the Table 5 you can see factors that we analyze for counting sale policy risks (Lapusta & Sharshukova 1998, Utkin 1998).

Supposed model lets us to optimize the goods distribution between existing sale channels. So we'll be able to select the optimal channels from the point of view of the price manager (Tereschenko & Madatova 2000).

3.3 The application of the developed approach

The application of the developed approach was carried out at one of the largest garment enterprises in Novosibirsk "Sinar".

One of the most common methods—the cost-based method—is used for pricing at "Sinar". So, this method suggests that the offered price is calculated as the sum of a cost price and a fixed additional value that is profits. One of the difficulties in its application is the fact that it is hard to determine the additional amount as there is no certain way or formula to calculate it. Everything changes depending on the product, season, and the state of competition. There is a danger that the amount that is added to the prime cost of a product or service seemed reasonable to a seller turns out to be unacceptable to a customer.

The markup level is dependent on the particular product, its consumer appeal. For example, S760 product that is cheap and focused on a wide range of customers is marked up 20%. A low markup is justified by a strong sales volume.

However, prestigious and expensive items are marked up between 50 and 110%. These products are focused on a wealthy customer and sales volumes of these exclusive models are weak.

Our task is to set the best price for a product for both a manufacturer and a consumer by means of cost, price and sales management.

Our analysis showed: only costs, percentage of profitability and VAT affect the end price, but neither competitor prices nor product worth to a customer is taken into account.

The given example shows the rigid cost approach with no possibility for other parameters to affect the price. In fact, the cost approach allows some adjustments to be made depending on the method used to account and allocate some costs. As a comparison, we consider the calculation of the product price using a full cost method and a direct cost method.

Price calculation using the full cost method enables to allocate fixed costs differently depending on the chosen base. The price can differ by more than twice depending on the method used to allocate fixed costs.

The use of the direct-cost method enables to deduct the price partially from the full product cost that comprises cost accounting on the whole.

This method is more acceptable because it enables to vary the price depending on the possible sales volume. However, neither the full cost method nor the direct cost method takes into account the buyer's opinion. Further pricing occurs in the sales and marketing department.

The attempt to build a parametric dependency between the price and the parameters that are significant to both the consumer and the manufacturer has resulted in identifying the relationship only with two factor features (out of more than 30 analyzed!) and with those that are not informative from the consumer perspective. Some of the parameters were dismissed even prior to building the model, as they affected pricing in no way, which was seen even with the naked eye (color, size, discounts, marketing circuit, and etc.). The results obtained confirm over again that pricing lacks in logic after the price is calculated using a cost-plus pricing method, then the process of pricing is rather random.

The suggested above tool for price calculation relying on product worth to a customer and competitor prices enables to create dependencies between the price and the parameters that are significant to the customer. The application of this method onward will make it possible to collect statistic data in order to build a parametric model that is more convenient, simple and justified to use for price calculation, price adjustment and price forecasting.

The further pricing is determined by the marketing circuit where the end product is found. At this stage, as it has already been noted, we propose to optimize the output programme taking into account the price policy of an enterprise.

In order to apply the approach proposed it is necessary to calculate price flexibility and sales policy risks.

We calculated the risks in reference to 188 customers of the sales department for the period of 3 years.

As there is a great amount of information to process it is impossible to give complete calculations, therefore, we elaborate on some results.

The risk priorities obtained with the method of expert evaluation are presented in the Table 6.

According to the classification developed this way the customers who have a stable turnover, small arrears (or they pay in advance), and the lowest percentage of the items returned are least risky. Only such customers are first of all entitled

Table 5. Analyzed factors for counting sale policy risks.

	Risk of loosing buyers	Rick of returning goods	Risk of unpayments
Input data for counting	Income	Value of returning	Debit debt
Factors for counting	Coefficient of variation	Value of returning for all period to income	The relation of Debit debt to income

to concessionary terms, such as: discounts, preliminary order for products, advantages of volume of purchase and assortment range.

Further application of the calculated risks is found in the optimization distribution model of the output programme relying on pricing policy.

In order to calculate the price flexibility, it is necessary to know the price at which the customer purchases the goods from the manufacturer, the costs of the manufacturer and the customer, and the break-even price of the customer. Price flexibility is defined a margin between a break-even price of the customer and a wholesale price of the manufacturer. Its economic essence is the potential of a manufacturer to increase a price to a break-even price.

The source data for optimizing the sales policy of one item within the existing marketing circuit of "Sinar" are presented in the Table 7.

Optimization Model Restrictions: the output volume – 10,000 pcs; "Sinar" company section buys at least 30% of the total output; "Sinar" has monetary assets of not more than 3,500,000 rubles; the merchandise market—no more than 1,500,000 rubles; the section in the store—2,500,000 rubles; the private enterprise—4,000,000 rubles; in order to increase sales volumes in the central department stores, the sections in the stores must purchase at least 70 per cent of the items purchased by private traders; sales volumes are not negative.

At the target function with the specified parameters and restrictions the distribution of the output volume between customers will as follows (see the Table 8):

Table 6. Evaluation of the risk priorities.

	Customer risk	Product return risk	Non-paymen risk	Total amount
Priority	1	2	3	
Weight	0.5	0.2	0.3	1

Table 7. Source data for optimization.

Marketing circuit	Sinar wholesale price, rub	Price flexibility, rub	Sinar profitability, %	Risk, %
Sinar manufacturer's outlet store	900	100	55	30
Merchandise market	920	95	56	20
Section in a store	950	100	60	45
Private enterprise	1000	120	65	50
Weighting factors		0.08	0.52	0.40

Table 8. Optimized output programme.

Marketing circuit	Sales volume	Turnover, rub
Sinar manufacturer's outlet store	3000	2,700,000
Merchandise market	609	560,301
Section in a store	2631	2,500,000
Private Enterprise	3759	3,759,398
Total amount	10,000	

This distribution of output is beneficial to the enterprise in terms of increasing its profitability, flexibility, risk reduction and the implementation of restrictions that are directed to support managerial decisions of an enterprise.

3 CONCLUSIONS

Given the model we will be able to calculate the price based on costs of producers and claims of buyers, explain the rise and fall of price, forecast the price trends, make decisions about new goods or reject old goods. So under current economic conditions this parametric model is beneficial for both producers and buyers.

But Russian enterprises are not able to collect so big statistic information to begin using new methods in short time. So we have to give the possibility for our enterprises to use both—old methods and collect data for new approaches. And just the parametric price model will allow us to combine old and new methods.

REFERENCES

Cochovich E. 1994 Financial mathematics. Moscow: Finance and Statistics.
Dragunova E. 2006. National features of Russian entrepreneurs; Int. J. Entrepreneurship and Small Business, 3(5): 607–620.
Lapusta M. & Sharshukova L. 1998. Risks in production. Moscow: Infra-M.
Madatova, A.M. 2002. Developing and Realization of Computer-Aided Pricing System Providing Efficient Price Management for Russian Enterprises; Proc. of The European Applied Business Research Conference, Venice 2003: 12–14.
Revetria, R. & Maxamadaminovich, U.K. 2014. Production advantages in textile and light industry and features of effective using of them in Uzbekistan; Proc. of the 8th international conference on management, marketing and finances (MMF'14) by WSEAS Press: 93–98. Greece.
Tereschenko O.V. & Madatova A.M. 2000. Optimization of pricing for industrial enterprises; Taxes and economics (10): 73–79
Utkin E. 1997. Price. Price policy. Moscow: Ekmos.
Utkin E. 1998. Risk-Management. Moscow: Ekmos.

Regulation of remaining deficit in subsystems of discrete manufacturing

M. Khayrullina & V. Mamonov
Faculty of Business, Novosibirsk State Technical University, Novosibirsk, Russian Federation

ABSTRACT: The article considers a production system consisting of subsystems ordered in a technological sequence. Each subsystem is a consumer of products of the previous subsystem and is a supplier of products for the subsequent subsystem. Delays in delivery of products in time are due to effects of random factors. The subsystem incurs reserve costs to compensate for the delay. These costs should be returned by the previous subsystem in the form of penalties. The article studies the dependence of the deficit value transferred on its value at the input.

1 INTRODUCTION

The manufacturing process at machine-building enterprises is multiple-stage, for this reason partial manufacturing operations are performed step-by-step by numerous production units. In discrete component manufacturing, production systems have a complicated, multisection structure with a significant amount of organizational and communication relations (Khayrullina et al. 2015, Fatkhutdinov 2008). Their stable and reliable functioning as a qualitative characteristic of the system operation depends on the functioning of its component elements (subsystems) (Lvov & Satanovsky 1984). The problem of product supply by subsystems "just in time" under actions of random factors upon the process throughout the manufacturing chain could be considered on the basis of introducing redundancy to subsystems and by using methods of economic regulation.

2 THE MAIN PROVISIONS

The following scheme of relationships between the production subsystems involved in the manufacturing process is discussed. According to the structure of manufacturing relationships, each subsystem upon completing its part of the manufacturing process supplies the products to the subsequent subsystem in the "supplier—consumer" chain on term. When random disturbances (both external and internal), which are random as related to an individual subsystem, affect the discrete component manufacturing, there is a difference between the actual and the scheduled dates of the products supply (Vatnik 1978). Due to the fact that in practice the products supply time could be set with the accuracy to a certain interval, so in the current system of the short-term production planning the "just in time" product supply is considered as a coordination interval, i.e. a certain period of time. This normative standard shows the accuracy (just in time) for production units to supply products for the allied production subsystems in the manufacturing chain. Thus, if production units in the "supplier–consumer" chain supply products within the coordination interval, so the system could be considered to meet the criteria of efficiency and reliability (Korovin 2006).

In the context of the structural and technological coherence of the subsystems by the sequence of the manufacturing processes performed, the product supply "just in time" is regulated by direct management methods of the manufacturing process. At the same time, with the aim of a better performance, the direct methods of production management are always completed with methods of economic regulation, which content and direction come down to taking into account economic interests of subsystems. The organizational form of such an account is a system including material incentives and compensation for damage (Ashimov 1986). Such a system ensures the stabilization of planning standards aimed at smooth carrying out of the manufacturing process by means of incorporating economic interests of the subsystems into the direct methods of management. With the efficient system of economic regulation, the effect of self-regulation is observed, i.e. the stable behavior of the manufacturing process is maintained by eliminating deviations from the "just in time" products supply between the subsystems. The system of economic regulation of the relationships between the subsystems supposes the interests of the subsystems in the elimination of deviations in products supply made by other subsystems at the input and corrected at the cost of its internal reserves, as well as in the minimization of deviations from the standard "just in time" products supply to other subsystems.

Due to the effect of random factors, the production of the subsystem might be delayed relative to the set interval of supply to a consumer subsystem. In this case, a supply time deficit appears. On the basis of a statistical model of the flow of subjects of labor over time throughout the "supplier—consumer" manufacturing chain, the instability in the subsystem operation can be described as average delays and time lags in production supply to the allied units (Mamonov & Poluektov 2005, Pervozvansky 1975).

Let the average duration of the input deficit into the subsystem throughout a set of production supplies be θ_0. In its turn, the subsystem as an intermediate link in the technological flow of the production phase functions with the output deficit θ. The value of the output deficit could be presented as a sum consisting of two components, i.e. the first component is an input deficit passed to the output (partially or completely); the second component is the output deficit added by the subsystem itself as a result of its instability due to the effect of random factors. Therefore the process of ensuring the "just in time" supply consists in the partial or complete elimination of the output deficit. The partial or complete elimination of the output deficit entails additional costs of the subsystem. Costs directed to a complete or partial elimination of the input deficit are in content costs for the regulation of the external instability Z. The value of these costs is a damage of the consumer subsystem and in accordance with the system of the coordination regulation of the process it should be compensated in the form of penalties W by the subsystem, which is the source of the deficit.

For the further study, it is necessary to present the character and content of the functions of costs and penalties.

The analysis of the process shows that the greater is the cost value necessary for the subsystem to eliminate the external instability in the form of the input deficit, the greater is its value. This can be explained by the fact that with the increase in the input deficit, the subsystem has less time for the "just in time" production supply to a consumer subsystem, which requires larger quantities of resources and reserves (Shamarina & Turovets 1981). That is why the higher are deviations from the "just in time" interval, the larger are the costs to compensate for the input deficit. On the other hand, the more are the regulation costs, the less is the value of the output deficit, which is a variable of the regulating process of the internal instability of the subsystem and which is transferred to a consumer subsystem.

So, the function characterizing the cost value of resources of the subsystem per each additional unit of the input deficit reduction has a shape of an increasing downward-convex function. Whereas the study includes two parameters, so it is quite sufficient to determine a function of costs for the external instability regulation as a function depending on the following two parameters: the value of the input deficit and output deficit.

The function of costs for the external instability regulation $Z(\theta_0, \theta)$ characterizes the use of resources and reserves of subsystems first of all in the procedures of ensuring redundancy to the subsystem. In practice, volume and time redundancy are widely used (Mamonov & Poluektov 2005). The general character of the dependence of costs on the values of the input and remaining (transferable) deficit, which are variables, shows that function of costs for the external instability regulation has a pronounced non-linear shape. The analysis of the $Z(\theta_0, \theta)$ function with the aim of its further approximation by simple functions shows that the most important is to determine interrelations between the growth of the function of regulation costs and the reduction of the remaining deficit. Let us approximate the function of regulation costs with the assumption of the consistency of the relative increase of costs per unit of the reduction of the output deficit:

$$[dZ(\theta_0,\theta) / Z(\theta_0,\theta)] / (d\theta/\theta) = -\alpha \qquad (1)$$

The solution of Equation 1 makes it possible to get the function of regulation costs depending on the values of the input deficit and the remaining deficit in the following way:

$$Z(\theta_0,\theta) = \vartheta\theta_0^{1+\alpha}(t+\theta)^{-\alpha}, \quad \alpha > 0 \qquad (2)$$

Obviously, the value of the costs directed for the regulation of the value of the output deficit and determined by Equation 1 takes into account the most important requirements. Firstly, the value of costs depends on the value of the input deficit and on the value of the output deficit; secondly, with the same values of the output deficit, the value of the regulation costs of the subsystem would be the greater, the greater is the value of the input deficit. The value of the output deficit $\theta = 0$ means, that the input deficit is completely compensated by the subsystem, i.e. the function of the regulation costs takes on the maximum value $Z(\theta_0, 0) = \max Z(\theta_0, \theta)$. The parameter ϑ, which is set exogenously and is equal to the value of costs for the full compensation per a unit of the input deficit, as well as the constant value t and the index of power α make it possible to flexibly change the shape of the function of costs and, thus, to ensure their agreement with the statistical data.

The use of linear rates in the penalty function in terms of the relationships of subsystems is proposed in work (Mamonov 2006). In this case, when the linear function of penalties for the output deficit is $W(\theta) = \beta\theta$, the minimum amount of the regulation costs and penalties is determined by the Equation 3 below:

$$W\left(\theta_0,\theta\right) = \min_{\theta}\left\{Z\left(\theta_0,\theta\right) + \beta\theta\right\} \quad at \quad \theta_0, \theta \geq 0 \quad (3)$$

At $\theta = 0$ the search for permissible values leads to the solving of the following equation:

$$\frac{dW\left(\theta_0,\theta\right)}{d\theta} = 0, \, at \quad \theta = 0 \quad (4)$$

We get:

$$-\alpha\vartheta\theta_0^{1+\alpha}\left(t+\theta\right)^{-\alpha-1} + \beta = 0, \quad \theta = 0 \quad (5)$$

from which:

$$t+\theta = \left(\frac{\alpha\vartheta}{\beta}\right)^{\frac{1}{1+\alpha}}\theta_0 = k\theta_0 \quad (6)$$

and with the remaining deficit equal to zero we get the value of the input deficit $\tilde{\theta}_0$ such that for all $\theta_0 \leq \tilde{\theta}_0$

$$\tilde{\theta}_0 = t/k \quad (7)$$

and the value of the function 3 is minimal.

For the values of the input deficit $\theta_0 > \tilde{\theta}_0$, the optimal value of the deficit θ_{ov} to be transferred to a consumer subsystem is found by solving the following Equation:

$$\frac{dW\left(\theta_0,\theta\right)}{d\theta} = 0, \quad at \quad \theta \geq 0, \theta_0 \geq \tilde{\theta}_0 \quad (8)$$

from which we find the value of the deficit at the output and which is equal to:

$$\theta_{ov} = k\theta_0 - t = k\left(\theta_0 - \tilde{\theta}_0\right). \quad (9)$$

Combining the results obtained for the two variation intervals of the value of the input deficit, we get finally the following Equation 10:

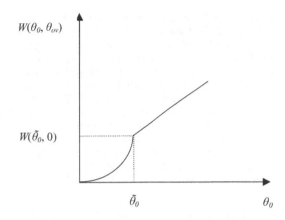

$$W(\theta_0, \theta_{ov})$$

$$W(\tilde{\theta}_0, 0)$$

$$\tilde{\theta}_0 \qquad \theta_0$$

Figure 1. Plotted function of costs for the external instability regulation of a production subsystem.

$$W\left(\theta_0,\theta\right) =$$
$$\begin{cases} (\beta/\alpha)\theta_0^{1+\alpha}\tilde{\theta}_0^{-\alpha}k = W\left(\theta_0,0\right), & \theta_0 \leq \tilde{\theta}_0 \\ (\beta/\alpha)\theta_0 k + \beta k\left(\theta_0 - \tilde{\theta}_0\right) = W\left(\theta_0,\theta_{ov}\right), & \theta_0 \geq \tilde{\theta}_0 \end{cases}$$
$$(10)$$

The general view of the function $W(\theta_0, \theta_{ov})$, which increases monotonically throughout, is given in Figure 1.

The optimal regulatory policy of the value of the remaining deficit is formulated on the basis of the result (Eq. 10): if the value of the input deficit in the subsystem is within the range of $0 < \theta_0 < \tilde{\theta}_0$, so it is economically advantageous for the subsystem to eliminate the deficit completely using its own resources and reserves. In this case, the amount of the reimbursement of the expenses for the redundancy to the subsystem in the form of penalties turns out to be less than the penalties, which would be brought to it in case of incurring of the remaining deficit.

In order to characterize the compensation share of the input deficit, it is proposed to apply the ratio of deficit reduction:

$$CRD = 1 - k\left[1 - \left(\tilde{\theta}_0/\theta_0\right)\right] =$$
$$1 - \left[k - \left(t/\theta_0\right)\right], \quad \theta_0 \geq \tilde{\theta}_0 \quad (11)$$

3 CONCLUSION

The obtained results can be used for considering the sequential scheme of a "supplier—consumer" manufacturing relationships or sequential fragments in general supply chains (Svensson 2002, 2003, Svensson et al. 2010, Chase et al. 2003). The functions of the regulation costs for the remaining deficit explicitly (Eq. 10) make it possible to use them in solving problems of the optimization of general costs for the redundancy in production subsystems and consequently in the system in general. The analysis of the publications allows us to hold out a hope that the introduction of the mechanism of the deficit reduction and compensation for damage as an element of planning management in supply chains will allow to increase their quality and performance (Battini et al. 2017, Egorov & Chernova 2015).

REFERENCES

Ashimov, A.A., et al. 1986. Soglasovannoe upravlenie aktivnymi proizvodstvennymi sistemami. Moscow: Nauka.

Battini, D. et al. 2017. Closed Loop Supply Chain (CLSC): *Economics, Modelling, Management and Control. International Journal of Production Economics* 183:319–321. DOI:10.1016/j.ijpe.2016.11.020.

Chase, R. et al. 2003. Production and Operations Management. Moscow: Publishing House "Williams".

Egorov, V.N. & Chernova, M.V. 2015. Supply Chain Management of Textile and Apparel Industry by PDM and ERP Software. *Proceedings of Higher Educational Institutions. Textile Industry Technology* 2(356): 5–10.

Fatkhutdinov, R.A. 2008. Proizvodstvennyi menedzhment. Saint-Petersburg: Piter.

Khayrullina, M. et al. 2015. Production systems continuous improvement modeling. *Quality innovation prosperity* 19(2): 73–86.

Korovin, D.I. 2006. Matematicheskie metody upravleniya ekonomicheskoi nadezhnostiyu proizvodstvennykh system: Author's Abstract of a Doctor of Sciences (Economics) Dissertation. Ivanovo.

Lvov, Y.A. & Satanovsky, R.L. 1984. Intensifikatsiya mashinostroitelnogo proizvodstva: organizatsiya i planirovanie. Leningrad: Mashinostroenie.

Mamonov, V.I. & Poluektov, V.A. 2005. Effektivnost vnutrennikh regulyatorov operativnogo upravleniya v obespechenii ustoichivosti funktsionirovaniya predmetno-zamknutykh uchastkov. *Proceedings of Higher Educational Institutions.* Machine Building 9: 71–76.

Mamonov, V.I. 2006. Regulirovanie velichiny ostatochnogo defitsita pri vzaimodeistvii proizvodstvennykh zveniev. *Polzunovsky Vestnik* 4–2: 255–259.

Pervozvansky, A.A. 1975. Matematicheskie modeli v upravlenii proizvodstvom. Moscow: Nauka.

Shamarina, L.I. & Turovets, O.G. 1981 Opredelenie optimalnogo urovnya rezervov dlya obespecheniya nadezhnosti proizvodstvennykh system. Teoriya i praktika organizatsii proizvodstva mashinostroitelnoi produktsii v sovremennykh usloviyakh: 119–123. Voronezh: VPI.

Svensson, G. 2002. The Theoretical Foundation of Supply Chain Management: A Functionalist Theory of Marketing. *International Journal of Physical Distribution & Logistics Management* 32(9): 734–754.

Svensson, G. 2003. Sub-Contractor and Customer Sourcing and the Occurrence of Disturbances in Firms' Inbound and Outbound Logistics Flows. Supply Chain Management: *An International Journal* 8(1): 41–56.

Svensson, G. et al. 2010. Balancing the Sequential Logic of Quality Constructs in Manufacturing (productive)-Supplier Relationships—Causes and Outcome. *Journal of Business Research* 63: 1209–1214.

Vatnik, P.A. 1978. Statisticheskie metody operativnogo upravleniya proizvodstvom. Moscow: Statistika.

Financial reporting for public sector entities by conceptual framework

M. Kicová
Faculty of Business Economics with seat in Košice, University of Economics in Bratislava,
Košice, Slovak Republic

ABSTRACT: The Conceptual Framework for general purpose financial reporting establishes the concepts that are to be applied in developing IPSAS standards applicable to the preparation and presentation of general purpose financial reports of public sector entities. The paper deals with requirements contained in the Conceptual Framework, namely role and authority of the Conceptual Framework, the objectives and users of general purpose financial reporting, qualitative characteristics of information in the general purpose financial reports, definition of the reporting entity, elements and recognition in the financial statements, measurement of assets and liabilities in the financial statements and presentation in the general purpose financial reports. The application of the requirements of the Conceptual Framework provides transparency of financial reporting by public sector entities. This means that the general purpose financial reporting of public sector entities provides transparent information that is useful for accountability and decision-making purposes.

1 INTRODUCTION

Globalization is a process which involves the increasingly interdependent markets of different countries. Globalization requires the breakdown of borders not only in economic, political, and social environment, but also requires the free flow of economic information. Whereas the main source of economic information is accounting and its harmonization is an important part of economic globalization. Harmonization of accounting is the process of convergence of accounting methods, accounting principles, accounting rules, but above all the contents of the financial statements to be comparable and understandable to the user in any country. Public sector accounting is an important part of the harmonization and regulation of accounting since 2000. As a part of the harmonization of accounting, the International Public Sector Accounting Standards Board (IPSASB) develops and issues set of recommended accounting standards for public sector entities referred to as International Public Sector Accounting Standards (IPSASs). The application of IPSAS standards helps to improve financial management, increase accountability in the public sector, and improve the quality and comparability of financial reporting around the world. An important step in the harmonization of public sector accounting was the release of the Conceptual Framework for General Purpose Financial Reporting by Public Sector Entities. The Conceptual Framework is a fundamental document that defines and explains the key characteristics of public sector

entities. It includes methods to be used in the creation of IPSAS standards and recommended principles to be applied in preparing the general purpose financial statements. The International Public Sector Accounting Standards Board has worked on development project Conceptual Framework since 2006, which was divided into four phases:

– Phase 1: History of the Conceptual Framework;
– Phase 2: Elements and Recognition in Financial Statements;
– Phase 3: Measurement of Assets and Liabilities in Financial Statements; and
– Phase 4: The Presentation of Information in General Purpose Financial Reports.

All phases of the development of the Conceptual Framework are interlinked. Phase 1 was completed in January 2013; the other phases were completed in October 2014. Phase 1: History of the Conceptual Framework includes four areas, which are divided into four chapters:

– Chapter 1: Role and Authority of the Conceptual Framework;
– Chapter 2: Objectives and Users of General Purpose Financial Reporting;
– Chapter 3: Qualitative Characteristics; and
– Chapter 4: Reporting Entity.

This paper includes detail knowledge of the areas related to the requirements contained in the Conceptual Framework, namely role and authority of the Conceptual Framework, the objectives and users of general purpose financial reporting, qualitative characteristics of information in the general purpose financial reports, definition of the reporting entity, elements and recognition in the financial statements, measurement of assets and liabilities in the financial statements and presentation in the general purpose financial reports.

2 LITERATURE REVIEW

The researched object that is financial reporting for public sector entities by the Conceptual Framework was chosen because of its timeliness and dynamic development. This topic was not a comprehensive solution until 2014 because there was not Conceptual Framework for public sector reporting. We got information about the researched object from book and magazine sources, conference proceedings and from our own previous knowledge of the research activities. We have worked with the current literature published in 2016, which was mainly in English. This literature is listed in the references section. The above issue deals mainly with Conceptual Framework for General Purpose Financial Reporting by Public Sector Entities that was first published in October 2014 on the IFAC website (IFAC 2014) and with provisions of relevant International Public Sector Accounting Standards that are listed in the Handbook of International Public Sector Accounting Pronouncements, which was published in 2016 on the IFAC website and was divided into two volumes (IFAC 2016a), (IFAC 2016b). We also used our knowledge of our research activities that is listed in monograph related to international harmonization of financial reporting in the general purpose financial reports of the public sector (Kršeková 2011) and articles in scientific journals and conference proceedings (Kicová 2006), (Kicová 2007), (Kršeková & Pakšiová 2014), (Kršeková & Pakšiová 2015a), (Kršeková & Pakšiová 2015b), (Kršeková 2016), (Moscalu 2011), (Pakšiová & Kubaščíková & Kršeková 2015) and (Pakšiová & Kršeková 2014). The knowledge gained forms the basis for the processing of results and conclusion.

3 DATA AND METHODOLOGY

The aim of this paper is to describe and analyze financial reporting for public sector entities by Conceptual Framework in terms of the requirements of the Conceptual Framework in the public sector. We applied epistemology as a basic method for researching this problem. Standard research methods, such as selection, analysis and synthesis, presenting basic methodical approach to paper processing are applied. We combined the obtained knowledge to form new, higher level of knowledge of research problems. In particular, ways of understanding and explaining the role and authority of Conceptual Framework, objectives and users of general purpose financial reporting, qualitative characteristics of information in the general purpose financial reports, definition of the reporting entity, elements and recognition in the financial statements, measurement of assets and liabilities in the financial statements and presentation in the general purpose financial reports, the inductive-deductive and analytic-synthetic logical scientific methods are used. In the conclusion, we stated opinions, in which we highlighted the importance of financial reporting for public sector entities by Conceptual Framework in a way that meets the objectives of financial reporting to provide information that is useful for accountability and decision-making purposes.

4 RESULTS AND DISCUSSION

The Conceptual Framework establishes and makes explicit the concepts that are basis for financial reporting by public sector entities prepared under the accrual basis of accounting. International Public Sector Accounting Standards Board applies these concepts in developing International Public Sector Accounting Standards (IPSASs) applicable to the preparation and presentation of general purpose financial reports (GPFRs) of public sector entities under the accrual bases of accounting. Financial statements prepared under the accrual basis of accounting inform users of those statements of past transactions involving the payment and receipt of cash during the reporting period, obligations to pay cash or sacrifice other resources of the public sector entity in the future, the resources of the public sector entity at the reporting date and changes in those obligations and resources during the reporting period. The Conceptual Framework does not establish authoritative requirements for financial reporting by public sector entities that adopt IPSAS standards, nor does it override the requirements of IPSAS standards. Authoritative requirements relating to the recognition, measurement and presentation of transactions and other events that are reported in general purpose financial reports are specified in IPSAS standards. The Conceptual Framework can provide guidance in dealing with financial reporting issues not dealt with by IPSAS standards. In these circumstances, preparers can refer to and consider the applicability of the definitions, recognition criteria, measurement principles, and other concepts identified in the Conceptual Framework. General purpose financial reports constitute a central element of transparent financial reporting by governments and other public sector entities. They are characterized as financial reports intended to meet the information needs of users who are unable to require the preparation of financial reports tailored to meet their specific information needs. General purpose financial reports contain multiple reports, each responding more directly to certain aspects of the objectives of financial reporting and matters included within the scope of financial reporting. General purpose financial reports encompass financial statements including their notes and the presentation of information that enhances complements and supplements the financial statements.

The objectives of financial reporting by public sector entities are to provide information about the public sector entity that is useful to users of general purpose financial reports for accountability and decision-making purposes. Financial reporting provides information useful to users of general purpose financial reports. The objectives of financial reporting are determined by reference to the users of general purpose financial reports and their information needs. Government and other public sector entities receive funding from taxpayers, donors, lenders and other of resources providers for use in the provision of services to citizens and other service recipients. These entities are accountable for their management and use of resources to those that provide them with resources, and to those that depend on them to use those resources to deliver necessary services. Entities that provide the resources and receive, or expect to receive, the services also require information as input for decision-making purposes. General purpose financial statements of public sector entities are developed primarily to respond to the information needs of service recipients and resource providers who do not possess the authority to require a public sector entity to disclose the information they need for accountability and decision-making purposes. The legislature and members of parliament are also primary users of general purpose financial reports and use of general purpose financial reports when acting in their capacity as representatives of the interests of service recipients and resource providers. The Conceptual Framework states that the primary users of general purpose financial reports are services recipients and their representatives and resource providers and their representatives. General purpose financial reports provide information about the financial results, financial performance and cash flows of the public sector entity during the reporting period. They also provide information about the assets, net assets/equity and liabilities at the reporting date, changes in the assets, net assets/equity and liabilities during the reporting period and the results achieved in the provision of services.

For accountability and decision-making purposes, service recipients and resource providers will need information that supports the assessments of the performance of the public sector entity during the reporting period, the liquidity and solvency of the public sector entity, the sustainability of the public sector entity's service delivery and other operations over the long term and changes therein as a result of the activities of the public sector entity during the reporting period and the capacity of the public sector entity to adapt to changing circumstances, whether changes in demographics or changes in domestic or global economic conditions which are likely to impact the nature or of the activities it undertakes and the services it provides.

General purpose financial reports present financial and non-financial information about economic and other phenomena. The qualitative characteristics of information included in general purpose financial reports are the attributes that make that information useful to users and support the achievement of the objectives of financial reporting to provide information useful for accountability and decision-making purposes. The qualitative characteristics of information included in general purpose financial reports of public sector entities are relevance, faithful representation, understandability, timeliness, comparability, and verifiability. Constraints on information included in general purpose financial reports are materiality, cost-benefit, and achieving an appropriate balance between the qualitative characteristics. Qualitative characteristics are applied to all financial and non-financial information reported in general purpose financial reports, including historic, prospective and explanatory information. The extent to which the qualitative characteristics can be achieved may differ depending on the degree of uncertainty and subjective assessment or opinion involved in compiling the financial and non-financial information.

The public sector reporting entity is characterized as a government or other public sector organization, program or identifiable area of activity that prepares general purpose financial reports. The public sector reporting entity may comprise two or more separate entities that present general purpose financial reports as if they are a single entity. In this case, the reporting entity is referred to as a group reporting entity.

Key characteristics of a public sector reporting entity are that (IFAC 2016a):

– It is an entity that raises resources from, or on behalf of, constituents and/or uses resources to undertake activities for the benefit of, or on behalf of, those constituents; and
– There are service recipients or resource providers dependent on GPFRs of the entity for information for accountability or decision-making purposes.

General purpose financial reports provide information useful to users for accountability and decision-making purposes. Service recipients and resource providers are the primary users of general purpose financial reports. As a result, a key characteristic of a reporting entity, including a group reporting entity, is the existence of service recipients or resource providers who are dependent on general purpose financial statements of that entity or group of entities for information for accountability or decision-making purposes.

The Conceptual Framework internationally regulate definition of the elements in the financial statements of public sector entities recognized in the statement of financial position and statement of financial performance. An asset under the Conceptual Framework is a resource presently controlled by the public sector entity as a result of a past event. A liability under the Conceptual Framework is a present obligation of the public sector entity for an outflow of resources that results from a past event. Net financial position under the Conceptual Framework is the difference between assets and liabilities after adding other resources and the deducting other obligations recognized in the statement of financial position. A net financial position can be a positive or negative residual value. Expense under the Conceptual Framework is a decrease in the net financial position of the public sector entity, other than decreases arising from ownership distributions. Ownership distributions under the Conceptual Framework are outflows of resources from the public sector entity, distributed to external parties in their capacity as owners, which return or reduce an interest in the net financial position of the public sector entity. Revenue under the Conceptual Framework is an increase in the net financial position of the public sector entity, other than increases arising from ownership contributions. Ownership contributions under the Conceptual Framework are inflows of resources to a public sector entity, contributed by external parties in their capacity as owners, which establish or increase an interest in the net financial position of the public sector entity. Surplus or deficit for the period under the Conceptual Framework is the difference between revenue and expense reported on the statement of financial performance.

The Conceptual Framework internationally regulate the recognition criteria for elements in the financial statements of public sector entities. Items in the financial statements satisfies the definition of an element in the Conceptual Framework and can be measured in a way that achieves the qualitative characteristics and considers constraints on information in the general purpose financial reports of public sector entities.

Measurement of assets and liabilities in the financial statements of public sector entities is a way of expressing assets and liabilities in monetary terms. Elements in the financial statements can be measured by requirements in Conceptual Framework. The objective of measurement is to select those measurement bases that most fairly reflect the cost of services, operational capacity and financial capacity of the public sector entity in a manner that is useful in holding the accountability and for decision-making purposes. Using the correct method of measurement means the selection of appropriate measurement bases of assets and liabilities providing a true and fair view of the assets and liabilities of public sector entities and determines whether the measurement of assets and liabilities achieves the qualitative characteristics and considers constraints on information in the general purpose financial reports of public sector entities.

The Conceptual Framework internationally regulate definition of measurement bases of assets and liabilities.

Historical cost for assets under the Conceptual Framework is the consideration given to acquire or develop an asset, which is the cash or cash equivalents or the value of the other consideration given, at the time of its acquisition or development. Market value for assets under the Conceptual Framework is the amount for which an asset could be exchanged between knowledgeable, willing parties in an arm's length transaction. Replacement cost for assets under the Conceptual Framework is the most economic cost required for the public sector entity to replace the service potential of an asset (including the amount that the public sector entity will receive from its disposal at the end of its useful life) at the reporting date. Net selling price for assets under the Conceptual Framework is the amount that the public sector entity can obtain from sale of the asset, after deducting the costs of sale. Value in use for assets under the Conceptual Framework is the present value to the public sector entity of the asset's remaining service potential or ability to generate economic benefits if it continues to be used, and of the net amount that the public sector entity will receive from its disposal at the end of its useful life.

Historical cost for liabilities under the Conceptual Framework is the consideration received to assume an obligation, which is the cash or cash equivalents, or the value of the other consideration received at the time the liability is incurred. Cost of fulfillment for liabilities under the Conceptual Framework is the costs that the public sector entity will incur in fulfilling the obligations represented by the liability, assuming that it does so in the least costly manner. Market value for liabilities under the Conceptual Framework is the amount for which a liability could be settled between knowledgeable, willing parties in an arm's length transaction. Cost of release for liabilities under the Conceptual Framework is the amount that either the creditor will accept in settlement of its claim, or a third party would charge to accept the transfer of the liability from the obligor. Assumption price for liabilities under the Conceptual Framework is the amount which the public sector entity would rationally be willing to accept in exchange for assuming an existing liability.

The basis of measurement of assets and liabilities is choosing the right measurement bases that are used for measurement of assets and liabilities. Choosing the right measurement bases affecting the explanatory power of all information reported in the financial statements of public sector entities and therefore has a significant impact on the financial situation and financial performance of public sector entities.

Presentation on information in general purpose financial reports is characterized as the selection, location and organization of information that is reported in the general purpose financial reports. The role of presentation is to provide information that meets the objective of financial reporting to provide information useful for accountability and decision-making purposes in compliance with the qualitative characteristics and considers constraints on information in the general purpose financial reports of public sector entities. Decisions on selection, location and organization of information are made in response to the needs of users for information about economic or other phenomena. Presentation decisions may result in the development of new general purpose financial reports, the movement of information between existing reports, the amalgamation of existing reports or be detailed decisions on information selection, location and organization within general purpose financial reports. The amount or type of information selected could have implications on whether it is included in a separate report or organized into tables or separate schedules. The result of the selection decision information is specifying the information to be disclosed in the financial statements or in general purpose financial reports outside the financial statements. Selection of information depends on its character and information has to be useful to users. Information is selected for display or disclosure in general purpose financial reports. Information selected for display communicates key messages in general purpose financial reports, while information selected for disclosure makes displayed information more useful by providing detail that will help users to understand the displayed information. Disclosure is not a substitute for display. General purpose financial reports contain key messages that are communicated, so

general purpose financial reports contain displayed information. Displayed information is kept to a concise, understandable level, so that users can focus on the key messages presented and not be distracted by detail that could otherwise obscure those messages. Displayed information is presented prominently, using appropriate presentation techniques such as clear labeling, borders, tables, and graphs. Disclosures and displayed information are necessary to achieve the objectives of financial reporting. That information necessary to meet the objectives of financial reporting is available for all public sector entities, while allowing information to be displayed in a manner that reflects the nature and operations of specific public sector entities. Decisions on information selection require continuing and critical review. Information identified for possible selection is reviewed as it is developed and considered for presentation, regarding its relevance, materiality and cost-benefit, although all the qualitative characteristics and constraints are applied to decisions on information selection. Decisions on information location are made about which report information is located within and which component of report information is located. The location of information has an impact on information's contribution to achievement of the objectives of financial reporting and the qualitative characteristics. Location may affect the way that users interpret information and the comparability of information. Location information helps to achieve the objectives of financial reporting and compliance with the qualitative characteristics considering the constraints of qualitative characteristics. Location information can affect the understand ability and the comparability of information. Suitable location information may be expressed relative importance of information, the nature and connection with other data. Location may be used to convey the relative importance of information and its connections with other items of information, convey the nature of information, link different items of information that combine to meet a user need and distinguish between information selected for display and information selected for disclosure. Information organization addresses the arrangement, grouping and ordering of information, which includes decisions on how information is arranged within general purpose financial reports and the overall structure of general purpose financial reports. Information organization involves a range of decisions including decisions on the use of cross-referencing, tables, graphs, headings, numbering, and the arrangement of items within a particular component of a report, including decisions on item order. How information is organized can affect its interpretation by users. Decisions about the organization of information consider important relationships between information and whether information is for display or disclosure. The information displayed within the financial statements are usually classified in the appropriate groups, including totals and subtotals, the arrangement provides a structured overview of the financial position, financial performance, and cash flows of the reporting entity.

5 CONCLUSION

The aim of this paper was to develop detailed knowledge of the specific problems of financial reporting by public sector entities, namely role and authority of the Conceptual Framework, the objectives and users of general purpose financial reporting, qualitative characteristics of information in the general purpose financial reports, definition of the reporting entity, elements and recognition in the financial statements, measurement of assets and liabilities in the financial statements and presentation in the general purpose financial reports. The Conceptual Framework for General Purpose Financial Reporting by Public Sector Entities and relevant International Public Sector Accounting Standards cover these issues. The result of examination of that topic is a comprehensive overview of knowledge about the requirements of the Conceptual Framework for General Purpose Financial Reporting by Public Sector Entities. The application of the requirements of the Conceptual Framework in general purpose financial reports by public sector entities provides transparent information in a way that meets the objectives of financial reporting to provide information that is useful to users for accountability and decision-making purposes. General purpose financial reports have the important impact on the future ensuring of public sector entities' finances and thereby allow increasing transparency of public sector entities' finances and ensure the transparency in financial reporting by governments and other public sector entities.

REFERENCES

International Federation of Accountants. 2014. *Conceptual Framework for General Purpose Financial Reporting by Public Sector Entities.* Retrieved 27th March 2017 from IFAC: http://www.ifac.org/system/files/publications/files/IPSASB-2014-Handbook-Vol-I.pdf.
International Federation of Accountants. 2016. *Handbook of International Public Sector Accounting Pronouncements. Volume I.* Retrieved 27th March 2017 from IFAC: http://www.ifac.org/system/files/publications/files/IPSASB-2016-Handbook-Vol-I.pdf.
International Federation of Accountants. 2016. *Handbook of International Public Sector Accounting Pronouncements. Volume II.* Retrieved 27th March 2017 from IFAC: http://www.ifac.org/system/files/publications/files/IPSASB-2016-Handbook-Vol-II.pdf.
Kicová, M. 2006. Medzinárodné účtovné štandardy pre verejný sektor (IPSAS) ako základ medzinárodnej harmonizácie účtovníctva vo verejnom sektore. In *AIESA 2006*: 1–7. Bratislava: EKONÓM.
Kicová, M. 2007. Medzinárodná harmonizácia účtovníctva verejného sektora. In *Mezinárodní Baťova doktorandská conference*: 1–14. Zlín: Univerzita Tomáše Bati ve Zlíně.
Kršeková, M. 2011. *Medzinárodné účtovné štandardy pre verejný sektor – IPSAS.* Bratislava: IURA EDITION.
Kršeková, M. & Pakšiová, R. 2014. Long-term fiscal sustainability of public sector entities' finances. In *Finance and risk 2014, 16th international scientific conference, 24th–25th November 2014*, (1): 115–124. Bratislava: EKONÓM.
Kršeková, M. & Pakšiová, R. 2015. Rámcová osnova pre finančné vykazovanie subjektov verejného sektora určené na všeobecný účel—Fáza 1. In *Ekonomika a informatika: vedecký časopis FHI EU v Bratislave a SSHI*, 14 (1): 59–73. Bratislava: EKONÓM. Retrieved 27th March 2017 from: http://www.fhi.sk/files/ekonomika_a_inform/Ekonomika_a_informatika_1_2015_ISSN.pdf.
Kršeková, M. & Pakšiová, R. 2015. Financial reporting on information about the financial position and financial performance in the financial statements of the public sector. In *Finance and risk 2015, 17th international scientific conference, 23rd–24th November 2015*, 14 (1): 136–145. Bratislava: EKONÓM.
Kršeková, M. 2016. Zmluvy o licenciách na poskytovanie služieb z pohľadu poskytovateľa—subjektu verejného sektora. In *Ekonomika a informatika: vedecký časopis FHI EU v Bratislave a SSHI*, 14 (1): 1–11. Bratislava: EKONÓM.
Moscalu, M. 2011. Financial integration in the euro area and SME's access to finance: evidence based on aggregate survey data. In *Financial Studies*, 19 (2): 51–66. Retrieved 27th March 2017 from: http://fs.icfm.ro/vol19i2p51-66.pdf.
Pakšiová, R. et al. 2015. Valuation of all fungible items' assets reduction and its influence on measuring companys' performance. In *Financial management of firms and financial institutions. 10th international scientific conference: 7th-8th September 2015*: 921–928. Ostrava: VŠB—Technical university of Ostrava.
Pakšiová, R. & Kršeková, M. 2014. Ethical Principles in Distribution of Recognised Gain in a Company. In *Ethics as an essential condition for sustainable economic development: proceedings of scientific papers*: 69–74. Bratislava: EKONÓM.

New Trends in Process Control and Production Management – Štofová & Szaryszová (Eds)
© 2018 Taylor & Francis Group, London, ISBN 978-1-138-05885-9

WIP inventory optimization in electrode production

O. Kislitsyna, A. Chuvaev & M. Khayrullina
Faculty of Business, Novosibirsk State Technical University, Novosibirsk, Russian Federation

ABSTRACT: Inventory management is an important element in the system of current asset management and production management. The world practice has developed methodological techniques of inventory management in a manufacturing enterprise, including various methods and approaches. The methods and tools of inventory management based on the principles of pull systems (Kanban, dynamic buffer management, in particular) could provide such examples. But in cases, when the production lead time exceeds the delivery time, the methods and tools mentioned above are to be oriented to form a high inventory level. The authors propose an optimization algorithm of works-in-progress inventory on the example of an electrode production enterprise. The solution is based on the combination of pull system tools and mathematical statistics methods. The application of the elaborated model makes it possible to reduce sufficiently the inventory level, to release current assets while maintaining the level of delivery reliability required by the consumer.

1 INTRODUCTION

The key problem of inventory management in a manufacturing enterprise is to ensure uninterrupted material flows, which are formed during the process of the transformation of raw materials into manufactured goods. The material flow as an object of industrial management forms a complex of the objects of labor used for production purposes from the source to the termination of the products manufacturing process within a certain time frame. This problem solution should meet the condition of the optimum inventory level to be accumulated in terms of financial expenses. On one hand, the inventory should be sufficient to ensure the uninterruptable material flow and that the manufacturing process is not stopped due to the absence of the objects of labor at the input, and the enterprise should carry out its obligations to clients as a reliable supplier. On the other hand, the inventory should not be excessive, because it leads to some problems, and the main problem is the freeze of financial resources, and, consequently, the decrease of the speed of current assets turnover.

The inventory level optimization is still an important problem of the modern production and operational management of an enterprise.

2 KEY THEORETICAL PRINCIPLES AND METHODS

The general-theoretical basics of inventory management are widely described in scientific literature.

Surely, theoretical advances contribute sufficiently to the inventory management science. At the same time, the application of theoretical models in practice requires their adaptation with the aim to elaborate concrete mechanisms and tools of inventory management in modern production systems.

Inventories can be classified in different ways depending on the aims of studies and analysis of problems.

The object of management in the proposed model is works-in-process inventory (WIP inventory). WIP inventory includes works and services which are completed, but not accepted by the client. WIP inventory includes also backorders and residual stock of semi-finished products of own manufacture. Materials and semi-finished products which are in process refer to WIP inventory under condition that they are already processed.

Approaches to WIP inventory management depend on the type of the logistic system used at the enterprise. For push systems, inventories are planned starting from the supply process to the process of sales of finished products. At the same time, the inventory level is calculated beforehand on the basis of the sales forecast or consolidated annual plans according to preliminary agreements. For pull systems, the formation of inventories is based on changes in demand for the products of the enterprise, and WIP inventory movements are determined by the sales process.

A logistic system type of the enterprise determines a system of inventory planning.

The Japanese system JIT (just-in-time) is a well-known system of inventory planning which involves the pull principle. Inventory management in the JIT system is carried out in the on-line mode, and it is based on the appearance of a need for a particular production. The tool which starts up the movement of objects of labor according to the arisen need for a particular product is a Kanban (signboards).

Another system elaborated within the framework of the theory of constraints (TOC) is based on inventory management by means of dynamic buffer management. Conceptual solutions referring to materials management by means of buffers differ depending on the type of manufacturing environments, i.e. make-to-stock (MTS), make-to-order (MTO) or make-to-availability (MTA).

Many factors influence the choice of a system of WIP inventory planning, including the following factors: demand fluctuations from consumers, specifics of technologies used, state of production facilities, fulfillment of obligations by suppliers of raw and other materials, etc.

In order to take into account the influence of these factors managers are to analyze a great deal of information, which results in a variety of solution options. The authors propose a solution which is based on the following basic provisions:

– WIP inventory management should take into consideration the specifics of material flows in a concrete enterprise dependent on the technology of production and organization of flows in space and in time;
– Specified inventory levels at stock locations should be determined taking into account profiles of the production facilities used at different stages of the manufacturing process;
– The information system should reflect up-to-date information of the required quality about the movement of objects of labor in the production process.

The determination of a type of the material flow is necessary in order to reveal problem points of the flow which arise in case of its convergence or divergence. From this point of view, the flows can be of the following types: A, V, I, T (Schragenheim & Dettmer 2000).

When production facilities are used unevenly, which is typical of a discrete component manufacturing, it is necessary to reveal that item of equipment or a group of resources which affect efficiency of the whole manufacturing process (performance of the production system), a bottleneck resource of the process. In other words, in the process of the operations ordered in a technological sequence, some production facilities can be underutilized, and some production facilities require a specific operating mode, which should be set in order to maximize their usage economically. In this connection, it is proposed to use in the model the prioritization of

the execution of production orders using the criterion of marginal revenue gained by the enterprise from individual orders. This provision is considered as a key principle for the formation of inventory before the bottleneck of the process.

The parameters of the data which form the information support to the process play an important role in the on-time management decision making. This process is described in detail in the paper (Khayrullina & Kislitsyna 2015, Pushkar & Dragunova 2014).

The proposed model combines pull principles and methods of mathematical statistics.

Many authors give similar definition of pull system. In general Pull system means producing only in response to a specific customer demand signal rather than making-to-forecast (Darlington et al. 2015). The downstream process/customer takes the product/service they need and 'pulls' it from the producer (Sundara et al. 2014). In other words in a pull system the succeeding stage demands and withdraws in-process units from the preceding stage only according to the rate and time the succeeding stage consumes the items (So & Pinault 1988). Each station can be viewed as an isolated station with its own supplier (the upstream station) and its own customer (the downstream station). When a customer order is placed, it will be fulfilled from the finished product inventory. As soon as the finished product is pulled from this inventory, a signal (or Kanban) is generated to trigger production of the upstream station in order to replenish the finished product inventory. Similar procedures take place until the first station, where it pulls raw material from the raw material storage (Ghrayeb et al. 2009). The structure of the information flow in pull system is local and decentralized (Marquès et al. 2012). One basic objective of a pull system is to minimize in-process inventory (So & Pinault 1988). The major justification for viewing inventory as undesirable is that it ties up capital that could be invested; and plants with large amounts of inventory tend to be less flexible, being bound to their current business through the sunk cost fallacy (Fienberg et al. 2016). However, the ideal pull system is not achievable in a real manufacturing environment and the main reasons are:

– Uncertainty in demand with medium to large demand variation. This can result in a significant backorders;
– Pull-type system often has longer delivery leadtime than that of the push-type system. And this leadtime can be longer than promised delivery time (PDT);
– Variation in processing times;
– Imbalance of workloads among stages and machine breakdowns.

Consequently, a substantial amount of safety stock may be required in each stage of the pull system in order to reduce the effects of these variations and machine breakdowns.

But so often, demand is not absolutely random in each period of time. Methods of mathematical statistics make it possible to distinguish between random and nonrandom components and thus to reduce a probable variation of demand uncertainty. The use of probability distributions makes it possible to determine what minimum inventory level ensures the required level of delivery reliability.

The combination of approaches of the pull system and mathematical statistics makes it possible to achieve a lower inventory level with other factors being equal.

3 WIP INVENTORY OPTIMIZATION ALGORITHM (ON THE EXAMPLE OF AN ELECTRODE PRODUCTION ENTERPRISE)

3.1 Algorithm of analysis of inventories at the enterprise in order to optimize their level

The chosen enterprise produces electrode productions (Novosibirsk Region). Among the whole range of production, the authors have chosen a product line group in order to test a hypothesis for the availability of excess finished goods and works-in-progress (WIP) inventory (hearth blocks) after the Annealing operation and before the shipment to clients. The general view of the production of hearth blocks is shown in Figure 1.

The authors have chosen one stock item for the results presentation, this stock item takes the largest volume in the product line group "hearth blocks".

The initial assumptions are as follows:

1. The preliminary shipment plan which is formed at the beginning of the year and contains only the total quantity of shipment for deliveries per corresponding months without any concrete quantities and dates.
2. The total quantity of the required production according to the amended shipment plan can vary from the preliminary plan within 10%.
3. The enterprise receives the amended monthly shipment plan from the customer on the fifth day of a month which precedes the deliveries. So, the enterprise is not able to correct its plan of annealing in accordance with the amended monthly delivery plan of the customer. This plan is the obligation of the enterprise for the customer. Further corrections referring to the increase in volumes and earlier deliveries are possible by mutual agreement of the parties

only, if there are free production facilities at the enterprise.
4. The production lead time:
   – Quality control—3 days;
   – Mechanical treatment—1 day;
   – The required date for the transfer of a workpiece to the mechanical treatment shop before the direct treatment is one day before the mechanical treatment;
   – The required date for the discharge from the furnace is one day before the transfer to the mechanical treatment shop;
   – Annealing—31 days;
   – The required date for the furnaces charge planning—3 days.

The maximum term from the beginning of the furnaces charge planning till the shipment of the finished products to the customer is 40 days.
5. The planning at the enterprise takes place once a month on the 20th–23rd day of a month.

So the production lead time (40 days) exceeds the time from the receiving of the amended data from the customer before the moment of the first shipment (25 days).

The current situation at the enterprise is presented in Figure 2.

When choosing a solution within the concept of pull systems for the MTO production environment it is foreseen that the client's order initiates the pro-

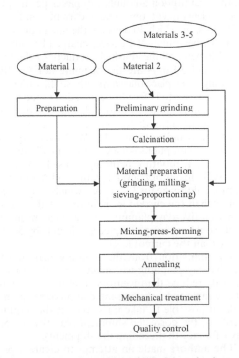

Figure 1.   Macrochart of hearth blocks production.

Figure 2. The current WIP inventory level referring to the selected stock item.

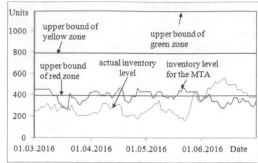

Figure 3. WIP inventory level for the MTA production environment, the replenishment time is 70 days.

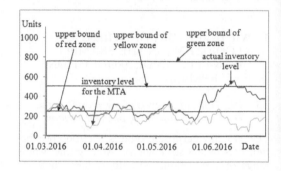

Figure 4. WIP inventory level for the MTA production environment, the replenishment time is 45 days.

duction, i.e. the delivery lead time is set taking into account the duration of the manufacturing process. For the given production, the duration of the manufacturing process is great enough, and the delivery lead time for the client's order from the moment of the order till the readiness, including the total duration of production lead time, becomes noncompetitive.

In case of the solution within the concept of pull systems for the MTA production environment, it is necessary to determine the following two indicators: maximum consumption and replenishment time. On the basis of the statistical data on consumption for the previous periods, as well as according to the predicted values, it is determined that the maximum consumption amounts 510 pieces per month. The replenishment time under current conditions is 70 days, of which 40 days are the maximum term from the beginning of furnaces charge planning till the shipment of the finished products to the customer, and 30 days are the planning horizon (the planning takes place once a month, so the order can wait for a month before to be included in the plan). If to organize a more flexible planning of the production (with a five-day interval), than the replenishment time could be 45 days. The WIP inventory level for these assumptions is shown in Figures 3 and 4.

In case a monthly planning is used, the average WIP inventory level for the MTA environment would make up 384 pieces, which is by 25.0% higher than the current level; in case of the transition to a five-day planning, the average inventory level would make up 194 pieces, which is by 37.0% lower than the current level.

The transition from the monthly planning to the five-day planning will make it possible to sufficiently reduce the inventory level in the system. At the same time, the presented approaches do not make it possible to take into account the requirements of the corrected shipment plan (by the 5th day of a month preceding the shipment).

The authors made an attempt to create a production planning algorithm which would take into account all the information available in order to minimize the WIP inventory level at the selected section of the production while maintaining a high delivery reliability of the enterprise as a supplier.

3.2 Production planning algorithm

1. The determination of the maximum potential volume of consumption per month, i.e. the volume according to the preliminary plan increased by 10% (according to the contract).
2. The determination of a possible distribution of shipments within a month according to the statistical information for the previous periods and in accordance with the amended shipments schedules under the following conditions:
   – Analyzed intervals—5 days;
   – The selection of distribution parameters according to the beta distribution, because this very distribution has a bounded interval and corresponds to the analyzed random value to a great extent.

As a result, the maximum possible accumulated consumption is determined according to the shipment schedules in each five-day period on the basis of the plotted distributions (the confidence level is 99%).

3. The elaboration of the schedule of furnaces charge at the Annealing operation section during a month using five-day periods in accordance with the potential monthly volume of production at the Annealing operation section and distribution of possible shipments. At the same time, possible maximum consumption per day is taken into account. For the analyzed stock item, it is 30 pieces in accordance with the amended shipment schedules.
4. The correction of the annealing treatment schedule by the value exceeding the annealing volume in comparison with the actual shipments of the current month and shipments delayed for the next month. The production at other stages proceeds according to the initial assumptions.
5. The correction of the shipments schedule and production schedule takes place on the 5th day, when the enterprise receives an amended shipments plan. If at this moment more products are put into production than it is needed according to the amended shipments plan, then the further production start-up is organized in order to ensure the minimum WIP inventory level under the current conditions in order to ensure the shipments according to the amended plan.
6. The production in the mechanical treatment shop proceeds according to the amended schedule dated the 5th day of the previous month (or the additional correction, but only in case of the availability of the sufficient quantities of the annealed workpieces).

The actually created plans for production and shipments are further implemented according to the rules of the pull system for the MTO production environment.

The results of the modeling of the WIP inventory level with the use of the proposed algorithm are shown in Figure 5.

The use of the proposed algorithm of planning makes it possible to reduce the average WIP inventory level for the analyzed period for the analyzed stock item to 106 pieces, which is by 65.3% lower than the actual inventory level.

4 CONCLUSIONS

As a result of the carried out study, the authors elaborated a model of WIP inventory optimization on the example of an electrode production enterprise. This model is based on the combination of the tools of pull systems and methods of mathematical statistics. The approbation of this model in practice at an electrode production enterprise in the Novosibirsk Region has proved that its application makes it possible to sufficiently reduce WIP inventory level, to release the current assets while maintaining the necessary level of shipments (delivery reliability of the enterprise as a supplier). An important condition of its application is the availability of the information system with up-to-date and complete information of the required quality.

REFERENCES

Darlington, J. et al. 2015. Design and implementation of a Drum-Buffer-Rope pull-system. Production Planning & Control 26(6): 489–504. DOI: 10.1080/09537287.2014.926409.
Fienberg, M.L. et al. 2016. Strategic Production Line Synchronisation. South African Journal of Industrial Engineering 27(2): 218–233.
Ghrayeb, O. et al. 2009. A hybrid push/pull system in assemble-to-order manufacturing environment. Journal of Intelligent Manufacturing 20:379–387. DOI 10.1007/s10845-008-0112-6.
Khayrullina, M. & Kislitsyna, O. 2015. Synchronization of the Information and Material Flows of Production Systems. International Conference on Advanced Management Science and Information Engineering (AMSIE): 656–662. Hong Kong.
Marquès, G. et. al. 2012. A supply chain performance analysis of a pull inspired supply strategy faced to demand uncertainties. Journal of Intelligent Manufacturing 23: 91–108. DOI 10.1007/s10845-009-0337-z.
Pushkar, D.I. & Dragunova, E.V. 2014. Development of information system of innovation evaluation at the meso level. 12th International Scientific-Technical on Actual Problems of Electronic Instrument Engineering, APEIE 2014 2–4 Oct. 2014: Proceedings: 651–655.
Schragenheim, E. & Dettmer, H.W. 2000. Manufacturing at Warp Speed: Optimizing Supply Chain Financial Performance. CRC Press Reference.
So K.C. & Pinault S.C. 1988. Allocating buffer storages in a pull system. International Journal of Production Research 26(12): 1959–1980. DOI: 10.1080/00207548808948008.
Sundara, R. et. al. 2014. Review on Lean Manufacturing Implementation Techniques. 12th Global Congress On Manufacturing And Management, GCMM 2014: Procedia Engineering 97: 1875–1885.

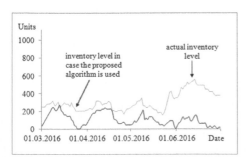

Figure 5. Inventory level with the use of the proposed algorithm.

New Trends in Process Control and Production Management – Štofová & Szaryszová (Eds)
© 2018 Taylor & Francis Group, London, ISBN 978-1-138-05885-9

Business performance management in small and medium enterprises

J. Klučka
Faculty of Security Engineering, University of Žilina, Žilina, Slovak Republic

ABSTRACT: Business performance management is a powerful tool to monitor effectiveness and sustainability of an enterprise. There is the software package NEO within the framework "Business Intelligence" aimed at small and medium sized companies. The objective of the paper is to clarify business strategy of a software company to successfully develop and implement this software. The obstacles are different: small and medium sized companies focus on "short term -" and "technological-" objectives; the objective for the software company is to provide added value—therefore not only attract a customer but also keep him/her should be the strategy objective. In the paper will be presented approaches for implementation in order to achieve business objectives.

1 INTRODUCTION

The software company Kros s.r.o. invented and implemented the software within the framework "NEO Bussiness Inteligence". The software solution is focused on small and medium enterprises (SME's). The current approach is based on assumptions—to create scheme of outputs for:

– Sales: Sales analysis, Sales forecast,
– Accounting: Accounting analysis, Accounting planning.

The structure of the software product (highest level) is defined according to activities/processes. The structure of the Sales analysis is split into modules: salesmen, products, warehouses, payment date, sales according to activities/depots/customers and customers. The module customers can be specified into: what your customers' buy, seasonality, regionalism and top customers. The module Top customers has parameters (y—axe): goods and services, warehouse, business cycle, businessmen, costs allocation and the last is company. The x-axe in this module are items: sales, average, discounts, profit from sale, payment date. The z-axe is time (year, month…). Further additional items that can be defined from roll—up menu are: type of document (e.g. invoice) and time period (a specific year or defined scope of time). All the items within each module can be defined based on the specific customers 'needs. The variety is huge but the ambition of the software company is to create a flexible tool that can be tailored to specific customers' needs.

The objective of the paper is straightforward as the mission of the software company:" We create and supply to our customers extraordinary sophisticated software to help them to be successful.

(Report, 2015) The scope will cover problem of performance measurement and its implementation within existing software.

The content of the paper is therefore divided to items:

– Why is it important for an entrepreneur/owner to analyze and forecast its business results?
– What approach to apply in order to get new customers and to keep existing ones—in other words—what approach should be applied to contributed added value to company's customers?
– Are there any other business fields where implementation of the software can small and medium company improve its competitiveness or long-term sustainability?

The content of the paper is based on: analysis of existing software, unstructured interview with head of the team responsible for development and implementation of the software, literature review to the subject business performance management and business analytics.

The reasons for monitoring/measuring business performance are expressed in (Pidd 2012):

– If you don't measure results, you can't tell success from failure.
– If you can't see success, you can't reward it.
– If you can't reward success, you're probably rewarding failure.
– If you can't see success, you can't learn from it.
– If you can't recognize failure, you can't correct it.
– If you can demonstrate results, you can win public support.

Therefore successful business decisions are based on relevant data, know-how and appropriate strategy. The proposed software modification can create the business solution viable, relevant and based not only on intuition but also on the quantitative and qualitative analysis.

2 BUSINESS ANALYTICS AND BUSINESS PERFORMANCE MANAGEMENT

Business performance management is being developing. At the beginning the stress was put on the financial ratios therefore called "Financial analysis". It covers financial figures with implicit statement "after all everything is based on financial results of a company". Constrains of this approach are focused only on financial data and on the past. To overcome a narrow focus was/is implemented Balanced Scorecard that adds to financial measures also dimensions like internal processes, customers and learning & growth. This scheme is based on the system approach and has an ambition to cover relations of all business activities with business strategy. Business analytics explicitly defines its relation to business strategy. The approach is based on the sequence: Business objectives—Key Success Factors—Key Performance Indicators.

Business objectives are formulated within strategy and created via SMART abbreviation (specific, measurable, agreed to, realistic, timely). Strategy is explicitly defined by its vision, mission and objectives. Critical success factors are specific subsets that should be targeted in order to achieve success. Key performance indicators (KPI) are specific measures that cover all relevant fields in a company. KPI's are quantitative measures of different but relevant aspects that determine strategy fulfillment. Therefore KPI's should be: relevant with strategy and critical success factors. Their thresholds need to be reasonable, realistic and achievable. They can be based on the experience from the past or as benchmarks when there are available data. Types of KPI's can be differentiated as: focused on process, input, output, leading, lagging, outcome, qualitative and quantitative.

A managerial decision consists of the sequence of following steps: problem formulation—data collection and their evaluation—analysis of alternatives—decision—feedback and evaluation. Any managerial decision is related to time and available resources (tangible, intangible). As nobody can change the past (only to analyze) the managerial decisions are focused on the real time and future. Therefore within business analytics there are three types of analytics: reporting (what happened in the past), descriptive analytics (why did it happen) and predictive analytics (what is going to happen).

Very specific but reflecting current trends is digital analytics—applied in monitoring web pages.

Diverse information sources are dealing with problems describing in the paper.

Pidd in his book (Pidd 2012) expresses principles of performance measurement, different uses for performance measurement and practical methods for performance measurement. The unique of the book is its focus to public services and application of performance measurement in public services. The comprehensive list of literature to the subject is provided in (Emsslin et al. 2017).

In the book (Brammertz et al. 2009) the financial analysis is explained in details. It is expressed its application generally and then in banks, life—and non –life insurance and in nonfinancial corporates. The book provides very specific detail analysis and description of input elements—like counterparty, behavior and also analysis from liquidation point of view. There are explained sensitivity analysis and methods for risk quantification.

Detailed explanation of financial analysis is expressed in the book (O'Regan 2007). As additional value can be stated the chapters dealing with creative accounting, corporate social responsibility, accounting standards with a case study. With the content of this paper is very relevant and challenging the chapter that deals with Future reporting.

Comments to their statistical research concerning performance management in 37 SME's can be found in the paper (Ates & Garengo 2013). It was found that less significance is set to strategy in SME's. "Short term priorities" and "technical excellence" are crucial—the same approach was identified in Slovak SME's. Efforts on internal orientation leads to short term performance. The approach to identify competitive capabilities and their evaluation in the model was the objective of the paper (Lorentz et al. 2016).

In the paper (Balfaqiha et al. 2016) are proposed the steps in implementation of performance management system. The framework is focused on the broad view. "Develop a preliminary PMS (Performance measurement system) by selecting suitable approach, technique, criteria, and metrics based on the defined SC's (supply chain) objectives and goals" was the objective of the paper. The approach—to identify KPI's in different enterprises and activities is presented in the paper (Popovič et al. 2016). There are presented benefit due to application of performance management in three companies.

Also in the research paper (Analytics & Data 2013) are presented results of study that dealt with Business Intelligence and performance management. From total 208 companies 30% were from UK and 23% from post-socialist countries; the rest was from European region. It seems very important that 63% of respondents reacted to the question "Has your organization set up a business intelligence competency center or another form of a centralized BI team" that they have or plan to establish competency center (deals with BI). This fact also articulates future potential of know-how and BI solutions in the Slovak republic.

3 PROBLEM FORMULATION AND PROPOSED SOLUTION

The business problem is related to customers and the software company. For customers are crucial added value and support. Added value reflects in tailored software solutions, on line help desk and functional software that is enough flexible and intuitive. From the software company's side it is to provide software solution that is attractive, flexible and financially viable. It should be eliminated short term orientation (see above) and technological/technical focus. Therefore the activities have to be diversified and oriented towards added value for customers—that mitigate strategy focus. The added value however does not have to be recognized because of lack of knowledge. The activities are supposed to be therefore oriented towards professional skills and their sustainable improvement.

Hence it can be adopted these two approaches:

– Current approach—to create different schemes that cover the key success factors for all existing and potential customers. These approach seems to be rather ambitious—because of different heterogeneous small and medium enterprises and their heads/owners requirements.
– Challenging approach—to create list of different possible measures (KPI's) and an end-user will choose the appropriate measures. This tailored made attitude seems to be more flexible and reflects possibility of specific customers' requirements. This approach together with educational/consultative support can be solution that links customers' needs and the software company declared mission. The proposed approach is described in the Figure 1.

Each relevant activity of a company will be monitored by specific KPI. For each KPI will be defined the set of data and responsible/owner of KPI. The results provided by the software package will be a tool to support managerial decisions.

The scope of the software solution can be broaden towards two new topics:

– Risk management,
– Future development, forecast.

The idea of risk management in a small and medium sized company can be covered with the existing software tool with identification Key risk indicators (KRI) measures that are related to spe-

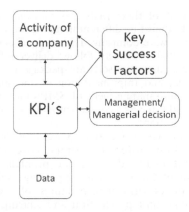

Figure 1. The framework of software package development.

cific risks company is faced. There also should be incorporated for each specific KRI minimum, maximum, demanded level of measures.

The future development can be expressed within the framework of predictive analytics. There can be applied different mathematical-statistical methods to evaluate future forecasts; to express future trends. The practical implication for the existing software is therefore—to divide analytics oriented towards history and forecast.

4 DISCUSSION AND CONCLUSION

The efficiency of software—its application in business decision making is determined also on factors:

– Business know-how—it covers not only knowledge of interior environment but also relations and identification of possible consequences of external environment to a company,
– Proficiency of executive officers/owners to be able to formulate key success factors,
– Preparedness and competency to interpret outputs/results,
– Availability of reliable data that are regularly updated; each data item must have its owner—responsible person for update and monitoring results,
– Proficiency and enthusiasm of staff is the key for software success in a company; therefore on-line consultation, educational activities and consultations tailored for specific needs of a customer should be within the core of marketing activities of the software company.

There can be identified two types of customers:

– Potential,
– Existing.

The core of their needs may be equal or very similar but must be different in marketing strategy of the software company. The range of activities for potential customers follow: deliver information concerning the software package (its functionalities) and implementation of the software package into management of SME's (advantages and added value as a consequence of the software implementation).

The activities focused to existing customers should be different. It should cover on line consultation (conduct ad hoc problems) and creation of the platform supporting exchange of practices (informal meetings of existing customers) and publication and sharing good practices.

In order to implement measurement in a company there must be exercised steps (Bergonzi 2016):

- Establish goals: define the objective and link them to desired performance quantifiers.
- Establish metrics: develop metrics to compare the actual performance to the desired one.
- Understand performance: study the gaps.
- Initiate improvement actions.

These steps should be reflected in the software package and supporting activities of the software company.

The content of business performance management (covered also in concept Business Intelligence and described in the software solution) can be powerful source for small and medium companies—their competitiveness. The software package is however a tool—therefore the key is end users preparedness.

The enthusiasm and professional know-how are fundamental conditions that can improve competitive advantage of a company. For the software company it means to intensify consultative and educational activities. The cooperation with university can be additional benefit for both sides.

ACKNOWLEDGEMENT

This publication was created in the frame of the project VEGA 1/0918/16 Risk management of SMEs in the context of clusters.

REFERENCES

Analytics & Data Warehousing Reader Survey—Europe. 2013. TechTarget, [on-line] http://docs.media.bitpipe.com/io10x/io_102267/item_835376/Analytics%20%20Data%20 Warehousing%20Reader%20Survey_2013_EU_UKcut.pdf

Ates, A. & Garengo, P. 2013. The development of SME managerial practice for effective performance. *Journal for Small Business and Enterprise Development.*

Bergonzi, L. 2016. eCall KPIs. EENA operations document, Brussels. [on-line] http://www.eena.org/download. asp?item_id=225

Blackburn, R. et al. 2013. Small business performance: business, strategy and owner-manager characteristics. *Journal for Small Business and Enterprise Development,* 20(1): 8–27.

Brammertz, W. et al. 2009. *Unified Financial Analysis.* Chichester: Wiley.

Ensslin, L. et al. 2017. BPM governance: a literature analysis of performance evaluation. *Business Process Management Journal,* 23(1): 71–86.

Hasan, B. et al. 2016. Review of supply chain performance measurement systems: 1998–2015. *Computers in Industry,* 82(1): 135–150.

Kucharčíková, A. et.al. 2011. *Efektivní výroba.* Brno: Computer Press.

Lorentz, H. et al. 2016. Cluster analysis application for understanding SME manufacturing strategies. *Expert Systems with Applications,* 66(1): 176–188.

O'Regan, P. 2007. *Financial Information Analysis.* Chichester: Wiley.

Pidd, M. 2012. *Measuring the Performance of Public Services.* Cambridge: Cambridge University Press.

Popovič, A. 2016. The impact of big data analytics on firm's high value business performance. *Information Systems Frontier.*

Report 2015. KROS, Žilina, [on-line] https://www.kros.sk/tmp/asset_cache/link/0000103584/Vyrocna_sprava_KROS_2015_update.pdf.

A new approach to public funds in the light of so-called anti-shell law

O. Kmeťová

*Faculty of Business Economics with seat in Košice, University of Economics in Bratislava,
Košice, Slovak Republic*

ABSTRACT: The path to a greater transparency in the management of public finances led to the adoption of the new Law no. 315/2016 Coll. on the register of public sector partners. This so-called anti-shell law introduces after 1st February 2017 a new register of public sector partners led by the Slovak Ministry of Justice. The register re-places the previous register of beneficial owners led by the Office of Public Procurement. The introduction of the register of public sector partners aims to define the legislative requirements for entities, with which it enters the state, resp. PSE's into legal relations, resp. in which a third party takes any transactions, including the sale of property to the state. The register should reveal the ownership structure of the companies. The aim of this subscription is to highlight the new legislation and to introduce rules, which the new law implements into practice.

1 INTRODUCTION

Transparency requirements of the management of public finances began by legislation of the mandatory disclosure of contracts of public bodies in the central register of contracts. The public may thus find out on the internet, which companies and under what conditions "conduct business with the state". In most cases it is not clear, which individuals are "hidden", behind such companies ultimately benefit from the business ("we know what, but we do not know who"). For such purpose exists a new "anti-shell-law". The term shell company can be a company, which has its seat mainly in so-called tax havens, does not carry out business activities in the country of residence, besides professional and rent-like "statutory", does not have regular employees, does not have technical or organizational facilities and in publicly available records it is not possible to get to the bottom of its true economic owner. It is basically a "mailbox".

Reports, which have emerged and originated in Slovakia, long resonated in the subconscious of people such as the "Notice-board Tender" and the "CT case" or "Váhostav" lead to realistic clues about the fact that behind companies conducting business with the state are different politicians, officials, sponsors of political parties or financial groups. Suspicions concerning conflict of interest, corruption and favoritism created a political will to require disclosure of ownership background of companies, which are managing with public finances.

2 THE PATH TO GREATER TRANSPARENCY

Initial legislative attempts to adjust the transparency of ownership structure by means of public procurement rules were not successful, since they failed to respect the fundamental European principles of non-discrimination of entrepreneurs from other Member States of the European Union and by illegal means limited the access to public contracts. Later changes, which introduced a register of end-user benefits for entrepreneurs in public procurement, on the contrary, did not meet the requirement for flatness, effectiveness of control and sanction enforcement.

After an extensive public and professional debate, a new legislative intent of legal regulation of state trade and the private sector reflected into the law on the register of public sector partners, so called anti-clipboard law approved by the National Council on 25th October 2016.

Although the basis for the law makers was the so called Fourth directive against money laundering (The Directive of the European Union and the European Council, No. 2015/849), the anti-shell law is not its implementation. While the Directive presents a minimum standard for uncovering the beneficial owner, the anti-clipboard law far exceeds its content.

The main idea behind the law is the principle that only those may conduct business with the State, who willingly and reliably uncovers its final owner, being one or more physical entities. This means that to accept money or non-monetary compliances from the public domain may only those private sector entities, which publish their final owner in the register.

According to Leontiev (2016), "those may conduct business with the state, who reveals its end-owner being a physical entity". The registration of the beneficial owner in the register is no longer enough, but written data will be verified prior to registration, consecutively, as well as on the basis of the initiative. The verifications will be carried out via a "professional", who will be guiding the register and review all date on the basis of complaints in the District Court of Žilina. The Register will be publicly available on the internet. Penalties for law breaking are:

– Contract withdrawal,
– Suspension of the contract due to contract compliance,
– Fines,
– Liability for fines,
– Withdrawal of economic benefits,
– Removal from the register,
– Registration of disqualified persons.

These policies shift public control of public funds in the European context on the highest level.

3 ASSIGNATION OF A PARTNER OF THE PUBLIC SECTOR

A key subjective in the new law is the partner of the public sector defined both positively or negatively. In a positive sense a partner of the public sector is a person, who in not subject of public administration and also meets one of the following criteria:

– Receives EU funds, investment incentives or funding of selected legal entities from his/her "daughters and granddaughters",
– Receives property or property rights, or rights to property from the "public sector" or health insurance,
– Is a supplier of public procurement,
– Is a health care provider,
– Is a claim-transferee against the "public sector",
– Is a qualified sub-contractor,
– Has public sector partner-status according to a separate law.

In the negative sense of the public sector there is no entity that receives funds exceeding one-off pay-

ment of EUR 100,000 and EUR 250,000 per year, or if assets acquired have a value of more than EUR 100,000 (i.e. de minimise exemption). It is also not a person, who operates in the non-profit sector, unless he is not a contractor in procurement a does not enter into "public property". A partner of the public sector is not even a bank, if it receives funds under the normal course of business, or a person receiving payment from an Embassy abroad or fulfilling in developing humanitarian aid, another State and its authorities and international organizations established under public international law and its institutions.

According to a special regulation a partner of a public sector is:

– Holder of mining authorization,
– Health insurance company,
– Creditor in bankruptcy proceedings, if his claim is over EUR 1000,000 against the bankrupt, who has been in the last five years a public sector partner and the lender wants to avoid a period of kin creditor,
– Holder of a geological authorization,
– Holder of an authorization to conduct business in the energy sector,
– Toll selector,
– Payment selector of highway signs.

Beneficial owner is a person who benefits from the activities of the partner of the public sector. This happens so that he

– Controls a legal person, physical entity—entrepreneur or property association or
– In favour of a legal person, physical entity—entrepreneur or property association carrying out business or trade,
– Together with another person is acting in agreement or concerted practices of a public sector partner.

End-user benefits include such physical entity that

– Directly or indirectly has a more than 25% share in the share capital or voting rights,
– Has the right to appoint authorities,
– Controls a legal person,
– Takes ownership benefits from him.

A physical entity entrepreneur is an end-user physical entity, which has from its business a more than 25% economic benefit. If a public sector partner and physical entity does not fulfil those conditions, as an end-user members of the top management are going to registered. Similarly, top management members are going to be assigned into the register in such case, when a public sector partner directly or indirectly, exclusively operated or controlled by the issuers of exchanges traded on a regulated market and no physical entity has no economic benefit of at least 25%.

3.1 Registration to the public sector partner register

The public sector partner register is an information system, whose manager and operator is the Ministry of Justice and is led by the District Court Žilina. When it comes to a system of forms, appeals, electronic statements, it is akin to the Commercial Register. The difference between them is that communication with the public sector partner registry will be carried out electronically via an authorized person using a qualified electronic signature. The register is publicly available on the internet. A public sector partner within must be registered with all the necessities at least for the duration of the contract, under which receiving public funds.

Enrolling of a public sector partner into the register and the verification of the end-user beneficial owner must be carried out by an authorized person, so called "professional". An authorized person may only be:

- An advocate,
- A notary,
- A bank, respectively foreign bank branch,
- An auditor,
- A tax advisor.

The authorized person must have a registered office or place of business in the Slovak republic and must have a written agreement with a public sector partner under the provisions of the Treaty of Control Activities. The authorized person during verification acts impartially, independently without instructions of a public sector partner, with professional care (appropriate to his role and status), after the procurement and assessment of available information (which he acquired or may acquire, which have or may have an impact on verification).

The subjects of registry arc:

- Public sector partner/name/business name/place of business/residence/ID),
- End-beneficial owners (name, address, date of birth, nationality, jurisdiction, whether he is a public official, respectively, whether a public official within the structure),
- Authorized person (name/business name, place of business/residence, ID).

The register also includes:

- Date of enrolment,
- Date of each verification,
- Date of changes,
- Date of deletion,
- Details of imposed sanctions,
- The verification of all documents.

Although there are many similarities in the registration into the Commercial Registry, in the case of public sector partner register the registration process is different. The claimant may only be an authorized person; a public sector partner only provides him with interoperability, while both are responsible for accuracy. The deletion of an authorized person on the proposal of the public sector partner must be submitted by a new beneficiary. Any information, respectively its change is recorded electronically, attached with a verification document, which has identified and validated end-user benefits.

Court acting is formal, thus the court examines whether the submitted documents agree with the proposal. The deadline for registration/modification/deletion of data is 5 working days (until the 31st of July 2017 is extended to 10 working days). The data are entered for an indefinite period. At registration-refusal, objections may be submitted in a 15 day period.

A public sector partner is obliged to inform an authorized person of any information change. The authorized person in such case shall apply for registration of changes in the 60 day fair period, i.e. since the change occurs. If the registrant body deletes the authorized person on its proposal, the public sector partner is obliged to ensure the entry of the new authorized person within 30 days after deletion. As a rule, a public sector partner may have registered only one eligible person, but the authorized person may be registered for a number of partners of the public sector.

The allegation of the end-user is written on the basis of a verified document. Within, the person entitled justifies, whereupon it verified end-user benefits, indicates ownership and control structure of the public sector partner, includes information concerning public officials and declares that the information in the documents correspond to the established conditions.

The verification, i.e. identifying or identification verifying of end-user benefits occur:

- During public sector partner register enrolment,
- During change in the end-user beneficial,
- During a change of the authorized person,
- On the 31st of December,
- At the termination of a contract or its modification,
- During the admission of a contract exceeding EUR 1 million.

4 OBLIGATIONS OF THE PERSONS CONCERNED AND THE CONSEQUENCES OF THEIR VIOLATION

A public sector partner is required via an authorized person, to which it provides interoperability and complete information, to enrol and at time update the register details of its end-user beneficial.

He is obliged to avoid conflicts of interest with a qualified person and in the case of an ex-post check bear the burden of proving the correctness of the enrolment.

Penalties for failure to fulfil obligations are both in the form of monetary fines and deletion from the register. The court imposes a fine in case of false or incomplete information recording of the end-user beneficial or public officials aloft economic benefits, if that cannot be determined, so in a lump sum of EUR 10,000 to 1 million. Register-deletion occurs when a public sector partner fails to pay the fine, further in the case of ex-post control will not be able to bear the burden of proof or if he used an authorized person, despite interest-conflict.

Obligations of an authorized person are in collaboration with public sector partners to write and enrol data in the register of end-user beneficial and verify end-user beneficial, independently and with due care (including the rejection of verification in the absence of reliable information). Similarly, as the public sector partner the authorized person shall avoid conflict of interest. If the statutory public sector partner is given a fine, the authorized person is liable for the payment, unless he demonstrates that he acted with due diligence. In case of breach of duty to avoid conflicts of interest the registration court shall impose a fine of EUR 10,000 to 100,000.

The amendment of Public Procurement (Act. No. 343/2015) imposes obligation and in case of infringements and sanctions, and the contracting authority. For breach of obligations shall be considered if a contracting authority, without justifiable reasoning, has not exercised the right to withdraw from a contract (if the a public sector partner has been deleted from the register, or could not bear the burden of proof, or if he has been fined for failure to update the end-user beneficial within the given period or in a conflict of interest or has not appointed in time a new authorized person), or if without justifiable reasoning has not exercised the right to suspend execution (if a public sector partner has not enrolled an end-user beneficial in the register within the deadline or he has not updated him during verification or not appointed in time a new person). When a contracting authority enters into a contract with a partner of the public sector, who does not have in the register an end-user beneficial registered, commits an offense (This a particular physical entity sanctions himself), for which it may be fined from EUR 1000 to 100,000.

If the end-user beneficial gets to know that he has become an end-user beneficial of the public sector, he is required within a 15 subjective period to notify the public sector partner, which notice he shall deliver to the beneficiary person registered in the register. For failure to do so, fines up to EUR 10,000 threatens the end-user beneficial.

5 TRANSVERSAL VERIFICATION OF DATA ENROLLED INTO THE REGISTER

Repeated verification of data enrolled into the register of public sector partners is an assurance of their maximum actuality. The data are being verified by "professionals", during registration and subsequently during each verification event (ex ante control). It can parallel data ex offo at any time or on the basis of a qualified initiative to check the registration court (ex post control).

Anyone may submit a qualified incentive for judicial review registration of the end-beneficial owner, however, must adduce to justify doubt concerning the veracity of such registration. The burden of proof in such case is transferred onto the partner of public sector, who is required to provide evidence about the truthfulness and completeness of data entered by an end-user beneficial. In case when a public sector partner will not sustain this burden of proof, the Court shall annul such person from the register and shall initiate proceedings to impose a fine. To the person who notifies, upon request, when having drawn attention to the incorrectness of the end-user beneficial at his employer, the person is entitled to protection as in the case of "whistleblowing".

6 CONCLUSION

The solving of complicated and years-persistent problems of money-laundering via so-called tax heavens, the Slovak Republic has offered Europe, in the form of progressive adjustments, so-called anti-shell laws. The chosen solution is particularly unique, especially in uncovering beneficial owners involved in creating shell companies and tax optimizing schemes. Slovakia, as the first country in the world, has begun to apply the highest standards to combat money laundering on its apparatus and in effecting its public finances, thus creating a new "role model", which may serve as a "compliance" model for other countries, as well as other entrepreneurs. The fact that it will be publicly known who and why trades with the state and uses public finances, will increase the transparency of the business environment, which will have positive impact on improving competition and removing market distortions in the form of handling state contracts unfairly via entrepreneurs.

REFERENCES

Leontiev, A. & Pala, R. 2016. Čo má spoločné Žitňanskej protischránkový zákon a Miklošova rovná daň? In *Verejné obstarávanie—právo a prax*. 6/2016. Bratislava: Wolters Kluwer. ISSN 1339-5963.

Leontiev, A. 2016. [retr. 2017–02–20]. Available online at: <http://www.ta3.com/clanok/1093212/host-vstudiu-a-leontiev--o-boji-proti-scrankam.html>.

The Directive of the European Union and the European Council, 2015/849 of 20 May 2015 on preventing the use of the financial system for the purpose of money laundering or terrorist financing on combating mailboxes.

Tkáč, J. & Griga, M. 2016. *Zákon o verejnom obstarávaní— veľký komentár.* Bratislava: Wolters Kluwer, 2016, 1260s. ISBN 978-80-8168-454-8.

Zákon č. 315/2016 Z. z. o registri partnerov verejného sektora a o zmene a doplnení niektorých zákonov.

Zákon č. 343/2015 Z. o verejnom obstarávaní a o zmene a doplnení niektorých zákonov. (Act. No. 343/2015 Coll. Public Procurement on amend-ments to certain laws as amended by Act No. 438/2015 Coll.).

Zemková, J. & Šingliarová, I. 2015. *Judikatúra vo veciach verejného obstarávania.* Bratislava: Wolters Kluwer 2015, 384 s. ISBN 978-80-8168-219-3.

New Trends in Process Control and Production Management – Štofová & Szaryszová (Eds)
© 2018 Taylor & Francis Group, London, ISBN 978-1-138-05885-9

Tourism information flows in destination management

G. Koľveková & E. Liptáková
Faculty of Economics, Technical University of Košice, Košice, Slovak Republic

C. Sidor & B. Kršák
Faculty of Mining, Ecology, Process Control and Geotechnology, Technical University of Košice, Košice, Slovak Republic

ABSTRACT: Regional Tourism Organizations in Slovakia play important role in destination management. They are understood here as decision makers with an influence upon achieving a tourism goal of Slovakia. The goal was set in Manifesto of the Government of the Slovak Republic 2016–2020. The goal includes apart from others encouraging destination management. The paper discusses information selection and their use in improving competitiveness of tourism sector as whole. The questions were answered by DMO located all over Slovakia and their answers allowed us to see behind a curtain of decision making and find out, which information are used preferably. The outcome of this paper is to provide an overview of how decision makers use information so far and suggest more possibilities for exploiting existing data in order to support destination management in Slovakia.

1 INTRODUCTION

1.1 *The act on the promotion of tourism in Slovakia*

The Act *No.* 91/2010 on the promotion of tourism provides support to tourism in terms of rights and obligations of individuals and legal persons engaged in tourism, creation of conceptual documents and funding for the development of tourism in the Slovak Republic. The performance of tasks to promote tourism laid down by this Act shall provide also regional and county destination management organizations (DMO). Therefore, the Act provides for the creation of rights and obligations, membership in the DMO, as well as the rights and obligations of the members of DMO.

1.2 *Regional destination management organizations*

The member of the regional DMO may be a municipality, natural or legal person, which is established or operates in DMO´s territory. The rights and obligations of the regional DMO also include the following:

– Support the activities of its members in developing and implementing the concept of tourism development in the area of responsibility;
– Create and execute marketing and promotion of tourism for its members and the community at home and abroad;
– Promote sustainable tourism development in order to preserve and maintain the environment in all its pages and to respect the way of life of the local population;
– Set up and implement the concept of tourism development, based on its own analyses, on the regional Concept of tourism development and national policy of tourism;
– Initiate or provide for the creation, management and presentation of tourism products in its territory;
– Update the integrated information system in its operation, including through the tourist information centre;
– Gather information on products, activities and values of destination in its territory in collaboration with municipalities, members of the regional organization and representatives of the expert community, etc.

1.3 *County destination management organizations*

County DMO encourages and creates the condition for development of tourism in the county. The members of the county DMO are the higher territorial unit and at least one regional DMO acting within its territory and established under The Act *No.* 91/2010.

1.4 *Manifest of the government of the Slovak Republic—field of tourism*

In this paper, the DMOs are understood as those which have the power of decision with influence in attaining the objectives of tourism in Slovakia.

Targets were set in The Manifest of the Slovak Republic for the years 2016–2020 (published by the Ministry of Transport and Construction as part of the Government Programme (Pompurová & Šimočková 2014).

Aims include, apart from the rest also the support of destination management. The paper examines the selection of information and its use when it comes to improvement of competitiveness of the tourism sector as a whole. DMOs in Slovakia were asked to respond to the questionnaire in order to find out their opinion on data collection in tourism sector.

Their answers allowed this paper to see behind the scene of the decision making and finding out which information are used as priority information. Result of this work is to provide an overview about how the decision-making unit in tourism sector uses information and eventually also suggest other options of using existing data in order to support destination management in Slovakia.

2 METHODS AND MATERIAL

2.1 Survey among DMOs in Slovakia

This paper was determined by survey that was realized among DMOs in Slovakia during months of September–November 2016. DMOs and its directors throughout Slovakia were kindly asked to fulfil the questionnaire focused on information that help DMOs in their functioning.

The return rate was 60%.

All questions in the questionnaire were aimed on finding out which information and in which way are collected and used by DMOs or in other words with what kind of problems and issues do DMOs struggle with in this process.

Questionnaire was divided in the following areas thematically:

1. Collecting information on tourists—real need versus practice.
2. Information in terms of offer/services—reality and issues when collecting information.
3. Collecting of feed-backs by providers of services or tourists respectively—which information and in which way they collet.
4. Monitoring of the tourists flows in the destination.
5. Monitoring of the infrastructure state in the destination.
6. Marketing of destination.
7. Activities of DMOs on Internet—web pages, social media.

Findings of the survey were analysed and evaluated sequentially. Analysis was taking into account

concepts used in information systems for intelligent destination management. For instance the concepts of image of the target place, locality, destination in the perspective of the advertisement or loyalty can be using continual three-dimensional approach or three-elements approach (Byun & Jang 2015, Zhang et al. 2014, Chung et al. 2015). Use of such analysis ensures the compatibility of results and creation for complete and complex information system for tourism sector. Orientation in the data collection of complex system can be pictured in simplified manner in Figure 1, where it is possible to see that primary data from a destination are available to be collected by means of surveys, questionnaires.

2.2 Scheme of information in tourism

In the course of questionnaire preparation there was a concept of the selection of the basic respondents' pool linked with several surveys, questionnaires. These questionnaires were aimed on DMOs, tourists—in general, events as well as for accommodation facilities (see the sketch in the Figure 1 in the middle). In the middle of the figure there are structured data that information system collects. Coming out from the title of the paper its content focuses on DMO survey. However, this inevitably joins together other data as the figure tries to point it out.

3 RESULTS

3.1 Types of information actually collected by DMOs

One of the questions was aiming on finding out *to which extent it is important to DMOs to know exact information about tourist and at the same time DMOs should answer question if they actually collect such information.* Respondents should select the importance of such information using Likert scale: 1 (the least important) till 5 (the most important). When answering question on actual collection of that information, respondents answered only yes/no. For picturing these answers on graph the yes answers were assigned the importance 5 and answers no importance of 1. Table 1 and radar graph on Figure 2 provide average values.

There exist a gap in information flow, as it is evident in Figure 2 and from the data of Table 1. The biggest gap arose in information on income, followed by information on type or size of the group and sex. The same gap was in information on age a visit purpose. The narrowest gaps were noticeable in the realm of visit satisfaction, visit duration

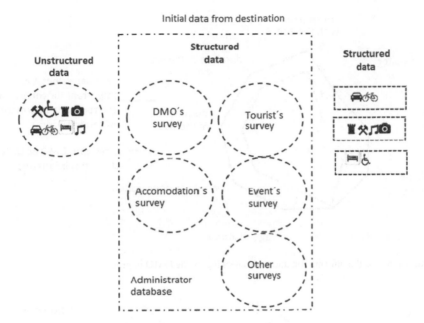

Figure 1. General scheme of information in tourism.

Table 1. Data specifying relationship of DMO to some type of information.

Information	Importance rate of information for DMO (1-not important, 5-the most important)	"Does your DMO actually collect this information?" (1 – no, 5 – yes)
Visit satisfaction	4,79	3,95
Visit duration	4,68	3,95
Visit purpose	4,63	3,74
Places/events visited	4,53	3,53
Age of visitor	4,11	3,11
Residence of visitor	4,05	3,95
Size/type of group	3,89	2,68
Income of visitor	3,58	1,21
Sex of visitor	3,21	2,05

and visit purpose. This information are collected in real practice as the easiest achievable and most frequent ones.

Results are therefore considered to be in line with assumptions.

With and exception of information on visit satisfaction all information that was the question asking for can be characterised as cognitive attributes of entities in an information system. An example of an entity are: tourist, DMO and other as it is shown in the scheme in Figure 3. Cognitive information is such an information that is clearly differentiable or

possible to distinguish from another one based on knowledge. On the other hand, affective information is such an information that is created based on emotions and senses. Missing affective attributes were the subject of other question in the same questionnaire. While the affective attributes were linked mainly with visitor's satisfaction and their needs, requests for service improvements.

Figure 3 shows ways via ones information are flowing between entities of system. In the perspective of this paper, the main entity were DMOs, which is why the figure utters only relations against this reality and neglects possible and existing relations among rest of entities. Tourist whose characteristics was already mentioned above can participate on various queries or can be questioned in an email. Tourist can use services of TIK—Tourist Information offices i.e. information are flowing in two ways. Similarly DMOs in their answer on the question of information flows and communication they revealed that they communicated also with the providers of services by means of queries and web page. DMOs use case studies that source is given by the ownership of product in question. Product can be for instance congress and various services linked with it (accommodation, food, relax). As it is evident in the practice the task of service provider can be dual. This means that the service provider can be the member of DMO. DMOs work with published statistics, analysis and databases of Statistical office of SR. It is prevail-

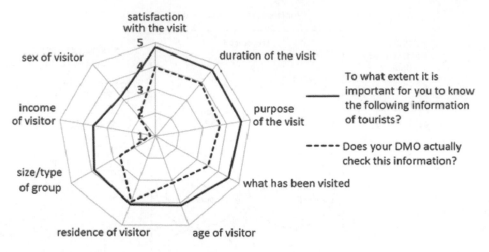

Figure 2.   Importance of the information and their checking by the DMO in practise.

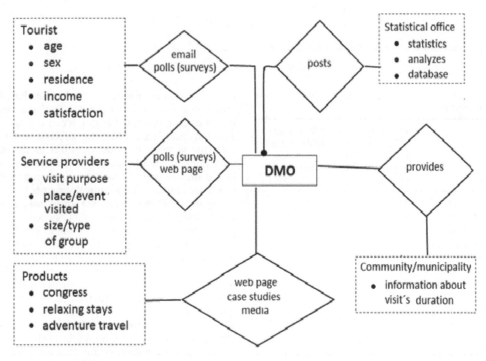

Figure 3.   Scheme of information flow and lines of communication between the DMO and other actors in tourism.

ingly one-way relation labelled by full dot in the figure (statistics published once a month, once a year and so forth). The rest of the relations are two-way and multiple. Finally yet importantly, DMOs communicate with municipal authorities in the region of their province. An example of useful information exchanged between the two entities is information of tourists visit duration.

3.2   *Importance of selected areas of information to DMOs*

In another question, the respondents were to match the rank according to importance of selected areas of information (1—the least important, 5—the most important). Results are shown as an average (arithmetic mean) for each one of areas (Table 2)

Table 2. Array of the information due to importance for DMO.

Areas of information	Importance rate of information for DMO (1 – not important, 5 – the most important)
Tourist profile (visit purpose/ duration, place/event visited, accommodation and transportation)	4,53
Quality of products/services	4,47
Extent of supply	4,42
Tourist profile (age, sex, residence)	3,68
Prices of products/services	3,67

and these were arrayed by the importance. While the cognitive attributes are within the tourist`s profile (age, sex, residence), prices of products/services and the extent of supply. Affective attributes were meant such as quality of products/services and tourist profile (visit purpose/duration, place/event visited, accommodation and transportation). Tourist profile must be in some part affective because the selection of transport means depends on emotions. e.g. tourist, who travels due to experience, can prefer train and meeting fellow travellers or playing guitar. Other tourist, who travels due to congress, seeks the fastest and the most effective means of transportation. This type of tourist will be emotionally indifferent between choices of sitting in a train or in a bus; in both cases he will be reading a book. It is self-evident that countless examples may be found and arose.

4 CONCLUSION

Thematic marketing is taking a turn from mono-thematic products towards production lines and these are reflecting attitudes of consumer, tourist at first in terms of affective perception, cognitive perception will often match implicitly. The creation of product line is possible by merge of the strengths ascribed to certain region such as in German region Rheinland-Pfalz the following strengths are being applied—tracking, biking, wine, wine county and health. The products are merged i.e. the selling of experiences in region that is also possible to apply in Slovakia for instance in region Tokai—biking and wine. Similarly popular among tourists is region in India, Goa that is having simple slogan: „Sun`n`sand".

The development and creation of products, visitor's service, creating and running a destination, all

afore mentioned can be the tasks of DMO. At the same time the advertising of destination for tourists is a component of overall DMO`s tasks aimed at achieving better competitiveness. Advertising of destination and loyalty of tourists towards destination are two elements, which one is at the foreground (on the stage) of the tourism process and the other on the background (behind the scene) of this very process. Loyalty can apart from other manifestations may demonstrate itself in repeated visits of tourist or recommendations to friends and acquaintances. Advertising as well as loyalty has its affective and cognitive attributes that were in this paper labeled as a type of information. Specific attribute is congruity (Zhang et al., 2014, Supak et al., 2015). Congruity of consumer, client, and tourist exists in the perception of the destination and at the same time this same congruity must have a reflection in advertisement. Only then it is possible to improve competitiveness of the tourisms as an industry as a whole. Concurrently, DMOs can create products and advertisements that use the language having both affective and cognitive meaning. This language, vocabulary is ascribed to given target group, to whom the DMO is oriented on as a market segment.

For some DMOs as respondents, the merge of preferences in terms of use of the type of information can be taken into account. DMO as a group would than has at its disposal particular preferences that are available to create by employing technique of oriented graphs (Ma, 2016). Therefore, the results acquired by questionnaire can be analysed this way and have an influence on the decision on data, which the group of DMOs will be preferring as a group. This will allow unifying the information system. It can unify preferences of CEOs of DMOs, which are focus on one given geographic area, region. This will make analysis more transparent and possibly easier.

ACKNOWLEDGEMENTS

This work was supported by the Slovak Research and Development Agency under the contract No. APVV-14-0797.

REFERENCES

Byun, J. & Jang, S. 2015. Effective destination advertising: Matching effect between advertising language and destination type. *Tourism Management*, 50, 31–40.
Chung, N. et al. 2015. The influence of tourism website on tourists' behavior to determine destination selection: A case study of creative economy in Korea. *Technological Forecasting and Social Change*, 96, 130–143.

Ma, L.C. 2016. A new group ranking approach for ordinal preferences based on group maximum consensus sequences. *European Journal of Operational Research,* 251, 171–181.

Manifest of the Government of the Slovak Republic 2016–2020 Tourism. Ministry of Transport and Construction of the Slovak Republic.

Pompurova, K., & Simockova, I. 2014. Destination attractiveness of Slovakia: perspectives of demand from major tourism source markets. *E & M Ekonomie and Management,* 17(3), 62–73.

Supak, S. et al. 2015. Geospatial analytics for federally managed tourism destinations and their demand markets. *Journal of Destination Marketing & Management,* 4, 173–186.

Zhang, H.M. et al. 2014. Destination image and tourist loyalty: A meta-analysis. *Tourism Management,* 40, 213–223.

Economic aspects of inspection of pipelines with unmanned aerial vehicles

C. Kovács, S. Szabo & S. Absolon
Faculty of Transportation Sciences, Czech Technical University in Prague, Prague, Czech Republic

ABSTRACT: The paper focuses on the economic benefits of the use of unmanned aerial vehicles compared to the existing monitoring system. Pipes, compressors and pumps are often found in difficult terrain, therefore their inspection is complicated. Progress in remote sensing solution, processing technology and image stabilization provides a basis for designing monitoring systems for pipelines using remote sensors and context-oriented image processing software. With an increasing age, pipelines are more susceptible to corrosion leading to the release of dangerous substances, which can have a negative impact on the environment, population, economy and the international oil market. The economic advantage of monitoring their operations is a matter of global importance. UAV's monitoring systems regularly provide measured data important for characterization of structural and functional qualities of pipes and that might help detect problems in a sufficient period of time allowing, if necessary, take appropriate safety precautions.

1 INTRODUCTION

Unmanned aerial vehicles allow safer manipulation. The decisive factor in this regard is primarily the absence of a pilot on board of UAVs, which reduces stress and fatigue of operator. UAV's pilot controls the direction of the drone from safe position on the ground. The UAV operator can control drone manually or set up autonomous flight according to the requirements of the inspection routes and operational possibilities. This fact provides the possibility of more regular inspections based on personal capacities of operating firm. In addition to safety, UAV has an advantage in the efficiency and accuracy of the flight. UAVs of smaller size may be navigated by RTK GPS with accuracy of a few centimeters and their suitable sensory equipment customized to a character of planned inspection allows more precise and accurate inspection of the surface. That's why UAVs bring more efficiency and lower costs with more accurate data to clients.

2 PIPELINE INSPECTIONS

There are approximately 1032 km (Transpetrol.sk 2017) of pipelines in the Slovak Republic, which need to be controlled in regular intervals. Helicopters are used mostly for these controls but their air operating costs are high. This way of control requires a high degree of safety and quality as well as highly qualified operators. Trends of rapid development of UAVs provided new market opportunities in the field of distribution valuable energy resources inspections (Fuchs et al. 2015).

2.1 Factors that affect the choice of UAV

When choosing a suitable UAV, it is important to consider next basic factors:

– Type of required information,
– Sensory equipment,
– Data processing,
– Payload,
– Terrain conditions,
– Legislation,
– Flight length,
– Platform.

2.1.1 Type of required information
The type of required information is mainly about direct observation of pipes and their transformation. The most common is hydrocarbon leak detection, soil characteristics or determination of phonological changes in plants affected by leaking. (Novák et al. 2016)

2.2 Sensory equipment

UAV used for detection of the leak in pipes should be equipped with sensors adapted to detection hydrocarbons and gas leaks. They should be easily identified regardless actual air conditions. The sensory equipment must be adapted to the platform of the UAV.

2.2.1 Data processing

Every measured data typically generates useful information, which consist of radiometric calibration and geometric correction. The geometrical correction ensures image adjustment in order to provide a precise data complex. Radiometric calibration is used to remove defects caused by external factors for example different position of the sun, changing atmospheric conditions, etc.

2.2.2 Payload

In case of UAV for monitoring pipelines there should be used a device that can carry at least sensory and ancillary equipment. The selection of equipment depends on client requirements primarily.

2.2.3 Terrain conditions

The most ideal route for inspecting by UAVs is the flat terrain with small elevation differences that ensures the constant scanning speed and altitude during convenient weather conditions.

An essential component of UAVs is a navigation system. The more accurate the system is, the more precise location data are sent to the operator in the case of failure.

2.2.4 Legislation

National law regulates the possibility of choosing UAVs. Generally, there are restrictions in the working frequency of UAVs to correspond to local requirements and conditions.

2.2.5 Fly length

The length of flight is dependent on the characteristics of the controlled distribution network for example pipes, pumping stations etc. Strategic proposal of the flight is the key to maximum applicability of capabilities of UAVs. The main condition is to minimize expenses (the operation) and maximize efficiency (the duration of the flight) (Říha et al. 2014).

3 UAV VS HELICOPTER

The comparison of all hourly expenditures of operating UAVs and helicopters is not always a straightforward matter, since there is no standardized protocol to calculate the cost per flight hour. It is important to consider the total cost including staff, transport, etc. Each operation has its own conditions of the comparison. When interpreting the advantages of UAV compared to helicopter systems, it is good to consider and compare the following facts:

- Prices of aerial work,
- Dimensional advantage,
- The regularity of inspections,
- Refueling,
- Maintenance and insurance,
- Sensory equipment.

3.1 Prices of aerial work

An hour of helicopter flight compared to the flight hour with UAVs is considerably more expensive, while the price of renting air service flight is usually determined by two parts:

- The cost of flight,
- Insurance.

Cheaper models of helicopters (for example Robinson R44) operates approximately for $800/hour, however, this is the lower price limit. The price of AS350 starts at $1,500/hour. Nevertheless, these prices are corresponding to the price per flight hour, not an hour of complex aerial work. Therefore, in the case of monitoring pipelines total expenses need to be increased by at least an hour or two for the transport. In addition, there must be reflected the rent of sensory technology which is also very expensive. The rent of Cineflex Gimbal together with sensors cost between 4,000 and 10,000 dollars per day. Using special sensory equipment leads to additional expenditures (Rising 2015).

On the contrary, drones are far more economical. For the price equivalent to an inspection by a helicopter at the duration of several days it is possible for a company to obtain a permit for the operation of air traffic, purchase the UAV and train their employees.

3.2 Dimensional advantage

Some situations require operations and monitoring of areas that are hard to reach. One example is a pumping apparatus located in an overgrown surrounding with the lack of transparency which would be scanned very difficult by a helicopter with risking the crew's safety. On the other hand, the UAVs can easily intervene in hard to reach places, they are able to fly alongside the buildings, trees and ridges of rocks (Voštová et al. 2014).

3.3 The regularity of inspections

Using UAV allows more regular inspection due to the fact that the cost of the operation is significantly lower. Drone equipped with sophisticated Thermal Vision costs about $85,000, while an inspection by helicopter costs about $3000/hour. The return on investment in an UAV is approximately 29 flight hours (Krishnamurthy 2013).

3.4 Problems with the refueling

Each air device needs a certain range of refueling for aerial work. This means finding a suitable place to land for refueling when using a helicopter at least and also in the case of pipelines inspection it can be required to transport the fuel in hardly accessible sites which increases the level of danger.

When using an UAV, it is enough either to add a small amount of fuel or to provide a replacement of its batteries. Thanks to the development in UAV, it is now possible to use the docks for refueling or battery replacement.

3.5 Maintenance and insurance

Maintenance together with insurance constitutes a significant part of operating expenses that largely affect the price of inspection control. Maintenance and insurance of UAV is much cheaper than in the case of helicopters. This positively affects the overall cost of inspections.

3.5.1 Operating costs of helicopter

Table 1. Cost comparison of helicopter (Helicopterforum 2010).

Cost/Year	Amount ($)	Notes
Insurance	7450	
Hangar	2400	Rent for a year
Machine control	1800	Annual fees
Inspection hours	1100	Planned hours 50–100 h
Fuel	4860	$h * l/h * p/l**$
Oil	127	
Total	17,737	
The purchase price	120,000–160,000	Depending on the age and number of hours flown

** h = hour, l = liters, p = price.

3.6 Operating costs of UAV

Insurance: valid in the European Union for commercial use is roughly $400 (CMMA 2017).

Hangar/year: $600, due to the functional dimensions of smaller airplanes, it is sufficient merely one third of the size of the hangar.

Annual fees for control: checking is provided by staff responsible for the operation of UAVs. The price is included in their salary +$100 to $200 for spare parts.

Planned inspection: included in the annual fees for monitoring.

Fuel: in the case of n-copter driven by electricity, the batteries costs about $50 per year (rough estimate).

In the case of combustion engines, the fuel consumption is 2 liters per flight hour. It is an estimate, which considers the payload of the UAV. The consumption for 120 hours is 240 liters. The price of fuel for conventional aerial vehicles is approximately $250 (for 120 flight hours) and it already includes a lubricating oil in the fuel.

The purchase price of drone is around $85,000.

Table 2. Cost comparison of UAV.

Costs/Year	Amount ($)	Notes
Insurance	400	
Hangar	600	Rent of hangar for 1 year
Machine control	150	The price is included in the employee's wage
Inspection hours	0	Planned hours 50–100 h
Fuel	50 / 250	Battery Combustion engine
Oil	0	Oil is included in the price of fuel
Total	1400	
Purchase price	85,000	Depending on the age and number of hours flown

3.7 Sensory equipment

During the last decade research and development of UAVs have progressed enormously. The same goes for sensory technology used during the flight. Nowadays, most of the development centers and companies dedicated to exploratory sensory devices of larger sizes which are part of the aircraft and helicopters gear focused on the development of these facilities with the possibility of using them for UAVs. These days, UAVs can be fully equipped with GNSS technology measurements and make use of laser scanning LIDAR technology. It should be noted that smaller devices are often less powerful than the larger dimensions devices, but UAVs can fly at lower altitudes, which largely compensates for this problem.

4 POSSIBLE SUBSTITUTION OF HELICOPTERS BY UAVS

Economically and technically efficient substitution of the helicopters by UAVs within existing legislation is possible by using two scenarios:

– N-copter.
– Fixed wing.

4.1 N-copter

N-copter qualifies for the survey of hard to reach places, it is also suitable for flight on shorter distances due to lower battery life, for local monitoring in case of leakage or the detailed observation of pumping stations etc. The operator should be provided with spare applicable batteries. The used N-copter should be provided with appropriate sensory; for example hyperspectral camera or Lidar together with the visible camera with high resolution. N-copter is able to effectively compensate for manual inspection activities by workers and manual inspection equipment (Hausamann et al. 2005).

Table 3. Parameters of n-copter.

Altitude	<50 m
Payload	<7 kg
Maximum endurance	<1 h
Platform	N—copter
Sensor	Nir cam + lidar, or 4K UHD + LIDAR

4.2 Fixed wing

This device is usable mainly for recording of longer routes within the legislation authorizes it and at lower altitudes to Class G <100 m. There is also a possibility of using it for regular control flights. Sensory technology should include at least the NIR camera and Lidar, however, the most appropriate would be a multisensor (Gómez et al. 2015).

Table 4. Parameters of fixed wing.

Altitude	<100 m
Payload	<25 kg
Maximum endurance	<5–6 h
Platform	Fixed wing
Sensor	Nir cam + lidar, or 4K UHD + LIDAR or multisensor

4.3 Combined operation

With the combination of UAVs in sections 4.1 and 4.2 it is possible to provide a full inspection of pipelines while using UAVs for the entire pipeline inspection route.

5 LIMITATIONS OF UAV'S OPERATION

The widespread use of developed UAVs is limited by legislation requirements of the states where the UAVs are used for operation. This fact indirectly has an impact on their development. The question is avoiding obstacles either organic or inorganic during manual or autonomous flight. Many states do not allow autonomous flights and if they do, the UAVs can be only used in the field of view of its operator. (Hospodka et al. 2015) Without resolving these legislative obstacles, the operation of UAVs for pipelines over long distances will be difficult. The solution could be using UAVs equipped with localization sensors which would fly in a dedicated flight altitude. The location of UAVs could be checked by operating pilots as well as operators of air traffic transport.

6 CONCLUSION

The use of unmanned aerial vehicles can be a suitable complement to conventional methods of monitoring pipelines carrying dangerous and valued materials. UAVs are able to reduce the amount of expenditures significantly by preventing leaking, while they are also far more efficient due to less expensive operation on the contrary to inspection by helicopters. This enables to undertake more frequent surveillance flights.

Within the inspection of equipment and pipelines, thanks to rapid research and development that has caused miniaturization of devices, the small-sized UAVs can be equipped with superior sensory technology, which was recently used for inspection by helicopters for the same purpose. Using UAVs eliminates the risk of security of the crew in emergency situations because of the absence of the crew and controlling it by ground-based operator in safe distance.

Companies carrying oil and gas have the opportunity to train their employees for operating unmanned aerial vehicles, which can lead to more effective manual controlling and eliminates time-consuming inspection in difficult terrain with handheld devices. This is the reason using UAVs can diminish unnecessary costs in the long term.

REFERENCES

Fuchs, P., et al. 2015. The Assessment of Critical Infrastructure in the Czech Republic. In: *Proceedings of 19th International Scientific Conference Transport Means*. Transport means 2015. Kaunas, 22.10.2015–23.10.2015. Kaunas: Technologija. 2015, pp. 418–424. ISSN 1822-296X.

Gómez, C. & Green, D.R. 2017. *Small-Scale Airborne Platforms for Oil and Gas Pipeline Monitoring and Mapping*. University Of Aberdeen. 2015 Report. [Online] [cit. 2017-03-26]. Available from: http://www.abdn.ac.uk/geosciences/documents/UAV_Report_Redwing_Final_Appendix_Update.pdf

Hausamann, D. et al. 2017. *Monitoring of gas pipelines— A civil UAV application*. Aircraft Engineering and Aerospace Technology. 2005, 77(5), 352–360. DOI: 10.1108/00022660510617077. ISSN 0002-2667.

Hospodka, J. et al. 2015. Influence of autonomous vehicles on logistics. International Review of Aerospace Engineering. 2015, 8(5), pp. 179–184. ISSN 1973-7459.

How much do helicopter cost to buy and run?: edspilot. 2017. In: *Helicopterforum* [online]. USA, 2010 [cit. 2017-03-26]. Available from: http://helicopterforum. verticalreference.com/topic/12708-how-much-do-helicopter-cost-to-buy-and-run/

Krishnamurthy, K. 2017. Feature-In Alaska's oilfields, drones count down to take off. In: *Reuters* [online]. 2013 [cit. 2017-03-26]. Available from: http://www.reuters. com/article/drones-oil-idUSL3N0E828H20130607

Novák, M. 2014. Implementation of the NDT into the Approved Maintenance Organization according to the Regulation (EU) No 1321/2014. In: *Ostaševičius*, V., ed. Proceedings of 20th International Conference Transport Means 2016. 20th International Conference Transport Means 2016. Juodkrante, 05.10.2016–07.10.2016. Kaunas: Kauno technologijos universitetas. 2016, pp. 180–184. ISSN 1822-296X.

Pojištění bezpilotních systémů č. 490 001 519. 2017. Pojištění odpovědnosti za škodu způsobenou provozem. *CMMA* [online]. [cit. 2017-03-26]. Available from: http://cmma.cz/pojisteni-bezpilotniho-letadla-2/

Rising, J. 2015. Drone Vs Helicopter—Part 1: *How Drones are Changing the Aerial Video Industry.* In: Http://flight-evolved.com [online]. 2015 [cit. 2017-03-26]. Available from: http://flight-evolved.com/drone-vs-helicopter/

Říha, Z. et al. 2014. Transportation and environment— Economic Research. In: *The 18th World Multi-Conference on Systemics, Cybernetics and Informatics*. The 18th World Multi-Conference on Systemics, Cybernetics and Informatic. Orlando, 15.07.2014–18.07.2014. Orlando, Florida: International Institute of Informatics and Systemics. 2014, pp. 212–217. ISBN 978-1-941763-05-6.

Ropovodná sieť v SR 2017. In *Traspetrol.sk* [online]. [cit. 2017-03-26]. Available from: http://www.transpetrol. sk/ropovodna-siet-v-sr/

Voštová, V. 2014. Conveyor new concept for parcel logistics in air transport [online]. In: *Applied Mechanics and Materials*. Conference on Research, Production and Use of Steel Ropes, Conveyors and Hoisting Machines. Vysoké Tatry—Podbanské, 23.09.2014–26.09.2014. Zurich: Trans tech publications ltd. 2014, pp. 114–118. ISSN 1660-9336. ISBN 978-3-03835-316-4.

New Trends in Process Control and Production Management – Štofová & Szaryszová (Eds)
© *2018 Taylor & Francis Group, London, ISBN 978-1-138-05885-9*

Corporate reputation in the automotive industry

M. Kozáková & J. Lukáč
Faculty of Business Management, University of Economics in Bratislava, Bratislava, Slovak Republic

ABSTRACT: Good corporate reputation is very important because of its potential for value creation and intangible characteristics which make replication by other firms more difficult. The concept of corporate reputation has gained widespread attention throughout the world because it is believed that a company's reputation, as experienced by various stakeholders, will ultimately influence commercial opportunities, sales, and profit and non-commercial (consumer trust, loyalty) outcomes. Corporate reputation can be a key contributor to an organization's success and it can just as easily be a contributing factor to an organization's failure. It is one of the most important intangible assets for maintaining and enhancing firms' competitiveness in the global marketplace. Corporate reputation encompasses the feelings of individuals toward a company, plays an important role in the growth of share value and also supports the company's ability to attract qualified staff. The aim of this paper is to evaluate and compare the corporate reputation in automotive industry.

1   INTRODUCTION

The current business environment shows that the company must be part of society and play an active role. Corporate reputation is built over time and it is the result of complex interactions and relationships between the company and its stakeholders. This means that the reputation is based on past actions, experiences and reactions of stakeholders, so the corporate reputation is based on the overall perception of stakeholders, for example customers, managers, employees etc. A good reputation is a valuable source that allows the company to achieve long-term profitability or sustainable superior financial performance. Positive effects of corporate reputation have a financial impact and lead to more profits, more stable and less vulnerable cash flows, and it influences on the firm's value, which is ultimately its market capitalization (shareholder value). Generally, customers buy from company they trust. Corporate reputation is the perception in time about a company. In this case trust is an important element that influences corporate reputation.

2   THE CURRENT STATE OF AUTOMOTIVE INDUSTRY IN THE WORLD

Today's economies are dramatically changing, triggered by development in emerging markets, the accelerated rise of new technologies, sustainability policies and changing consumer preferences around ownership. Digitalization, increasing automation and new business models have revolutionized other industries, and automotive will be no exception. These forces are giving rise to four disruptive technology-driven trends in the automotive sector: diverse mobility, autonomous driving, electrification and connectivity (McKinsey 2016).

Long product cycles and deep capital investments make planning in the automotive industry a complex endeavor. For the past 10 years, OEMs and suppliers have generally chased global sales growth while hoping to improve margins by leveraging automobile platforms in multiple regions and striving for scale wherever possible. The results of this strategy have been decidedly mixed. In 2015, global economic conditions worsened. This trend makes any new commitment to invest in a country or region a risky one that must be deliberately crafted using a clear-eyed assessment of market conditions (Pwc 2016).

2.1   *European union*

Sales have improved in the European Union since the financial downturn, but the EU automotive industry is held hostage by local economies that are teetering on the edge of recession. In 2015, new car registrations in the EU rose 9.3% year-on-year to 12,6 million units. But that is well below the record year of 2007, when more than 18 million vehicles were sold in the region. Carmakers in some EU countries (France, Greece, Italy, Portugal and

Spain) struggling to grow their economies—they face losses or low profits, fragmented markets and the inefficiencies of model proliferation. The EU automotive industry must figure out ways to better match production capacity to market demand, while simultaneously investing in new potentially strong product areas (f. e. small SUVs and crossovers) and in new automobile technologies.

2.2 Emerging nations

Perhaps the biggest downward macroeconomic force in the automotive industry today is the underperformance of emerging markets, which not too long ago represented a big opportunity for major gains in the global auto sector. While sales in India remained roughly flat in 2015, China's year-over-year growth slowed to 7.3% from a 10% gain in 2014 and 16% gain in 2013. New vehicle ownership restrictions in China's largest cities will further curtail sales in the coming years. Russia had its second straight year of decline in 2015—sales were almost 50 percent below the peak in 2012. Sales in Brazil fell by nearly 1,3 million units (or 30%) from its record high in 2012—this drop was larger than the entire Mexican car market (Pwc 2016).

2.3 Middle east and Africa

Over the next five years, the Middle East and Africa, which is relatively unmotorized region, will likely see strong and consistent automobile sales growth. The biggest improvements are expected in Iran, Egypt, South Africa, and Nigeria. Along with this growth, carmaker factory activity in the region will increase significantly. Nearly 3 million cars will be produced yearly in the region by 2021, which constitute increase of about 50%. Substantial factory capacity improvements are likely in Algeria, Nigeria, Egypt, and Iran. Given the diversity of this region—there are more than 50 distinct markets—carmakers face the obstacle of satisfying multiple unique local requirements. Among them are domestic assembly quotas, import and export tariffs and duties for parts and vehicles, gas or diesel preferences, and local customs that may dictate the design of interior and exterior features.

2.4 North America

U.S. markets are peaking at historic levels, setting a sales record of just under 17,5 million vehicles in 2015, up 5.7% from the year before. U.S. sales are likely to be relatively flat in the next two years and may face a moderate downturn in 2018, the victim of economic cycles, higher car loan interest rates as the Federal Reserve raises overnight

Table 1. The 10 TOP selling car brands worldwide in 2015.

Position	Brand	Cars sold	Market share
1.	GM	3,082,366	17.6%
2.	Ford	2,603,082	14.9%
3.	Toyota	2,499,313	14.3%
4.	Chrysler	2,200,834	12.6%
5.	Honda	1,586,551	9.1%
6.	Nissan	1,484,918	8.5%
7.	Hyundai	761,710	4.4%
8.	Kia	625,818	3.6%
9.	Subaru	582,675	3.3%
10.	Mercedes-Benz	372,977	2.1%

Source: Own processing according to Carophile's data.

rates and an expected flood of vehicles into the used car market. Mexican car sales outpaced forecasts in 2015—jumping 19% to more than 1,3 million units. It is expected to surpass 1,5 million by 2021 (Pwc 2016). Investments in new car factories in Mexico are surging as well; installed capacity is likely to grow more than 50% over the next five years. These conditions compel carmakers and suppliers to manage supply chains and factory usage cautiously in the U.S., while continuing to expand in Mexico. Table 1 summarizes the top 10 selling ca brands around the world in 2015.

2.5 Automotive industry in Slovakia

The automobile industry is very important for the Slovak Republic and is one of the traditional drivers of the Slovak economy, not only because it employs a lot of people, has a long-term tradition and regularly produces more than a million vehicles, but also because of its position within Europe. Even despite the relatively small size of our country and economy compared to other member states, the Slovakia's developed automobile industry is not losing pace, but is even remaining ahead, according to some indicators. Recently, Slovakia has grown to become one of the leading car producers in world, primarily thanks to the presence of three world-class automotive companies being established in Slovakia: Volkswagen Slovakia in Bratislava (since 1991), PSA Peugeot Citroën Slovakia in Trnava (since 2003) and Kia Motors Slovakia in Žilina (since 2004). The automotive industry has a strong tradition in Slovakia and became the most important sector and driving force of the Slovak economy. Over the past 20 years it has been an important source of foreign direct investment. Almost 80 000 people are

Table 2. The 10 TOP selling car brands in Slovakia in 2015.

Pos.	Brand	Cars sold	Market share
1.	Škoda	16,570	21.3%
2.	Volkswagen	7616	9.8%
3.	Hyundai	6302	8.1%
4.	Kia	5603	7.2%
5.	Peugeot	4472	5.7%
6.	Opel	4220	5.4%
7.	Dacia	3313	4.2%
8.	Renault	3008	3.9%
9.	Citroen	2674	3.4%
10.	Suzuki	2638	3.4%

Source: Own processing according to Trend Top 2016 data.

employed directly by the 3 car producers. 200,000 people are employed directly and indirectly by the automotive industry in Slovak Republic and produced 2,5 BN € of value added (Sario 2016). The 10 TOP selling car brands in Slovakia in 2015 are listed below in Table 2.

3 COPORATE REPUTATION AND AUTOMOTIVE INDUSTRY

3.1 Concept of corporate reputation

Corporate reputation is important in today's highly competitive environment. Companies have traditionally been able to sustain high profit margins, facilitated by strategies of product differentiation and market segmentation. However, today, companies are having a hard time differentiating their products, even with large research and development budgets. Any innovation could be quickly copied by the competition. That is the reason, why corporate reputation should be consider an important strategic asset that helps a company to differentiate itself from others companies.

Corporate reputation is still relatively new as an academic subject. It is becoming a paradigm in its own right, a coherent way of looking at organizations and business performance, but it is still dogged by its origins in a number of separate disciplines. There are lots of different definitions of what corporate reputation is and every discussed discipline brings a different view and approach to corporate reputation. It has been studied under several disciplines: institutional theory, financial theory, economic theory, organizational behavior theory, etc. and from various perspectives: from explaining its role and up to researching the processes of establishment

and analyzing a company's abilities to shape its reputation.

Fombrun et al. (2000) define corporate reputation as a collective assessment of a company's ability to provide valued outcomes to a representative group of stakeholders. He considers corporate reputation an intangible multi-stakeholder concept resulting from the perceptions that diverse stakeholder groups—such as customers and suppliers. Gotsi & Wilson (2001) also define the corporate reputation as a stake holder's overall evaluation of a company over time. This evaluation is based on the stake holder's direct experiences with the company, any other form of communication and symbolism that provides information about the firm's actions and/or a comparison with the actions of other leading rivals.

Alternatively, Gray & Ballmer (1998) define corporate reputation as a valuation of a company's attributes, performed by the stakeholders, what would almost completely exclude affective components. Hall (1992) combines cognitive und affective components by formulating that a company's reputation consists of the knowledge and the emotions held by individuals. Fombrun & van Riel (1997) have defined corporate reputation from the perspective of six distinct academic subject areas (Table 3).

Table 3. Categorization of corporate reputation literatures..

Discipline	Categorization of reputation
Accountancy	Reputation seen as an intangible asset and one that can or should be given financial worth.
Economics	Reputation viewed as traits or signals. Perception held of the organization by an organization's external stakeholders.
Marketing	Viewed from the customer or end-user's perspective and concentrating on the manner in which reputations are formed.
Organizational behaviour	Viewed as the sense-making experiences of employees or the perception of the organization held by an organization's internal stakeholders.
Sociology	Viewed as an aggregate assessment of a firm's performance relative to expectation and norms in an institutional context.
Strategy	Reputation viewed as assets and mobility barriers. Since reputations are based on perception, they are difficult to manage.

Source: Fombrun and van Riel (1997).

3.2 Why is corporate reputation important?

Favorable reputations are accepted as intangible assets that have been related to positive outcomes for an organization (Coombs 2007).

It helps companies differentiate themselves, in a positive way, from their peers; therefore it provides competitive edges (Surroca et al. 2010). It can provide a variety of benefits, including reduced financing, advertising and supplier costs; increased access to new strategic opportunities and partnerships; greater ease in recruiting talented employees, increased employee morale, productivity; and greater good will with stakeholders when something goes wrong (Barnett & Pollock 2012). Reputation has been identified as playing a significant role in improving firm value (Fombrun & Shanley 1990). It can enhance consumer perceptions of product quality (Milgrom & Roberts 1986) and permitting access to cheaper capital. Favorable corporate reputation is considered intangible asset that offers a strategic competitive advantage to companies and leads to value creation. Corporate reputation is able to bring higher sales, revenues and financial benefits the company, so there is connection between corporate reputation and financial performance of the company (Rose & Thompsen 2004).

Companies with a bad reputation on the other hand can have more difficulties in drawing attention of investors and receiving funding (Kiambi & Shafer 2015).

Concluding, a positive reputation is of importance for an organization and can even be seen as the single most valued organizational asset (Gibson et al. 2006).

3.3 RepTrak Pulse model

The Reputational Institute determines a model called RepTrak Pulse to measure the corporate reputation and to develop the ranking of the most reputable companies all over the world. The RepTrak Pulse is a measurement for the public opinion of companies detected by an annually conducted global survey by the Reputation Institute. The rating describes how much consumers trust, like and admire a company (Fombrun 2015). The survey analyzes people's perception about company and asks the public to rate company in a terms of its products and services, innovation, workplace, citizenship, governance, leadership and performance.

The most reputable companies from automotive industry all over the world in 2016 were BMW Group and Daimler Mercedes Benz. Table 4 shows the reputation coefficient determined for companies in automotive industry by Reputational Institute 2013 to 2016.

Figure 1. Rep trak pulse model.
Source: 2016 Global RepTrak®100- The World's Most Reputable Companies. [online] https://www.rankingthebrands.com/PDF/Global%20RepTrak%20100%20Report%202016,%20Reputation%20Institute.pdf.

Table 4. Reputation coefficient.

Company	2016	2015	2014	2013
BMW group	77.9	78.98	77.2	78.39
Daimler (Mercedes Benz)	77.7	77.85	75.4	76.58
Rolls Royce Aerospace	75.8	74.49	73.2	–
Michelin	75.7	74.07	74.2	72.49
Toyota	73.2	71.65	71.6	70.49
Honda motor	71.4	69.89	70.9	70.93
Volvo group	70.5	71.24	70.2	70.05
Ford motor	69.7	69.43	67.7	68.12
General motors	67	65.59	-	65.85
Volkswagen	61.3	75.02	74.9	74.38
Nissan motor	–	–	66.9	65.82
Suzuki	–	–	66.2	65.53
PSA Peugeot-Citroen	–	–	65.7	65.04

Source: Own processing.

The industry average in 2016 drops to a 70.5 from a 72.5. The only carmaker that sees a drastic drop is Volkswagen. Since 2015, Volkswagen's reputation dropped by 13.7 points globally. Across the 7 dimensions of reputation, VW saw an average drop of 10.9 points with biggest drops in Governance (–17 p), Citizenship (–15 p) and Leadership (–11 p). VW's reputation dropped significantly across 14 markets. Due to its drop in reputation, VW saw a drop across all supportive behaviors. Figure 2 represents the reputation coefficient and its development over time.

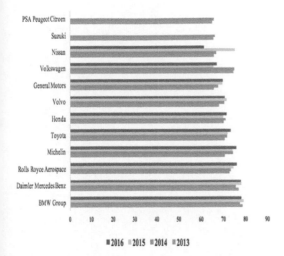

PSA Peugeot Citroen
Suzuki
Nissan
Volkswagen
General Motors
Volvo
Honda
Toyota
Michelin
Rolls Royce Aerospace
Daimler Mercedes Benz
BMW Group

0   10   20   30   40   50   60   70   80   90

■2016 ■2015 ■2014 ■2013

Figure 2.   Development of coefficient over time.
Source: Own processing.

4   CONCLUSION

Companies are increasingly becoming aware that intangible assets provide more competitive advantages than product-related sources. Numerous authors identify corporate reputation as one of the most important intangible assets playing an increasingly important role in terms of firms' propensity to influence important stakeholder groups, such as financial analysts, employees, and customers in global markets. That is the reason, why companies should carefully manage corporate reputation and try to understand the potential factors that can enhance corporate reputation.

ACKNOWLEDGEMENT

This article is one of the partial outputs of the currently solved research grant **APVV** no. **APVV-15–0511** entitled *"Research of the issue of Online Reputation Management (ORM) of subjects from automotive industry"*.

REFERENCES

Carophile.com. The Top selling car brands. [online] http://www.carophile.com/the-top-selling-car-brands/
Coombs, W.T. 2007. Protecting organization reputations during a crisis: The development and application of situational crisis communication theory. *Corporate Reputation Review,* 10(3):163–176.
Fombrun, C. & Shanley, M. 1990. What's in a Name? Reputation Building and Corporate Strategy. *Academy of Management Journal,* 33: 233–258.
Fombrun, C.J. & van Riel, C.B.M. 1997. The reputational landscape. *Corporate Reputation Review,* 1(1/2): 6–13.
Gibson, D. et al. 2006. The importance of reputation and the role of public relations. *Public Relations Quarterly,* 51(3): 15–18.
Gotsi, M. & Wilson, A.M. 2001. Corporate reputation: Seeking a definition. *Corporate Communications,* 6(1): 24–30.
Gray, E. & Ballmer, J.M.T. 1998. Managing Corporate Image and Corporate Reputation. *Long Range Planning,* 31: 695–702.
Hall, R. 1992. The Strategic Analysis of Intangible Resources. *Strategic Management Journal,* 13: 135–144.
Kiambi, D.M. & Shafer, A. 2015. Corporate Crisis Communication: Examining the Interplay of Reputation and Crisis Response Strategies. *Mass Communication and Society,* 1–22.
KPMG. 2017. Global Automotive Executive Survey 2017 Bratislava: KPMG International Cooperative. [online] https://assets.kpmg.com/content/dam/kpmg/xx/pdf/2017/01/global-automotive-executive-survey-2017.pdf.
Mckinsey.com: Disruptive trends that will transform the auto industry. [online] http://www.mckinsey.com/industries/automotive-and-assembly/our-insights/disruptive-trends-that-will-transform-the-auto-industry.
Milgrom, P. & J. Roberts. 1986. Price and Advertising Signals of Product Quality. *Journal of Political Economy,* 94: 796–821.
Reputation Institute. 2016. *Global RepTrak®100—The World's Most Reputable Companies.* [online] https://www.rankingthebrands.com/PDF/Global%20RepTrak%20100%20Report%202016,%20Reputation%20Institute.pdf.
Rose, C. & Thomsen, S. 2004. The impact of corporate reputation on performance: some Danish evidence. *European Management Journal,* 22(2): 201–210.
Sario. 2016. Automotive sector in Slovakia. Bratislava: Slovak Investment and Trade Development Agency. [online] http://www.sario.sk/sites/default/files/content/files/sario-automotive-sector-in-slovakia-2017-03-02.pdf.
Strategyand.pwc.com: 2016. Auto Industry Trends. [online] http://www.strategyand.pwc.com/trends/2016-auto-industry-trends.
Surroca, J. et al. 2010. Corporate Responsibility and Financial Performances: The Role of Intangible Resources. *Strategic Management Journal,* 31: 463–490.
TREND. 2016. *Slovenský trh osobných automobilov. Trend TOP 2016.* Bratislava: News and Media Holding.

New Trends in Process Control and Production Management – Štofová & Szaryszová (Eds)
© *2018 Taylor & Francis Group, London, ISBN 978-1-138-05885-9*

Use of modern technologies in the ergonomics application in logistics

M. Kramárová, Ľ. Dulina, I. Čechová & M. Krajčovič

Faculty of Mechanical Engineering, University of Zilina, Žilina, Slovak Republic

ABSTRACT: The article describes the dependence between the operator's load parameter during his control of a forklift during dangerous logistic activities and subjective reaction of operator to this load in the form of strain. Evaluation of the operator's strain is based on the force that is exerted on the intervertebral segment L4/L5 and muscles in the spine by Low back analysis at Digital Factory. It also provides information about values of operator's awkward postures differing from the neutral working posture during a specific working dangerous activity. Subsequently, the article defines the parameters for an operator, during control of a forklift, when he doesn't enter into non-physiological working postures associated with discomfort. On the following factors and outputs are proposed solutions for their elimination.

1 INTRODUCTION

In the long term, when a spine is loaded repeatedly in the same way, the feeling of pain can turn into a chronic cervical or lumbar back injury. Forklift operators spend most of the shift in the seated position associated with frequent lateral and axial spine rotations. After the eight—hour work shift feeling of pain or eventually stiff muscles can be expected in many operators. Improvement of ergonomic conditions and therefore elimination of the load on the operator is a factor that can be primarily influenced by the producer of the forklift and secondary by the right choice of the forklift type, for the given activity. Then the conditions of work for the forklift operators may be adapted by improving the organization of work, and by keeping the operators informed about the ergonomic principles. The main aim should be to provide the operator with such work conditions, that his productivity in the first and last hours of the shift would be qual. (Forklift Operator Ergonomics 2012)

2 FACTORS AFFECTING THE OPERATORS OF FORKLIFTS

Two of the most important factors that cause forklift operators back injuries are repetitive and non-physiological working postures. Range of impact on the physical load is enhanced by vibration and shock. These can be transmitted to the operator's body through the seat or steering wheel of forklift.

Other factors affecting on forklift operator load are whole body vibration, dimensional solution of forklift cabin, view cone and visibility, configuration of foot and hand controls, configuration of control panels and displays and fatigue.

The operators, whose substantial part of work is a forklift operating during work shift, are exposed to increased risk of pain or spine diseases. Specifically, hernia or bulging intervertebral disc, scoliosis, chronic back injury, and others. The most common form of hernia is the disc herniation between vertebrae L4/L5. It is the intervertebral disc, which carries the most weight and pressure of whole spine. It is part that is analyzing during load the most. It is the most susceptible to injury and cure of it is for long time. (Forklift Operator Ergonomics 2012, Gajšek & Duklic 2015, Gregor et al. 2016)

Operation of forklift requires from operator the works mainly with static load in the position of the seats, in certain types of forklift stand, associated with large amounts of non-physiological working position. However, this position is directly dependent on the type of forklift. Except of static load, the operators are during operating of forklift loaded by non-physiological working postures as extension of the cervical spine, axial or lateral rotation of the trunk, combination of lateral and axial rotation at same time and other (Fig. 1). (Forklift Operator Ergonomics 2012, Gajšek & Duklic 2015)

In Figure 1 are defined selected the most common non-physiological working postures of forklift operators in logistics. Each of these working postures have defined limits of unacceptable postures in the law of Slovak, which can't be exceeded

Figure 1. Risky logistic activities with limits of deflection from neutral postures.

Table 1. The table of results from the Nordic Questionnaire.

Location of symptoms	The number of forklift operators with difficulty, last 12 months	The number of forklift operators with difficulty, last 7 days cervical spine
Cervical spine	9	6
Shoulders	9	9
Lumbar and Sacral spine	12	7
Thoracic spine	4	3
Hip/thighs/buttocks	4	3
Wrists/hands	4	4
Ankles/legs	5	4
Knees	7	3

for more than 30 minutes per one shift. For extension of worker's cervical spine, for example during loading and unloading of pallet to/from racks (Fig. 1a), is posture unacceptable without whole head support during static holding. During operating of forklift where is lateral rotation of the trunk for better vision (Fig. 1c, d) is unacceptable postures defined by law of Slovakia no. 542/2007 when lateral is more than 20° or when is more than 20° with frequency of move more or equal like 2 times per minute. When there is axial rotation of the torso during reversing with forklift (Fig. 1b), this posture is unacceptable when act more than 20° or when is axial rotation more than 20° with frequency of move more or equal like 2 time per minute.

Working postures when operating forklift were chosen on based of information about the problem of inclusion these activities to a higher category of hazardous work. This is the reason why were examined and verified limits of risks and value of the operator load in the specific job position. However, on the range of operator load during operating the forklift does not affect only the working posture but also other factors which influence each other and increase the total load.

2.1 Nordic Questionnaire as a tool for assessment of the subjective reaction of forklift operator on load

Vibrations, non-physiological working postures, shocks and static loads for longer time spent in the seat of forklift are factors, that effect on the formation of musculoskeletal system disorders. The most common problems of musculoskeletal system diseases include diseases of the cervical and lumbar spine, shoulders and forearms. Discomfort, pain and

decreased mobility are just some symptoms of the start form these difficulties. (Capacity electric stand-up end control forklifts 2016, Štefánik et al. 2003)

From survey realized by Nordic questionnaire was asked 16 forklifts operator's different types. It identified a subjective reaction to the load (Table 1) in the form of difficulties in the lumbar and sacral spine, shoulder and neck pain. And based of this result, was further reason to solve the impact of working position the operator on the size of the load. Specifically, Influence of trunk angle rotation on the final risk level and load on the intervertebral disc L4/L5, as the weakest part on the spine.

3 FORKLIFTS OPERATORS' EXCESSIVE STRAIN AS A RESULT OF NON-PHYSIOLOGICAL WORKING POSTURES

One of the factors that increase the physical load of operators at work is also working posture. When we talk about postures at work, we divide them into physiological and non-physiological working postures. In case of keeping healthy workers, reduction of absence and diseases, the workers should keep working only at physiological working postures. It means, that they should be, the least deflected from the neutral body position. (Gregor et al. 2015, Matuszek et al. 2009)

3.1 Ergonomic analysis OWAS for evaluating operator load rate

In context of operator's load during forklift operating and with analysis the level of risk to health due to the working position was made ergonomic analysis OWAS (The Ovako Working Posture

Analysis System). For the three types of work postures with different types of angles during extension of cervical spine, lateral and/or axial rotation (Fig. 1).

From the analysis OWAS we get information about operator load by working posture, specifically if there is low no. 1, middle no. 2, increased no. 3 or high no. 4 risk to health (Table 2).

Just OWAS analysis is usable for determine the risk level of health harm caused by working postures. OWAS analysis compare with the law of Slovak has different results in the case of cervical spine extension and lateral rotation of the trunk. While in the law of Slovak is defined extension of cervical spine as unacceptable positions (Table 3), an analysis OWAS assessment this position with low level of risk. Reason is that the analysis OWAS doesn't take account deflection of the cervical spine from a neutral position but only the deflection of the spine as a whole.

In the case of lateral rotation of the trunk, the results of analysis OWAS is almost identical to the law of Slovak. Law of Slovak defines lateral rotation greater than 20° as non-physiological working posture. OWAS analysis assessment lateral rotations of spine greater than 20° as the posture with middle risk. However, with regards to time factor is a significant difference in the results. On the basis of which it can be concluded that the Law of Slovak is stricter than OWAS analysis. And the reason is that law of Slovak defined as unacceptable posture in the lateral rotation of more than 20° for more than 30 minutes per one shift. Analysis

OWAS evaluate these type of posture as unacceptable for up to 4 hours.

3.2 Forklift operator's load of intervertebral disc L4/L5 during operating the forklift

Load of human due to working postures during operating the forklift can be analyzed through the several factors that act on him immediately. One of them is even force that is expended on L4/L5 segment. At this point is analyzed dependence of intervertebral disc L4/L5 load on the size of the deflection from the spine neutral position, also on the gender and the operator anthropometric characteristics that are in our article, height and weight.

The smallest height was selected on the basis of the smallest height of women at anthropometric atlas in Tecnomatix Jack. The highest height was selected on the basis of the highest man at anthropometric atlas in Tecnomatix Jack.

For each height in the range from 150–195 cm with the spacing values 3 cm were analyzed for intervertebral disc three weight variants. Underweight, ideal weight and slightly overweight that are different for man and woman. To determine the size of the intervertebral disc L4/L5 load with given parameters was used ergonomic analysis LBA (Lower Back Analysis) in Digital factory Tecnomatix Jack.

Even though the fact that with an increasing extension of the man and woman cervical spine is reducing the pressure on the intervertebral discs L4/L5 (Figs. 2, 3) is this posture unacceptable from view of the static hold. Based on law of Slovak is

Table 2. Possible results from OWAS analysis.

OWAS (Result of the risk rate)

1 Low risk—not necessary corrective actions
2 Middle risk—need corrective actions in the near future
3 Increased risk—need corrective actions asap
4 High risk—need immediate corrective action

Table 3. Analysis OWAS for assessment of man and woman forklift operator's posture.

Extension of the cervical spine		Lateral rotation of the trunk	Axial rotation of the torso
0°	1	1	1
10°	1	1	2
20°	1	2	3
25°	1	2	3
30°	1	2	3
35°	1	2	4

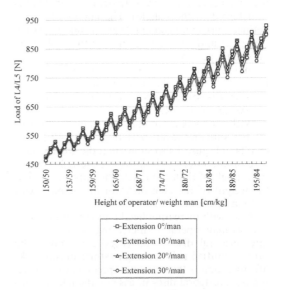

Figure 2. Dependence of intervertebral disc L4/L5 load from anthropometric characteristics of man operator and the extension of the cervical spine.

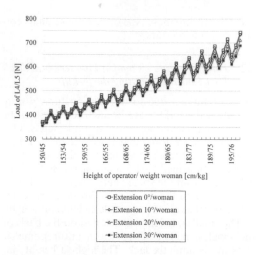

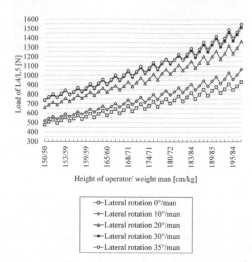

Figure 3. Dependence of intervertebral disc L4/L5 load from anthropometric characteristics of woman operator and the extension of the cervical spine.

Figure 4. Dependence of intervertebral disc L4/L5 load from anthropometric characteristics of forklift man operator and the lateral rotation of spine.

not allowed exposes the operators to static hold of head and neck while extension without the support of the whole head.

It is recommended only in the context of rehabilitation exercises in a very short intervals and a very short time. Reason is that extension of head is one of the rehabilitation positions with disc problems, that the get disks to the original position. Alternatively, if there is head extension, the muscles of the spine are contractions and around the disk acts a lower pressure because part load is transferred to the muscles. The law of Slovak is defined maximum acceptable extension of cervical spine 15° and less than 2.min-1 during a 4-hours maximum per one shift.

In the case of exposure of operators to the lateral rotation of the spine during forklift operating can be observed increasing pressure on the intervertebral discs L4/L5 (Figs. 4, 5).

With the increase of spine deflection, increases the pressure on the intervertebral discs L4/L5. The results in Figure 4 also shows that the greater pressure on the intervertebral discs is exerted with the same angle of deflection, with the same height and the weight of the man operator in compare with woman operator.

Also, with increasing height and weight of man and woman operators increasing a load on the intervertebral disc L4/L5 (Figs. 2–9). From the three identified risk working postures during forklift operating is the maximum pressure exerted on the intervertebral discs in case of the lateral rotation of the spine (Figs. 4, 5).

In the case of axial rotation of the spine occurs the similar situation of intervertebral disc L4/L5

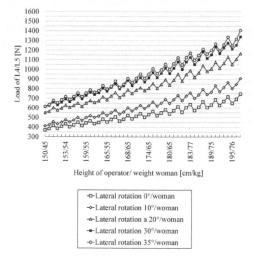

Figure 5. Dependence of intervertebral disc L4/L5 load from anthropometric characteristics of forklift woman operator and the lateral rotation of spine.

load like in case of the lateral rotation. The reason is that, with increasing angle of rotation rising the pressure on the intervertebral discs, which is at the same angle of rotation, the same height and weight greater in man workers than for women workers. (Figs. 7, 8) But only in relation to the exerted pressure on the intervertebral discs is less load on the worker in case of axial rotation (Figs. 2–9).

When during the operator working posture is worker also exposed by lateral and axial rotation of spine there is not significantly increasing pressure

290

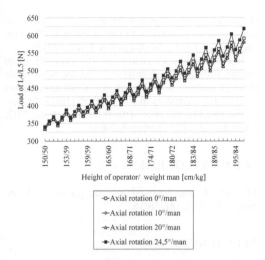

Figure 6. Dependence of intervertebral disc L4/L5 load from anthropometric characteristics of forklift man operator and the axial rotation of spine.

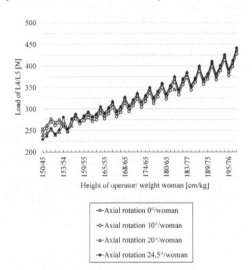

Figure 7. Dependence of intervertebral disc L4/L5 load from anthropometric characteristics of forklift woman operator and the axial rotation of spine.

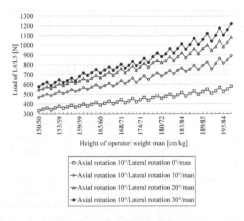

Figure 8. Dependence of intervertebral disc L4/L5 load from anthropometric characteristics of forklift man operator and the combination of lateral and 10° axial rotation of spine.

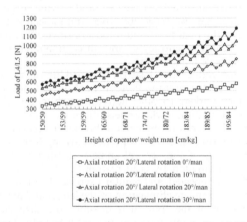

Figure 9. Dependence of intervertebral disc L4/L5 load from anthropometric characteristics of forklift man operator and the combination of lateral and 20° axial rotation of spine.

on the intervertebral disc L4/L5 compared with the situation in which there is only lateral rotation. (Figure 8–9).

3.3 *Forklift operator's strain of muscles during operating the forklift*

Muscle tension on the spine was analyzed for the same working postures in relations to changes of the same parameter as in the case of analyzing intervertebral disc L4/L5 load in the software Tecnomatix Jack. Increase of muscle tension on the

spine were identified only in case of the Erector spinae muscles.

In the case of spinal axial rotation of woman operator and a man operator were tension in the muscles, almost identical in all cases. During the working posture seats with extension of the cervical spine were analyzed minimal muscle tension on the spine, only at significantly higher operators. The reason is that, during extension of the cervical spine respectively during bent backward of head occurs to muscles retraction on the spine thus to reduce tensions. In working position with lateral rotation of the spine, was analyzed the highest tension of Erector spine muscle. However, even at these identified values it isn't assumption of formation any difficulties in relation to the muscle-skeletal system, because values are minimal.

4 SOLUTIONS FOR ELIMINATING THE RATE OF RISK OF FORKLIFTS ACTIVITIES

Over the years, occurred several changes in the area of forklifts innovations in the meaning to smaller load of operators, but it is still a great challenge to adjusted forklift with less load on operator's spine. In relation to ergonomics and forklifts was detected that, 67% mentioned work characteristics, 33% forklift design and 8% human-machine systems, personal protective equipment, prevention, workspace as causes for all kind of injuries related to work with forklift. (Gajšek & Dukic 2015, Capacity electric stand-up end control forklifts 2016, Ďurica et al. 2015)

Results from the study indicated that ergonomic features in trucks increase productivity. Ergonomics has two basic objectives and indicators to assess the quality of solutions: a positive effect on the health of employees and an economic effect. (Gajšek & Dukic 2015)

For reduce the load on the forklift operator through the organization of work is suitable to rotate workers to other activities such as forklift operating during whole one shift, or allow to operators longer or more frequent breaks. If necessary, eliminating the load of operators arising from the extension of the cervical spine is a potential for risk elimination by head restraint in forklifts, by PPE or repeated and regular exercise. The reducing or complete elimination of the lateral and axial rotation of the spine, it is possible to realize the technical actions in connection with forklifts. For example, forklift with rotating cabin (Fig. 10). The rotation of cabin allows to turn of 30° to the left and 180° to the right. Alternatively, use at workplace type of forklift, allowing to operate always face to the direction work respectively driving

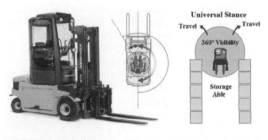

Figure 10. Forklift with rotating cabin (left) and forklift with universal standing posture (right). (Gajšek & Dukic 2015).

(Fig. 11). What allows working in the forklift in neutral stand posture with upper limbs holding on the sides of the body. (Gajšek & Dukic 2015)

Removing the extension of the cervical spine during unit insertion into the shelves, is possible through camera and display at forklift that allows monitoring activities in the field of vision directly in the forklift cab (Fig. 10). (Štefánik et al. 2003, Ďurica et al. 2015, Bubeník 2004)

ACKNOWLEDGEMENTS

This article was created for the purpose of the project VEGA 1/0936/16.

REFERENCES

Bubeník, P. 2004. Scheduling system for minimizing the costs of production. In: Journal of mechanical engineering. Vol. 50, no. 5 (2004), p. 291–297. ISSN 0039-2480.

Capacity electric stand-up end control forklifts. 2016. [Online]. 2016, [cit. 2016-12-01]. Available on the Internet: http://www.mcfa.com/en/jungheinrich/forklifts-pallet-trucks/Jungheinrich/ClassI/Stand-Up-End-Control-Trucks/ETG-230–355.

Ďurica, L. et al. 2015. Manufacturing multi-agent system with bio-inspired techniques: CODESA-Prime. In: MM science journal. ISSN 1803-1269.—December 2015. p. 829–837, on-line ISSN 1805-0476.

Forklift Operator Ergonomics 2012. [Online]. 2012, [cit. 2016-11-26]. Available on the Internet: http://www.aalhysterforklifts.com.au/index.php/about/blog-post/operator_ergonomics.

Gajšek, B. & Dukic, G. 2015. Review of ergonomics solutions to protect from injuries of lower back in case of forklifts. In Researchgate. [Online]. 2015. [cit. 2016-11-28]. Available on the Internet: <https://www.b2match.eu/alimentariabrokerage2016/participants/729>.

Gregor, M. et al. 2016. Knowledge in healthcare. In: Annals of DAAAM. Proceedings of the 26-th DAAAM International symposium on intelligent manufacturing and automation. ISSN 1726-9679. Vienna: DAAAM International Vienna, 2016.—ISBN 978-3-902734-07-5-online, p. 1115–1121.

Gregor, M. et al. 2015. The factory of the future production system research. In: ICAC 2015 Proceedings of the 21-st International conference on automation and computing. Glasgow, UK, September 11–12, 2015. [S.l.]: IEEE, 2015. ISBN 978-0-9926801-0-7.

Matuszek, J. et al. 2009. Estimating prime costs of producing machine elements at the stage of production processes design. In: Modelling ang designing in production engineering/ed. A. Świć, J. Lipski; Lubelskie Towarzystwo Naukowe, Lublin 2009, Vol. 5 No. 2, s. 80–92, p-ISBN: 978-80-89333-15-8.

Štefánik, A. et al. 2003. Tools for Continual Process Improvement—Simulation and Benchmarking. In: Annals of DAAM for 2003 & Proc. of the 14th Intern. DAAAM Symposium: Intelligent manufacturing & automation: Focus on reconstruction and development, 2003, p. 443–444, ISBN 978-3-901509-34-6.

New Trends in Process Control and Production Management – Štofová & Szaryszová (Eds)
© 2018 Taylor & Francis Group, London, ISBN 978-1-138-05885-9

Safety, quality and reliability in operations and production

A. Lališ, P. Vittek & J. Kraus
Department of Air Transport, Czech Technical University in Prague, Prague, Czech Republic

ABSTRACT: This paper focuses on the distinction between safety, quality and reliability in the context of operations and production management. Integrated management systems are readily available but the drivers for their implementation are often market or profit based whilst the actual contribution of proper management with regard to each of the areas is frequently matter of confusion, especially with regard to safety. In the first part, theoretical background with the highlights of distinctions and overlapping areas between safety, quality and reliability are identified and analyzed. Subsequently, a case study from the domain of transportation demonstrates application issues as well as opportunities to benefit from synergy effects. Conclusions provide a rationale for correct distinction between respective management processes and the long-term advantage of their effective integration.

1 INTRODUCTION

Operation and production are typically subjected to variety of factors, which determine the final product or service to be delivered. From customer's perspective, most of them are hidden or not explicitly perceived as important; their preferences are driven by conscious or subconscious perception of final product quality, safety and reliability, as applicable, supporting importance of proper management to assure these qualities. Even though they seem to be clear for managers, there are several issues. With regard to quality, for example, one of the existing definitions by Noriaki Kano (Kano et al. 1984) and others divides it into "must-be quality" and "attractive quality". Whilst the former regards more what professionals understand as product reliability, the latter is what matters more when momentary decision to purchase product or service comes to the fore. Manufacturers and service suppliers usually understand this and put an adequate emphasis on marketing their products or services. However, this means that quality may be split into different aspects of what is to be managed by quality management systems and it may be solely decision of respective company or software system manufacturer, which one will gain more focus.

Not only is this a consequence of ambiguous interpretations, such as those concerning quality, but also of market competition, dynamic environment and, after all, the customer's behaviour. For management, the key objective is to maintain the business, putting constraints on production or operations. Managing such intangible properties as safety or quality are, does not clearly fit economic stability as it normally entails costs and investments rather than profit. Sometimes, safety of operations and production is assured only due to applicable regulations. Furthermore, customer's behaviour dictates requirements, which may not necessary comply with what is supposed to be quality, safe and/or reliable product.

Whilst companies are busy dealing with everyday concerns and working on their strategies, integrated management systems reflect the problem with ambiguous interpretations and limited motivation to benefit from proper management systems. Integrated systems are perceived rather as an opportunity to improve image, competitiveness and reduce operational costs related for example to quality or safety management systems using common technology platform. However, there are no motivations associated with process optimization, where proper quality management may seriously affect reliability or safety of the final product/service (Asif et al. 2008, Salomone 2008).

This paper aims at addressing the problem with flawed interpretations of safety, quality and reliability. Making clear distinctions and developing proper framework for implementing integrated safety, quality and reliability management is a fundamental step towards more effective and efficient integrated management systems. There are many definitions available to date, which serve as a basis for various systems, but none of them provides understanding comprehensive enough to build effective integrated systems.

The directions in this paper serve also as an academic discourse to support further domain research, which may find its application in various conceptualizations and dedicated solutions from software engineering. The ideas are materialized in transportation domain based on research projects executed in the domain of Czech aviation safety environment.

2 METHODS

2.1 *Definitions*

To date, there are no unified definitions available for safety, quality and reliability. Even though these terms are used world-wide and for quite some time, their interpretations vary upon the domain of application. This is probably a natural outcome of specialization, which is necessary as part of their implementation. Empirically, more accepted seem the definitions used by authorities in each domain. Following subchapters provide an overview for each of the terms interpretations.

2.2 *Safety*

Safety is normally emphasized and put in the first place. One of the oldest definitions dates back to 1966, which was derived by U.S. Department of Defense as "freedom from all conditions that cause injury of any kind or death of a person, or damage to or loss of equipment or property" (Malasky 1974). This definition was a success because of its genericity and domain independence and it is likely the reason, why it is still used by many agencies. Today's interpretations allow variations such as "freedom from danger, risk or threat" (Crutchfield & Roughton 2013) or more in-depth contemplations such as the one by World Health Organisation (WHO), which divides safety into two dimensions based on external (objective) or internal (subjective) criteria (Nilsen et al. 2004).

Quite interesting point is that some rulemaking agencies issuing standard safety documentations, such as International Civil Aviation Organization (ICAO), avoid base definition of safety and already expect their stakeholders to know what is referred to (ICAO 2013). This underpins the fact that despite the lack of unified definition of safety, it is not an ambiguous term.

Its extensions in form of interpretation variations allow better approach to research and development in safety by specializing the term. For example, using the term "risk" rather than "condition that cause injury" leads to the suggestion that safety may be approached via statistics and probabilistic methods, which is the case in modern safety engineering. WHO distinguishes between what it is to "feel safe" and to "be safe" as it has implications in the domain of medicine and health care where the typical "noninjury" understanding of safety limits the application.

Despite the domain variations, there are no serious inconsistencies among the existing definitions and thus any of them may be used when referring to integrated management systems. However, it is worth to note that the saying "If you can measure it, you can manage it" (Deming 2000) supports rather those definitions which allow for quantification, i.e. those based on the concept of risk.

2.3 *Quality*

For the term quality, there are many definitions, which are more inconsistent compared to safety. This is caused by the genesis of the term which dates to ancient Greek and which relates to object properties. Modern business definitions are still property-related, even though they encompass broader sense of the term in order to allow its enterprise manageability.

According to International Organization for Standardization (ISO), quality is "degree to which a set of inherent characteristics of an object fulfils requirements" (ISO 2015) and it is one of the best generic definitions available to date. American Society for Quality extends this definition by dividing it into two separate concepts, namely 1) "the characteristics of a product or service that bear on its ability to satisfy stated or implied needs" and 2) "a product or service free of deficiencies" (ASQ 2017). Technical (measurable) variation is derived in Six Sigma as "number of defects per million opportunities" (Taguchi 1993). Definitions that are more philosophical describe quality as "value" to some person (Weinberg 1991) or "fitness for use" (Juran & Godfrey 1998) and this is where the aforementioned definition by Kano of "must-be quality" and "attractive quality" fits in.

From practical point of view, the concept of quality is elusive and different people tend to have different interpretations (Ross & Perry 1999). This can be experienced in everyday life as regular people often do not know what a "quality product" means, even though they frequently use the term. Similarly, managers can easily have different understanding of what is a quality product or service their companies should deliver. This implies that practical applications deal with the philosophical definitions too, interpreting quality more in terms how it is perceived by customer and not only what it technically means within the manufacturing process. In some cases, the former may attract much more attention than the latter, what is definitely not an ideal approach.

For integrated management systems, more appropriate seem the definitions allowing measurability, i.e. counting number of defects as by Six Sigma or confronting the final product or service with the original requirements and identifying the extent of discrepancies.

2.4 Reliability

This term is the newest out of the three discussed in this paper as it originates from 19th century (Saleh & Marais 2006). It emerged as a result of need for a technical term referring to the possibility to repeat a test or process under controlled conditions. Even though it relates much more to products than services, its application in science and engineering suggests some implications also for operations.

Because of its novelty and technical background, the term consistency and interpretations are uniform and typically vary only slightly based on domain of application. The most common definition today is that "reliability is the probability of success or the probability that the system will perform its intended function under specified design limits" (Pham 2006), or in other words, it is the probability that a product will be working for a specific time period under given conditions, i.e. it will exhibit desired performance. Less measurable interpretations include "resistance to failure of an item over time" (Anderson & Neri 1990) or "capacity of an item to perform as required over time" (Hamada et al. 2008).

The term relates to operations with regard to the technology that is used to run them. It is important to account for lifetime cycles, maintenance and set proper redundancy in order to assure that the technology will not fail so that it would seriously affect the ability to operate. For example, this is extremely important in transportation where timing is frequently the focus and inadequate handling of technology reliability leads to delays, cancelled connections and penalties.

Concerning integrated management systems, reliability, as defined today, perfectly suits the management systems. Experience from production and operation provides sound bases for its implementation and integration.

2.5 Integrated framework for safety, quality and reliability

As soon as these qualities are considered in the most suitable way for integrated management systems, i.e. in quantifiable form, common points and distinctions between them become clearer. All of them represent an effort to control certain properties in production and operations and they use the same principles to achieve it.

They all have in common that, due to practical reasons, none is being measured directly as there are no sensors which could serve this purpose. Obviously, there are requirements to assure safety, quality and reliability of products and services, but they stem from long-term experience and best practices. However, there is no way how to categorically declare a production, operation and their resulting product or service as safe, quality or reliable if not by an absence of safety occurrences, discrepancies from customer requirements and failures over time, respectively. Dedicated management systems aim to assure the absence of this type of undesired outcomes by monitoring and investigating their occurrences. In many cases, one occurrence may be both safety, quality and reliability relevant. This is crucial to realize with respect to the synergy, which may be achieved by appropriate integration.

On the other hand, the core difference lies with the property they aim to assure. Safety is a criterion by itself and it may pose a disjoint set of requirements as opposed to those for quality and reliability. In extreme case, safety may contradict some of the other requirements and then, optimal trade-off between the criteria should be identified.

Unlike reliability, quality requirements typically include no time frame. Also, they are more focused on what is important for trading the final product and services, whilst reliability regards exclusively technical domain. However, in real world it is only rarely the case, that reliability requirements could be omitted whilst safety and quality requirements are achieved at the same time.

Theoretically, it is admissible to consider both extreme cases, namely where a) safety, quality and reliability completely overlap and b) all properties are entirely disjoint. However, it is rather the case that the requirements overlap to certain extent than not. Purchaser requirements often implicitly or explicitly stipulate safety and reliability because it is natural to demand certain level of these qualities regardless of application domain. Further, the more advanced the product and service, the more likely will the requirements overlap. It is almost impossible to produce complex systems, which could entirely omit any of the properties.

This property is expressed by the following notation:

$$S \cap Q \wedge S \cap R \wedge Q \cap R = \{X_1, X_2, ..., X_n\}, \qquad (1)$$

where S, Q and R represent sets of safety, quality and reliability requirements, respectively, or related occurrences (discrepancies, failures) and X are the elements, which bear relevant information at least for two of the qualities. It is possible to express both requirements and occurrences as data patterns and as such they can be stored and integrated in common database.

As already mentioned, it is rarely the case, that the expression (1) has an empty set of elements on the right side. With respect to this, almost every production and operation has to deal with overlapping elements $(X_1, X_2, ..., X_n)$ and if the elements are addressed separately, i.e. without storing their context with respect to all qualities, some of the information is lost. This is often the issue and the very elements determine the need of proper integration of all qualities in terms of their management. Not only it allows inferences, which could otherwise not be identified, but it also provides sound basis for overall approach to process optimization.

Figure 1 exemplifies the outlined approach to the common framework. Managerial systems are based on deriving knowledge (intelligence) out of their data. If the data are put in single system and database, robust dependencies could be identified. Moreover, additional internal qualities, such as efficiency, can become part of the integrated systems to complete the circle (Szabo et al. 2013, Pudło & Szabo 2014). Managing efficiency regards process optimization that can justify the whole concept and lead to continuous motivation to improve.

From technical standpoint, this implies collecting information from production and operation in common database and common format. To this end, various solutions may be used, such as conceptual modelling and structural models (Guizzardi 2005). It is important to assure that the data content is integrated properly and that no information is lost as well as no bias is generated.

As additional layer to the database, knowledge (intelligence) module is to be established. This module may simply depict key trends or, in more advanced systems, compute complex relations using mathematical modelling. Various decision criteria can be incorporated as soon as strategy goals are confronted with the achieved results.

If part of the integrated data comprise economic data (efficiency-related), these may be interrelated with data from safety, quality and reliability and identified dependencies may serve the decision process in the future to assure achieving targets set in respective company.

3 CASE STUDY

In transportation domain, all three qualities are of high importance. The most advanced means of transport—aviation—uses the most mature managerial systems because it is also the most susceptible to significant negative consequences if these qualities are not maintained. Therefore, the case study focuses specifically on aviation.

In the aviation, safety is dominant and always emphasized. This is because the risks in aviation operation can lead to disasters. It is also reflected by the fact that the most advanced aviation organizations are Air Navigation Service Providers (ANSPs). The role of these companies is to ensure safety and continuous flow of air traffic, i.e. meeting the needs of different airspace users whilst ensuring their safety. The quality of ANSPs operation and provided service is assessed with regard to the level of safety achieved in an airspace they control.

3.1 *Air navigation service providers*

ANSPs work with highly sophisticated and expensive technology (communication, navigation, surveillance equipment and facilities), which is known for demanding requirements on its reliability. Many redundancies or highly reliable elements constitute the equipment, ensuring that the organizations would be able to operate continuously under any conditions. Unserviceability of vast majority of elements comprising the equipment and facilities implies higher risk and generally lower quality of provided service. Therefore, it is possible to pronounce safety and quality as highly dependent on technology reliability, thus there are significant overlaps. Additionally, because part of the service provided by these organizations is safety of air operations, quality of the service captures safety so that it is in fact a subset of quality. Figure 2 exemplifies the case.

Figure 1. Knowledge engineering in production and operation.

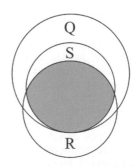

Figure 2. ANSPs and safety (S), quality (Q) and reliability (R) in their operations.

From the figure, it is apparent that the resulting overlap area (in grey) is significant and should be considered for integration. In Europe, this is the case in recent initiatives to integrate safety and quality for this type of organizations but there are no achieved results yet. This initiative also implicitly comprises reliability as there is no reliability management system in the aviation. Reliability is normally part of technology maintenance. Due to this, the grey zone in Figure 2 as well as the overlap zones between quality and reliability should be carefully analyzed prior to any integration. In addition, these qualities are not yet considered for relation with financial data to generate overall intelligence as in Figure 1. This is mainly because ANSPs are state-owned enterprises with no competition what leads to reduced efforts to operate them efficiently, at least compared to other aviation organizations. However, this may change in the future so it is worth to research the potential for further optimization.

3.2 Airlines

Airlines have different role in the aviation. Spe-cifically, they have to compete on the market and meet customers' requirements (Rozenberg & Heralová 2015, Endrizalová & Némec 2014). This shifts the quality management towards requirements, which are less coupled with safety. On the other hand, technology reliability plays as significant role as it does for ANSPs. Aircraft and other technology is subject of intensive and thorough maintenance due to assuring safety but also due to making it possible to fly. Therefore, the overlap between reliability and the other two qualities is of the same significance.

Figure 3 exemplifies the case for airlines. Safety is not a subset of quality anymore and because airlines are subject of market competition, their main driver is overall efficiency, which determines the

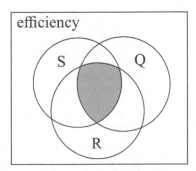

Figure 3. Airlines and safety (S), quality (Q), reliability (R) and efficiency in their operations.

approach to all the qualities. The resulting overlap area is smaller than for ANSPs, but still significant. Its importance is underlined by the fact that overall efficiency can be optimized further if these three qualities are integrated appropriately, i.e. with the goal to maximize derived knowledge out of the existing data. There exists great potential for improvements in this domain as there is no such integrated management system available to date, which would address the distinctions, overlap areas and which provides an overview to allow efficiency optimization.

This case is more frequent than the one for ANSPs. Majority of companies have to compete and meet different customer's requirements. At the same time, they are operating technology, which must be reliable to certain extent in order to allow for operation or production. Safety standards are naturally present in any industry and domain.

4 RESULTS

According to the theoretical knowledge gathered and the definitions identified, there are overlaps between safety, quality and reliability in production or operation and how there are managed.

The definitions themselves provide possibility to identify distinctions only if considered in quantified form, i.e. the form, which allows their actual control. For all of them, this is based on occurrences, which indicate lack of the qualities.

According to the case study performed, there are significant overlapping zones between the three qualities, typically in larger and complex companies, and proper distinction between them with subsequent adequate integration can provide their overall optimization. Additionally, they can be related to other criteria, such as efficiency of production or operation and used to optimize them as well.

5 DISCUSSION

The case study exemplified two common types of organizations, which provided realistic look how safety, quality and reliability overlap and why it is important to consider their integration.

From theoretical point of view, none of them represents extremes, i.e. completely overlapping or entirely disjoint sets of requirements and data. This is because extreme situations are rare if not almost impossible. However, the study showed that when shifting the attention to regulatory bodies or authorities, which usually deal with safety assurance, the three qualities tend to overlap more. ANSPs are imposing constraints on air traffic and cooperate closely with national supervisory authorities. In some cases, air navigation services may be even provided directly by those authorities as is the case for instance in the USA.

On the other hand, when shifting the attention towards standard market companies, overlapping zones will be smaller. This is due to customers, which often have no direct knowledge about safety and reliability, and whose requirements can easily contradict them. In addition, the less advanced technology is used for production or operation, the less requirements for reliability will there be. With regard to this, it only makes sense to integrate advanced production and operation processes, which are based on technology that is more complex.

6 CONCLUSIONS

This paper provided analysis of available definitions for safety, quality and reliability. Definitions variations and proper distinctions were identified. Case study exemplified the importance of such distinctions for the purpose of their proper integration in advanced managerial systems. This can positively influence not only on the very management of these three qualities, but their optimization and long-term efficiency improvements of particular production or operation.

This paper is an initial overview based on authors experience from safety-domain research projects. It provides rationale for correct distinctions and subsequent academic discourse on how these qualities work in real environment. It does not provide any suggestions on the most correct definitions of safety, quality and reliability so as it does not intends to materialize any technical solutions to the problem.

Further potential exists with regard to deeper analysis of the integration issue in particular environment, i.e. particular production and operation. Practical application will need to be specialized per each case but the general knowledge on safety, quality and reliability integration could be reused.

There are efforts to integrate respective managerial systems but to truly motivate companies to responsibly approach the issue, the integration will need to be considered with economic aspects. They can lead to comprehensive quantification of the long-term benefits, which could be achieved by such systems not only because of common technology platform but also due to correct distinctions leading to extended intelligence about the system controlled.

ACKNOWLEDGMENT

This work was supported by the Czech Technical University in Prague, junior research grant No. SGS16/188/OHK2/2T/16.

REFERENCES

Anderson, R.T. & Neri, L., 1990. *Reliability-Centered Maintenance: Management and Engineering Methods*. Springer.

Asif, M., de Bruijn, E. & Fisscher, O.A., 2008. Corporate motivation for integrated management system implementation: why do firms engage in integration of management systems: a literature review & research agenda. In *The 16th Annual High Technology Small Firms Conference: May 22–23, 2008 + May 21 Doctoral Workshop, University of Twente, Enschede, The Netherlands*. Enschede: University of Twente, NIKOS.

ASQ, 2017. Glossary – Entry: Quality. American Society for Quality (ASQ).

Crutchfield, N. & Roughton, J., 2013. *Safety Culture: An Innovative Leadership Approach*. Butterworth-Heinemann.

Deming, W.E., 2000. *The New Economics for Industry, Government, Education (MIT Press)*. The MIT Press.

Endrizalová, E. & Němec, V., 2014. The costs of airline service - the short-haul problem. *MAD - Magazine of Aviation Development* 2(8): 14.

Guizzardi, G., 2005. *Ontological foundations for structural conceptual models*. Enschede Enschede: Centre for Telematics and Information Technology Telematica Instituut.

Hamada, M.S., Wilson, A., Reese, C.S. & Martz, H., 2008. *Bayesian Reliability (Springer Series in Statistics)*. Springer.

ICAO, 2013. *Safety Management Manual (SMM): Doc 9859 AN/474*. Montreal, Quebec: International Civil Aviation Organization (ICAO), 3 ed.

ISO, 2015. ISO 9000:2015, Quality management systems – Fundamentals and vocabulary. International Organization for Standardization (ISO).

Juran, J. & Godfrey, A.B., 1998. *Juran's Quality Handbook*. McGraw-Hill Professional.

Kano, N., Seraku, N., Takahashi, F. & Tsuji, S., 1984. Attractive quality and must-be quality. *The Journal of the Japanese Society for Quality Control*.

Malasky, S.W., 1974. *System safety: planning/engineering/management*. Spartan Books; [distributed by] Hayden Book Co., Rochelle Park, N.J.

Nilsen, P., Hudson, D.S., Kullberg, A., Timpka, T., Ekman, R. & Lindqvist, K., 2004. Making sense of safety. *Injury Prevention* 10(2): 71–73.

Pham, H., 2006. *System Software Reliability (Springer Series in Reliability Engineering)*. Springer London.

Pudło, P. & Szabo, S., 2014. Logistic costs of quality and their impact on degree of operation level. *Journal of Applied Economic Sciences* 9(3): 469–475.

Ross, J.E. & Perry, S., 1999. *Total Quality Management: Text, Cases, and Readings, Third Edition*. CRC Press.

Rozenberg, R. & Heralová, D., 2015. Quality in air transport process of LOT polish airlines. *MAD - Magazine of Aviation Development* 3(16): 20.

Saleh, J. & Marais, K., 2006. Highlights from the early (and pre-) history of reliability engineering. *Reliability Engineering & System Safety* 91(2): 249–256.

Salomone, R., 2008. Integrated management systems: experiences in italian organizations. *Journal of Cleaner Production* 16(16): 1786–1806.

Szabo, S., Ferencz, V. & Pucihar, A., 2013. Trust, innovation and prosperity. *Quality Innovation Prosperity* 17(2): 1–8.

Taguchi, G., 1993. *Taguchi on Robust Technology Development*. ASME International.

Weinberg, G.M., 1991. *Quality Software Management: Systems Thinking*. Dorset House.

New Trends in Process Control and Production Management – Štofová & Szaryszová (Eds)
© 2018 Taylor & Francis Group, London, ISBN 978-1-138-05885-9

Application of computer aided technology in production system planning

Š. Václav, P. Košťál, Š. Lecký & D. Michal
Faculty of Materials Science and Technology in Trnava, Slovak University of Technology in Bratislava, Trnava, Slovak Republic

ABSTRACT: Production system planning in production processes and it's subsystems like assembly etc. Computer aided technology play huge role when company want to be competitive. Optimization by CA systems can improve whole production process and cut manufacturing or assembly times. This kind of systems give to company an advantage in market. Tool which can eliminate mistakes in production planning even before production starts is very powerful. If CA technology is used correctly in production planning it can safe financial resources and lot of time spent for errors fixing. This paper deals with application of computer aided technology specifically in production planning. Software Tecnomatix Plant Simulation from company SIEMENS will be used for creation of examples. Tecnomatix Plant Simulation was applicated into experiments which are shown in this paper. This software is aimed for planning and optimizing of production systems with high level of complexity.

1 INTRODUCTION

We can see that pressure on improving effectiveness of production systems is increasing and that is based purely on competition in international production network. Logistic concept is needed for international logistic networks. For managing, these needs are made tools like digital factory in context of product life cycle management.

Connection to logistic concept is needed for international logistic networks. "Simulation of complete material flow with all important activities like production, storage and transport activities is key component of digital factory in industry. Lowering the storage capacities by 20–60% and increasing of throughput of existing production system by 15–20% is possible in real life projects. The purpose of running simulations varies from strategic to tactical up to operational goals" (Václav & Lecký 2017).

Users answers the questions like which production plants in which countries are the best for future production of new product with regard on factors like logistic solutions, working efficiency, downtimes, flexibility, storage resources etc. This is considered by users for upcoming years. User evaluate flexibility of production system or sub-system. In present statistic data is a topic which is important.

Computer aided technology is more advanced from one year to another.

"Products and service variants offered by different competitors often resemble so much, that the delivery time and price become the crucial competition criteria" (Consiglio et al. 2007).

Figure 1. Product lifecycle management.

In production system planning area is a lot of software that can be used. Companies that have the biggest market share have in their portfolio quite modern and advanced software. One of software like these is Tecnomatix Plant Simulation and Tecnomatix Process Simulate from company SIEMENS.

2 LITERATURE ANALYSIS

"Market demands require manufacturers to bring products to market faster than ever, even though the products themselves have become more complex. This puts pressure on machine builders and manufacturers. They must work faster to create new, more flexible equipment while continuing to meet customers' cost and quality requirements. Product lifecycle management (PLM) software provides a tool-3D process simulation-that can help alleviate the pressure (Control Engineering, 2008)".

We can say for sure that greater customer choice, faster delivery and reliable product performance are key ingredients to the success of a modern manufacturing enterprise (Brooks 2000).

In this concept simulation plays huge role and ensuring the competitiveness of company.

2.2 Simulation in production system planning

Simulation technology in field of production system planning is an important tool. It is especially important in implementing of complex technical systems like production or assembly systems or sub-systems etc. Obviously, trends in economy are shortening of planning cycles and that involves (Bangsow 2011):

– Iincrease of product complexity,
– Increase of quality,
– Shorter product lifecycles
– Lower storage capacities,
– Increase of competitiveness.

The most used simulation software in market (Siderska 2016):

– WITNESS (Lanner),
– Tecnomatix Plant Simulation (SIEMENS),
– Quest (Delmia),
– Enterprise Dynamics (Incontrol Enterprise Dynamics),
– ProModel.

Advantages in simulations:

– Replacing the real system experiments,
– Use in cases where there is no possibility of analytical solutions,
– The use of a large number of random effects,
– Possibility of modeling time,
– Possibility of verification solutions, which was acquired by another route,
– Better understanding of the real system (Trebuňa et al. 2014).

Disadvantages in simulations:

– Each simulation is individual,
– Correctness of the design needs to be studied,
– Simulations are costly means to study systems,
– Creation of simulation model is often lengthy,
– Simulations are the numerical methods, which means that for any change of parameters are

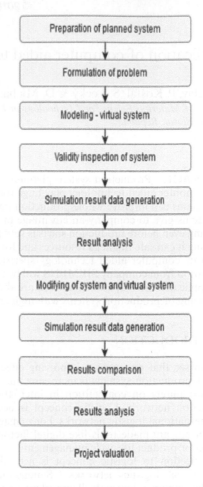

Figure 2. Simulation process of planned system (Václav & Lecký 2017).

required for new solutions (Trebuňa et al. 2014).

2.3 Tecnomatix plant simulation

"Tecnomatix is a complete digital manufacturing solutions portfolio realized by Siemens company, that offers innovative solutions by correlating all manufacturing methods for one product—starting from the design process, simulation of the process, validation and ending with the manufacturing process. The simulator helps to create digital models of logistic production systems in order to explore the systems' characteristics and to optimize their performance (Petrila et al. 2010)."

Advantages of Tecnomatix Plant Simulation (https://www.plm.automation.siemens.com/):

- Testing of innovative strategies in no risk virtual environment,
- Maximal usage of production resources,
- Lowering the investment risk with help of quick simulation,
- Size optimization of systems and storage place,
- Quick identification of problems in logistics and production,
- Lowering of storage capacities by 20–60% based on size of system,
- Lowering the investment costs for new system by 5–20%,
- Lowering the employees' capacities and manipulation technology,
- Quick accomplishment of positive results and identification of effect.

2.4 *Tecnomatix process simulation*

Process Simulation is the solution to minimize the risk of changes in the production and launch of new production system. It allows to verify the plans from design concept to start of production. It helps allay these risks. Ability to use 3D data makes easier virtual validation, optimization and commissioning of production processes. This results in faster launches and better production quality. Tecnomatix—Process Simulation can verify the feasibility of the assembly process by verifying reachability of the robot or human, and collisions between moving devices or between man and machine. This is done by simulating complete assembly sequence of the products and their working tools (Rusnák 2006). Tools such as measurement and detection of collisions allows detailed control and optimization of assembly processes. The software is fully integrated with the platform Teamcenter. Technology can be reused and can verify the production processes. Makes easier simulation of assembly processes, human operations and mechanical methods of tools, devices and robots (https://www.plm.automation.siemens.com/).

Main functions:

- 3D simulation,
- Static and dynamic detection of collisions,
- 2D and 3D view,
- 3D measurement,
- Scanning operations,
- Planned assembly operation,
- 3D geometry and kinematics.

Advantages:

- Reduce the risk in the production system,
- Shortening the planning of new production systems,
- Reduce the cost of change thanks to the early detection of errors,
- Analysis of ergonomic process.
- Choice of the best production variant.

3 TECNOMATIX PLANT SIMULATION— EXAMPLE 1

Before we start to create virtual production system we need do define what questions we want to answer. What is the purpose of simulation and if simulation is even needed.

Research question for example number one:

Can be production system optimized for better throughput and storage?

In this section, we introduce production system with six stage assembly line. This model is designed in Tecnomatix Plant Simulation software from SIEMENS company.

Production workplace consist of these processes:

- Handling,
- Six assembly stations,
- Control.

Assembly line relates to input, handling, control and output by conveyors. Simulation time is ten days of continuous work. Every process or workstation must have some time management assigned by user.

Failure for each station except control and parallel process was programmed as constant. That's based on experiences from real life production system. Working times and blocked times are calculated from time management of process.

After running of simulation, we get starting information like throughput (9588), production (65.15%), transport (24.19%) or storage (10.66%).

After replacement of conveyors by pick & place machines and after combination of two processes into one parallel process of two we get improved results.

Simulation results of improved model were slightly better in throughput (9593) but transport (0.85%), production (99.15%) and storage (0.00%) were decreased significantly.

Table 1. Time management of processes.

Handling	30 seconds
Assembly 1	70 seconds
Assembly 2	60 seconds
Assembly 3	60 seconds
Assembly 4	60 seconds
Assembly 5	60 seconds
Assembly 6	60 seconds
Control	60 seconds

Figure 3. Initial model of production system.

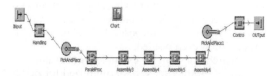

Figure 4. Improved model of production system.

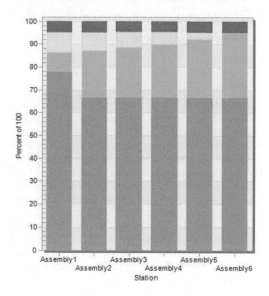

Figure 5. Resource statistic of assembly line in base model.

Figure 6. Legend for resource statistics graph.

As you can see in figure number five and six resource statistics of assembly line of base model are around 70% working time and resource statistics of improved model are also around 70% working time. But in improved model is assembly line less blocked and that makes production system more flexible.

Research question was how can be production system optimized for better throughput and storage. Based on showed virtual models, simulation

Table 2. Simulation statistics.

	Production	Throughput	Transport	Storage
Base model	65.15%	9588	24.19%	10.66%
Improved model	99.15%	9593	0.85%	0.00%

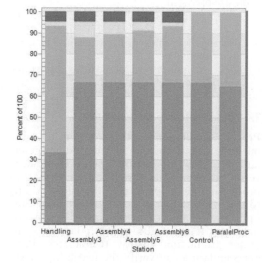

Figure 7. Resource statistic of assembly line in improved model.

and statistics we can say that production system can be improved by combination of 2 processes into one parallel process of two and by switching of conveyors for pick and place machines.

4 TECNOMATIX PROCESS SIMULATION—EXAMPLE 2

For example of using simulation software was designed simplified production system, which includes a number of components in the form of CAD data in a virtual environment. It includes NC machines, industrial robots, conveyors etc.

The simulation software enables work to create assembly or production system or device in digital form in advance of implementation. It is possible to verify the changes, test the reachability handling points and detecting conflict situations. You can see it in Figure 9. Reachability of industrial robotic arm (on left) and collision warning of industrial robotic arm with CNC milling machine doors.

Use of simulation software found that the industrial robot does not meet the desired parameters and is unfit for this role because of unreachable

Figure 8.   Designed production system.

Figure 9.   Reachability of industrial robotic arm (left) and collision warning of industrial robotic arm with CNC milling machine doors (right).

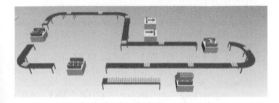

Figure 10.   Visualization of production system and layout in tecnomatix plant simulation.

workpiece through handling (insert and remove) of the workpiece in the CNC machine. It is necessary to change the requirement for the selection of an industrial robot or change the type of CNC machine.

Collision occurred at the wrong time of movements between industrial robots and the doors of CNC device. A collision occurred while closing the door of CNC device while industrial robot moved out of the area of CNC device. These software functions allow the user to search not only the optimal deployment of equipment in the production area but also the optimal timing of movements of elements in the system.

Simulation is suitable tool for experimenting with the structure of work and modernization of the production system. Simulation allows detection and the subsequently reducing collision situations in the design of the production system in the digital space and the help to find unfeasible and dangerous situation.

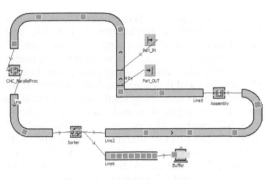

Figure 11.   Throughput simulation of production system in tecnomatix plant simulation.

4.2   *Tecnomatix plant simulation—cooperation with tecnomatix process simulate in planning*

Tecnomatix plant simulation was used for whole plant visualization, layout valuation and throughput simulation of previous example number 2.

Throughput of production system was 1483 (in 20 days) with one minute working time on every station. This throughput was acceptable so no addition changes were made to simulation model.

5   CONCLUSION

Computer simulations with IT tools are nowadays necessary activity which support design of new production and logistic systems or even already existing systems (Siderska 2016). Simulation methods are used for evaluation different aspects of production systems or subsystems. Repeatability is important and basic attribute of computer simulation. Because of exact values and parameters which have their own values assigned to them can be the same process executed many times. In real life, this is not possible (Klosowski 2011).

Application of innovative design methods is one factor that positively impacted the process of fast modern assembly systems implementation (Madarász et al. 2008).

This paper deals with application of Tecnomatix Plant Simulation software and Tecnomatix Process Simulate software from company SIEMENS into computer designing and planning of production process. Simulated production systems that were made as examples showed that if production system had some flaws bottlenecks or small imperfections, it can be improved by simulation. Simulation experiment can be tested with different properties and different kinds of adjustments of model. It is user's choice if it is necessary to change base model or just few properties of production model.

With software like Tecnomatix Process Simulate, movement of mechanism or machine can be controlled by time sequencing. Simulation of workstations and machines is helping engineers to prove if kinematic and dynamic properties of mechanisms and machines are right. Information's exported from simulation results are influenced by precision of input data and geometric precision of CAD models. Use of simulation methods of different processes in manufacturing, logistics or planning of new production systems is element of Industry 4.0. It makes production system planning and optimization of existing systems easier.

ACKNOWLEDGEMENTS

The article was written as part of the VEGA 1/0477/14 project "Research of influence of selected characteristics of machining process on achieved quality of machined surface and problem free assembly using high Technologies" supported by the scientific grant agency of the Ministry of Education of the Slovak Republic and of Slovak Academy of Sciences.

REFERENCES

Bangsow, S. 2011. *Tecnomatix Plant Simulation: Modeling and programming by means of examples.* England: London. ISBN 978-3-319-19502-5.

Brooks, B. 2000. *Software planning of assembly lines.* Emerald Group Publishing, limited. Hong Kong: Bedford. ISSN 01445154. Internet source: http://www.scopus.com/.

Consiglio, S. et al. 2007. *Development of hybrid assembly workplaces.* Internet source: https://www.scopus.com/.

Control Engineering, 2008. *PLM: 3D process simulation delivers.* Internet source: https://www.scopus.com/.

Klosowski, G. 2011. *Zastosowanie symulacji komputerowej w sterowaniu przepływem produkcji mebli,* Zarządzanie Przedsiebiorst-wem/Polskie Towarzystwo Zarządzania Produkcją 2: 29–37.

Madarász, L. et al. 2008. *Selection of assembly system type.* SAMI 2008 6th International Symposium on Applied Machine Intelligence and Informatics—Proceedings, art. no. 4469196, pp. 44–47. ISBN 978-1-4244-2105-3

Petrila, S. et al. 2016. *The use of Tecnomatix software to simulate the manufacturing flows in an industrial enterprise producing hydrostatic components.* Institute of Physics Publishing. Romania: Bacau. ISSN 17578981. Internet source: https://www.scopus.com/.

Plant Simulation: *Product overview.* SIEMENS, ©2016 [quotation 2016-11-11]. Internet source: https://www.plm.automation.siemens.com/.

Rusnák, F. 2006. *Possibilities of simulation applications in assembly.* Trnava: STU.

Siderska, J. 2016. *Application of Tecnomatix Plant Simulation for modeling production and logistics processes.* VGTU, ©2016 [quotation 2016-11-10]. Internet source: http://www.bme.vgtu.lt/index.php/bme/article/view/316.

Trebuňa, P. et al. 2014. *Creation of simulation model of expansion of production in manufacturing companies.* Publisher: Elsevier Ltd. ISSN 18777058. Internet source: http://www.scopus.com/.

Václav, Š. & Lecký, Š. 2017. *Impact of computer aided assembly technologies and simulation in production planning.* Scientific technical union of mechanical engineering conference—"INDUSTRY 4.0". Bulgaria: Borovets. ISSN 2543-8582.

Creative quality management

A. Linczényi & R. Nováková
Faculty of Materials Science and Technology in Trnava, Slovak University of Technology in Bratislava, Trnava, Slovak Republic

ABSTRACT: Nowadays, the issue of quality management is focused on the application of ISO standards. If a business obtains a certificate of the quality management system, the issue of quality management is more or less resolved. However, we forget that ISO standards are focused only on the internal processes and in any case, they do not resolve the whole issue of the complex quality management. External processes are decisive for the efficiency of management. The authors of this paper identify such a state a crisis of the quality management and they propose so-called creative quality management as a starting point. The article will contain the proposal of the quality management model in the context of new ideas and trends.

1 INTRODUCTION

ISO standards related to quality management were not designed as an instrument for building the quality management system. The purpose of the first revision of these standards, which was adopted in 1987 and which was actually the best one, was to obtain the client's trust and declare that the delivered goods will comply with the conditions defined in the contract. For this reason, this standard was focused only on the internal processes because the customer was not interested at all in the processes, such as inputs in the quality management and on ensuring pre-production stages, verifying of the suppliers, post-production and after-sale processes and the quality economy. According to classics of quality management, quality management is aimed to ensure that a business will produce the products that will be feasible and that will ensure the business profitability. Satisfying the customer's needs was only the instrument to reach this objective. Quality management is not substantiated without quality management.

Focusing only on internal processes is typical for all the following versions of ISO standards. The application of these standards became the subject of Q-Business while several minor modifications were implemented into each subsequent revision of the ISO standards in order to justify re-certification of already certified quality management system. If someone studied the genesis of the standards, he/she must have derived such a conclusion. The current economy of businesses imposes completely different requirements on quality management.

2 CRISIS OF QUALITY MANAGEMENT

The introduction focuses on the purpose of the first revision of ISO standards, as well as on the fact that the subsequent versions were focused towards the mere necessity of the re-certification of already certified management systems of a production business. According to the first version of ISO standards, suppliers were evaluated on the basis of the quality of the products produced, on the basis of meeting the deadlines, ensuring the service etc. After carrying out the certification process, suppliers were evaluated on the basis of the compliance of the documentation of internal process of the producer with the requirements of ISO standards. This is true also nowadays. During certification, the certifying authorities are not interested at all in the quality of the products produced. A business whose documentation complies with the requirements of the ISO standards is awarded a certificate regardless whether it supplies high quality products or not. Thus, we can see frequent cases when a certified business supplies products where "ppm" in the supply is 50,000 or more. Hence, the certification from the point of view of the selection of suppliers lost sense. Moreover, the certification of quality management system negates a basic principle of quality management, i.e. continuous improvement. It can be possible to discuss about benefits of different amendments of ISO standards, however, the scope of this article does not allow it.

At the international conferences about quality we observe the increasing number of articles criticizing the current state of quality management including the applications of ISO standards. In our environment, such a declaration provokes very irritated reactions, especially from the part of advisory companies, and references mainly to the fact that these standards, although non-mandatory, are supported by the EU bodies. The authors of this article agree with these critical comments on the current approach to quality management and they claim that quality management is in a crisis these days. The crisis of quality management should not be only the subject of expert discussions and opinion exchange but in practice it can have very negative impacts on the sustainability of business profitability.

The crisis of quality management can be characterized by the following crucial factors:

- Existence and influence of myths in quality management,
- Normative attitude to quality management,
- Reluctant attitude of ISO standards to innovation and to its implementation,
- Ignoring quality economy by ISO standards.

The article will contain only brief characteristics of these influences and each reader can form its own opinion on whether the declarations in the article are true or not.

The first mentioned factor is the influence of myths in quality management. The term "myth" is related to the Ezop's fables in which we can find supernatural creatures with supernatural characteristics. People believed in the existence of these creatures. Nowadays, the term "myth" is understood as a false declaration that something exists but it is not true or that an influence is positive but it is not true. Nowadays, it is very modern to use the term "post-truth" declarations. It means that people believe in rumours, and not in arguments based on facts. These declarations are, however, mentioned in literature, taught at universities and they are used also in the economy practice due to the activities of advisory companies.

These declarations include, without limitation, the following declarations:

- Declaration that a certificate of quality management system according to ISO standards is a document about the efficiency of this system. This declaration is not true because ISO standards do not cover at all the whole area of quality management but only internal processes, but the most important processes of today are external processes. Moreover, during certification the certifying organisations are not interested at all in whether a producer produces and supplies high quality products but only whether he has the documentation which complies with the requirements of a standard.
- Another myth is a declaration that costs of quality is expressed by a PAF model (prevention costs, appraisal costs, failure costs), i.e. that costs of quality consist of prevention costs, appraisal costs and failure costs resulting from low-quality production. This model had its meaning in the conditions of the unsaturated market when it was important to produce as many products as possible and the loss result-

ing from faulty pieces was covered by the profit from increased production. Today, identifying the loss from low-quality productions which is the wastage of material, energy and labour force in the production process as the cost of quality means the misunderstanding of the content of the term "cost" and the content of the term "product quality". Indeed, the loss of low-quality production must be monitored because it diminishes the economic result of a business but it does not have in any case the nature of costs of quality. On the contrary, they express low quality and wastage.

- Another myth is the declaration that statistical methods of quality control are the most efficient form of quality control. Also, this method was created around 1930. It was created because producers wanted to produce as many products as possible, and therefore the 100% control was replaced by the selection control based on the application of statistical methods. The main part of these methods was the statistical regulation. Nowadays, various methods of the 100% control are used (Zero Defect, Zero Control etc.). The statistical regulation gives the answer to the question whether there is a significant influence in the production process or not but it is completely inefficient as a method of control.
- Another myth is the declaration used by producers that the main objective of producers is to satisfy customer's requirements. This declaration is an express lie. The main objective of each producer to produce under conditions ensuring profitability of invested capital and satisfying the customer's requirements is only an instrument to achieve this objective. If a product does not satisfy customer's requirements, it is unsaleable and the producer is affected by losses. In the American literature, there is a range of examples when the producer was focused on satisfying customers. As a result of this, its costs and subsequently the prices increased. Products became unsaleable and the producer bankrupted.

We can indicate more similar post-truth declarations which occur at the universities and in the industrial practice. Such myths appear because of the fact that production conditions, conditions of product realisation but also the customer requirements are subject to changes. Neither the theory nor the practice respond to such changes. The role of quality management is to respond to such changes instantly. This role pertains to top management and not to quality management units. It is forgotten that quality management is the function of top management and quality management units are only the executive body whose role is only to realise the decisions of top management.

The second factor characterizing the current crisis of quality management is the normative approach to quality management. In this case, the ideas of classics saying that there is no type solution of complex quality management and that complex quality management must be based on specific conditions of individual businesses. The normative approach to quality management started to be promoted by adopting ISO standards and by significant support of these standards by EU governing bodies. Although ISO standards for quality management systems are not mandatory, they are de facto mandatory because they are promoted by EU bodies and subsequently by national authorities of member states. The current USA literature says that one of the causes of the leg of EU countries behind countries like USA, Japan, Korea, China in the economic development is the application of ISO standards in quality management. This idea is confirmed by the fact that if we express such an indicator, such as proportion of GDP per inhabitant, the leaders in such a ranking are the countries where these standards are not applied and the EU legs behind these countries significantly. The normative approach to quality management has 2 basic shortcomings:

- ISO standards are conceived so widely that are same almost for all businesses regardless whether they produce yoghurts or railway wagons,
- The normative approach ignores the basic requirement of quality management, i.e. continuous improvement, and adaptation to market conditions.

As far as the first shortcoming is concerned, it is necessary to stress what has been already said above that complex quality management must take into consideration conditions which are specific for each company, i.e. production programme, technological equipment of the business, staff and the level of automatization (this is important mainly with regards to Industry 4.0). In this context, it is important to emphasize that in addition to ISO standards for quality management there are many field and product standards that have an absolute priority in the realisation of products. This can lead, metaphorically speaking, to a kind of "schizophrenia" in ensuring quality.

The second shortcoming of the normative approach is that it conserves the state in the area of quality management for a certain period (3-year validity of the certificate). It is interesting that the requirement of continuous improvement has been somehow forgotten.

The crisis of quality management is significantly reflected also by the approach of the current system of quality management to innovation. It is important to stress in this context that innovation without quality management cannot exist and quality without innovation cannot exist, either. The basis of quality development is innovation.

The last cause indicating the current crisis of quality management is complete ignorance of quality economy in ISO standards. It is forgotten that quality management was created as an instrument for ensuring the production of the products that will ensure profitability of the invested capital and as a result, they lead to sustainability of the organisation on the market. Taking into account this fact, quality economy is the most important process of quality management.

3 CREATIVE QUALITY MANAGEMENT

The reply to the above-mentioned crisis of quality management is, according to the authors of this article, the substitution of a normative approach in quality management by the creative quality management.

For the sake of completeness, it should be added that until the certificate about the quality management system will be required in tenders as a proof of the efficiency of quality management system, businesses will be forced to apply ISO standards in order to obtain such a certificate. On the other hand, managements of businesses must realize that the content of the notion "quality management" goes far beyond the requirements of ISO standards for the quality management system. If the business wants to have truly efficient quality management, it must replace the current approach by the creative quality management.

The basic role of creative quality management is to create conditions for the smooth and efficient realization of product innovations in industrial businesses and at the same time the implementation of efficient tools of economic management in quality management. This requires a new way of defining costs of quality and indicators of quality profitability.

The notion "creativity" is known in the current professional terminologies. There are many definitions of creativity and we do not dare to judge which one is the best. In general, the notion "creativity" can be defined as a set of capabilities and knowledge which permit to realize new ideas as well as overall progress of the society on the basis of scientific and other creative activity. In the industry, creative approaches are attributed to the activities in the area of innovations. The authors of this article rely on the idea that creativity in the area of innovations requires the creative approach to quality management. It would be an approach which is not rigid but is based on possibilities of continuous improvement not only in the area of innovations but also in the area of production and quality management. In this article, we refer to product innovations but on the other hand it cannot be excluded that certain product innovations can lead also to process innovations. In particular, in relation to the application of 4.0.

The prerequisite of successful realization of product innovations in the industrial businesses is the integration of main areas of the value process. Creative quality management is considered the instrument of such successful realization of innovations.

In line with magic trio of creativity, creative quality management can be defined as the triad of the following elements:

- Focusing business activity on the quality development by applying the innovation strategy.
- Using modern methods of quality assurance oriented on using IT, especially in the area of production control.
- Focusing of the top management of the business on regular monitoring and assessing of the development of quality of products produced.

The content of creative quality management can be understood as a set of activities carried out in the individual areas of the innovation development of quality of products produced including the potential change of the production programme which will guarantee that products will have characteristics identical to constantly changing requirements of customers, all delivery conditions will be fulfilled and products will be sold for prices which are acceptable for customers and will ensure profitability of invested capital for a producer.

Such a definition of quality management reflects the current needs in the area of quality management and in the area of the production of industrial products. Indeed, quality management understood in this way requires a new approach to what is defined in ISO standards as a system of quality management.

We avoid the term "quality management system" intentionally and we return to the term introduced by classics—complex quality management. Complex quality management cannot be burdened by errors about which we spoke in relation to the system of quality management and cannot be addressed by the further amendment of ISO standards. This would oppress the creative approach to quality management. The authors of the article suggest the model of creative quality management which they identified by the acronym "Model RIQP". This name of the model expresses the basic areas of creative quality management and at the same time expresses the substance of the integration of the whole value process "Research—Innovation—Quality—Profitability". We come across the term "integration" also nowadays where integration integrates the processes for which the ISO standards are approved. Such integration is not very beneficial. Only documentation is integrated.

The substance of the RIQP model is to elaborate specific activities for the individual areas of the value process regardless the nature of the business, its production programme, technical equipment and staff etc. The business then will choose the processes and activities which correspond to the character of the business while this choice depending on the constantly changing conditions can be changed anytime. For the sake of completeness, it is necessary to add that some businesses do not have their own research (although the development usually exists in these businesses). The model can be adapted to such a state without problem.

4 CONCLUSIONS

In the authors' view, the current state of quality management was marked by the range of shortcomings and does not correspond at all to challenges to which quality management faces today and which will have to deal with in the near future. The main shortcoming is in the first place the insufficient approach to innovations and the absence of quality economy in the system of quality management built under the ISO standards. The quality economy is the main process and objective of quality management. Equally, the development is not possible without innovations. The authors emphasize the shortcomings of the current approach and they suggest creative quality management as a solution. In the article, creative management is defined and the authors also proposed "Model RIQP" as a replacement for the system of quality management built on the ISO standards.

REFERENCES

Hudec, O. et al. 2014. Regional decision-making criteria: Strategic investment in the central Europe. In *Theoretical and Empirical Researches in Urban Management* 9 (2): 104–117. ISO Standards 9000.

Linczényi, A. & Nováková, R. 2013. Communication with customers and its impact on the profitability quality. In *IBER* 4(1): 260–266.

Linczényi, A. & Nováková R. 2011. Management of quality or quality of management. In *II. Journal of International Scientific Publications: Economy & Business* 5: 7.

Linczényi, A. & Nováková, R. 2011. Returns on Quality—ROQ Model. *55th EOQ Congress World Quality Congress: "Navigating Global Quality in a New Era".*

Nováková, R. 2011. Scientific-Research Cluster as a form of Knowledge Transfer. In *Drvna industrija.* 62(4) 291–300.

Nováková, R. 2014. The innovations and quality assurance support for SMEs as a smart policy tool of regional public administration. In *Proceedings of the International Scientific Conference Quality and Leading Innovation.* 168–176 [CD-ROM]. Slovakia: Košice.

Sedliačiková, M. et al. 2014. Factoring and Forfeiting in Slovakia and Possibilities of its Application in Wood-Working Industry. In *Drvna industrija.* 65 (1): 53–59.

New Trends in Process Control and Production Management – Štofová & Szaryszová (Eds)
© *2018 Taylor & Francis Group, London, ISBN 978-1-138-05885-9*

Higher education of Ukraine and the needs of consumers

P. Maciaszczyk
Faculty of Social and Humanities Sciences, State Higher Vocational School Memorial of Prof. Stanislaw Tarnowski in Tarnobrzeg, Tarnobrzeg, Poland

I. Chaika
Faculty of Management, Poltava University of Economics and Trade, Poltava, Ukraine

I. Britchenko
Faculty of Technical and Economic Sciences, State Higher Vocational School Memorial of Prof. Stanislaw Tarnowski in Tarnobrzeg, Tarnobrzeg, Poland

ABSTRACT: The aim of the paper is the development of theoretical approaches and practical recommendations for the formation of a system to respond to the needs of consumers in the company that provides the service. Theoretical and methodological foundation work was to study and rethink works of Ukrainian and foreign scientists on improving marketing activities, services based on the formation of a system to respond to the needs of consumers. Based on systematic methods, structural analysis, causal analysis, synthesis and argumentation proposed system respond to the needs of consumers in higher education (universities). The importance of formation of the system to respond to the needs of consumers for Ukrainian universities and determined its structural components; classified obstacles to implementation in Ukrainian university system to respond to the needs of consumers and the ways to overcome them; Chart systematization process designed event involving consumers and roadmap event (for example, of "Career Day").

1 INTRODUCTION

Modern scholars point out many aspects of higher education marketing in Ukraine and the need for cooperation between higher education institutions and different partners and stakeholders. Company Teletov O.S. in the work "Marketing in Educational Activities" proves that "in the area of administrative reform, it is intended to deepen cooperation between scientific and educational institutions, business entities and governmental agencies at national and regional level" (Teletov 2015). Vasylkova in her work "Strategic Marketing Planning in Higher Education" presents the directions for optimizing the university management processes by improving the strategic and tactical processes of marketing planning that were identified on the basis of the study on the experience of strategic planning in some foreign universities (Vasylkova 2015). Illyashenko in his work "Analysis of Factors Influencing Consumer Choice in the Education Market" shows, on the basis of the analysis, the differentiation of the marketing activities within the educational services of the higher education institutions in order to increase their effectiveness (Illyashenko 2015). Drawing the attention of academics to the issues of university marketing points to their importance, and it also presents many aspects that need improvement, one of which is the process of cooperation of educational institutions, similar to the cooperation of the company with its customers. The systematic work in the context of the development of the process of communication with consumers can significantly improve individual processes and university programs, together with the quality of service, and ultimately increase the effectiveness of the higher education system.

Previous publications have already described who are the consumers of educational services presented by universities (Tchaika 2013).

This article attempts to develop a consumer response system in higher education, including students and potential employers. This system should encourage every university employee to take into account all types of occupational activity within the mission and resources of higher education institutions "through the eyes of the consumer", which means taking into account their vision of the educational process.

2 RESEARCH MATERIAL

Introducing a consumer response system requires the development of a number of programs that will allow for understanding why something is preferred by consumers and how to respond timely to their needs. For students and employers, the implementation of this program can mean:

- Continuous improvement of the quality of education, practical training, student internships in companies based on the program created by students and employers,
- Stimulation of the workforce of higher education institutions, students and employers to work together to achieve better curriculum outcomes and expand their "personal horizons",
- Systematic collection and processing of information as well as quantitative and qualitative data on projects involving students and employers, with mandatory analysis of the achieved outcomes and prospects for further improvement,
- Monitoring of students' and employers' complaints about university procedures, quick and flexible response to them,
- Formulating procedure quality standards and database of objective data on their practical planning and implementation in the context of higher education for the purpose of their comparison with the norm and conducting analyzes,
- Business ethics practice.

The leaders of higher education institutions, managers and employees of business units who are directly linked to cooperation with employers (practice units, Deans, Faculties, etc.) should participate in the development of programs. It is very important to eliminate the barriers between these units that duplicate some functions and overemphasize others, which makes it difficult to implement appropriate quality procedures and reduces customer satisfaction. For example, all Faculties have the same purpose, that is the invitation of employers to a Career Day organized by the university. Faculties perform tasks conscientiously, but without prior internal information about the interrelated activities that target the same groups. Such a way of communicating with the external environment is an unfavorable demonstration of the promotion activities of the university. Preparing next events, based on previous experience, the faculty can not invite all popular employers, because there is a possibility that other faculties will invite the same stakeholders.

Program planning is possible in practice by modernizing activities using the cause-and-effect diagram proposed in the early 1950s by K. Ishikawa, often referred to as the fishbone diagram (Ishikawa 1988). In the field of "consequences", an "effect" is put and in the place where traditionally "causes" are presented, "means to achieve results" are formulated. The subject of the analysis focuses on standardizing the "fish head" quality concerning the carrying out of the Career Day (Fig. 1).

It describes qualitative and quantitative indicators that can be measured and analyzed. For instance:

- 100% of invited employers participated in this event,
- 100% are students who attended the event and talked with employers,

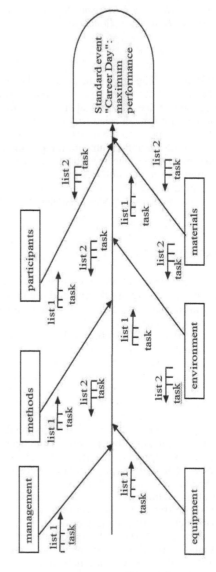

Figure 1. The systematization process diagram of conducting the Career Day.

Table 1. The Career Day roadmap.

	Participants	Methods	Management	Environment	Equipment	Materials	
list 1	Enterprises and students: – list of issues to be agreed with employers, – including student wishes concerning the event	Persons who will cooperate with employers before and after the event: – phone calls, – postal distribution, – creating and updating the employers' information database for the university	Responsible for the analysis of the previous employer and student questionnaire: – developing corrective actions to carry out the Career Day, – informing employers and students of the change	Persons responsible for carrying out the questionnaire: – initial employer questionnaire, – initial student questionnaire, – final results of the employer questionnaire, – final results of the student questionnaire	Agreed with employers: – furniture, – office equipment, – coffeemakers, – dishes (if necessary)	Materials which must be obtained from employers: – list of vacancies and requirements for jobseekers, – information on company, – information about people who are at the event; – employers' wishes concerning the event	work with customers
list 2	Responsible structural units of the University that organized the event: – formation of the team, – defining the purpose of the event, – identifying additional specialists who will be involved, – allocation of duties, – defining time and resource constraints	Persons who develop the questionnaire for students and employers: – development of the previous student questionnaire, – development of the previous employer questionnaire, – development of the final student questionnaire, – development of the final employer questionnaire	Responsible for the final analysis of the employer and student survey: – analysis of problems that occurred during the execution, – development of corrective actions for the future	Room: – for the meeting of the invited, – for direct carrying out the event, – for coffee breaks (if planned)	Persons responsible for the presence and operation of equipment: – a coordinated action plan to ensure timely synchronization	Materials which should be prepared by the higher education institution: – questionnaire (described above), – signposts to direct the invited, – materials for the project of event registration	internal functioning

- The result of the activities and talks is confirmation of the intention to continue to cooperate with the University by 100% of invited employers,
- The result of the activities and talks is confirmation by 100% of students and participants that they have been invited to take up a job or attend an interview,
- Employers and students rated the results of the event at the highest level (based on the results of the study).

In turn, the manager responsible for conducting the event collects feedback from all faculties involved in its organization, makes corrections (side lines) in its key areas: people and contractors (participants), equipment and technology, and methods, among others (from three to six). The first row factors that are the most significant for achieving the target are determined. Then the deepened verification of factors from the first level should be conducted. When analyzing it, it is important to gather information about all possible factors, even meaningless. A potential design variant of the process of event preparation and handling is presented below (Table 1).

As a result of the analyzes, a roadmap of actions is obtained, in which one's actions should be compared. It should be emphasized that every participant is involved in its organization and execution. It is important that such a map with a list of names of people responsible for the process is available to every university representative. This is to avoid duplication of functions and provide flexibility as well as enable meeting the needs reported by participants/customers.

After such a meeting/event, an analysis of bottlenecks should be performed, if such exist, based on one of the classic methods of constructing a cause-and-effect diagram (Fig. 2). In the core box of the diagram, the so-called "fish head", a problem that occurred in practice during the implementation of the event is defined, and in the so-called "bones" – the causes of the problem.

The diagram should be compiled based on the collected final results of the analysis of the questionnaire conducted among event participants. The reasons for incomplete customer satisfaction may be deeper than it seems during the overall verification of the event. At the same time, event organizers are able to observe the passivity of the students, and the survey will diagnose the true cause of this behavior, for example, their lack of interest in some of the proposed job offers. Any shortcomings that hindered the maximum event effectiveness should be considered and reflected in the new modified roadmap for the next future project called the Career Day.

Consequently, the consumer response system in the scope of higher education consists of four subsystems (Fig. 3), which, according to the principles of holistic marketing, are as important as their objectives, which are as follows: continuous

Figure 3. Customer response system in the enterprise.

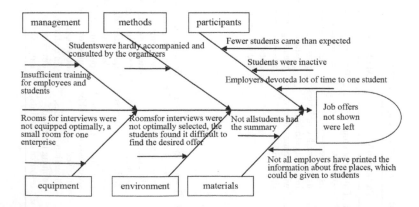

Figure 2. The cause-and-effect diagram of the analysis of the problem of "Unnecessary job offers".

Table 2. Obstacles related to the introduction of a customer response system in the scope of higher education.

Process	Obstacles	Corrective actions aimed at reducing the recognized obstacles
Consumer needs and satisfaction survey	– lack of resources in higher education: money, experts, qualifications, – focus on other types of work, – barriers in the flow of information between departments	– development of public relations procedures targeted at internal and external audiences, – improvement of procedures for the transfer of information between departments, – appointment of a person responsible for survey, – development of material and moral programs that encourage employees
Employee training	– rigid hierarchy, – impersonal policy, – the scope of responsibilities for every employee, – low salaries, employee burnout	– formation of operational standards for responding to consumer needs, – giving employees wider powers, – encouraging employees to make suggestions, – introduction of a broad employee training program, – transparent procedures for remunerating employees
Complaints analysis	– routine approach to work, – symptoms of openness, interest, – lack of willingness to respond to specific situations beyond the scope of responsibilities	– proposing procedural innovations concerning dealing with complaints, – statistical approach to complaints analysis, – developing and implementing parameters of complaints handling performance, – reduction of time to make a complaint decision

Table 3. Measures to encourage consumers to express various suggestions about the functioning of the institution and the educational services provided and offered.

Condition	Measures of implementation
The variety and convenience of the technology for submitting proposals	– a separate phone number, – mailboxes at the University, – e-mail address, – discussion in specialized groups on social networks
Building consumer confidence in the process of university improvement	– procedures for quick notification about accepting proposals for consideration, – operational responses to proposals submitted according to needs, – creating an online database in the form of an offer for consumers describing the results achieved by the University, – procedures for consumer access to the online database of results and proposals.

improvement through transparency in decision-making, decentralization of management, establishing and maintaining feedback, and free participation of every participant in this process.

Assumptions regarding possible obstacles to the implementation of the national university system in the context of the response system for all customer groups and prepared corrective actions are presented below (Table 2).

Taking into account the fact that not only the external environment is changing for the university, but also the rapidly changing internal environment (for example, new students every year), consumer response systems cannot be fixed once and for all. "Improving quality and increasing customer satisfaction is a long process, not a single event" (Kotler et al. 2011). The higher education institution which aims to create an effective system for responding to the needs of consumers should encourage the latter to improve its functioning and services.

Dissatisfied customers who made a complaint provide valuable information that should be used to improve the effectiveness of the adopted system and the level of provision of educational services in the future. In order to gain the necessary knowledge about the effectiveness and quality of some of the implemented procedures, universities are verifying them among all categories of their beneficiaries: students, employees, parents, students, employers, etc. Universities should be interested in obtaining such information for further improvement. Many people, consumers, who have valuable suggestions may not express them if the university does not take further steps to simplify the process of submitting such proposals. In order to encourage consumers, proposals should be made to create the right conditions (Table 3).

Universities should focus on developing appropriate procedures aimed at informing consumers about the technology for making proposals and accessing the Internet database to encourage consumers to submit their proposals. Some results of the complaints analysis can also be submitted to the Internet database, which makes the University's activities more transparent to consumers and helps to improve its image. It is very important that consumers not only have the opportunity to submit a complaint, but, above all, to propose changes to the University or get a reasoned response as to why some proposals are impossible to implement. Universities must be able not only to improve their activities but also inform consumers about it, make changes that are obvious to them, and teach consumers to use innovation.

3 CONCLUSIONS

In conclusion, it can be noticed that creating a system in response to the needs of consumers in the scope of higher education requires significant changes in the approach to communication both with customers and between employees. Efforts should be made to improve and enhance the value of educational services provided. The prospect for further research should be the development of high-quality education concepts: what is quality, and how and who decides to purchase and establish a quality system for higher education.

REFERENCES

Chaika, I.P. 2013. *Holistic Marketing and its Importance in the Functioning of Higher Educational Institutions.* Marketing and management of innovations (3): 53–60.

Ilyashenko, S. M., Konopelko G. M., Nedilko M. O., Starkov I. L. 2015. *Analysis of factors influencing the consumer choice of entrants in the market of educational services.* Economy and entrepreneurship (1) 34–35: 32–39.

Ishikawa, K. 1988. *Japanese methods of quality management.* Moscow: Economics.

Kotler Ph. & Karen F. A. Fox 2011. *Strategic Marketing for Educational Institutions.* Kiev: UAM, Khimzhest Publishing House.

Teletov, O.S. 2015. *Marketing in educational activities.* Economy and Entrepreneurship (1), 34–35: 78–86.

Vasylkova, N.W. 2015. *Strategic marketing planning Strategic Marketing Planning at Higher Education Institution.* Economy and entrepreneurship (1) 34–35: 5–14.

Municipal companies

H. Marcin Jakubíková
Faculty of Economic Informatics, University of Business Economics, Bratislava, Slovak Republic

ABSTRACT: The city as a legal entity may use public funds from its budget to establish a company or use them for acquisition of equity stakes in companies. The reason should be the need to provide specific services for the citizens who live there, such as addressing transport, waste, heat management, water management, social services, culture and art. This city thereby becomes an employer which provides citizens with an opportunity for employment. The aim of this paper is to map municipal companies regional capitals of the Slovak Republic include as a consolidated accounting entity of public administration. These are the eight regional capitals, namely the capital city of Bratislava, Košice, Prešov, Banská Bystrica, Žilina, Nitra, Trenčín and Trnava. A municipal company in consolidated whole of public administration and may have the status of a subsidiary accounting entity of public administration or a joint or affiliated accounting entity of public administration.

1 INTRODUCTION

The city is a separate territorial unit; it joins the people permanently residing in its territory. The territory of a town is a territorial unit, who may form one or more cadastral territories.

The city is managed independently with its own assets and its own income. The city is required to manage the property for the benefit of the development of the municipality and its citizens. The assets should provide enhancement and protection for the city. The city can also provide repayable financial assistance from its budget of the company, which has established, however, cannot use the refundable financing sources.

When the city is the founder of the company, it may decide to give a part of the budget to provide a subsidy for a task, an action in the public interest or for the benefit of the development of the city.

The bodies of the city are the city council and mayor. The city as a legal entity is listed in the registration of organizations kept by the Statistical Office of the Slovak Republic for local government.

1.1 Competences of city council

The city council decides about important questions in the life of the city. It is a representative body composed of members who have been directly elected by citizens for four years in direct elections.

2 PROCESS OF DECISION MAKING IN THE CITY COUNCIL

2.1 Establishment, termination of a commercial company, acquiring equity stakes in a legal entity or selling them

The city council approves a deposit into the owners' equity of a company or approves the establishment or termination of a legal entity. It also approves the representatives of the municipality in their governing bodies. They proceed in accordance with the accepted principles of management or statute.

The city budget in any given financial year should include funds for the payment of a financial deposit into the company when establishing a company or acquiring an ownership interest in a business entity. Changes to the budget during the financial year are also approved by the city council in these cases. The sale of securities and stakes in legal entities which are owned by the city are decided by the city council, which has the choice of three options for implementing a change in ownership of property. The choice is between tender, voluntary auction or direct sale for the general value of the property determined by experts in the sector and field.

The city must publish sale by tender in the usual way 15 days before the deadline for proposals. It also publishes the intention for direct sale of municipal property itself and the direct sale itself is also published at least 15 days prior to the approval by the city council. It is approved twice, specifically the intention to sell municipal property directly and the actual direct sale of municipal property. In the case of direct sale of municipal property its value is assessed using an expert opinion. If the value of municipal assets is higher than EUR 40 000 the city can sell the property by tender or auction but cannot sell it directly.

2.2 Sources of information—city accounting

If a citizen of the city wants to find out what businesses his city has established or what ownership interests the city has in legal entities, the information can be found in the separate financial statements of the municipality or the consolidated financial statements of the accounting entity of public administration in the case where the city has an obligation to prepare one under the Law on Accounting. The financial statements of the accounting entity of public administration (city) and the consolidated financial statements of the entity of public administration (city) are publicly available financial records published in the register of financial statements.

From consolidated financial statements of the entity of public administration i.e. the city, the citizen learns how many subsidiary entities (e.g. information about budgetary organizations, contributory organizations, commercial companies), joint entities (commercial companies) or affiliated entities (commercial companies), including the parent entity of public administration, the city, are included in the consolidated entity of public administration for the accounting period up to 31.12. This information is stated by the parent entity of public administration, the city, in the notes of the consolidated financial statements of the accounting entity of public administration.

Budgetary organizations and contributory organizations are always subsidiaries accounting units of public administration, since they are directly dependent on their parent, the city, through the budget. The aim of these consolidated financial statements is to provide information about the consolidated entity as a single economic unit through the findings and subsequent cancelling out of mutual claims, liabilities, costs, revenues and other relations between the organizations of consolidated unit (Kašiarová & Majorová 2008).

2.3 Selection of objects of research

The objects of research were the eight regional capitals of the Slovak Republic. This is Bratislava—capital of the Slovak Republic, Košice, Prešov, Banská Bystrica, Žilina, Nitra, Trenčín, Trnava. I researched their published financial statements, consolidated financial statements as entities of public administration, annual reports, consoli-

dated annual reports, the commercial register. A common feature is that they are regional capitals that have established companies or have an equity stake in legal entities, they produce consolidated financial statements of an accounting entity of public administration for a group of entities, regardless of their registered office.

2.4 Regional capitals

– Capital City Bratislava, the head of office: Primaciálne nám. 1, P. O. Box 192, identification number (later referred to as ID no.) 00603481.

The capital of the Slovak Republic, it is situated in the foothills of the Lesser Carpathians and extends on both sides of the River Danube. The area of the city is 367 km^2 and the altitude is 126 metres above sea level, the population density is 1 268 inhabitants/km^2 and it has 413 000 inhabitants. The capital city, Bratislava, has 17 city boroughs.

(Capital City Bratislava, Summary annual report for 2015).

– Košice, the head of office: Trieda SNP 48/A, 040 01 Košice, ID no 00691135.

Košice is the second largest city in Slovakia. It lies in the valley of the River Hornád in the Košice Basin. From the east it is surrounded by the Slanské Hills of volcanic origin. The city of Košice is divided into 22 boroughs. The area of the city is 243.73 km^2, the altitude is 255 metres above sea level, the population density is 981.95 inhabitants / km^2 and it has 239 200 inhabitants in 2015.

(Košice, Summary annual report for 2015).

– Prešov, the head of office: Hlavná 73, Prešov, ID no 00327646.

The City of Prešov is surrounded by the Košice Basin, Slánske Hills and Highlands. It is the third largest city in Slovakia and also the capital of the largest region. The area of the city is 70.43 km^2 and an altitude of 255 metres above sea level, population density is 1 277.28 inhabitants / km^2 and it has 89,959 inhabitants in 2015.

(Prešov, Summary annual report for 2015).

– Banská Bystrica, the head of office: Československá armáda 26, ID no 00313271.

Banská Bystrica is located in the centre of Slovakia, in the valley of the River Hron between the Kremnica Hills, Starohorské Mountains and Polana. Banská Bystrica has 16 boroughs. The area of the city is 103.37 km^2 and the altitude is 362 metres above sea level, population density is 761.31 inhabitants / km^2 and it has 78, 758 inhabitants in 2015 (Banská Bystrica, Summary annual report for 2015).

– Žilina, the head of office: Námestie obetí komunizmu, Žilina, ID no: 00321796.

The city of Žilina is situated in Žilina Basin, at the confluence of three rivers: the Váh, Kysuce and Rajčanka. The area of the city is 80 km², the altitude is 378 m asl., the population density is 950 inhabitants/km² and in 2015 it had 83,570 inhabitants in 2015.
(Žilina, Summary annual report for 2015).
- Nitra, the head of office: Stefanikova trieda 60, Nitra, ID no. 00308307.
The City of Nitra is situated in the territory extending between Zobor massif and hills which can be considered part of the Tribeč Mountains, separated from the main massif by the River Nitra.
The area of the city's territory is 4083 ha, of which the built-up area covers 194 ha. The elevation ranges from 138 to 587 metres above sea level. As of 1.1.2015, Nitra had a population of 80 524.
(Nitra, Summary annual report for 2015).
- Trnava, the head of office: Hlavná 1, Trnava, ID no. 00313114.
The City of Trnava lies in the Danube Basin on the Trnava Platform, which is part of the Danube Hills and to the northwest are the Lesser Carpathians. The city is located 50 km east of Bratislava. Near Trnava flows the longest Slovak river, the Vah and its tributaries the Dudváh, Parná and Trnávka.
The city's territory is 4083 ha, of which the the built-up area is 194 ha. The altitude is 147 metres above sea level. As of 1.1.2015 Trnava had a population of 64 439. The territory of the city of Trnava is divided into two cadastral areas: Trnava and Modranka.
(Trenčín, Summary annual report for 2015).
- Trenčín, the head of office: Mierové námestie 2, Trenčín, ID no: 00312037.

Trenčín is located in the western part of Slovakia. The flat, Trenčín Basin, which bends down along the river Vah, is closed in the east by the massifs of mountain ranges Považský Inovec and the Strážovské Hills, to the west by the foothills of White Carpathians.
The area of the city is 8 199.80 hectares and the altitude is between 204–210 metres above sea level. As of 31.12.2015, Trenčín had a population of 55 155.

3 CONSOLIDATED ENTITY OF PUBLIC ADMINISTRATION—REGIONAL CAPITALS

By examining these consolidated entities of public administration, regional capitals, for 2015 I found that the city of Žilina in 2015 had the largest number of commercial companies included in its consolidated entity as subsidiaries, followed by the cities of Prešov and Košice. The largest number of affiliated accounting entities of public administration in the consolidated whole of public administration in 2015 was in Banská Bystrica.
A joint accounting entity—trading company was included in the consolidated entity for Trenčín.
By examining these consolidated entities of public administration for 2014 I found that the city of Žilina in 2014 had the largest number of companies included in its consolidated entity as subsidiaries, followed by the cities of Prešov and Košice.
The largest number of affiliated accounting entities of public administration in the consolidated entity of public administration in 2014 was in Banská Bystrica.

Table 1. 2015—Consolidated entity of public administration—regional capital.

Consolidated entity of public administration	Bratislava	Košice	Prešov	Banská Bystrica	Žilina	Nitra	Trenčín	Trnava	Total
Parent accounting entity									
City	1	1	1	1	1	1	1	1	8
Subsidiary accounting entity									
Budgetary organization	25	62	32	14	38	15	14	12	212
Contributory organization	10	5	1	2	1	3	0	3	25
Commercial Company	4	5	6	3	8	4	1	2	33
Total:	39	72	39	19	47	22	15	17	270
Joint accounting entity	0	0	0	0	0	0	1	0	1
Affiliate accounting entity	5	2	1	7	2	0	0	3	20
Total for consolidation entity	44	74	40	26	49	22	16	20	291

Source: prepared by the author, the consolidated financial statements of accounting entities of public administration, register of financial statements.

Table 2. 2014—Consolidated entity of public administration—regional capital.

Consolidated entity of public administration	Bratislava	Košice	Prešov	Banská Bystrica	Žilina	Nitra	Trenčín	Trnava	Total
Parent accounting entity									
City	1	1	1	1	1	1	1	1	8
Subsidiary accounting entity									
Budgetary organization	25	61	32	14	38	15	14	12	211
Contributory organization	10	5	1	2	1	4	0	3	26
Commercial Company	4	5	4	3	8	4	1	2	31
Total:	39	71	37	19	47	23	15	17	268
Joint accounting entity	0	0	0	0	0	0	1	0	1
Affiliate accounting entity	0	1	1	7	2	0	0	3	14
Total for consolidated entity	39	72	38	26	49	23	16	20	283

Source: prepared by the author, the consolidated financial statements of accounting entities of public administration, register of financial statements.

3.1 Number of employees of accounting entities included in the consolidated unit of public administration

The consolidated financial statements of public administration should contain information about the number of employees of accounting entities included in the consolidated entity of public administration. This number is mentioned in the notes of the consolidated financial statements of public administration. Based on this information, I prepared an overview of the development of number of employees of entities included in the consolidated general government—regional cities for the years 2014 and 2015, which I state in Tables 3 and 4.

As can be seen from the processed data in Tables 3 and 4, the greatest number of employees who work directly or indirectly for the city is in the capital city, Bratislava, followed by Košice. While in 2015 there was a loss of employees from the capital city Bratislava and Košice relative to 2014. The city of Prešov did know show data on employees for 2014 in the notes to the consolidated financial statements of the accounting entity of public administration, only in the individual financial statements of the accounting entity of public administration. Bratislava, in its notes to the consolidated financial statements of the accounting entity of public administration also did not state the number of management employees. This information is also given in the notes to individual financial statements of the City of Bratislava. This information is not disclosed in the notes to individual financial statements of Trenčín.

Table 3. Number of employees stated in notes to the individual financial statements of the accounting unit of public administration as of 31.12.2015 and in the notes to the consolidated financial statements of the accounting unit of public administration up to 31.12.2015.

Monitored item	Number of employees	Of which management employees
Bratislava	6428	374
Košice	5547	289
Prešov	2451	*
Banská Bystrica	1788	178
Žilina	2484	114
Nitra	2145	149
Trenčín	1252	100
Trnava	2098	163
Total	24,193	1367

Source: prepared by the author, notes to the individual financial statements up to 31.12.2015, notes to consolidated financial statements of public administration up to 31.12.2015, register of financial statements,* unknown data.

3.2 Economic activity of accounting entities including in the consolidated entity of public administration

On the basis of the consolidated unit of public administration I have identified extra recurring economic activities according to SK NACE Statistical Classification of Economic Activities, carried out by business companies established by the city or in which the city has an equity stake. These are business companies in the consolidated entity of public

Table 4. Number of employees stated in notes to the individual financial statements of the accounting unit of public administration as of 31.12.2014 and in the notes to the consolidated financial statements of the accounting unit of public administration up to 31.12.2014.

Monitored item	Number of employees	Of which management employees
Bratislava	6579	383
Košice	5803	497
Prešov	2431	*
Banská Bystrica	1758	170
Žilina	2484	114
Nitra	277	168
Trenčín	1139	121
Trnava	2060	161
Total	21,899	1231

Source: notes to the individual financial statements, notes to consolidated financial statements, register of financial statements,* unknown data.

administration for Bratislava, Košice, Prešov, Banská Bystrica.

In this examination, I discovered that their cities used public funds to establish companies or acquire equity interests for economic activities relating to:

- Transport,
- Supply of electricity, gas, steam and cold air,
- Supply of water,
- Treatment and removal of waste water,
- Waste and services for removing waste,
- Activities in the area of real estate,
- Professional, scientific and technical activities, art, entertainment and recreation.

3.3 Economic activities

I list the processed overview of business companies included in the consolidated entity of public administration—regional capital by economic activity

SK NACE – 49.31.0 Urban and suburban passenger land transport.
Dopravný podnik Mesto Bratislava, Joint Stock Company, ID no: 00192 736.
Dopravný podnik Mesto Košice, Joint Stock Company, ID no: 3170914.
Dopravný podnik Mesta Prešov, Joint Stock Company ID no: 31718922.
Dopravný podnik mesta Banská Bystrica, Joint Stock Company, ID no: 36016411.
Dopravný podnik mesta Žilina, Limited liability Company, ID no: 3607099.

SK NACE – 52.29.0 Other transportation support activities.
Bratislavská integrovaná doprava, Joint Stock Company, ID no: 35949473.
SK NACE – 52.21.0 Activities incidental to land transportation.
Mestský parkovací systéme spol. Limited liability Company in liquidation, ID no: 3578880.
SK NACE – 50.40.0 Inland freight water transport.
Slovenská plavba a prístavy, Joint Stock Company ID no: 35705671.
SK NACE – 35.30.0 Supply of steam and cool air.
Tepelné hospodárstvo, Limited liability Company. Košice, ID no: 31679692.
SPRAVBYTKOMFORT Joint Stock Company Prešov, ID no: 31718523.
STEFE Banská Bystrica Joint Stock Company ID no: 36024473.
SK NACE – 36.00.1 Collection, treatment and supply of drinking and utility water.
Bratislavská vodárenská spoločnosť, Joint Stock Company, ID no: 35850370.
Východoslovenská vodárenská spoločnosť, Joint Stock Company, ID no: 36570460.
SK NACE – 38.11.0 Collection of non-hazardous waste.
Odvoz a likvidácia odpadu, Joint Stock Company.
ID no: 00681300.
Technické služby mesta Prešov, Joint Stock Company.
ID no: 31718914.
SK NACE – 38.21.0 Treatment and disposal of non-hazardous waste.
Kosit Joint Stock Company.
ID no: 00681300.
SK NACE – 02.10.0 Silviculture and other forestry activities.
Mestské lesy Košice, Joint Stock Company.
ID no: 31672981.
Mestské lesy Limited liability Company.
ID no: 31642365.
SK NACE – 68.20.0 Renting and operating of own or leased real estate.
Halbart—Slovakia Joint Stock Company.
ID no: 31404693.
STREDOSLOVENSKÁ VODÁRENSKÁ SPOLOČNOSŤ Joint Stock Company.
ID no: 36056006.
MBB Joint Stock Company.
ID no: 36039225.
BIC Banská Bystrica, Limited liability Company.
ID no: 31609465.
SK NACE – 68.31.0 Real estate agencies.
Zámocká spoločnosť Joint Stock Company in liquidation, ID no: 35687533.
SK NACE – 68.32.0 Managing real estate.

Bytový podnik mesta Košice Limited liability Company, ID no: 44518584.
PREŠOV REAL Limited liability Company, ID no: 31722814.
BPM Limited liability Company, ID no: 31598056.
SK NACE – 71.12.9 Other engineering activities and related technical consultancy.
Metro Bratislava Joint Stock Company. ID no: 35732881.
SK NACE – 81.29.0 Other cleaning activities.
Správa a zimná údržba prešovských ciest, Limited liability Company.
SK NACE – 82.30.0 Organization of conventions and trade shows.
Incheba Joint Stock Company, ID no:
SK NACE – 93.11.0 Operation of sports facilities.
Národné tenisové centrum Joint Stock Company ID no: 35853891.
Košická Futbalová arena Joint Stock Company ID no: 47845660.
FC TATRAN Joint Stock Company. ID no: 36503975.
SK NACE – 11.07.0 Production of soft drinks.
MINERÁLNE VODY Joint Stock Company. ID no: 31711464.

4 CONCLUSION

Cities play an important role in economic and financial development of local self-government. They perform a fundamental role in terms of care of the overall development of their territory and making it possible to meet the needs of the population. To carry out some tasks, the city has the opportunity to set up a company or to take an equity stake in a legal entity, if there are available public funds. The city council and mayor have an important role because the vote to decide about this. I have found information on business companies by examining selected individual consolidated entities of public administration in the regional capitals: Bratislava, Košice, Prešov, Žilina, Banská Bystrica, Nitra, Trnava and Trenčín. I found that the large number of economic activities that these trading companies performed related to transport services, supply of electricity, gas, steam and cold air, water supply, treatment and removal of wastewater, waste management and treatment activities, real estate activities, professional, scientific and technical activities, arts, entertainment and recreation. The cities are also important employers in their territories. The most employees are in Bratislava and then Košice.

REFERENCES

Kašiarová, L. & Majorová, M. 2008. A practical guide to consolidation of municipalities—part 1. Publisher Andra vzdelávacie centrum n.o., p.11.
Law no. 138/1991 on municipal property and on the amendment of certain laws.
Law no. 369/1990 on municipal subsidiaries and the amendment of certain laws.
Law no. 431/2002 on Accounting and on the amendment of certain laws.
Law no. 523/2004 on budgetary rules for public administration and on the amendment of certain laws.
Law no. 540/2001 on State statics.
Law no. 583/2004 on budgetary rules for local government and on the amendment of certain laws.
MF Decree no. MF/27526/2008-31, which sets out the details of the methods and procedures of consolidation in the public sector and details of the structure and identification of the items in consolidated financial statements in the public sector as amended.
Statistical Classification of Economic Activities.

Legislative changes concerning VAT in the context of e-commerce

S. Martinková & A. Bánociová
Faculty of Economics, Technical University of Košice, Košice, Slovak Republic

ABSTRACT: E-commerce is the sale or purchase of goods and services, whether between businesses, consumers, government and other organization, which is run via the Internet. From an economic perspective e-commerce brings mainly savings in the information flows of business logistics and acceleration of business processes. E-commerce nowadays attracts in particular the attention of tax administrators and with that are also related the changes in legislation. The paper clarifies the legislative changes in the taxation of VAT in EU countries under the influence of e-commerce expansion. The paper, based on the literature, highlights selected weaknesses of VAT in the view of its development in the context of business practice, further describes the VAT issue from the perspective of a modern e-commerce expansion, as well as options of VAT regulation.

1 INTRODUCTION

The value added tax represents a major source of national budgets, and at the same time, it is a key element of tax systems of individual countries. Revenues from this tax formed about 7% of GDP and about 17% of the total tax revenues of the EU member states in 2016 (Eurostat 2017).

The current technological progress in selling goods and services provides new types of trading that are of electronic, digital, respectively of intangible character (Brid 2005). Challenges to address the issues of e-commerce are mainly associated with optimal taxation of these transactions in order to maintain their neutrality and fairness.

The EU as a community of 28 countries significantly simplifies the free movement of persons, goods, services and capital between the member states. The policies of the EU member states, under the influence of globalization, adapt to legislative changes in the area of tax rules applicable within this Community.

At the time of VAT harmonization, individual countries report a number of changes in their tax systems. The countries are trying to adjust their tax systems in line with the EU directives, which focus on the issues of VAT harmonization. Additionally, the usage of the single currency within the Eurozone countries significantly expands the regular and stable cross-border trade between these countries.

Significant progress in the development of technologies in recent years, as well as the digitalization of cross-border trade of goods and services, have developed an interest of the international organization, political authorities and economists in the taxation of e-commerce, in which the individual transactions are carried out online. The development of e-commerce is associated with efforts to minimise administrative costs of traders. However, it causes considerable problems related mainly to the legislation of this type of business, as well as the issues of e-commerce in the context of the VAT application (EU 1997, Agrawal & Fox 2016).

This paper discusses the e-commerce taxation in the context of value added tax during the period 2004–2016. Based on the empirical and theoretical literature, the aim of this article is to provide a comprehensive view of the situation in economic practice, as well as of the current process of VAT application according to harmonised tax policies in the EU and possible changes and trends in this tax development.

In line with the set objective, the paper is divided into three parts. The first part focuses on characteristics and understanding of VAT in terms of economic literature and the view of its development within the EU. The second part describes the VAT issue from the perspective of a recent e-commerce expansion, as well as the options of VAT regulation in the context of e-commerce, characterises the development changes related to e-commerce taxation in terms of the European directives and new rules created to harmonize these taxes with the EU. The last part of the paper provides an overview of present empirical studies that analyse VAT and its impact on e-commerce in the EU.

2 VALUE-ADDED TAXES IN THE EUROPEAN UNION

The VAT is a percentage value levied on the price of goods turnover. There are Slovak and Czech (Bánociová 2009, Kubátová 2010, Mihóková et al. 2016, Andrejovská & Martinková 2016) as well as foreign literature (Van der Merwe 2004, Ene 2011) that inclines to such definition. This tax was in its basic form included in legislative for the first time in 1910–1921 as a sales tax. The length of this period results from the German and American writings (Adams 1921). The reason for the introduction of this tax was a particular need for tax revenues in order to improve tax administration of individual countries. In the first form of this tax, the offered goods were a subject to taxation within indirect taxes several times, and thus the rates of this tax were relatively low.

The current form of this tax derives from the basic idea of refund of business inputs, which creates revolutionary changes in the tax policy that was based on a cascade method of taxation (Kubátová 2010, James 2011, Široký & Střílková 2015).

2.1 Investigating VAT from the perspective of economists

From an economic point of view, the VAT on product represents a sum of incomes from used production factors, namely wages, profits, rents and interests. Such determination of VAT is defined as *an income method*. It represents an investigation of consumers' behaviour, expected economic growth, changes in employment, respectively in the impact on international trade. Several authors were studying the individual components of the income approach, e.g. examination of unexpected changes in consumers' incomes as one of the key determinant affecting VAT was studied by Attanasio & Weber (2010), employment by Metcalf (1995) and international trade by Desai & Hines (2005).

Currently, all countries that use VAT apply so called *the indirect differential method*. This approach identifies the added value as a difference between the output and input of a taxpayer, where the input tax is deducted from the output tax. Both are adjusted by the respective rates of a tax period (Široký et al. 2014). Such determination of VAT underlines its neutrality because the tax share on the final consumer price is the same for all goods and services with the same rate.

Despite many questions related to the use and the optimal determination of this tax, currently, the positive views on its existence prevail. The highlighted is mainly the neutrality of VAT towards both, consumers and producers, the possibility of taxing the services, preferential taxation of international transactions, in particular arising from the exemption of tax on exports and reliability of tax revenues for the state. Of course, the existence of VAT is also connected to some disadvantages, such as the difficulty of the transition to the new tax, costly administration of its existence, the macroeconomic consequences associated with its existence mainly in the context of increasing inflation.

The arguments supporting the existence of VAT are still a subject of experts' discussions, and opinions of economists and politicians are not always uniform. The strength of arguments varies in individual countries also due to trends of other macroeconomic and tax factors.

The importance and role of the VAT are apparent mainly in the economy of a country and public budget and its status and role can be determined through measurable tax indicators. In order to study the current situation and application of VAT in the EU, it is important to monitor the tax indicators that are reflected in the short term as well as in the long term in form of measurable indicators, which are:

– Standard VAT rate in %;
– Reduced VAT rate in %;
– Tax quota VAT as % of GDP;
– VAT revenues as % of total tax revenues.

The changes in the rate of value added tax depend not only on the timing but also on the macroeconomic indicators that affect them. The average VAT rate was in the EU member states in the range 17–27% in 2016. The lowest VAT rate was reported in LUX and the highest in HU (Table 1).

Ramona et al. (2011) assume that increasing VAT rate negatively impacts the purchasing power of the population. The changes in VAT rate thus significantly influence also a number of real wages. The studies of previous authors suggest that the increase of base VAT rate by one percentage point has a negative proportional character on the development of revenues from this tax in the budget and the existence of a reduced VAT rate causes a regressive impact of this tax.

Most countries used both standard and reduced VAT rate. Among the countries using only a single VAT rate and one reduced rate belong BG, DE, EE, LV, NL, SL, SK, GB. DK is the only country in the EU where goods and services are taxed at a single rate, without the existence of reduces rates.

From the perspective of studying the changes in the development of revenues from VAT in the EU countries during period 2004–2015, we see a growing trend of this indicator. The figure below (Fig. 1)

Table 1. Standard VAT rates in the EU 2016 (EU 2016).

LU	17%
MT	18%
CY, DE	19%
AT, BG, FR, SK, RO, UK	20%
BE, CZ, ES, NL, LV, LT	21%
IT, SI	22%
EL, IE, PL, PT	23%
FI	24%
DK, HR, SE	25%
HU	27%

illustrates the revenues from this tax shown as a% of GDP (scale on the right) and as a% of total tax revenues (scale on the left) within the EU28 average.

2.2 *VAT in the view of the EU*

The VAT is a fundamental element of tax systems of individual EU countries and is a major source of revenues for government budgets of the EU countries. In order to avoid loopholes in the definition of tax base in each country, it is required to spend more on administration and collection of these taxes.

The introduction of the VAT in the EU its current form is associated with the 70 s of the 20th century, when the VAT replaced the previously existing turn-over taxes, general taxes on consumption. The VAT was introduced within the EEC for the first time in the 1st Directive in 1973, first in France and later in other countries DK, GB, GE, NL or SE (James 2011, Široký & Střílková 2015).

The rise of the VAT introduction in Western Europe was accelerated by the implementation of the 2nd, 3th and 4th Directive of the EEC for the member states in the 80 s of the 20th century, in order to harmonize the VAT after the accession of the countries to the EC. The VAT harmonization in the EC is currently based on the 6th Directive, while the rules for the tax base as well as the tax rate of this tax have been revised and changed several times.

A harmonized VAT tax base is every sale of goods and services, excluding financial and legal services and capital goods, for which it is difficult to determine the tax base, respectively it is not clear what exactly the tax base is. Exempt from taxation is also the health care, education and other goods under state protection. In order to establish uniform rules for the EC during the period 1979–2004, this form of VAT was adopted by EL, PT, AU, FI, SE

and ES. A mass introduction of this tax occurred mainly in the countries with developed market economies as well as in post-communist countries in the 90 s of the 20th century.

With the expanding number of the EU countries and an increasing extent of transactions, there also emerges relatively weak point of VAT, which is the possibility of tax evasion. The 5th enlargement of the EC, in 2004–2006, is the first to define so-called principle of carousel fraud thoroughly (Široký & Střílková 2015, p. 54). Less developed countries replaced the valid sales tax later, because the deterrent element was particularly the unpreparedness of financial authorities, demanding accounting and administration of these taxes (Kubátová 2010). Under the influence of adoption the 5th Directive of the EU, several countries adjusted their currently valid tax laws in the context of VAT, in order to harmonize the tax legislation. Specifically, it was CY, CZ, EE, HU, LV, LT, MT, PL, SK and SL.

The 6th Directive has undergone numerous amendments since its creation. Therefore its text became confusing and outdated. During the adoption of this Directive in 2007–2012, the member states BG, RO and HR transformed their tax laws in the context of VAT. From the previous directives and their amendments it is obvious that since the introduction of the single market in 1993, the EU member countries work under the transition regime. The harmonized rules for determining the VAT tax rate where:

– Standard tax rate may be only one and at the minimum level of 15%;
– Further, there may be no more than two reduced tax rates and at minimum level of 5%;
– Increase tax rates are not allowed.

However, the introduction of the uniform VAT still faces the issues of existing national rules of the Member states, which differ significantly mainly in matters of territorial competence of taxation, taxable transactions, VAT deductions or registration obligations, where and when the provided services should be performed, what is the indicator for taxation of the services at export or at import (Terra & Kajus 1992). The activity of the EU bodies to harmonize VAT is significant. However, the results are modest. This is the best demonstrated by the variation range of the base rate, which varies in the Member States within the range 17–27% (the EU countries 2017).

3 E-COMMERCE IN THE CONTENT OF VAT

In general, it can be stated, that the period of formation and development of the digital age, as well as the development of e-commerce, would not be feasible without the formation of the web and graphic web browser Mosaic in 1993, as until then the internet was not considered to be "user

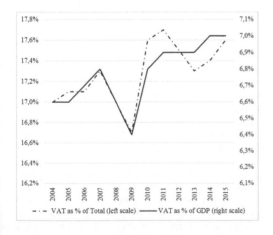

Figure 1. Selected indicators of VAT determines as the EU28 average in period 2002–2015 in % (EU 2016).

friendly" (Poole 2005). E-commerce provides several positives to merchants, such as the ability of suppliers to include the service delivery without difficulties to ensure a quick cross border trade.

Despite the fact that the EU does not define the e-commerce in its Directives, the methods to approach the e-commerce are further specified within the "A European Initiative in Electronic Commerce" (COM(97)). E-commerce can be understood as e.g. electronic share trading or public producement, inline sourcing, but in general, it is doing business electronically, which implies processing and transmitting data electronically (COM(97)).

E-commerce is a global phenomenon in the area of providing goods and services. It is still questionable whether it is important to distinguish between goods and services that are taxed at a single rate in individual jurisdictions. Such a distinction could cause distortion effects of this tax, and therefore it is important that a global approach comply with the classification of digital products.

The introduction of e-commerce allowed to several suppliers outside of Europe to avoid application of VAT on supplies in the EU. Their position, compared to suppliers from the EU whose supplies were subject to VAT, was in their favour. Such behaviour of suppliers was endangering the behaviour of e-commerce businesses, while the EU's effort was to preserve the individual market shares of these companies and not to create additional distortion effects of this tax (Van der Merwe 2004).

The EU did not pay any special attention to the e-commerce topic until 2002. It was in the mentioned year when the EU was forced to adopt a series of guidelines and additional rules governing the EEC Directive on VAT and services provided through e-commerce.

The original proposal of the Directive 77/388/ECC has been replaced by a new proposal on the regulation of e-commerce 2002/388/EEC, which:

– Removed competitive tax disadvantage;
– Amended the legislative base of VAT;
– Took into account a wider international cooperation between countries;
– Ensured the taxation of cross border e-commerce transactions (2006/112/EC).

Currently, there are three valid adopted documents governing the e-commerce, namely:

– Regulation 282/2011, which clarifies the concepts arising from the intracommunity transactions in the context of e-commerce;
– Directive 2008/08/EC;
– Basic VAT Directive, which defines the concepts and legal aspects of VAT for goods and services in electronically supplied services (ESS).

The harmonization of VAT with the EU regulations and directives is currently at a high level. Although the individual EU member states respect the EU legislative measures adopted in the context of VAT harmonization at the supranational level, their individual laws can in some cases and some VAT tax rate differs from each other.

The e-commerce creates the serious issues in the taxation of goods ordered electronically. The question remains where and when the offer of these goods arises, what is the value of such offer if the offer is created by a taxpayer who is a subject to VAT and how is the VAT involved in such offer.

3.1 VAT regulation in the EU

COM(97) defines both forms of e-commerce, direct and indirect. Within the direct e-commerce, the goods and services are supplied exclusively through e-networks. Within ESS, the electronic goods and services are considered to be those, which are traded through internet or e-network and the nature of which renders their supply essentially automated and involving minimum human intervention, and impossible to ensure in the absence of information technology (EC 2011/282, art.7). In the case of e-commerce in the context of VAT, we distinguish between trading done as business-to-business (B2B) and business-to-customer (B2C). In most cases, a consumer is considered to be a taxpayer and supplier is a taxable person. Within the VAT Directive applies that individual goods are taxed at the place of delivery and services are taxable at the place of their execution, depending on their use. During the adoption of the VAT Directive, there is no difference between supplies from the EU Member States and non-Member States. Until 2003 there was no need for the intracommunity trade regulation. The transactions from non-European suppliers were exempt from VAT (within B2C), and in addition, the transactions with the EU were subject to VAT taxation regardless of the transaction's destination. Purchases from non-European suppliers were preferable because their prices were without VAT and therefore the same goods were cheaper. Moreover, the EU suppliers were at disadvantage also at the international level, because they had to pay VAT from all their supplies. In the context of e-commerce several goods were taxed at the place of their origin (Table 2).

The adoption of the Directive 2002/38/EC, which came into force after 2003, the EU revised the deliveries of non-European suppliers into the EU. Individual e-commerce deliveries from EU suppliers to non-European countries are not subject to VAT taxation. VAT should be levied for the purchase of goods if it is a product from non-European country and of VAT rates applicable in a country where the goods were delivered. The same rules apply to

Table 2. The location of taxation for ESS before 2003 (Authors' elaboration).

Supplier	B (EU)		B (non-EU)	
Customer	B	B	B	B
	(EU)	(non-EU)	(EU)	(non-EU)
Principle	Origin	Origin	Origin	Origin
Customer	C	C	C	C
	(EU)	(non-EU)	(EU)	(non-EU)
Principle	Origin	Origin	Origin	Origin

Table 3. The location of taxation for ESS after 2003 (Authors' elaboration).

Supplier	B (EU)		B (non-EU)	
Customer	B	B	B	B
	(EU)	(non-EU)	(EU)	(non-EU)
Principle	Destination	Destination	Destination	Destination
Customer	C	C	C	C
	(EU)	(non-EU)	(EU)	(non-EU)
Principle	Origin*	Destination	Destination	Destination

* after 2015 is B2C (EU-EU) on destination principle too.

services. From 2003, all types of transactions were taxed according to the destination principle.

Such guidance based on the destination principle (Table 3) faces several criticisms, as individual companies prefer to move their headquarters to countries with low VAT and EU suppliers established in a country with high VAT are at competitive tax disadvantage (Mc Lure 2003).

4 EMPIRICAL RESEARCH

4.1 The tax implications of VAT regulation in the context of e-commerce

Several authors (Rendahl 2009, Hellerstein 2008) dedicated to the fulfilment of the basic function of VAT in context of e-commerce. Krensel (2004), Praxity (2013), Merkx (2015) or EU (2016) observe neutrality when introducing Mini One Stop Shop, and as a problematic category of taxation, they still consider digital products, especially e-services in the case of B2C. Another part of the literature focuses on the place of taxation of these digitized products. In the EU, several authors prefer a principle of destination. This is due to the settlement of the marginal cost of production (Mc Lure 2003, Agrawal 2015). In contrast, the drawbacks of origin principle are property interests between countries, intensification of the competition and a decline in VAT revenues.

New alternative approach to the taxation of e-commerce is dealt by for example Ivinson (2003), who addresses the question whether there should be tax at this business industry and what are the positive aspects of taxation at origin. New technological possibilities of taxation are focused on by Lamensch (2012), who proposes the solutions based on real-time VAT, RTVAT.

Some authors are considering a zero or reduced VAT rate on e-commerce. According to Slemrod & Wilson (2009) or Johannesen (2010), the existence of a preferential VAT on e-commerce is dangerous especially because of the expansion of tax heavens. In the opposite position there stand Keen & Wildasin (2004) or Keen & Lahiri (1998), who point out at importance to establish a definite VAT rate as well as to assess the existence of imperfect competition.

4.2 The macroeconomic impacts of VAT regulations in the context of e-commerce

Issues of optimal taxation of e-commerce were also monitored by Hatta (1986), who assumed that "the optimal tax rules generally do not indicate identical commodity tax rates for either final or intermediate goods, though uniform taxes are optimal under certain conditions". Zodrow (2006) adds that preferential taxation of e-commerce also affects the changes in the labour supply. According to Bruce & Fox (2013), the majority of e-commerce is produced by SMEs, which make up to 57% of B2C e-commerce. A significant element influencing e-commerce is the internet (Keen, Lahiri & Raimondos-Møller 2002, Karake-Shalhoub & Qasimi 2006, Gillies 2008, Agrawal 2015). The influence of the internet on tax competition has demonstrated that option of online shopping leads to downward the pressure on the tax rates.

5 CONCLUSION

Taxation of e-commerce in the EU is a complicated and complex issue affected by several fiscal and macroeconomic factors. Challenges to solve the issues of e-commerce are mainly associated with the optimal taxation of these transactions in such a way that preserve the role of neutrality and fairness of VAT. VAT is a key element of tax systems and is a major source of revenue for government budgets of EU countries. The activity of the authorities of the EU to harmonize VAT in EU countries is currently significant too, as evidenced by a number of modifications of final adoption. Despite this, there are several legal and practical issues to alleviate the problems associated with VAT, cross-border cooperation in e-commerce and digital products.

REFERENCES

Adams, T.S. 1921. Fundamental Problems of Federal Income Taxation. *The Quarterly Journal of Economics* 35(4): 527–556.

Agrawal, D.R. & Fox, W.F. 2016. Taxes in an e-commerce generation. *Int Tax Public Finance*: 1–24.

Agrawal, D.R. 2015. *The Internet as a tax haven? The effect of the Internet on tax competition* (February 1, 2017). Available at SSRN: <http://dx.doi.org/10.2139/ssrn.2328479> [Accessed 27 March 2017].

Andrejovská, A. & Martinková, S. 2016. The Impact of Value-Added Taxes Rates on the Economy of the European Union Countries using Data Mining Approach. *JAES* 6(44): 1085–1097.

Attanasio, O.P. & Weber, G. 2010. Consumption and saving: models of intertemporal allocation and their implications for public policy. *Journal of Econ. Lit., American Econ. Association* 48(3): 693–751.

Bánociová, A. 2009. Analýza dane z pridanej hodnoty v SR. *Economics and Management*, 4: 104–115.

Bird, R.M. 2005. Taxing Electronic Commerce: The End of the Beginning? *Bulletin for International Fiscal Documentation* 59(4): 130–140.

Bruce, D. & Fox, W.F. 2013. An analysis of Internet sales taxation and the small seller exemption. *SBA Research Summary* 416: 4.

Communication from the Commission to the Council, the European Parliament, the Economic and Social Committee and the Committee of the Regions—A European Initiative in Electronic Commerce/COM(97)/16 April 1997.

Council Directive 2002/38/EC of 7 May 2002 amending and amending temporarily Directive 77/388/EEC as regards the value added tax arrangements applicable to radio and television broadcasting services and certain electronically supplied services.

Council Directive 2006/112/EC of 28 November 2006 on the common system of value added tax.

Council Directive 77/388/EEC of 17 May 1977 on the harmonization of the laws of the Member States relating to turnover taxes Common system of value added tax: uniform basis of assessment.

Council Implementing Regulation (EU) No 282/2011 of 15 March 2011 laying down implementing measures for Directive 2006/112/EC on the common system of value added tax.

Desai, M.A. & Hines, J.J. 2005. *Value-Added Taxes and International Trades: The Evidence*. Michigen: the University of Michigan.

Ene, C.M. 2011. Tax Evasion-Between legal and illegal mechanisms of the underground economy. *Internal Auditing & Risk Management* 24(4): 12–20.

European Union 2015. *Taxation trends in the European Union*. Luxembourg: Publications Office of the European Union.

European Union 2016. *VAT Aspects of cross-border e-commerce—Options for modernisation*. Luxembourg: Publications Office of the European Union.

Gillies, L. 2008. *Electronic Commerce and Private International Law*. Routledge: Ashgate Publishing.

Hatta, T. 1986. Welfare effects of changing commodity tax rates toward uniformity. *Journal of Pub. Econ.* 29: 99–112.

Hellerstein, W. & Gillis, T.H. 2010. The VAT in the European Union. *Tax Notes* 127: 461–471.

Ivinson, J. 2003. *Why the EU VAT and E-commerce Directive Does not Work*. London: International Tax Review.

James, K. 2011. Exploring the Origins and Global Rise of VAT. *Tax Analysts*: 15–22.

Johannesen, N. 2010. Imperfect tax competition for profits, asymmetric equilibrium, and beneficial tax havens. *Journal of International Economics* 81(2): 253–264.

Karake-Shalhoub, Z. & Al Qasimi, S.L. 2006. *The Diffusion of E-commerce in Developing Economies: A Resource-based Approach*. Cheltenham: Edward Elgar.

Keen, M. & Lahiri, S. 1998. The comparison between destination and origin principles under imperfect competition. *Journal of Internat. Economics* 45: 323–350.

Keen, M. & Wildasin, D. 2004. Pareto-efficient international taxation. *American Economic Review* 94(1): 259–275.

Keen, M., Lahiri, S. & Raimondos-Møller, P. 2002. Tax principles and tax harmonization under imperfect competition: A cautionary example. *European Economic Review* 46: 1559–1568.

Krensel, A. 2004. *VAT Taxation of E-Commerce*. Ph. D. Law at the University of Cape Town.

Kubátová, K. 2010. *Daňová teorie a politika*. Praha: Wolters Kluwer.

Lamensch, M. 2012. Unsuitable EU VAT Place of Supply Rules for Electronic Services-Proposal for an Alternative Approach. *World Tax Journal* 4(1): 77–91.

Mc Lure, C.E. 2003. The value added tax on electronic commerce in the European Union. *International Tax and Public Finance* 10(6): 753–762.

Merkx, M. 2012. Fixed Establishments and VAT Liabilities Under EU VAT-Between Illusion and Reality. *International VAT Monitor* 23(1): 22–26.

Metcalf, G.E. 1995. Value Added Taxation: A Tax Whose Time has Come? *The Journal of Economic Perspectives* 9(1): 121–140.

Mihóková, L. et al. 2016. Estimation of vat gap in the Slovak Republic. *APE* 6(180): 327–336.

Poole, W. et al. 2005. *The Internet: a historical encyclopaedia*. Santa Barbara: ABCCLIO.

Praxity 2013. *European Value Added Tax (VAT). European VAT—Business and Taxation Guide*. Rotterdam: Mazars Paardekooper Hoffman N.V.

Ramona, M.E. et al. 2011. Value Added Tax In The Economic Crisis Context. *The Journal of the Faculty of Economics—Economic* 1(1): 389–395.

Rendahl, P. 2009. Cross-border consumption taxation of digital supplies. [e-book] *IBFD Tax Portal*. Available through: IBFD Tax Portal website <https://www.ibfd.org/> [Accessed 30. January 2017].

Široký, J. & Střílková, R. 2015. *Trend, Development, Role and Importance of VAT in the EU*. Brno: CERM.

Široký, J. et al. 2014. Reflection of the change in VAT rates on selected household expenditures in the CZ and the SR (2007–2013). *Acta Un. Agric. et Silvic. Mend.* 62(6): 1465–1474.

Slemrod, J. & Wilson, J.D. 2009. Tax competition with parasitic tax havens. Journal of Pub. Econ. 93(11–12): 1261–1270.

Terra, B.J.M. & Kajus, J. 1992. *Introduction to Value Added Tax in the EC after 1992*. Boston: Kluwer Law and Taxation Publishers.

Van der Merwe, B.A. 2004. VAT in the European Union and Electronically Supplied Services to Final Consumers. *South African Mercantile Law Journal* 16: 577–589.

Zodrow, G.R. 2006. Optimal commodity taxation of traditional and electronic commerce. *National Tax Journal* 59: 7–31.

New Trends in Process Control and Production Management – Štofová & Szaryszová (Eds)
© *2018 Taylor & Francis Group, London, ISBN 978-1-138-05885-9*

The sector of biotechnology in Poland

P. Marzec & G. Krawczyk
The John Paul II Catholic University of Lublin, Lublin, Poland

ABSTRACT: The article discusses the activities in the field of biotechnology in Poland on the basis of selected indicators, namely: the number of scientific institutions engaged in research and development activities, spending on research and development in the field of biotechnology scientific units, the areas of activity in the field of biotechnology indicated by scientific institutions, people working in research and development activities in the field of biotechnology research units, companies operating in the field of biotechnology, expenditures on activity connected with biotechnology in enterprises, people working in the field of biotechnology in companies. The analysis showed that biotechnology is one of the fast-growing sectors in the country and it is still an emerging sector. In the coming years we can expect further dynamic development of the domestic market of biotechnology.

1 INTRODUCTION

The dynamics of the changing world imposes on all organizations actions concerning designing and implementing changes leading to balance with the environment. These changes have to be systemic, fulfilling certain technological, economic and social criteria. The changes having these characteristics take the form of innovations which are focused on products (goods or services), processes, organization (structural and procedural) or marketing. The necessity of systemic use of innovations treated as the primary factor in the development of organizations, regions, entire economies, as well as the increase in their competitiveness is becoming more and more important in the awareness of managers (Baruk 2016). Poland, with the adoption of the Lisbon Strategy from 2000, agreed to stimulate research and development activities, which are the undisputed source of innovative and competitive economy (Gardocka-Jałowiec 2012). Giving innovation a key role in the processes determining economic phenomena is connected with a growing demand for knowledge necessary for their development and implementation (Grzybowska 2013). An enterprise needs innovation to develop. Companies which fail to notice it are in danger of stagnation and, as a consequence, elimination from the market (Prystrom 2012). Innovation is therefore one of the most important economic issues, both from a theoretical and political perspective. A lot of governments make efforts in order to transform the state into an economy based on knowledge (Truskolaski 2013). Innovativeness of the Polish economy against the background of European Union Member States is still low. According to

the report of Innovation Union Scoreboard 2015, Poland, reaching the synthetic result of innovation of 0.313, was placed among the so called moderate innovators. It was classified as the 24th in the ranking, one position higher in relation to the rating of 2014, and ahead of Romania (0.204), Bulgaria (0.229), Latvia (0.272) and Lithuania (0.283) (MG, 20.5).It is extremely important, from the point of view of the trends of contemporary development, to strive to develop a wide range in regional economy concerning modern fields of engineering and technology, and above all: information technology and telecommunications, biotechnology, environmentally friendly technologies in the chemical industry, energy-saving technologies of consumer products as well as renewable and alternative energy.

For over a decade, biotechnology has been one of the fastest developing fields in the world. Currently around 22 million people work in European bioeconomy, which constitutes 9% of the overall employment in the EU. Changes within the Union take place very rapidly and they influence not only a single user but also the whole chain of values or even whole communities. New solutions are not formed spontaneously, and their creation is often extremely expensive (Woźniak-Malczewska 2015).

This article attempts to assess the activities in the field of biotechnology in Poland. It has been hypothesized that the size of the expenditure on actions related to the industry of biotechnology is crucial for the development of the sector, which has the opportunity for further growth.

Financing innovation in biotechnology is not a simple task due to the long waiting period for the return of the investment. Investments are necessary to ensure the growth and development of the sector because within this sector the research and development activity is expensive and the changes are often caused by radical inventions. Implementation and launching of products on the market is important, but providing and financing adequate R&D which is more important because it allows long-term planning (Woźniak-Malczewska 2015).

The analysis used Local Data Bank maintained by the Central Statistical Office and available on its websites at www.stat.gov.pl.

2 BIOTECHNOLOGY IN POLAND

Biotechnology is an interdisciplinary field of science and technology dealing with the change of animate and inanimate matter through the use of living organisms, their parts or products derived from them, as well as models of biological processes for the creation of knowledge, goods and services (Central Statistical Office—GUS, 2017).

In the analyzed period, the number of scientific units conducting research and development activities has increased. We can observe an increase in both the government and private sectors as well as in the higher education sector (Fig. 1).

Government sector includes all departments, offices and other bodies that provide public services to all citizens, and also entities who are responsible for the administration of the State and the economic and social policies in a given society as well as non-profit institutions controlled and financed mainly by authorities but not administrated by the higher education sector. Public companies are assigned to the corporate sector whereas entities directly related to higher education are placed in the category of higher education sector. The sector of private non-profit institutions includes non-market private non-profit institutions serving households (ie. the general public) and private individuals as well as households. The higher education sector includes all universities, technical colleges and other institutions offering education higher than A-levels or final exams, regardless of their funding sources or legal status. This also includes all research institutes, experimental stations and clinics operating under the direct control of higher education institutions administrated by these institutions or affiliated with them (CSO-GUS, 2017).

The highest increase in the number of units according to the research they conduct can be observed within the basic research and applied and industrial research. Development work shows a decrease (Fig. 2).

Basic research is theoretical and experimental work undertaken most of all in order to acquire or expand knowledge about the causes of phenomena and facts, undirected, in principle, to obtain specific practical applications. Applied research (including industrial research) is undertaken to acquire new knowledge having particular practical applications. The results of applied research are test models of products, processes or methods. Industrial research aims at acquiring new knowledge and skills in order to develop new products, processes and services or implementation of significant improvements into existing products, processes and services. Development work is construction work, technological and design as well as experimental, involving the use of existing knowledge gained thanks to research work or as a result of practical experience, to develop new or significantly improve existing materials, devices, products, processes, systems or services, including the preparation of prototypes and experimental pilot installations (CSO-GUS 2017).

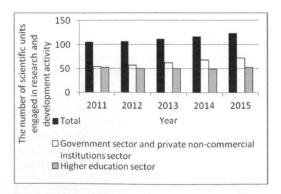

Figure 1. The number of scientific units engaged in research and development in the field of biotechnology.

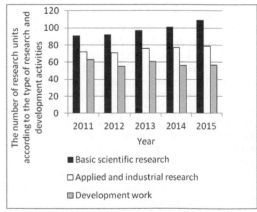

Figure 2. The number of research units according to the type of research and development activities.

Analyzing the current internal expenditure on research and development activities in scientific units according to their spending directions we find an increase in expenditure on: current expenditure, basic research, applied and industrial research as well as development work (Fig. 3).

Current expenditures on R&D—personal expenditures as well as costs of use of materials, perishable items and energy, the costs of external services (other than R&D) including: outside processing, transport services, repair, banking, postal, telecommunications, information technology, publishing, utilities, etc., business travel expenses and other current costs, including in particular taxes and fees charged to operating costs and profits, property insurance and cash equivalents to employees in the part relating to R&D activities. Total current expenditures do not include depreciation of fixed assets, as well as VAT (CSO-GUS 2017).

Analyzing internal expenditures on research and development in the field of biotechnology we can observe an increase in both total expenditure and current expenditure (Fig. 4).

Investment or capital expenditures on fixed assets show fluctuations in the studied period. Internal expenditure is all expenditures on R&D activities performed during the given period within the particular statistical unit regardless of the source from which the funds are disbursed. Capital expenditure is gross expenditure spent on the acquisition of fixed assets used in R&D programs by a particular statistical entity per annum.

Analyzing internal expenditures on research and development according to the origin of the funds, an increase in budgetary resources can be observed as well as fluctuations in funds coming from enterprises, funds from abroad and other. (Fig. 5).

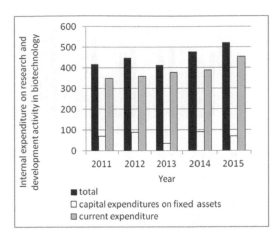

Figure 4. Internal expenditure on research and development activity in biotechnology.

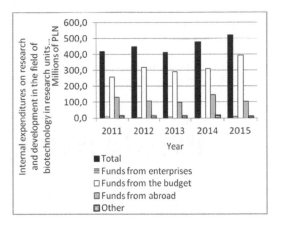

Figure 5. Internal expenditures on research and development in the field of biotechnology in research units according to the origin of funds.

Internal expenditure on R&D in biotechnology in 2015 reached 522.6 million of PLN. Resources from enterprises are funds received from companies and being their own funds. Budgetary resources are resources for R&D from the budget (Ministry of Science and Higher Education and other ministries and local government units). Funds from abroad are resources received for R&D activity from the European Union, international organizations and foreign institutions. Other: own funds and funds received from academic institutions and private non-profit institutions (which are own funds of these institutions) (CSO-GUS 2017).

Analyzing the areas of business activity in the field of biotechnology, the biggest number of enterprises operate in the fields of health and the

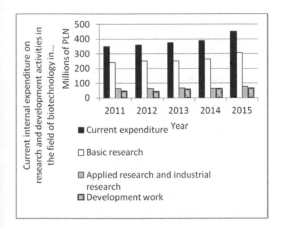

Figure 3. Current internal expenditure on research and development activities in the field of biotechnology in scientific units according to spending directions.

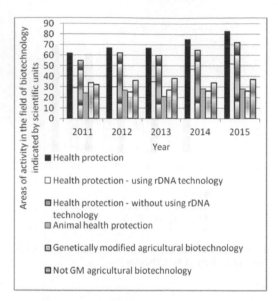

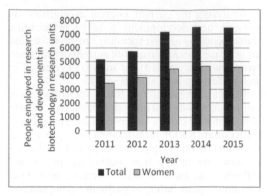

Figure 8. People employed in research and development in biotechnology in research units.

Figure 6. Areas of activity in the field of biotechnology indicated by scientific units.

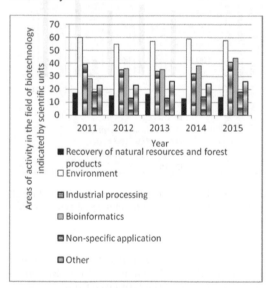

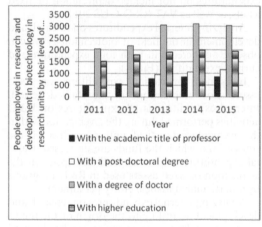

Figure 9. People employed in research and development in biotechnology in research units by their level of qualification.

Figure 7. Areas of activity in the field of biotechnology indicated by scientific entities.

environmental protection. The lowest interest lies in the fields of nonspecific use and recovery of natural resources and forest products (Figs. 6, 7).

Analyzing the employment in the studied period, we notice an increase in the number of people employed in the field of biotechnology in research units. Women constitute a majority here. Analyzing the employment from the level of qualifications, the majority has a PhD while the fewest are professors (Figs. 8, 9).

In 2015 research and development in biotechnology was conducted by 230 entities, of which 160 were units belonging to the business sector. In the analyzed period we can observe a steady increase in companies operating in the field of biotechnology, companies conducting R&D and employing 49 and fewer persons as well as employing 50 or more people (Fig. 10).

A company operating in the field of biotechnology (biotechnology company) is a company that uses one or more techniques of biotechnology (in accordance with definitions of biotechnology) for the production of goods or services and/or to conduct systematic research and development (R&D) in order to increase resources of knowledge and its economic applications (CSO-GUS 2017).

The number of biotech companies is a widely used indicator of the involvement of a country in the use of biotechnology, mainly due to the ease of

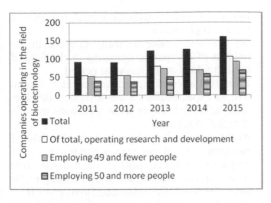

Figure 10. Companies operating in the field of biotechnology.

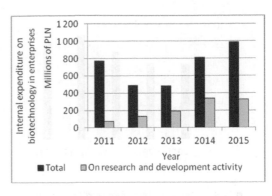

Figure 11. Internal expenditure on biotechnology in enterprises.

obtaining it. The disadvantage of this indicator is limited comparability resulting from a large variety of companies operating in the field of biotechnology—mostly in the scale and nature of their involvement in biotechnology activity, as well as according to other criteria (such as, among others, the size of the entity or type of activity) (CSO-GUS 2015).

In 2015, 106 (66.3%) of companies led research and development work in the field of biotechnology. This is a category of companies featured in the analyses of the OECD as enterprises conducting R&D activities in the field of biotechnology. Among them, 58 companies dealt with only R & D in the field of biotechnology while 48 blended research and development with biotechnological production. 54 enterprises dealt with biotechnological production only. 92 companies were small (employing up to 49 people) while 68 (26,9%) of them were medium or big companies (CSO-GUS 2015).

In 2015 internal expenditures in enterprises to operate in the field of biotechnology reached 989.8 million PLN, ie. 22.4% more than the year before. Among the biotech companies an increase was observed in internal expenditure on activities involving the production of biotechnology—it rose by 192.4 million PLN (40.9%). The lowest level of expenditures was reported in 2013. On the other hand, expenditure on research and development dropped by 11.1 million PLN (3.3%). Both internal total expenditure and expenditure on research and development in biotechnology in the analyzed period is growing (Fig. 11).

Analyzing internal expenditure in the field of biotechnology in enterprises according to the origin of the funds, most of them are their own resources which were increasing in the last three years. The remaining funds are those which come from abroad and they show an upward trend during the period and the budgetary resources which are also growing (Fig. 12).

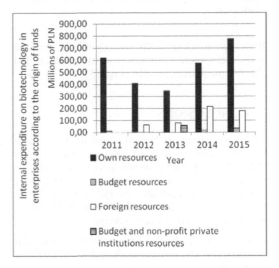

Figure 12. Internal expenditure on biotechnology in enterprises according to the origin of funds.

Resources of the companies are the funds received from businesses and being their own funds. The budgetary resources are resources for R&D work from the budget (Ministry of Science and Higher Education and other ministries and local government units). Funds from abroad are the funds received for R&D from the European Union, international organizations and foreign institutions.

Employment in enterprises in the field of biotechnology in the studied period shows an upward trend. 10016 people were involved in the business of biotechnology in Poland in 2015. 2534 of the people worked in the business sector, by 8.4% (about 231 people) fewer than the year before. In the analyzed period we observe a growth in both total employment, women employees and people employed in research and development field.

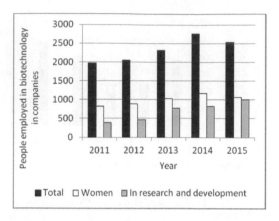

Figure 13. People employed in biotechnology in companies.

1008 people dealt with biotechnology research and development activities in companies, i.e. 181 (21.9%) more than in 2014 (Fig. 13).

3 CONCLUSIONS

The article presents the situation in the field of biotechnology in Poland. Recent years (2011–2015) depict variable trends in figures describing the condition in this area. Periodic fluctuations do not change the favorable overall situation in the field of biotechnology. In this summary, therefore, we should pay attention to several issues:

- In the analyzed period the number of scientific entities conducting research and development is growing.
- While examining the current internal expenditure on research and development in scientific units according to the directions of spending, we find an increase in expenditure on: current expenditure, basic research, applied and industrial research as well as development work.
- Analyzing internal expenditures on research and development in the field of bio-technology, we can observe an increase both in total expenditure and in current expenditure.
- Capital (investment) expenditures on fixed assets depict a fluctuation in the studied period
- Analyzing internal expenditures on research and development according to the origin of resources, it can be observed that budgetary resources have increased. At the same time there are fluctuations of the resources from companies, funds from abroad and other.
- As far as employment in the studied period is concerned, we observe an increase in the number of people employed in the field of biotechnol-

ogy in scientific units with a predominance of female employment.
- We can also notice a steady increase in the number of companies operating in the field of biotechnology in the analyzed period.
- Both total internal expenditure in enterprises as well as expenditures on research and development in biotechnology show a growing trend in the studied period.
- Analyzing internal expenditure in the field of biotechnology in enterprises according to the origin of resources, most of them derive from their own resources which are increasing in the last three years. The remaining funds come from abroad and they show an upward trend during the period and the budget resources are also growing.
- Employment in enterprises in the field of bio-technology in the examined period tends to grow.

On the basis of this analysis it has to be stated that biotechnology is one of the fast-growing sectors in the country and it is still an emerging sector. In the coming years we can expect further dynamic development of the domestic market of biotechnology, which is stimulated both by the implementation of innovative research projects conducted by Polish biotech companies and research institutions, as well as by the inflow of foreign investment into this sector.

REFERENCES

Baruk, J. 2016. Innowacyjność przedsiębiorstw w państwach Unii Europejskiej. *Wiadomości Statystyczne* 8/2016: 64–78.

Gardocka-Jałowiec, A. 2012. Nakłady na działalność badawczo rozwojową a innowacyjność polskiej gospodarki. *Ekonomista* 1/2012: 79–99.

Grzybowska, B. 2013. Wiedza i innowacje jako współczesne czynniki wzrostu gospodarczego. *Ekonomista* 4/2013: 521–532.

GUS 2015. Główny Urząd Statystyczny. Biotechnologia i nanotechnologia w Polsce w 2015 r. Opracowanie sygnalne.

GUS 2017. Główny Urząd Statystyczny. Słownik pojęć. www.stat.gov.pl.

MG 2015, Ministerstwo Gospodarki. Polska 2015 Raport o stanie gospodarki 2015, Warszawa 2015.

Prystrom, J. 2012. Narodowy system innowacji jako czynnik rozwoju gospodarczego na przykładzie Szwecji. *Ekonomista* 4/2012: 499–513.

Truskolaski, S. 2013. Działalność innowacyjna inwestorów zagranicznych i przedsiębiorstw polskich. *Ekonomista* 4/2013: 533–552.

Woźniak-Malczewska, M. 2015. Finansowanie innowacji w polskim przemyśle biotechnologicznym. *Zagadnienia Naukoznawstwa*. 1/2015: 45–57.

New Trends in Process Control and Production Management – Štofová & Szaryszová (Eds)
© *2018 Taylor & Francis Group, London, ISBN 978-1-138-05885-9*

Shared services centers and their task at enterprise financial management

M. Mateášová & J. Meluchová
Faculty of Economic Informatics, University of Economics in Bratislava, Bratislava, Slovak Republic

ABSTRACT: The businesses began using shared service centers and outsourcing to improve back-office efficiency more than two decades ago. Finance led the way, followed by IT. The Shared Services Centers (SSC) seeks to achieve more efficient processes and cost savings itself. Shared services are the result of globalization of economic activities within the company and are and often a spin-off of the corporate services to separate all operational type of tasks from the corporate headquarters, which has to focus on a leadership and corporate governance. Target of the article is to analyze outstanding debts with the focus on cash management, credit risk and bad debt provision. The article contributes to evaluate the optimal outstanding debt management in a specific international company. Financial management of outstanding debts is a key element to provide financial stability and responsibility of a company from solvency point of view.

1 INTRODUCTION

1.1 *Development of shared service centers*

According to the methodology developed by Galgoci (2015), emerging shared services centres (hereinafter referred to as "SSC") go through four development stages. The first stage "start-up" (simple and manual processes with a low level of standardization is moving into SSC). The second stage "growth" (SSC has established standardized processes, controls processes and tools to support customer service begin to gradually implement). The third stage "expansion" (SSC has in addition to internal also external clients, automated business processes, system control processes and operational risk management). The fourth stage "centre of excellence" (SSC already acts as a sovereign organization that generates income, provides highly professional and strategic services for internal and external customers).

Currently, through shared service centres there have been carrying out the purchase of inputs, customer support so called call centres, science and research, and IT services. They have become second-generation centres at the level of business partnership with its parent company. The organization Smižanská & Patoprstá (2016) conducted a survey of services provided by SSC: financial services and accounting (89%), IT services (68%), customer service (63%), human resources (52%), sale and orders processing (47%), purchase (42%) and others (10%). This article focuses on optimizing receivables management, which is provided through SSC. It examines instruments by which it is possible to monitor rating of customers, and thus eliminate credit risk Tumpach & Baštincová (2014).

1.2 *Possibilities for streamline operations through SSC*

Currently, the majority of SSC in Slovakia is in the third development stage of "expansion" with the ambition to move to the fourth stage of "excellence", when SSC should ensure the provision of services with the highest added value. According to our own survey we conducted in 2016 in one multinational company, designated for the purposes of the article as "XYX", SSC, established by parent company, deals with increasingly sophisticated services, which provides all companies in the group of XYX, which are more than 100. The uniform chart setup allows transparent monitoring process and eliminates operational risks such as human error, updating processes in changing legislation preventing disclosure of information, and so on. All events in the parent company will subsequently be reflected in accounting operations and transactions and will affect the reporting of financial results in the financial statements by which is the company presented to the external environment (i.e. user). The information presented must meet certain quality characteristics that present a true and fair view of the assets, liabilities and financial position of the company, and thus could be a subject to further analysis, for example, what we can see in work (Tumpach et al. 2014) or Juhászová & Domaracká (2015).

Increasing demands on professional services of SSC are also related to the growing technological capabilities, which will automate several activities, and thus the proportion of manual work decreases, thereby reducing operational risk. The basis for prosperity of each company is doing business i.e. closing deals and sells. Therefore, the most important area of business is the automation of the ordering process, which begins by reviewing a new customer by limiting the volume of production, which will be the subject of business, followed by exposure to customer orders, delivery of services or products, exposure to invoice the customer for the collected volume, pairing invoice with the order, posting transactions, financial settlement process to recover outstanding debts. With regard to possible risks related to payment discipline of business partners, the effort of XYX is to eliminate credit risk and verify the credibility of business partners before entering into business partnerships in order to avoid future secondary insolvency. Many activities are within the SSC consistently set by parent company XYX for the entire group of companies. The aim is to streamline processes and achieve cost savings Dvořáček & Tyll (2010).

1.3 Financial management of business risks with an emphasis on claims

Most important is to verify a new business partner before starting the business Ondrušová & Parajka (2014). For this, there are used services of external bodies, such as international information offices that provide information and assessment on the reviewed companies (business partners). SSC has established a department, which provides the following services to all companies in the group of XYX by exploiting various externally processed databases, as well as publicly available information, such as the Business Register, Financial Report of the SR, data from the Social Insurance Agency and health insurance companies, data obtained from the Register of Financial Statements, official journals, statistics produced by the Statistical Office of the Slovak Republic, etc. All information and other resources allow checking a business partner, but this is time consuming. Offering analytical outputs is generally not sufficient and does not have a uniform data format which complicates further processing. In this case, it is preferable to use paid services of international information offices that collect highly relevant marketing and financial information from various relevant sources that are processed in a single database which the group of XYX uses.

A survey of the Slovak Compliance Circle in 2015 shows that many companies solve prevention of insolvency issues of business partners. Up to 77% of respondents confirmed that the reputation of the trade partner, its stability and ability to meet contractual obligations, has a serious impact on the status, business success and results of the company. Most serious potential impacts may be risks associated with operations, sales and reputation of the company. Based on the findings and our own research, it is possible to summarize the benefits of screening business partners before entering into a business relationship. These are: minimizing the impact of financial risks (such as poor payment discipline), minimizing the impact of legal risks on our trading company, avoiding jeopardizing the reputation of our company, early detection of unethical practices and specifications are subject to entities doing business unethically. However, a trading company cannot underestimate verification of business partner and it is necessary on a weekly or monthly basis to examine existing partners whether there is no significant change in its financial situation. More than half of respondents as well as our company XYX confirmed that the verification of business partners is performed regularly and the findings of the risk profile are incorporated into its data-

Table 1. International Information Agencies with jurisdiction in the SR and their services.

Company	Coface	Creditre form	Bonnier-bisnode AB	Kompass
Country	France	Germany	Sweden	USA
Brand name	Coface	Creditre form	Dun & Bradstreet	Kompass
Output data	Business inform.	Economic inform.	Economic governance	Business inform.
Rating	Credit rating	Index credibility	D&B Rating	–
	Solvency	Solvency	D&B Paydex	–
Recommended resp.	maximum credit limit of business			
Other services	Country RiskLine report			Demand public contract
	Analysis of insolvency			

* Own source.

bases which have an impact on future business contracts.

1.4 Financial management of receivables

After examining a new business partner, company identifies all the risks. SSC records identified risk profile in the internal databases (investigated Group of XYX works in SAP) to create customer master data that are constantly updated. Based on the evaluation, it is recommended to limit the maximum volume of the contract which will be a subject of trade. For small companies it may not be able to find out all the relevant data. If it fails to comprehensively examine the potential partner, then the risk is eliminated by reducing turnover of business. Another potential tool for reducing the risk is requesting financial assistance in advance or bank guarantee. XYX trading company in order to eliminate the credit risk of the customer implements concrete measures. In order to reduce risk, including political risk, the following risk mitigation measures should be used as appropriate: Advance payment and progress payments, Cash On Delivery, Guarantees or promissory notes, Taking security (lien, collateral, mortgage, etc), Surety Bonds, Stand-by or confirmed letters of credit/bankers acceptances, External credit insurance, Export credit agency risk insurance.

Subsequently, there is closed a business in volume which has come out a review of the specific business partner. Within the subsequent implementation of trade it is necessary to continuously monitor compliance with the terms of trade, especially debt maturity (invoices issued to the customer) and at least on a monthly basis. For company XYX Credit Control department assessment carries out monitoring of receivables through the Country Credit Committee (CCC). The Commission meets monthly and evaluates existing debts, delay of debt maturity and by the time of delay takes measures for provisioning, provides rescheduling of the debtors, sets a souvenir process, enters a call centre to call for amounts owed, etc. Forms of customer assessment - for each new customer it is necessary to assess the credit risk before any order can be booked. Assessment of the customer is required during creation in SAP. Possible forms of customer credit assessment based on requested credit limit amount is showing Table 2. This table is for guidance only. The final credit rating and limit is assigned based on the Credit Control department assessment.

Followed by the implementation of the trade, accounting for a transaction in SAP system where the procedure of individual work is established. The first circuit is referred to "the process of claim settlement" in which there are monitored, for

Table 2. Possible Forms of customer assessment based on requested credit limit amount.

Credit limit request	Assessment of customer existence	D&B rating + D&B "more credit limit"	Credit control inter. assessment payment history
Up to basic credit limit	Possible	xx	xx
Up to 100 kUSD	xx	Possible	Possible
Up to 1 MUSD	xx	Possible*	Possible*
Up to 5 MUSD	xx	Possible*	Possible*
Up to 50 MUSD	xx	xx	Possible*
Above 50 MUSD	xx	xx	Possible*

* The Credit control department has to consider other relevant information where appropriate.

Table 3. Basic credit limit which is assigned to all customers where normal business is allowed.

Country	Basic credit limit
DE	8000 USD
CZ	6000 USD
SK	6000 USD
HU	500 USD
UA	1200 USD

* Own source in XYX company.

example, recording the payment of debt, setting recovery (a souvenir process) to external customers in the event of late payment, keeping correspondence with external customers regarding setting new terms for payment, issuing and accounting penalty invoice, setting of default interest, fines and penalties, monitoring of debts, creation of provisions for bad debts, evaluation of bad debts, write-off receivables, continued cooperation with the CCC in the event of further steps. All claims against business partners are assessed for impairment, thus carrying the potential risk arising from non-payment of debts to the cost of doing business. For accounting purposes, impairment losses are recognized, if there is any event that indicates a negative impact on future cash flow from the customer, which reflects the degree of credit risk. Each claim is considered individually, and this activity is the responsibility of the Credit Control Department assessment. Provisions are not formed for receivables that are insured or secured by a bank guarantee or secured by a letter of credit. Provisions for bad debts are cancelled (reversed) at the moment when the reasons for their creation disappear (in the case of payment, write-off, assignment of debt or its sale).

Setting the customer's credit limit, assessment and subsequent re-evaluation of credit risks, control of the hedging instruments (such as a bank guarantee), communication with business partners, if agreed schedule of payments and so on are reviewed within the Credit Control. All these activities have an impact on secondary insolvency, which could arise, if a company does not properly manage their claims. Therefore, follow-up activity is cash flow management, where SSC provides the XYX group with setting activities, such as cleaning and checking bank accounts, management of payments and receipts through the banking system, entering payments, and bank master data management, archiving records, communication with the bank.

2 CONCLUSION

Successful debt management gives the company a chance to stay solvent and successfully manage their cash flow. An important tool to eliminate the risk of insolvency is to set up its own system of evaluation of business partners using databases and services of external entities. The incorporation of customers in rating categories, according to the risk of insolvency, is a useful tool that is currently used mainly by multinational companies with large numbers of customers. Creating a rating scheme is an intensive process for date and comprehensive data. Well-developed internal rating of the creditworthiness of customers simplifies work with orders and accounts receivable, speeds up decision making and helps to draw attention to an increased risk of insolvency. The more is the rating process automatized and linkages with other components of the information system, the larger are the benefits for a company. In the context of improving the effectiveness it is advantageous to set the global rating scheme for the needs of the entire enterprise groups, as is the company XYX. The advantages are time and cost savings, but above all, the complexity of mapping risks, especially for those companies that are moving the global market and entrepreneurship within individual countries, would want to mask their unethical business practices. Rating requirements, each company adapts

to their needs. For existing subscribers it is possible to incorporate into the rating an important source of information e.g. historical customer's payment discipline that extends the possibilities of its comprehensive assessment.

Calculating the creditworthiness of the customer is usually based on non-financial information (such as historical payment behaviour, industry risk, country risk, testimonials from business partners, reports on execution proceedings, lawsuits, etc.) and related databases from external entities such as Coface, Creditre form Dun & Bradstreet. Important information may also provide credit insurance company, if the company cooperates with them. Another source of information is financial indicator based on an analysis of the accounts of customers providing information on liquidity, profitability and indebtedness of trading partner. Verification of business partner prior to the conclusion of the contract should be the basis for setting fair and long-term partnerships.

REFERENCES

Dvořáček, J. & Tyll, L. 2010. *Outsourcing a offshoring podnikatelských činností.* 183 p. Praha: C.H. Beck.

Galgoci, G. 2015. Business Service Center Forum. Online, *BSC-2015_flyer_sk.pdf*: 36 p. Bratislava: BSCF.

Juhászová, Z. & Domaracká, D. 2015. Premiums earned in the financial statements. *In Financial management of firms and financial institutions: proceedings: 10th international scientific conference. 7–8 September, 2014.* 479–483. Ostrava: VŠB.

Ondrušová, L. & Parajka, B. 2014. The revaluation of assets and liabilities at fair value in merger. *In 7th International Scientific Conference on Managing and Modelling of Financial Risks. 8–9 September, 2014.* 577–581. Ostrava: VŠB.

Smižanská, M. & Patoprstá, J. 2016. *Slovak service centers are on the way to first division* In. Trend. Issue: 16. 32–38. Bratislava: Trend Holding.

Tumpach, M. & Baštincová, A. 2014. Cost and Benefit of Accounting Information in Slovakia: Do We Need to Redefine Relevance? *11th International Scientific Conference on European Financial Systems 12–13 Jún, 2014.* 655–661.Brno: Masaryk University.

Tumpach, M. et al. 2014. *Relevance of System of National Financial Reporting from the Point of View of Creditors as Non-privileged Users.* In. Economic Journal, Vol. 62. Issue: 5. 495–507. Bratislava: Slovak Academy of Science.

New Trends in Process Control and Production Management – Štofová & Szaryszová (Eds)
© *2018 Taylor & Francis Group, London, ISBN 978-1-138-05885-9*

Endogenous and exogenous effects of R&D in enterprise innovations activity

M. Matušovič & D. Hrušovská
Faculty of Business Management, University of Economics in Bratislava, Bratislava, Slovak Republic

ABSTRACT: The strategic importance of intellectual property and innovation, their support and their maximum use in specific applications today absolutely accents. Constantly accelerating innovation activity subjects brings both positive and negative effects. The current focus on competitiveness and innovation accentuates ensure a minimum threshold of social status of man. Confrontation of theory and practice reveals new insights on the social status of the people. These facts led us to the decision to diagnose selected processes in the implementation of innovations. The starting point is known theoretical knowledge and practical experience in this area. We used the available data and knowledge. We have also implemented their own data collection. Generalized results of this analysis, we consider it useful incentives for the development of theory and practice in the field of social innovation and implementation in real practice.

1 INTRODUCTION

In this article, we have focused on selected innovation factors. When considering the essence of innovation, we have met with several opinions of the authors. Different views of authors are minimal. We do not see them as fundamental. These are rather minor cosmetic treatments. From our point of view, however, we see in the last decades the existence of significant social processes (challenges, phenomena) that theory and practice treat as innovation. Doubts as to whether or not the classical contemporary concept of innovation in general sufficiently reflects these new phenomena has resulted in the recognition of social innovation as a tool, as a product, as a product, such as innovation and others. In their opinions on innovation, the authors preferred structure, ownership, income, value, yield, return or competitive advantage. The views of the authors are probably theoretical and practical. We need to complement the views of the authors in some parts about our attitudes that originate from the studied literature, but also from the results of the research conducted by the respondents' opinions, especially in the area of perception of substance/content/selected concepts. We examined the various views of respondents and the views of authors in the literature on the content of selected concepts of innovation and social innovation. The survey focused on the field of innovation of all types, including social innovations. This article is elaborated in VEGA project No. 1/0784/15

"Knowledge of results from the implementation of" Examining the relationship between social innovation and business economy in order to increase the competitiveness of the entrepreneurial subject "Agenda 2030 for Sustainable Development Objective 9. Build solid infrastructure, promote inclusive and sustainable industrialization and strengthen innovation".

2 STATE OF KNOWLEDGE

The basic stone on which the object and subject of our study is built is an extensive literary research. However, the limited scope of the post allows us to summarize the current state only very briefly. We are therefore only giving an incomplete selection of authors. Innovative activity and innovation are often a topic to be published. (Valenta 2001, Molnár & Dupaľ 2008, Barták 2008, Caulier-Grice & Gul 2005, Corrado et al. 2005; Hilman & Pulley 2005, Romadoni 2006, Adairn 2004, Allen 199,; Bačišin et al. 2006, Nicholls & Murdock 2012, Deiglmeier & Miller 2008). The issue of innovation is an integral part of publications on corporate governance, but also on the sustainability of economic growth. The literature lacks an examination of specific factors for the growth of innovation activity and their interrelations at national and global level.

3 METHODOLOGY

The aim of the innovation research was to define the content of the selected concepts, to map current knowledge and models, to examine the selected processes that concrete entities implemented in real practice. We have been examining innovation issues, including social innovations. Linking to the most comprehensive review of the current state of innovation issues. The intention is to define selected concepts used in the field of innovation, to identify and examine selected problems of selected processes and models. We will ensure the achievement of the following objectives: summarize, organize and analyze the theoretical and practical knowledge of the selected concepts of each type of innovation, specify some selected problems at all stages of the life cycle (from idea generation to implementation and impact assessment); Design and test correlations in innovation (including social innovations). At this stage of the study, we formulated the following objectives:

- We assume that there is ambiguous understanding of the content and use of selected terms in the area of social innovation in business practice,
- We assume that the theoretical opinions elaborated in the literature on the content of the concept of
- Social innovation are inconsistent,
- We assume that research costs show a minimal mild correlation with selected parameters (indicators).

The basis for the formulation of objectives was the study of professional literature, innovations analyzes and other individually obtained or generally accessible information dealing with the given issue.

Selected issues have been explored through the following methodological steps: to search and study available related literature and data, to formulate a goal, to gather information on the subject under discussion by participating in various professional events, to contact practitioners and to obtain information on the status of the subject of the targeted interviews, to search for and analyze Metadata in the field of research, conducting own research, identifying problem areas, selecting appropriate tools and methods of investigation, identifying specific target research objects, creating a database for analysis processing, analyzing collected data, evaluating results and processing conclusions, proposing solutions and recommendations for theory and practice.

Achieving the goals was achieved through the use of recognized general scientific methods. In addition to the study of literature, empirical methods, especially observation methods and managed interviews, were used to obtain information. In this context, a comparison method has been used. The system approach was used to process the information obtained. We have exploited the synergy of several scientific methods. Analysis of selected indicators of innovation performance, factor and statistical analysis. Total, line and residual squared sum:

$$S_T = \sum_{i=1}^{I} \sum_{j=1}^{m} \left(Y_{ij} - y_{..} \right)^2 = \sum_{i=1}^{I} \sum_{j=1}^{n} Y_n^2 - \frac{Y_{..}^2}{n} \tag{1}$$

$$S_A = \sum_{i=1}^{I} n_i \left(y_i - y_{..} \right)^2 = \sum_{i=1}^{I} \frac{Y_i^2}{n_i} - \frac{Y_{..}^2}{n} \tag{2}$$

$$S_e = S_T - S_A \tag{3}$$

In addition to the methods already mentioned, other generally accepted methods of scientific research have been used, particularly the basic and most extensive methodological procedures such as analysis and synthesis, deduction and induction. The survey method used the purpose analysis (questionnaire survey), which was focused on a selected sample of Slovak enterprises. The data were obtained through a survey of respondents' opinions representing selected companies operating in the Slovak Republic. Several questions have been formulated in the questionnaire in order to obtain the background to achieving the research goal. The questions were formulated with the intention of characterizing the respondents, characterizing the individual objects of the survey and the actual state of innovation in the corporate sphere, as well as the opinions of the respondents on the actual innovation structure. First we tried to look for data that somebody had collected and analyzed before. So we did not find such a range of data that would fulfill our stated goals. For these reasons, we collected our own data. We have formulated 33 questions that we logically organized in the questionnaire. The first set of questions asked us to provide information about the respondent—gender, age, educational attainment and job placement. The second set of questions asked us to provide information about the business—the object of the survey from which the respondent drew information on answering questions. In cases where the respondent did not find the information in the given enterprise, he provided expert opinion or opinion. The business information included the legal form, the share of foreign capital, the seat of the enterprise, the business year, the main business, the sector and the size in terms of number of employees and realized turnover. The third group of questions should provide us with information on the structure of the assets of the undertaking concerned—total volume, share of intangible assets, use of the main subject of business, linking the existence of assets to business records and identifying specific components of non-registered and used intangible assets. The fourth group of questions should provide us with information on the structure of the assets of the enterprise concerned in terms of substance, value, evidence and protection of intangible property. In this group of questions, the respondent also expresses the perception of the content of the selected concepts of property and innovation. We responded to the respondent by asking questions open (the respondent could indicate or add—to clarify the opinion stated) and closed (the respondent could indicate one or more of the taxative options listed). The questionnaire was compiled in accordance with published author's current opinions in available literature and in accordance with current legislation, including accounting practices in the Slovak Republic at the time of its preparation. Due to the complexity and novelty of the problem under consideration, as well as the considerable scope and complexity of the questions, we have distributed the questionnaire also in electronic form with the possibility of online opening of commentary and explanation on its filling in for reasons of clarity and minimization of filling time. Any ambiguities could be consulted if necessary in person. We selected the list of addressed companies from existing databases. 238 respondents were actively involved in the research. The object of the survey was a set of companies, whose structure is dominated by limited liability companies, followed by joint stock companies and tradesmen. Only 36% of the companies surveyed had partners with foreign capital participation of 51% or more. Almost 80% of enterprises are represented in the Bratislava Region and are active from 2 to 17 years with an average life of 11 years. The average age of all surveyed businesses is just over 14 years. The main subject of the activities of the surveyed enterprises is mainly focused on services and industrial production. Enterprises with up to 100 employees are 79% with an average of 46 employees. The average number of employees in the entire group of surveyed enterprises is 36. Some questions offered respondents a choice of possible responses, others asked for additional information or respondent's opinion. We have processed open questions based on individually tailored methodologies.

Given the limited scope of the article, we will narrow our attention to investigating the company's research activities in relation to the effects of the innovation activities carried out.

4 DISCUSSION

According to Jaffe, there is a model for generating new innovations that is conceptually based on existing patented technologies that are publicly available. Innovations are implemented on a generic principle (Jaffe 2011). The distinctive feature of Tassey—Gallaher—Petrus' conceptual framework is that it contains impulses to develop business performance, market development and risk reduction. Over the past decade, it has been highlighted to explore the linking of investment and innovation to productivity factors. The growing importance of the accounting framework of the innovation system is focused on the use of methods to measure more accurately the value of individual components of intangible assets (Corrado et al. 2005, Haskel & Wallis 2009).

The theoretical foundations of innovative activity measurement indicators are extensive. A global agreement is needed to measure the main innovation activities (measuring capabilities and trends in human capital, research and development and innovation) that would be globally and historically comparable. Indicators are compiled by National Statistical Offices, UNESCO, OECD, Eurostat, NCSES has released human capital and R&D indicators. In 2010, the Agency began publishing statistics on research, development and innovation: the monitoring of innovative business activities has been particularly sophisticated. Simple monitoring of domestic R&D expenditure by a small number of large manufacturers has to be replaced by a much more complex monitoring of R&D, innovation and innovation spending across the world and across sectors, collecting, analyzing and evaluating the available information for monitoring innovation, research, development, science, technology, engineering, and mathematics (STEM) of the workforce has also been complicated. In the past, the statistical authorities have reliably collected consistent and objective and relatively comparable information. At present, however, the amount of raw (incomparable) data that is readily available online has risen sharply. These data sources, though more detailed, are problematic to process them into objective and relatively comparable information—there are doubts as to whether the statistical activities of the relevant institutions are properly focused on the production of information that politicians, scientists and businesses need. These issues become particularly acute in view of the current development of national economies, the global economy and the role (role) of innovation. In recent years, we have seen a comprehensive review of relevant science, technology, and innovation (STI) data collection methods and indicators from different countries, including Japan, China, India, countries in Europe, Latin America and Africa.

An important impetus for the design of our analysis was the structure of the generally valid accounting records of the company in the Slovak Republic. From the point of view of evidence of intellectual property, it is mainly based on generally binding, internal binding standards and external sources of information in this field. We also need to take into account the specific fact that research potential is not only the property of an enterprise but also, in particular, the personality and property rights of a particular entity (originator or owner). Only property rights can be transferred to third parties. Therefore, the originator is not always directly associated with the business. Business research activities can be partially identified in the inventory of intangible assets. Purchased results of external research activities are also problematic in the company's costs. This question was for the respondents who agreed that the company records the component of intangible assets "activated development costs". Those respondents were 94. 2.5% responded to question 21 although they did not allow for the existence of activated development costs in the intangible asset structure in question 19. These respondents estimate the share of the activated development costs to 65% on average. Respondents prefer the design and operation of the selected alternative. The shares of the components of the activated development costs are between 18% and 46% in the companies surveyed. The structure is similar in the case of reality and estimation. The highest share is "Making and Running the Selected Option" (41% to 46%) and "Making and Testing Prototypes and Models" (33% to 38%). Respondents unequivocally (just over 3%) prefer "Making and running a selected alternative" and "Others" (24% of respondents). The other components of the activated development costs have a 20% preference. It is clear from the respondents' opinion that R&D is only limited in the surveyed enterprises.

The small – 1% – difference in reality and the estimation of the respondents' opinion indicates that they do not see the problem in insufficient evidence. We suppose that in the company records we find items in the area of outsourced services, wages or costs of entire centers, or organizational units that deal mainly or exclusively with research. However, we did not get data that would allow us to assess the reality of our assumptions. Freely available business listing information is in the structure that did not allow us to filter data. In our opinion, however, the real share of research is higher than respondents said. The reason for our claim, apart from complex evidence, lies in the existence of multinational corporations, which include almost 40% of the enterprises surveyed by us. Research management may not be within their competence, but it is implemented by the headquarters. However, this does not mean that they do not use the results of research, it is not reflected in the evidence of their intangible assets. We have reviewed this fact. The share of research in enterprises with foreign capital is higher by 1% (in case of fact and estimation). In the case of a more detailed survey of enterprises with any foreign capital participation, our assumption has been confirmed. From the point of view of the surveyed respondents, we mainly observed the opinion of the owners and managers. Both groups have options to influence the status. Owners assumed 10% of the research and 15% managers. Owners do not consider the share estimate to be substantial and managers think, the reality of research evidence is lower by 5%. So they declare the inconsistency of evidence with reality.

The average value of the research activity scale is $M = 7.492$; Standard deviation $SD = 10.9$. Minimum value $Min = 0$, maximum $Max = 50$. The scale was filled by 238 respondents. We also attempted to correlate the selected parameters in order to identify the correct parameters (indicators) to measure the effects of the research results. We have identified a correlation between spending on research, employment, turnover, wages and salaries:

Table 1.   Correlation between research expenditure and selected business indicators.

	zam_s	obr_s	mzd_s	mzd_o	odm_s	odm_o	VYS_s	VYS_o
zam_s	1							
obr_s	0,711	1						
mzd_s	0,071	0,004	1					
mzd_o	0,071	0,129	0,549	1				
odm_s	0,065	0,07	0,362	0,167	1			
odm_o	0,064	0,068	0,155	0,335	0,59	1		
VYS_s	−0,004	−0,053	0,261	0,122	0,069	0,0006	1	
VYS_o	0,067	0,198	0,156	0,234	0,05	0,0884	0,646	1

We have not shown a correlation between research spending and selected indicators (turnover, employment, wages and salaries).

5 CONCLUSIONS

We have encountered ambiguous understanding and use of concepts when examining social innovations. We have identified fundamental differences in the perception of social innovation in a search of available literature. The differences of opinion between the authors are significant. We also found opinions in businesses. We collected the data via a questionnaire. Defining social innovations is also ambiguous from an enterprise perspective. Accounting provides sufficient space to record the individual research components that business uses in business. However, in the records of enterprises, we do not find the individual components of intellectual property related to research, despite the fact that, according to experts, their value represents a notable share of 40–65%. The problem is to introduce their value to the accounting because we do not recognize it. Certain research components are created only for the business of a particular subject and must be classified or protected by registration in the databases of the Industrial Property Office of the Slovak Republic or other relevant competent institutions. However, this kind of protection is particularly problematic in the case of small businesses. We also see the problem in regular revaluation in accounting or other records. This is unfeasible in view of the specific nature of some of the components of intellectual property and the respect of the precautionary principle. Research expenditure in Slovakia does not show a minimum correlation rate in relation to selected indicators. It can be said that the indicators and therefore the effects of innovation are due to the transfer of intellectual property. We propose to modify analytical accounting records.

REFERENCES

Adair J. 2004. *Efektívní Inovace*. Alfa Publishing, 2004, 240 s., ISBN 80-86851-04-4.

Aggarwal, V.A. & Hsu, D.H. 2014. Entrepreneurial exits and innovation In: *Management Science*, 60 (4), pp. 867–887.

Allen J.C. 1994. *Inovačné podnikanie*, Elita, 1994, 276 s. ISBN 80-85323-70-2.

Allred, B.B. & Park, W.G. 2007. Patent rights and innovative activity: Evidence from national and firm-level data. In: *Journal of International Business Studies*, 38 (6), pp. 878–900.

Bačišin, V. 2006. Inovácie sú hnacím motorom ekonomického rastu. In *Dialógy o ekonomike a riadení*: odborno-informačný časopis pre členov Klubu ekonómov Ekonomickej univerzity a jej absolventov.—Bratislava: Klub ekonómov Ekonomickej univerzity v Bratislave, 2006. ISSN 1335-4582, marec 2006, roč. 8, č. 24, s. 35–42

Bačišin, V. 2010. *Financovanie inovácií*. Dizertačná práca. Prognostický ústav SAV. 109 s. 2010.

Baker A., 2002. The Alchemy of Innovation: Perspective from leading edge, Spiro [Press, 2002, 224 p. ISBN 1 904298 01

Baláž V. 2004. *Prechod na poznatkovú ekonomiku. Priority ľudských a finančných zdrojov a hodnotenie výskumu a vývoja v SR*. Prognostický ústav SAV. 89 s. ISSN 0682–91–37.

Baláž v. et al. 2007. *Inštitúcie a ekonomická transformácia*. Veda,. 2007, 233 s.

Barták J. 2008. *Od znalostí k inovacím: tvorba, rozvíjení a využívaní znalosti v organizacích*. Alfa Nakladatelství, 2008, ISBN 978-870-87197-03-5.

Borovský, J. & Gál, P. 2005. *Inovácie a transfer technológií*, Eurounion, 2005, 78 s. ISBN 80-88984-86-6.

Brown J.R. et al. 2017, What promotes R&D? Comparative evidence from around the world. In: *Research Policy*, 2017, vol. 46, issue 2, 447–462.

Čarnický, Š. & Mesároš, P. 2005. Manažment znalostí s pohľadu teórie a praxe, In *Acta Oecomica Cassoviensia*, Košice 2005, s. 50–507, ISBN 80-225-2038-1.

Caulier-Grice G.M. 2010, *The Open Book of Social Innovation*, NESTA, 2010.

Chesbrough H. 2003. *Open Innovation: The New Imperative for Creating and Profiting form Technology*, Harvard Business School Press, 333 p. 2003, ISBN 1-57851-837-7

Christensen, C.M. – Rayner, M. 2003 *The Innovators´s Solution*, Harvard Business School Press, 2003, 289 p. ISBN 1-57851-852-0.

Christensen, C.M. 1997. *The Innovator's Dilemma: When New Technologies Cause Great Firms to Fail*. Boston: Harvard Business School Press, 1997, ISBN-10: 0060521996.

Čimo, J. & Mariaš M. 1993. Inovačná stratégia firmy, Elita 1993. ISBN 80-85323-47-7.

Čimo, J. & Mariaš, M. 1998. *Inovácie vo firemnej stratégii*, Sprint, ISBN 80-88848-29-6.

Corrado, C. et al. 2005. "Measuring Capital and Technology: An expanded framework", in C. Corrado, J. Haltiwanger and D. Sichel (eds), *Measuring Capital in the New Economy*, National Bureau of Economic Research, Studies in Income and Wealth,Vol. 65, Chicago, IL: University Chicago Press, pp. 11–45.

Davila, T. et al. 2006. *Making Innovation Work*. Wharton School Publishing. 2006. 320 p. ISBN 0-13-149786-3.

Deiglmeier & Miller. 2006. Rediscovering Social Innovation. In: *Stanford Social Innovation Review*. 2008, Vol. 6, Issue 4.

Drucker, P.F. 1985. *Innovation and Eneterprepreneurship.*, 1985, ISBN 0-06-091360-6.

Duman, P. et al. 2009. *Klastre na podporu rozvoja inovácii—analytická štúdia*, Slovenská inovačná a energetická agentúra, 2009. 35 s.

Dundon, E.A. 2002. *The Seeds of Innovation: Cultivating The Synergy That Fosters New Ideas*, AMACOM 2002., ISBN 0-8144-7146-3.

Dvořák, I. – Procházka. P. 1998. *Rizikový a rozvojový kapitál.* Praha: Management Press. 170 s. 1998. ISBN 80-85943-74-3.

Gorfinkeľ, J.V. 2009. *Ekonomika innovacij,* Izdateľstvo Vuzovskij učebnik, Moskva 2009, 416 c., ISBN 978-5-9558-0110.

Hargadon, A. 2003. *How Breakthroughts Happen: The Surprising Truth About How Companies Innovative,* Harvard Business School Press, 2003, 291 p. ISBN 1-57851-904.

Harris, M. & Albury, D. 2010, Collabortive innovation in the public sector.

Hilman, H. & Romadoni, A. 2006. *Managing and Protecting Inteliectual Property Assets,* SanFrancisco: Berrett-Koehler Pubvlishers, 2006. s. 394. ISBN 978-0-74944-052-7

Jaffe, A. 2011. Analysis of public research, industrial R&D, and commercial innovation: Measurement issues underlying the science of science policy. In K.H. Fealing, J. Lane, J.H. Marburger, III, and S. Shipp, eds., *The Science of Science Policy*: A Handbook. Stanford, CA: Stanford University Press.

Janek, B. et al. 2007. *Inovácie ako zdroj rozvojového potenciálu firmy,* zborník prednášok—Európska konferencia produktivity EPC 2007 a 10.národné fórum produktivity, str.179 -183, SLCP, Žilina 2007, ISBN 978-80-969391-7-6.

Joly, A. 2010. *The Innovation Handbook.* New York: Kogan, 2010. 336 s. ISBN: 97-8-07494-568-70.

Kelemen, J. et al. 2008. *Kapitoly o znalostnej spoločnosti.* Iura Edition, 2008, 295 s. ISBN 978-80-8078-209-2.

Kisslingerová E. et al. 2008. *Inovace nástroju ekonomiky a managementu organizací.* Nakladatelství C.H. Beck, 1. vydání. 2008, 293 s. ISBN 978-80-7179-882-8.

Klas A. Et Al. 2005. *Technologický a inovačný rozvoj v Slovenskej republike,* Bratislava 2005. Ústav slovenskej a svetovej ekonomiky, 390 s. ISBN 80-7144-147-3.

Kline, S.J. – Rosenberg, N. 1996. *An overview of innovation.* In: Landau, R. The positive sum strategy. Washington: National Academy Press, 1996, 656 p. ISBN-10: 0309036305.

Kolesárová, L. a kol. 2012, Innovation activity of enterprisees in the Slovak republic 2008–2010, Štatistický úrad Slovenskej republiky, 2012, ISBN 978-80-8121-156-0.

Košturiak, J. – Chaľ, J. 2008. *Inovace vaše konkurenční výhoda* !, Computer Press, Brno, 2008, 164 s. ISBN 78-80-251-1929-7.

Litan, R.E. et al. 2012 *Improving Measures of Science, Technology, and Innovation: Interim Report,* Washington, DC: The National Academies Press. 78 s. ISBN 978-0-309-25389-5.

Mihok J. Et Al. 2010. *Podpora Inovácií,* Centrum inovácií a technického rozvoja, Košice 2010, 296 s., ISBN 978-80-970320-0-5.

Mitsauki, Sh. 2004. *Innovative Business Paradigm,* Japan, Tokyo, 2004, ISBN 4-64.

Moczala, A.. 2005. *Zarandzanie Innowacijami,* Wydawnictwo Akademii Techniczno humanisticznej v Bielsku Bialej. 193 s. 2005. ISBN 83-89086-3-1.

Molnár, P. – Dupaľ, A. 2008. *Manažment Inovácií,* Ekonóm, 170 s. 2008. ISBN 978-8.

Mozga, J. 2000. O inovacích—cast I. *E+M Ekonomie a Management,* 2000, roc. 3, c. 3, s.30–33. ISSN 1212–3609

Nicholls, A. A Murdock, A. 2012, *Defining Social Innovation The Young Foundation.*

Perlaki, I. 1977. *Inovácie v organizácii,* Alfa 1977, 154 s., 302 07 24, 63 – 070 – 77.

Porter, M. 2000. Location, Competition and Economic Development: Local Clusters in a Global Economy, *Economic Development Quarterly* 14 (1), s. 15–34, Sage ublications.).

Riegel, K. 1985. *Inovace ve výrobní organizaci* SNTL—Nakladatelství technické literatury, 1985, 184 s

Rothwell, R. 1994. Towards the Fifth-generation Innovation Process, *International Marketing Review*, Vol. 11 No. 1, 1994, pp. 7–31.

Rughuram, G.R. – Zingales, L. 2003. *Saving Capitalism from the Capitalists, Crown Business,* 2003, 486 p., ISBN 0-609-61070-8.

Schmookler, J. 1966. *Invention and Economic Growth.* Hravard University Press.1966. 348 p. ISBN 10: 0674464001

Schumpeter, J.A. 1934. *Theory of Economic Development.* [1st ed. 1911]. Cambridge, MA: Harvard University Press.

Schumpeter, J.A. 1987. *Teória hospodárskeho vývoja: analýza podnikateľského zisku, kapitálu, úveru, úroku a kapitalistického cyklu (z nemeckého originálu:* Theorie der wirtschaftlichen Entwicklung). Bratislava: Pravda, 1987, ISBN 80-7325-044-6.

Stangova, N. & Vighova, A. 2016. Possibilities of creative accounting avoidance in the Slovak Republic. In: *Economic Annals-XXI,* Volume 158, Issue 3–4, 21 June 2016, Pages 97–100.

Tureková, H. & Mičieta, B. 2003. *Inovačný manažment,* 2003, EDIS—Vydavateľstvo Žilinskej univerzity, 169 s., ISBN 80-8070-055-9.

Valenta, F. 1969. *Tvorivá aktivita—inovace—efekty,* Praha, 1969, Svoboda, 262 s.

Valenta, F. 2001. *Inovace v manažérské praxi.* Nakladatelství velryba ®, s. r. o. 2001, 151 s. ISBN 80-85860-11-2.

Vodáček, L. – Dvořák V. 1989. Podnikatelské *řízení inovačních procesu,* Institut řízení, 78 s. 1989, ISBN 80-7014-021-6.

New Trends in Process Control and Production Management – Štofová & Szaryszová (Eds)
© *2018 Taylor & Francis Group, London, ISBN 978-1-138-05885-9*

Risk resources of decision-making in small and medium enterprises

V.T. Míka & M. Hudáková
Faculty of Security Engineering, University of Žilina, Žilina, Slovak Republic

ABSTRACT: The making of the decisions is the core of management and a result of conscious activities which is under way in a particular environment and concrete conditions. The managers decide about the goals, procedures and about the methods how to respond to the changes and to the problems which developed. Their decisions affect the effectiveness, quality, economy and the overall successfulness in every organisation. In spite of this fact they do not pay sufficient attention to the individual steps of the decision-making process. They emphasise more how to cope with the individual methods and techniques of making decisions and forget about the way how to cope with analysing the problem or assessing the individual solution variants. In many cases the underestimating of the analytical phase can lead to an incorrect assessment of the problem and this can then negatively influence its further solution.

1 INTRODUCTION

It is important for every manager to make efforts to take such decisions which reflect the basic conditions, states and development of the environment. The small and medium-sized enterprises (SMEs) are more sensitive to the changes of the external conditions and also to the problems of the internal environment.

In general, it is also valid for the SMEs that the biggest risks result from the character of the market environment, the entrepreneurial environment in the line of business and with the economic and legal conditions in the given country. However, the SMEs can be also threatened by the personal risks, the risks resulting from the insufficient qualification of the employees, low motivation and by violating discipline and safety measures as well as other rules and standards. The preparedness of the management, its management skills and experience which are missing in many cases and are replaced by professional knowledge and routine, play a special role.

2 RISKS IN THE DECISION-MAKING PROCESS OF THE SMEs

More companies are aware of the need and importance of the risk management in their enterprises. The global research concerning the implementation of the risk management shows that in spite of certain problems the companies try to implement the risk management into the managing processes (Merna 2007). The top managers are usually involved in the monitoring and control of the risks. Afterwards they link the risk management with strategic planning of the enterprise. On the other hand, only half of the companies believe the risk management system is an important competitive tool of the enterprise. The support for an early identification of the risks in the company is missing; the trainings and consultancy in this area are not sufficiently provided.

The European enterprises and we can assume the majority of the large Slovak companies pay the biggest attention to managing the risks of the foreign currency exchange rate, of the credits and those which can result from the failures of the counterparties, e.g. insolvency of the business partners. It is interesting that the European enterprises perceive some risk more strongly than e.g. the risk of the loss of reputation, the risk in the internal systems or the risks caused by the failure of the business partners. Although mostly large enterprises took part in this research, the practice shows that also small and medium enterprises perceive the market and financial risks as the most important ones.

The situation regarding the implementation of the risk management in the Slovak companies is not positive. The approach to the risk management is in many companies (compared with the advanced countries) less systematic and is implemented with certain reserves. The companies are missing an overall framework of the risk management which is sufficiently interconnected and works without any link to the company strategy (Belás 2015).

The majority of the companies are aware of the occurrence of the entrepreneurial risk, however, its importance is perceived with a different intensity. The solutions are often limited to an informal risk assessment. In practice in Slovakia it is not common that the risk management is a natural part of the managerial decisions. The 2009 financial crisis aroused interest in the risk management in the Slovak enterprises and strengthened the positions of the financial managers, however, by far not to an extent as it is usual in the foreign countries (Vodák 2014).

The situation in the SMEs where are several objective and subjective barriers which prevent the companies to manage the risks effectively is more complicated. There are several causes as follows:

– The management of the SMEs (subject to exceptions) has a worse access to information necessary for analyses, assessment and risk management,
– The management, mainly of the small enterprises, is on the one hand created by experts of the given area, however, on the other hand frequently by the owners or partners who are often missing economic and management knowledge and experience,
– Creating a specialised position for an analyst or specialist in the area of the risk management would inadequately increase the costs and moreover, there is no space for such a position in the simple organisational company structure,
– The business experience of the owners and managers or the current successful development and especially the missing strategic analysis of the environment conditions can significantly suppress the need of implementing the risk management in the company.

In 2013 we realised a statistical research of the SMEs' entrepreneurial risks in the region of Žilina in the framework of the project FaME/2013/MSPRISK—Entrepreneurial Risks of the Small and Medium Enterprises in the Turbulent Economic Environment (Hudáková 2015). Out of the total number of 164 SMEs 30% of the entrepreneurs identified the market risk as the most serious one, for 22% it was the financial risk, for 14% the human resources risk, for 14% the legal risk, for 12% the security and safety risk and 8% of the entrepreneurs consider the operation risk to be the

least serious one. The figure one shows the percentage of the SMEs' identified risks.

The personal interviews show that the entrepreneurs consider the insufficient market research, insufficient advertising, incorrect strategy of the selling price, prices of the competitors' products but also the language barriers to be the main sources of the market risk.

Except for the market and financial risks it is very important to orient also on the *human resources risks* which are perceived as very important by the SMEs, however, they do not take any measures for their reduction and elimination or even they are not simply carried out. But it is important to realise that some causes and phenomena of the human resources risks are connected closely with the effectiveness of the measures for reducing other entrepreneurial risks. We consider the risks linked with the owners and managers to be the most underestimated sources of the human resources risks in the SMEs. They are those risks to which the managers (in several investigations) do not draw attention because they concern just them. It is also important to realise that the majority of the *shortages and risk sources are closely connected with the decision-making processes, i.e. with the key management activities*. Therefore the next part will deal just with the risk sources and causes of failures during the decision-making process.

3 OBJECTIVE AND SUBJECTIVE RISK RESOURCES

The management processes and especially the decision-making processes in the area of the small and medium entrepreneurship are affected by a whole range of factors which have both an objective and subjective character. The characters of the environment where the SMEs realise their function belong to the *objective factors*. While the conditions of the local or regional market the SMEs operate in are relatively foreseeable, the information for a strategic analysis of a broader market, economic, technological and social environment can be acquired harder.

Therefore the *subjective factors* which are frequently closely connected with the possible risks of erroneous decisions come to the foreground. There are especially the following facts:

– The decision-making process of the managers in the SMEs is based on until that time successful and more or less intuitive decisions based more on experience and clear relations with the business partners than on any deeper analyses of the situation,
– Although they sometimes make decisions based on a rational decision-making procedurethe

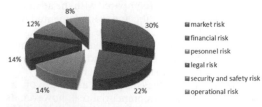

Figure 1.   Percentage of identified risks of the SMEs in the region of Žilina in 2013 (Hudáková 2015).

mangers (or the owners) are missing any larger knowledge and experience resulting from identifying, analysing and assessing the risks of the possible decisions including the knowledge and experience from the implementation of the necessary methods,
– The decision-making process is affected by the decision-maker's personality, i.e. the owners or managers, their qualification, the economic and managerial experience, analytical abilities, the level of their strategic thinking as well as their personal qualities, communication and interpersonal capabilities and several psychical processes (fatigue, stress, influence of an illness, etc.).

These subjective factors create a certain summary comprising the possible causes of errors during making decisions which—in connection with the existence of the aforementioned objective factors—can be perceived as two groups of the risk sources during the decision-making process:

1. The underestimation of the risk analysis in the individual phases of the decision-making process, i.e. wc underestimate the implementation of suitable methods of the risk management into the individual phases of the rationally understood decision-making process.
2. The list of the so called cognitive failures which result from the deformed perception of information, sometimes caused by routine erroneous patterns of processing the information. The cognitive failures are often the causes of the incorrect intuitive decision-making.

Based on the aforementioned facts it is possible to say that the risk appears on two levels during the decision-making process:

a. The risk related to the environment conditions—the risk as an uncertain situation resulting from insufficient or incomplete information about the state of the relevant factors of the environment as a probability of their important properties—in this case we can speak about the situation of a risk.
b. The risk related to the subject's activity, its goal (a risk resulting from our purposeful activity)—the risk as a possibility of a result which differs from the determined goal.

The decision-making processes as the key processes of the company management, especially in their rational form, are closely linked with implementing the risk management tools. On the other hand the individual phases of the decision-making process itself include elements of the risk management. This is depicted in the Figure 2.

In the phase of *identifications and analyses of the problem* it is important to recognise the causes of

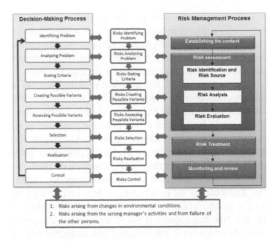

Figure 2. Risk management elements in the individual phases of the decision-making process.

the rise of the problem and to estimate which risks would result from its solution and on the other hand it is necessary to count on the fact that solving the problem can bring new risks. Unfortunately, this phase is not fully respected in practice and concentrates only on formulating the problem.

One of the reasons can be that the management is loaded by a number of unambiguous solutions (at the first sight) which do not require any analysis but an immediate and most frequently intuitive solution. The managers have a feeling (and it is necessary to say that a justified one) they recognise what type of a problem it is and therefore they concentrate on the realisation phases of the decision-making process.

However, underestimating the analytical phase can mean that the problem is assessed incorrectly and this fact affects its further solution negatively. In this case the team search and knowledge of the causes of the arisen problem can make its real solution more effective, not to speak about a situation when it prevents the development of another problem. In our opinion it is suitable to utilise the participation approaches and to create a good communication atmosphere especially in the phase of analysing the problem. The management often responds to the situation when the problem is transparent but its cause remain unknown and the problem can emerge in other connections and frequently with other symptoms.

The methods of the risk analysis find their place in the problem analysis understood in this way. They enable assessing the risk rate and to choose a suitable procedure of the further steps in the decision-making process.

The process of *determining the criteria* for selecting a suitable solution is understood as a routine activity, however, also in this phase the management can make a mistake. In spite of respecting the basic principles—the purposefulness, effectiveness, feasibility and lawfulness, an incorrectly determined criterion can cause a risk decision.

The key phase of the decision-making process is the *creation and assessment (or testing) the possible solution variants*. The utilisation of the risk management tools helps reveal the possible negative consequences of this solution variant which suits the stated criteria in the best way. In the case of a SME management it concerns e.g. the selection of the business partners, selection of an advantageous credit, the decision about purchasing a new technology or a new development investment. The implementation of the risk management tools will enable the company to consider if the detected risk is acceptable or if it will be better to find another solution.

The textbooks of management and risk management do not bring much information about the risks of the already chosen decision, however, in spite of this the realisation phase, *the phase of implementation,* hides a lot of risks. The risk is hidden in wrong communication between the management and the realisers—they need not understand each other and due to this fact an inappropriate implementation can develop.

The preparedness of the realisers—their knowledge, experience, their relation to the decision being made—substantially influences the final effect of the decision. It is also important how the implementation is organised—we forget the selection and formulation of an appropriate decision is only the assumption of an effective solution of the problem and only its fast and well coordinatedrealisation can show if the selected decision was correct.

4 MANAGER'S ERRORS DURING THE DECISION-MAKING PROCESS AND PROPOSALS FOR THEIR REMOVAL

It is obvious that making appropriate decision is affected by several more or less influenceable factors creating the decision-making system (Hammond 2006). We have already mentioned that the source of the risks in making decisions of a manager or owner of a SME have both objective and subjective character.

The decision-maker personality is of key importance. In spite of the fact that the knowledge, experience and the aforementioned comprehensive preparedness to solve complicated situation in the decision-making processes are the basic prerequisites, making mistakes cannot be excluded. Some of them result from the shortage of relevant information or from taking unsuitable or incorrect decision-making methods; some of them can be caused

by the defective human perception (Frankovský 2015).

The errors in the decision-making process and proposals for their removalor reduction:

1st – attaching too big importance to the first information.

It is something like "anchoring" the first information, impressions, the first estimations, significant experience from the past in the thinking of the decision-makers.

The proposals for removing this error consist especially in:

– a thorough analysis of the problem, assessing the problem from several aspects and broader interconnections, searching for relevant information and opinions,
– utilising a combination of methods for ealizati the problem and searching for solution variants, if the situation allows it.

2nd – the effort to maintain the status quo.
The new situation is not perceived as a stimulus for a necessary change. The effort to prolong the current situation is much stronger than the need to work out new measures.

The proposals for removing this error require especially:

– Rejecting the status quo as the only alternative, to consider rationally the advantages and limitations of the possible solutions in relation to the current situation,
– Realising that the individual factors of the current state, i.e. also the sources of the risks and crises develop with time and taking adequate decisions reduces the probability and strength of the possible crises.

3rd – the effort of "framing" the problem to a known scheme complies with the general and quite appropriate tendency to ealiza the acquired experience. It can cause the unwillingness to look for next information (anyway, everything is clear; we know what the matter is).
The proposals for removing this error are:

– To pay attention to the correct identification of the problem and its thorough analysis before the "framing" process,
– Due to the frequent time stress during solving an acute crisis it is sometimes inevitable to adopt a certain framework, to compare it with an already solved problem but when it is only a little possible, to try looking for other schemes for solving the problem, to compare our perception of the problem with the colleagues' opinions.

4rd – preferring information which confirms our estimations are partially connected with the previous errors.Especially the relatively experienced managers have a tendency to rely on their estimations and prognoses. In the complicated situations this tendency is strengthened by the fact we do not possess enough necessary information and also the time for their thorough analysis.

The proposals for removing this error are restricting these influences does not mean that we cast doubt upon the experience and intuitions. It could also prevent assessing the information before we are able to analyse it, at least partially.

During a reaction to the possible risks the accompanying phenomenon of the rational measures can be that we rely on the decision-maker's intuition. The complicated situations of various characters place heavy demands on the subjects that decide. The manager has to have not only professional knowledge, but also the knowledge of the decision-making methods and techniques. He/she has to be prepared to bear the consequences of incorrect decisions. This requires certain experience and psychical and mental competencies.

The improvement of the mangers' training in the SMEs with an emphasis on implementing the selected risk management tools, on the capability to analyse the situation correctly, to make the necessary conclusions for choosing the selection methods to pick the best solution variant is only one side of improving their competencies (Urbancová 2015). Another and also very important area is to increase their psychical and mental abilities to perceive the situation correctly, to analyse the information rationally, to decide wisely and to improve their abilities to implement the adopted decision successfully and to manage their realization effectively

5 CONCLUSION

This paper concentrates on that aspect of the management competencies of the leading employees of the SMEs which are linked with errors during making decisions. The incorrect selection of the goals, erroneous assessment of the situation, underestimation of the importance of monitoring the development of the important indicators, an inadequate reaction to the changes of the conditions are the most frequent errors during the decision-making process. Although the causes for incorrect decision can have also a less objective character, e.g. the relevant information is not accessible or there is a need to decide under the time stress, the majority of erroneous decisions can have a subjective character. We think the unpreparedness of the management and the existence of the causes for making wrong decisions belong to important sources of the entrepreneurial risks in the conditions of the small and medium enterprises. Therefore in the risk management process of the SMEs, especially in the phase of identifying the risks, it is necessary to give more weight to the risks resulting from the character of the staff and in the case of the management to their possible erroneous decisions.

ACKNOWLEDGEMENT

Publication of this paper was supported by the Scientific Grant Agency of the Ministry of Education, Science, Research and Sport of the Slovak Republic—VEGA No. 1/0560/16.

REFERENCES

Belás, J. et al. 2014. Significant Attributes of the Business Environment in Small and Meduim-Sized Enterprises. *Economics and Sociology* 7 (3): 22–39.

Frankovský, M. 2015. Kognitívne omyly a rozhodovanie manažérov. *Zborník príspevkov Manažérske rozhodovanie v kontexte situačných a osobnostných charakteristík*. Prešov: FM PU.

Hammond, S. et al. 2006. The Hidden Traps in Decision Making. *Harvard Business Review*: Reprinted in 1/2006: 118–125.

Hudakova, M. et al. 2015. Analysis of the risks of small and medium-sized enterprises in the Zilina region. *Communications: scientific letters of the University of Žilina* 17 (1): 34–39.

Kral, P. & Kliestik, T. 2015. Estimation of the level of risk based on the selected theoretical probability distributions. *Proceedings of the 10th International Scientific Conference on Financial Management of Firms and Financial Institutions Location*. Ostrava: VŠB.

Merna, T. & Faisal, F. Al-Thani. 2016. *Risk management*, Brno: Computer Press.

Vodak, J. et al. 2014. Cooperation management in Slovak enterprises. Procedia Social and Behavioral Sciences. *Proceedings of 2nd world conference on business, economics and management*. Istanbul: Elsevier.

Zibrinová, Ľ. & Birknerová, Z. 2015. *Myslenie v kontexte kognitívnych omylov*. Prešov: Bookman.

Urbancová, H. & Hudáková, M. 2015.Employee Development inSmall and Medium Enterprise in the Light of Demographic Evolution. *Acta Universitatis Agriculturaeet Silviculturae Mendelianae Brunensis* 63 (3): 1043–1050.

New Trends in Process Control and Production Management – Štofová & Szaryszová (Eds)
© 2018 Taylor & Francis Group, London, ISBN 978-1-138-05885-9

Economic conditionality of law and responsibilities associated with it

T. Mikhaylina
Vasyl' Stus Donetsk National University, Vinnytsya, Ukraine

D. Palaščáková
Faculty of Economics, Technical University of Košice, Košice, Slovak Republic

ABSTRACT: The paper is devoted to the economic conditionality of rights through the use of the dialectical method. Disclosed the process of formation of the law since its inception in the actual public relations prior to the official implementation in the legislation. It justifies the thesis that a significant consideration of social (including economic) conditionality of rights improves the quality of the regulatory material and the effectiveness of its actions. On the basis of the comparative law method are examined specific examples of economic conditionality of law and the consequences of including or ignoring this provisions by the legislator. Given that research is conducted in the framework of National Scholarship Programme of the SR SAIA, this paper summarizes the results of complex interdisciplinary study in the fields of Economics and Law, as well as comparisons of Slovakia and Ukraine's economic and legal systems.

1 INTRODUCTION

The interaction between law and economy was probably the most significant problem both at the theoretical and at a practical level. Its value is determined by the fact that there are spheres of public life, which would not have been affected by the economy. And quality of economic relations, in turn, is determined by the effectiveness of their regulation.

In fact, the law and economy do not occur independently and as a result of the mutual influence on each other in different spheres of the social system. That's why the most problems in law and economy cannot be solved without taking into account the impact of other areas and social phenomena. Inasmuch economic and social systems are relatively independent but still interconnected subsystems of the social system, only interdisciplinary research at the present stage can ensure the implementation of comprehensive social systems studies. Only this approach can ensure the improvement of economic system and system of law, as well as help to find the solutions of urgent challenges of our time.

Louis Kaplow and Steven Shavell in their fundamental work "Economic Analysis of Law" (Kaplow & Shavell 1999), mention of Bentham, who still in the 19th century addressed the problem of interaction of law, morality and economy (Bentham, 1970, Coase 1960, Becker, 1968, Calabresi & Melamed 1972). But their own research is taking us in the plane on the economic characteristics of private law institutions: the right of ownership, litigation, criminal liability, contracts, claims, legal procedures and more (Kaplow & Shavell 1999).

From contemporary researchers on the economic aspects of law it is necessary to remember Posner, whose stunning, comprehensive works reflect the logic of the economic manifestations of the right and Vice versa (Posner 2014, 2016).

However, in order to realize the deep connection of these phenomena it is reasonable to refer to the origins of the interdisciplinary economic-legal researches. Such is certainly "Das Kapital" by Karl Marx (1867). He initially identified the economy as the basis for all other social relations, which he called the superstructure. And it should be noted that his economic theory is still quite popular in the post-soviet states, although in recent years it has been criticized.

This is due to the fact that this theory, despite the generally faithful proposition, however, has a tinge of etatism, and claims the leading role of the state. At the same time, the mass of economic relations does not require the intervention of the state through law that will be discussed below.

And yet, obviously, the interaction of economy and other social phenomena is more complicated than just the basis and superstructure. The conceptual approach is the recognition of the permanent, continuous, and systematic of infraconnection of economic and other social subsystems. This fully applies to the right.

2 THE PURPOSE OF SCIENTIFIC RESEARCH AND METHODOLOGY

Given that scientific research is conducted in the framework of National Scholarship Programme of the Slovak Republic for the Support of Mobility of Students, PhD. Students, University Teachers, Researchers and Artists (SAIA), this article summarizes the results of complex interdisciplinary study in the fields of Economics and Law, as well as comparisons of Slovakia and Ukraine's economic and legal systems.

The purpose of this paper in this regard is the development and substantiation of the approaches to economic stability using legal means in various legal systems (including the countries of Romano-Germanic and Post-Soviet law), in connection with how economic relations affect the process of law formation.

Scientific research is primarily based on dialectical and structured systems analysis, since it is considered a constant in time correlation of many phenomena of the relatively independent subsystems of the social system: law and economy. Also, to identify the mutual influence of the individual elements, first of all, it's necessary to structure the integrated systems for distribution opportunities and the degree of their interpenetration. Significantly is applied an extrapolation method, inasmuch, firstly, it allows to extrapolate the most general knowledge of one scientific industry to interconnected one. And secondly it will give an opportunity to use the experience of economically developed countries (on an example of Slovakia) for positive changes in the developing countries' economic systems. Helpfully would be to use the statistical method to confirm positive or negative consequences for the economic system as a result of the application of certain legal mechanisms. Finally, it is advisable to use the method of comparative law in light of the fact that success or failure in the regulation of economic relations is determined by different approaches and legal institutions. In connection with this it is necessary to reveal their specificity and the ability to use them in other economic-legal system.

3 RESULTS

Economic theory performs methodological, practical, cognitive, predictive, educational and ideological functions. Also highlight the ideological function, however, we must remember that excessive indoctrination may obstruct the establishment of the objective laws of economic development. The theoretical model of the economic system is so important as the mutual exchange of the latest developments enriches both theory and practice. Practical problems for deeper and more comprehensive understanding is raised at the theoretical level, returning to practice after rethinking, and that "circulation" of knowledge ensures identification of General patterns and laws of development.

The idea of selecting rights through the social relations reflected, for example, in Hayek's theory of law, influenced by the common-law experience and sees law (like morals) as evolved abstract rules that have been selected through a lengthy historical process of cultural evolution, where the advantageous rules have been filtered through group selection (Hayek 1988).

So, to build a theoretical model of the interaction of law and economy go back to the original thesis that the economy is primary. It's not good or bad, it is naturally. And this is confirmed by a number of factors, including the fact that initially most of the model of relations is formed in the economy. For example, buying and selling, or rent, or newer legal structures (for instance, outstaffing), first, always appear not in the legislation. They are born in the actual relationship, when the idea of using some kind of new economic model comes to mind some people (individuals or representatives of companies). If the new model is unsuccessful, it is used a single time and dies naturally. While its effectiveness has been confirmed, it is spreading rapidly in society. And the subjects of these social relations develop rules of conduct in such situations, which are called actual (factual, real) norms. And only after a sufficient spread in society such a relationship and the actual rules governing there, the legislator receives a kind of signal that the structure of this relation should be enshrined in legislation. The task of the legislator in this process is not to create a rule from scratch, and to develop a legal model matching reality and to fix it in the law of adequate means of legal technique. That is, ideally, the right does not appear out of nothing, it grows up from the actual public (including economic) relations. Therefore, in most cases, the right slightly behind the economic relations, what can explain the presence of a certain number of gaps in the law. This means the formation of law is not only law-making, which is its final stage, it starts much earlier in society.

As defined, the right is always secondary to economic relations. But what is its role in that connection? It organizes economic relations, but its effect can be both positive and negative, depending on whether the right of the actual economic situation and the laws of economic development. If the legislation reflected important economic preconditions of the modern stage, the right promotes economic moving forward, helps to effectively exit the crisis situations.

According to Schneider and Colin (2013), one of the purposes of government is to provide the legal framework within which economic activity takes place; and societies with good institutions prosper. But when it comes to shadow work those legal institutions are bypassed: contracts can often be unenforceable; economic relationships can become marred by violence; and it can become very difficult for businesses to expand because they then come to the attention of the authorities (Schneider & Colin, 2013).

In case if the economic conditionality of the right is low, legal rules are created in favor of someone's narrow interests, sociological studies of the economy are not carried out at all or carried out with violation of the basic methodology, and then the right only inhibits economic relations, deforms them. This is due to the fact that the persons of law at the level of their legal consciousness perceive that the specific legislative decisions violate their rights. And instead of directing their efforts towards achievement of economic effect from activity, they try to avoid unfair from their point of view legal requirements.

This leads to several negative consequences: efficiency of the economic system is reduced which leads to slower growth of the economy, the unwanted realignment of the tax system, the increase in the number of economic offences, the decrease in the level of legal order in society, the negative deformation of citizens' legal consciousness, and finally reducing the overall prestige of law.

All the described situation in full measure characterizes Ukraine as a country with a very low level of economic conditionality of rights. For an example, in Ukraine practically are not conducted sociological researches in Economics, not carried out calculations of economic feasibility, for example, the tax burden on enterprises of various sectors. As a result, small businesses are at a disadvantage and after raising taxes and unified social contribution for in 2016, in late 2016 – early 2017 in our state were closed 330 thousand private entrepreneurs (http://biz.censor.net.ua/news). Also, Ukraine has extremely illogical system of charging and declaring VAT, so combined with the selective refund of this tax it is not surprising that VAT in Ukraine has become the subject of corruption.

In contrast, Slovakia has a fairly transparent system of taxation and commensurate with the type of economic activity tax rates, which contributes to its economic growth (Čulková et al. 2015).

It is also no secret that one of the foundations of the stability of the economic system is the flow of investment. It is the economic component. But from a legal point of view, the investors in Ukraine are not protected so, additionally, very little is guaranteed private property, what makes the risk of losing your investment is very high. This is a clear reason of absolute nonsystematic investment flows to Ukraine from abroad. According to the Ministry of Finance of Ukraine there isn't any systematic character of the income of foreign direct investment in 2002–2016. In author's words, it is most similar to "tropical fever" (http://112.ua/ekonomika), because the data lack any pattern, are characterized by spontaneous "jumps" or "drops" from year to year. And the indicator of foreign direct investment in 9 months of 2016 is grown up only from Russia (http://index.minfin.com.ua). Such evidence suggests the impossibility of prediction in the investment field that deprives of sense and forecasting the overall rate of economic growth.

Another important aspect of economic relations is the price formation. But the government, through legal rules may participate in this process. Thus, prices in 15 categories of goods had been regulated by government since 1996 in Ukraine. They are flour, bread and bakery products, pasta, cereals, rice, some types of meat, boiled sausages, milk, butter and sunflower, cheese, eggs, sugar and vegetables. This does not mean that the government has set prices for them, but it was possible to control the maximum size of the margin. In 2016, the government as an experiment abandoned the legal regulation of the prices of these foods. But it is worth noting that in conditions of rapid inflation and impoverishment of the Ukrainian people, the abolition of state regulation should be accompanied by targeted support to vulnerable categories of the population. However, no targeted assistance programs were not suggested. At the present stage, the Government of Ukraine plans to completely abandon its involvement in pricing (http://newsrbk.ru/news), not taking into account that a significant part of the population is below the poverty line. The most people will not able to buy even basic foodstuffs in consequence of application of the aforementioned measures from the government. So, according to the latest figures, one of the products which has become more expensive during this time, was bread. That gives the opportunity to ascertain the absolute inefficiency of legal regulation of the economic sphere in Ukraine, as the relationship requiring in fact greater economic freedom is regulated excessively. Conversely, economic problems requiring legal regulation, in the law is not fixed or fixed not in accordance with economic prerequisite.

Turning to the instrumental nature of the right it would be logical creation of it given the economic condition of the legal norms in full accordance with the formation law which is described above. In addition, regulation should be applied legal means corresponding the actual state of economic relations.

According to Trebilcock (1997), with respect to positive economic analysis of legal issues, the analyst tends to ask the following kind of question: if this (legal) policy is adopted, what predictions can we make as to the likely economic impacts, allocative (the pattern of economic activities) and distributive (winners and losers), of the policy, given the ways in which people are likely to respond to the particular incentives or disincentives created by the policy? And since the current political course is reflected in the legislation each political force coming to power, should initially be focused on the creation of economic incentives and disincentives. While enshrined in legislation, legal means will not cause opposition from the society, they should be perceived by most people as right and fair specifically for them. The legal norm being a model of legal behaviour exists and operates only in terms of its realization within legal relationships, and for that it is necessary for this norm to obtain a corresponding character in sense of justice for both a law user and an ordinary citizen (Permyakov 2003). Consequently, statistically high level of social conditionality of the legislation should be the basis of its effectiveness.

To achieve a positive assessment of the legal regulations by subjects of economic relations, it is necessary to ensure the smooth mechanism of their formation, starting with really true needs assessment in the regulation of economic relations (for which it is advisable to use sociological methods of data collection) and ending with the clever use of legal techniques, including the legal terms and constructs (Mikhaylina & Zlobin 2015). However, by itself, the application of sociological methods to identify the economic prerequisites for the establishment of legal norms is clearly not enough. The research should be conducted in full compliance with the methodological principles and be unbiased from a political point of view. As for Ukraine, the failure to comply with these basic provisions leads to the unreliability of sociological research in the field of economy, the result of which always corresponds to the interests of the state or large companies. But most often such research is generally not carried out replacing exclusively by the statistical analysis of economic data that demonstrates the consequences of certain actions of the government, but did not accurately reflect social causes.

It should be mentioned, that the legal norm, though perfectly prescribed in terms of the legal technique and being socially conditioned still may not be deprived of the certain implementation problems. It is true that a corruption constituent in any case dwindles down to nothing all the efforts of a legislator. Hasty organizational mechanisms, bureaucratic in taking administrative decisions, the permissive system which is rather complicated in Ukraine, absence of the really operating system of defending the right for ownership do not assist to the development of positive forms of sense of justice, and as a result it is not possible to observe formation of the socially oriented market economy. Above-stated facts give the ability to speak with confidence about the formation of law as a holistic process, where a wrong assessment or use of any element leads to the ineffectiveness of legal regulation.

4 CONCLUSIONS

The strategic objective of any modern democratic state should be to ensure the stability of its economic system, which can be achieved by itself economical or sociological, political, institutional, legal means. The importance of the latter determined by its formalization and the provision of state coercion means.

Therefore, in the process of creation can be nothing insignificant, every detail plays an important role. It is necessary to distinguish the category "formation of law" and "law-making" ("legislating"). The first of them refers to all the stages of the emergence of legal norms, considering their emergence in society, including in economic relations. Meanwhile law-making represents the final stage of the formation of law, its official recognition and approval of state authorities.

The formation of law is constantly and continuously simply because social life and social relations also are continuous, and one of the most dynamic are economic relations as the fundamentals for the development of the social system. Specifically, in this plane should be sought economic conditionality of the legal norms. Legal regulation may have a high degree of efficiency only in the case when it comes from the actual rules of conduct, which in one form or another are formed in society before institutionalizing them. This does not mean that they exist in structuralizing and formalized form, but in essence they are clear, reasonable, and fairly common and recognized by many members of society.

Economic, social conditionality of rights may lie not only in the actual relationship and norms, typical is its search through the interests, desires, and emotions of the subjects of law, their understanding of legal norms, the attitude towards them, that is, through the sense of justice. Thus, if in the bulk the attitude to particular legal requirements is negative (as in the case of tax law in Ukraine), so they are characterized by their low social conditioning and it likely will affect the effectiveness of their actions.

Extrapolation of the positive experience of foreign developed countries on the economic and legal system of developing countries (in particular Ukraine) should be based on similarities of the historical, cultural, economic way of their formation, or be accompanied by comprehensive research of the terms and methods of achievement of positive result, that unfortunately is ignored by the government of Ukraine, therefore reforms are spontaneous and situational. Thus, it is necessary to specify that the economic environment, the actual conditions of doing business are no less important than the applicable legal mechanisms, what is especially important to take into account of developing countries.

REFERENCES

Becker, G. 1968. Crime and Punishment: An Economic Approach. *Journal of Political Economy,* 76(2): 169–217. [on-line] www.jstor.org/stable/1830482

Bentham, J. 1970. *An introduction to the principles of moral and legislations.* London.

Calabresi, G. & Melamed A.D. 1972. Property Rules, Liability Rules, and Inalienability: One View of the Cathedral. Faculty *Scholarship Series. Paper 1983.* [on-line] http://digitalcommons.law.yale.edu/fss_papers/1983

Coase, R. 1960. The Problem of Social Cost. The Problem of Social Cost. *Journal of Law and Economics*, 3: 1–44. [on-line] www.jstor.org/stable/724810

Čulková, K. 2015. Development of Risk Payment Index in Slovakia Comparing with Chosen EU Countries. *Polish Journal of Management Studies,* 12(1): 37–47.

Food prices: to let go forever or control completely? 2017. [on-line] http://newsrbk.ru/news/4181550-c%B3ni-na-produkti-v%B3dpustiti-nazavzhdi-chi-kontrolyuvati-povn%B3 styu.html

Foreign direct investment in Ukraine in 9 months of 2016, grew by 6.2%. 2016. [on-line] http://112.ua/ekonomika/obemy-pryamyh-inostrannyh-investiciy-v-ukrainu-za-9-mesyacev-2016-goda-vyrosli-na-62-gosstat–352909.html

Foreign direct investment Ukraine from 2002 to January 2017. 2017. [on-line] http://index.minfin.com.ua/index/fdi/

From the end of 2016, 20 February 2017, their activities ceased 330 thousand private entrepreneurs. 2017. [on-line] http://biz.censor.net.ua/news/3021279/s_kontsa_proshlogogoda_zakrylis_330_tysyach_fizlit-spredprinimateleyi_minfin

Hayek, F. 1988. *The Fatal Conceit. The Errors of Socialism, Chicago.* University of Chicago Press.

Kaplow, L. & Shavell, S. 1999. *Economic Analysis of Law.* [on-line] www.law.harvard.edu/programs/olin_center/papers/pdf/251

Luts, L.A. 2005. *European Intergovernmental Legal Systems: General Description.* Extended abstract of candidate's thesis. Kyiv.

Marx, K. 1867. Das Kapital. Kritik der politischen Ekonomie. Hamburg.

Mikhaylina, T. & Zlobin, I. 2015. The formation of public awareness as the key to sustainable economic development. Economics of sustainable development: theoretical approaches and practical recommendations. In *Proceedings International scientific and practical conference,* Košice, Slovak Republic, 13–16 September 2015.

Permyakov, J.E. 2003. *Grounds of Law.* Samara.

Posner, R.A. 2014. *Economic Analysis of Law.* Wolters Kluwer Law & Business.

Posner, R.A. 2016. *Divergent Paths: The Academy and the Judiciary.* Harvard University Press.

Schneider, F. & Colin W.C. 2013. *The Shadow Economy. London: The Institute of Economic Affairs.* [on-line] https://iea.org.uk/wp-content/uploads/2016/07/IEA%20Shadow%20Economy%20web%20 rev%20 7.6.13.pdf

Trebilcock, M.J. 1997. An Introduction to Law and Economics. *Monash University Law Review*, 23(1): *123–158.*

New Trends in Process Control and Production Management – Štofová & Szaryszová (Eds)
© 2018 Taylor & Francis Group, London, ISBN 978-1-138-05885-9

Identification of start-ups through financial indicators: A case of the ICT branches

S. Mildeová
University of Finance and Administration, Prague, Czech Republic

ABSTRACT: The aim of the paper is to provide new research knowledge of the issue of start-ups; to search for related key entities in the process of starting and functioning of start-ups; and in particular, to explore through financial indicators the metrics for defining start-ups in information and communication technologies and systems. The research questions are tested through research activity, evaluation of the situation in the field, and through an analysis of the "hard data" of organisations in the ICT field acquired from the Albertina database for business and marketing. The subject of investigation consists of small enterprises. There is some evidence that an examination of small enterprises requires systems approach and this requirement is even more significant in the case of start-ups. The results of the paper should in part contribute towards the development of a theory of corporate science and a theory of applied informatics.

1 INTRODUCTION

Small enterprises, i.e. small organization is a legal or natural person who employs at least one employee and no more than 25 employees, are an important part of the Czech economy (Breckova 2015). Similarly small and medium-sized enterprises play a key role in shaping the Slovak economy, as mentioned by Stricik & Mehes (2012). They are a source of economic growth and they play an important role in employment. In this context, start-ups are a recent phenomenon, typically newly established, rapidly evolving and changing young enterprises with innovative potential. Start-ups are often linked to the use of modern information and communication technologies and systems and their connection with business. A separate category is the emergence of start-ups directly in the fields of information and communications technologies and systems.

The goal of the paper is to provide new research knowledge of the issue of start-ups in information and communication technologies and systems. The research questions that the paper will seek answer to are the following:

- What are the metrics for defining a start-up in information and communication technologies and systems?
- What are the global innovation trends in information and communication technologies and systems?
- It is possible, on the basis of accounting data, to infer a start-up in information and communication technologies and systems?
- How to search patterns of behaviour for sustainability of start-ups (small enterprises)?

The research questions are tested through research activity, evaluation of the situation in the field, and through an analysis of the "hard data" (including financial statements) of organisations, including companies in the Czech Republic in the ICT field acquired from the Albertina database for business and marketing. The subject of investigation consists of small enterprises in the fields of information and communication technologies and systems chosen according to the CZ-NACE classification of economic activities. The analysis is conducted in relation to the life cycle of an enterprise. The author is based on (Fiala & Hedija 2015) from the assumption that smaller enterprises in the Czech Republic grow faster than larger enterprises. For an illustrative case study, Software publishing was chosen as a typical area within ICT.

2 MATERIALS AND METHODS

2.1 *Start-up*

The essence of knowledge economics is a permanent innovation process, say Mihola et al. (2015). Innovations are an important part of business and are often also a key factor that determines the company's success in the market. The term innovation is closely related to the concept of start-up, that is, a company at the beginning of its business journey with a new innovative idea. Start-ups are enjoying interest and support. The Start-up Map, which works at the StartupJobs server, currently notes 1320 start-ups in the Czech Republic. Seznam.cz is also building a database of Czech start-ups in cooperation with the StartupYard accelerator; the data will be gathered through its service Firmy.cz.

In the Czech Republic, small business and enterprise innovation policy are institutionally established within the Ministry of Industry and Trade. The stimulation of innovation was significantly supported through the European Regional Development Fund co-financed by the Operational Program Enterprise and Innovation 2007–2013 (OPEI). The SME support concept for 2014–2020 also sets specific measures aimed at effective functioning and overall development of small and medium-sized enterprises with an emphasis on the creation and diffusion of innovation.

As has been said, start-ups are a recent phenomenon, but they are also among high-risk businesses. However, it should be added that while any failure of the startup business plan may have a fatal impact on the stakeholders, i.e. the founders and the investors, unlike large organizations, it will not have fatal consequences for the whole economy or society, associated with the number of jobs and other consequences, due to the decreased production volume.

2.2 *Start-up in ICT fields*

Industry 4.0 is the current trend of industry development that is causally linked with technologies and information: automation and data exchange, cyber-physical systems, the Internet of things, cloud computing, service oriented architecture, etc. (Lasi et al. 2014).

Informatics is a dynamic field that pervades all domains of our activities, not only those specifically economic. However, due to economic reasons, information and communication technologies are experiencing an unprecedented boom in the form of start-ups, meaning newly established or emerging economic organisations with innovative potential, often with a technological focus, which are rapidly evolving and changing. The economic reasons include low barriers of entry in the industry and the easy distribution of the start-up services (products) via the internet, especially through wireless connection, as stated by Ortin et al. (2016) or Hulsink & Elfring (2003). All of this, assuming success of the start-up and its services, leads to low unit costs of the product (service). Globally, we can talk about the transition to the knowledge economy (Almakenzi et al. 2015), which further underlines the importance of ICT start-ups in the economy.

The above-mentioned economic reasons lead to high competitiveness on one hand and the existence of global and international ICT companies on the other hand. Demands on services (start-up products) are directly connected to the innovative trends in ICT.

A study by Grilli (2013) notes the importance of human capital in the success of start-ups. Bilkova (2013) shows a significant wage differentiation for individual sectors in the Czech-Slovak economy. From this point of view, requirements for successful startups in the field of human resources are neccesery to consider—in particular, the length of experience and skills as programming, knowledge of specific skills ICT (Maryska & Doucek 2012). A specific issue is the matter of unemployment and income inequality among ICT professionals (Potuzakova & Mildeova 2015).

Factors leading to the internationalization of services (products) of Czech start-ups are also important in the studied area in the Czech Republic, consistently with studies by Cannone and Ughetto 2014, and analysis of key factors of success (Buchegger 2014) and performance (Botha et al. 2015).

It is necessary to also discuss the role of investors, start-up accelerators and incubators in the system framework or generally speaking innovation methods and their financing in small and medium enterprises. It means identifying the main obstacles to the development and the main positive factors of foundation and development of start-ups created by the state's economic policy in its complexity (Hajduova et al. 2015). The international context and the processes (Bartuskova & Nemcova 2014) must be, of course, also considered (e.g. Europe 2020, cohesion policy sub-projects). Dlaskova & Havlicek (2013) states, that companies that operate in the branches of new technologies can expect the participation of foreign investors.

Last but not least an important role has the studies dealing with the definition of startups: "developed a system for improving the completeness and accuracy of the start-ups identification process. The application of our method to three sample counties in North Carolina illustrates how a consistent, systematic approach to tracking business start-ups can significantly improve the reliability of data." (Luger & Koo 2005, p. 24).

2.3 *Methods of investigation*

To ensure a true and fair view of the property and financial situation of the company, it is important to identify individual items of assets, liabilities, revenues and costs correctly and to correct to value them. Financial situation is determined by analyzing the valuation of the assets of the entity at a certain date and the financial situation is described by financial analysis of indicators from the financial statements, as mentioned by Paksiova et al. (2015).

Financial statement analysis is used to identify and evaluate the financial situation of the analyzed enterprises. It is a monitoring of financial performance, an assessment of viability, stability and profitability of a business. Financial statement analysis is a method that uses data that capture business activities predominantly in monetary units. Financial statement analysis is based on outputs of the accounting information system. Its importance for small (and medium-sized) businesses is mentioned by Mares (2016).

The most typical methods of financial statement analysis are the Analysis of Extensive Indicators (status, absolute) and within that Horizontal Analysis (Analysis of Trends) and Vertical (Percentage) Analysis; furthermore, Analysis of Funds and Analysis of Ratio Indicators. The ratio indicators include Profitability Returns, which quantify profitability, Solvency indicators that show how the company finances its assets and whether it is able to repay its long-term liabilities in the long-term, Liquidity indicators that demonstrate the company's ability/inability to meet its commitments in the course of the year, and Stability indicators which capture the ability to remain in business in the long run. Ivanickova et al. (2016) draw attention to the need not only to monitor current financial health and discuss the forecasting of financial health of a company by different models of prediction. The examination of the relationship between financial performance indicators and other indicators in quantitative analysis has also, by Mura et al. (2015), strong ability to communicate.

The Albertina database is used to map enterprises operating in the area investigated and their financial statement analysis. Information from the Albertina database records economic data for individual businesses, based on annual accounting reports (Bisnode/Albertina 2017). The Albertina database contains more than 2.8 million records of a total of 1543 organizations. When the data is correctly filled in, it amounts to about 250 data points per year per enterprise. This information serves to evaluate the economic success of enterprises.

The analysis is conducted in relation to the life cycle of an enterprise according to the Miller & Friesen (1984) general lifecycle model. The model is based on the fact that every company is going through some development during its existence. The lifecycle model describes the life of a business as a sequence of stages that (usually) consist of emergence, growth, stabilization, crisis, and downfall (Fig. 1). Fractal analogy is applied to describe the lifecycle model of a general successful company in terms of the most significant time-dependent phenomena observed on the researched sample of Czech and Slovak enterprises (Pawliczek 2015).

Depending on the type of data acquired, the methods used for the data processing will be primarily statistical methods and analyses of time series trends from 2003 till 2015.

Systems approach is applied to the study of start-ups in information and communication technologies in a holistic way.

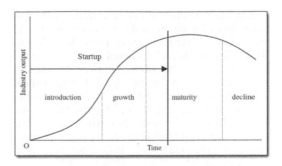

Figure 1. Lifecycle model (source: author according to value based management.net. 2016).

3 IDENTIFICATION OF START-UPS THROUGH FINANCIAL INDICATORS

3.1 Results

Let's now focus on the metrics for identifying a start-up. The definition of a start-up shows that the organization must address at least one innovation trend. The innovative trends in ICT are social informatics, 3D printing, decentralized services, NFC chips, clouds, the Internet of things, virtual reality, big data, data security and privacy, and, with the bitcoin growth, security and encryption of virtual currency payments (Bhandari & Regina 2014), (Kamilaris & Pitsillides 2016), Lansky (2016) or (Smutny 2016).

The definition of a start-up also states that it is an emerging organization, that is, an enterprise in the first developmental phases of a company's existence. These phases include the preparatory seed phase, the start of the business, its growth, and, in part, stabilization. From the perspective of a start-up, the primary phase is the stage of growth with an increase in production volume and penetration into new markets. However, once an enterprise has achieved a truly stable result, i.e. it is in the state of achieving optimal size considering the market opportunities and regularly generates a profit, the start-up label is no longer appropriate (Fig. 1).

The choice of financial indicators for characterization of corporate performance depends on the purpose of the financial analysis. From the point of view of the start-ups, the appropriate liquidity indicator is the Interval Rate which by comparing assets with current expenses indicates for how many days the business has enough liquid assets without acquiring additional cash. Alternative to liquidity, a more synthetic indicator is profitability as the ratio of the profit and the capital invested. Here it is also worth mentioning the frequently used profit category modifications such as EBIT (Earnings before Interests and Taxes), EBT (Earnings before tax), and EAT (Earnings after tax), known as the profit for the year. In the case of start-ups, the Return on

Equity (ROE) indicator can be considered the primary indicator for profitability since it shows the profitability of the investors' deposits. According to the DuPont model, ROE is heavily dependent on the return on total capital, ROA (Return on Assets), and the interest rate on foreign capital. Foreign capital is also reflected in the indicator of financial stability Total indebtedness. The indicator testifies to the level of financing of assets by foreign sources. The above-mentioned DuPont model shows that a more intensive representation of foreign sources leads to higher profitability of equity (Penman 2010).

Let's now go to our own analysis of the data from the Albertina database. The subject of investigation consists of small enterprises. The basic set of investigated enterprises is defined by:

- territorial aspects – enterprises based in the Czech Republic,
- sectoral aspects – Enterprises belonging to ICT fields according to the CZ-NACE classification,
- size – enterprises with a staff of 1 to 24,
- legal form – without distinction,
- time – from 2003 till 2015.

On the basis of a data analysis from the Albertina database, the fields for investigation have been chosen according to the CZ-NACE classification of economic activities according to the Czech Statistical Office (Table 1).

When selecting enterprises from the Albertina database, it was necessary to take into account the extent and quality of the economic information available in this database. (Some enterprise information in the database is incomplete to such an extent that it was not usable for subsequent analysis). The selection criterion was not to achieve the maximum representativeness of the sample. The aim was to identify enterprises whose available data would have the capability to relate to the purpose of the survey.

Based on the availability, complexity and quality of the accounting information in the Albertina database, a subsequent data analysis was performed. For an illustrative case study, Software publishing as a typical area within ICT was chosen from ICT fields shown in Table 1. Organizations with 1 to 24 employees are analyzed for the purposes of our research. In the context of the Software publishing (and Computer programming activities) case, there are 51 organizations. The records obtained are subjected to a more detailed analysis in order to get an answer to the main research question. The analysis of these organizations according to the field of business shows that there are very few companies which as the subject of activity mention only Software publishing and/or Computer programming activities. Many companies have a wider range of activities, for example, apart from Software publishing (and Computer programming) activities, there is also Retail or Wholesale.

Table 1. Fields according to the CZ-NACE classification Fields.

| Wholesale of computers, computer peripheral equipment and software |
| Retail sale of computers, peripheral units and software in specialised stores |
| Software publishing |
| Publishing of computer games |
| Other software publishing |
| Computer programming, consultancy and related activities |
| Computer programming activities |
| Computer consultancy activities |
| Computer facilities management activities |
| Other information technology and computer service activities |
| Data processing, hosting and related activities; web portals |
| Data processing, hosting and related activities |
| Web portals |

Source: List of CZ-NACE.

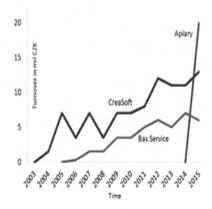

Figure 2. Model 1 and 2.

Predetermination of the Albertina database structure led to the fact that the starting point of our financial analysis is primarily the time series of turnover. In this way, four typical behavioral patterns (model situations) are found among small businesses in the field of Software publishing and Computer programming activities. The first, most optimal model, is the unicorn start-up, represented here by the company Apiary and the business rocket launch. It is a technology start-up, providing tools for interfaces between Internet programs. In the second model of growth behavior, when it is no longer a unicorn, the organization grows slowly and this growth is often accompanied by downward fluctuations. This model can be characterized as the most frequented in the area of Software publishing and Computer programming activities. Bas Service, a web design company, and CreaSoft,

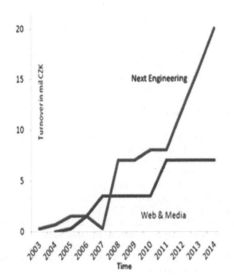

Figure 3. Model 3.

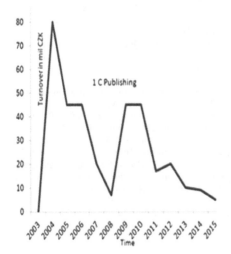

Figure 4. Model 4.

specializing in the financial market, are such cases (Fig. 2). Jumping turnover development, illustrated by the firm Web & Media and Next Engineering, is suggested as the third model of behavior (Fig. 3). The fourth model documents that even in such an "innovative" field as ICT, the initial startup behavior can also lead to an immediate decline of the turnover, as seen in 1C Publishing (Fig. 4). Noteworthy is the wavelike pattern of development.

3.2 Discussion

An examination of small (and medium-sized) enterprises requires system approach and this requirement is even more significant with the SMEs start-up because of the high degree of complexity of the problem and the need to take into account the dynamics of their development. The results of the article can be considered as the basis for further analytical work and for the topic of further complex research.

4 CONCLUSIONS

Start-ups are one of the most-discussed topics in today's modern economy and newly established starting enterprises with innovation potential are the hope for tomorrow. In Czech Republic the reality is of course that 50% of start-ups close down within 5 years, and only a fraction of them survive over the long term beyond the initial boom. Start-ups in information and communication technologies and systems are no exception. This paper focused on these enterprises.

The author tried to search the metrics for defining start-ups in information and communication technologies and systems with the emphasis on financial indicators. First, innovation trends were defined. Through an illustrative case study four typical behavioral patterns of recently formed organizations in the field of Software publishing and Computer programming activities were introduced. These models have confirmed that within the lifecycle model it makes sense to examine typical behaviors and occurrence of similar patterns.

The importance of such research is due to both the share of revenues from this field in the state budget, and primarily the cross-sectional and key role of information and communication technologies and systems in growth in practically all other economic and non-economic sectors. Due to their focus on a certain innovation, startups have the significant advantage of being built from the beginning with a specific purpose, and even in the need for a change in strategy, due to their smaller size, they are more flexible. The results of the paper should in part contribute towards the development of a theory of corporate science and a theory of applied informatics.

These results should provide recommendations in the sense of the initialisation of start-ups and the emergence of unicors in information and communication technologies and systems. For the analysis of small enterprises in the fields of information and communications technologies and systems it was drawn primarily from historical financial data from the accounting closures. The ex post analysis approach is particularly suited for already existing enterprises. From the very beginning of the start-up, the fact that it is a start-up organization is, of course, rather an ex ante analysis. Here we can see the ambivalence of our research.

REFERENCES

Almakenzi, S. et al. 2015. A Survivability model for Saudi ICT startups. *International Journal of Computer Science & Information Technology*, 7(2): 145–157.

Bartuskova, H. & Nemcova, I. 2015. Convergence and Integration of the Central Europe. In *Proceedings of the 12th International Scientific Conference: Economic Policy in the European Union Member Countries* (pp. 45–50). Karviná: Silesian University in Opava.

Bhandari, S. & Regina, B. 2014. 3D Printing and Its Applications. *International Journal of Computer Science and* Information, 2(2): 378–380.

Bilkova, D. 2013. Modeling of wage distribution in recent years in the Czech Republic using l-moments and the prediction of wage distribution by industry. *E a M: Ekonomie a Management*, 16(4): 42–54.

Bisnode A. 2017. Retrieved from http://www.albertina.cz/.

Botha, M. et al.. 2015. An integrated entrepreneurial performance model focusing on the importance and proficiency of competencies for start-up and established SMEs. *South African Journal of Business Management*, 46(3): 55–65.

Breckova, P. et al. 2015. Prosperity and Stability of SME Segment by Industry in the Czech Republic. In Loster, T. & Pavelka, T. (eds.), *The 9th International Days of Statistics and Economics* (220–230), Prague: Melandrium.

Buchegger, T. 2014. *How To Start Up A Software Business Within A Cloud Computing Environment: An Evaluation Of Aspects From A Business Development Perspective.* e-book, Anchor: Hamburg.

Cannone, G. & Ughetto, E. 2014. Born globals: A cross-country survey on high-tech start-ups. *International Business Review*, 23(1): 272–283.

Dlaskova, G. & Havlicek, K. 2013. Approach to Valuation of Assets in Czech Accounting, Comparison to IFRS and Impacts on Controlling Process of SMEs. In Doucek, P., Chroust, G. & Oskrdal, V. (eds.). *Proceedings of the 21sh Interdisciplinary Information Management Talks (IDIMT-2013)* (pp. 147–156). Linz: Trauner Verlag universität.

Fiala R. & Hedija, V. 2015. The Relationship Between Firm Size and Firm Growth: The Case of the Czech Republic. *Acta Universitatis Agriculturae et Silviculturae Mendelianae Brunensis*, 63(5): 1639–1644.

Hajduova, Z. et al. 2015. Case study in the field of innovation in selected companies in Slovak Republic. *Analele Stiintifice ale Universitatii Al I Cuza din Iasi—Sectiunea Stiinte Economice* [online], 62(1): 103–119. DOI: 10.1515/aicue-2015–0008.

Hulsink, W. & Elfring, T. 2003. Network effects on Entrepreneurial Processes: Start-ups in the Dutch ICT Industry 1990–2000. *ERIM Report Series Research in Management.* Retrieved from https://repub.eur.nl/pub/976/.

Grilli, L. 2013. *Experience and high-tech start-up survival during an industry crisis.* Retrieved from http://cei.ier.hit-u.ac.jp/Japanese/database/documents/WP2008-22000.pdf.

Ivanickova, M. et al. 2016. Assessment of companies' financial health: Comparison of the selected prediction models. *Actual Problems of Economics*, 180(6): 383–391.

Kamilaris, A. & Pitsillides, A. 2016. Computing and the Internet of Things: A Survey. *IEEE Internet of Things Journal*, 3(6): 885–898.

Lansky, J. 2016. Analysis of Cryptocurrencies Price Development. *Acta Informatica Pragensia*, 5(2): 118–137. DOI: 10.18267/j.aip.89.

Lasi, H. et. al. 2014. Industry 4.0. *Business & Information Systems Engineering*, 4(6): 239–242.

List of CZ-NACE. 2017. Retrieved from http://wwwinfo.mfcr.cz/ares/nace/ares_nace.html.en

Luger, M.I. & Koo, J. 2005. Defining and Tracking Business Start-Ups. *Small Business Economics: An Entrepreneurship Journal*, 24(1): 17–28. doi:10.1007/s11187-005-8598-1.

Mares, D. 2016. Accounting and Controlling Business Management System. *ACTA VŠFS—Economic Studies and Analyses*, 10(2): 126–141.

Maryska, M. & Doucek, P. 2012. ICT Speicalist Skills and Knowledge—Business Requirements and Education. *Journal on Efficiency and Responsibility in Education and Science*, 5(3): 157–172, DOI:10.7160/eriesj.2012.050305.

Mihola, J. et al. 2015. Is the most innovative firm in the world really innovative? *International Advances in Economic Research*, 21(1): 41–54.

Miller, D. & Friesen, P.H. 1984. A Longitudinal Study of the Corporate Life Cycle. *Management Science*, 30(10): 1161–1183.

Mura, L. et al. 2015. Quantitative financial analysis of small and medium food enterprises in a developing country. *Transformations in Business and Economics*, 14(1): 212–224.

Ortin, J., et al. 2016. Optimal configuration of a resource-on-demand 802.11 WLAN with non-zero start-up times. *Computer Communications*, 96: 99–108.

Paksiova, R. et al. 2015. Valuation of all fungible items' assets reduction and its influence on measuring companys' performance. In *Proceedings of 10th international scientific conference Financial management of firms and financial institutions* (pp. 921–928). Ostrava: VŠB—Technical university of Ostrava.

Pawliczek, A. 2015. Lifecycle of Enterprises and its Dynamics: Using Fractal Analogy Model in Empirical Study of Czech and Slovak Enterprises. *Procedia-Social and Behavioral Sciences*. 181: 331–341.

Penman, S.H. 2010. *Financial Statement Analysis and Security Valuation: International Edition.* 4. ed. New York: McGraw Hill.

Potuzakova, Z. & Mildeova, S. 2015. Analysis of causes and consequences of the youth unemployment in the EU]. *Politická ekonomie*, 63(7): 877–894. DOI:10.18267/j.polek.1043.

Smutny, Z. 2016. Social informatics as a concept: Widening the discourse. *Journal of Information Science*. 42(5): 681–710.

StartupMap. 2017. Retrieved from http://www.startupmap.cz/

Stricik, M. & Mehes, M. 2012. *Vybrané kapitoly z podnikania v malých a stredných podnikoch.* Bratislava: Vydavateľstvo EKONÓM.

Value Based management.net. 2016. Retrieved from http://www.valuebasedmanagement.net/methods_product_life_cycle.html.

New Trends in Process Control and Production Management – Štofová & Szaryszová (Eds)
© 2018 Taylor & Francis Group, London, ISBN 978-1-138-05885-9

Sustainable development of region in decline with the support of tourism

L. Mixtaj, A. Csikósová, E. Weiss & R. Weiss
Faculty of Mining, Ecology, Process Control and Geotechnologies, Technical University of Košice, Košice, Slovak Republic

ABSTRACT: The contribution is focused on the inquiry into the potentials for sustainable development in region of Upper Gemer, which in the past was a mining centre of the wider region. Based on a survey of real geographical, social a demographic background features of the territory, it is tourism that appears to be the determining factor for sustainable development. The resulting proposal to the solution aiming at sustainable development in the selected area by way of systematic and long-term management of tourism while making use of the tools defined by the legislative framework of the Law on support of tourism at all levels of public administration, not omit the private sector. The final discussion is focused on comparing and identifying the importance of current weight of tourism among the rest of the existing branches of national economy as regards the area of our concern in the context of sustainable regional development.

1 INTRODUCTION

Regional development is currently is an important component of further development and stronger integration of the member states into the European Union. A crucial problem is the unequal development and the lagging behind of the peripheral regions, mainly the countries of the former Eastern block. The European Commission does have tools developed to support these regions, which, in long-term perspective, meet the requirements for drawings of funds accumulated in the financial reserves of the EU. As a rule, the regions are lagging behind not only from economic, social and technological points of view, but they also are suffering from devastated living environment as a result of inconsiderate industrial development in the era of socialism. The role of competent institutions at the appropriate level consists in continuous search for possibilities and active approach to the elimination of regional disparities in the areas mentioned above. One of possible solutions could be in targeted support to tourism in those regions, which in the past have been known for their mining activities, and today are noted as disadvantaged or directly backward when compared to the average of the European Union. The negative impacts of those activities currently represent a social burden, which in via a thoughtful and targeted support of the competent bodies can be turned into attractive areas of primary offer for geo- an mountain-tourisms.

2 LITERATURE REVIEW

The theme of sustainable development is currently on the agenda of a fair number of authors both local and from abroad. Their research involves the general theoretical model of sustainable development in various areas of the society, such as economics, ecology, industry and demography. Another group of authors is dealing with sustainable development from the point of tourism and sustainable geotourism (Radwanek-Bak et al. 2015, Meakin 2011, Kiernan 2013, Amorfini et al. 2015). Sustainable tourism has been defined by the United Nations World Tourism Organization (WTO) and United Nations Environment Programme as "Tourism that takes full account of its current and future economic, social and environmental impacts, addressing the needs of visitors, the industry, the environment and host communities" (2005). Burlando et al. (2009) are pointing out the importance of natural features and cultural heritage of the investigated area, which is considered by them as a basis for developing plans and strategies in support of the sustainable development of the area. Organizations such as the UNESCO, the European network of geo-parks and others provide tools, according to which concrete measures and activities at lower levels of state administration or destination management are taken.

Although modern geotourism, as a form of sustainable geoheritage tourism, was only recognized as such in the 1990s, its roots lie in the seventeenth century and the Grand Tour with its domestic equivalents. The development of the Grand Tour and the landscape aesthetic movements, the various influential institutions, key personalities and locations are considered insofar as they provide an overview of the background to historical geotourism (Hose 2016).

As per Citiroglu et al. (2016) a territory, which comprises a lot of geologically interesting objects such as caves, geological disorders, discovery sites of fossils, as well as old mining assets, are of not only scientific but educational and economic importance. These element are important from the point of sustainable development, to which tourism can be the driving force.

Olafsdóttir and Dowling (2014) state that geotourism is partially an answer to the need for minimizing the negative effects of mass-scale tourism in geologically and geographically interesting destinations, which, at the same time, appear to be a catalitic agent for the sustainable development of rural areas.

Authors Walliss and Kok (2012) explore how the application of more experiential landscape driven interpretative strategies might shape the development of two Australian mining sites which offer potentials for Geotourism: the operational mining landscape of Queenstown, Tasmania and the coal mining sites of Yallourn North in Victoria's La Trobe valley. Through the case studies, authors highlight how the championing of the experience of landscape itself as a mode of interpretation and education (as distinct from textual interpretations) offer valuable techniques for engaging with biotic and abiotic characteristics, aesthetic qualities, ecological processes, cultural histories and sustainable futures.

Post-industrial tourism presents an opportunity for rescuing industrial heritage and protecting the natural environment. Kruczek Z. and Kruczek M. (2016) concentrate on the possibilities of using oil routes for post-industrial tourism as a means to revitalise post-industrial sites and ensure sustainable development for the region. Examples of educational and tourist reclamation and revitalization of the post-mining terrains carried out authors Poros and Sobczyk (2014).

2 METHODOLOGY OF GEMER REGION RESEARCH

Gemer is the name of a historical area, geographically bounded by the headspring the river Hron in the north, through the Oždianske Hills in the west, Cerová Uplands in the south and, finally, the Soroška Pass in the east. Along with the territory of the Malohont country, they both constitute the historical County of Gemer-Malohont, one of the oldest inhabited localities in the present Slovakia. From administrative point of view, the Gemer region is spreader over the territories of two Self-administration counties, that of Banská Bystrica (districts of Revúca and Rimavská Sobota) and that of Košice (district of Rožňava). From historical point of view, its neighbors with the Regions of Spiš and Upper Hron in the north, Novohrad from the west, Abov from the east and the national borders of Hungary (www.gemer.sk).

2.1 Geo-morphology of the territory

The area of the Gemer region is mostly hilly. From morphological point of view, it belongs to the sub-province of the Internal Western Carpathians. Within it, almost the entire surface is made up of the geo-morphological units of the Slovak Ore Mountains. Heights above sea level ranges between its highest point, Stolica peak (1476 a.s. l.), and the lowest at the flat land of the Turňa river leaving the territory of Slovakia further to the Republic of Hungary. The natural environment is fairly varied—the northern part offering high-mountain setting, whereas the river walleyes are typical for advanced level of agriculture. In view of the substantial differences in elevations, the regions are situated in three climatic areas, from warm up to cold ones.

2.2 The basic macro-economic characteristics

The Gemer region is home to three district centers, cities of Revúca, Rožňava and Rimavská Sobota and 211 residential areas of which 8 enjoy the privileges of cities. Presented in the Table 1. below are the basic data of the districts. Permanent stay is recorded for as much as 188.083 inhabitants, of which 40.237 living in the Districts of Revúca, 63.082 in that of Rožňava and 84.764 in that of Rimavská Sobota (Dzianová 2015).

In all the district centers of the region there is a decreasing tendency as for the natural increase in the number of inhabitants. In year of 2013 more people died than were born in each of the cities mentioned. More to it, people tend to leave the region for other cities, causing the number of population to decrease and grow older.

Districts of Rimavská Sobota, Revúca and Rožňava show a long-term record of highest unemployment rate within Slovakia as a whole. The word is about the districts, which relatively tightly interconnected, not only in terms of geographical location. They are firmly interwoven with a common history, culture and lots of cultural and demographic characteristics involving the first place in the rating in unfavorable social characteristics mostly in terms of rate of unemployed. In view of the current administrative classification, they belong to different regions of self-administration

Table 1. District centres.

District	Area (km²)	Number of villages	Cities included	Density of population
Revúca	730	42	3	55 inh. per km²
Rožňava	1 173	62	2	54 inh. per km²
Rimavská Sobota	1 471	107	3	58 inh. per km²

(www.upsvar.sk), see Table 2. Splitting this traditional region can represent a barrier to more efficient solution of the issue of unemployment, and the related economic and social effects. A compelling issue is the one of the structure of the local industry, which has been built in accordance with natural dispositions of the region, concentrating on mining and processing of raw material resources. The recent two decades, however, have witnessed the cutback mining, mostly due to economic inefficiency.

2.3 Turistical amenities

They are made up of facilities providing services of accommodation, food, transportation and mediation. They form an important component of potential for the development of residential tourism. Their function is to offer visitors accommodation in the selected destination by satisfying various needs and wishes depending primarily on the type of tourism they are participating in.

Illustrated in Figures 1 and 2 are the numbers of facilities offering accommodation and that of

Table 2. Unemployment in the Gemer region.

Indicator /District	Revúca	Rožňava	Rimavská Sobota
Rate of recorded unemployment	26.46%	24.09%	29.71%
Economically active inhabitants	19 028	30 562	40 693
Available number of applicants for employments	5 053	7 361	12 089
Number of the those applying for employment	6 519	8 537	13 984

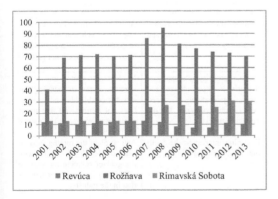

Figure 1. Number of facilities providing accommodation in the Gemer region.

the visitors of the Gemer region by districts (www. datacube.statistics.sk).

2.4 SWOT analysis regarding the development of tourism in the Gemer region

As it follows from the SWOT analysis (Fig. 3), primary offer is the strongest feature of tourism in the Gemer region.

Also suitable are its natural conditions added with rich cultural and historical heritage. The attractiveness of the region is multiplied by the fact that the area is neigh boring with the Republic of Hungary. The weakest point, on the contrary, is in its secondary offer. There is a fair availability of facilities providing food and accommodation, though, their level of services fall short of the standards

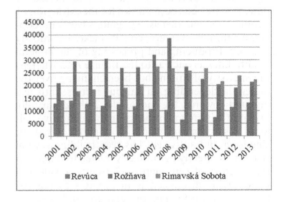

Figure 2. Number of visitors to the accommodation facilities in the Gemer region.

S – Strengths	W – Weaknesses
- Attractive primary offer for the development of various forms of tourism - Multilingual and multinational features - Geographical position of the region - Culture, technical heritage, traditions	– Weak secondary offer – Lack of managerial organization of tourism – At some places, catastrophical status of transport infrastructure unemployment
O – Opportunities	**T- Threats**
– Cross-border cooperation with Hungary – Changing preferences in the holiday-making destinations and customer behaviour of tourists – Rich heritage from the great past of mining	– Chaotic organization and uncoordinated activities within tourism – More active regions of the „competition" – Decline in culture and tradition – Lack of interest in tourism on the part of local public administration

Figure 3. SWOT analysis of the Gemer region.

required. Simultaneously, transport infrastructure is insufficient. Advertisement of the region is another disadvantage, which at the same time offers the best opportunity to make the region more visible. Long-term unemployment is another problem resulting in poverty. The region is known to be a backward one, a fact that tourists may find distracting. The greatest opportunity for the development of tourism consists in acquiring funds, mostly those of the EU.

3 CONDITIONS FOR THE DEVELOPMENT IN THE AREA UPPER GEMER

Thanks to its unique geo-physical structure rich in siderite and sulphitic dikes, the region of the Upper Gemer belongs to the oldest and best known mining areas of Europe. For quite a long period of time, this area has been known for the extraction of various precious metals, e.g. gold, silver, and copper, antimony, mercury, cobalt, nickel but mostly iron ore. Producing and processing of the mineral wealth dates as back as to the Bronze Age, whereas the intensity of mining has been changing over the millennia depending on various factors until terminated towards the end of the 20th century 20, bringing mining and processing to a definitive halt in all segments and plants of mining companies. The long and active era has seen building several structures directly or indirectly related to mining in the territory of our interest. Currently, a wide and society-wide discussion is in progress as regards the need for protection, preservation and development of these objects. Thorough mapping, evaluation and cataloguing are prerequisites of further support to be extended to the new forms of tourism, i.e. geo- and mountain-tourisms.

Currently the territory of the Gemer region belongs to the poorest ones in Slovakia, with the rate of unemployment reaching a record level of almost 30% in the three districts. On the other hand, it is this region that along with the regions of Orava and Spiš belongs to the most beautiful tourist destinations, another reason for the region to be developed.

Institutions supporting the development of the Gemer region are next:

– *Agency for the Development of Gemer*—the agency for regional development is part of the Integrated network of regional agencies for development of the IS RRA, providing services in clasuse in §14 and Law No. 539/2008 Coll. on supporting regional development. Established in 2008, it is a voluntary association of legal entities residing in the city of Hnúšťa. The responsibilities of the agency are formulated in its official documents.

– Agency for regional development Rožňava—the agency for regional development Rožňava was established on 01 March, 1996, in compliance with the Decision of the Slovak Government No. 96/95 and No. 548/95 as of a 25 July, 1995. Its purpose is to provide support to the economic development of the districts of Rožňava and Revúca. The agency for the regional development was formed as an association of legal entities from the public, financial and private sectors in compliance with the Civil Code. Defined in its articles are the areas of activities in support of the regional development.

4 RESULTS

At this time, the development of the region depends on global trends resulting from both the threats and opportunities as well. One of them is tourism, for sure. Recently Slovakia has recorded a rapid growth in the performances bound to tourism also having made a sensitive impact upon the macro-economic indicators of the whole country.

Another important factor is the switch in the holiday-making preferences of the bulk of inhabitants in the EU. The changes can be related to political and security concerns as regards the countries formerly belonging to their favourite destinations for summer holidays. The worries can be under-

Barriers to development	Suggestions to solutions
In-transit region for most of the visitors.	Developing adequate products of tourisms for the appropriate type of tourism.
Uncoordinated activities and relations between the public and the private sectors, lack of interest on the part of local administration.	Fast-lane for the establishing of managerial organization of tourism within the existing legal framework for tourism.
Separatism and unhealthy competitions among the private competitors as providers of secondary offers-services.	Coordination of management or marketing activities among organizations engaged in tourism.
Desolate state as for the majority of objects used in mining and metallurgy in the past.	Accelerated efforts in reconstruction of the technical sights, plans of gradual establishing open-air museum and educational paths of mining.
Unclear property relations particularly in the forest-covered areas, association of forest owners, church-owned forests, municipal forests.	Involvement owners and administrators of the surrounding forests for establishing educational tracks of mining and forestry or open-air museums on the related territories.
Declining interest in the traditions and legacy of mining.	Activation of the mining associations of the Gemer region with the goal of reviving the interest of children and younger generation in the industry.

Figure 4. Comparing tourism and other branches of the national economy.

stood as unique chances for Slovakia disposing of a wealth of primary offer and a strong potential of its regions becoming favourite destinations for both summer and winter holidays. The Gemer Region disposes of an extraordinary rich primary offer made up of its attractiveness for geo- and mountain tourisms.

Identifying the barriers and potential solutions in the area of geo- and mountain tourisms in the Gemer region shows Figure 4.

5 CONCLUSION

Compared to other segments of the national local economy, local tourism in the region appears to be rather underdeveloped. Evidently, the region does dispose of unique primary offer both of natural and anthropogenic, by origin. The question where it got stuck is to be answered if to handle the acute backlog in the development of tourism compared to other regions. A unique negative feature typical for smaller units of administration is marked with the lack in the availability of relevant of sources as for primary and secondary offer in tourism, which are needed for real planning of further strategic development with the involvement of all the interest groups. It can be stated that the development of geo- and mountain tourism in the Gemer Region is hindered by organizational and managerial deficiencies that can be eliminated by way of changing the attitudes and preferences of the local interest groups applying marketing- and management-oriented principles to the development of tourist destinations.

ACKNOWLEDGMENT

Contribution is partial result of project solving VEGA No 1/0310/16 and VEGA No 1/0316/16.

REFERENCES

Amorfini, A. et al. 2015. Enhancing the Geological Heritage of the Apuan Alps Geopark (Italy). In *Geoheritage to geoparks: Case studies from Africa and beyond, Book Series: Geoheritage Geoparks and Geotourism.*

Burlando, M. et al. 2011. From geoheritage to sustainable development: Strategies and perspectives in the Beigua Geopark (Italy). *Geoheritage*, 3(2): 63–72.

Citiroglu, H. et al. 2016. Utilizing the Geological Diversity for Sustainable Regional Development, a Case Study-Zonguldak (NW Turkey). *Geoheritage*, 1–13.

Csikósová, A. et al. 2014. *Strategic aspects of business in context of regional development.* Ostrava: VŠB-TU.

Dzianová, O. 2014. *Štatistická ročenka regiónov Slovenska 2014.* Bratislava: ŠÚSR.

Gergeľová, M. et al. 2013. A GIS based assessment of hydropower potential in Hornád basin. *Acta Montanistica Slovaca*, 18(2): 91–100.

Hose, T.A. 2016. Three centuries (1670–1970) of appreciating physical landscapes. In *Appreciating physical landscapes: Three hundred years of geotourism. Book Series:* Geological Society Special Publication.

Kiernan, K. 2013. The Nature Conservation, Geotourism and Poverty Reduction Nexus in Developing Countries: A Case Study from the Lao PDR. *Geoheritage*, 5(3): 207–225.

Kruczek, Z. & Kruczek, M. 2016. Post-Industrial Tourism as a Means to Revitalize the Environment of the Former Oil Basin in the Polish Carpathian Mountains. *Polish journal of environmental studies*, 25(2): 895–902.

Labant, S. et al. 2014. Graphical interpretation deformation analysis of stability area using of strain analysis. *Acta Montanistica Slovaca*, 19(1): 31–40.

Meakin, S. 2011. Geodiversity of the Lightning Ridge Area and Implications for Geotourism. In *Proceedings of the linnean society of New South Wales.*

Mihalčová, B. et al. 2014. Tourism in the condition of small family firms of a selected region in Slovakia. In: *17. mezinárodní kolokvium o regionálních vědách.* Brno: Masarykova univerzita.

Ólafsdóttir, R. & Dowling, R. 2014. Geotourism and Geoparks-A Tool for Geoconservation and Rural Development in Vulnerable Environments: A Case Study from Iceland. *Geoheritage*, 6(1): 71–87.

Poros, M. & Sobczyk, W. 2014. Reclamation Modes of the Post-Mining Terrains in the Checiny-Kielce Area in the Context of Its Use in an Active Geological Education. *Rocznik ochrona srodowiska*, 1(16): 386–403.

Radwanek-Bak, B. et al. 2015. New Polish Geopark proposal Wislok rivervalley—the Polish Texas. In *SGEM 2015, Proceedings International Multidisciplinary Scientific GeoConference. Albena, Bulgaria.*

Región Gemer. [on-line] www.gemer.sk.

Seňová, A. et al. 2011. Influence of economical and marketing environment to the effectiveness of investment of mining business. In *SGEM 2011, Proceedings 11th International Multidisciplinary Scientific Geoconference and EXPO, 20.–25.6.2011, Varna, Bulgaria.*

Štatistické údaje. [on-line] www.datacube.statistics.sk/ TM1 WebSK/ TM1 WebLogin.aspx.

UNEP and UNWTO 2005. Making Tourism More Sustainable—a Guide for Policy Makers, United Nations Environment Programme, Nairobi.

ÚPSVR. [on-line] www.upsvar.sk/statistiky/ nezamestnanost-mesacne.

Walliss, J. & Kok, K. 2012. New interpretative strategies for geotourism: an exploration of two Australian mining sites. *Journal of tourism and cultural change*, 12(1): 33–49.

New Trends in Process Control and Production Management – Štofová & Szaryszová (Eds)
© 2018 Taylor & Francis Group, London, ISBN 978-1-138-05885-9

The fee policy and the financing of air navigation services in the Czech & Slovak Republic

Ž. Miženková, E. Jenčová, A. Tobisová & J. Vagner
Faculty of Aeronautics, Technical University of Košice, Košice, Slovak Republic

ABSTRACT: The aim of this article is practically point to the fee policy and financing of air navigation services and the main economic aspects, operational indicators of companies providing air traffic services. The aim is also to define and explain notions, describe principles of economic regulation of such companies, samples of the general framework for the financial evaluation and the way of its design. It also describes information sources, which are drawn from data required for financial assessment.

1 INTRODUCTION

Air traffic control is a fundamental prerequisite of the air transport existence all over the world. It is also the phenomenon significantly affecting its safety and effectiveness. The methodology of air traffic services is a discipline which is solved by international rules and recommendations.

The general economic principle of entities providing air traffic services is not maximizing profits, but especially the profitability of the provided services. They must be managed so that they return the costs incurred in the provision of services i.e. that they are finally unprofitable. It is the economic principle of all enterprises in the market economy. Possible profit from the operations is primarily used to improve the quality of services with emphasis on safety and fluidity of air traffic (Říha 2014).

2 POSITION AND ECONOMIC ASPECTS OF AIR TRAFFIC MANAGEMENT BUSINESS

The main task of the enterprises of air traffic control is to ensure a safety environment for air traffic in accordance with the provisions of the relevant ICAO documents. In practice, this means to provide a number of different services to the users of the airspace and airports of the state territory (Vagner 2014). It is necessary to take care on the international character of air transport, in accordance with international standards, while observing maximum safety of operation and to perform it at the same time financially and capacity effective manner (Pudło 2014).

2.1 Economic status of air traffic management enterprises from a historical perspective

The basic rules and historical background of an economic position of air traffic management companies result from the different development of an organisation and an ownership situation in the aviation industry. In this paper, due to practical reasons, we deal with the issues of air traffic management from the European perspective, i.e. when the airlines pay the services and also pay the specified fees. Their amount depends on the level of ANSP costs, the density of traffic, the route length and aircraft weight. An alternative way of funding these services is used, for example, in the U.S., where air traffic control services are financed by a special tax on aviation fuel.

In the framework of the provision of air traffic services—again in terms of economy—a route control service was separated from the terminal control services. This development is reflected also the ICAO document entitled as "Policies on Charges for Airports and Air Navigation Services", published under the name of Doc 9082/7. It describes additional rules that providers of air traffic services are obliged to follow—the rule of gradualism, the rule coverage the costs and a reasonable profit, the rule of transparency and the rule of consultation with service users (Vittek 2015).

The provider of air traffic services has, in own segment of the airspace, a monopoly position and must be a subject of regulation. This regulation, as well as in other areas of civil aviation is performed at the national and international level. In the conditions of a monopoly the minimum competition must therefore necessarily enter the economic regulation. The economic regulator with executive powers in the EU and at a global scale has not existed yet. The regulatory process is kept rather in the form of recommendations (recommended practices) ICAO, benchmarking, productivity, and prices (Eurocontrol), lobbying, negative/positive publicity of the expensive/the cheap providers (IATA, AEA and other associations of carriers) and political pressure. For an efficient economic regulation in EÚ the individual states are responsible. The problem with this solution is that in many cases, the national regulatory authorities absent from the professional experience in the field of air transport economy, and so regulation may be limited to a simple comparison of the price levels applied in other countries.

In the Czech Republic the role of economic regulator is divided between the Ministry of Transport (maintenance of international treaties under the Civil Aviation Act) and the Civil Aviation Authority (financial position of the air traffic services in accordance with EC Regulation no. 550/2004). A similar situation exists in Slovakia. The national regulator should be in the control of its national air traffic control services to monitor and evaluate the following at least:

– The user's compliance with the rules of non-discrimination,
– Covering the corresponding costs,
– Achieving a reasonable profit,
– Compliance with other recommendations of ICAO.

3 COST BASE AND PRINCIPLES FOR PRICING OF AIR TRAFFIC CONTROL

Prices for air traffic services are determined on a net cost-base (cost relationships). Title principles for determining the cost-base of both air traffic services types are described in ICAO document "Policies on Charges for Airports and Air Navigation Services" – Doc. 9082/7 and also in EC Regulation No. 1794/2006. Clear rules defining the form and structure determination of cost base are summarized in the Eurocontrol document "Principles for Establishing Cost Base for Route Facility Charges and the Calculation of the Unit Rates" which is binding for all member countries of the organization, including the Czech Republic and Slovakia. This is a document, targeting solely to route services (ICAO 2001).

ANSP cost base for calculating en-route and approach navigation services is the sum of all costs invested in these services in a given calendar year. To calculate the cost base in a given calendar year, in accordance with the rules of Eurocontrol, it applies also a very important rule called an over/under recovery principle. This rule basically means that for smaller/larger than anticipated sales in year

N – 2 (e.g.: 2017, included under/over recovery in 2015), this difference accrues or is deducted from calculating a base cost of the planned (future) year. The cost base of the national air traffic control services consists of national costs, the cost of Eurocontrol (in the case of en-route), over/under recovery previous years and profit margins. The price in terms of air traffic services is calculated as a proportion cost base, and the total number of traffic units anticipated in the future, which is the next calendar year. Providers of air traffic services, by this price, cover own costs which are associated with the services, and also ensure a reasonable profit (EUROCONTROL 2014). The price is calculated individually for each of the types of services, i.e. for en-route and aerodrome and approach flights control service to which corresponding costs are assigned. Each price is charged separately, has different traffic unit and different approval procedures. An essential element of calculation rates charged to air carriers for air traffic services are performance units. For a track service, the performance unit is called a service unit. The fees for service unit flights depend on the maximum take-off aircraft weight (MTOW) and the length of the flight:

$$\sqrt{\frac{MTOW}{50}} \times D(100 \ km) \tag{1}$$

where: $MTOW$ = maximum take-off aircraft weight; D = length of the flight.

Among the users themselves and between users and providers the track services are held heated discussions about whether this method of calculation corresponds to the rule of bond prices to the actual costs. Operators of larger aircraft types advocating the philosophy that the important factor is not the weight, but the time spent in the airspace of the ANSP and that the jets with a higher MTOW are usually faster, time they spent in the system is lower should therefore pay lower prices for services.

4 ANALYSIS AND COMPARISON OF THE ECONOMIC SITUATION OF ENTERPRISES PROVIDING AIR TRAFFIC SERVICES IN SLOVAKIA AND THE CZECH REPUBLIC

For companies providing air traffic services charges for en-route navigation services are a key source of income, as well as charges for approach and aerodrome services. It is therefore important to monitor the statistics and aircraft movements in the airspace of both countries. The movements are expressed either in the performance unit (s) or in overflight units (Maximum take-off weight/

MTOW) dependent on the weight of the aircraft (ICAO 2007).

4.1 Movements in FIR Bratislava

In 2015 in comparison with 2014 recorded an overall increase in en-route air traffic by 6.9% (ATS SR 2015). This development contributed significantly events in Ukraine, which resulted in a complete circumvention of its airspace. Despite this significant increase in traffic a high level of safety of services has been maintained. Due to the volume of controlled flights at ACC Bratislava 5 sectors were simultaneously activated during the peak hours. Despite the challenging operational conditions the priority given to the training of new ATCO was not reduced. In the provision of air traffic control some delays were recorded, but did not exceed the limits set for the Slovak airspace.

4.2 Movements in FIR Prague

Totally, in 2015, 782,552 aircraft movements took place in the airspace of the Czech Republic. The increased number of movements for the year amounted to 7.2% in comparison with 2014. This result is generally assessed as positive, despite the negative developments at the end of the year caused by strong entry of still lasting global crisis of the aviation industry caused by economic recession in Europe, leading to a significant limitation in some cases up to the end of the operation of many air carriers. Still a significant proportion of the volume of air traffic has kept the segment of low-cost air transport. The airspace of the Czech Republic remained in 2015 a transit area for long-haul flights, directed mainly from Europe to the Middle East and Southeast Asia, which represents equally outstanding share of the total traffic.

Upto 31st December 2015, LPS SR had total revenues amounting to € 65,684,970, representing 99.10% of the Annual Plan. This result was largely influenced by revenues for en-route navigation

services (en-route charges), invoiced in the amount of € 59,379,460. In comparison with 2014, the revenues for en-route services decreased by 4.46%. ANS CR, s. p. in 2014 reported revenues for en-route navigation services (en-route charges), invoiced in the amount of 2,978,624,000 CZK, representing 80.03 % of total company revenues (ATS CZ 2015).

For a comparison the table above is displayed, which shows calculation of the profit per unit of outputs (movement of a plane). It's clear, that despite the greater volume of air traffic in the airspace of the Czech Republic, the amount of income per unit is almost on a par with Slovakia.

4.3 Charges for air traffic services

Charges for air traffic services are generally divided into:

– Charges for en-route navigation services,
– Charges for approach and aerodrome control,
– Services,
– Fees for training (training) flights.

Revenues from these charges represent the most important part of company revenues; we can call them traffic revenues. A summary table is below, where we find the distribution of revenues from charges for air traffic service.

Due to the incompleteness of the data that we had available, the following table has been given for only distribution of traffic revenues of ANS CR and total transportation revenues of LPS SR. The largest proportions of sales are from en-route charges.

4.4 Fees in the Slovak Republic

Revenues from sales of services, namely revenues from en-route charges, have a decisive impact on the total generation of revenues. They are dependent on the number of controlled service units and

Table 1. Recalculation of the yield per performance unit.

	LPS SR, š.p. (in €)	ŘLP CZR, s.p. (in CZK)
Revenues for track services and the proportion of the total profit	59,379,460 90.04%	2,978,624,000 (110,233,670 €) 80.03%
Number of aircraft movements	454,342	782,552
Revenues for a performance unit	130.69	3806.29 (140.86 €)

Table 2. Distribution of traffic revenues.

	LPS SR, š.p. (in €)	ŘLP CZR, s.p. (in CZK)
Revenues for en-route services	59,379,460	2,978,624 (110,233.67 €)
Revenues for approach and aerodrome control services	3,981,090	506,701 (18,752.12 €)
	-	
Fees for training		2358 (87.3 €)
All revenues together	65,684,970	3,707,482 (137,212.51 €)

the amount of the national unit rate for en-route charges.

In 2015, 1,071,382 service units were controlled, representing an increase by 27,039 service units, i.e. by 2.58%, compared to 2014, when 1,044,343 service units were controlled.

The amount of the national unit rates for en-route charges was established according to the rules laid down in "Principles for Establishing the Cost-Base and the Calculation of Unit Rates EUROCONTROL DOC 13.60.01". Effective from 1 January 2015, it was set at € 55.38, representing a decrease by € 5.55 when compared to 2014; in percentage it was a decrease by 9.10%. Since its Implementing Decision (EU) 2015/347 of 2 March 2015 the European Commission has not approved the "FAB CE Performance Plan" In May 2015 a reviewed FAB CE Performance Plan was submitted. Based on the above-stated and in accordance with the Article 17(2) of the Commission Implementing Decision (EU) No. 391/2013 as amended, a unit rate for en-route charges was modified based on the reviewed FAB CE Performance Plan to EUR 54.99, effective from 1 September 2015 (ATS SR 2015)..

4.5 *Fees in the Czech Republic*

The amount of the national unit rate for en-route charges for the year 2014 was set at 1064 CZK. The year 2015 was the first year of the Plan's performance for the second reference period (2015–2019), which is applied to a performance scheme for air navigation services (ATS CZ 2015). Under this sys-tem, each Member State must contribute to the achievement of Pan-European performance targets through the national performance plan approved by the European Commission. A key performance indicator for the cost-effectiveness was to determine the national unit rate for en-route control service for the entire reference period.

5 CONCLUSION

Companies providing air traffic services that are a subject to economic regulations must carefully consider their economic decisions. For air carriers, when choosing an air traffic service nowadays the main decisive factor is the quality and price of services. These factors, ultimately, air traffic service providers are unable to achieve because of two reasons. On one side there is reducing of the total costs of business that did not increase prices for services, on the other side investing in improving the quality of services. Although these are two contradictory measures they are closely related and mean success and further development for the companies.

To achieve the above results is not easy and also time consuming. It is necessary to continuously monitor various indicators and their development trends. Companies must be able to take appropriate measures to improve and streamline their activities. Some factors, however, are heavily influenced, for example by geographical location or background of the company, which is also dependent on the length of exposure time in the environment of air traffic services, etc.

REFERENCES

ATS CR 2015. Annual Report 2015, Řízení Letového Provozu ČR, s. p. Available on internet: http://www.rlp.cz/generate_page.php?page_id = 127.

ATS SR 2015. Annual Report 2015, Letové prevádzkové služby SR, š. p. Available on internet: http://www.lps.sk/docs/vyrsprava2014sk.pdf.

EUROCONTROL 2014. Principles for establishing the cost-base for en-route charges and the calculation of the unit rates, *Doc. N° 07.60.01*, Available on internet: http://www.eurocontrol.int/crco/gallery/content/public/docs/circulars/princ_doc_10_60_01_en_mar2014.pdf.

ICAO 2001. Policies on Charges for Airports and Air Navigation Services, *Doc. 9082*, Eight edition Available on internet: http://www.icao.int/icaonet/dcs/9082/9082_8ed_en.pdf.

ICAO 2007. Manual on Air Navigation Services Economics, *Doc. 9161*, Fourth Edition, Available on internet: http://www.icao.int/icaonet/dcs/9161/9161_en.pdf.

Pudło, P. & Szabo, S. 2014. Logistic costs of Quality and Their Impact on Degree of Operation Leverage. In: *Journal of Applied Economic Sciences*. Vol. 9, no. 3 (29), p. 470–476. ISSN 1843-6110.

Říha, Z. et al. 2014. Transportation and environment-economic research, In: *WMSCI 2014–18th World Multi-Conference on Systemics, Cybernetics and Informatics, Proceedings*.

Vagner, J. & Jenčová, E. 2014. Comparison of Radar Simulator for Air Traffic Control, In: *Our Sea*, Vol. 61 (1–2), p. 31–35. ISSN 0469-6255.

Vittek, P. & Lališ, A. 2015. Safety Key Performance Indicators system for Air Navigation Services of the Czech Republic, In: *Transport Means—Proceedings of the International Conference*, p. 537–542.

New Trends in Process Control and Production Management – Štofová & Szaryszová (Eds)
© 2018 Taylor & Francis Group, London, ISBN 978-1-138-05885-9

Transition from functional to process management of the selected company

Z. Nižníková, E. Manová, J. Lukáč & J. Simonidesová
Faculty of Business Economics with seat in Košice, University of Economics in Bratislava, Košice, Slovak Republic

ABSTRACT: The aim of paper "Transition from functional to process management" is business process mapping and creation of a process map in an optimized form. The first part is intended to provide theoretical analysis of given problems. The main emphasis is put on the distinction between functional and process management, process analysis and mapping and approaches to process improvement. The next part deals with analysis of processes in the company AAA, a.s. and the design of process maps. The closing part of the paper summarizes achieved and expected costs and benefits linked with the implementation of process management in the company.

1 INTRODUCTION

Financial management or in other words, financial management plays in corporate life a very important role. Without sound financial management company in the competitive environment it is very difficult to survive. Level management of the enterprise to its success and prosperity characterizes financial situation. Managers draw up its decisions based on the company's financial situation, both in operational and strategic roles of business, with particular focus must be on quality and solvency of the company. Their role is also in the design of distributed profits and, at the same time in a possible loss and the sources of its coverage., "There are currently many investment and financial products that can be used for capitalizing of savings." (Feranecová et al. 2016).

1.1 *Financial policy and financial management*

"Corporate activities consist of three parts: supply, production and sales performance. The process may be carried out only if the available funds in a certain amount, structure, and time." (Majdúchová &Neumannová 2008).

"The transformation process in enterprises such entities employed a kind and in cash page. Factually it begins attracting factors of production, followed by their involvement in productive activity that results in goods and services for the market. The whole process has a second—financial page. His movement mediate the money is going chaining countless acts of purchase and sale. This enables companies to get into the multilateral monetary relation to its economic hinterland: the suppliers, customers, its employees, banks, insurance companies, the tax authority and other entities. This financial page of the transformation process will ensure the financial business." (Sedlák 2010).

"Corporate Finances Display the movement of money, capital and financial resources for which the company receives in the various quantitative and qualitative monetary relationships with other businesses, employees and the state." (Kupkovič 2002).

1.2 *Categories used in corporate financial management*

"Financial management has in its target orientation seek ways which best meet their financial goals. From the assessment of needs through the search and evaluation of alternatives to the adoption of specific financial decisions occur Permanent evaluation of basic categories, namely profit, cash flow, interest rates, time value of money, risk and the like. knowledge of these concepts is essential for the effective work of the financier." (Fetisovová 2009).

2 TOOLS AND METHODS FOR MAPPING PROCESSES

There are currently on the market many software tools developed specifically for the mapping process. Most of these instruments describe the process and its activities using graphic symbols. For each process or activity can be linked to their characteristics. Tools for mapping processes can be divided into three main groups:

– Tools representations flow—This is the drawing tools that are at the lowest level and help describe the migration process verbal description into graphic symbols.

– CASE tools—provide a conceptual framework for modeling processes and their hierarchy description. Usually they provide the possibility of linear, deterministic and statistical analysis.
– Simulation Tools—provides in-depth analysis of the dynamic of continuous or discrete data. Allows you to view, as a customer, or any other object passes through the system. Simulation tools are mostly better part of CASE tools.

2.1 Transition to management process

For creating a process model, the company decided to purchase information system EISOD. Process maps are portrayed in this system using the EPC diagrams. As the company started to use EPC diagrams in the transition to process management, I decided I also all the procedural maps illustrated by the EPC diagrams.

EISOD (electronic ISO documentation) is a software product from the Institute of Industrial Management, Ltd. This system is the product of an electronic network management and proven competence Quality Management System according to ISO, TS, etc. EISOD system contains separate modules, of which the company bought the following four:

– Module ORYX QPM—process modeling: allows the creation of complex enterprise process model—Process organization and integration of quality management and operational structure
– Module Records management SMJ (Quality Management System)—is intended for complex administration and management documentation Quality Management System ensures controlled circulation of documents between users, regardless of where the application originated, or what application to include.
– Module Management audits, non-compliance, action—is designed to manage internal and external audits, it also includes tools for quick and easy assessment of the state of quality management in the form of summary tables and graphs.
– Module the report gauges and devices—is designed to manage the instruments and equipment restricted, individual items are listed in the registration card, the monitoring of their history and planned other events.

Each user of the system is exactly specified type of access granted access rights and range of activities that can take place in the individual modules.

Objects that are displayed in proposed process maps are as follows (Fig. 1).

2.2 Marketing

The initiation process status is defined marketing strategy. For strategy development is responsible

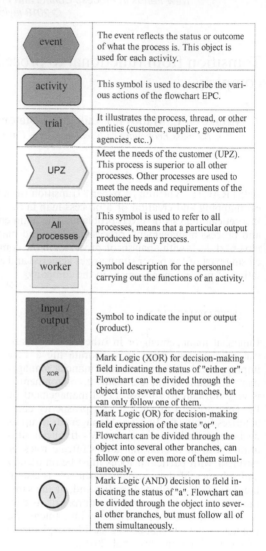

event	The event reflects the status or outcome of what the process is. This object is used for each activity.
activity	This symbol is used to describe the various actions of the flowchart EPC.
trial	It illustrates the process, thread, or other entities (customer, supplier, government agencies, etc..)
UPZ	Meet the needs of the customer (UPZ). This process is superior to all other processes. Other processes are used to meet the needs and requirements of the customer.
All processes	This symbol is used to refer to all processes, means that a particular output produced by any process.
worker	Symbol description for the personnel carrying out the functions of an activity.
Input / output	Symbol to indicate the input or output (product).
XOR	Mark Logic (XOR) for decision-making field indicating the status of "either or". Flowchart can be divided through the object into several other branches, but can only follow one of them.
V	Mark Logic (OR) for decision-making field expression of the state "or". Flowchart can be divided through the object into several other branches, can follow one or even more of them simultaneously.
∧	Mark Logic (AND) decision to field indicating the status of "a". Flowchart can be divided through the object into several other branches, but must follow all of them simultaneously.

Figure 1. Objects used in flowchart.

CEO and this document is subject to approval by the Board of Directors.

3 DESIGN AND DEVELOPMENT PROCESS

Design and development process begins with pent-up demand, and ends with the creation of complete documents for pricing. During this process they are identified specific customer requirements and if the customer so desires, it also created a prototype of the product. Design and development process consists of two processes: finding and developing customer requirements.

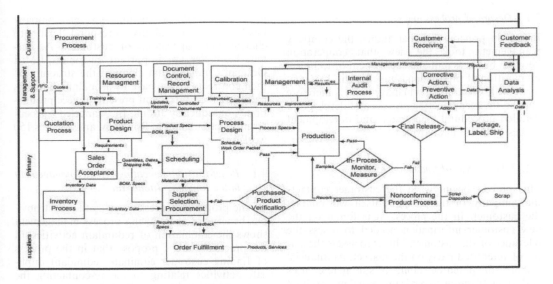

Figure 2.   Process map core processes in the company.

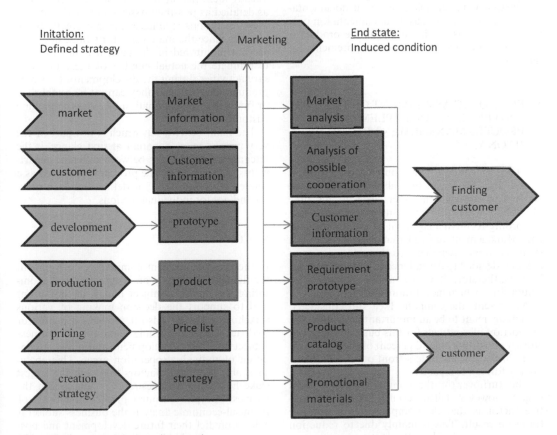

Figure 3.   Process marketing—input and output.

3.1 Design of process maps

Based on previous internal audits, the company management took the view that cooperation between business unit and department of technical preparation of production is insufficient. In agreement with the company management, I therefore applied the reengineering process and developed a detailed procedural map design and development process, which I have divided.

3.2 The process of finding customers

This process begins when the customer contacts the company, or the company itself will contact the customer. In the process, it is found on the new customer information needed to assess the relevance of the customer. This is to assess the status of interested party on the market, its financial stability, the potential volume of supply references and so on. If the customer is interested in a product that business yet producing, followed by an assessment in terms of strategic goals. Product committees assess whether the new product is in line with our product and strategic profiling of the race. If the customer has complete technical documentation for the product or has no documentation only preliminary ideas on the future of the product is necessary to define the technical requirements for the product.

4 ECONOMIC ASSESSMENT OF THE COST AND BENEFITS OF IMPLEMENTING PROCESS MANAGEMENT COMPANY JITONA, A.S

The need for changes in the company management Jitona, Inc. He began to realize this was before 2003. The strategic customers was at this time of IKEA, where the company was Jitona, Inc. According to the management of wholly dependent. Management realized that dependence on a single customer represents to the enterprise great risk and decided to dependence from Ikea reduced. In 2003, therefore, it was to buy the company Tusculum, Inc., which meant more customers.

At present, the company Ikea still considered by management to be an important customer. The proportion of contracts for this business, however, already constitute only 10 percent of the total turnover, 90 percent are orders from other customers, mainly foreign companies.

The turnover of the company and its sales began, however, fallen recently. Another negative factor is that the company has a negative financial result. This is mainly due to reduction of the volume of orders. Production has increas-

ingly nature-off production, so the company has become a vital ability to respond flexibly to changes and suggestions of customers. As current Examples are important customers LMiE German company that changed the structure of their contracts. In the past it was mainly the series nature of the contract, at present, a piece production. Consequently Jitona, Inc. canceled Carpenter production plant in Rousínov, and releases 150 workers.

4.1 Benefits of implementing process management in the company Jitona, Inc

Of the proposed process maps as the most obvious benefits of introducing process management shows the elimination of redundant activities. In the practical part, I propose that in the process of finding customer eliminate redundant duplicate activities relating to the re-contacting the customer to supplement the information on the required product characteristics, product prototype respectively. It is a useless activity, that would not be necessary if the customer requirements already identified at the start of the negotiations as detailed as possible. Assessment of the cost of these operations is unnecessary but greatly problematic. Since the management of these activities are not monitored in the past, it is not possible to estimate the actual number of these unnecessary activities during the development of a single product. On this basis they cannot be established or funds that the elimination of excess contacting customers save.

The main benefit, but which is not part of the very process maps apparent at first glance, is the integration of activities between different departments in the process. The process is perceived as a whole and responsibility is defined for each process and not for individual sections.

5 CONCLUSIONS

In the present turbulent competitive environment the financial analysis is an essential part of monitoring the business subject and is an important tool to support the decision making of various stakeholder groups. Also it provides a picture or feedback about the whole condition of business subject and their development and about a condition of individual operation areas. This analysis is able to identify factors that with the largest stake have caused undesirable results within the business subject. Through prediction models of financial-economic analysis the business subject is able to predict their future development and possible option for bankruptcy. Among the benefits of this article belongs financial-economic analysis focused on the business subject's ratio indicators of activity, profitability, liquidity and indebtedness itself. Related proposals mentioned in this article for weaknesses elimination which were found by financial analysis are focused on practical use in the business subject's experience. Introduction of process management in the company Jitona, inc. In my opinion will rebuild the benefits are mainly in the possibility of a more rapid response to change and increase market competitiveness. It should be noted, however, that this is not only a technical change, but the success of the launch process management is dependent on the worker of the enterprise. After the implementation of process management will work with processes is far from over. Processes should pass a permanent improvement and process maps should be amended based on the needs of businesses and consumers.

ACKNOWLEDGEMENT

"This paper is part of Project for young teacher, researchers and PhD students, no. I-17-105-00, 2017: Evaluation of the economic and financial performance of SMEs in the V4 countries."

REFERENCES

Feranecová, A. et al. 2016. Selecting the savings account in the Slovak Republic., In *Problems and perspectives in management* 14(4), pp. 8–16, 2016, ISSN 1727-7051.

Fetisovová, E. et al. 2009. *Podnikové financie: praktické aplikácie a zbierka príkladov*. 1. vyd. Bratislava: Iura Edition., 2009. s. 7. ISBN: 978-80-8078-259-7.

Fiala, J. & Ministr, J. 2003. *Průvodce analýzou a modelováním procesů*. 1. vyd. Ostrava: Vysoká škola báňská—Technická univerzita, 2003. 109 s. ISBN 80-248-0500-6.

Hammer, M. & Champy, J. 2000. *Reengineering—radikální proměna firmy: Manifest revoluce v podnikání*. 3. vyd. Praha: Management Press, 2000. 212 s. ISBN 80-7261-028-7.

Hromková, L. 2001. *Teorie průmyslových podnikatelských systémů I*. 1. vyd. Zlín: Univerzita Tomáše Bati ve Zlíně, 2001. 118 s. ISBN 80-7318-038-3.

Kovář, F. et al. 2004. *Teorie průmyslových podnikatelských systémů II*. 1. vyd. Zlín: Univerzita Tomáše Bati ve Zlíně, 2004. 250 s. ISBN 80-7318-189-4.

Kupkovič, M. et al. 2002. *Podnikové hospodárstvo*. Bratislava: Sprit vfra, 2002. s. 424. ISBN: 80-88848-93-8.

Majdúchová, H. & Neumannová, A. 2008. *Podnikové hospodárstvo pre manažérov*. 1. vyd. Bratislava: Iura Edition, 2008. s. 213. ISBN: 978-80-8078-200-9.

Sedlák, M. et al. 2010. *Podnikové hospodárstvo*. 1. vyd.. Bratislava: Iura Edition, 2010. s 303. ISBN: 978-808-8078-317-4.

New Trends in Process Control and Production Management – Štofová & Szaryszová (Eds)
© 2018 Taylor & Francis Group, London, ISBN 978-1-138-05885-9

Types of unmanned aerial vehicles for package transport

M. Novák & D. Hůlek
Faculty of Transportation Sciences, Czech Technical University in Prague, Prague, Czech Republic

ABSTRACT: Current logistics is dynamic and it requires using of modern technologies. It is ensured that packages are delivered in time and for a price that is as low as possible thanks to these technologies. Unmanned aerial vehicles are the modern technology that is capable to fulfil these requirements. This article deals with possibilities of using the unmanned vehicles at the modern logistics focusing on a choosing of suitable type of the unmanned vehicle. The first part of the article summarizes a current state of using the unmanned vehicles at the logistics. Suitable types of the vehicles are chosen in the next part of the article. The choice of the suitable type depends on several factors. The last part of the article is about a model situation and its economic assessment. The package transport by the unmanned vehicle is compared with a common way of the transport using a surface transport.

1 INTRODUCTION

An innovation in an area of the logistics has come with the advent of the new millennium. New modern technologies have improved and still improving a process of the packages transportation. The process are becoming fast and reliable. This trend is still here and the modern technologies are still implemented into the process. Thanks to a massive expansion of a one technology—unmanned aerial vehicles (UAV), it gets into the process too. Although the package transport by the unmanned aerial vehicles is still at the beginning, some tests have already been done. There are some companies listed below that have started to experiment in an area of an unmanned aviation. The companies are Amazon, DHL or DPD. The most companies use x-copters. The most used are quad-copters, hexacopters or octocopters. (Amazon. com 2017 & DPD 2017) The DHL Company has an unmanned vehicle that can tilt wings. Thanks to this the unmanned vehicle with a fixed lifting surface becomes the vehicle with rotating lifting surface (tilt-wing). (DHL 2017) The question is if the x-copters are the best structure for this. The goals of this article are to assess advantages and disadvantages of each type of the unmanned aerial vehicle used for logistics and try to answer to the said question (Voštová et al. 2014).

2 TYPES OF UNMANNED AERIAL VEHICLE STRUCTURES

There are many types of structures used for the unmanned aerial vehicles and aircraft models in the world. But not all of them are suitable for using in the area of the logistics. There are the most common types listed below that are suitable for the area of the logistics too. These types are:

– Airplanes
– Helicopters
– x-copters
– Tilt-wings
– Airships

2.1 Airplanes

The airplanes are aircraft with a fixed lifting surface (wings). A control is provided by aerodynamic rudders. Advantages of this structure are a high transport speed and possibility to use all types of drive units (for all types of drive units see the next chapter). Thanks to the second advantage the airplanes can have a long flight time. The disadvantages are a need of a long runway and problems with a weight and dimensions of the package. The second disadvantage is only for smaller airplanes. It is not suitable to use the airplanes for the package transportation to an addressee. The autonomous unmanned vehicles are good for the package transportation between logistic centres. The logistics centres have space for the runways. Bigger airplanes (span 1.5 m and more) are more suitable. They can carry bigger and heavier packages.

2.2 Helicopters

The helicopters are the aircraft with a rotating lifting surface. The helicopters are controlled by tilting the rotating lifting surface and by changing an angle of attack of rotating blades. The most helicopters have a tail rotor too. The helicopters have several advantages. The advantages are vertical take-off and landing, hovering and possibility of using all types of the drive units. It means long flight time. The helicopters can be easily modified for big packages. The disadvantages are lower flight speed, lower safety resulting from rotor dimensions and complexity of the structure. It is suitable to use the helicopters for the package transportation between logistics centres and also to the addressee because of the helicopters attributes. It is better to use them for the transportation to the addressee because of the vertical take-off and landing and hovering.

2.3 *X-copters*

The x-copters are the aircraft with the rotating lifting surface. The x-copters have more lifting rotors and they are controlled by changing engines rpm. The advantages of the x-copters are vertical take-off and landing, hovering and simply structure. The disadvantage is at a used drive units and thus at the flight time. Nowadays, the electromotor is only drive units that is used for the x-copters. Thanks to this the x-copters have smaller range than aircraft with another types of the drive units. More about the drive units see the next chapter. The x-copters are suitable for the transportation of smaller and lighter packages for a short distance because of the used drive units. The x-copters are suitable for transportation from the logistics centre to the addressee.

2.4 *Tilt-wings*

This type of the unmanned aerial vehicles combines the aircraft with the fixed lifting surface and the rotating lifting surface. Thanks to this it combines their advantages. The tilt-wings can take-off and land vertically and they can hover. The flight speed is high. It is because of the fixed lifting surface configuration. The disadvantage is the structure complexity. The tilt-wings can be used for the transportation between the logistics centres and to the addressee too. However the authors prefer to use another type of the structure for the transportation to the addressee.

2.5 *Airships*

The airships are aircraft lighter than an air. The airships can fly thanks to a gaseous medium which is lighter than the air and it uphold all structure. It is controlled by propellers and aerodynamic rudders. The advantages are vertical take-off and landing, hovering, safety, possibility to transport big and heavy load, structure sim-

plicity and long flight time. The disadvantages are poor controllability, very low flight speed and dependence on suitable meteorological conditions. (Frei et al. 2016) The airships are not suitable for the package transportation because of their disadvantages. If the airships are used, it is better to use them for the transportation between logistic centres.

3 TYPES OF DRIVE UNITS

There are the three most used types of drive units for the unmanned aerial vehicles and also for the aircraft models listed below:

− Electromotor
− Piston engines
− Turbine engines

3.1 *Electromotor*

The electromotor can be used for all mentioned types of structures. Its advantages are good reliability and structural simplicity. The disadvantage of the electromotor is small power supply. The energy for the electromotor is gained from batteries. Nowadays, the most used batteries are Li-Po, Li-Ion, NiMh or NiCd. A capacity of the best batteries enables to work approximately 45 minutes. But it depends on many factors (for instance: engine type, UAV weight or payload). A common flight time is something about 20 minutes. There is a dependence of the flight time on the payload in the Graph 1. The dependence is for the helicopter LaHeli 700 powered by the electromotor. However the research in the area of the batteries is still in progress and the results show that the capacity of the batteries could be better. This would mean longer flight time. (Reichl 2015 & Novinky.cz 2017) It is better to use the electromotor for lighter packages transported to shorter distances because of the mentioned things. An example can be the package transportation to the addressee by the helicopter or x-copter powered by the electromotor.

3.2 *Piston engines*

The piston engines have more complex structure than the electromotor. They are heavier too. But the flight time is extended from tens of minutes to hours. There is a dependence of the flight time on the payload plus quantity of fuel in the Graph 2. The Graph is for helicopter Radikal G30 V2 which is similar to the helicopter LaHeli 700. The piston engines are suitable for the transportation of heavier packages for longer distances because of the flight time. It is suitable to use them for the transportation between the logistics centres.

3.3 *Turbine engines*

The turbine engines have the most complex structure. On the other hand the power from the turbine engine is incomparably bigger than the power gained from the electromotor or piston engine. The turbine engines can be used for big unmanned aerial vehicles that have dimensions similar to manned aircraft. These vehicles can transport extremely big packages for very long distances between logistic centres.

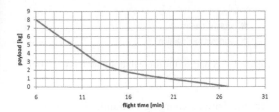

Figure 1. The dependence of the flight time on the payload for the helicopter LaHeli 700 powered by the electromotor. (Brázda 2015).

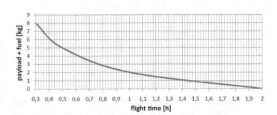

Figure 2. The dependence of the flight time on the payload plus quantity of fuel for the helicopter Radikal G30 V2. (Brázda 2015).

4 DEFINING A MODEL SITUATION

The transportation from the logistics centre to the 10 addressees by a car and unmanned aerial vehicle will be compared within this model situation. All addressees are in a radius of 3 km from the logistics centre. The model situation should clear up in what conditions is suitable to use the unmanned vehicle instead of the car. The conditions are costs per 1 km of the transportation and time needed for delivering all packages. The packages are letters or packs up to 1 kg. There is a scheme of the addressees and logistic centre placement in the Figure 3. Table 1 is a matrix with distances between all points.

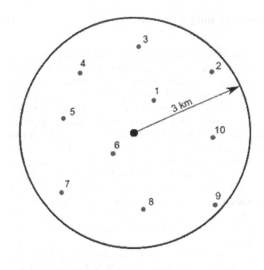

● logistic centre

• addrerssee location

Figure 3. The scheme of the 10 addressees and logistic centre placement.

Table 1. The matrix with distances between all points (see also Figure 3).

	centre	1	2	3	4	5	6	7	8	9	10
centre	0	1010	2560	2280	2130	1920	760	2490	2020	2860	2060
1	1010	0	1660	1480	2100	2460	1770	3470	2860	3180	1810
2	2560	1660	0	2020	3470	4080	3360	5060	4010	3470	1730
3	2280	1480	2020	0	1730	2750	2890	4340	4270	4590	3070
4	2130	2100	3470	1730	0	1270	2280	3140	3940	4950	3900
5	1920	2460	4080	2750	1270	0	1630	1950	3180	4630	3980
6	760	1770	3360	2890	2280	1630	0	1730	1660	3040	2670
7	2490	3470	5060	4340	3140	1950	1730	0	2240	4050	4270
8	2020	2860	4010	4270	3940	3180	1660	2240	0	1840	2600
9	2860	3180	3470	4590	4950	4630	3040	4050	1840	0	1770
10	2060	1810	1730	3070	3900	3980	2670	4270	2600	1770	0

5 COMPARISON OF TWO TRANSPORT TYPES

The transportation by the car are going to be described as the first one. There is a part of a route that the car with the packages has to go throw in the Figure 2. It is assumed that the transportation without waiting at the addressees will take approximately 1 hour. The courier (car driver) will be late at each addressee 5 minutes on average. The courier will load all packages at the logistic centre and his route will be logistic centre – 6 – 5 – 4 – 3 – 1 – 2 – 10 – 9 – 8 – 7 – logistic centre. This route has been chosen based of the Bellman´s equation (1). The equation says that the shortest route through a graph is composed of the shortest routes between each two points.

$$u(x,y) = \min_{x \neq y}(u(x,z) + a(z,y)) \qquad (1)$$

where $u(x, y)$ is a route length from point x to point y and $a(z, y)$ is a distance between points z and y. (Neckář 2016) It is obvious from the table 1 that a total length of the route is approximately 27 km. Depreciations per 1 km are 3.90 Kč. Kč means Czech Crown and 27 Kč is approximately 1 €. A price for fuel is 29.50 Kč per 1 l. A fuel consumption is about 9 l per 100 km. (Hlavsa 2017) A courier´s salary is an average couriers salary in the Czech Republic – 20 000 Kč per month. (Platy. cz 2017) Total cost for the transportation of all packages by the car are calculated in the Table 2.

A part of the transportation by the unmanned vehicle is in the Figure 4. The unmanned vehicle always transports only one package to the addressee and back. Shown route is only illustrative. It doesn't depend on which addressee will be first, second and so on. The total length of the route is approximately 48.2 km. Assumed average flight speed is 35 km/h. The price 1.95 Kč per 1 km has been chosen because of that the depreciations for the unmanned vehicles doesn´t exist. This price has been chosen by this way: It is assumed that a price for a new car is approximately 300 000 Kč. The depreciations are 3.90 Kč. If an assumed price for an unmanned vehicle is something about 150 000 Kč, the assumed depreciations are half too. (Pudlo et al. 2014) The unmanned vehicle is powered by the electromotor. The price

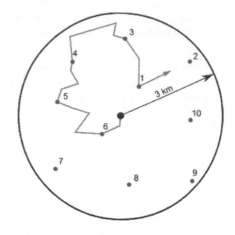

● logistic centre

● addrerssee location

Figure 4.   A part of the route that the car has to go throw.

for one battery charging is hundredths of Czech Crowns (Kč). The fuel price for the unmanned vehicle can be neglect because of this. But the price for batteries and their wearing out cannot be neglect. It is assumed that the battery will be replaced with a charged one after every single package delivery. 10 batteries will be needed. The price for one battery is approximately 1 000 Kč and it can be recharged 1 000 times. It is obvious that the one recharging costs 1 Kč. If 10 batteries for one unmanned aerial vehicle are used, the costs are 10 Kč. The autonomous unmanned aerial vehicle has to have 2 operators. The first operator controls a flight and state of the UAV and the second operator changes the batteries and packages. The salary for both operator has been chosen by the authors. The first operator´s salary is 700 Kč per hour and second operator´s salary is 300 Kč per hour. Both operators can work with and control 10 vehicles. The hour rate for one unmanned vehicle is 70 and 30 Kč per hour because of this. Total cost for the transportation of all packages by the car are calculated in the Table 2 (Hospodka 2014).

It is obvious from the Table 2 that the unmanned aerial vehicle's operations is cheaper. The car will go throw the rout in 1 hour and 50 minutes. The unmanned vehicle transports all packages in approximately 1 hour and 45 minutes. It takes almost the same time. So it can be said that it doesn´t matter which type of transport will be chosen. Unmanned aerial vehicle´s social benefits are that it can deliver the package to the exact place (for instance balcony). The unmanned vehicle is friendlier to the environment because it produces less emissions. However it is very difficult to economically assess which type of the transport is

Table 2.   Total costs table for the car and unmanned aerial vehicle.

item	car [Kč]	UAV [Kč]
depreciation	105,57	93,99
fuel	71,87	10,00
salary	229,17	142,58
total	**406,61**	**246,57**

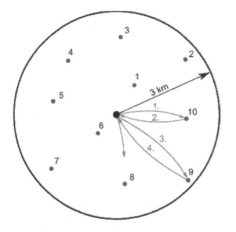

● logistic centre

● addrerssee location

Figure 5.   A part of the transportation by the unmanned vehicle.

better because of many estimates and inaccuracies during the calculations (Říha et al. 2014).

6 CONCLUSION

The goal of this article isn't to determine exact and only one way how and with what type do the unmanned courier services. It only shows some predictions of actual data and experiences how this area will probably evolve. The structures like airplanes or tilt-wings will be probably used for the transportation between the logistic centres. The most suitable vehicles for transportations to the distance up to 3 km will be x-copters. The most suitable vehicles for transportations to the medium distances could be the helicopters with one main rotor or coaxial helicopters. But this depends on several things like a shape of the terrain or package weight.

A one part of the article is about the model situation of the final transportation by the car and the x-copter. This model is very simplified and it is based on many estimates that cannot be verify now. A significant deviation from a future reality can be expected. Nevertheless this model indicates that the use of the autonomous unmanned aerial vehicles for the courier services can be interesting.

REFERENCES

Amazon Prime Air. 2017. *Amazon.com* [online]. USA, 2017 [retr. 2017-03-23]. Available from: <https://www.amazon.com/Amazon-Prime-Air/b?node = 8037720011>.

Brázda, S. 2015. *Tracking Unmanned Aerial Vehicle CTU FTS: Construction of Airframe.* CZE, 2015, 74 s. Diploma thesis. CTU in Prague Faculty of Transportation Sciences. Thesis leader Ing. Martin Novák, Ph.D.
Delivery drones. 2017. *DPD* [online]. Germany, 2017 [retr. 2017-03-23]. Available from: <https://www.dpd.com/home/insights/delivery_drones>.
DHL. Press Release. 2017. *DHL* [online]. Germany, 2017 [retr. 2017-03-23]. Available from: <http://www.dhl.com/en/press/releases/releases_2016/all/parcel_ecommerce/successful_trial_integration_dhl_parcelcopter_logistics_chain.html>.
Frei, J. et al. 2016. Use of Mode-S Radars for Extracting of Meteorological Data from Aircraft. In *International Review of Aerospace Engineering.* 2016, 9(4), pp. 99–106. ISSN 1973-7459.
Hlavsa, T. 2017. Náhrady za použití soukromého auta. In *Založení s.r.o. naklíč* [online]. CZE, 2017 [retr. 2017-03-23]. Available from: <http://www.help1.cz/clanek_vydaje_na_auto_pri_podnikani-nahrady-za-soukrome-vozidlo.html>.
Hospodka, J. 2014. Cost-benefit analysis of electric taxi systems for aircraft. In *Journal of Air Transport Management.* 2014, 20(39), pp. 81–88. ISSN 0969-6997. Available from: <http://www.sciencedirect.com/science/article/pii/S0969699714000532>.
Kurýr. 2017. *Platy.cz* [online]. CZE, 2017 [retr. 2017-03-23]. Available from: <http://www.platy.cz/platy/doprava-spedice-logistika/kuryr>.
Nečkář, J. 2016. Problém nejdelší cesty. *Algoritmus* [online]. CZE, 2016 [retr. 2017-03-23]. Available from: <https://www.algoritmy.net/article/36597/Nejkratsi-cesta>.
Průlom? Vynálezce lithium-iontových baterií přišel s novým akumulátorem. 2017. *Novinky.cz.* [retr. 2017-03-23]. Available from: <http://oenergetice.cz/technologie/elektroenergetika/zpusobi-nova-hlinikova-baterie-revoluci-na-trhu/>.
Pudlo, P. & Szabo, S. 2014. Logistic Costs of Quality and their Impact on Degree of Operaton Level [online]. In *Journal of Applied Economic Sciences.* 2014, 9(3), pp. 469–475. ISSN 1843-6110. Available from: <http://www.cesmaa.eu/journals/jaes/files/JAES_2014_Fall.pdf>.
Reichl, T. 2015. Způsobí nova hliníková baterie revoluci natrhu? In *OEnergetice.cz* [online]. CZE, 2015 [retr. 2017-03-23]. Available from: <http://oenergetice.cz/technologie/elektroenergetika/zpusobi-nova-hlinikova-baterie-revoluci-na-trhu/>.
Říha, Z. et al. 2014. Transportation and environment—Economic Research. In *The 18th World Multi-Conference on Systemics, Cybernetics and Informatics.* The 18th World Multi-Conference on Systemics, Cybernetics and Informatic. Orlando, 15.07.2014 –18.07.2014. Orlando, Florida: International Institute of Informatics and Systemics. 2014, pp. 212–217. ISBN 978-1-941763-05-6.
Voštová, V. et al. V. 2014. Conveyor new concept for parcel logistics in air transport [online]. In *Applied Mechanics and Materials.* Conference on Research, Production and Use of Steel Ropes, Conveyors and Hoisting Machines. Vysoké Tatry—Podbanské, 23.09.2014 – 26.09.2014. Zurich: Trans Tech Publications Ltd. 2014, pp. 114–118. ISSN 1660-9336. ISBN 978-3-03835-316-4.

New Trends in Process Control and Production Management – Štofová & Szaryszová (Eds)
© 2018 Taylor & Francis Group, London, ISBN 978-1-138-05885-9

Towards alliance's cooperation of low-cost carriers?

A. Novák Sedláčková & A. Tomová
Faculty of Operation and Economics of Transport and Communications, University of Žilina,
Žilina, Slovak Republic

ABSTRACT: While horizontal cooperation of air carriers in the form of international alliances was historically typical of full service network carriers, on the present, the trend towards closer horizontal cooperation in the form of alliances is observed in the segment of low-cost carriers, too. In the paper, we discuss motives and potential impacts of horizontal cooperation among allied low-cost carriers using the case of Value Alliance, the international alliance established by eight Asian low-cost airlines in 2016. The newly allied partners in Value Alliance are investigated with regard to the attributes of business models and geographical patterns of their operation. Based on the findings, we can assume that intensifying mutual horizontal cooperation of airlines, including low-cost ones, will be among the leading phenomena in the air transport industry. This will contribute to even higher complexity of the industry what will shape significantly the global competition of airlines.

1 INTRODUCTION

Since the first modern alliance of airlines was created in 1975, alliance agreements among airlines accelerated (Li 2000). To be allied (or not) seemed to be one of the crucial strategic decisions of airlines for the last years as was predicted by Gudmunsson & Rhoades (2001). Historically, the pace of alliances was dominantly captured by traditional carriers and different forms of alliances among traditional carrier emerged (and also terminated), confirming thus the anticipated growth of airlines alliances as one the most significant developments in the airline industry (Evans 2001). Finally, the process of alliance cooperation among airlines led to the creation of the triad of global alliances (Oneworld Alliance, Star Alliance and Sky Team Alliance) with intensive level of horizontal cooperation among their members. Such configuration in the up-stream part of the air transport industry with airlines globally allied and airlines not globally allied persisted for about ten years. The question whether this status quo will be overcome in the future and if, then how, was not fully answered by academic research. Evans (2001) even emphasized that airlines positioned out of Triad countries will be increasingly interested in the alliance cooperation with the established alliances and that the formation of new substantial alliances might be difficult.

The establishment of Value Alliance in 2016 by eight Asian carriers, which declare themselves as low-cost airlines, is evidence that international alliances have penetrated into the low-cost part of the market. Moreover, global alliances formed specific collaborative platforms for low-cost air-

lines[1] recently and this indicates the growing role of low-cost carriers within the air transport industry and simultaneously growing interest of traditional carriers towards low—cost ones. Whether these changes in the global alliances which offer a bridge to cooperate with low-cost carriers within global alliances have been driven defensively or offensively (Tomová & Materna 2017) is not fully crystalized at the present, however, without any doubt, international alliances of airlines are in a state of flux.

This paper will be focused on Value Alliance as a new grouping in the air transport industry. We shall investigate the member airlines of the new international alliance with regard to the attributes of business models and geographical patterns of their operation, discussing potential motives for such alliances' cooperation and potential scenarios of future development. We shall confront the conclusions of scientific literature in the field. Our focus on Value Alliance's airlines in this paper will go through the following research questions:

– Is the newly formed alliance an equity based alliance?
– Are there any (equity or cooperative) linkages of the new alliance to the existing triad of global alliances?
– Are business models of partners in Value Alliance similar to each other and do they coincide

---

1. "The Oneworld alliance, which includes British Airways and American Airlines, is reviewing the case for seeking members from the ranks of discount carriers while exploring measures to fill gaps in emerging markets." (cited from Transport, Airlines, February 10, 2017).

with the archetypical attributes of a pure low-cost business model?
– What are probable scenarios of future developments in the case of international airlines' alliances, including low-cost ones?

2 ALLIANCES OF AIRLINES: WHAT DO WE KNOW?

International alliances of airlines as specific collaborative arrangements among airlines based on extensive horizontal cooperative agreements were studied intensively in the past and even at the present, researchers focus on airlines' cooperations within alliances. Morrish & Hamilton (2002) reviewed major studies of airline alliances aimed at the relationship between alliance membership and the performance of allied airlines. Using the DEA approach, Min & Joo (2016) examined the performance of three alliance groups and a non-alliance's group of airlines. Later, the effect of code-sharing which is a core cooperative scheme within airlines' alliances on airline profitability was investigated by Zou & Chen (2017). The above-mentioned studies indicated that to be allied with others might have impact on an airline's performance and profitability, however, this impact might depend on many factors, including the scope and efficiency of cooperative agreements among airlines as well as many external circumstances. In the context of the newly formed Value Alliance, scientific contributions to the issues of alliances' stability, duration and survival are also relevant. Li (2000) predicted that alliances based only on code-sharing without serious financial tie-ups were predisposed to fail. Gudmunsson & Rhoades (2001) brought typology of alliances stemming from the levels of complexity and resource commitment and found that equity alliances were expected to be the most unstable and vulnerable due to their complexity. Kleymann (2005) discussed the multilateral allying among airlines as a comprehensive process among independent miscellaneous firms and, for smaller airlines, he stated that stabilization could be achieved through cooperation with partners of equal size that cover a complementary geographical network. Agusdinata & de Klein (2002) worked with four categories of airlines and discussed drivers that are important for the stability of alliances. They stated that airline alliances primarily did not comprise niche players in the budget and leisure markets with air services.

According to our current knowledge, none of the studies have explicitly anticipated the formation of international alliances comprising low cost-carriers in the close future. This is at least interesting in the light of willingness to create

such an alliance also in Europe.[2] This is in a sharp contrast with what Agusdinata & de Klein (2002) expected: *"In view of market size, it is believed that what will mainly determine the number of major airline groupings will be the number of powerful American airlines."* Taking into account low-cost carrier alliances' projects, whether existed[3] or planned, the issue of "low-cost airline alliances" needs to be researched more to better understand a changing patchwork in the global airline industry.

3 VALUE ALLIANCE: WHAT DO WE KNOW?

Value Alliance declares itself to be the world's first pan-regional *low-cost* carrier alliance comprising eight Asia and Pacific market champions. Indeed, among the member airlines of the new alliance, Cebu Pacific from the Southeast Asia region and Jeju Air from the Northeast Asia region are found. Zhang et al. (2008) mentioned just these airlines among main low-cost players in Asia. Besides them, NokAir, Scoot, Nok Scoot, and Vanilla Air allied. Tigerair Singapore and Tigerair Australia complete the list of members.

Within Value Alliance, older airlines like Cebu, but also recently established airlines like Scoot, Vanilla Air and Nok Scoot are allied. Many of them declare themselves as low-cost or budget airlines, Scoot is even a low-cost long-haul carrier.

3.1 *Equity's linkages*

As the composition of an international alliance may undermine its success or failure, we bring a simplified scheme in Figure 1 which illustrates the complexity of linkages among Value Alliance's airlines. The scheme shows not only main equities' links among the allied airlines but also subsequent links to global alliances which raises questions about the complexity of relationships within the alliance as well as the motivation to be allied.

In Value Alliance, comprehensive equities' connections exist among the members. We revealed NokScoot as a joint venture founded by Scoot and Nok Air, both members of Value Alliance. Scoot is, however, a subsidiary of Singapore Airlines, which is a member of Star Alliance. A merger between Tigerair and Scoot with Tigerair is planned to come into the Scoot brand and move to a single operating license.

2. "The head of budget airline Norwegian has revealed plans to join forces with the likes of Ryanair and easyJet, in a bid to create a powerful "low-cost alliance" (The Telegraph, February 2, 2017).
3. U-fly Alliance was the world's first low-cost carrier alliance founded in 2016. Within the alliance, HK Express, Lucky Air, Urumqui Air and West Air allied.

Table 1. Value alliance memberships—basic information.

Allied airline	Country of residence	Year of establishment	Number of destinations
Cebu Pacific	Philippines	1996	66
Jeju Air	South Korea	2005	34
Nok Air	Thailand	2004	29
NokScoot	Thailand	2014	9
Scoot	Singapore	2011	24
Tigerair Singapore	Singapore	2004	42
Tigerair Australia	Australia	2007	14
Vanilla Air	Japan	2013	11

Source: (websites of the airlines, March 2017).

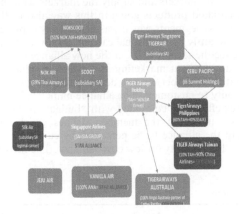

Figure 1. The scheme of revealed equity's linkages of the allied airlines (as February, 2017)[4].

The merger ought to be completed at the end of 2017. Tigerair Australia is now a fully owned subsidiary of Virgin Australia Holdings. Vanilla Air is owned fully by All Nippon Airways, which is a member of Star Alliance. Although not disposing of all equities' linkages among the allied airlines, we could label the new alliance as equity based, or at least as an entity with equity linkages which are not omissible. The links to the existent global alliances (although indirect in several cases) allude to the fact that the global alliances' know-how can be exploited by Value Alliance.

3.2 Fleet's composition and routes' coverage

The analyses of the fleet's size, structure and commonality together with the analyses of routes' coverage may help to understand capability of the new

4. Member airlines of Vallue Alliance are drawn as blue

grouping to supply markets with air services within the Asia-Pacific region and out of the region as well. The main fleet's indicators are contained in Table 2.

The analysis of fleets suggests that although the airlines declare themselves as low-cost, only five airlines within the grouping operate a uniform fleet which is just an attribute typical for low-cost carriers. The values of HHI for Cebu Pacific, Nok Air and Scoot indicate that business model of these airlines is hybridized. As a whole, the grouping has the fleet composed of aircraft coming from different aircraft families, which are represented by aircraft of different types with different flight range. This can be interpreted as capability to serve different O-D markets interlinked intensively through the alliance's cooperation. This ambitions is expressed at the official website: *"Passengers will be able to book flights offered by any Value Alliance partner at the most competitive fare when they visit any member website … Value alliance will offer guests multi-destination options at one go."*

With regard to the supply of routes, Value Alliance offers a large scale of domestic and international destinations in many countries. To detect definitely whether the alliance is (or will be) more parallel or complementary would require a deeper analysis, although, several destinations are wholly represented only by one airline within the alliance. This points to at least some complementary

Table 2. Value alliance fleet's data and indicators.

Allied airline	Aircraft numbers according to types	% of the most populous aircraft type	HHI[5]
Cebu Pacific	36 A320–200 7 A330–300 8 ATR72–500 3 ATR72–600	66.6	4914
Jeju Air	25 B 737–800	100	10,000
Nok Air	3 B777–200 21 B 737–800 6 Q400 NextGen	70	5400
NokScoot	3 B737–200	100	10,000
Scoot	6 B787–800 6 B 787–900	50	5000
Tigerair Singapore	23 A320	100	10,000
Tigerair Australia	14 A 320	100	10,000
Vanilla Air	12 A320–200	100	10,000

Source: (Authors using data available at websites of the airlines, March 2017).

5. Herfindahl-Hirschmann Index.

features of the new alliance (Greece—Scoot, Middle East—Cebu Pacific). Similarly, airlines within the grouping such as Tigerair Australia (owned by Virgin Australia) which operates more domestic Australian air services than international may enforce complementarity within the alliance.

The data in Table 3 shows that routes of the airlines allied in Value Alliance cover miscellaneous countries and territories in five world regions which at least suggests potential of the alliance to be more than a pan-regional Asian–Pacific player in the global airline industry. The long-haul low-cost product in the supply of Scoot together with planned merger of Tigerair may promote it.

3.3 Product policies

All airlines within the alliance also conduct business with cargo transportation. The business in cargo segments is, however, more typical to traditional carriers than to low-cost carriers. With regard to travel classes in passenger segments, all allied airlines utilize a multi-class product policy.

The offer of different products to different segments of passengers, the implementation of frequent flyer programs and the involvement in cargo transportation file the member airlines of Values Alliance among hybridized carriers more than among low-cost ones. The frequent flyer programs with cooperative linkages even out of the alliance

and cooperative linkages at least among the frequent flyer programs of several member airlines foreshadow perspectives to be integrated more in the future, acquire and keep loyal customers and, finally, survive as an alliance (Li 2000). Our investigation also shows that according to the distributional channels, the airline within Value Alliance may be grouped in two pools: the first pool includes the member airlines using only websites (Jeju Air, NokAir, Tigerair Singapore, Tigerair Australia and Vanilla Air) which is symptomatic to low-cost carriers and the second one includes the member airlines using diversified channels (Cebu Pacific, NokSccot, Scoot) which is symptomatic to traditional carriers. This equally points to the hybridization of Value Alliance's airlines and the alliance as a whole. On the other hand, these characteristics also mean that Value Alliance wants to operate on different markets and supply the markets with differentiated products going in this way also to the business segments which was historically captured dominantly by traditional carriers. In this context, we see the term "value" in the new alliance's name as an excellent marketing idea which emphasizes that the member airlines deliver products of high quality to customers. The alliance name "value" is thus efficiently combined with "budget" or "low-cost" image of the member airlines which can help the alliance to be perceived by customers as an "high quality for low cost" alliance. The impor-

Table 3. Value alliance countries and territories served, destinations.

Allied airline	Countries (territories) served	No of destinations	International/ domestic routes
Cebu Pacific	Philippines, Taiwan, Kuwait, Qatar, Saudi Arabia Emirates, Singapore, Australia, South Korea, Japan, Hawaii, Thailand, Brunei, Cambodia, China, Hong Kong, Vietnam, Malaysia	66	29/37
Jeju Air	South Korea, Japan, Hawaii, Thailand, Philippines, Vietnam, Taiwan, China, Guam	34	27/7
Nok Air	Thailand, Vietnam	29	24/5
NokScoot	Thailand, China, Singapore	9	3/6
Scoot	Singapore, Australia, Greece, Thailand, Taiwan, Hong Kong, China, Japan, South Korea, India	24	23/1
Tigerair Singapore	Singapore, Bangladesh, China, Hong Kong, India, Indonesia, Macau, Malaysia, Maldives, Myanmar, Philippines, Taiwan, Thailand, Vietnam	42	41/1
Tigerair Australia	Australia, Malaysia, Bali	14	1/13
Vanilla Air	Japan, Vietnam, Philippines, Taiwan Hong Kong	11	5/6

Source: (websites of the airlines, March 2017).

Table 4. Value alliance—product policies of member airlines.

Allied airline	Classes	Frequent flyer programme
Cebu Pacific	Fly Only Fly + Baggage Fly + Baggage + Meal	GETGO
Jeju Air	Saver Fare Special Fare Regular Fare	Refreshed Points = Bonus Tickets
Nok Air	Promotion NokEco NokFlexi	Nok Fan Club = Nok Miles
NokScoot	Fly FlyBag FlyBagEat ScootBiz	N/A
Scoot	Economy Scoot Biz	Plus Perks (selected routes) Earn Krisflyer Miles
Tigerair Singapore	Economy Scoot Biz	PlusPerks—Earn Krisflyer Miles
Tigerair Australia	Light Express	Velocity Points/Points + Pay
Vanilla Air	Campaign Simple Inclusive	Use ANA Miles to Fly Vanilla Air = Vanilla Air Flight Award

Source: (websites of the airlines, March 2017).

tance of customers' perception in this context was shown by Wang (2014) who confirmed a positive relationship between the membership of airlines in the global airline alliances and brand equity on one side and purchase decisions made by passengers on the other side, albeit using only two airlines from Taiwan in his analysis.

4 CONCLUDING REMARKS

Without any doubt, the emergence of Value Alliance (and further existing or planned low-cost alliances) is a new phenomenon in the provision of air services. Alliance's cooperation among traditional carriers in the form of global alliances helped them to compete against expanding, but fragmented (i.e. not allied) low-cost airlines, exploiting thus economies of scale, density, scope, and positive network externalities. At the present, the competitive status of global alliances may be threatened by new emerging alliance's groupings of low-cost airlines which also "discovered" advantages stemming from being allied.

The future evolution in the global airline industry in terms of alliances and competition is not easily to be forecasted and several scenarios may occur. New international pan-regional alliances of low-cost carriers may rival more effectively with traditional carriers allied in global alliances,

strengthening thus competition of global nature on markets with long-haul air services. Due to this, the global airline industry dominated by three global alliances will be changed and may acquire a more polycentric fashion.

The new international pan-regional alliances of low-cost airlines can also strengthen competition among low-cost rivals in the respective world regions. Airlines from three ASEAN member countries (Singapore, Thailand and Philippines) participate in Value Alliance, while Air Asia—a further main market player in the region—is an airline of Malaysia which is also a member country of ASEAN. The effort to create a single aviation market in ASEAN and standing liberalization in the region (Tan 2013) may explain partially willingness to be allied. Thus, to be allied or not may be among the crucial factors influencing the survival of low-cost players on markets with air services within world regions in which liberalization is in progress.

In the long term, one cannot exclude the emergence of a global low-cost alliance built upon the emerging pan-regional international low-cost airlines' alliances.

A further scenario is seen by us in the creation of a mega alliance comprised of a global alliance including traditional carriers and an international pan-regional alliance of low-cost carriers. New collaborative platforms for low-cost carriers, which are being created at the present, reveal an appetite of traditional carriers to cooperate with low-cost carriers which were not previously perceived by traditional carriers as candidates for cooperation. On the other hand, when alliances of low-cost carriers are created, this will increase their bargaining capability with global alliances to cooperate.

Likewise, the thinning of the existing global alliances due to the willingness of several of their members to be allied differently, i.e. within an alliance of low-cost carriers (with their own low-cost daughters or joint-ventures and regional partners) is a possible scenario for future developments.

In all these scenarios, the survival of emerging alliances of low-cost carriers has been assumed, however, whether the emerging low-cost alliances will be among lasting alliances is not fully clear now because current research has not answered the question completely. Therefore, Value Alliance and further alliances of low-cost carriers ought to be thoroughly investigated. As the most perspective research track we see the impact of being allied on the low-cost airlines' performance and competitiveness against rivals—not allied low cost carriers and global airlines' alliances as well.

REFERENCES

Agusdinata, B. & de Klein, W. 2002. The dynamics of airlines alliances. In: Journal of Air Transport Management 8, pp. 201–211.

Evans, N. 2001. Collaborative strategy: an analysis of the changing world of international airline alliances. In: Tourism Management 22, pp. 229–243.

Gudmunsson, S. V. & Rhoades, D. L. 2001. Airline alliance survival analysis: typology, strategy and duration. In: Transport Policy 8, pp. 209–218.

Kleymann, B. 2005. The dynamics of multilateral allying: a process perspective on airline alliances. In: Journal of Air Transport Management 11, pp. 135–147.

Li, M. Z. F. 2000. Distinct features of lasting and non-lasting airline alliances. In: Journal of Air Transport Management 6, pp. 65–73.

Min, H. & Joo, S.-J. 2016. A comparative performance analysis of airline strategic alliances using data envelopment analysis. In: Journal of Air Transport Management 52, pp. 99–110.

Morrish, S. C. & Hamilton, R. T. 2002. Airline alliances—who benefits? In: Journal of Air Transport Management 8, pp. 401–407.

TAN, A. K.-J. 2013. Toward a Single Aviation Market in ASEAN: Regulatory Reform and Industry Challenges. ERIA Discussion Paper Series ERIA-DP-2013-22. Available at: http://www.eria.org/ERIA-DP-2013-22.pdf.

Tomová, A. & Materna, M. 2017. The directions of on-going air carriers' hybridization: Towards peerless business models? TRANSCOM 2017: International scientific conference on sustainable, modern and safe transport, in press.

Wang, S. W. 2014. Do global alliances influence the passenger's purchase decision? In: Journal of Air Transport Management 37, pp. 53–59.

Zhang, A. et al. 2008. Low-cost carriers in Asia: Deregulation, regional liberalization and secondary airports. In: Research in Transportation Economics 24(1), pp. 36–50.

Zou, L. & Chen, X. 2017. The effect of code-sharing alliances on airlines profitability. In: Journal of Air Transport Management 58, pp. 50–57.

New Trends in Process Control and Production Management – Štofová & Szaryszová (Eds)
© 2018 Taylor & Francis Group, London, ISBN 978-1-138-05885-9

Effect of personality traits (BFI-10) and gender on self-perceived innovativeness

C. Olexová
Faculty of Business Economics in Košice, University of Economics in Bratislava, Košice, Slovak Republic

F. Sudzina
Faculty of Social Sciences, Aalborg University, Aalborg, Denmark

ABSTRACT: The aim of the paper is to investigate if gender and personality traits influence self-perceived innovativeness. There are two versions of the dependent variable used—innovativeness in the eyes of others, and innovativeness in one's own opinion. Big Five Inventory-10 is used to measure personality traits: openness to experience, conscientiousness, agreeableness, extraversion, and neuroticism. The research was conducted in Slovakia using a paper-based questionnaire which contained 10 statements. This paper is a replication of a previous study conducted in Denmark. According to the findings, neuroticism and openness to experience have impact on self-perceived innovativeness in the eyes of others, while in previous study, conscientiousness influences self-perceived innovativeness in the eyes of others. Openness to experience influences self-perceived innovativeness in one's own opinion in both researches. In this research, agreeableness and neuroticism have significant influence, too.

1 INTRODUCTION

Innovativeness is a notable trait that is examined from different points of view nowadays (Feist 1998, Fursov et al. 2017, Rothmann & Coetzer 2003, Stock et al. 2016, Sung & Choi 2009, Wolfradt & Pretz 2001).

The importance of user-innovation has largely been argued through efficiency of product development (Hienerth et al. 2014, Majerník et al. 2015). Innovativeness is emphasised in business practice at present, to invent new products and services, to improve, upgrade or streamline current products, services or processes. Therefore, the pressure on people to be innovative is gaining strength. One of the important factors for innovativeness identified by Conradie et al. (2016) is dissatisfaction. Dissatisfied people make effort to ameliorate products or processes to be pleased.

Innovativeness attracts attention of many researchers who examine the factors influencing this trait. The aim of this paper is to investigate whether personality traits and/or gender influence innovativeness.

The impact of the Big Five Inventory personality traits that include five factors as the basic dimensions of individual differences: neuroticism, extraversion, conscientiousness, agreeableness, and openness to experience (Costa & McCrae 1992), on consumer innovation success was

investigated by Stock et al. (2016). A literature preview supporting the hypotheses why innovativeness should be influenced by personality traits is also provided in this paper (Stock et al. 2016). Although the findings regarding the personality traits of user-innovators are still rather rudimental, this field holds opportunities for future research (Fursov et al. 2017).

Creativity is positively linked to two personality traits, openness to experience and extraversion (Feist, 1998, Rothmann & Coetzer 2003, Sung & Choi, 2009, Wolfradt & Pretz 2001). With regards to conscientious, a positive link is found by Rothmann & Coetzer (2003) but a negative relation is found by e.g. George & Zhou (2001). In some studies, there was found a negative link between creativity and neuroticism (Rothmann & Coetzer 2003) and agreeableness (King et al. 1996). Number of solutions generated by individual subjects was significantly correlated to openness to experience (Stock et. al 2016) and problem solving is examined in more detailed way in further research of Stock et al. (2017).

The results of testing multiple models by Stock et al. (2016) are as follows:

- ideation (1st stage) - was influenced by openness to experience and gender;
- prototyping (2nd stage) - was influence by extraversion, conscientiousness and gender;

- a) peer-to-peer diffusion and b) commercial diffusion (3rd stage) - were both influenced by conscientiousness.

Based on the literature, it is realistic to expect that openness to experience, extraversion, conscientiousness and gender may probably turn out to be significant also in the analysis presented in this paper. Overall, the research presented in this paper can be considered as a replication of a part of Stock et al. (2016) model, and a full replication of (Sudzina 2016). The goal is to see whether the identified relationships hold even if fewer items are used to measure the Big Five Inventory than Stock et al. (2016) used and innovativeness is measured differently from Stock et al. (2016).

The rest of the paper is organised as follows: data and methodology contains the description of the questionnaire and the analysis, the following section contains results of the research, and finally, the summary of the findings and discussion is provided.

2 DATA AND METHODOLOGY

Data were collected in the February 2017 by using a broader questionnaire dealing with personality traits. Respondents were students of the University of Economics in Bratislava, Slovakia. In total, 136 students (of whom 44 were male and 92 female) answered all relevant questions. Moreover, there was one respondent who did not provide information on gender but filled in all other answers. This additional, 137th respondent will be used in streamlined models, which do not contain gender.

The research presented in this paper measures innovativeness using two statements based on (Gimpel et al. 2014, Gimpel et al. 2016); they used it to measure innovativeness as a part of a self-identity construct. The instruction was "Please indicate to what degree you agree with the following statements":

- "People consider me as somebody with an innovative mind";
- "I consider myself as somebody with an innovative mind".

A 1–5 Likert scale was used where 1 meant strongly disagrees and 5 stood for strongly agree. Despite both answers are self-reported, they provide an insight in how respondents perceive their innovativeness in the eyes of others and in their own opinion.

Stock et al. (2016) used Costa & McCrae's (1992) instrument to measure the Big Five Inventory; the instrument contains 50 statements. The research presented in this paper is based on the

newer version of the questionnaire (Rammstedt & John 2007) which contains 10 statements. The aim is to test whether the instrument with one fifth of questions compared to Costa & McCrae's (1992) questionnaire for the Big Five Inventory can lead to significant results. The instruction was to rate "How well do the following statements describe your personality" with statements "I see myself as someone who..."

1. ... is reserved;
2. ... is generally trusting;
3. ... tends to be lazy;
4. ... is relaxed, handles stress well;
5. ... has few artistic interests;
6. ... is outgoing, sociable;
7. ... tends to find fault with others;
8. ... does a thorough job;
9. ... gets nervous easily;
10. ... has an active imagination.

On a 1–5 Likert scale where 1 meant strongly disagrees and 5 stood for strongly agree. Extraversion was calculated as an average of the 1st (reversed-scored) and the 6th answer, agreeableness as an average of the 2nd and the 7th (reversed-scored) answer, conscientiousness as an average of the 3rd (reversed-scored) and the 8th answer, neuroticism as an average of the 4th (reversed-scored) and the 9th answer, and openness to experience as an average of the 5th (reversed-scored) and the 10th answer. Cronbach alphas for personality traits will not be reported since the Big Five Inventory-10 (Rammstedt & John 2007) was not constructed with this statistics in mind.

The Slovak translation of the questions was used according to the Slovak translation of BFI-2 items by Halama & Kohút (John & Soto 2015), as it was published on official web site of the authors of BFI-2.

A generalized linear model (GLM) was used to analyse impact of gender and of five personality traits (extraversion, agreeableness, conscientiousness, neuroticism, openness to experience) in three models where the dependent variables were:

1. innovativeness in the eyes of others ("People consider me as somebody with an innovative mind");
2. innovativeness in one's own opinion ("I consider myself as somebody with an innovative mind");
3. innovativeness in the eyes of others minus innovativeness in one's own opinion.

A multivariate approach to testing was used. Parameter estimates tables will be provided (instead of ANOVA-style tables) in order to be able to see signs of parameter estimates (not only p-values). The results should be equivalent to a multiple linear regression model estimates in case the dummy variable is set to 1 for male and to 0 for female. R^2 and R^2_{adj} are provided in order to be transparent about how much a model explains though it may be significant.

To measure correlation between answers for statements "People consider me as somebody with an innovative mind" and "I consider myself as somebody with an innovative mind", Pearson product-moment correlation coefficient is used. To test a difference between these two variables, a paired samples t-test was used. SPSS software was used for all the tests.

3 RESULTS

Parameter estimates for the generalized linear model analysing impact of gender and of personality traits on self-perceived innovativeness in the eyes of others are provided in Table 1.

With regards to the explanatory power, $R^2 = 0.218$, $R^2_{adj} = 0.175$, p-value < 0.001. Neuroticism and openness to experience have significant impact on self-perceived innovativeness in the eyes of others. In (Sudzina, 2016), the model per se was borderline significant (p-value = .064), $R^2 = 0.069$, $R^2_{ad} = 0.035$ and conscientiousness was the only significant variable.

Submodels were tested to see whether omissions of certain independent variables could improve p-values. Parameter estimates for the best submodel are provided in Table 2.

In tested models with agreeableness and/or conscientiousness on top of neuroticism and openness to experience, p-values for agreeableness and/or conscientiousness were well over .1. With regards

to the explanatory power of the streamlined model from Table 2, $R^2 = 0.186$, $R^2_{adj} = 0.174$, p-value < 0.001. In (Sudzina 2016), the streamlined model containing only conscientiousness was significant (p-value = 0.003), $R^2 = 0.052$, $R^2_{adj} = 0.046$.

Parameter estimates for the generalized linear model analysing impact of gender and of personality traits on self-perceived innovativeness in one's own opinion are provided in Table 3.

With regards to the explanatory power, $R^2 = 0.399$, $R^2_{adj} = 0.371$, p-value < 0.001. Agreeableness, neuroticism, and openness to experience have significant impact on self-perceived innovativeness in one's own opinion, and extraversion has a borderline significant impact. In (Sudzina, 2016), the model per se was not significant (p-value = 0.247), $R^2 = 0.046$, $R^2_{adj} = 0.011$, while openness to experience and agreeableness had the lowest p-values, and when a bivariate test was used, only openness to experience was borderline significant (p-value = 0.071), $R^2 = 0.019$, $R^2_{adj} = 0.013$.

Submodels were tested to see whether omissions of certain independent variables could improve p-values. Parameter estimates for the submodel without conscientiousness and gender are provided in Table 4.

With regards to the explanatory power, $R^2 = 0.385$, $R^2_{adj} = 0.366$, p-value < 0.001. In general, nothing changed in a sense that agreeableness, neuroticism, and openness to experience are still significant, and extraversion is borderline significant.

Table 1. Parameter estimates for model 1.

Parameter	B	Std. Error	t	Sig.
Intercept	3.188	0.553	5.769	0.000
Extraversion	0.027	0.086	0.316	0.753
Agreeableness	−0.119	0.083	−1.439	0.153
Conscientiousness	0.123	0.075	1.643	0.103
Neuroticism	−0.269	0.072	−3.762	0.000
Openness to experience	0.170	0.069	2.454	0.015
Gender	−0.361	0.727	−0.497	0.620

Table 2. Parameter estimates for streamlined model 1.

Parameter	B	Std. Error	t	Sig.
Intercept	3.299	0.322	10.244	0.000
Neuroticism	−0.310	0.066	−4.728	0.000
Openness to experience	0.174	0.069	2.532	0.013

Table 3. Parameter estimates for model 2.

Parameter	B	Std. Error	t	Sig.
Intercept	3.184	0.586	5.430	0.000
Extraversion	0.162	0.091	1.780	0.077
Agreeableness	−0.298	0.088	−3.388	0.001
Conscientiousness	0.114	0.080	1.432	0.155
Neuroticism	−0.373	0.076	−4.909	0.000
Openness to experience	0.356	0.073	4.855	0.000
Gender	−0.242	0.152	−1.593	0.114

Table 4. Parameter estimates for the first streamlined model 2.

Parameter	B	Std. Error	t	Sig.
Intercept	3.503	0.531	6.601	0.000
Extraversion	0.178	0.090	1.965	0.051
Agreeableness	−0.306	0.086	−3.555	0.001
Neuroticism	−0.423	0.071	−5.933	0.000
Openness to experience	0.366	0.073	4.995	0.000

But p-value for extraversion decreased from *.077* to *.051*; with a few more respondents, this could possibly drop below *.05*. Excluding also extraversion from the model leads to all independent variables being significant, as it can be been in Table 5.

With regards to the explanatory power, $R^2 = 0.367$, $R^2_{adj} = 0.352$, p-value < 0.001. Regression coefficients changed only marginally.

The correlation coefficient for innovativeness in the eyes of others and innovativeness in one's own opinion is *0.746 (p-value < 0.001)*. In Sudzina (2016), it was *0.596 (p-value < 0.001)*. The correlation coefficient of *.746* translates into Cronbach's alpha of *0.844*, i.e. higher than Nunnally's (1978) threshold of *0.7*.

On average, innovativeness in the eyes of others was *3.02* and innovativeness in one's own opinion was *3.16*; the difference *0.14* is significant (*p-value = 0.009*). In (Sudzina 2016), the averages were *3.29* and *3.38*, i.e. respondents rated their innovativeness in their own opinion higher than in the eyes of other; the difference of *0.092* was not significant (*p-value = .113*).

Parameter estimates for the generalized linear model analysing impact of gender and of personality traits on the difference between self-perceived innovativeness in the eyes of others and in one's own opinion are provided in Table 6.

With regards to the explanatory power, $R^2 = 0.172$, $R^2_{adj} = 0.134$, p-value < 0.001. In (Sudzina 2016), the model per se was not significant (*p-value = 0.316*), $R^2 = 0.041$, $R^2_{adj} = 0.007$, and conscientiousness had the lowest p-value.

Table 5. Parameter estimates for the second streamlined model 2.

Parameter	B	Std. Error	t	Sig.
Intercept	4.059	0.454	8.945	0.000
Agreeableness	−0.269	0.085	−3.169	0.002
Neuroticism	−0.451	0.071	−6.379	0.000
Openness to experience	0.371	0.074	5.004	0.000

Table 6. Parameter estimates for model 3.

Parameter	B	Std. Error	T	Sig.
Intercept	0.004	0.462	0.008	0.993
Extraversion	−0.135	0.072	−1.881	0.062
Agreeableness	0.179	0.069	2.577	0.011
Conscientiousness	0.009	0.063	0.148	0.883
Neuroticism	0.104	0.060	1.730	0.086
Openness to experience	−0.186	0.058	−3.224	0.002
Gender	0.076	0.120	0.634	0.527

Table 7. Parameter estimates for streamlined model 3.

Parameter	B	Std. Error	T	Sig.
Intercept	0.055	0.414	0.132	0.895
Extraversion	−0.144	0.071	−2.043	0.043
Agreeableness	0.194	0.067	2.887	0.005
Neuroticism	0.112	0.056	2.014	0.046
Openness to experience	−0.191	0.057	−3.328	0.001

Submodels were tested to see whether omissions of certain independent variables could improve p-values. Parameter estimates for the best submodel are provided in Table 7.

With regards to the explanatory power, $R^2 = 0.175$, $R^2_{adj} = 0.150$, p-value < 0.001. (The reason why R^2 of the streamlined model is higher than R^2 of the full model is inclusion of the respondent, who did not provide information on gender, in the streamlined model.) In (Sudzin 2016), the model including only conscientiousness was significant (*p-value = 0.049*), $R^2 = 0.023$, $R^2_{adj} = 0.017$.

4 CONCLUSION

The aim of the paper was to analyse impact of gender and of personality traits on self-perceived innovativeness. There were two versions of the dependent variable used—innovativeness in the eyes of others, and innovativeness in one's own opinion.

Neuroticism and openness to experience significantly influenced self-perceived innovativeness in the eyes of others. Agreeableness, neuroticism, and openness to experience significantly influenced self-perceived innovativeness in one's own opinion; the significance of extraversion was borderline. Conscientiousness was expected to be also significant, but it was not confirmed in the research. Gender was not found to be significant in any of the models.

Innovativeness in the eyes of others and innovativeness in one's own opinion correlate and can be used together as a construct due to reasonable value of Cronbach's alpha. There is a significant difference between averages of the two variables.

These findings could be strengthened by examination of the relations between facet scales and innovativeness in future research, to reveal whether there is a need to examine the relations in more detailed way. For this reason, 60-item version of the questionnaire for the Big Five (Soto & John 2016) should be applied.

Conradie et al. (2016) identified also academic degree as significant. Although it was not examined in this research due to the same education of all respondents in time where the data were collected in this research, the education as a control variable could be used in future research.

REFERENCES

Conradie, P.D. et al. 2016. Product Ideation by Persons with Disabilities: An Analysis of Lead User Characteristics. In *DSAI 2016, Proceedings of the 7th International Conference on Software Development and Technologies for Enhancing Accessibility and Fighting Info-exclusion* 2016: 69–76. Portugal.

Costa, P.T. & McCrae, R.R. 1992. *Revised NEO Personality Inventory (NEO-PI-R) and NEO Five-Factor Inventory (NEO-FFI) Professional Manual*. Odessa: Psychological Assessment Resources.

Feist, G.J., 1998. A meta-analysis of personality in scientific and artistic creativity. *Personality and Social Psychology Review* 2(4): 290–309.

Fursov, K. et al. 2017. What user-innovators do that others don't: A study of daily practices. *Technological Forecasting & Social Change* 118: 153–160.

George, J.M. & Zhou, J. 2001. When openness to experience and conscientiousness are related to creative behavior: an interactional approach. *Journal of Applied Psychology* 86(3): 513–524.

Gimpel, G. et al. 2014. Mobile ICT acceptance in late adopter countries. In *Proceedings of 13th International Conference on Mobile Business 2014*. London.

Gimpel, G. et al. K. 2016. Mobile ICT use in early adopter vs. late majority countries. *International Journal of Mobile Communications* 14(6): 610–631.

Hienerth, C. et al. User community vs. producer innovation development efficiency: a first empirical study. Res. Policy 43(1): 190–201.

John, O.P. & Soto, Ch. J. 2015. BFI-2 Slovak self-report form and scoring key. <http://www.colby.edu/psych/wp-content/uploads/sites/50/2013/08/bfi2-form-slovak.pdf>.

King, L.A. et al. 1996. Creativity and the five-factor model. *Journal of Research in Personality* 30(2): 189–203.

Majerník, M. et al. 2015. Process innovations and quality measurement in automotive manufacturing. In *Production Management and Engineering Sciences: Proceedings of the International Conference on Engineering Science and Production Management (ESPM 2015)*, Tatranská Štrba, High Tatras Mountains, Slovak Republic, 16th-17th April 2015 (p. 179). CRC Press.

Nunnally, J.C. 1978. *Psychometric theory*, 2nd edition. New York, NY: McGraw-Hill.

Rammstedt, B. & John, O.P. 2007. Measuring personality in one minute or less: A 10-item short version of the Big Five Inventory in English and German. *Journal of Research in Personality* 41(1): 203–212.

Rothmann, S. & Coetzer, E.P. 2003. The Big Five personality dimensions and job performance. *South African Journal of Industrial Psychology* 29(1): 68–74.

Soto, C.J. & John, O.P. 2016. The next Big Five Inventory (BFI-2): Developing and assessing a hierarchical model with 15 facets to enhance bandwidth, fidelity, and predictive power. *Journal of Personality and Social Psychology* (in press).

Stock, R.M et al. 2017. Problem Solving Without Problem Formulation: Discovering Need-Solution Pairs in a Laboratory Setting. Available at SSRN: https://ssrn.com/abstract=2902117 or http://dx.doi.org/10.2139/ssrn.2902117.

Stock, R.M. et al. 2016. Impacts of personality traits on consumer innovation success. *Research Policy* 45(4): 757–769.

Sudzina, F. 2016. Do gender and personality traits (BFI-10) influence self-perceived innovativeness? In *Proceedings of the International Scientific Conference of Business Economics Management and Marketing (ISCOBEMM) 2016*. Brno: Masaryk University, 23–29.

Sung, S.Y. & Choi, J.N. 2009. Do Big Five personality factors affect individual creativity? The moderating role of extrinsic motivation. *Social Behaviour and Personality* 37(7): 941–956.

Wolfradt, U. & Pretz, J.E. 2001. Individual differences in creativity: Personality, storywriting, and hobbies. *European Journal of Personality* 15(4): 297–310.

New Trends in Process Control and Production Management – Štofová & Szaryszová (Eds)
© 2018 Taylor & Francis Group, London, ISBN 978-1-138-05885-9

The significance of emergency planning in crisis management

M. Ostrowska & S. Mazur
Krakowska Akademia im. Andrzeja Frycza Modrzewskiego, Kraków, Poland

ABSTRACT: It is fairly common for crisis events to occur unexpectedly and to bring consequences whose negative nature goes well beyond the means available to local rescue services. Therefore, measures taken at the first stage of the crisis are usually chaotic, especially in terms of organisation. These seem to be the most essential factors to be considered in the entire crisis management process. The article describes the role and significance of planning in the crisis management system because the plan, in envisaging all types of potential threat, makes it possible to assume the implementation of mitigating measures from a relatively safe point, but developing a good-quality crisis management plan is neither easy nor straightforward. Several difficulties are encountered in the process. These are related not only to work organisation, but also to the lack of sufficient information and delays in the crisis management team work.

1 INTRODUCTION

Management can be considered as being the act of efficiently coordinating human actions to achieve a certain objective (Gontkovičová et al. 2014). It can also be defined as performing the managerial duties which entail identifying and pursuing specific objectives by way of adequately utilising information resources and processes under specific circumstances (cultural, legal, social, economic, etc.), (Walas-Trębacz & Ziarko 2010). Of note, the use of any such resources and processes must be efficient and effective, and consistent with the rationality of the duties performed. In line with a different definition, management involves using certain resources to obtain specific benefits, controlling diversity and transforming potential conflicts into cooperation (Penc 1998).

From a system-oriented perspective, management enables the employment of all organisational resources available within fundamental sub-systems, i.e. human resources, technologies, organisational structures and goals (Sienkiewicz-Małyjurek 2010).

2 CRISIS MANAGEMENT

Crisis management forms an essential part of national security. Its role cannot be overestimated in providing effective solutions to every kind of security-related issues, and in combating, and preparing for, the many threats which may occur. Essentially, it is the process of maintaining and restoring stability. Crisis management is characterised by the reasonable activity undertaken by government bodies at all State organisation levels, and involving both specialised organisations, such as security and inspection services, and the general public (Więcek & Bienick 2014). However, treating crisis management as merely being a series of attempts to undertake mitigating measures in an emergency situation provides a very limited picture. Nowadays, crisis management consists of numerous actions involving an array of interacting elements. The notion is very broad, and its framework comprises organisational structures and measures aimed at: reducing the likelihood of a severe crisis impact, efficiently keeping the conflict under control, reducing crisis consequences, mitigating the consequences and restoring the previous state, enabling the authorities to take adequate steps to make the crisis evolve in a desirable direction, and to develop and put in place optimum solutions.

Crisis management involves the activities performed by State administration bodies within the area of national security management. (Grocki 2012). It involves preventing the occurrence of emergency situations, being able to control these by way of well-planned actions, adequately responding to emergency situations, removing their consequences and restoring critical resources and infrastructure (Ordinance No. 86 of the President of the Council of Ministers of 14 August 2008 on the Organisation and Working Mode of the Government Crisis Management Team http://isap.sejm.gov.pl/).

Certain elementary principles of crisis management can also be distinguished, including in particular:

- The principle of territorial division supremacy—which is considered to form the underlying structure of the territorial division model of the country, with the industrial division acting as a supplementary structure (Mihalčová et al. 2014).
- One-person management—which implies that one person takes and is held accountable for all critical decisions.
- The principle of accountability of public authorities—which means that these bodies assume responsibility for taking decisions in emergency cases.
- The unification principle—according to which the administrative authorities are granted the principal powers ensuring the adequate performance of duties entrusted to these bodies.
- The threat categorisation principle—consisting of threat assignment grouped by type and impact scale, for which many specific legal, organisational and financial solutions are identified.
- The principle of generality—which means that crisis management is to be organised by public authorities, in cooperation with existing specialised institutions and organisations, as well as the public (Sobolewski 2011).

The principal objective of crisis management is to ensure the security of people and to establish the conditions conductive to the development of all types of entities in each area. Crisis management covers (Sienkiewicz-Małyjurek 2015).

- The following three types of threats: natural and non-natural, technical, warfare-related,
- All management levels (corresponding to authority levels): local, provincial, central.

Following an analysis of the literature and normative acts concerning this subject matter, it can be inferred that crisis management is comprised of several stages:

- Prevention, i.e. activities aimed at eliminating or reducing the likelihood of a catastrophic event, or mitigating its consequences through: analysis (threat categorisation): assessment of the public sensitivity to threat, legal regulations, rationally planned spatial development, budget compliance and planning, assessment of the number of casualties, as well as property and infrastructure losses caused by the catastrophic event, the devising of a plan for preventive measures, identification of control and supervision principles and methods,
- Preparation, i.e. indicating the ways of reacting and measures to be taken with the aim of increasing the resources and means necessary to adequately react to a crisis event, through:

development of an appropriate crisis management plan, establishment of crisis management centres, formulating fundamental principles of communication, identification of the in-place monitoring systems, identification of alert and warning systems, the putting into place of effective procedures for requesting and providing assistance, definition of the principles of legal coercion as applicable in respect of individuals, non-governmental organisations and private sector entities, establishment of warehousing bases and databases to enable the ready-supply of the necessary resources and materials, development of databases, promotion of public awareness, competence-raising of rescue and first-response services, the seeking of social acceptance of potential incurred costs, the continuous updating of preparation elements.
- Response, i.e. a set of measures taken immediately after the occurrence of a catastrophic event. More specifically, response aims at providing assistance to the injured, and limiting the scale of secondary damage or loses, through: launch of continual information processes (information management), establishment of contact points (information provision), operation of alert and warning systems, generation of immediate appropriate response within the affected local community, launch of adequate emergency response procedures, activation of rescue measures, launch of evacuation processes, neutralisation of threat focal points, organisation of community self-assistance, provision of operational support to armed forces, involvement of social and humanitarian organisations, the provision of psychological care to casualties and survivors, the establishment of appropriate temporary survival conditions for the affected.
- Restoration, i.e. measures aimed at restoring the previous condition and making the infrastructure less sensitive to subsequent catastrophic events, through: estimation of the scale of damage, provision of assistance to the affected population, provision of treatment and rehabilitation to the injured, compensation to the affected, the injured and displaced, provision of information on the rights and obligations of the affected, re-creation and supplementation of necessary supplies, restoration of rescue services readiness, restoration of ecological balance and security, the rebuilding and restoration of infrastructure efficiency, the establishment of legislative initiatives, restoration of efficient administration, settlement of the response costs (liabilities), summation of unfolded events and the drawing of appropriate conclusions, modification and updating of response plans, documentation of event (reporting) (Szczupaczyński 2002).

The management, i.e. planning, analysing and steering, covers such aspects as:

- Objectives – plans and tasks
- Activities – disposing, guiding, planning and controlling
- Authorities – managers and supervisors,
- Resources – information,
- Vehicles – organisation, incentives and needs.

The Act on Crisis Management directly obliges local-government and central administration bodies to plan various alternatives and methods of responding to all types of incidents and adversities. Under this Act and as indicated in the literature, emergency planning forms part of the management process. This involves taking decisions concerning the rational use of the available means and resources to overcome a given crisis (Act of 26 April 2007 on Crisis Management).

3 CRISIS PLANNING

The essence of planning lies in forecasting future events and determining, by way of a document referred to as "a plan", the tasks and resources necessary for the efficient performance of future activities by a given unit, aimed at accomplishing a certain (public) objective in relation to strategic, tactical and operational duties.

Prior to commencing the planning process, it appears indispensable to learn about the scope of the plan, its structure, content and purpose. In crisis management theory, three types of plans are distinguished:

- Strategic plans – long-term general plans containing decisions on the allocation of resources, along with priorities and measures necessary to accomplish certain strategic objectives;
- Tactical plans – mid-term plans oriented towards accomplishing certain tactical objectives, drawn up for implementing strategic plan elements;
- Operational plans – short-term plans oriented towards implementing tactical plans, with a view to accomplishing certain operational objectives (Grodzki 2012).

The planning process can be divided into: military planning aimed at countering enemy activity, civil planning, the purpose of which is to save human lives and health.

In the national commanding process, three stages have been recognised, i.e.

- Determining the location and scope of the disaster at hand,
- Planning and delegating duties,
- Controlling (Bieniok et al. 1999).

The duties envisaged as part of the civil planning include:

- Identifying threats that are likely to occur within the activity area,
- Identifying the means to overcome: threat to people's health and well-being, threat to equipment, civil material threat,
- Establishing dedicated teams and structures, and making them ready to perform rescue actions,
- Laying down the principles of cooperation with: supervisors (district and provincial bodies, as well as government bodies), neighbours (commune, district and provincial bodies), other countries.

Emergency plans are public documents which set the objectives, legal bases and objectives to be pursued. They should be developed at all management levels, considering that an obligation to provide for emergency measures is binding on the authorities of all levels. The content of an emergency plan defines the duties to be assumed by the authorities at different levels in the event of crisis. It is also worth noting that commune and district emergency plans should focus on elementary measures to protect the local population. These mainly include providing adequate information and warnings, making the population aware of the evacuation methods and procedures, and sheltering possibilities for evacuated individuals.

Moreover, district emergency plans should be oriented towards exploiting the available means and resources to combat the emerging threat. The generated plans contain foreseen threat characteristics, along with assessments of risk of occurrence, information on the hardiness of critical infrastructure, as well as risk and threat assessment maps. Herein, critical infrastructure is deemed both whole systems and interrelated system constituents, such as buildings, equipment, installations and services that are crucial to national security and citizen safety, as well as services aimed at ensuring the efficient functioning of various entities, institutions or enterprises.

Emergency plans are directly used to determine the scope of preparatory measures, while also providing the basis for the organisation of training and drill sessions. Training sessions provide an opportunity for the crisis management staff not only to familiarise themselves with their duties, but also to acquire the necessary skills to adequately perform their functions. Drills, in turn, serve as a chance to verify the emergency plan effectiveness and to test the response procedures. They also provide the means of verifying the actual skills of the crisis management staff.

It is the crisis management staff's duty to provide a safety net understood as a set of threats to be addressed, included in the catalogue, with an indication of the managing entity and cooperating bodies, as well as measures intended to target a given type of threat (Ivaničková et al. 2015). A set of means and resources to be used in emergency situations should include the resources that could be directly exploited within certain periods of time, under specific crisis circumstances. It should take the form of a continually updated database (Act of 26 April 2007 on Crisis Management).

Summing up, the master crisis management plan deals with arranging for emergency measures in the community, along with specifying the legal bases for emergency operations and the circumstances in which they should be undertaken. It also explains the overall action concept, providing a list of duties related to emergency measures (Pružinský 2015). This also concerns the obligation to develop standard operational procedures for individual measures.

Individual plans should also be consistent with the solutions provided for in the National Crisis Management Plan, and they should comprise measures to be implemented at all crisis management stages (Ordinance No. 23 of the Minister of Administration and Digitisation of 13 December 2013 on the Guidelines to Provincial Crisis Management Plans) (i.e. prevention, preparation, response and restoration).

The structure and content of the plan should be consistent with the duties and objectives to be implemented. It serves the purpose of:

- Ensuring the uniformity of the performance principles for various types of crisis management services and organisations,
- Establishing the principles of the co-existence of various administrative levels, depending on the scale and type of threat or crisis circumstances,
- Identifying the necessary planning documents for government and local-government administration bodies, combined and special services, as well as entities in charge of taking measures in emergency situations,
- Identifying the potential types of threat and functions of individual entities to ensure the proper performance of measures, mainly as regards communications, medical care, water supply, food, social assistance, evacuation, transportation, energy, alert and warning systems, media cooperation, financing, public order, law or critical infrastructure protection (Grodzki 2012).

The crisis response plan, developed by taking a functional approach, comprises: the master plan and functional annexes, the latter containing information on specific events, a list of organisations entrusted with specific tasks under the general plan or functional annexes, as well as implementation schemes of the crisis response plan.

The master plan covers:

- Threat specifications and risk assessments,
- Duties and responsibilities of the crisis management staff,
- A set of means and resources to be used in emergency situations.
- The master plan should be accompanied by functional annexes containing specific procedures and guidelines. These should include (Act of 26 April 2007 on Crisis Management):
- The procedures for the organisation of crisis management tasks,
- The organisation of communication between various entities performing crisis management tasks,
- The organisation of threat monitoring, warning and alert systems,
- The principles of disseminating information on threat and threat management methods,
- The principles of organising evacuation from the threat impact area,
- The organisation of rescue and medical services, as well as social and psychological assistance,
- The organisation of protection against province-specific types of threat,
- A list of contracts and agreements concluded directly for the purpose of implementing the tasks envisaged in the crisis management plan,
- The principles and mode of damage assessment and documentation,
- The procedures for launching strategic reserves,
- A list of critical infrastructure elements.

Emergency plans are directly used to determine the scope of preparatory measures, while also providing the basis for the organisation of training and drill sessions. Training sessions provide an opportunity for the crisis management staff not only to familiarise themselves with the duties, but also to acquire the necessary skills to adequately perform their functions (Pružinský & Mihalčová 2015). Drills, in turn, serve as an opportunity to verify the emergency plan effectiveness and to test the response procedures. They also provide the means of verifying the actual skills of the crisis management staff.

Emergency planning also facilitates response and short-term restoration. If the plan is flexible enough and can be used in the event of any threat, it provides a solid basis for crisis management, thanks to which the community has an opportunity to continue, in good faith, the activities aimed at long-term prevention oriented towards specific types of threat.

4 CONCLUSION

Summing up, the plan, in envisaging all types of potential threat, makes it possible to assume the implementation of mitigating measures from a relatively safe point, but developing a good-quality crisis management plan is neither easy nor straightforward. Many difficulties are encountered in the process. These are related not only to work organisation, but also to the lack of sufficient information and delays in the crisis management team work.

The cooperation in such a diverse team of experts (in the field of communications, medical care, water supply, food, social assistance, evacuation, transportation, energy, alert and warning systems, media cooperation, financing, public order, law or critical infrastructure protection) requires a good organisational preparation, i.e. developing an action plan and proposed concepts for a plan containing a list of issues to be covered and their due consideration. Before commencing the implementation, it appears indispensable to assess the available resources, both human and material. It also seems advisable to indicate a managing unit in charge of the plan development, to point out cooperating bodies, to appoint a planning team, and to establish a local coalition for the plan implementation purpose.

REFERENCES

Act of 26 April 2007 on Crisis Management (*Journal Laws* of 2013, item 1166).

Act of 26 April 2007 on Crisis Management *(Journal of Laws* of 2007, No. 89, item 590 as amended).

Bieniok, H. et al. 1999, Efficient action methods, planning, organising, motivating and controlling, Placet, pp. 57–58.

Gontkovičová, B. et al. 2014. Youth unemployment—current trend in the labour market? In *Global conference on business, economics, management and tourism*, Procedia economics and finance: 2nd Global conference on business, economics, management and tourism, 30–31 October 2014, Prague, Czech Republic. Web of Science. The Netherlands: Elsevier B.V., 2015. SCOPUS ISSN 2212–5671, 2015, vol. 23, pp. 1680–1685. VEGA 1/0708/14. Available at: ftp://193.87.31.84/0207088/1-s2.0-S2212567115005547-main.pdf.

Grocki, R., 2012, Crisis management. Good practices, Difin, Warsaw, pp. 105–115.

Ivaničková, M et al. 2015. Intercultural differences in the Visegrad group: the Hofstede model application. SCOPUS. In *Actual problems of economics: scientific economic journal.* - Kyiv: National Academy of Management, 2015. ISSN 1993-6788, 2015, no. 6, p. 284–291.

Mihalčová, B. et al. 2014. Brand positioning of a Slovak company [Pozicioniranje marke Slovačkog poduzeća].

In *Management (Croatia)*. Split: University of Split—Faculty of Economics. ISSN 1331-0194. Volume 19, Issue 2, 2014, SCOPUS. Pages 197–204.

Ordinance No. 23 of the Minister of Administration and Digitisation of 13 December 2013 on Guidelines to Provincial Crisis Management Plans—Under Article 14, Par. 3 of the Act of 26 April 2007 on Crisis Management (*Journal Laws* of 2013, item 1166).

Ordinance No. 86 of the President of the Council of Ministers of 14 August 2008 on the organisation and Working Mode of the Government Crisis Management Team.

Penc, J. 2015. *Management for the Future. Creative corporate management*, Professional Business School, pp. 25–29.

Pružinský, M. & Mihalčová, B. 2015. Material flow in logistics management. (Conference Paper) *International Conference on Engineering Science and Production Management*, ESPM 2015; Tatranská Štrba, High Tatras Mountains; Slo-vakia; 16 April 2015 through 17 April 2015; Code 154039. In Production Management and Engineering Sciences—Sci-entific Publication of the International Conference on Engi-neering Science and Production Management, ESPM 2015. Editorial: Majerník, M., Daneshjo, N. & Bosák, M., Lon-don: Taylor & Francis Group, 2015, 610 p. ISBN 978-1-138-02859-2 (Hbk); SCOPUS, ISBN: 978-1-315-67379-0 (eBook PDF) pp. 523–528.2016, Pages 523–528.

Pružinský, M. 2015. Knowledge management in the marketing mix of small food businesses. In Smart City 360°, Volume 166 of the series Lecture Notes of the Institute for Computer Sciences, *Social Informatics and Telecommuni-cations Engineering*. Editors: Akan, O., Bellavista, P., Cao, J., Coulson, G., Dressler, F., Ferrari, D., Gerla, M., Koba-yashi, H., Palazzo, S., Sahni, S., Shen, X.S., Stan, M., Xiaohua, J., Zomaya, A.Y. Belgium: ISSN: 1867-8211 pp 541–550. Available online: http://link.springer.com/bookseries/8197.

Sienkiewicz-Małyjurek, K, 2015, *Efficient crisis management (revised)*, Difin, Warsaw, pp.11–18.

Sobolewski, G. (ed.), 2011, *Emergency threats*, National Defence University, Warsaw, pp. 67.

Szczupaczyński, J,. 2002, *Organisation management anatomy,* International Management School. Warsaw.

Walas-Trębacz, J. & Ziarko J., 2010, *Crisis management basics, part 1: Crisis management in public administration,* Andrzej Frycz Modrzewski Krakow University, Kraków 2010, pp. 10–11.

Więcek, W. & Bieniek J., 2014, *Crisis management basics and training scenarios,* National Defence University, Warsaw, p. 38 of 13 December 2013 on Guidelines to Provincial Crisis Management Plans—Under Article 14, Par. 3 of the Act of 26 April 2007 on Crisis Management (Journal Laws of 2013, item 1166) of 13 December 2013 on Guidelines to Provincial Crisis Management Plans—Under Article 14, Par. 3 of the Act of 26 April 2007 on Crisis Management (Journal Laws of 2013, item 1166).

New Trends in Process Control and Production Management – Štofová & Szaryszová (Eds)
© 2018 Taylor & Francis Group, London, ISBN 978-1-138-05885-9

Reporting of sustainable development of enterprises in Slovakia

R. Pakšiová, D. Oriskóová & K. Lovciová
Faculty of Economic Informatics, University of Economics in Bratislava, Bratislava, Slovak Republic

ABSTRACT: The aim of this paper is to analyse reporting of sustainable development in the Slovak Republic in accordance with the transposition of EU Directives to the national legislation. New requirements arising from international harmonization of reporting have a significant impact on disclosure in financial statements, as well as in annual reports. Reporting is a key to assessment of sustainable development, which has an impact on evaluation of the achievement of business activity goals. Different scope of reporting of financial and partially non-financial information according to size groups of enterprises is a key aspect for the assessment of sustainable development, which has its advantages and disadvantages. The advantage is lower administrative burden for micro-enterprises, but disadvantage, on the other hand, is that the aggregated information is not sufficient. An important supplementary source of information for sustainable development assessment is the annual report because stakeholders can more easily asses the comprehensive information.

1 INTRODUCTION

The concept of sustainable development of the United Nations allows to fulfil present needs without compromising the needs of future generations. Sustainable development is a basic principle of development in the European Union and Slovak Republic. Slovakia participates in translating the principles of sustainable development into daily life. Evaluation of the approach to sustainable development is made through measurable indicators, many of which are published by enterprise by various means, such as on the Internet and by publication of a statement on social responsibility. In the document "The National Strategy of Sustainable Development of the Slovak Republic" these main objectives were declared: reducing environmental burden and the proportion of use of non-renewable resources, and supporting the use of renewable resources. The strategy of the sustainable development is translated into national and multinational legislation to move enterprise towards sustainable development. Application of social responsibility as a prerequisite for sustainable development is shaping the social, institutional, environmental, and economic environment. Information arising out of the obligation of disclosure of information on corporate social responsibility are reported in the annual report, which includes the financial statements. The annual report discloses that financial and non-financial information, which is an important resource in assessing the degree of implementation of sustainable development.

2 THE AIM AND METHODOLOGY

The aim of this paper is an analysis of disclosure requirements of the sustainable development of enterprises in the Slovak Republic. International harmonization has a significant impact on financial and nonfinancial information disclosure. To achieve the objective, various scientific methods of examination were used. The first step in examination was analysis of literature including international and national reporting regulations. The selected significant findings were further processed using deduction to summarise the requirements for financial and nonfinancial information disclosure of enterprises in the Slovak Republic. This disclosed information is information about the reporter's sustainable development. Financial information is information which can be quantified in monetary terms. Information which cannot be quantified in monetary terms represents non-financial information. Using the comparison method, differences in national legislation before and after transposition of Directive 2013/34/EU of the European Parliament and of the Council were examined. Attention was given to alternatives offered by Directive 2013/34/EU of the European Parliament and of the Council, and to the extent of their transposition into national legislation. In conclusion, the synthesis method summarized findings.

There was a need for unification of the various synonyms of enterprise, accounting unit, legal person used in various regulations, and in this paper the term enterprise will be used.

3 DISCLOSURE OF FINANCIAL AND NONFINANCIAL INFORMATION

One of the key aims of any business activity is making a profit and ensuring sustainable development. An elementary accounting principle is the going concern principle, which is in relation to fact that business activity should prosper and continue for an indefinite period. Sustainable development is a kind of extension of the going concern principle, because it indicates the enhancement, expansion, and the flourishing of business activities. The sustainable development of the enterprise can be evaluated through its financial and nonfinancial information.

3.1 Disclosure of financial information in financial statements

Financial information suitable for analysis of sustainable development is included in financial statements, which in the Slovak Republic consist of a balance sheet, profit and loss statement, and notes to the financial statements. Evaluation of financial situation, financial performance and changes in financial situation based on information disclosed in financial statements is a keystone for users' decision making.

Financial performance is reflected in the financial situation and allows analysis of changes in financial situation as well as analysis of business development. One of the significant financial performance indicators is profit or loss.

A milestone in financial information disclosure in financial statements was the transposition into national legislation of Directive 2013/34/EU of the European Parliament and of the Council of June 26, 2013 on the annual financial statements, consolidated financial statements and related reports of certain types of undertakings, amending Directive 2006/43/EC of the European Parliament and of the Council and repealing Council Directives 78/660/EEC and 83/349/EEC. The Directive was created on the "think small first" principle, aimed at decreasing the administrative burden and improving the business environment mainly for small and medium sized enterprises. According to the Directive, enterprises are categorized based on criteria such as balance sheet total, net turnover, and average number of employees during the financial year. The Directive distinguishes between these types of enterprises: micro-undertakings, small, medium-sized, and large undertakings. Significant changes in disclosure in the Slovak Republic according to Directive 2013/34/EU of the European Parliament and of the Council arose by adoption of an amendment to Act No. 431/2002 on Accounting, as in effect (hereinafter the "Act on Accounting") in 2015 (Kosovska et al. 2015).

Because the Directive allows individual states a variety of alternatives in transposing the Directive, Slovak legislation provides the following size classes: micro, small, and large accounting unit. The enterprise shall be classified to the corresponding size category if it does not exceed the limits of two categories. Table 1 shows the comparison of size criteria according to multinational and national legislation.

According to Act on Accounting, the enterprise also assesses the immediately preceding accounting period for purpose of classification in size classes.

Table 1. Criteria for categorization according to Directive 2013/34/EU of the European Parliament and of the Council and Act No. 431/2002 on Accounting, as in effect.

Categories of enterprises	Directive 2013/34/EU of the European Parliament and of the Council	Act No. 431/2002 on Accounting, as in effect
Micro-undertakings	a) balance sheet total: €350,000; b) net turnover exceeds: €700,000; c) average number of employees during the financial year: 10.	a) balance sheet total: €350,000; b) net turnover exceeds: €700,000; c) average number of employees during the financial year: 10.
Small undertakings	a) balance sheet total: €4,000,000; b) net turnover exceeds: €8,000,000; c) average number of employees during the financial year: 50.	a) balance sheet total: €350,000–€4,000,000; b) net turnover exceeds: €700,000–€8,000,000; c) average number of employees during the financial year: 10–50.
Medium-sized undertakings	a) balance sheet total: €20,000,000; b) net turnover exceeds: €40,000,000; c) average number of employees during the financial year: 250	not defined
Large undertakings	a) balance sheet total exceeds €20,000,000; b) net turnover exceeds €40,000,000; c) average number of employees during the financial year exceeds 250.	a) balance sheet total exceeds: €4,000,000; b) net turnover exceeds €8,000,000; c) average number of employees during the financial year exceeds 50.

An enterprise changes its classification starting with the next accounting period after it becomes aware it has exceeded or no longer fulfils the conditions of the size group in the two immediately preceding accounting periods.

Requirements for disclosure of financial information increase proportionally with the size of the enterprise. Large enterprises face the greatest requirements for disclosure. The Directive allows member states to adopt exceptions to certain regulations for small enterprises, occasionally for medium-sized enterprises as well.

The Slovak Republic has applied the possibility of a simplified disclosure structure for micro-enterprises, while requirements for small and large enterprises are the same. Simplification can be identified in the financial statements of micro-enterprises, for example on the balance sheet. Due to the aggregation of some items of assets, liabilities and capital disclosed individually for small and large enterprises on 145 lines, in micro enterprises the disclosure is on only 45 lines. These are mainly the disclosure details concerning inventories, long-term and short-term receivables, financial instruments, and financial accounts (cash, bank accounts).

Measurement of the fair value of assets and liabilities contributes to a true and fair view of the enterprise's financial position. (Ondrusova & Parajka 2014) Micro-enterprises do not revaluate to fair value as of the date of the financial statements, which has a negative impact on the true and fair view of enterprise's financial situation, yet a true and fair view of the enterprise, as well as a proper understanding thereof, is important for financial decision-making. (Juhaszova et al. 2014) The reporting of relevant information about an enterprise should be assessed not only in terms of the significance of the information, but also in terms of the expenses incurred for obtaining it in relation to benefits arising from the disclosure (Tumpach & Bastincova 2014).

The basis for assessment of sustainable development from the perspective of financial information is the profit and loss statement as one of the parts of the financial statements. The structure of the profit and loss statement explains the creation of profit or loss for the accounting period. The profit and loss statement presentsthe amounts contributed by the individual revenues and expenses in the creation of profit or loss. (Slosarova et al. 2016) According to the transposition of Directive 2013/34/EU of the European Parliament and of the Council to the Act on Accounting, the reporting of extraordinary revenues and extraordinary expenses was repealed, and now only revenues from operating activities and from financial activities, and expenses from operating activities and from financial activities are reported.

Directive 2013/34/EU of the European Parliament and of the Council offers two different forms of profit and loss statement. If both structures of profit and loss statement are adopted into national legislation, states may allow enterprises to choose one or the other alternative. In the Slovak Republic, the structure of the profit and loss statement is adjusted for a specific group of enterprises in accordance with the Measure of the Ministry of Finance of the Slovak Republic (Measure MF SR No. MF/15464/2013-74, MF/23378/2014-74, MF/23377/2014-74) laying down the details of the arrangement, marking, and content specification of items in individual financial statements and the extent of data determined for publication from individual financial statements for entrepreneurs using double entry bookkeeping. Directive 2013/34/EU of the European Parliament and of the Council offers simplification for micro, small and medium-sized enterprises through preparation of a condensed profit and loss statement, but in the Slovak Republic financial statements for small and large enterprises are the same. Micro enterprises have the possibility of preparing a condensed profit and loss statement, which may have an impact on the relevance of the disclosed information. The difference between condensed financial statements and normal financial statement is also in the extent of the profit and loss statement. The profit and loss statement for micro enterprises consists of 38 lines, while for small and large enterprises it has 61 lines. Based on information disclosed in the financial statements of micro–enterprises, it is possible to determine key financial indicators such as leverage, profitability, and liquidity. In practice some users, for example creditors, need more specific information not disclosed in the financial statements of a micro-enterprise, which if required can be sought from the enterprise. (Parajka 2016) Correct and understandable disclosure of information in financial statements is the basis for the financial analysis of an enterprise (Kubascikova & Juhaszova 2016).

Complementary financial information outside the financial statements, as well as non-financial information included in the annual report, plays an important role in reporting on sustainable development.

3.2 Disclosure of financial and non-financial information in annual reports

International harmonization of reporting information also had a substantial impact on the reporting of information in the annual report. The annual report should include, in addition to financial information, non-financial information that is key for the assessment of sustainable development.

On 22 October 2014, Directive 2013/95/EU of the European Parliament and of the Council amending Directive 2013/34/EU about disclosure of non-financial and diversity information by certain large undertakings and groups was adopted.

The aim of Directive 2014/95/EU is to improve the consistency and comparability of non-financial information throughout the European Union by including in their annual reports (or management reports) non-financial statements. Non-financial statements should contain at least information regarding the environmental, social and employee-related matters, human rights, anti-corruption and bribery matters. It should also include a description of policies, their results and risks associated with these matters.

Where relevant, a company's non-financial statements should include information on its supply and subcontracting chains to identify current and potential negative impact and prevent or mitigate it.

Companies should provide, among other things, sufficient information about issues that could lead to risks with serious consequences and should be assessed according to their extent and gravity. The risk of negative impact can arise simply from the business activities of an enterprise, and where relevant and appropriate from its products, services, or business relationships, including their supply and subcontracting chains.

On 6 May 2015, the Slovak Parliament approved the Act No 130/2015 amending and supplementing the Accounting Act. Certain provisions relating to the change in the content of the annual report and the related reports shall take effect 1 January 2016 and some will come into force on 1 January 2017.

Information in the annual report can be separated into financial and non-financial information according to the nature of the information. In some cases, it is not possible to clearly identify whether it is financial or non-financial information, as such information has the characteristics of both types of information.

Information that meets the definition of financial information in the annual report pursuant to the Accounting Act from January 1, 2017 is disclosed in:

– Financial statements for the accounting period for which the annual report is prepared,
– Auditor's report on those financial statements, unless special legislation provides otherwise.

Other information in the annual report can be about:

– Costs associated with research and development,
– Acquisition of own shares, temporary certificates, business interest and shares, interim certificates and shares of the parent entity,

– Proposal for the distribution of profit or settlement of loss.
– The non-financial information in the annual report is mainly information about:
– Development of the enterprise, the position and the significant risks and uncertainties to which the entity is exposed, including information on the impact of the enterprise on the environment and employment, regarding data in the financial statements;
– Expected future development of the enterprise.

Information that meets the definition of both financial and non-financial information is information about:

– Any important events that have occurred after the end of the accounting period for which the annual report is prepared,
– Expected future development of the enterprise,
– The information required by special regulations,
– Whether the enterprise has a branch abroad.

By stating non-financial information in the annual report, the company will fulfil the obligation to indicate non-financial information on the impact of its activities on the environment and employment, regarding data in the financial statements.

Beginning 1 January 2017, public-interest enterprises are required to provide in their annual report information regarding the development, performance, position, and effect of the company on the environmental, social and employment issues, information regarding the respecting of human rights and information concerning anti-corruption and bribery.

The annual report of these enterprises must contain non-financial information to include mainly:

a. Brief description of the business model,
b. A description of the main risks with impact on social responsibility,
c. A description and the results of the social responsibility policy employed by the enterprise,
d. A reference to the sums shown in the financial statements and an explanation of such sums as regards their impact on social responsibility, if appropriate,
e. Significant non-financial information about the individual activities of the entity.

Trade companies and public-interest enterprises operating in the mining industry or in natural forest grubbing compile and publish an annual report on payments to public authorities (hereinafter referred to as the "report on payments"). Payments indicated in the report on payments are broken down by the countries in which the entity operating mineral extraction or grubbing operations in natural forests.

Payments by type are broken down as follows:

a. Payment for mining rights,
b. Corporate income tax and similar tax,
c. Dividends,
d. Premium for signing of a contract for finding and mining,
e. Payment for the improvement of infrastructure,
f. License fee, entrance fee and other consideration ensuing from the lease agreement.

The Accounting Act defines exactly what the report on payments shall include:

a. The total sum of payments provided to a single public authority classified according to payment type,
b. Payments relating to a project if they can be assigned to the project, the total sum, and the total sum according to the particular type of payment; payments which the accounting unit is obligated to pay and which cannot be assigned to the project, need not be assigned to the project and can be recognised separately,
c. Payments made in-kind, including specification of their monetary value, while the content and extent of such a payment in-kind must also be disclosed and described and it is necessary to explain how the monetary value expression of such a payment was determined,
d. Declaration of the director of an enterprise signed by an authorized person and indicating that the report on payments provides a true representation of the payments.

4 CONCLUSIONS

The concept of sustainable development is to fulfil present needs without compromising future generations. Nowadays, regulations concerning disclosure of financial and non-financial information reflect user requirements in accordance with the concept of sustainable development. The breakdown of enterprises by size and different requirements as to the scope of information disclosed has both advantages and disadvantages. The advantage is that disclosure by size groups reduces administrative burdens for micro-enterprises. On the other hand, simplified disclosure of information provides aggregated information to users which is not sufficient for assessment of financial situation and financial performance. The range of activities of a microenter prise that meets the lowest criteria concerning number of employees, net turnover and balance sheet total indicates that the enterprise probably does not account for securities and other types of long-term or short-term financial instruments and does not perform capi-tal-intensive activities. This implies that the micro enterprise does not have relevant information for several items on the balance sheet and profit and loss statement designed for small and large enterprises. Use of simplified statements is not required by the Act on Accounting and enterprises use their own judgment as to whether to prepare financial statements in the structure for small and large enterprises.

In addition to information from financial statements, the annual report also provides information for users, as in addition to financial information it also contains non-financial information due to the increased requirements of users. Currently the prevailing effort is to improve the consistency and comparability of non-financial information throughout the European Union by including non-financial statements in annual reports or management reports. Non-financial statements should contain at least the information regarding the environmental and employment matters and social responsibility as a precondition of sustainable development. By including information concerning sustainable development in the annual report or management report, an enterprise can easily assess whether or not there is a distortion of sustainable development.

ACKNOWLEDGEMENT

This article is an output of the project of the Scientific Grant Agency of the Ministry of Culture of the Slovak Republic and Slovak Academy of Sciences (VEGA) no. 1/0512/16 (2016–2018) *Retention and Development of Business Property of Company as a Relevant Tool for Sustainable Development.*

REFERENCES

Act 431/2002 Coll. on Accounting as amended.
Act 130/2015 amending the Act on Accounting.
Directive 2013/34/EU of the European Parliament and of the Council of 26 June 2013 on the annual financial statements consolidated financial statements and related reports of certain types of undertakings, amending Directive 2006/43/EC of the European Parliament and of the Council and repeating council Directives 78/660/EEC and 83/449/EEC.
Directive 2014/95/EU of the European Parliament and of the council of 22 October 2014 amending Directive 2013/34/EU as regards disclosure of non-financial and diversity information by certain large undertakings and groups.
Juhaszova, Z. et al. 2014. Fair Value and its Importance for Financial Decision—making. In *IFRS: Global Rules & Local Use, 2nd International Scientific Conference on IFRS—Global Rules and Local,*

Prague, 10 October 2014. Prague: Anglo-American University.

Kosovska, I. et al. 2015. Amendments in the accounting of entrepreneurs specifics of measurement, bookkeeping and reporting in a micro accounting entity in the Slovak republic. In *Financial Management of Firms and Financial Institutions, 10th International Scientific Conference, PTS I-IV.10th International Scientific Conference on Financial Management of Firms and Financial Institutions, Ostrava, 07–08 September, 2015*. Ostrava: VSB—Technical University of Ostrava.

Kubascikova, Z. & Juhaszova, Z. 2016. Analysis of Financial Statements Focusing on Detection of Ponzi Schemes Using XBRL. In *Europena Financial System 2016: Proceedings of the 13th International Scientific conference, 13th International Scientific Conference of the European Financial Systems Location, Brno, 27–28 Jun 2016.* Brno: Masaryk University.

Ondrusova, L. & Parajka, B. 2014. The revaluation of assets and liabilities at fair value in merger. In *Managing and modelling of financial risks: 7th International Scientific Conference, PTS I-III, 7th International Scientific Conference on Managing and Modelling of Financial Risks Location, Ostrava, 08–09 September 2014.* Ostrava: VSB—Technical University of Ostrava.

Measure of the Ministry of Finance of the Slovak Republic No. MF/15464/2013-74 defining details of the arrangement, marking, and content specification of items of an individual financial statement and extent of data determined for publication from an individual financial statement for entrepreneurs using double entry bookkeeping for micro accounting units.

Measure of the Ministry of Finance of the Slovak Republic from 3. December 2014 No. MF/23378/2014-74 defining details of the arrangement, marking, and content specification of items of an individual financial statement and extent of data determined for publication from an individual financial statement for entrepreneurs using double entry bookkeeping for small accounting units.

Measure of the Ministry of Finance of the Slovak Republic from 3. December 2014 No. MF/23377/2014-74 defining details of the arrangement, marking, and content specification of items of an individual financial statement and extent of data determined for publication from an individual financial statement for entrepreneurs using double entry bookkeeping for large accounting units and public interest entity.

Parajka, B. 2016. Micro Accounting entities in the Slovak Republic—A Year After an Introduction. In *Strategic Management: international journal of strategic management and decision support systems in strategic management.* 21 (3): 43–48.

Slosarova, A. et al. 2016. *Účtovníctvo.* Bratislava: Wolters Kluwer.

Tumpach, M. & Bastincova, A. 2014. Cost and Benefit of Accounting Information in Slovakia: Do We Need to Redefine Relevance? In *Europen financial systems, 11th International Scientific Conference on European Financial Systems, Lednice, 12–13 Jun 2014.* Brno: Masaryk University.

New Trends in Process Control and Production Management – Štofová & Szaryszová (Eds)
© *2018 Taylor & Francis Group, London, ISBN 978-1-138-05885-9*

Comparison of the fee tools in the area of waste management in the EU

D. Palaščáková & M. Janošková
Faculty of Mining, Ecology, Process Control and Geotechnologies, Technical University of Košice, Košice, Slovakia

ABSTRACT: Currently, the problem of waste management often belongs to disputable topics at European as well as global level. People are still more aware that the resources of our planet are exhaustible and it is necessary to save them more economically. One of the areas, which significantly contribute in the conditions of our environment, is the waste management in all of its forms, whereas every country has different solutions to the problem. Waste management is too broad term, authors of the article focused on the comparison of selected fee tools in the area of waste management in Slovakia and the countries of the European Union. The goal of this article is to draw an attention to alternatives, which can be applied when improving the waste management in Slovakia and suggest the possible solutions in accordance with a sustainable development and costs necessary for its realization.

1 INTRODUCTION

Recently, Europe faces not only economical, but even political and environmental changes. According to the strategy Europe 2020, EU should change into intelligent, sustainable and inclusive economy. These are the priorities, which complete each other and cannot be separated. Intelligent growth means to create the economy, which is based on knowledge and innovations. Sustainable growth talks about the support of ecology and competitiveness in the field of economy. Inclusive growth helps to create social and regional unity with the help of providing a high rate of employment.

The quality of our life is, to certain level, influenced by natural resources, which are an unconditional presumption to European and even global economy operation. According to the European Commission, there is an expected growth of global population in 30% till 2050, which means it won't be able to use natural resources in the same way as now and it influences mostly the waste management a lot. The minimization of waste production, reduction of production inputs and optimization of production process will be necessary (European Commission 2011).

The use of opportunities brought by environmental value networks, mostly by more efficient use of resources, reducing the amount of waste and transformation of waste into new products and services requires environmental innovations, new mediator or mediator services. Small and middle businesses and businessmen need beneficial environment to be able to join the new industry relationships, which enable them the transition to circulation economy (Sturgeont 2013).

Environmental taxes presented a huge problem in the past. The main problem was definitely the lack of literature, laws, rules, but even disinterest of people and politicians to solve the problem of environment. Today, there is a lot of literature on the potential of environmental taxes towards the increase of the effectiveness of environmental politics. So-called ecological tax reforms, which consist of the increase of the rate of environmental taxes and the reduction of other taxes, play an important role regarding this.

The main goal of environmental taxation is to fill the purpose of the reduction of environment damaging. The significance of the problem of economical tools, particularly environmental taxes, proves the fact that one of the goals of the Ministry of Environment of the Slovak Republic in the area of environmental politics is to issue the vision of sustainable development of society, which we understand as its natural part.

Slovak legislation defines waste management within the Law 79/2015 Coll. on waste, which is defined in the Rule of European Parliament and the Council 2006/12/ES from April 5, 2006 on waste. According to this law, the waste management is collection, transport, recovery and disposal of waste including the supervision of all of these activities and following treatment of the places of disposal and includes even the acting of businessman and mediator.

Among the fee tools in the area of waste management belong taxes and fees in various forms. Such as environmental taxes, landfill fees and waste disposal fees.

2 ENVIRONMENTAL TAXES

Environmental challenges increase the pressure on the governments of particular countries in order to find the way how to reduce the damaging of environment and not to limit an economic growth at the same time. Governments dispose the whole set of tools including regulations, innovative politics, information programs, donations in the field of environment and environmental taxes. Taxes are dominant key part of this set of tools.

Environmental taxes hold several important advantages, such as environmental efficiency, economic effectiveness, and ability to grow the public incomes or transparency. These taxes are used to answer the wide scale of questions such as waste disposal, water and air pollution. Without the intervention of the state, there is no market stimulus for companies and households to consider the environment pollution. Therefore, the protection of environment generally requires a collective acting, usually led by government (OECD 2011).

Legal definition of environmental taxes can differ in particular countries and can distinguish from the definition used in national accounts. Both are relevant for statistics of environmental tax. We distinguish four categories according to the rate of environmental tax on the protection of environment (European Commission 2001).

- *Energy taxes* – taxes for energetic products used for transportation and stationary purposes. The most important energetic products for transportation purposes are gas and diesel oil. Heating oils, natural gas, coal and electricity belong to energetic products for stationary use.
- *Transportation taxes* – taxes related to the ownership and the use of vehicles. Taxes for the use of other means of transport, such as airplanes and other related transport services are included among transportation taxes just in case they meet general definition of environmental taxes.
- *Pollution taxes* – taxes for measured or estimated emissions of air or water, solid waste management and noise. CO2 taxes do not belong to the category of pollution taxes, but energetic taxes.
- *Natural resources' taxes* – the income from these taxes makes only the minor part of the overall incomes from environmental taxes. There are different opinions, whether that extraction of energetic resources is or is not harmful to environment.

3 LANDFILL TAX

Currently, there are two ways of how to charge landfilling in EU countries. We talk about the landfill tax or the fee of landfilling. Bakeš (2003) defines landfill tax as a payment tool, which is set by law and is set to acquire incomes to satisfy social needs. We don't need to provide an equivalent consideration to tax subjects. We cannot guarantee the equivalency. Tax is regularly repeated payment set by law, which is non-refundable, ineffective and non-equivalent.

Non-equivalency and non-refundability are, according to Dienstbier (2006), the condition for distinguishing the tax from the fines. Fine is defined as obligatory payment, which is, compared to a tax, related to the consideration of receiver. According to the definition of taxes, it is obvious that one of their primary functions is fiscal task, i.e. fund-raising to cover the expanses of public budgets. Beside of the fiscal task, taxes perform motivational and allocation functions. They are even of environmental scope, because the protection of environment belongs among the basic state politics funded by state budget.

Specifically, in case of landfill tax, we talk about the type of financial reimbursement for landfilling or the financial reimbursement fined for landfill waste disposal. The term "landfill tax" is often replaced by the term "fine". These two terms have different meaning, i.e. landfill tax is related to soil, construction waste, remains of mechanic-biological treatment, dangerous waste and combustible waste (www.odpady-portal.sk).

In Figure 1. we present the comparison of selected European countries related to this tax. Every country presents, whether particular type of waste is charged by landfill tax, free of tax, partially taxed, or there is the prohibition of landfilling of

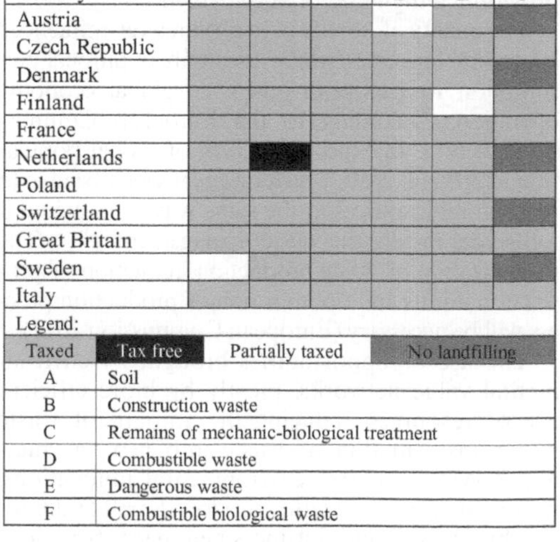

Country	A	B	C	D	E	F
Austria						
Czech Republic						
Denmark						
Finland						
France						
Netherlands						
Poland						
Switzerland						
Great Britain						
Sweden						
Italy						

Legend:			
Taxed	Tax free	Partially taxed	No landfilling
A	Soil		
B	Construction waste		
C	Remains of mechanic-biological treatment		
D	Combustible waste		
E	Dangerous waste		
F	Combustible biological waste		

Figure 1. Landfill tax in selected EU countries and Switzerland.
Source: Overview of the use of landfill taxes in Europe.

such a waste in this country. Generally, we can see that there is certain form of landfill tax in every single country. In some countries, e.g. Austria, Denmark or Netherlands, it's prohibited to landfill combustible biological waste. There is no landfill tax in Slovakia at all, there is just a fine for landfill waste disposal, and therefore it's not presented in Figure 1.

4 FEE FOR LANDFILL WASTE DISPOSAL

According to Kačmárová & Kaprová (2012), the fees for landfill waste disposal are the payments, which should fine the environment pollution and endangering the health, or human lives, but even vegetation as a consequence of human activity. These fines are set with the goal to motivate the originator/causer of pollution to reduce negative impact of his activity on environment. Another reason of the fees is that they contribute to the projection of negative externalities to the expenses of the polluter of environment.

Dienstbier (2006) marks this fee as an obligatory payment, which is related to the consideration of receiver. Consideration in the field of environment protection is the most frequently understood as the allowance to use economic functions of environment. He also states that neither fees nor taxes present the overall price for landfill waste disposal, they are just one of the elements.

These are named as "gate fees" or "entrance fees". According to the final report of the European Commission (2012), landfill operator will receive these fees for providing the services for waste disposal. They are set to cover the costs and benefit the landfill operator. They depend on more aspects, e.g. landfill capacity, sort of waste, market variations, etc.

5 ANALYSIS OF LANDFILL TAX IN SELECTED EU COUNTRIES

We used the observation of tariff rates of landfill tax in selected EU countries for the purpose of the analysis of landfill tax. There is no landfill tax introduced in the Slovak Republic, there is "the fee for landfill waste disposal", which is a certain form of landfill tax. We found out that there are too low landfill waste disposal fees applied in the Slovak Republic. This means that citizens do not have any reason to separate or recover the waste in other ways and it is easier and especially cheaper for them to dispose the waste at landfill "without any consequences". Just to compare with Slovakia, we have selected 22 European countries making a diversified specimen. In Table 1. and Table 2., there

are countries of various areas, populations, locations or their GDP production, ordered ascending from the lowest tariff rates to the highest ones. We suppose that such a variable specimen of the countries is a necessity for better complexity and the quality of analysis. In Table 1, we presented countries with the only one tariff rate of landfill tax for solid municipal waste (hereinafter just "SMW"). In Table 2, there are other selected countries with the interval of the amount of landfill tax stated in their source data.

Countries of Western and Northern Europe dispose with higher landfill taxes, their waste management is managed in a different way than in Slovakia. We did not present Germany in Table 1. and Table 2., because there was the prohibition of landfill disposal of untreated communal waste adopted in 2005. It

Table 1. Tariff rate of landfill tax for solid municipal waste in selected EU countries in 2015.

Country	Tariff rate of landfill tax (Eur/t)
Lithuania	0
Romania	0
Portugal	5
Latvia	9.96
Netherlands	17
Bulgaria	18
Czech Republic	20
Poland	26.60
Estonia	30
Luxemburg	30
Hungary	35
Norway	37.40
Greece	40
Sweden	45
Finland	60
Denmark	63
Ireland	75
Austria	87
Great Britain	97.30

Source: Confederation of European Waste-to-Energy-Plants.

Table 2. Tariff rate of landfill tax for solid municipal waste in selected EU countries in 2015.

Country	Lower tariff rate limit (Eur/t)	Higher tariff rate limit (Eur/t)
Slovenia	2.20	11
Italy	10	25
Belgium (Walloon)	25	65
Belgium (Flanders)	31.70	84.89

Source: Confederation of European Waste-to-Energy-Plants.

helped to reduce the landfill disposal to the minimum, it reduced the number of waste landfills and helped to increase the recycling rate, which is used as the form of waste recovery. The highest fees per one ton of landfill disposed waste, up to 97.3 EUR, from the selected countries are in Great Britain. There are also countries, which did not introduce any landfill tax, e.g. Lithuania and Romania.

In Table 2. we presented other countries with set tariff rate of landfill tax depending on the type of solid municipal waste. Belgium was divided into two regions (Walloon and Flanders) because of different interval limits.

Table 3. consists of the fee information, which are set in Slovakia in 2016. We presented even data from previous two years for the purpose of comparison. According to Law no. 17/2004 Coll. on fees for waste disposal, the amount of fees changes depending on the average annual inflation rate of previous calendar year. As the inflation rate for selected period of three years scaled around zero, the fees rate hasn't changed.

If we compare the tariff rate for municipal waste, which are not sort out with other selected countries in Table 1 and Table 2, Slovakia would be a country with very low fees. It would specifically rank a shared 4th place of the lowest fees together with Latvia. Standard of living in Austria and even Great Britain is higher than in Slovakia, but these countries have more than 8-times higher fees than Slovakia. SMW disposal in the Czech Republic is more than twice higher than in Slovakia.

The positive is that when we sort out the elements of SMW, tariff rate of fees is decreasing, however, this tendency is just half in case of sorting five elements out. In order to increase the motivation of society to separate the waste, we would need to

increase the fee for unseparated SMW. To have even bigger motivation, more significant decrease of tariff rate for separated waste would help at the same time. Germany deals with the problem of environment pollution differently, it had prohibited landfilling of unseparated SMW in 2005 and achieved a significant success in terms of recycling.

In Table 3. we present the review of the fee situation in Slovakia, namely tariff rates for inert, dangerous and other waste. For instance, people pay 2.2 EUR per ton for inert waste in Slovenia. If we consider the highest fee, which they have set for SMW (i.e. 11 EUR per ton), it's similar to Slovakia, but the tariff rate for inert waste is 7-times higher over there.

According to official report Landfill Tax Bulletin (2016), there is the fee for landfill inert waste disposal of approximately 3 EUR after conversion per ton in Great Britain. If we compare it with Slovakia (with the tariff rate of 0.33 EUR per ton), it makes 9-times higher difference. Based on the comparisons, we can state that the solution in the field of waste disposal is offered even here, i.e. in increasing the fees not only in SMW, but even in all sorts of waste.

6 ANALYSIS OF ENVIRONMENTAL TAX IN EUROPEAN UNION COUNTRIES

According to the press release of Eurostat, environmental taxes in 2015 held the share of 6,3% of overall tax incomes of the whole EU. Hence, environmental taxes present major item for waste management.

The second tool for the analysis of fees tools in the field of waste management was the comparison of the amount of environmental tax by EU countries to set the overall ranking of the countries in this index and determine the actual position of Slovakia. Data in Figure 2. are ordered ascending, i.e. beginning with the countries with the lowest rate of environmental taxes per GDP to the countries with the highest rate.

In order to compare all EU countries, we couldn't use the indices of the amount of absolute incomes of environmental taxes for particular countries. Therefore, we have chosen the index relating the incomes from environmental taxes to overall produced GDP of these countries. We have made the ranking in Fig. 8 in this way. Slovakia is the second lowest rate of incomes (1,79%) from environmental taxes per GDP in this index. Lower incomes are shown only by Lithuania (1,7%). Denmark, as the only country, passed the rate of 4%.

We also compared the percentage rate of environmental taxes on GDP for EU countries during the period of 2004–2014, when we found out that, during the observed period, these values still

Table 3.   Fees for waste disposal in Slovakia (in EUR/t).

Waste item	2014	2015	2016
Inert waste, replaced construction waste, soil and gravel not consisting of dangerous materials, waste from landfill remediation	0.33	0.33	0.33
Other waste	6.64	6.64	6.64
Communal waste after separation of less than four elements	9.96	9.96	9.96
Communal waste after separation of four elements	5.98	5.98	5.98
Communal waste after separation of five elements	4.98	4.98	4.98
Dangerous waste	33.19	33.19	33.19

Source: processed according to the Law no. 17/2004 Coll.

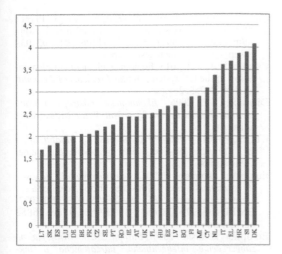

Figure 2. The amount of environmental taxes in the rate of GDP for EU countries (in %).
Source: processed by Eurostat.

had decreasing tendency in Slovakia. Since Slovakia joined EU in 2004, with the rate of 2,45%, its amount has been decreasing up to the rate 1,79% in 2014. All the categories of environmental tax are related to the sustainable development and play an important role in the question of environment pollution. We improve the state of environment by increasing the taxes of any category, because higher fees will motivate the subjects to strive for, e.g. less landfilling, less usage of unsustainable resources of energy and more care about our environment. We have proved that Slovakia faces the problems of waste management. Hence, environmental taxes should serve as other tool, which would solve actual undesired state.

7 CONCLUSION

"Globalization of value orientations is needed by present rulers whom a simplified mechanism of the choice of values makes their ruling easier."

Environmental values are gradually succeeding in the present globalized world, in most of the value systems on Earth. In spite of this, that the processes of globalization possess negative signs even in the relationship to axiological questions, they even possess positive signs and the expansion and the establishment of environmental values in a global measure is one of them (Brožík 2000). The rate of probability is increasing, as environmental values will be included into all value systems.

New opportunities are rising even in relation to a present redevelopment of environmental value

chains. Their development must be considered by all countries. Sooner or later, the companies of different countries will join these chains, if coordinator of this chain is interested in it, or if the company finds the chances to join itself. The only thing that matters is how effective will this integration be and how their national-economic contribution will be. New demands on the state policy are rising.

Some member countries of EU achieved the revenue share of environmental tax on the overall tax incomes of the amount of 10% in the field of the reforms of environmental tax, whereas their fiscal incomes were kept and they improved the competitiveness and energetic efficiency. It's proved by the fact that it's possible to move the taxation to environmentally harmful activities in the context of responsible economic scope. On behalf of more effective measuring of the transfer to price signals necessary for the support of greater amount of investments to more effective use of resources, another index can be needed, e.g. effective tariff rate of the tax from environmental pollution or the use of resources (European Commission 2011).

The level of fees for the services in waste management is set too low in Slovakia. Other countries achieved more effective treatment with waste and the improvement of environment by legislative repeal of landfilling of particular sorts of waste or by higher waste disposal fees. In accordance with a sustainable development and expanses necessary for its implementation, we suggest implementing such steps even in Slovakia supposing the similar change will happen.

Following previous sections, we present some facts, which we need to focus on when setting the amount of fees for landfilling waste disposal:

If landfilling fees are set too low, it won't mean a substitution effect on other ways of waste landfilling, but it rather advantages subjects, which are willing to dispose the waste by landfilling.
– If landfilling fees are set higher, neighboring countries with higher expenses on waste management (e.g. Austria or Hungary) won't have the reason to dispose the waste in the territory of Slovakia and keep their waste in their own territory.
– When setting the amount of fees, it's necessary to get inspired by the countries of Western Europe, which manage the waste better than Slovakia does.
– Too rapid increase in the fees will result to a rising number of illegal landfills. Therefore, it's advisable to increase the fees gradually and in certain limits.
– Beside of the increasing of fees for landfilling, it's necessary to increase even the sanctions for illegal landfill waste disposal. If the sanctions

were not increased, it would lead to a worse state of environment.

- Increased fees will create new resources, which will support environmentally economical waste management.
- The fees limit should extend the prices, which offer competitive technologies of waste management (e.g. material assessing of waste) and which are more economical for environment.

Last but not least, it is necessary to consider even overall incomes of population, which does not consider the increase of fees as suitable, but the purpose of landfill restriction will be filled. Educational activities are very important, because the state can create the conditions to increase public awareness in relation to waste management by the policy in the field of education, granting the research and cooperation of universities (research institutions) with corporation sphere. Promotional activities are important part of waste management, because the results in this field are difficult to achieve without public awareness.

The goal of the European Commission is to achieve a shift from taxation of work to environmental taxation till 2020, what will cause a serious increase in the share of environmental taxes on public incomes in accordance with proved processes in member EU countries.

The fact that the impact of the application of environmental taxes on entrepreneurial sphere is generally different, it supposes to apply suitable correctional mechanisms, e.g. compensations. It is possible to synchronize economic and environmental goals by their application towards the support of economic growth, but even the process of adaptation of the structure of economy on behalf of sustainable development.

ACKNOWLEDGMENT

This paper was written in connection with scientific project VEGA no. 1/0961/16, financial support from this Ministry of Education's scheme is also gratefully acknowledged.

REFERENCES

Bakeš, M. et al. 2009. *Finanční právo*. Praha: C.H. Beck.

Brožík, V. 2000. *Hodnotové orientácie*. Nitra, 2000.

COM. 2014. Zelený akčný plán pre MSP. [on-line] http://eurlex.europa.eu/legalcontent/SK/TXT/?uri=CELEX:52014DC0440

Confederation of European Waste-to-Energy Plants. 2015. *Landfill taxes and bans*. [on-line] http://cewep.eu/media/cewep.eu/org/med_557/1406_2015-02-03_cewep_-_landfill_inctaxesbans.pdf.

Csikósová, A et al. 2015. Ecological elements implementation in the tax system. *Actual Problems of Economics,* 164(2): 253–261.

Dienstbier, F. 2006. *Ekonomické nástroje ochrany životního prostředí—otázky právní*. Brno: Masarykova univerzita.

ETC/SCP. 2012. *Overview of the use of landfill taxes in Europe*. [on-line] http://scp.eionet.europa.eu/publications/

Európska komisia. 2001. *Environmental taxes—A statistical guide*. [on-line] http://ec.europa.eu/eurostat/documents/3859598/5854253/KS-39-01-077-EN.PDF/5c97b328-6539-4290-9bca-97dea7b882bd?version=1.0.

Európska komisia. 2010. *Európa 2020: Stratégia na zabezbečenie inteligentného, udržateľného a inkluzívneho rastu*. [on-line] http://eur—lex.europa.eu/LexUriServ/LexUriServ.do?uri=COM:2010:2020:FIN:SK:PDF.

Európska komisia. 2011. *Oznámenie komisie európskemu parlamentu, rade, európskemu hospodárskemu a sociálnemu výboru a výboru regiónov*. Plán pre Európu efektívne využívajúcu zdroje. SEK (2011) 1067, SEK (2011) 1068.

Európska komisia. 2012. *Use of economic instruments and waste management performances*. [on-line] http://ec.europa.eu/environment/waste/pdf/final_report_10042012.pdf.

Eurostat. 2016. *Environmental taxes in the EU*. [on-line] http://ec.europa.eu/eurostat/documents/2995521/7236510/8-22042016-BP-EN.pdf/b910e804-e410-4b9c-b9ab-1893398e2a2d.

HM Revenue & Customs. 2016. *Lanfill Tax Bulletin*. [on-line] https://www.uktradeinfo.com/Statistics/Pages/.

Kačmárová, P. & Kaprová, K. 2012. *Databáze OECD/EEA ekonomických nástrojů ochrany životního prostředí: Poplatky a daně*. [on-line] http://www1.cenia.cz/www/sites/default/files/Databaze%20OECD.pdf.

Maciaszczyk, M. & Maciaszczyk, P. 2016. CSR and disbaled consumers. *Procedia Economics and Finance,* 39(2016): 855–860.

OECD. 2011. Environmental Taxation—A Guide for Policy Makers. [on-line] https://www.oecd.org/env/tools-evaluation/48164926.pdf

Pružinský, M. & Mihalčová, B. 2015. Material flow in logistics management. In *Proceedings International Conference on Engineering Science and Production Management, ESPM 2015*.

Sturgeont, T.J. 2013. *Global Value Chains and Economic Globalization—Towards a New Measurement Framework*. Report to EUROSTAT.

TaxAndDutybulletins.aspx WP2012_1/wp/WP2012_1

Zákon č. 17/2004 Z. z. o poplatkoch za uloženie odpadov.

New Trends in Process Control and Production Management – Štofová & Szaryszová (Eds)
© 2018 Taylor & Francis Group, London, ISBN 978-1-138-05885-9

Consumer credit relations in the Slovak Republic

L. Pelegrinová
Faculty of Law, Pavol Jozef Šafárik University in Košice, Košice, Slovak Republic

L. Štofová & P. Szaryszová
Faculty of Business Economics with seat in Košice, University of Economics in Bratislava, Košice, Slovak Republic

ABSTRACT: The aim of the paper is to identify key problematic aspects of consumer credit relations on the basis of analysis of individual features in researched field. In this case is devoted to the current legislation of consumer credit relations in which related sources of consumer rights in the Slovak Republic and the European Union are identified. Consequently, authors in the paper define subjects of consumer credit relations as consumer, creditor, as well as supervisors. There are described the requirements of consumer credit agreement under the Consumer Credit Act, consequences for not complying with requirements of consumer credit agreement and breach of consumer protection laws by creditors. Authors are also discussing about a sanction mechanism in the field of consumer credit, selected aspects of consumer credit relationship and referring to problems within the application of consumer legislation.

1 INTRODUCTION

Identifying the initial impulses of consumer protection policy is quite challenging. Controlling business practices and products, and the link with buyer protection is not a modern concept. Significant development of the concept of protection the consumer can, however, be attributed to a successful European integration.

The European Communities have begun to address the issue of consumer protection in the 1970 in the context of the establishment of the internal market and customs union. However, it took more than a decade until the idea of consumer protection emerged in Primary law of the European Communities. Since the introduction of consumer protection into the founding Treaties through the Single European Act (1986), the concept of consumer protection has evolved into comprehensive adjustment of consumer relations, which are currently affecting all aspects of the company's economic life.

Due to the complexity of the legislation, it is now possible to distinguish between the general and specific protection of the consumer's rights. Under the general protection of consumer rights, we understand legislative adjustments to contractual relationships that apply to all types of transactions in which the consumer is in a disadvantaged position. These include, for example, consumer co tract terms, unfair terms in consumer contracts, or misleading advertising.

The specific protection of consumers' rights constitutes a partial adjustment of legal relations, sale of consumer goods, credit agreements, rental properties, travel purchase contracts, street and mail sales, etc.

2 REQUIREMENTS OF CONSUMER CREDIT AGREEMENT AND UNFAIR COMMERCIAL PRACTICES

2.1 Requirements of consumer credit contracts

The current legislation on consumer law in Slovakia is mainly the result of the application of the legal norms of the law of the European Communities and of the European Union (hereinafter referred to as "EC" and "EU") to which the Slovak Republic committed itself by the Treaty of Accession of the Czech Republic, Estonia, Cyprus, Latvia, Lithuania, Hungary, Malta, Poland, Slovenia and Slovakia, signed in 2003, which entered into force on the first May of 2004. In the European Union law we find consumer law standards as in both primary and secondary law. Primary law is based on the concept of consumer protection under Title XV of the Treaty on the Functioning of the European Union. Art. 169 on consumer protection: "In order to promote the interests of consumers and to ensure a high level of consumer protection, the Union shall contribute to the protection of consumers' health, safety and economic interests, as well as to the promotion of their rights to information, education and the formation of associations to protect their interests" (Vojčík et al. 2012).

2.2 Breach of the creditor's obligation in consumer credit relationships

In the case that the supplier of the credit fails to comply with the conditions laid down by the laws governing consumer credit agreements, this may result in the invalidity of the contract as a whole, or other consequences, such as consumer credit default, fines and other sanctions from supervisors. Consumer legislation is sufficiently reprehensible to creditors in cases of breaches of consumer protection law. The Consumer Credit Act clearly and comprehensibly lays down the terms and conditions for the offering and provision of consumer credit, as well as the consumer loan contract, the breach of which combines specific legal consequences (Strémy 2015).

For a consumer loan agreement, it is necessary to realize that all the transactions the creditor plans to obtain for a consumer credit must be part of the consumer credit agreement. It is not possible for the creditor to demand from the consumer interest, fees or other performance of any nature not provided for by law or included in the consumer credit agreement (Mazák & Jánošíková 2011).

If a creditor tried to avoid consumer credit provisions in order not to comply with the terms of consumer credit, it is also a lawful act. According to § 9 of the Consumer Credit Act, If the lender has taken advantage of the consumer's error and has used it the contractual terms and conditions excluding the application of provisions relating to consumer credit, the contract is deemed to be a consumer credit contract unless the creditor proves that he did not intend to circumvent this law. The credit agreement will be free of charge and free of charge even if the creditor has grossly breached his obligation and, prior to concluding a consumer credit agreement or before changing the consumer credit agreement, has not assessed the consumer's ability to repay the consumer credit taking into account the period of time consumer credit, consumer income and, where appropriate, the purpose of consumer credit.

According to § 7 par. 1 of the Consumer Credit Act, The creditor is not entitled to require from the consumer a one-off repayment of consumer credit. The gross breach of the obligation to assess the creditworthiness of the consumer under § 11 par. 2 also considers the assessment of the consumer's ability to repay the loan without any data on the income, expenditure and family status of the consumer or without taking account of data from the relevant database or register in order to assess the consumer's ability to repay the consumer credit. The Consumer Credit Act is the consequence of the invalidity of the credit agreement under § 11 par. 3, for example, to grant unauthorized consumer credit without authorization. If, in such a case, the unauthorized person had provided the service and the consumer would have incurred the obligation to issue the financial benefit provided by the invalidity of the contract, the unauthorized person in the invalid a consumer contract labeled as a creditor would be obliged to allow the consumer to pay only the actually paid financial compensation in installments and a period which, however, may not be shorter than the period within which the consumer should return the financial compensation if there is no reason to invalidate the credit agreement, unless agreed otherwise. This is without prejudice to the consumer's right to reimburse the financial benefit provided at the same time. This means that such a contract would become void and would also be gratuitous provision of funds.

The consequence of invalidating the consumer credit agreement is also connected with the violation of the provisions of the Civil Code. Under Article 53c of the Civil Code, a Consumer Credit Agreement is also invalid if the provisions of the Consumer Contract as well as the provisions contained in the General Terms of Business or any other contract documents related to the Consumer Agreement are given to the consumer in an inacceptable and minor font, implementing regulation. The Consumer Credit Agreement, which contains an unacceptable contractual condition set out in Section 53 of the Civil Code, as stated in the verdict of the court ruling and its conclusion was reached with the use of unfair commercial practices or insults, is invalid.

2.3 Sanction mechanism

The sanction mechanism of breaches of consumer rights and interests in credit relationships is sufficiently rigorous to providers. Given the National Bank of Slovakia (NBS)'s position within the system of supervision and supervision of creditors, the NBS has the right to impose sanctions on creditors as well as on individuals in special cases. The NBS is competent against the creditors for the breach to impose sanctions ranging from financial sanctions to the withdrawal of a license or permit to conduct business.

Pursuant to § 23 par. 2 of the Consumer Credit Act is relevant to the NBS:

- Impose on the creditor the obligation to take measures to eliminate and remedy deficiencies identified.
- Impose a fine of up to EUR 3000 on a creditor who is a natural person and a repeated failure or

serious shortfall of up to EUR 7000; to a creditor who is a legal person, a fine of up to EUR 150,000 and a repeated failure or serious shortfall of up to EUR 500,000.

- Impose on the creditor the obligation to submit specific statements, reports and other information.
- Restrict or suspend the creditor's performance of the consumer lending activity.
- Withdraw the creditor's authorization.

NBS has the right to sanction an individual as a creditor employee up to twelve times the monthly apart from the possibility to sanction the creditor as a legal entity or a natural person providing a consumer loan of the total earnings from the lender for the previous year.

The head of the organizational unit may be fined up to a maximum of 50% twelve times the monthly average of the total income from the lender for the previous year. The person who has become an untrustworthy person by the effective imposition of the fine is obliged to withdraw from the post without delay (Zákon č. 129/2010 Z.z.).

3 SELECTED ASPECTS OF CONSUMER CREDIT RELATIONS

Consumer contracts in general, as well as consumer credit agreements, are governed by a special regime as regards their negotiation, closure, requirements as well as other parameters. Significant consumer preference in legal relationships often appears to be unjustified, especially from the point of view of creditors. The new financial consumer protection legislation imposes on the creditors obligations that increase the administrative, technical, legal and financial demands of consumer credit (Deville 2016, Dettling & Hsu 2017).

This is inherently threatening that the consumer will be discriminated against as compared to a legal entity or a natural person to entrepreneurs in providing funding to both bank and non-bank entities. Especially non-bank entities, which are their turnover, structure, human resources, and so on. At a level other than banking, feel the provision of consumer credit as a non-profitable cost item and a high risk of loss and business is just an activity based on profit making.

Providing funding for a legal entity or a natural person to an entrepreneur is simpler, does not require additional costs for accessing different databases, does not restrict the creditor to offer any interest rates taking into account the market mechanism, allows sufficient coverage of the claim for the creditor in a satisfactory manner, for example as a result of early repayment or withdrawal.

On the other hand, consumer law has its importance precisely because financial services booming financial services has come to the forefront of financial service providers who have failed to respect basic moral values, abuse and circumvent legislation. The most common victim of such practices were consumers, particularly those with a lack of information or misuse.

3.1 Withdrawal from the consumer credit agreement

The consumer's right to withdraw from a consumer credit contract is based on Section 13 of the Consumer Credit Act. This right is granted to the consumer under § 13 par. 1 within the statutory time limit of 14 days from the conclusion of the consumer credit or from the day when the contractual terms of the contract pursuant to Article 9 or 10 are delivered to the consumer if that day follows the day of conclusion of the consumer credit agreement. If the consumer credit agreement does not contain contractual terms under § 9 or § 10, the withdrawal period begins from the moment the consumer receives these terms and conditions.

If the consumer exercises his right to withdraw from the contract, according to § 13 par. 3 shall be obliged to pay to the creditor the installment and the aliquot part of the interest for the period of drawing of the consumer loan without delay, no later than 30 calendar days after sending the notice of withdrawal to the creditor. Interest shall be calculated on the basis of the agreed interest rate on the consumer credit. The creditor loses the right to other compensation from the consumer, except for the non-refundable fees the lender has paid to the public authority for this consumer credit. The Consumer Credit Act does not, however, regulate the consumer's entitlement to certain transactions which may have arisen in the context of the contract for the benefit of the consumer, for example, various payments on behalf of the insurer, or consumer credit processing fees incurred by the creditor in connection with the consumer's creditworthiness test. The law remains unsettled when the consumer, within the statutory time-limit, withdraws from consumer credit after taking over the subject of financing at which time the damage will cause damage to the creditor or other person.

3.2 Repayment of consumer credit before maturity

Consumer rights and creditor claims for early repayment of consumer credit are governed by Section 16 of the Consumer Credit Act. The consumer is entitled to full or partial repayment of the loan even before the agreed maturity, and the creditor is obliged to immediately provide the consumer with information about the amount of the lender's claim. In this case, the consumer pays only the creditor's costs and the interest accruing from the granting of the loan to his repayment. If the lender should strictly abide by the provisions of the law, it would be necessary to clearly determine the due date of the loan for the purpose of calculating such costs that the consumer would have committed to repay the loan on that day. In fact, the creditor's costs can be factually indicative only and may vary depending on when the consumer actually pays the consumer credit.

Another problem that arises from the provision of § 16 par. 3 is the statutory limitation of the creditor's cost of repayment that is set at a maximum of 1% of the amount of consumer credit repaid before maturity, if the period between repayment of the consumer credit before maturity and the agreed date of termination of the consumer credit agreement exceeds one year or 0.5% of the amount of consumer credit repaid, if this period does not exceed one year. The reimbursement of the creditor's costs is also limited to the amount of interest that the consumer would have paid during the borrowing period until his repayment.

It is therefore clear that if the consumer chooses to repay consumer credit prematurely, the borrower gets a significant disadvantage with regard to the expected profitability. The creditor, as a finance provider, must also receive these funds for a certain interest on the financial market. Revenues from repayments repaid by borrowers in the creditor's portfolio are far from equilibrium with spending on the provision of finance. At the same time, the lender has fixed costs with the company's operations, from paying rent, employees, legal and accounting services, to the cost of technical and information equipment. We cannot omit the variable costs associated with paying commissions to intermediaries, accessing databases, and meeting the legal requirements imposed on it by changing legislation. The processing of a business case involving a consumer contract requires personnel, administrative and technical costs. Considering the vastness of the contractual documentation for the consumer that needs to be provided to him prior to the conclusion of the consumer contract, costs are incurred by printing the contractual documentation or the cost of another durable medium. None of these Costs will take into account the Consumer Credit Act.

The fact that the legislator does not recognize the different specifics of financial services is also considered to be imperfect. For example, when leasing is offered, the funded item becomes the property of the leasing company, the consumer becomes the holder who pays the installment for the use of the item. The basis of the lease is the transfer of ownership right after the agreed time to the consumer under a purchase contract. It is a legitimate assumption that the sales contract will include, among other things, the sales price. In the case of a lease, the standard sale price is determined symbolically, but the provisions of the Consumer Credit Act do not allow the consumer to demand the prepaid price at the pre-reimbursement.

However, to the detriment of the consumer there is a tendency for creditors to provide consumer loans "without any increase". In such a case, the consumer will pay a fee for processing a consumer credit contract at the time of the conclusion of the contract, which is predominantly the amount of the prepaid consumer credit costs for the whole agreed period. Repayment of this charge at the outset will not damage the creditor's early termination. On the other hand, the consumer is thus harmed by virtually repaying interest until the end of the contract, and the Consumer Credit Act does not impose an obligation on the lender to repay or set off an aliquot part of the fee.

3.3 Inappropriate retaliation and annual percentage rate of charge

One of the most frequent reasons for which the court decides on matters in the detriment of the creditor is precisely the deficiencies in informing the consumer of the pay and the Annual Percentage Rate (APR) as well as their incorrect calculation or exceeding the legal limits. Consumer information on pay, APR and average APR must be part of the consumer credit contract as well as standard consumer credit information. It must also include information on the assumptions used to calculate them. The APR calculation method does not take into account the discontinuity of some transactions. Thus, if the consumer pays 100 EUR fee for a loan, this fee becomes part of the APR calculation. Not as a one-time performance but as a repeat performance of the contract (Babčák et al. 2012).

Another problematic aspect that needs to be pointed out is that some services are included in the APR calculation, for which the lender is only the intermediate financial flow. Such a situation arises, for example, when the subject is insured by a creditor who is also a separate financial agent for the provision of insurance. The legal regulation does not distinguish between different types of insurance. As an example, Collective insurance, i.e. insurance provided by the creditor directly to the consumer. The subject matter of which the consumer is the holder is insured within the scope of the subject matter of the creditor and the consumer pays the premium directly to the creditor. In such a case, the inclusion of the insurance in the calculation of the APR would be justified.

However, more frequent are the direct debit insurance where the creditor merely "will collect" insurance and in the full amount it transfers directly to the insurer. This is the insurance relationship where the creditor only mediated the insurance, the full amount of the payment from the consumer is passed on to the insurer, even if the consumer is late with repayments or the loan does not pay. In such a case, it is doubtful whether the amount of the premium is to form part of the APR, since it is not a payment intended for the creditor as an end user.

In disputes between the creditor and the consumer, the amount of the APR and its calculation becomes a regular part of the decision not only of the national courts of the EU Member States but also of the European Court of Justice. Inappropriateness of retribution and APR appeared, for example, in the judgment of the District Court of Žilina in case 8C/158/2013 of 18/11/2013, when the court ruled on consumer credit risk because of the disproportionate amount of APR despite the fulfillment of other formalities of the consumer credit agreement.

3.4 Assessment of consumer credit repayment

Consumer crediting in the consumer credit process is currently an important milestone in increasing consumer protection. Along with legislation, however, there are also various application problems. The amendment to the Consumer Credit Act, which was submitted to the National Council of the Slovak Republic in October 2014, introduced a new process for obtaining a consumer credit authorization as well as a process for verifying the consumer's ability to repay consumer credit. In the previous version of the law, it was the duty of the consumer to provide the creditor with the information necessary to assess his ability to pay. After the adoption of the amendment to Act no. 35/2015 Coll., amending Act no. 129/2010 Coll. on consumer credit and other consumer credit and loans and on the amendment of certain laws as amended, and amending certain laws, the responsibility for the assessment of the repayment capacity has been fully transferred to the creditor.

The amendment supplements Section 7 of the Consumer Credit Act to paragraphs 3 to 17. Under the new legal framework, the creditor is obliged to provide consumer credit data on at least one electronic consumer credit data record for the purpose of providing consumer credit and

is required to use professional care for the purposes of assessing the consumer's ability to repay consumer credit to obtain and expeditiously use consumer credit data. The creditor is required, without the consent of the consumer, to provide information on each consumer with whom he has entered into a consumer credit agreement within one month of the conclusion of the consumer credit agreement, being responsible for the correctness, completeness and timeliness of the data provided in the register.

The creditor is legally obliged to proceed to provide consumer credit in a manner that does not harm consumers and with professional care. And just professional care is obligatory and credible proof. Based on statutory assumptions, the creditor is required to create and maintain a system of assessing the consumer's ability to repay consumer credit.

However, due to the fact that it is a software application of law, it is impossible to rule out a certain error. The error is attributable solely to the creditor who could submit incorrect data about the consumer or consumer credit to the system.

However, due to the fact that consumer credit is almost exclusively a financial institution and not a software company, other parties are also involved in the creation and maintenance of the system. In application practice, the process of transferring information from the lender through a software solution company that subsequently transfers information to a system that was created and maintained by another software company on behalf of the lender.

4 CONCLUSIONS

Since the accession to the European Communities/European Union, the protection of consumers' rights has also been given special attention in the Slovak legal system, where this issue did not play a significant role until the approximation of law in the accession process. The European Commission has been and continues to be the main initiator and supervisor of consumer legislation and the harmonization of its legislation in the Member States.

We can now say that consumer protection has become a comprehensive agenda that governs the wide range of legal relations and is a major concern in most areas of economic life. Although the European Commission is trying to harmonize legislation, there is no consumer protection at the same level in all Member States, or we may find variations, particularly when applying directives which give the Member States room for customizing some of their aspects.

We have come to the conclusion that problems with the practical application of the Consumer Credit Act in Slovak conditions stem from the transposition of EU law, exempting imperfect language translations. Slovak legislation is the strictest version of the directive on consumer credit agreements. We also consider the enforcement of the creditors' right in the context of permanent controls by public authorities not only in accordance with the Consumer Credit Act but also under other legislation (control of personal data protection, financial intermediation control, etc.). Too much emphasis on fines rather than help and guidance in legal uncertainty results in the country's almost bullying behavior towards creditors. Such an approach may, however, become indirect discrimination against consumers, given the difference in complexity and riskiness of the provision of funds to entrepreneurs.

REFERENCES

Babčák, V. et al. 2012. *Finančné právo na Slovensku a v Európskej Únii.* Bratislava: Eurokódex. ISBN 978-80-89447-86-2.

Dettling, L.J. & Hsu, J.W. 2017. Minimum Wages and Consumer Credit: *Impacts on Access to Credit and Traditional and High-Cost Borrowing.*

Deville, J. 2016. Debtor publics: tracking the participatory politics of consumer credit. *Consumption Markets & Culture,* 19.1 (2016): 38–55.

Mazák, J. & Jánošíková, M. 2011. *Lisabonská zmluva: Ústavný systém a súdna ochrana.* Bratislava: Iura Edition, spol.s.r.o. ISBN 978-80-8078-416-4.

Strémy, J. 2015. *Ochrana finančného spotrebiteľa.* Praha: Leges. ISBN 978-80-7502-089-5.

Vojčík, P. et al. 2012. *Občianske právo hmotné.* Plzeň: Aleš Čeněk. ISBN 978-80-7380-402-2.

Zákon č. 129/2010 Z.z. o spotrebiteľských úveroch a o iných úveroch a pôžičkách pre spotrebiteľov a o zmene a doplnení niektorých zákonov §23 ods.

New Trends in Process Control and Production Management – Štofová & Szaryszová (Eds)
© *2018 Taylor & Francis Group, London, ISBN 978-1-138-05885-9*

Going worldwide with a local strategy—international business

P. Poór
Faculty of Manufacturing Technologies with a seat in Prešov, Technical University of Košice, Prešov, Slovak Republic

ABSTRACT: Brands play a critical role in firms international expansion. A coherent international brand architecture is a key component of firms overall international marketing strategy as it provides a structure to leverage brands into other markets, assimilate acquired brands, and rationalize firms international branding strategy. Strong brands command customer loyalty. With a strong brand, customers will stick with it even in challenging situations such as bad customer experiences or defective products. Strong brand facilitates firm's long-term growth, it can leverage the value of its brand, to more easily expand its product line and achieve an immediate customer acceptance of the new products. Also expand into new markets based on the strong brand. Presented article deals with expanding worldwide with a local strategy using the techniques of international business. First it theoretically describes the problems and situation in international business, then gives practical view of positives and negatives of going worldwide.

1 INTRODUCTION

International business refers to the performance of trade and investment activities by firms across national borders. Firms organize, source, manufacture, market, and conduct other value-adding activities on an international scale. They seek foreign customers and engage in collaborative relationships with foreign business partners. While international business is performed mainly by individual firms, governments and international agencies also undertake international business activities (businessculture.org).

International trade is the exchange of goods and services across international borders and is also known as exports and imports. Exports are goods and services produced by a firm in one country and then sent to another country. Imports are goods and services produced in one country and brought in by another country. International trade refers to exchange of products and services across national borders, typically through exporting and importing. Exporting is the sale of products or services to customers located abroad, from a base in the home country or a third country. Importing or global sourcing refers to procurement of products or services from foreign suppliers for consumption in the home country or a third country (Štofová & Szaryszová 2016). International investment refers to international transfer or acquisition of ownership in assets. Foreign direct investment is an internationalization strategy in which the firm establishes a physical presence abroad through acquisition of productive assets such as capital, technology, labor, land, plant, and equipment.

Simply international market is the application of marketing principles to more than one country. According to the American Marketing Association the "international marketing is a multinational process of planning and executing the conception, put price, promotion and distribution of ideas, goods and services to create exchanges that satisfy individual and organizational objectives."

2 WHAT DOES "GOING WORLDWIDE" MEAN?

Going international in any field of endeavour provides organizations with exciting new opportunities for growth as well as challenges;

- Cost management strategy must be of concern for going international. Moving into new territories require new organizational structures, which decentralizes operations. Training of new staff to abreast them with organizational objectives, investing in technology to ensure business operations stay on schedule all requires effective cost management.
- Trade legalities and regulations by local government however would put measures in place to protect the interest of the indigenous industries. Local government want to protect the interests of their local industries and so may enact laws to make foreign companies behave in their economy's interest and not only get enticed with making profit and repatriating their resources.
- Low pricing and quality control strategy lowers cost of product yet ensures quality; and equally

compete with local companies. However, high logistics cost, operation, human and raw material as well as capital invested among other costs could increase product price tag when not combined efficiently and effectively.

- Change management and marketing strategy had to be adopted by coke to achieve its stated plan for growth in the global market by replacing Beecham and Grand Metropolitan as national bottlers with Cadbury Schweppes. The appointment of a new marketing manager and replacement of complacent local franchisers with a shift from carbonated drinks to bottled water were all adopted.
- Companies operating internationally must not ignore social corporate responsibilities. Thus, to be responsive to local needs, reaching out to local communities and getting involved in civic and charitable activities.
- Monitoring and Evaluation: It is important to continuously and efficiently monitor and evaluate franchisers in terms of profitability, demand for products, cost control and pricing strategies that takes into account segments. Ability to pay, market conditions, competitor actions and input costs among others.
- Sales Promotion Strategy: A company should take its eye off the "ball". It must always continue the processes that took them to the top in terms of sales and promotions strategies by increasing its impact and reducing risks and costs as much as possible.
- Adopt Strategies to keep it competitive: A company should always endeavour to adopt strategies that that will keep it competitive in the market. Choosing a low pricing strategy for a product to make it popular and later selling it to customers by giving discounts.

Simply international market is the application of marketing principles to more than one country. According to the American Marketing Association the "international marketing is a multinational process of planning and executing the conception, put price, promotion and distribution of ideas, goods and services to create exchanges that satisfy individual and organizational objectives."

2.1 These are the types a company can learn lessons from involvement in the international market

1. Efficient monitoring and evaluation of franchisers in terms of profitability, demand for product, cost control. It always important to review franchisers. This would enable the company to know whether it is doing well. For example it can change the franchise ownership leading to increase in sales and profitability.

Figure 1. International business.

2. The need to recognize each market as unique and develop tailored and personalized offerings to suit them: for example by lowering the prices backed by intense advertising to boost sales in one market as compared to their focus on expanding distribution networks in another one.
3. Always recruit efficient and effective managers. Due diligence in employing managers. Lazy workers must be replaced. Managers must be well motivated and compensated for their work. Appointing a new manager could lead to increase in sales.
4. Outlining strategies and measures to give u competitive edge. Competitive pricing, building a better distribution network, excellent social and corporate responsibility gives competitive advantage. Low cost and rapid delivery were the key strategic drive for coca cola success.
5. Consumer recognition is built with a strong brand. Consumers, associate some level of credibility and quality with brand names, which becomes easy for them to operate internationally.
6. Increased sales margins will entreat them to operate internationally since consumers would accept a higher price for a strong brand.
7. Consumer loyalty would boost sales irrespective of the price tag. This desire would interest them to go global since they will not sacrifice a good brand for another.
8. The addition of other product line brings about economies of scale; facilitate long-term growth, and reduces marketing cost. Suppliers, vendors and distributors, would all be enticed to provide continuous supplies and services.
9. A good brand has competitive edge/advantage over others and this boosts the desire to expand operations to new regional markets.

2.2 Barriers to competition that obstructs the search for global competitive advantage

International trade is the most important and most profitable business nowadays but there are some barriers to international trade that could be faced as a result of local competition. For desiring to enter into international trade, the company faces some obstacles.

Entering new markets is never easy thing to do. That move always comes with its issues.

However going international is relatively easier for a strong brand in the sense that the customer awareness is high due to the strength of the brand and the savoir-faire of the company can be replicated elsewhere or adapted if conditions are not the same. Also with a realistic PESTEL (political, economic, social, technological, environmental and legal) and SWOT (strength, weakness, opportunities and threats) analysis the organisation can make an informed decision before entering the market.

The global firm needs to recognize the crucial role of foreign subsidiary strategy in building global competitive advantage and emphasize corporate control of important strategy elements. In particular, there is a need to examine the role of foreign subsidiary strategy in the firm's efforts to break through local barriers to competition that obstructs the search for global competitive advantage.

– Distribution
– Promotion
– Product
  Physical evidence
– Advertising
– Plannnig
– Organising
– Sfaffing People

A product is the physical presentation of what a company sells. It need to be branded in a way that will suit a particular country. The international company should know the existing product in a country that will be a possible substitute to its product.

The distributing channels that will be used by the company should also be taken into consideration. Whether it should be through franchising or distributing van delivery the product to retailers or middle men.

The price is the amount charged for the product, since there will be local competition, the price must not be charged too high or too low since some countries perceived a low price of goods to be inferior goods.

Promotion and advertising should be the hall mark of every company.

However, the challenge of international laws on cross border trade are also a barrier. Where the law protect the local companies, international companies may have favorable and attractive consumer demand like lower price. Also an instance of giving discount to distributors who were stocking only their product even though it could be another way of penetrating the local market.

The of European Union eliminating all internal tariffs in Europe made gained more profit due to easy transportation of soft drinks from one European country to the other. This made focused more on low cost and rapid delivery as the key strategy to successful business.

Another step is to implement principles which to gain local citizens trust in their product thereby introducing the "think local, act local" strategy, high involvement of local manager and focus of brand promotion at the regional level to eliminate the popular local beverage was made possible. This possibility was largely also due to the partnership coke went into with Schweppes and San Pellegrino. Schweppes and San Pellegrino had already a trusted brand on the local market and the partnership made the coke brand easily acceptable by the local market.

2.3 Risks in internationalization

Internationalizing firms are routinely exposed to four major types of risk, cross-cultural risk, country risk, currency risk, and commercial risk. The firm must manage these risks to avoid financial loss or product failures.

Cross-cultural risk – A situation or event where a cultural misunderstanding puts some human value at stake. Risks arise from differences in language, lifestyle, attitudes, customs, and religion, where a cultural miscommunication jeopardizes a culturally valued mindset or behavior. Negotiations are required in many types of business transactions; e.g., Mexicans are friendly and emphasize social relations, whereas Americans are assertive and get down to business quickly. Standards of right and wrong vary considerably around the world. For example, bribery is relatively acceptable in some countries in Africa, but is generally unacceptable in Sweden.

Country risk – Exposure to potential loss or adverse effects on company operations and profitability caused by developments in a country's political and/or legal environments. For example the U.S. imposes high tariffs on imports of sugar and other agricultural products. Doing business in Russia often requires paying bribes to government officials. Venezuela's government has interfered much with the operations of foreign firms. Argentina has suffered high inflation and other economic turmoil.

Currency risk – Potential harm that arises from changes in the price of one currency relative to another. Risk that exchange rate fluctuations will adversely affect the value of the firm's assets and liabilities. High inflation, common in many countries, complicates business planning and the pricing of inputs and finished goods.

Commercial risk – Firm's potential loss or failure from poorly developed or executed business

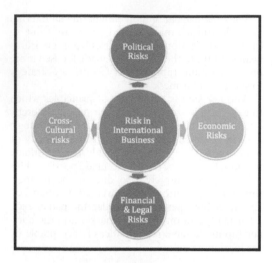

Figure 2. Risks in internationalization.

strategies, tactics, or procedures. General commercial risks such as weak partner, operational problems, timing of entry, competitive intensity and poor execution of strategy lead to sub-optimal formulation and implementation of the firm's international value-chain activities.

No matter how strong a brand is internationally, it faces the issue of cross cultural risk where it is exposed to changes in tastes and preferences due to religious beliefs, culture, and lifestyles. This gives local firms a certain advantage when it comes to meeting the direct needs of consumers. So even though the international companies come to increase competition in a particular market because of low cost of production and in effect low prices, the local firms can outcompete international ones by coming up with products that are specific to that market's tastes and preferences.

Consumer base created by local competition could be a barrier. Market penetration by an international company could be a challenge and a drain on resources due to high advertising costs. How to get consumers to switch their taste to products offered by the foreign company can also be a great barrier to growth.

Depending on the brand any foreign citizen is a potential customer. But customers living in international market have preference difference buying habit and priorities. There should be research and cultural survey. Though strong brand but without this invites troubles. For example, Muslims do not take in alcohol so selling soft drink in Middle Eastern countries will succeed. For instance, all though strong brand but without international marketing plan, budgets and expertise, it is difficult to establish partnership and net-works with other companies, review of international marketing strategies may not help product foot-hold for success.

Wherever a business thrives, competition evolves and a multimillion company could be displaced to a position where it could barely break even. Competitors are necessary evils; they keep businesses on their toes and equally drive them out of operation.

– Acquisition and mergers of local companies by international companies could pose as competition to local markets. They dominate the market and call the shots, which can threaten the survival of other local companies.
– Regulations and trade policies by international companies over ride local markets due to their anti-competitive nature.

According to 2012 survey by World Federation Advertisers

– 95% of WFA members believe future growth will from outside USA.
– Over 71% think best advert campaigns are being developed abroad.
– 75% of marketers think brand can lean lot from foreign campaign.

This means strong brand may help operate internationally but without research, understanding cultural believes proper planning, establishments of partnership and net-works, companies will face many difficulties.

Often than not there are multiple forms of goods and services available on the market prior to entry of international companies. In simple terms this can be referred to as local competition. The effect of local competition could be used as a stepping stone to excel in business, that is, for new entrants seize these shortfalls and improve on or offer better and alternative solutions.

3 CONCLUSION

International companies need to partner or establish relationships with distributors and located in the countries whose markets they are seeking to enter. Where International firm have no prior experience in that country, finding partners who are honest and experienced can be a challenge. Time difference and distance, laws and legal procedures protect the local companies and culture having a great influence on consumer purchase, local companies are always at the advantage and international companies need to push harder.

In a very strictly localized market, penetration could be very difficult if the inhabitants have a relationship with the brand and product especially where they have the sense of ownership.

Also penetration strategy adopted by the company will be crucial for them to integrate to the local environment. This can be through offering scholarships to the inhabitants, sponsoring festivals and community development projects. If the growth of the foreign firm is made local in the hands of the indigenes, that is, the management and operations of the firm is trusted in the locals it is a sure way to beat competition from the local competition

ACKNOWLEDGEMENT

In conclusion, we would like to express thanks for the support of projects VEGA No. 1/0492/16 "Investigation of possibilities of elimination of deformations of thin-walled parts using high-speed machining" and ITMS project 26220220125 "Research and implementation of experimental simulation methods to optimize processes on technological workplaces."

REFERENCES

Bartlett, Ch., A. & Ghoshal, S. 2000. *"Going global: lessons from late movers."* Reading 1.3 (2000).

Gasper, K. & Gerald L.C. 2002. "Attending to the big picture: Mood and global versus local processing of visual information." *Psychological science* 13.1 (2002): 34–40. http://businessculture.org/business-culture/what-is-international-business/.

Kearney, M. 1995. "The local and the global: The anthropology of globalization and transnationalism." *Annual review of anthropology* 24.1 (1995): 547–565.

Kreheľ, R. & Gregová, L. 2006. The measurement of forces in turning operations / - 2006. In: *Scientific Bulletin.* Vol. 20, serie C (2006), p. 209–212. - ISSN 1224-3264.

Novotný, L. et al. 2016. Simulations in multi-pass welds using low transformation temperature filler material. *Science and Technology of Welding and Joining,* 21 (8), pp. 680–687, 2016, DOI: 10.1080/13621718.2016.1177989

Reade, C. 2003. "Going the extra mile: Local managers and global effort." *Journal of Managerial Psychology* 18.3 (2003): 208–228.

Rhiannon, L. & Cockrill, A. "Going global—remaining local: The impact of e-commerce on small retail firms in Wales." *International Journal of Information Management* 22.3 (2002): 195–209.

Roth, K. & David M. 1991. Schweiger, and Allen J. Morrison. "Global strategy implementation at the business unit level: Operational capabilities and administrative mechanisms." *Journal of International Business Studies* 22.3 (1991): 369–402.

Rugman, A. & Hodgetts, R. 2001. "The end of global strategy." *European Management Journal* 19.4 (2001): 333–343.

S. Cavusgil, T. et al. 2014. *International Business,* Pearson Australia, Aug 1, 2014 - Business & Economics - 657 pages

Simon-Miller, F. 1986. "World marketing: Going global or acting local? Five expert viewpoints." *Journal of Consumer Marketing* 3.2 (1986): 5–7.

Stofová, L. & Szaryszová, P. 2016. Environmental Criteria of Public Procurement as a Tool of Development Sustainability. *Calitatea,* 17(152), 67.

Deficits in social disasters' management and tasks for future research

D. Procházková & S. Szabo
Faculty of Transportation Sciences, Czech Technical University in Prague, Prague, Czech Republic

ABSTRACT: The paper characterizes disasters the cause of which is in human body and in human society. On the basis of specific investigation of data on social disasters it gives their impacts on public assets and evaluates their severity. It also gives results of expert judgement of level of social disasters' management in the EU. At the end it gives set of adverse problems that might be solved and which need the support from research.

1 INTRODUCTION

One of the main components of the human system is the social system in which there are also inherent the phenomena that damaged both, the human society and the whole human system (UN, 1994). Because the current aim of the EU population is to live in a safe area, which has development potential, the EU and its Member States have a primary function to provide protection and a development of the human society (Šoltés & Gavurová 2014, 2015). Unfavourable phenomena that are connected with human body and human society are: diseases, human errors and wrongful appropriation of property; killing a human being; bullying; religious and other intolerance; criminal acts such as: vandalism and illegal business, robberies and muggings, illegal entries, wrongful use of property and services, thefts and frauds, intimidations and extortions, destructions and sabotages, terror against an individual, terrorist attacks, local and other armed conflicts.

The security policy is a set of measures and steps for the prevention and elimination of harmful effects in the human society, or at least for mitigating the large impacts at the realization of risks that outcome from them with target to ensure internal and external safety, defence and protection of citizens and the state. It is done by a foreign defence and economic policy and by a policy in domain of internal safety and by public awareness. For this purpose the European Union and its Member States formulate or already have formulated the basic challenges, principles and legislative measures, which should serve or serve as the structure stones for the building, preparation and application of the relevant elements of the security system.

The questions of the EU safety are viewed from the perspective of the internal safety and simultaneously also from the view of external one, which are mutually overlapping, complementing and conditioning, because the aim of the management is the integral safety that includes both partial safeties (Procházková 2011a). The present work assesses the level of the EU governance of disaster management from the view of the management of disasters, which have cause in the human society. Based on the identification made in the FOCUS project (Procházková 2012) the work deals with disasters, such as: are evolved gradually. Throughout history, in accordance with development of knowledge, it is created and transformed the system of relationships the state of which is called public welfare is a state in a system of relations (in human society) which provides relative security and relative enough for the vast majority of individuals in a society. The rate (measure) of public welfare depends on how the basic needs of individuals are satisfied in the system and how the acceptable life standard is provided, how except lives and health protection is provided the certainty and safety, protection of property, the standard rights of individuals and also the conditions for their standard development. The development of this system of relations depends of course on the system management, on maintenance of public welfare through moral, ethical and legal rules. It is the managers who represent the very first persons to be capable of identifying changes in the environment and communicate the appropriate signals of changes to the rest of the employees by way of influencing their working activities (Gallo & Mihalčová 2016). In fact, throughout history in the subject area there were and now are many conflicts of different significance and importance which are solved by the governments, interest groups and individuals. However, it appears that the human system development was always more obvious and faster when there were fewer conflicts and when they were not so deep, when the feeling of security and public welfare was prevailing.

2 DATA USED FOR RESEARCH

To assess the severity of the impacts of social disasters there were used data about individual emergency situations which caused selected social disasters. In the research that is in detail described in study (EU 2013) there were considered 13 disasters. The method of investigation is based on expert evaluation of scenarios and impacts of disasters by using specific method What, If (Procházková 2011b).

For the assessment of the level of disaster management in the EU in the case of disasters, which have a cause in the human society there was used data from professional domains that comes from:

- Web sites (official EU documents and the Czech Republic documents (the archives of the Ministry of Interior of the Czech Republic) - strategy, agreements, conventions, programs, laws, reports, papers, etc.), the documents are enumerated in (EU 2013),
- Documents and statements of other international, national, governmental and non-governmental organizations (Ministries of Interior, Ministries of Defence, Ministries of Foreign Affairs, Europol, Interpol, ENISA, GRECO, Freedom House Europe, Transparency International, etc.),
- Printed materials (books, handbooks, periodical reports, professional journals, newspapers),
- Electronic databases (mostly on CD and DVD media).

The assessment of the level of disaster management in the EU was based on Questionnaire (Procházková 2012). Its evaluation and proposals' processing were done by data and criteria comparison on the basis of professional knowledge and occupational experiences with the relevant issues on the Ministry of Interior, using the professional consultations mutual improper behaviour of an individual or groups of individuals: wrongful appropriation of property; killing a human; bullying; religious and other intolerance; criminal acts such as: vandalism and illegal business, robbery and attacking, illegal entry, unauthorized use of property or services, theft and fraud, intimidation and blackmail, sabotage and destruction, terror against individuals, terrorist attacks; local and other armed conflicts:

- Intentional disuse of technologies, such as: improper application of CBRNE substances; data mining from social networks and other cyber networks used for psychological pressure on a human individual,
- Incorrect governance of public affairs: corruption; abuse of authority; and the disintegration of human society into intolerant communities.

3 BEHAVIOUR OF A HUMAN AND HUMAN SOCIETY

According to the evolutionary history, human is yet the last link in a development range of living nature. Human was created and lived there, where there were basic conditions for his existence. The human is a subject to all laws that are applicable to living organism. Monitoring the evolutionary path from an anthropoid to today's man lasting more than million years helps us to understand the modern knowledge of heredity and of human system dynamics. According to data in the expert literature and from practical experience about human behaviour in different situations, human responses to external (and internal) stimuli are very diverse. They can take the form of unconditional response, such as "automatic", innate ways of responding to stimuli (recoiling when there is an unpleasant stimulus), conditioned responses (e.g. in the form of habits), or the purposeful, will-conducted behaviour.

Human is an inherent part of the human system, which he needs for his life, and therefore he can manage only what he does. This means that a human, to provide his security and sustainable development in the range of possibilities, carries out arrangements and actions, he protects himself with and provides life in a way not to cause irreversible processes in the human system, which are undesirable for him, because they do not give a possibility to security and development.

Human as a social creature lives in a society. The human society is a social organization of people the structure and functioning process of which in institutions: Educational Faculty of Palacky University in Olomouc, the National Headquarters for Combating Drugs SKPV, and the Unit for Combating Organised Crime SKPV PCR. There were also used data on legislation and control mechanisms in the EU (EU 2017).

The surveillance object is the investigation of phenomena that disturb the human security and security of the whole human society and that are caused in human society. List of observed phenomena is given exhaustively in the introduction.

4 IMPACTS OF SOCIAL DISASTERS ON HUMANS AND HUMAN SOCIETY

The selected set of 13 social disasters, which includes:

- Incurable diseases: aids, cholera, swine flu, bird flu, SARS,
- Terrorist attack: the attack with nerve paralytic gas sarin on Tokyo subway in 1995, and Anders Breivik extremist attack in Oslo and on Utoya island in 2011,

- Abduction/Children disappearance – the disappearance of English girls Madeleine McCann from the Portuguese resort in 2006,
- Vandalism: damage of the copy of the Ishtar Gate and other acts of vandalism in Babylon done by American army in 2003–2004; damage of the spring of the mineral water Ida in Náchod in 2011, and riots at football,
- Murder/Murder attempt/Assassination – deadly dioxin poisoning of presidential candidate Viktor Yushchenko in Ukraine in 2004,
- Cybercrime or computer crime – fraud on the Internet committed by D. B. Sundin and his accomplices, who gained more than $ 100 million in 2006–2008,
- Local conflicts – a coup in Egypt in 2011, was investigated by help of What, If method in-depth (EU 2013) one after the other. The summary of outputs shows that all incurred emergency situations have a great impact on the nearest surroundings. Relatively large number of victims, namely not only direct, but also those who suffered from the psychological harm, connects these emergency situations.

Generally, the observed disasters cause fear, mistrust and also strong disappointment. A number of processed events are local issues that came by influence of media into public awareness of the world. Impacts on people can be divided minimally into three groups:

- Direct/primary impacts – people affected by severe disease, victims of crimes (kidnapped girl, poisoned presidential candidate, the people who lost their money on the Internet, people who died or were injured in the revolution in Egypt … etc.),
- Secondary impacts – close people and relatives of patients and victims (especially psychical shock, depression …),
- Tertiary impacts – people who have to solve the consequences of disasters (doctors, policemen, rescue workers, investigators, and media).

The most critical situation occurs when public administration fails, because there is no force that would organize a targeted response.

We cannot say that the disease or social disaster, i.e. deliberate attacks on people or their property, would avoid some of the world's areas. Generally, that everywhere where people live, there are diseases, crimes and criminal activity. The most serious consequences of processed disasters are losses of lives, injuries, damage of property or damage and destruction of monuments and historical sites in the event of vandalism. The psychological impact plays an important role on a narrow, but of course in muted form, on a wider range of people in surrounding of affected people and the victims.

The consequence of losses of lives, injuries, losses and damages of property and psychical damage are fear, distrust and uncertainty, i.e. loss of safety.

For judgement of level of social disasters' management the monitored adverse social disasters effects were put together for reasons of internal relations to the following groups:

- Subsequent crime and other offences. The group includes: vandalism and illegal risk behaviour, robbery raids and attacks, property crime, killing and rioting,
- Tax fraud and fraud. The group includes: tax fraud, fraud,
- Damage to the customs laws, including: customs fraud, smuggling of prohibited goods,
- Illegal access to any information systems. The group includes: data theft or data changes, espionage, partly fraud—forgery of documents, partially terrorist attack, data mining from social networks leading to the psychological pressure on people,
- Corruption and serious economic crime, including money laundering, extortion and humiliation. The group includes: corruption, abuse of authority,
- Society disintegration into the intolerant groups. The group includes: religious and other intolerances.

There were not considered the following phenomena due to lack of data: child labour, sabotages, infringement of law by government agencies, maritime piracy, severe negligence with criminal responsibility, misuse of postal services, an anonymous notice of alarming information, environmental crime including pollution, and violations of security regulations.

The experts' investigation showed that:

- The impacts of mutual improper behaviour of an individual or groups of individuals are: violation of basic human rights and freedoms (violation of religious freedoms and freedoms of religion, freedom of movement and expression, political opinion, affiliation to national or ethnic minority, discrimination based on sex, race, colour, language …), organized criminal community (trafficking with humans and children for commercial sexual exploitation and forced labour, production and distribution of drugs, illegal arms sales, extortion, terrorist attacks, sabotages, local and other armed conflicts …), violence against individuals and groups of people with the threat to health or life, financial and property harms (theft, robbery, burglary, credit and financial fraud, identity theft), social unrest, migration including illegal migration, extremism, xenophobia, etc.,

- The impacts of the intentional disuse of technologies are: incorrect application of CBRNE substances, illegally obtaining information from social networks and other cyber networks for psychological pressure on the human. This involves activities such as flaming (on-line attacks and communication), hacking/cracking/phishing (theft and destruction of data, misuse of personal data), cyber grooming/cybersex (manipulative behaviour), SMS spoofing/malware (malicious codes) spamming/hoax (unsolicited and alarm messages), cyberbullying/cyberstalking (humiliating and offensive behaviour) and of course cyberterrorism (i.e. attacks against critical infrastructure). Such as conventional terrorism, cyber terrorism impact is both economic and political,

- Incorrect governance management impacts are: disruption of internal security and public order, the increase in the specific crime (organized crime, corruption, financial fraud and illegal financial transactions), tax evasion, illegal financing of political parties and everything else that followed these activities,

- Impacts of corruption: disruption of the institutions and values of legal state; infiltration into government authorities that then do not decide in favour of public interest; threat to public security; undermining the society foundations by actions

Question	Answer (sentence + reasons for)
Does the list of followed disasters given above contain all disasters possible in the EU territory?	It should be added: - illegal production and distribution of narcotics and psychotropic substances, - illegal migration, - proliferation of the weapons of mass destruction.
Which disasters from the followed one are the most horrible for the EU territory? Give / Put them in order according to your own knowledge and experiences.	- illegal access to information systems, cybercrime - terrorist attacks - corruption in government and public administration, including the political scene - serious economic crime, including money laundering, tax evasion - trafficking with human beings and illegal migration - illegal production and distribution of psychotropic substances - extremism, all forms - discrimination and intolerance
For which followed disasters the EU does not systematically perform prevention? Is the prevention level sufficient? What is the situation in the CR? What is necessary to improve?	Prevention is not systematically carried out for any of the above given disasters. Prevention is often declared by signed treaties, conventions, treaties or bilateral / multilateral agreements. The level of prevention is not sufficient. The situation needs to be improved: - strict compliance with European and national legislation, consistent prosecution and punishment for breaking the law as an effective preventive tool - equipment by the most effective technical means for the detection of these forms of crime, - close interdisciplinary cooperation of all parties involved at national level and consistency with other central institutions within the EU states, - sharing good practice, continuing education and training of experts responsible at the pan-European level.
For which followed disasters the EU does not systematically perform preparedness? Is the preparedness level sufficient? Is the preparedness performed by all important society components (including public) sufficient? What is the situation in the CR? What is necessary to improve?	The EU does not systematically ensure the preparedness for coping with the above given disasters. The preparedness depends on the level of the crisis management of individual EU countries, i.e. on the level of detection (intelligence services, technical means, and the level of experts ...). The preparedness is the most well established the best on a theoretical level, the level of practice is greatly affected by the economic stability of a particular Member State. The CR level is the same as level of other EU countries. The impact of individual disasters that affect a large area has in all cases of the pan-European dimension; therefore, it is necessary to ensure consistent pan-European co-operation.
For which followed disasters the EU does not systematically prepare qualified response? Is this response level sufficient? Is response prepared by all important society components (including public) sufficient? What is the situation in the CR? What is necessary to improve?	In view of the above given information it follows that the responses to disasters are prepared on a theoretical basis, but it is difficult to estimate the real situation. The response depends on the type of a disaster, the situation is different in case of a terrorist attack, for which there are still many unknowns; satisfying situation on a European scale is in the detection of production and distribution of psychotropic substances. There is necessary to improve the consistency of trans national components at the detection and combat of individual types of disasters and it is necessary the unification of key legislations.
For which followed disasters the EU does not systematically prepare qualified renovation (renewal)? Is this renovation level sufficient? What is the situation in the CR? What is necessary to improve?	In view of the above given information it follows that qualified renovation for followed disasters recovery is not on a uniform level within the EU. Highly unacceptable impact on the current situation in EU countries they have long-term consequences of an economic crisis. The impact of the economic crisis very affects or causes the outflow of experienced professionals from the components of law enforcement (withdrawal into the private sphere) and it influences the level of technical support that is necessary for detecting and combating crime. A similar situation exists in the judiciary and prison system. It is necessary to improve the staff situation, qualification of experts, technical background and continuous current analysis of the potential impact of individual disasters.

Figure 1. (*Continued*).

Which followed disaster can cause the critical situations in the EU? Which followed disaster can cause the critical situations in the CR?	Terrorism, especially cyberterrorism and intentional abuse of technologies leading to disruption of critical infrastructure. The Czech Republic is not a primary target country, but within the EU it would be a major form of (cyber) terrorism with such fatal consequences.
Which followed disaster can cause the crisis situations in the EU? Which followed disaster can cause the crisis situations in the CR?	Terrorism, especially cyberterrorism and intentional abuse of technologies leading to disruption of critical infrastructure. The Czech Republic is not a primary target country, but within the EU it would be a major form of (cyber) terrorism with such fatal consequences.
For which crisis situations caused by followed disasters in the EU the level of crisis management is not sufficient? For which crisis situations caused by followed disasters in the CR is the level of crisis management not sufficient?	Terrorism, especially cyberterrorism and intentional abuse of technologies leading to disruption of critical infrastructure. This applies in both, the EU and the EU Member States.
Where the vulnerabilities of human society in the EU can cause a change of a critical situation into the extreme situation? Where vulnerabilities of human society in the CR can cause the change of a critical situation into the extreme situation?	The greatest vulnerability lies in the possibility of easy ideological abuse of the internet - bullying people, an attack on the infrastructure of critical systems, intervention to information and telecommunication infrastructure, providing the ability to co-ordinate terrorist activities in the EU / worldwide.
Do we have reliable methods for the determination of the scenarios of all disasters expected in the EU? Do we have reliable methods for the determination of the scenarios of all disasters expected in your country?	There are various methods, guidelines or legislative measures for determination of the object scenarios in both, the EU and the individual countries or regions. It is missing, however, effective mutual consultation and co-ordination of procedures and their flexible adaptation to the rapidly evolving global (transnational) conditions that bring new threat scenarios, and therefore, they require new more reliable methods determining new reliable scenarios.
Do we know for all followed disasters given above successful preventive, mitigation, response and renovation measures and activities? Which weaknesses are in knowledge on preventive, mitigation, response and renovation measures and activities?	Prevention and preparedness are conditioned by the greatest possible synergy between the domain of followed disaster risk reduction and the domain of adaptation to global changes with aim in order that financial support to prevention activities may increase resilience to future crises. In this aspect, the EU and individual Member States have substantial reserves.
What is necessary to improve?	- cooperation in the security research, - the implementation of existing directives and legislation, - strengthening the individual response tools of the EU to appurtenant disasters.
What research is the most effective for the improvement of safety management of the EU? What research is the most effective for the improvement of safety management of the CR?	Comprehensive research of safety in all relevant domains, which are in such research to be linked, and therefore, it is possible to achieve synergistic effect and outcome.
What principles, legislation and co-operation rules in the EU are necessary for security and sustainable development of humans?	The EU legislation is in fact sufficient. E.g., the Lisbon Treaty provides the opportunity to build powerful, comprehensive, co-ordinated and effective response capacities for the response to appropriate disasters in the European Union. However, it is necessary to continually ensure and oversee the effective use and application of all the existing EU legislation.
Can you propose measures for averting the social crises in the EU?	Due to the impact of the economic crisis on the European scale it is not possible to eliminate the spread of social unrests, which are already taking place in the most affected countries (Greece, Spain, Portugal, Italy ...). To the calm of situation it may contribute consistent and exemplary combat against phenomena such as corruption of politicians and officials and confiscation of the proceeds of crime - are examples of efforts to remedy the system.

Figure 1.   The results of the assessment of safety management level in the social domain.

of dictate, client or corrupt networks (the result can be the loss of public confidence in honesty and impartiality of public institutions, distortion of market relations, economic decline and the state destabilization). Unclear boundary between political and criminal motivation, fuelled by corruption, often leads to the linking of the structures of organized crime with terrorist networks. Corruption has been long seen as an internal national problem. Over time, it has become increasingly apparent that it also has its cross-border implications, and therefore, the successful fight against it cannot be performed in isolation. Cross-border organized crime is often resorted to bribery as a means suitable for penetration into the legal system. Necessary laws to tackle corruption have been largely accepted and the institutions relevant for this purpose more or less established, but the process of implementing anti-corruption policies across the EU is still unbalanced. The Stockholm Programme invites the Commission to develop indicators for assessing Member States' efforts in the fight against corruption.

5   SURVEY RESULTS

From the analysis and expert assessment of the efficiency of existing EU documents, the list of which is in study (EU 2013) they follow the relevant facts, e.g.:

– According to the current EU strategy (EU 2010) there are processed: the EU Action Plans that may prevent access to the chemical, biological, radiological and nuclear (CBRN) substances; legislative and non-legislative measures that enable barriers in the purchase of components for the explosives manufacturing, monitoring network, etc. By the implementation of the Action Plan into practice, the Commission has undertaken to give up to 100 million EUR till 2013 according to its strategical meaning,

- The EU deals with human society in relation to the possible disintegration in the level of the efforts on cooperation and understanding between individual countries, respectively among states and EU nations as a whole. In December 2006 the European Parliament and the EU Board established an action program "Europe for Citizens" as promotion of quality European citizenship,
- Negatively it is perceived illegal migration and its possible consequences, such as connections to the organized crime,
- The EU's efforts to achieve life quality and security of all children,
- Reduction of corruption (it is estimated that annual losses in the EU are 120 billion, i.e. 1% of EU GDP). Although the nature and extent of corruption are varying, the state impairs all EU Member States and the Union as a whole. Therefore, in the EU there is valid the Framework Decision 2003/568/SVV. Other international anti-corruption instruments: several EU Member States have ratified all the existing international anti-corruption instruments or their majority; three Member States (Italy, Germany, Austria) has not yet ratified the Criminal Law Convention on Corruption of the Europe Board; twelve states (Czech Republic, Estonia, Finland, Italy, Lithuania, Austria, Hungary, Malta, Germany, Poland, Portugal, Spain) haven't ratified its additional Protocol; and seven states (Denmark, Ireland, Italy, Luxembourg, Germany, Portugal and the UK) haven't ratified the Civil Law Convention on Corruption. Three Member States (Czech Republic, Ireland and Germany) have not ratified the UN Convention against Corruption. At the present time in the Czech Republic there is prepared the government material to the application for access to the UN Convention against corruption, submitted by the Minister of Justice,

- Five EU Member States haven't ratified the OECD Convention on Combating Bribery (Cyprus, Latvia, Lithuania, Malta, and Romania). These Member States are not OECD members. The only Member State that is not a member of the OECD and ratified this convention is Bulgaria. In some corruption and fraud cases the indictment passing has been delayed for many years from the first report to the national judicial authorities by OLAF. The reasons for these delays arise mainly from time-consuming process procedures for mutual legal assistance and the lack of leadership of prosecutions at EU level.

Evaluation of the questions from questionnaire 1, which is adapted to the field of observation, is shown in Figure 1.

Based on data in Figure 1 there is evaluated the level of EU governance in terms of social disasters´ management and there are identified fundamental deficiencies connected with management of appropriate disasters and there are identified domains in which there is necessary to accept measures (Fig. 2). The result implies that weaknesses are many. It is caused by the fact that in management there is missing the focus on priority issues.

Disaster	List of gaps	Type of measures and activities for remove of gaps				
		Legislation	Specific management	Research	Education	Other
Mutual inappropriate behaviour of an individual or groups of individuals	Unauthorized appropriation of property; killing a human being; bullying; religious and other intolerance; criminal acts such as: vandalism and illegal business, robbery and attacking, illegal entry, unauthorized use of property or services, theft and fraud, intimidation and extortion, destruction and sabotage, terror to individuals; terrorist attacks; local and other armed conflicts.	Yes	Yes	Yes	Yes	M
Intentional abuse of technologies	Improper application of CBRNE substances; mining information from social networks and other cyber networks for psychological pressure on the human being.	Yes	Yes	Yes	Yes	M
Incorrect governance of public affairs	Corruption, abuse of power, and the disintegration of human society as an intolerant community.	Yes	Yes	Yes	Yes	TM

Figure 2. Proposal for the solution of identified deficiencies. Bold letters denote domains that need special attention to real problem solving; in the "other" column M indicates the necessity to systematically carry out monitoring and TM refers to monitoring and tough sanctions.

6 SPECIFICATION OF PROBLEMS ASSOCIATED WITH MANAGEMENT OF ADVERSE EFFECTS IN HUMAN SOCIETY

In the first we show the general results and after we concentrate to the Czech Republic strategy for internal security.

6.1 *Problems associated with management of adverse effects*

Identified gaps or deficiencies listed above clearly indicate the need to continue efforts that would be directed to ensuring the even greater level of protection and security of all EU citizens. Existing legislation, strategies, programs and other crucial documents that have already been put into practice, provide an appropriate basis, which should further improve and develop integrated tools and policy. It is necessary to focus on specific steps, procedures and principles, which mainly include:

1. The EU safety is a tool that ensures the security and development of the EU population, i.e. their protection in all directions.
2. The safety of the EU and all its Member States must be understood as a tool incorporating a number of measures that have both, the horizontal (system) and the vertical (geographic) character. Similarly, it is necessary to understand the nature of co-operation among the individual EU bodies and components, and thus to ensure that it may be coordinated, integrated and efficient. One of the most necessary steps in a given sense is ensuring the effective co-operation and assistance of the judicial authorities and law enforcement authorities of the Member States.
3. In the context of the EU safety management it is necessary to emphasize the prevention and to create new or to improve existing preventive mechanisms. IT is equally important to engage in preventive activities the law enforcement agencies and professionals, and the academic and educational institutions, other relevant public and private sector bodies and / or civil societies as such.
4. It is also necessary to create comprehensive information-sharing tools that include all relevant EU databases and that allows access to them if necessary. Naturally, in this case they must ensure adherence of the right to privacy and personal data protection.
5. To ensure professional and systemic co-operation in the development and introduction of new technologies, namely in both, the public and the private sector.
6. Emphasis must be put on the continued development of integrated EU's external borders, which would ultimately be used to address issues such as trafficking with human beings and also had an important position in maintaining internal and external safety of the EU and individual Member States.
7. Based on the existing results of research projects that were implemented under the joint research and development program, the EU should determine which aspects of EU security policy need to be intensively pursued in the frame of intensifying the safety concept in individual Member States and in the EU as a whole.

Based on the information, conclusions and recommendations presented in the tables above it can thus be stated that the safety of the EU and all its Member States must be based on a commitment to adhere principles such as consensus, co-operation, democracy and rule of law with regard to adherence of the principles of the Charter of Fundamental Rights of the European Union. Therefore, it is necessary to have clearly defined rules, which must be applied both, horizontally and vertically. Subsequent evaluation of the effectiveness of the rules and spending means (cost effectiveness) enables to respond flexibly to changing circumstances (such as new criminal phenomena related to EU accession to the Schengen area and the consequent right of free movement of persons), to adapt to them, thus ensuring the EU inhabitants the highest possible level of security.

Because management of social domain means influencing the human behaviour, one of the most difficult tasks to achieve is he necessary changes in behaviour, particularly with regard to the fact that the rapid development of social networks is not accompanied by equally rapid increase of the user awareness of the consequences of providing huge amounts of data, the credibility of which is not often guaranteed. Mandatory notification at each internet service in principle may be positive, but it could cause problems for businesses. They should consider the information protocols according to the category of service—internet shop, internet service providers, search engines, social networks, etc. (Koščák & Kolesár 2015). Using the internet for the sales of new drugs and rapid exchange of information on new drugs through social networks are the new challenges faced by current drug policy and the traditional methods of prevention.

6.2 *Comparing the strategies for internal security in the European Union and the Czech Republic*

The main risks and threats related to crime which Europe faces today, such as terrorism, serious and organized crime, drug trafficking, cybercrime, human trafficking, sexual exploitation of minors and child pornography, economic crime and corruption, illicit arms trafficking and cross-border crime, particularly rapidly adapt to changes in science and technology.

Therefore, it is necessary to create a unified model of security (internal safety) and safety as such, that will be based on the integration of strategies and approaches that would confirm the Stockholm framework programme. This programme strengthens both, the civil law and the EU security. For this reason, an "Internal Security Strategy of the European Union—Towards a European model of security", was established and it was approved by the Europe Board in March 2010, it establishes common threats and challenges, a common EU policy on internal safety (security) and the principles on which this policy is based on and it defines the model of European security.

To cope with the above given phenomena there are the EU Member States´ (including the Czech Republic) own national security policy and strategy, and bilateral or multilateral agreements and regional cooperation. However, these efforts in many cases are not enough to prevent and combat crime, there is definitely needed a common approach at the EU level.

An important move towards achieving this goal was done by reality that the European Commission issued "Internal Security Strategy of the European Union: Five steps towards a safer Europe." This document builds on the challenges, principles and guidelines The Internal Security Strategy Council and suggests, how the EU should in the next four years, effectively prevent the serious crime and organized crime, terrorism and cybercrime, strengthen the management of EU external borders and increase resistance to natural and man- made disasters (Piľa et al. 2016)

The above two documents are indeed essential, but not the only tools to ensure security within the EU. From evaluation and analysis of all essential documents that were created for this purpose it follows that we still greatly need to improve or create tools that facilitate co-operation among the Member States.

7 CONCLUSION

Based on the above given facts it is clear that the EU and CR are aware of the importance of the management of social disasters and they make a lot of measures and activities. The performed analysis shows that the current policies, measures and activities in the given sector are partial; there is no uniform system tool for the entire social domain.

With regard to present situation we concentrate our research to security of humans at communities. Namely, we concentrate to workplaces, companies and public outdoors

REFERENCES

Antošová, M. et al. 2014. Assessment of the balanced scorecard system functionality in Slovak companies. *Journal of Applied Economic Sciences,* 9(1), 15–25.

EU 2010. The EU Internal Security Strategy in Action: Five steps towards a more secure Europe. [on-line] http://ec.europa.eu/

EU 2013. *FOCUS project study*-FOCUS. [on-line] http://www.focusproject.eu/documents/14976/-5d763378–1198–4dc9–86ff-c46959712 f8a

EU 2017. [on-line] http://eur-lex.europa.eu

Gallo, P. & Mihalčová, B. 2016. Models of Evaluation of Managing People in Companies. *Quality – Access to Success,* 17(155): 116–119.

Grinčová, A. et al. 2015. Meas-uring and comparative analysis of the interaction between the dynamic impact loading of the conveyor belt and the support-ing system. *Measurement,* 59 (2015): 184–191.

Grinčová, A. et al. 2016. Fail-ure analysis of conveyor belt in terms of impact loading by means of the damping coefficient. *Engineering Failure Analsis,* 68(2016): 210–221.

Hovanec, M. et al. 2015. Tecnomatix for successful application in the area of simulation manufacturing and ergonomics. In: International Multidisciplinary Scientific GeoConference Surveying Geology and Mining Ecology Management, SGEM 2015. STEF92 Technology Ltd, 347–352.

Koščák, P. & Kolesár, J. 2015. Safety of Airport Operation, Aeronautica 15. Lublin: Lublin University of Technology, 2015, 193–204.

Koščák, P. et al. 2015. State and Airports Safety Management, MOSATT 2015. Košice: Perpetis, 107–112.

Pavolová, H. & Tobisová, A. 2013. The model of supplier quality management in a transport company. *Nase More,* 60(5–6): 123–126.

Piľa, J. et al. 2016. New technologies in aircraft maintenance versus safety, Zarzadzanie bezpiec-zenstwem panstwa - wyzwania i ryzyka. Warszawa: NWP, 553–563.

Procházková, D. 2011a. Strategic Management of Territory and Organisation. Praha: ČVUT.

Procházková, D. 2011b. Methods, Tools and Techniques for Risk Engineering. Praha: ČVUT.

Prochazková, D. 2012. Questionnaire for special investigation. FOCUS 2011. [on-line] www.focus.eu

Rozenberg, R. et al. 2016. Critical elements in piloting techniques in aerobatic teams. In: Transport Means 2016. Juodkrante: Kansas University of Technology, 444–449.

Socha, V. et al. 2016. Training of pilots using flight simulator and its impact on piloting precision. In: Transport Means 2016. Juodkrante: Kansas University of Technology, 374–379.

Šoltés, V. & Gavurová, B. 2014. Innovation policy as the main accelerator of increasing the competitiveness of small and medium-sized enterprises in Slovakia. *Procedia Economics and Finance,* 15(2014): 1478–1485.

Šoltés, V. & Gavurová, B. 2015. Modification of Performance Measurement System in the intentions of Globalization Trends. *Polish Journal of Management Studies,* 11(2): 160–170.

Tkáč, M. et al. 2016. Analysis of request for proposals in construction industry. *Quality Innovation Prosperity,* 20, (1): 104–117.

UN 1994. *Human Development Report.* New York: UN. [on-line] www. un.org

New Trends in Process Control and Production Management – Štofová & Szaryszová (Eds)
© 2018 Taylor & Francis Group, London, ISBN 978-1-138-05885-9

Customers' motivation through personal sales with good grooming

M. Pružinský
Faculty of Business Economics with seat in Košice, University of Economics in Bratislava, Košice, Slovak Republic

ABSTRACT: In article, we discus effort within personal sales. We focus on the use of strategies in motivation, approaches and good grooming. These are in high use especially in personal selling of cosmetic products. We use descriptive-purposive method of research such as One Way Analysis of Variances statistic interpreted the correlates of personal selling strategies on motivation, approaches and good grooming. We may assume that motivation strategies are the tools in providing information from sellers to customers. This information is send to both the sellers and customers from manufacturer as well. Both the producer and seller have to keep in mind, if they want to sell, there is necessary to observe, and at the same time, obtain demonstrates their influence on effective managing of the buying/selling objections of potential buyers. Based on an example we recommend some practices that help in today's very important part of communication mix.

1 INTRODUCTION

Sellers should show the use of the product, emphasizing the benefits for consumers, demonstrate courtesy, patience, honesty, and use the results of surveys of sales, especially in questions that raised by the potential buyers of the product. Perfect clothes and use perfume are good signals towards buyers. Sellers implicate them in both the personal sales strategies and personal sales. Motivation, attitudes and good care are very important for the above segmentation strategy used in personal selling. (Ručinský & Ručinská 2007). Segmentation strategies are an essential component and an essential part of the sale, and thanks to them increases not only the quantity of products sold but also the quality of sellers who through this process better and become celebrities. To give full answer on good grooming is difficult. More and less the term is understanding on people grooming habits. Potential employers look more favorably on people who are well-groomed because good grooming suggests the interviewee is meticulous and organized. Poor grooming, on the other hand, we sometimes interpret as a sign of laziness. Therefore, it is important for interviewees to come to interviews well-groomed and in appropriate business attire. Psychologists also evaluate a patient's grooming. For example, having unkempt clothes and hair is a sign of depression, while scars on the back of the hand suggest eating disorders like bulimia. Overall, being well-groomed is often interpreted as a marker of self-confidence and self-worth. Generally, we may say that following 10 tips of good grooming should be of hands in personal sales:

– Stay clean! Bathe or shower at least every other day (if you have exerted yourself at all, shower DAILY), and on the off days, sponge bath the "naughty bits and pits" and wash your face and feet.
– Wash your hair at least 3 times a week, daily if it is greasy.
– Wear deodorant, fresh every day.
– Change your underwear every day, and wear outer clothes only twice and rewear only if they look and smell PERFECT. Check your clothes frequently during the day for dandruff, spills, messes and change as needed.
– Shave as needed daily to twice a wcck depending on hair growth. Pluck stray hairs from inside nose, chin, cheeks, and eyebrows.
– Brush your teeth at least morning and night. Brush your tongue with toothpaste too. Floss daily.
– Comb or brush your hair daily
– Wash your hands frequently. With soap. Wash them every time after you visit the bathroom, before you work in the kitchen, after you work in the kitchen or garden, after you change a diaper, after you pet an animal.
– Keep your clothing in good repair. Don't wear clothes that are torn where they aren't supposed to be.
– Keep your shoes clean.

Personal seller should wear the right dress, presents the sleek good behavior towards his clients. Effective is immediate physical testing of the product sold directly to each other, highlighting the advantages of purchase that the customer will buy, highlighting the utility value and price comparisons with other dealers or similar competitive products. In every situation, he appears positive and responsive to questions posed by potential consumers.

1.1 Personal selling

Personal selling or salesmanship are synonymous terms; with the only difference that the former term is of recent origin, while the latter term has been traditionally in usage, in the commercial world. Since a salesman, in persuading a prospect to buy a certain product, follows a personal approach; salesmanship, in the present-day-times in often popularly called as personal selling. Personal selling (or salesmanship) is the most traditional method, devised by manufactures, for promotion of the sales of their products. Prior to the development of the advertising technique, personal selling used to be the only method used by manufacturers for promotion of sales. It is, in fact, the forerunner of advertising and other sales promotion devices. Personal selling is a face-to-face contact between the salesman and the prospect; through which the salesman persuades the prospect, to appreciate the need for the product canvassed by him—with the expectation of a sales-transaction, being eventually materialized as we see in Figure 1.

Personal selling gives the best strategy that captures a target consumer. "Sales force" is group of people who persuade and convince a prospect to accept a product or a service. The aggressive and direct approach to a buyer is a strategy where the salesperson extends a lot of motivation, approaches and good grooming (Garcia & Villanueva 2009). This personal encounter of a salesman relies on guts of a skilled marketing communicator making the most challenging part of personal selling (Helbig 2011). In the language of sales and marketing "Personal selling" singles out those situations relates to communicating values and attitudes (Arante & Gomez 2000). A sales strategy

in personal selling is something that requires hitting the sales quota, realizing the sales presentation by addressing obstacles, and making sure that of a sales voucher (Gilleland 2004). The demonstrated message of motivation is the key to successful strategy. This is within selling group. The most complex activity that a salesman does is to create a connection thru effective communication. It leads to a favorable action of making a sale, requiring a strategy that takes persistence, energy and focus (Phopal 2009). This sales strategy sets out in details on how the salesperson delivers the objectives of personal selling using some approaches. This takes an edge in communicating his values caught from the way he looks, and carry himself during the sales presentation. (Ocon & Alvarez 2013). Good grooming is essential part of the salesman's passion for salesmanship. His strategies and approaches come along the way to successful communication that motivates and confirms the ethics of personal selling.

2 ANALYTICAL FRAMEWORK

In Figure 2 we present relations and synergy effect of personal selling strategies.

In this study, we investigate the implication of personal selling on the use of strategies in motivation, approaches and good grooming. We followed three objectives:

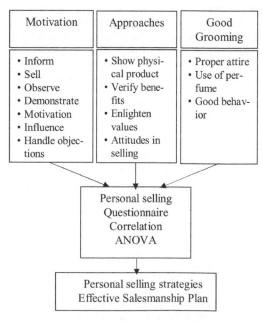

Figure 2.   Three personal selling strategies model.

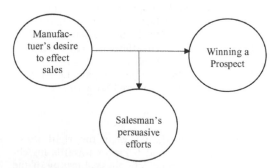

Figure 1.   Salesman's persuasive effort.

- To identify the needs of personal selling strategies in motivation, approaches and good grooming;
- To know the significant correlation between the strategies on personal selling variables: motivation, approaches, and good grooming;
- To Propose a Strategic Personal Selling Model for Cosmetic Products.

The pattern shows that there are three personal selling strategies used by salesmen and resellers. These are motivation, approaches, and good grooming. The focus of motivation strategies is to inform, sell, observe, demonstrate and influence the consumer by showing the physical product, its benefits, values, and price comparison; the focal points in approaches' strategy are attitudes in selling and handling objectives during the sales presentation; while in good grooming, its edge is proper attire and use of perfumes or cologne. Generally, a Salesmanship Model Plan has a concept for the benefit of sellers and owners of retail businesses. (Ocon & Alvarez, 2014).

3 METHODOLOGY OF RESEARCH

We used the descriptive-purposive method of research in order to investigate the most common personal selling strategies by the seller respondents. We utilized the mean, ranks, frequency and percentage that were in the analysis of descriptive data while the one-way Analysis of Variances (ANOVA) realized the significant correlation between motivation, approaches and good grooming strategies in personal selling. Before we validated the floating of the questionnaire in content analysis and split, half reliability was computing at 0.68. We also asked the cluster respondents to suggest and add their statements to complete the final content of the questionnaire before floating to the referees. There were questioned of 59 seller referees of the five outlets of cosmetic products in investigated region comprise the total population of this study. We conducted interviews in order to supplement the additional data computed in a master table analyzed and interpreted through a Micro Statistics.

4 RESULTS AND DISCUSSION

4.1 *Most common personal selling strategies*

Sellers used as the most common personal selling strategies "Good Grooming" (3.55) as we may see in Table1. Second ones are the strategies on motivation (3.40) and the least were approaches strategies (3.10). In the motivation strategies, it takes

Table 1.   Common sales strategies used by the salesmen.

Common sales strategies	Mean	Qualitative equivalent	Rank
Strategies on motivation	3.40	Often	2
Approaches strategies	3.10	Often	3
Good grooming strategies	3.55	Always	1

Table 2.   Motivation strategy in personal selling.

Motivation strategy	Mean	Qualitative equivalent	Rank
Sell the benefits of the physical products	3.77	Always	1
Demonstrate the uses of the product	3,69	Always	2
Influence the consumers citing importance of the product	3.51	Always	3
Observe the consumer's interests and desires	3,38	Often	4
Inform consumers about pricing and its comparison with others	3,14	Often	6
Ask question about the product that leads to getting attention	3,34	Often	5
Influence negative vibrations of consumers to affirmative	2,95	Often	7
Overall Mean	3.40	Often	

the seller to inform, sell, observe, demonstrate, influence and handle effectively the sales objections of the potential buyers of cosmetic products. In good grooming where result discloses the highest rank, includes proper attire, use of behavior and good behavior towards his clients. Lastly, the approaches posted the following: showing the physical product to the buyer, verifying the benefits when in case the client will buy the product, enlighten their values, comparing the price, and showing good attitudes when the consumer asks questions.

4.2 *Needs of personal selling strategies in motivation, approaches and good grooming*

To identify the needs of personal selling strategies in motivation, approaches and good grooming, it needs description to the specific items that covered the content of questionnaire in the clustered categories on strategies in personal selling. (Dzurová et al, 2007). Table 2 we show evaluated attributes within Motivation strategy in personal selling.

Table 3. Showing attitudes approaches.

A. Showing attitudes approaches	Mean	Qualitative equivalent	Rank
With courtesy like saying "hi", "hello" or "good morning"	3.41	Always	1
Patient in introducing the product	3.15	Always	2
Provide verification on the value of the product	2.90	Often	3
Caring attitude during the actual selling	3.03	Often	4
Industrious in prospecting potential consumers	3,21	Always	6
Agreeable attitude while explaining the features of the product	3.03	Often	5
Reliable attitude	2.99	Often	7
Overall Mean	3.10	Often	

B. Negotiating objections	Mean	Qualitative equivalent	Rank
Agreeable about objections		Often	5
Understand fully the objections		Often	4
lead the consumer to answer his own objections		Often	3
Answer only what is being asked		Always	2
Answer as honestly the best you can		Always	1
Confirm and verify the just concluded explanation		Often	6
Overall Mean		Often	

Good grooming strategy	Mean	Qualitative equivalent	Rank
Wearing appropriate attire		Always	1
Using cologne or perfume		Always	2
Proper hair style		Often	4
Wearing simple accessories		Often	3
Wearing correct combination of attire		Often	5

4.3 Approaches strategies on showing attitudes and negotiating objections

There are two approaches used in personal selling strategies. These are showing attitudes and negotiating objections. Helbig (2011) mentioned six guiding principles that should be written and shared to everyone in the selling business. These guides expected to contribute efforts in making communication more effective. This indicated the reflection of stakeholder's principles, processes, powers, and purposes (Connick 2013). Perfecting the plans was cited as to strategize by using the 4 P's (Riley 2012) that empower and reflect the purposes and objectives of a sales plan. (Gilleland 2004). It can be developed by whoever uses the strategies to their best advantage. Result of showing attitude approach showed a higher mean than the negotiating objectives. This disclosed the strategic personal plan to have its focus on showing courtesy, being patient and industrious and answering honestly the questions asked by the client. Wearing proper attire and using perfume or cologne was confirmed of need in a strategic selling plan. In Table 3 we provide values of attitudes approaches.

4.4 Analysis of variances in motivation, approaches and good grooming

In the Table 4, it shows that selling the benefits of the physical product ranks 1, followed by the demonstration on the use of the product, influence by citing the importance, observe the consumers and ask questions that lead to attention of the client to continue listening to the oral presentation of the salesman.

The data in Table 4 demonstrate that motivation, approaches and good grooming are significantly correlated: The critical value of 2.78 at 0.05 degrees of confidence level and 4.17 at 0.10 level of confidence rejected the null hypothesis. It is concluded that motivation, showing attitudes, negotiating objections, and good grooming are significantly correlated. Those that described selling benefits, demonstrate the use of product, citing the benefits to the consumer, courtesy, industry in prospecting consumers, patience, honesty and answering only those that are asked, wearing of appropriate attire and using cologne or perfume were implied in personal selling's very important in the inclusion of strategic personal selling in motivation, approaches and good grooming.

Table 4. ANOVA on relationship between motivation, approaches and good grooming.

Source	df	ss	Variance EST	F Ratio	Critical ratio	Decision
Between the group	9	1,180, 327,923	393,442, 641	42.3778	2.78	Reject
Within the group	50	5,162, 003,385	9,284, 178.75		4.17	
Total	59					

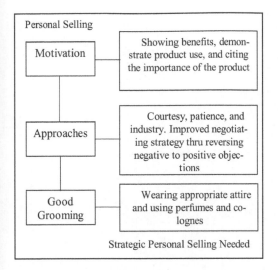

Personal Selling

| Motivation | → | Showing benefits, demonstrate product use, and citing the importance of the product |

| Approaches | → | Courtesy, patience, and industry. Improved negotiating strategy thru reversing negative to positive objections |

| Good Grooming | → | Wearing appropriate attire and using perfumes and colognes |

Strategic Personal Selling Needed

Figure 3.   Personal selling model framework plan.

5   PROPOSED STRATEGIC PERSONAL SELLING MODEL FRAMEWORK PLAN

In Figure 3 we explore personal selling model plan that is the most common variant for Cosmetic Products.

6   CONCLUSIONS

The results of our research show that the use of strategies in motivation, approaches, and good grooming are significantly correlated. For Good Grooming strategy, we must prepare ourselves, so that we achieve the highest selling price. Within the articles on good grooming we may find various opinions on it. In a perfect world, appearances don't have an impact on business, but in the real world, unfortunately, looks seem to matter. Studies have shown that people considered to be attractive earn more money and generate higher sales results (Burdett 2013). The studies aren't necessarily conclusive, but this from being in sales for many years: buyers do judge—appearances matter in sales and despite any conscious efforts to be objective, this judging is happening at least some of the time if not most of the time.

REFERENCES

Arante, L. & Gomez, J. 2000. Published by National Bookstore, Phillipine Copy Right, Printed by Printing Co., Inc-. Retrieved from www.nationalbookstore.com.

Burdett, E. 2013. Do Looks and Appearances Matter in Sales? In *Sales Management & Leadership*. Publisher: peaksales recruiting. Retrieved from http://www.peaksalesrecruiting.com/how-important-are-looks-for-sales-professionals/

Connick, W. 2013. 8 *Tips for productive sales call*. Retrieved from http://sales.about.com/.

Dzurová, M. & Lieskovská, V. & Fridrich, B. & Vokounová, D. & Kubiš, J. 2007. *Spotrebiteľské teórie a reálie*. Bratislava: Vydavateľstvo EKONÓM, 2007.

Garcia, A. & Villanueva, M. 2009. *Professional salesmanship*. Philippines: Rex Booklore Publishing Corporation.

Gilleland, J. 2004. *Eight sales strategy tactics for new business sales of P&C insurance*. Retrieved from http://www.profitableunderwriting.com/8 sales tactics.html.

Helbig, D. 2011. 5 *Keys to successful sales strategies*. Retrieved from http://smallbitztrends.com/2011/11/5-keys-successful-salesstrategies.html.

Ocon, J.A. & Alvarez, M.G. 2014. The Implication of Personal Selling Strategies in Motivation, Approaches and Good Grooming. In *Procedia—Social and Behavioral Sciences* 155 (2014) pp. 53–57.

Phopal, L. 2009. *15 strategies to help increase your sales and boost your bottom line*. LLC—Retrieved from www.stratcommunications.com.

Riley, J. 2012. *Promotional mix—personal selling*. Retrieved from http://www.tutor2u.net/business/marketing/promotion_personalselling.asp.

Ručinský, R. & Ručinská, S. 2007. *Factors of regional competitiveness*. Košice: Technická univerzita v Košiciach.

New Trends in Process Control and Production Management – Štofová & Szaryszová (Eds)
© 2018 Taylor & Francis Group, London, ISBN 978-1-138-05885-9

Changes in the attitudes towards consumers entailed by marketing 3.0

M. Rzemieniak & M. Maciaszczyk
Faculty of Management, Lublin University of Technology, Lublin, Poland

ABSTRACT: The aim of this article is to present changes in the attitude towards consumers entailed by the concept of marketing 3.0. Studies conducted for the purposes of the article included literature analyses and desk research on secondary sources. The article makes reference to reports for 2016 and the authors' own research conducted for the purposes of the following publications: "Management of Corporate Intangible Assets" (Rzemieniak 2013) and "Modern Consumer and Organisation in the Context of Contemporary Market Trends" (Rzemieniak & Maciaszczyk 2015), as well as "Consumer Behaviour of Persons with Mobility Impairments—Study Report" (Maciaszczyk 2014). In the past, marketing evolved from the initial focus on the product (marketing 1.0) towards the focus on the consumer (marketing 2.0). Nowadays, companies continue to expand their perspective to include, apart from products and customers, also more general social concerns.

1 INTRODUCTION

Nowadays, a business enterprise is no longer perceived as a lone, self-reliant actor in a competitive environment, but rather as an entity cooperating with a network of loyal partners—employees, distributors, sales force, suppliers. Under the concept of marketing 3.0, companies treat the consumer as a strategic starting point and are prepared to embrace them in all their humanity, to balance profitability with corporate responsibility (Kotler et al. 2010). If a company selects its partners by matching their goals to its own aspirations, it gains an opportunity to secure a competitive advantage and strengthen its market position (Matysewicz 2014). That, however, is only possible when it is prepared to share its mission, vision, and values with other team members within its respective cooperation network.

Marketing 3.0 is the stage at which the focal point of marketing is no longer the customer as such, but rather a human being and the values he or she represents. Marketing is very strongly correlated with macroeconomic phenomena. When an element of the macroenvironment changes, so does consumer behaviour, which in turn necessitates a change in marketing strategies. A business enterprise is an entity engaged in continued cooperation with a specific network of partners—employees, distributors, sales force, and suppliers who, if selected with due deliberation, can be a source of considerable competitive advantage. Businesses generate revenues by providing high quality offerings to their customers and the rest of their milieu. Under the concept of marketing 3.0,

companies treat the consumer as a strategic starting point and are prepared to embrace them in all their humanity, to balance profitability with corporate responsibility.

2 SOCIAL MILIEU IN MARKETING 3.0

Creative society plays an increasingly significant role in the modern world. In such a society, the use of the right hemisphere is considerably more prevalent that it used to be in the past. This fact has a significant impact on the directions of development that define contemporary marketing. Our everyday lives are increasingly affected by factors that are constantly evolving and therefore require a far greater capacity for creative thinking. The spheres of science and art are now closer to everyday concerns than ever before, a trend that is particularly evident in the sector of professional services. The success of the most developed countries as well as those boasting the fastest rates of economic growth is rooted in the creativity of individuals. This relates to the conclusions formulated by A. Maslow in the evening of his life, with regard to his hierarchy of needs. He claimed that the pyramid ought to be reversed as the most important human needs are in fact those related to self-realisation and spirituality (Romański 2014).

Indeed, in creative communities, the fulfilment of material aspirations has become the least pressing concern, a consequence of self-realisation, rather than a goal in itself. Psychological and spiritual needs are the ones that matter the most, their fulfilment can potentially become the most important element in allowing a brand to stand out on the market. A consumer is a human being who adheres to certain core values, one that can be treated as a real partner. For considerable time now, many companies have based their PR strategies on values and pro-social initiatives, but where conscious marketing 3.0 comes into play, values must lie at the very core of the organisation. The defined values should not be merely articulated but ought to permeate the entire enterprise and its culture.

3 KEY TRENDS IN MARKETING 3.0 MACRO-ENVIRONMENT

The mass development of technology is a factor facilitating the attitude of an enterprise aiming to "make the world a better place". The trend emerged from the widespread availability of cheap computers and cell phones, the decreasing costs of internet access and mobile data transfer, as well as easy access to software. These solutions allow individuals and groups to quickly connect in the global environment.

Modern consumers are well equipped to verify any declarations made in the mission and vision, as well as those expressed at the level of PR, against the actual conduct of the enterprise. This is possible mainly due to current technological advances in social communication. This heightened awareness of market actors requires that particular focus be placed on spirituality, culture and intensified cooperation with consumers (Sznajder 2014).

The above factors are determined by two types of social media outlets. Expressive social media such as blogs, YouTube, Facebook, Twitter and other similar services allow their users to express and share their emotions with others. Whereas cooperative social media such as Wikipedia are oriented towards user cooperation.

The impact of expressive social media is evident worldwide. It caused fundamental changes on the market and rebalancing chances in the process of values sharing. Consumers then have a lot of possibilities of gaining needed products and services (Wyrwisz 2014). Certain Twitter users can boast higher popularity ratings than CNN. Actor Ashton Kutcher reportedly hit the 1 million followers mark on Twitter, beating out even CNN (Kotler et al. 2010). Moreover, expressive social media are most commonly personal in character, with authors disseminate their opinions and ideas among a select group of readers. Some comment on world events, express their views or publish short opinions on random topics. Other social media users proceed to share their views on selected companies and review their products. For instance, a disgruntled

blogger with a sizeable following could discourage a considerable number of consumers from cooperating with a given company. Also video materials from individual content creators sharing their experiences with other users are increasingly popular these days.

With the increasing communicativeness of social media, active customers continue to gain ever greater influence on their fellow consumers. Shared opinions, feelings, conjectures, or experiences tend to be more convincing than corporate advertising. Moreover, less and less consumers actually watch commercials these days (a permanent trend, it seems) because they find themselves more absorbed by other concerns, including those activating social media activity.

In turn, cooperative social media utilise open-source software and offer users the possibility to participate in its development. For instance, the content of Wikipedia is the work of a great number of contributors devoting their personal time to create entries on a variety of topics. Similar services offering small ads for different products exert considerable competitive pressure on newspapers and magazines selling advertising space (Tapscot & Wiliams 2008).

The trend towards consumer cooperation is widely felt in business as it entails a partial loss of control over the brands. Modern companies have no choice but to cooperate with their customers.

4 GLOBALISATION AND CULTURE-ORIENTED MARKETING

Globalisation is yet another driving force that shapes the new consumer. It is intrinsically tied to technology, and modern technology allows information exchange that transcends the borders of countries, companies or entities the word over. Transport technology facilitates commerce and physical distribution within global value chains. Similarly to technology, globalisation is prevalent, affecting virtually every corner of the globe, creating an interlinked economy of multi-level co-dependencies (Kotler et al. 2010).

The phenomenon creates polarisation manifested in two contradictory standpoints: namely that of globalisation and nationalisation. Some authors claim that the world is now bordeless and the transfer of goods, services and people can take place efficiently and without obstacles thanks to the availability of cheap transport and information technology (Friedman 2006). Then there are those who observe that borders between countries or communities (e.g. the European Union) will never truly disappear because they are a direct product of politics and psychology. On the one hand, globalisation processes level the playing field for countries all over the world, but on the other, they create tangible dangers. Many countries make efforts to protect their local markets from the impact of globalisation, which in a way suggests that globalisation in fact aggravates nationalism. Indeed, globalisation itself is full of contradictions. China, a state far removed from the standards of democracy, is currently gaining alarming global prominence. On the other hand, although globalisation requires integration it does not create "equal" economies (Stiglizt 2004). Besides, globalisation processes contribute to cultural diversity and strengthen local culture. The intensifying processes of integration are met with a strong sentiments towards tribality. Still, Ch. Handy believes that rather than focus on rectifying said paradoxes, one ought to find a way to come to terms with the environment defined by those "inconsistencies" (Handy 1995). The ability to cope with such situations is a challenge faced by today's marketers.

The main market consequence of the described paradoxes is the fact that companies must now compete to be perceived as entities committed to continued cooperation and providing a sense of community. The mentioned trend results in a situation where marketers no longer have complete control over their brands. In order to be able to shape it to their advantage, they must collaborate with engaged, supportive groups while at the same time competing against other groups whose interest in the brand may or may not be keen but who perpetuate a less desirable image of the same. The solution is to be provided by cultural brands capable of resolving paradoxes in society (Kotler et al. 2010). Brands that will overcome social, economic, environmental, and any other problems encountered in the social context. Cultural brands must be characterised by high dynamism, which is why managing them requires constant monitoring of trends evidencing themselves in the microenvironment (Fig. 1).

Under the concept of marketing 3.0 the customer, until now perceived as prone to either rely on reason or be manipulated by emotional stimuli, becomes a person with specific needs emerging at a certain individual, spiritual level, a human being who is guided by particular values. Those values are the key to a well prepared marketing campaign. The starting point for any plans ought to be the mission and vision of the given company. Preparing a product offer with due consideration of its functional, emotional, as well as spiritual aspects is becoming a necessary condition of securing competitive advantage.

The concept of marketing 3.0 is based on cooperating, building a community, and developing brand personality. Any form of social media

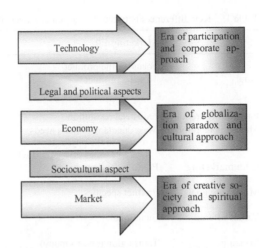

Figure 1. Key macro-environment changes leading to the emergence of marketing 3.0.

that effectively alters the specificity of marketing endeavours can prove a useful tool in achieving the goals of new marketing (Sadowski 2013). Communities now play an increasingly important role as early as at the stage of research and product development. The emergence of the same relates to the concept of tribality and assumes that consumers will spontaneously come together in a community focused around specific brands or products in order to exchange information and voice their opinions. Companies deliberately help these communities to take shape and actively support their members.

5 CREATIVE SOCIETY AND MARKETING 3.0

Members of a creative society are the people who take advantage of their right hemispheres, people active in the spheres of science, art, and professional services. In the opinion of D. Pink, it is an effect centuries-long human evolution whose main driving force is technology (Pink 2005). Studies unambiguously demonstrate that regardless of the fact that the working class majority greatly outnumbers creative individuals, the significance of the latter is now becoming ever more ubiquitous (Robbins 2008). Their social and creative contributions are often expressed by assuming the roles of innovators, inventors, developers of technologies and other solutions, as well as the keen users of such solutions.

As a result of the described social processes, consumers are not only searching for products and services that will satisfy their expectations and needs, but also appraise them against personal

Table 1. Key differences between the particular stages of marketing development.

	Marketing 1.0	Marketing 2.0	Marketing 3.0
Goal	Focus on the product To sell a product	Focus on the customer To satisfy the consumer's expectations and secure return business	Focus on values To make the world a better place
Facilitating trends	Industrial revolution	Technological revolution	New wave technology
Market perception	Mass clients with mass needs related to the physical characteristics of the product	Intelligent consumers guided by their intellect and heart	A human being following their reason, heart and spirit
Core marketing concept	Product development	Market distinction	Values
Marketing goals	Product specifics	Product and company positioning	Mission, vision and corporate values
Values offered	Functional	Functional and emotional	Functional, emotional and spiritual
Consumer interaction	Transaction (one-to-many)	Relationship (one-to-one)	Cooperation (many-to-many)

experiences and business models that they find spiritually satisfying. A business model based on complementing the basic market offer with broadly understood "meaning" manifested through values lies at the very core of marketing 3.0 (Table 1).

Researchers now believe that the factors that motivate humans in life can be successfully incorporated into corporate missions, visions and values (Barnett 2011). Naturally, this does not pertain to entities that take advantage of these ideals out of calculation, without actually believing or living up to what they declare. For such companies, the civic values they allegedly subscribe to are merely means of communicatively "adulating" consumer communities. Marketing 3.0 is not about creating artificial images (Berżer 2013). What matters is genuine incorporation of important values into the actual corporate culture and capitalisation on the customer's appreciation of this fact.

6 DIFFERENT MARKETING APPROACHES TO THE CONSUMER

Modern marketing practices are largely shaped by the constantly evolving customer behaviour and attitude. The necessity to shift corporate focus towards the customer requires more elaborate forms of activity that are rooted in cooperation and the emphasis on culture and spirituality. Technology allows consumers to cooperate with each other—they can exchange information, share ideas, and disseminate opinions (Table 2).

Technology is the driving force behind globalisation processes taking place in a variety of contexts (political, legal, economic, social), one that also facilitates and activates the creative society.

The future of marketing will always be shaped predominantly by current events and only to a considerably lesser extent by long-term influences. Trust is manifested in lateral relations. Consumers place greater trust in each other than they do in corporations. The emergence of social media is, in a way, a reflection of the migrating consumer trust—see Table 3 (Kotler et al. 2010).

The departure from vertical relations will define the future of marketing. In order to regain consumer trust, companies must come to terms with the evolution of the "consumer trust system". The next step is to develop marketing strategies around the concepts of co-creation, building communities, and shaping brand personality. The co-creation refers to new ways of developing products and generating related experiences by relying on the cooperation between enterprises, consumers, suppliers, distribution channel partners, and interlinked innovation networks (Prahaland & Krishnan 2008).

For a brand to be truly able to appeal to consumers, it must develop an authentic DNA, i.e. its Distinctiveness, Novelty, and unique Attributes. These components ought to be characteristic enough to distinguish the brand on the market (Borland & Lindgreen 2012). The word personality is used to signify that a brand, much like a person, will continue to develop over the entire course of its life, thus contributing to its unique (personal) character. Analogically, similarly to the way people are perceived, the market perception of a brand is positively correlated to the authenticity of its personality.

Table 2. Components of marketing 3.0.

Components What to offer?		
Content	Corporately oriented marketing	STIMULUS—era of participation
Context	Culturally oriented marketing	PROBLEM—era of the globalisation paradox
How to make the offer?	Spiritually oriented marketing	SOLUTION—era of creativity

Table 3. The future of marketing.

Marketing disciplines	Modern concept of marketing	Future concept of marketing
Product management	Product, price, distribution, promotion	Co-creation
Customer management	Segmentation, targeting, positioning	Building a community
Brand management	Developing a brand	Building brand personality

7 CONCLUSIONS

Marketing 3.0 is based on values which at the same time constitute elements of external organisational culture and the foundations for initiatives undertaken internally. Tools such as corporate vision, mission and set of core values must also refer to reason, emotions and spiritual concerns. Thus understood marketing in not merely a tool for generating demand and supporting sales, indeed, it determines the future of the company in terms of strategy, while at the same time helping to inspire trust in customers and other stakeholders.

In order to succeed, companies must understand that modern consumers put ever greater stock in co-creation, community, and social heroes. Market managers should be aware of that fact at all times.

REFERENCES

Barrett, R. 2011. Liberating the Corporate Soul: Building a Visionary Organisation. London: Routledge Taylor & Francis Group.

Berżer, J. 2013. Efekt wirusowy w biznesie: dlaczego pewne produkty i usługi zdobywają rynek. Warszawa: MT Biznes.

Borland, H., Lindgreen, A. 2012. Sustainability, Epistemology, Ecocentric Business and Marketing Strategy. Ideology, Reality and Vision. Journal of Business Ethics. 10/2012.

Friedman, T.L. 2006. Świat jest płaski. Krótka historia XXI wieku. Poznań: Wydawnictwo Rebis.

Handy, Ch. 1995. The Age of Paradox. Harvard Business School Press.

Kotler, Ph. et al. 2010. Marketing 3.0. Warszawa: Wydawnictwo MT Biznes.

Maciaszczyk, M. 2014. Zachowania konsumenckie osób niepełnosprawnych ruchowo-raport z badań. Lublin: Wydawnictwo Politechniki Lubelskiej.

Matysewicz, J. 2014. Usługi profesjonalne w globalnej gospodarce. Warszawa: Wydawnictwo Placet.

Pink, D. 2005. A Whole New Mind. New York: A Member of Penguin Group Inc.

Prahaland, C.K. & Krishnan, M.S. 2008. The New Age of Innovation. The McGraw-Hill Companies.

Robbins, R.H. 2008. Globalne problemy a kultura kapitalizmu. Poznań: Wydawnictwo Pro Publico.

Rzemieniak, M. & Maciaszczyk, M. 2015. Współczesny konsument i organizacja w świetle nowych trendów rynkowych, Tarnobrzeg: Wydawnictwo Państwowej Wyższej Szkoły Zawodowej im. Profesora Stanisława Tarnowskiego w Tarnobrzegu.

Rzemieniak, M. 2013. Zarządzanie niematerialnymi wartościami przedsiębiorstw. Toruń: Wydawnictwo Dom Organizatora TNOiK.

Sadowski, M. 2013. Rewolucja social media. Gliwice: Wydawnictwo Helion.

Stiglizt, J. 2004. Globalizacja. Warszawa: Wydawnictwo Naukowe PWN.

Sznajder, A. 2014. Technologie mobilne w marketingu. Warszawa: Oficyna Wolters Kluwer Business.

Szromnik, A. 2013. Marketingowa koncepcja kreowania i rozpowszechniania idei-zarys strategii marketingowej (cz. I). Marketing i Rynek, 4/2013.

Tapscot, D. & Wiliams, A. 2008. Wikonomia. O globalnej współpracy, która zmienia wszystko. Warszawa: Wydawnictwa Akademickie i Profesjonalne.

Romański, J. 2017. PortalMarketera.pl after: Marketing 3.0. From Products to Customers to the Human Spirit, http://portalmarketera.pl/publikacje/widok/marketing-30/, [retrieved on: 21.03.2017].

Wyrwisz, J. 2014. Rola komunikacji marketingowej w dobie marketingu 3.0, [in:] Przedsiębiorczość i wiedza w kreowaniu rozwoju regionalnego. Lublin: Wydawnictwo Politechniki Lubelskiej.

Innovative character of the contemporary enterprise and determinants of innovation

A. Rzepka
Lublin University of Technology, Lublin, Poland

ABSTRACT: Innovation is an inherent feature of progress and modernity which largely contributes to the economic development and serves as its most important driving force. Companies that operate on the market of today's world, which is developing dynamically, in order to survive, must be characterized by innovation.

An enterprise that aims to achieve market success should strive to implement innovation and thus be counted as being innovative. The purpose of this article is to specify determinants of innovation and assess whether the companies of the Podkarpackie region in Poland are of an innovative character. To accomplish this objective a research has been carried out among the entrepreneurs belonging to the SME sector operating in the above mentioned region. The goal of the research was to analyze the level of innovation of these enterprises.

1 INNOVATION AND INNOVATIVE CHARACTER OF A MODERN ENTERPRISE

Meeting the demands of contemporary enterprises requires an efforts to produce modern products. Goods manufactures in a traditional way when compared to the products manufactured based on advanced technologies often cannot be compared in terms of their quality. The necessity to be innovative is a feature of the XXI business (Olesiński et al. 2016).

Both terms *innovation* and *innovative character* are linked with one another (Cho & Pucik 2005). In management science, innovation is considered an important attribute of the company which is manifested in the ability of a given enterprise to implement cutting edge solutions. Innovation can, therefore, be associated with a skill of offering new solutions or with a skill of manufacturing state-of the art goods (Hilami 2010). The process of innovation of an enterprise thus also includes the ability of launching new products to markets thus expanding the scope of business (Danneels 2000).

It is, therefore, the competence of a given enterprise to use information, imagination and initiative for the purpose of translating an idea or invention into a good or service that creates value. Innovation also synonymous with risk-taking because organizations that create revolutionary products or technologies undertake often risky decisions and unproven solutions to be first on the market (Kotler & Keller 2006).

J. A. Schumpeter, the precursor of the theory of innovation, emphasized the role of entrepreneurship and the seeking out of opportunities for novel value-generating activities which would expand and transform the circular flow of income. These activities include (Bełz & Barbasz 2014):

– introducing new products or improving the existing products;
– introducing new or improved production technologies;
– opening new sales market or distribution of production or supply;
– using raw materials or semi-finished products;
– implementing changes in the organization of production.

2 THE CONDITIONS FOR IMPLEMENTING INNOVATION IN ENTERPRISES

Organizations are constantly faced with the challenge of gaining competitive advantage over competitors. It is generally known that any innovation activities make the business better and more competitive. As it has been emphasized by P. F. Drucker, a company that wants to be innovative has to "cultivate such organizational climate that fosters innovation, track the effectiveness of actions taken, modify organizational structure, and develop a system of incentives for change" (Drucer1992).

The process of becoming innovative depends on the presence of different mediating structures that can influence the interactions, behaviors and beliefs of employees, as a source of creating and successfully applying new ideas within an organization. Then, the process involves integrated and coordinated actions that would enable to turn these new ideas and concepts into something that will create commercial, social or organizational value. It may be done through making improvements in existing products or services, introducing new processes, changing marketing or business models, enhancing existing products or services, and introducing technological innovation, applying an innovative approach to improve operational processes in an enterprise and many more. There are certain factors, that may have an impact on the ability of an organization to become innovative.

Innovation of a given enterprise, as it was presented above, depends on a whole range of factors. High-risk and high-potential innovation ideas require well coordinated innovation management and innovation capacity building. An enterprise is able to become innovative if is market-oriented, has research and development focused strategy, promotes growth, monitors changes in business environment, cooperates with customers thus creates high returns of investment. Not only resources of an enterprise matters, but, above all, its organizational structure and attributes. This all combined form the culture of innovation.

There is a close relationship between innovation of a given enterprise and its attributes. Type, size of a company, intensity of competition, market structure, factors governing the production of knowledge such as appropriability, technological opportunities resource endowment are the main determinants of a firm's innovative activity.

What is more, labor productivity depends on physical and human capital as well as on new knowledge and innovation, so there is also a quantitative relationship between innovation and economic performance.

The tendency of an enterprise to implement innovation means a willingness to become innovative. Both the capacity and tendency are being shaped by the external influences, inter-organizational relations, internal resources, the size of a company, its market share and diversification, demand-pull and technology-push indicators. If knowledge is used together with technological potential, the innovation effectiveness increases.

Pichlak (2012) enumerates attributes of an innovative enterprise. According to the author they include:

– the tendency to generate innovation;
– the ability to implement innovation;
– the willingness to take a risk, that is inseparable to innovation.

It is worth to notice that these attributes also characterize an agile enterprise that is fast moving, flexible and robust and which gives an adequate response to market changes. Actions which facilitate the speed and change, and which aim at achieving competitive advantage in serving its customer require implementing innovative solutions

3 INNOVATION OF COMPANIES OF PODPARPACKIE REGION IN POLAND— AN EMPIRICAL APPROACH

Author conducted a research to measure the level of innovation of companies belonging to SME sector operating in the Podkarpackie region.

The study was designed to investigate whether companies are willing to undertake innovative activities and have attributes of an innovative organizations. Efforts were also made to determine the nature of the innovation introduced by the surveyed entrepreneurs. The aim of the study was also to determine the form of adoption of innovative activities and plans for the future also associated with innovation.

The research method was based on a survey among 100 enterprises with a random sampling. All of the enterprises belonged to the sector of small and medium sized companies, however, the research outcome should not be treated as a representative and generalize as to to assess the level of innovation of the whole group of companies of SME sector in Poland. However, the research results describe and constitute an attempt to assess the innovation of SME companies operating in Podkarpackie region.

Questions were of a varied nature, the majority of which focused on the form of conducting business activity, its scope, the market share all of which with a purpose to assess the attributes of a given enterprise, its resources and actions undertaken which may be called innovative.

The questions enabled the researchers to present a detailed characteristics of companies which operate in SME sector in Podkarpackie region in Poland as they verified the size of an enterprise, business sector, period of functioning on the market, number of people being employed, and the presence on the market (local, domestic or foreign).

More than half of the respondents are micro enterprises, ie. companies that employ up to 10 people (54% of the total number of the surveyed companies), 28% of respondents are small enterprises that employ not more than 50 people, and 18% of respondents are classified as medium-sized enterprises (up to 250 employees) (Table 1).

The largest number of companies operated in the service industry (48% of respondents), as well as in the trade industry (28%). Companies that function in processing, mining, transport and construction industries together accounted for 14% of the research group. 10% of surveyed companies declared trading in other sector than mentioned in the survey (Fig. 1).

Among the 100 examined companies, 38% of them functioned on the market for no longer than 3 years. The largest group of enterprises, 24%, conducted their business activity in a period from 4 to 7 years. The smallest group of companies included the longest-operating in the market, i.e. from 12 to 15 years (Fig. 2).

The biggest group of the respondents conduct business which is limited only to a local market (48% of the examined companies). The second place takes a group of enterprises operating on a

Table 1. The size of the examined enterprise.

The size of an enterprise	The number of respondents	Percentage share of respondents that provided an answer
A micro-enterprise	54	54%
A small enterprise	28	28%
A medium sized enterprise	18	18%
Total	100	100%

Podział przedsiębiorstw ze względu na kierunek prowadzonej działalności

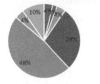

- Przemysł przetwórczy
- Przemysł wydobywczy
- Przemysł transportowy
- Przemysł handlowy
- Przemysł usługowy
- Przemysł budowlany

Figure 1. Industries in which companies conduct their business activity.

Okres istnienia na rynku

- do 3 lat
- od 4 do 7 lat
- od 8 do 11 lat
- od 12 do 15 lat
- 16 i więcej

Figure 2. The period of conducting business activity by the surveyed organizations.

Table 2. The type of market on which an enterprise conducts its business activity.

Type of market	The number of respondents	Percentage share of respondents that provided an answer
Local market	48	48%
Domestic market	46	46%
Foreign Eastern market	2	2%
Foreign Western market	4	4%

domestic market (46%), and the third place takes a group of organizations that operate on foreign markets, including 2% on the Eastern market, and 4% on the Western market (Table 2).

To summarize, the above data show that the majority of companies involved in the research are micro-entrepreneurs who employ up to ten people. The largest number of entities operate in the service sector. As for the period of functioning on the market, the largest number of respondents are entrepreneurs conducting business for no longer than three years and also the biggest number of entities are on innovative activities undertaken by the examined companies, both in the past, present and planned to be realized in future.

Very interesting was the outcome of an answer about undertaking innovative activities by companies in the past three years. More than 2/3 of the respondents, i.e. 68% of companies provided an affirmative answer, only 1/3 of the respondents, i.e. 16% provided a negative answer. This proves a growing level of awareness of the need to introduce innovative activities among enterprises from the SME sector.

As far as the type of innovative activities is concerned, the respondents most often declared implementing marketing innovation, i.e such that is based on the implementation of new marketing strategies (37% of the respondents). The second largest group of answers was about implementing such innovation as launching new type of product or service (34%). The third type of innovation which obtained the greatest number of answers concerned the innovation of an organizational nature, that is, implementing for example new methods of business management (19% of the respondents).

Among the entrepreneurs of the SME sector in Podkarpackie region the smallest number of companies introduced innovations related to the process innovation which is the implementation of a new or significantly improved production or delivery method (including significant changes in techniques, equipment and/or software) or pioneering methods for the production process (Fig. 3).

There is a vast number of companies that is planning to take innovative actions in the nearest future. For 38% of the surveyed enterprises such activities will be taken for the first time, while for 34% of them for a next time. 2% of respondents admits that they are not planning to implement any innovations in the nearest future while more than a quarter of the surveyed entrepreneurs have not decided yet whether they intend to take such action. It is worth to emphasize that among the surveyed companies, which have already implemented innovative activities in the past none of them denied planning similar activities in future (Table 3).

As it was mentioned earlier, innovation is the search for, and the discovery, development, improvement, adoption and commercialization of new processes, new products, and new organi-

zational structures and procedures. It involves uncertainty, risk taking, probing and reprobing, experimenting, and testing. It is an activity that is unlikely for a company to be carried out without cooperation inside and outside the company. Many research proved the difficulty of single-firm innovation, the faster innovation created by cooperation and specialization between joint participants, the pursuit of innovation in new fields by utilizing shared technology and know-how and the enhancement of the technological level of each participant through the interchange of technology.

Unfortunately, 64% of the examined enterprises did not cooperated with other entities for the purpose of implementing innovative solutions. Only 36% of them expressed a willingness to undertake such cooperation, but primarily with institutions operating in the same industry. Among this group of respondents, only two enterprises cooperate with organizations conducting research and technological centres, universities, and four with science and technology park. Three companies declared their willingness to cooperate with organizations from abroad. The largest group of companies, 32 of the surveyed entrepreneurs is working with organizations operating in the same field, 14 of the surveyed enterprises declared their willingness to cooperate with companies operating in a different industry.

Podział innowacji ze względu na ich charakter

- Innowacje o charakterze procesowym
- Innowacje o charakterze produktowym
- Innowacje o charakterze marketingowym
- Innowacje o charakterze organizacyjnym

Figure 3. The nature of innovation implemented by entrepreneurs of the Podkarpackie region.

Table 3. Entrepreneurs who are planning to undertake innovative activities or implement next.

An answer	The number of respondents	Percentage share of respondents that provided an answer
Yes, I am planning to undertake an innovative action	38	38%
No, I am not planning to undertake an innovative action	2	2%
Yes, I am planning to implement next innovative actions	34	34%
No, I am not planning to implement next innovative actions	0	0%
I do not know, it has not been decided yet	26	26%
Respondents in total	100	100%

4 CONCLUSION

The research with the use of a survey was an important tool that allowed to become more aware of the nature of enterprises of SME sector operating in Podkarpackie region. The outcome allowed to form the following conclusions:

1. Companies of SME sector believe that innovation is something that should be considered very carefully;
2. Poor innovative experience of companies of SME sector may be connected with a more general problem of Polish economy. It is being observed that companies have a limited financial capital and they fear of giving it over unproven and risky solutions. Moreover, companies cannot overcome the problem of bureaucracy in administration sector. Too many complications and problems with obtaining funding for innovative development cause that a small number of companies decide to go for broke. Some companies also simply lack of proper management with R & D skills and knowledge.
3. The drawback of Polish companies is also the reluctance to cooperate. Not many companies implement join ventures or simply form a group of companies that participate in a common activity. The majority of enterprises work on

their own, are unwilling to take up risk and is unduly reserved.

4. The research proved that there is a group of companies that is willing to undertake innovative activities. Interviews carried out with the management may be treated as a positive conclusion as there are companies that have all attributes of innovative enterprises. They have resources, people and features that will enable them to implement innovation in future and have the tendency (mentioned in Fig. 1) to implement it.

The success of an innovative enterprises on the market, as well as the realization of its objectives depends on the development and diffusion of innovation. An innovation to be implemented must be supported by the entire organization.

Enterprises operating in Podkarpackie region have the potential to become innovative but it is not exploited enough. The survey proved that many SME companies are aware of the need to implement new solutions and invent new products and services as it is a necessity of the contemporary economy. If all innovative activities are implemented by companies of the Podkarpackie region (as declared while the interviews), the economic development of it over the next decade may be faster than ever, especially in the service sector.

The most important thing is, however, to raise the awareness of the importance of innovation among companies of Podkarpackie region in Poland, help them to become more aware of their potential, more willing to take up risk, to search for cooperation. There is a potential in SME sector and with a little guidance it may increase the attractiveness of the medium-sized enterprises of the Podkarpackie region, both domestically and internationally and lead to changes and profits for organizations, region and the whole economy in future.

REFERENCES

Bełz G. & Barbasz A. 2014. *Research Papers. Management Forum 4.* Wydawnictwo Uniwersytetu Ekonomicznego we Wrocławiu, Wrocław 2014.

Cho H.J. & Pucik V. 2005. Relationship between Innovativeness, Quality, Growth, Profitability, and Market Value. *"Strategic Management Journal"* 2005, no. 26.

Danneels E. & Kleinschmidt E.J. 2000. *Product Innovativeness from the Firm's Perspective: Its Dimensions and their Impact on Project Selection and Performance.* Institute for the Study of Business Markets, The Pennsylvania State University, ISBM Report 2000.

Drucker P.F. 1992. *Innowacja i przedsiębiorczość*, Praktyka i zasady. PWE, Warszawa 1992.

Hilami M.F. et al. 2010. Product and Process Innovativeness: Evidence from Malaysian SMEs. *"European Journal of Social Science"* 2010, no 16 (4).

Kotler, Ph. & Keller K.L. 2006. Marketing Management. Prentice Hall, New Jersey 2006.

Olesiński Z. et al. 2016. *Międzyorganizacyjne sieci współpracy gospodarczej na przykładzie* Polski, Kanady i Gruzji. Texter, 2016.

Pichlak M. 2012. *Uwarunkowania innowacyjności organizacji. Studium teoretyczne i wyniki badań.* Wydawnictwo Difin Warszawa 2012.

New Trends in Process Control and Production Management – Štofová & Szaryszová (Eds)
© *2018 Taylor & Francis Group, London, ISBN 978-1-138-05885-9*

Accounting statements in the financial decision making

J. Sedláček & R. Skalický
Faculty of Economics and Administration, Masaryk University in Brno, Brno, Czech Republic

ABSTRACT: The paper is devoted to the comparison of accounting statements according to International accounting standards and Czech accounting standards and impact assessment of identified differences on the decisions of investors and owners of corporations. The first objective is to identify accounting methods and practices that lead to differences in reporting of financial position and performance of corporations, analyze their causes and examine their impact on the statements. The second objective is to assess the impact of these differences using methods of financial analysis on data specific company. For this purpose, is created parallel system of ratio indicators. The case study has proven that despite the ongoing harmonization efforts, there are still significant differences in both approaches and should be taken into account when making decisions. Similar results were also found when compared with the financial statements according to Indian accounting standards.

1 INTRODUCTION

1.1 *International Financial Reporting Standards (IFRS)*

Accounting is generally considered a reliable instrument of financial management, which accurately shows assets and liabilities of trade corporations and their net income. It is based on principles and rules which are universally recognised and accepted all over the world. And it is the understanding of these principles and their implementation at international and national levels which causes the differences in valuation, ultimately leading to different reporting of performance of businesses (Brealey et al. 2006), which is usually measured by the profitability and liquidity indicators. The environment in which accounting develops and functions plays a key role. It is therefore logical that accounting is influenced by the economic, legal, social, political and cultural environment in the given jurisdiction. Also, the application of accounting principles in national rules (particularly for valuation) is different in each country. This occurs despite the ongoing efforts to harmonise accounting globally in the form of the International Accounting Standards (IAS) and the IFRS, including the interpretations of SIC (Standing Interpretations Committee) and IFRIC (International Financial Reporting Interpretations Committee). Furthermore, a problem arises from the fact that the individual principles contradict each other and it is up to the will of businesses to decide which principle is more important when valuing their assets and liabilities. The selected method of valuation then directly affects the of reported assets and liabilities of a trade corporation, it has an impact on the amount of costs (as asset consumption expressed in monetary terms) and revenues (as accrued assets expressed in monetary terms) as well as on the amount of reported net income of a trade corporation.

According to Pacter (2015) from the IFRS Foundation 130 countries have publicly pledged to adopt the IFRS as a single set of global accounting standards, 114 countries require their use in all or most public companies and their use in other countries is possible. The ongoing convergence of the US GAAP, which are oriented to the needs of corporations financed through financial markets (Bohušová 2011), also contributes to the global nature of the IFRS. Accounting is thus becoming more and more international and national accounting is more and more pushed to the sphere of small and medium-sized trade corporations, in which it plays the role of recording for the tax purposes or commercial law. The International Accounting Standards Board (IASB) seeks to establish a coherent global system of accounting standards, but the way of their implementation in individual countries is still within the authority of national governments or economic groups. In the Czech Republic, being a member country of the EU, only those trade corporations which are traded on the public capital markets are obliged to submit financial statements according to the IFRS from 1st January 2005 onwards. Other corporations follow the Czech accounting standards, which leads to inconsistent reporting of their financial and revenue situations on an international scale. It is clear that the differences in reporting according to the international and national accounting standards (Oxelheim 2003) will be particularly affected by the economic and legal environment of individual countries.

1.2 Czech accounting legislation and standards (CAS)

The Czech national accounting system is primarily based on rules which are codified in legislation. As a member country, it is a subject to the legislation of the European Union and therefore it has to carry out its obligations. The main pillar of the Czech accounting legislation in its broader concept is the Accounting Act, which is a basic, generally valid legislative norm with state-wide force, containing adjustment of accounting methods and reporting for all business units in the country ranging from the smallest to the largest (also multinational) ones, whose scope of business and purpose of foundation are fundamentally different. The form and content of the act are determined not only by the rules and the content of the European legislation, but also by the Czech legislative rules and the requirement for full compliance (both factual and terminological) with the other regulations of the Czech legal order. Considering the fact, that the act is also designed for very small business units, e.g. the self-employed who cannot be expected to have broad theoretical knowledge in accounting and related fields, it is necessary to make the text as much comprehensible and clear as possible. The national accounting system is also influenced by the tax requirements, because the accounting net income is at the same time the basis for the assessment of corporate income tax (Sedláček 2007). As a result in practice the management makes a lot of estimates when preparing accounting statements with regards to the potential tax implications of the particular accounting procedure. This may lead to adoption of an accounting stance based on tax counts and not on considerations about how it is to be expressed in fair and true view of a transaction in its essence. Although the Czech accounting standards (below CAS) has gradually taken over a wide range of procedures from the IFRS and it is continuously updated in the form of amendments, there are still differences resulting from the different priorities and principles. Identification of the differences between the two accounting systems and examination of their causes is the subject of many research studies, see e.g. papers by audit companies Ernst and Young (EY 2013) and Pricewaterhouse-Coopers (PwC 2013).

2 PROBLEM DEFINITION, DATA AND METHODOLOGY

The aims of our research are to compare the accounting methods and procedures for the preparation of financial statements in compliance with the IFRS and the CAS and to identify the main differences of both approaches which affect the reporting of the financial position and performance of corporations; to assess the degree of the influence, or the significance, of these differences using the method of financial analysis. We use a parallel system of ratio indicators, which are applied to a specific corporation. The identification of the main differences will arise from the comparison of the two approaches as regards the recognition of the basic items of financial statements, their measurement methods, and the procedures used for their booking (Ding et al. 2007).

IFRS define in the conceptual framework precisely essential elements (assets, debts and other liabilities, costs, revenues and equity), along with basic rules for their recognition and measurement in Czech legislation general definition are missing.

2.1 The recognition of the basic items of financial statements

Asset—the IFRS characterize an asset as a future economic benefit, which will very likely occur and is controlled by the corporation (i.e., it can only use the benefits following from the asset in its own favour). Unlike in the case of the CAS, the ownership is not important for the recognition of the asset. This affects the reporting of the balance sheet total if, e.g., the corporation uses financial leasing.

Liabilities and other debits—represent the obligation of the corporation to settle a liability (debt), which occurred in consequence of past situations, and its settlement will result in reduced economic benefit (reduced assets or increased liabilities). CAS does not provide a definition of liabilities or other debts or criteria for their recognition.

Equity—is defined as the remaining proportion of the corporation's assets after deducting the liabilities. The definition is the same in the case of the CAS; however, due to the different definition of costs and incomes and other liabilities, the equity will be reported with a different value.

Costs—reduce the economic benefit of the corporation during the accounting period by using assets or creating liabilities, which leads to reduced equity in a different way than allocation to owners. The IFRS distinguish two types of costs: expenses (i.e., costs related to common activities of the corporation) and losses (the opposite of gains).

Incomes—represent an increase in the corporation's economic benefit by increasing the value of assets or reducing liabilities. They increase the equity in a different way than owners' deposits. Unlike the CAS, the IFRS distinguish two types of incomes—revenues (i.e., the income from the common corporation's activities) and gains (i.e., the income from other than common activities, which are reported in compensation with the related costs—losses).

2.2 The measurement methods for assets, liabilities and other debits

Valuation in accounting should generally be derived from the benefit which an asset or liability will bring to the owners. It is basically the right choice between two extreme approaches based on the one hand on historical costs, and on the other hand on fair prices (Sedláček 2010). Both approaches have their pros and cons and that is why the choice of an appropriate valuation variant is always problematic and ambiguous. The measurement method used will naturally affect the value of the reported assets and liabilities, equity and the amount of the costs, which are reflected in the activation and the profit or loss account for the period given (Christensen & Nikolaev 2013).

As a basic model presents IFRS: historical cost, current cost, realizable value, present value, fair value or nominal value. The standards also allow using other measurement models and techniques that provide useful information for economic decisions. They are gradually turning away from an accounting model based on displaying the ability of the corporation to reproduce historically expressed costs or retain nominally expressed financial capital towards models responding to the behaviour of markets. Within production, measurement relates to the market in the form of purchase prices (costs) of the assets which have been consumed during the production. The factor of time is reflected in the measurement of assets and liabilities by the method of the effective interest rate. When testing the assets as regards the value depreciation, the recoverable amount is determined together with the value in use.

The models presented in the CAS are basically consistent with the IFRS with exception of the measurement by equivalence which is applied for long-term financial assets. The key factor for determining the differences resulting from measurement are the different approaches used at the stages of the acquisition (creation) of the assets and liabilities, in course of holding the assets and the existence of liabilities, and after the disposal of assets or settlement of liabilities. The issues related to measurement are not only choosing the right model, but also differently defined requirements as regards the level of the data entering the model.

2.3 Accounting procedures and reporting

The differences in this area originate in the concept of accounting methods and procedures, which are based on defined principles in the IFRS, while in the CAS they mainly take the form of detailed rules (Epstein & Jermakovicz 2010). Out of the many differences that may significantly affect financial statements, we mention: the liabilities following from financial leasing, the cost of the disposal of fixed assets and restoration of the original state, methods of depreciation of fixed assets, spare parts, results of research and development, intangible assets with indeterminate period of use, exceptional items, corrections of significant errors, building contracts, employee benefits, real estate investments, reserves, etc. The possible alternative solutions do not contribute to comparability, and neither do the problems which are not addressed at all in the Czech legislation in contrast to the IFRS (Bohušová & Svoboda 2013).

An increased risk of errors and inaccuracies in the course of a financial statement creation may occur in the case where a corporation creates statements in accordance with the IFRS, but they need to know the result in compliance with the Czech legislation in order to determine the tax base. It is ideal to do dual accounting, which is more expensive, however, especially during the preparation and implementation of the two systems. Therefore, it is more popular in practice to create statements in compliance with the CAS and then transform them into statements in compliance with the IFRS using transformation rules (Hýblová 2014). This procedure is always inaccurate, and the result is in proportion to the cost the corporation considers acceptable.

2.4 Parallel system of ratio indicators

A system of indicators is chosen to cover four basic areas of the financial analysis: liquidity, debt ratios, equity, and profitability of corporations. In total, seven ratio indicators are selected: which are described by the equations listed in Table 1.

The calculations of the indicators are performed using data of a real international company, which operates in the building industry and is categorized

Table 1. Ratio indicators.

Ratio	The definition of ratio indicator
CR	Current Ratio = Current Assets/Current Liabilities
QR	Quick Ratio = Current Assets—(Inventory + Prepaid Expenses)/Current Liabilities
DER	Debt Equity Ratio = Total Liabilities/Stockholders Equity
ICR	Interest Coverage Ratio = EBIT/Interest Expense
PR	Proprietary Ratio = Stockholders Equity/Total Assets-Intangibles
ROS	Return on Sales = Earnings after Tax/Net Sales
ROE	Return on Equity = Net Income/ Stockholders Equity

Source: Brealey et al. 2006, authors.

as a large accounting unit by the Act on accounting (Act no. 563 1991). The development of the differences between the indicators in compliance with the IFRS and the CAS is measured by inter annual indexes expressed in percentage points (Valouch et al. 2015). The time interval covers six immediately preceding years.

The size of the difference and its direction are the basis for the interpretation of the causes of the development of the individual variables entering the indicator.

3 RESEARCH RESULTS AND DISCUSSION OF THE ISSUES

Data for the calculation of ratio indicators are taken from the financial statements (balance sheet and profit and loss account) of the company drawn up in compliance with the IFRS and concurrently the CAS in individual years of the researched period. Independent variables are shown in Table 2 including the resulting values of financial indicators calculated on data reported by corporations under IFRS and CAS.

The closest relationship between the two systems was observed in the time series of Interest Coverage Ratio Return on Sales ry = 0.9997 as measured by the Pearson correlation coefficient; followed by Return on Equity and Return on Sales. On the contrary, the lowest of the measured correlations were found in the time series Proprietary Ratio (ry = 0.8095). No negative correlation was found. The size of the difference (D_t) between the indicators compliant to the IFRS (R_I) and the CAL (R_C), expressed in percentage is calculated for the individual years (t) in accordance with the equation:

Table 2. Development of financial ratios under CAS and IFRS.

Ratios	System	2009	2010	2011	2012	2013	2014
CR	CAS	1.62	1.58	1.77	1.93	1.75	1.89
	IFRS	1.59	1.57	1.78	1.90	1.70	1.88
QR	CAS	1.58	1.57	1.75	1.89	1.60	1.82
	IFRS	1.53	1.52	1.65	1.76	1.51	1.76
DER	CAS	1.76	1.98	1.63	1.62	1.90	1.49
	IFRS	1.78	2.04	1.68	1.70	2.02	1.59
ICR	CAS	26.83	34.11	28.67	33.70	25.80	18.12
	IFRS	27.78	35.60	29.41	35.00	26.60	18.54
PR	CAS	0.32	0.29	0.32	0.33	0.30	0.33
	IFRS	0.31	0.29	0.33	0.35	0.31	0.33
ROS	CAS	0.03	0.03	0.04	0.04	0.01	0.02
	IFRS	0.03	0.03	0.03	0.04	0.01	0.02
ROE	CAS	0.14	0.13	0.14	0.14	0.05	0.07
	IFRS	0.15	0.13	0.14	0.14	0.05	0.08

Source: Authors' work and calculations.

$$D_t = [(R_I - R_C)/R_C] \times 100 \qquad (1)$$

The calculated values of the differences are entered in the graph in Figure 1, which also shows the development in the period 2009–2014. As regards the differences that have a negative progress throughout the observed period, higher values of ratio indicators were calculated compliant to the CAS than the IFRS. This concerns indicator Return on Sales, which is interannually influenced in particular by the efficiency of the production and the costs. The difference ranges around 6.4% and the causes lie in a different approach to the recognition of revenues. Additionally, liquidity measured using the data reported in compliance with the IFRS achieves lower values than when reported compliant with the CAS. It is a consequence of the different recognition and measurement of current assets and short-term debts. Spare parts are always included in reserves in compliance with the Czech standards, including reserves from the company's own activities and all related costs enter the measurement. The curve of the indicator Quick Ratio difference values is under the curve of Current Ratio, i.e. the differences for Quick Ratio are larger. This is caused by the different structure of short-term debts, which are reported in a lower value compliant to the CAS compared to the IFRS. Although the corporation achieves a relatively high level of liquidity and there are no large fluctuations in between the years, the distortion of the reported value may reach up to 6.7%, as occurred in 2012. The lowest volatility in the course of the observed period was measured for the Current Ratio difference, which ranged from – 2.9% to + 0.6%.

The corporation has a relatively low Debt Equity Ratio, which even declined during the period observed. It is a good signal for investors. The corporation manifests higher values in compliance with the IFRS compared to the CAS, as confirmed

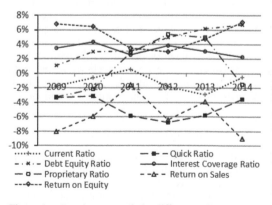

Figure 1. Development of the differences between the financial ratios under IFRS and CAS.

454

by the positive progress of the differences. The difference reached its peak in 2014. The cause can be particularly found in the different accounting concerning leasing. Interest Coverage Ratio difference curve suggests a very low distortion between the two approaches to reporting, which reached a maximum of 4.4%. The ability of the corporation to pay the interests exceeds the recommended values several times. However in reality, the indicator declined due to the declining values of EBIT.

Completely unprecedented development of the differences was observed in the Proprietary Ratio, which reflects the corporation's endowment with equity. The difference increased from negative values to positive ones in the years 2011 to 2013, and then again returned to a minimum value of – 1.5%. A greater degree of influence can be attributed to the item intangible assets, which are reported with a substantially lower value by the CAS than by the IFRS. The corporation did not reach a very high value of profitability, which had a downward trend throughout the observed period. The volatility of Return on Equity differences ranged between 3% and 7.1%. The higher values of the indicators reported under the IFRS were most likely influenced by the accounting methods for deferred taxes, depreciation, and valuation by fair value, which were reflected in the reported higher net profit.

4 COMPARISON WITH INDIAN NATIONAL STANDARDS

Unlike the Czech legislation, which requires using the IFRS in case of companies traded on the public market, India has accepted the IFRS as a single set of world accounting standards. In order, to reveal differences between international and national standards (the IGAAP), a dataset is employed. This dataset concerns a large Indian company which specializes in services in the area of IT (Information Technology) and which is traded on world stock markets (Kumar 2015). According to the same methodology, financial indicators are computed (see Table 3) and the differences between these indicators are measured. The development of these differences of the reporting period is shown in Figure 2.

Both liquidity indicators imply that the company creates sufficient level of current assets during the reporting period, and that the difference between statements has the identical process during the whole reporting period, except for the last year during which the quick ratio value was in accordance with the IFRS 4.4% lower than in case of the IGAAP. The better position of liquidity reported under the IFRS rules is likely to be a consequence of higher assets evaluation and a con-

Table 3. Development of financial ratios under IGAAP and IFRS.

Ratios	System	2010	2011	2012	2013	2014
CR	IGAAP	2.26	2.27	1.99	1.82	1.89
	IFRS	1.90	2.31	2.32	2.12	2.18
QR	IGAAP	2.16	2.16	1.92	1.80	1.82
	IFRS	1.83	2.21	2.32	2.10	2.02
DER	IGAAP	0.78	0.63	0.60	0.54	0.69
	IFRS	0.68	0.55	0.52	0.54	0.59
ICR	IGAAP	46.72	81.35	21.28	28.19	28.12
	IFRS	42.85	33.61	20.98	30.19	25.54
PR	IGAAP	0.67	0.72	0.74	0.70	0.69
	IFRS	0.72	0.77	0.79	0.74	0.73
ROS	IGAAP	0.17	0.17	0.15	0.16	0.16
	IFRS	0.17	0.17	0.15	0.18	0.17
ROE	IGAAP	0.25	0.23	0.21	0.23	0.24
	IFRS	0.23	0.22	0.20	0.22	0.22

Source: Authors' work and calculations.

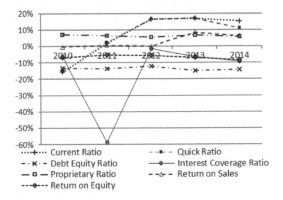

Figure 2. Development of the differences between the financial ratios under IFRS and IGAAP.

sequence of rent reporting manner. The negative course of the debt equity and the interest coverage ratios mean that in accordance with the IGAAP are reported higher values, comparing to the IFRS. The reasons for this fact (stated by Kumar) are different accounting procedures related to reserves and loans stated in foreign currencies. With regards to the interest coverage, the highest difference had been spotted in 2011 before this indicator was stabilized (the rate was 58.7%). The positive result of the proprietary ratio proves better stability of the company and lower creditor's risk in accordance with the IFRS view. Such conclusion arises from higher owner's equity, if this value is compared with the value reported in accordance with the IGAAP. The development of the difference between both the systems (in terms of owner's equity) is relatively constant, having the average value 6.4% during the whole reporting period. The indicator Return on sales was developing more, or less in the same way in both systems, except for the year 2013 when certain branches were spun-off. At the same time, the reserves were reported in accordance with the IFRS rules. The difference between both systems increased to 8.46%.

The reported return on owner's equity has the negative course during the whole reporting period. It is, as well as in the previous case—the proprietary ratio—caused by the way of net profit reporting procedure. The differences between both systems accounted for 6.61% (using median as a measure). The biggest spread of the values (57.2%) was measured in case of the Interest coverage ratio, while the lowest volatility was revealed in case of the Proprietary ration (1.6%) and Debt equity ratio (2.5%). The computed medians related to differences of individual indicators computed by the IGAAP were between −13.72% and 11.15%, while in accordance with the CAS only between −6.16% and 5.63%. It is, therefore, possible to claim that the CAS methods have already got closer to the IFRS, which indicates that these procedures (the CAS) should not significantly influence the decision-making process of a company, as it is in case of the IGAAP.

5 CONCLUSIONS

The world economy globalization requires that adequate information are provided for users' economic decisions. This requirement is met by the IFRS, which are developed by an independent institution; due to their qualities, they are gradually adopted (completely or partially) by economic groups and individual jurisdictions across the continents. The Czech Republic has adopted the IFRS for selected corporations; the others are bound by the national accounting legislation. In spite of the amendments to the CAS, which follow the IFRS updates and in fact absorb their new amendments, some differences remain, which in effect reduce the mutual comparability of the reported data.

The results of the case study conducted have confirmed the differences in reporting, which are mainly caused by different approaches to the recognition of assets and liabilities, their measurement and depreciation, which results in increased volumes of long-term assets and debts, equity and profit/loss in the statements prepared under the IFRS. Financial statements then show a different balance sheet total or other profit or loss of the corporation, depending on whether they are created in compliance with the IFRS or CAC.

The introduction of uniform standards will facilitate a better comparison of the economic position and performance of corporations, regardless of national borders, encourage the free flow of transnational capital, and increase the credibility of financial statements. Permanent innovation of global accounting standards with respect to user needs will increase the reliability and facilitate use of the financial statements, which will be reflected in the quality of the decisions based on them. However, as demonstrated by our study, in the conditions of the Czech economy, investors, lenders, managers, and other users when making their decisions should take into account the correction factor arising from the differences between the existing national and multinational financial reporting systems.

REFERENCES

Act no. 563/1991 Coll., on accounting, as amended.
Bohušová, H. & Svoboda, P. 2013. The evaluation of new methodological approaches to lease reporting on the side of lessor. *Acta univ. agric. et silvic. Mendel. Brun.* 61(4): 881–891.
Bohušová, H. 2011. General Approach to the IFRS and US GAAP Convergence. *Acta univ. agric. et silvic. Mendel. Brun.* 54(4): 27–36.
Brealey, R.A. et al. 2006. *Principles of Corporate Finance.* Boston: McGraw-Hill.
Christensen, H.B. & Nikolaev, V.V. 2013. Does fair value accounting for non-financial assets pass the market test? *Review of Accounting Studies* 18(3): 734–775.
Ding, Y. et al. 2007. Differences between domestic accounting standards and IAS: Measurement, determinants and implications. *Journal of Accounting and Public Policy* 26(1): 1–38.
Epstein, B.J. & Jermakovicz, E.K. 2010. *Interpretation and Application of International Financial Reporting Standards.* New Jersey: John Wiley & sons.
EY 2013. EY publication. http://www.ey.com/CZ/cs/home/library
Hýblová, E. 2014: Analysis of mergers in Czech agricultural companies. *Agricultural Economics.* 60(10): 441–448.
Kumar, K.A. 2015. International Financial Reporting Standards (IFRS) Adoption on Financial Decisions. *Journal of Accounting and Marketing* 4(3): 141–146.
Oxelheim, L. 2003. Macroeconomic Variables and Corporate Performance. *Financial Analysts Journal* 59(4): 36–50.
Pacter, P. 2015. IFRS as global standards: a pocket guide. http://shop.ifrs.org/Libraries/File_Library/Sample_Pages_from_Pocket_Guide.sflb.ashx.
PwC 2012. IFRS publication. http://www.pwc.com/cz/cs/ucetnictvi/ifrs-publikace.html.
Sedláček, J. 2007. Analysis of the development of financial efficiency of enterprises in the Czech Republic. *Journal of Economics* 55(1): 3–18.
Sedláček, J. 2010. The methods of valuation in agricultural accounting. *Agricultural Economics* 56(2): 59–66.
Valouch, P. et al. 2015. Impact of Mergers of Czech Companies on their Profitability and Returns, *Journal of Economics* 63(4): 410–430.

New Trends in Process Control and Production Management – Štofová & Szaryszová (Eds)
© 2018 Taylor & Francis Group, London, ISBN 978-1-138-05885-9

Process of supplier's evaluation and selection through software support

A. Seňová, A. Csikósová, M. Janošková & K. Čulková
*Faculty of Mining, Ecology, Process Control and Geotechnologies, Technical University of Košice,
Košice, Slovak Republic*

ABSTRACT: The paper deals with solution of the process of evaluation and selection of suppliers through software support SAS in JMP environment. The intention of the authors is to outline modern tools, methods and trends in procurement. For evaluation of suppliers, we used several methods, by which we compared individual criteria facilitating the selection of the final supplier, such as price, quality, type of material, reliability, supplier's location. Suppliers of the selected company were evaluated by the scoring model and selection of suppliers was done in the JMP program, based on single criterion and as well on multi-criteria decision-making. The program allows to filter data based on various criteria and to create new tables with the search results, which contributes to ensuring the continuity of business operations. The benefit is a created database of suppliers in the JMP programming environment, containing all relevant data and information needed to support decision-making.

1 INTRODUCTION

As regards supplier-customer relationships, entrepreneurs perceive their business partners via their financial situation, which largely influences the conditions of their business connections. Suppliers check the financial situation of those partners, with whom they do business in order to ensure the lowest risk possible in case that the debtor will not be able to service their debts i.e. will not pay them out. The reasons may vary. It can be insolvency, unwillingness to pay and administrative barriers in the debtor's country. Thanks to their creativity, the small and medium-size enterprises in Slovakia are able to react on the latest trends to a considerable extent and promptly meet the expectations of their customers. On the other hand, the main reasons for their wind-up are high contribution payments and tax burden to be paid to the state or changing legislation. Business entities determine prices of their products and discounts and, at the same time, they have to try to reach a profit while also paying attention to the optimization of all the processes. Purchasing represents a key tool for company profit increase or savings.

We are standing before new paradigms, in which the suppliers will turn into customers. Thus the customer acting as a supplier can try to gain us as a business partner and geographical barriers disappear.

2 LITERATURE REVIEW

In past time there is increasing importance of purchase as a tool for profit increasing, resp. as a tool of saving source. Companies inverted their attention from former effectiveness of sale, reinforcing of pressure to client by the way of aggressive marketing campaign, business techniques and improving of sale services to production optimizing that is realized by orientation to increasing of production quality as well as all processes of the company (Antošová ct al. 2014). Procurement in a company mostly represents a group of activities, which are connected with the demand for material sources to be used for the performance of business activities related to transport, procurement, acceptance, distribution, stock management, inspection and complaints concerning poor-quality inputs. The basic function of purchasing is to effectively provide the expected progress of main, auxiliary and operational production and non-production processes by raw and other materials, products and services, all that in the required amount, range, quality, time and place (Oreský 2011).

According Kita (2010), the purchasing process consists of 8 phases:

1. Issue occurrence;
2. General characteristics of the need;
3. Specification of the product subject to purchase;

4. Survey of potential procurement sources;
5. Submission of quotations;
6. Supplier Selection;
7. Making an order;
8. Evaluation of the delivery and the supplying organization.

According STN EN ISO 9001:2008 "an organization has to evaluate and select its suppliers based on their ability to deliver a product as per organization's requirements. The criteria for evaluation and re-evaluation criteria selection have to be defined. The records made of evaluation results and any inevitable activities incurred due to the evaluation must be maintained".

Authors Benková & Hudymačová (2009) present factors, influencing purchasing decision are as follows:

- *Conditions of supply*—supply is realized in certain supplementation and payment conditions.
- *Quality*—purchase of material or product for acceptable price, for determination of most acceptable variant there is necessary to use value analysis, and systematic study of purchased components.
- *Volume*—determination of necessary number of material or product.
- *Price*—the best price does not mean always the lowest purchasing price, goal is to provide goods and services in demanded quality at the lowest purchasing price.
- *Time*—determination of time, when to purchase material or services.
- *Supplier*—choice of supplier is one of key factors, which are assumption of proper purchase.

Examples buying process and factors collaboration with suppliers carried out authors Bowersox et al. (1996) and Cavinato (2000). According to authors (Kačmáry et al. 2010, Weiss et al. 2013, Seňová 2015) last step is import of suppliers' results to database of the company and resulting backward relation to supplier. Ideal platforms of database functionality is *MS Access,* or *JMP SAS*, which find out wanted suppliers, according various criteria, for example according type of material, period, in which evaluation had been realized, achieved results, etc.

The medium and large-size enterprises are also paying more and more attention to electronic procurement (e-procurement). Under the term 'electronic procurement', more than just an on-line procurement system is understood, as it connects the company with its suppliers and ensures all the activities related to purchasing. The modules ensure the mutual correspondence between customers and suppliers; register the entire agenda of risk and condition assessment, price comparison, price evaluation, purchasing agreement management

and complaint procedures etc. (Pekarčíková & Trebuňa 2011, Kačmáry & Malindžák 2010).

3 METHODOLOGY AND INVESTIGATION METHODS

The presented paper focuses on the preparation of a database and evaluation of suppliers in the JMP environment of SAS. The selection of a suitable supplier or suppliers is a complicated process, in which it is necessary to take into consideration a lot of evaluation criteria. When evaluating suppliers, various methods used to compare individual criteria (on the basis of a single criterion or on the basis of multi-criteria decision-making), which make it easier to select the final supplier. We pay attention to the scoring model, namely to the method of pairwise comparisons.

The following belong to the fundamental factors considered to be initial for the cooperation with suppliers:

- Quality management and quality assurance for each of the deliveries;
- Price level;
- Payment terms and conditions and payment due dates;
- Supplier's technical and technological degree;
- Flexibility related to supplies and capacity adjustment;
- Geographical and political risks.

In order to evaluate suppliers based on the criteria defined it is most common to use a credit-scoring model in practice. It is a scorecard system of evaluation, which depends on the number of suppliers or evaluation criteria and can be of the following form:

- Potential suppliers (A, B, C etc.) are stated in the table heading;
- Evaluation criteria to be met by the given supplier are stated in columns (price, costs, payment terms and conditions etc.);
- Priority criteria are given a high weighed value;
- Criteria of lower significance are assigned lower weighed values;
- Each supplier is given a score for each of the criteria given on the basis of a score scale, which might be of, for example, 1 to 5, 1 to 10 or some other scale;
- Scores are then multiplied by a relevant weighed value and the supplier with the highest weighed score will come first.

The method of pairwise comparisons is used to determine preference relations of pairs of criteria. The purpose of the method is to form a matrix, in which the weights of individual criteria are determined using pairwise comparisons. When comparing two criteria the above means that the most significant criterion is assigned the value 1 and the least significant criterion is assigned the value 0 from the perspective of our preferences. There might be a case when the compared criteria are considered equally significant. If so, the value 0.5 is assigned to each of them. There are several modifications to the method given; the difference between individual modifications lies in the way preferences are evaluated and weights are given to individual criteria (Pekarčíková & Trebuňa 2011, Nydick & Hill 2009).

4 CONDITIONS FOR SUPPLIER EVALUATION AND SELECTION SUPPORTED BY SOFTWARE

The given matter was reviewed in a company dealing with the sales of electrical appliances. As regards its business activities and purchasing relationship management, the company subject to our analysis used the following scheme showing individual hierarchical steps of suitable supplier selection.

We based our evaluation on a credit scoring model. According to Lambert (1998), we used seven criteria to analyse and evaluate all of the suppliers of white and black appliances):

– Procurement value;
– Goods quality;
– Supply cycle;
– Invoice due dates;
– Supplier's reliability—Reliability of supplies;
– Supplier's reliability—Price;
– Supplier's reliability—Quality of supplies.

Figure 1, shows the database of suppliers and the credits their scored on the basis of the scoring

model together with all the information and data necessary for decision-making and selection of the most suitable supplier or suppliers, all that depending on a particular situation.

The last step following database generation and evaluation is the selection of suppliers. The created program allows to make decisions and select suppliers either on the basis of a single criterion (e.g. a particular product, procurement price) or several criteria (e.g. a particular product in price category No. 3 with a 24-hour delivery). It is possible to use the 'Data Filter' functionality in order to make decisions and select suppliers, see Figure 2.

When our decision-making process is based on a single criterion, the selection is performed as follows, in our case based on a particular product:

1. Company employees need to replenish their stock levels e.g. of dishwashers, and need to check, which suppliers offer the given product. They have two options:
   a. The first one is to search for the given product by simply scrolling through the product menu. This option is, however, time-consuming considering the range of products offered.
   b. The second option is to use the above-mentioned 'Data Filter' function.
2. Once you open the 'Product' folder, a particular product is then searched for, in this case a dishwasher. As individual items are listed in their alphabetical order, it will not take much time to find the product of interest.
3. The last step is to create a new table using the 'Show Subset' option located under the red triangle of the data filter.

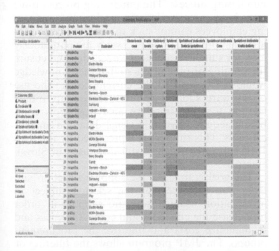

Figure 1. Supplier evaluation in the JMP program.

Figure 2. Data filter in the JMP program.

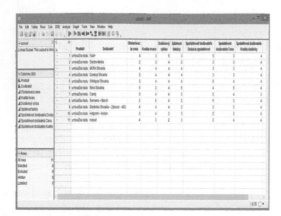

Figure 3. Final table No. 1.

Figure 4. Final table No. 2.

As seen in the Figure 3., we can see a final table containing only filtered data; in this case all the suppliers offering the product subject to search, i.e. dishwashers (Fast+, Electro-Media, MORA Slovakia, Gorenje Slovakia etc.). The employees may further decide e. g. on the basis of product prices or quality using the same procedure as the above.

Through its 'Data Filter' functionality, the JMP software offers an option of multi-criteria decision-making. It allows to filter suppliers based on several criteria. The procedure is as follows:

1. The company employees search for the supplier meeting the following requirements:
   - Product: microwave;
   - Procurement value: 4—means that the price is 10% below the average price;
   - Supply cycle: 4—means that the supplier is able to deliver the given goods in 2–3 work days;
   - Supplier's reliability considering the price: minimum 3—it means that the supplier does not change the defined prices or offers products for lower prices than defined.
2. Another step to take is to gradually enter the criteria in the data filter:
   - Once the first criterion, which is a particular product, i.e. a microwave, is entered, the program will automatically mark all the suppliers offering the goods given.
   - The second criterion is the Level 4 price— once the criterion given is entered the number of suppliers able to deliver the given product in the amount below the average price was reduced to 4.
   - The third criterion is the level 4 promptness of delivery. There are 3 suppliers able to deliver the given product in the level 4 price and in 2–3 work days.

- The last requirement of the company related to its suppliers is the reliability related to a, at least, level 3 price.

The last step is to prepare a new table containing the result corresponding with the criteria entered.

As seen in the Figure 4., the suppliers meeting all the required criteria are MORA Slovakia, Electrolux Slovakia—Zanussi—AEG and Amica. The given suppliers are able to deliver the product for a price lower than the prices offered by other suppliers in 2–3 work days and, in each of the cases, they follow the predefined prices or offer even lower prices than those agreed.

5 RESULTS

It is currently impossible to imagine that purchasing activities would be managed without using any up-to-date information systems and technologies. They have therefore become an integral support of decision-making activities within the purchasing process. The purchasing process topic is such an extensive issue nowadays that managing employees cannot stick to outdated methodological approaches. It is necessary to continuously look for suitable sources of knowledge and ensure the training of purchasers so that their knowledge would correspond to the current needs. The use of software support seems to be an up-to-date method that uses modern analytical tools combining acquired dynamic visual data and analytical tools. This support is represented by the SAS JMP software, which is still little used in Slovak enterprises.

The goal is to show the supplier selection and evaluation processes for the selected subject to investigation using a software support in order to increase the efficiency and competitiveness and to minimize costs and improve the procurement process. The JMP program allows the filtering of data on the basis of different criteria and generating new tables showing search results; this contributes to the faster and more fluent work of company employees.

6 CONCLUSION

When selecting a supplier the purchasing strategy is of substantial significance for the decision-making process. The assignment of suppliers to the selected groups makes sense if trying to reach a certain target, e.g. once a year suppliers are evaluated to monitor the level of business relations with business partners. The suppliers were evaluated on the basis of a database created in the SAS JMP program; thanks to this database it is possible to make the purchasing process more effective and to show the need of implementing such a database in practice. The established database is fully applicable in real life of other companies in Slovakia or abroad. The above-stated shows that it is necessary to check progressive approaches, procedures, new methods and trends in the field of purchasing processes and international purchases, which are able to interconnect information flows within and outside the company.

ACKNOWLEDGMENT

Contribution is partial result of project solving VEGA No 1/0310/16.

REFERENCES

Antošová, M. et al. 2014. System of supplier's selection and evaluation for increasing of buying process effectiveness in the company. *Applied Mechanics and Materials, Logistics Development, CLC 2014*, 708: 257–262.
Benková, M. & Hudymáčová, M. 2009. Suggestion of multicriteria methods for choice of relevant supplier's of company. [on-line] www.katedry.fmmi.vsb.cz/639/qmag/mj64-cz.pdf
Bowersox, D.J. & Closs, D.J. 1996. *Logistical management: The Integrated Supply Chain Process*. New York: The McGraw-Hill Comp. Inc.
Cavinato, J.L. & Kaufman, R.G. 2000. *The purchasing Handbook: A Guide for the Purchasing and Supply Profesional*. New York: McGraw—Hill Professional.
Kačmáry, P. & Malindžák, D. 2010. Trade and production prognosis in condition of dynamic changes of market conditions. *Acta montanistica Slovaca*, 15(1): 53–60.
Kita, J. 2010. Nákup a predaj: základné obchodné funkcie výrobného podniku. Bratislava: Iura Edition.
Lambert, D. et al. 1998. *Fundamentals of Logistics Management*. USA: McGraw-Hill Higher Education.
Nydick, R.L. & Hill, R.P. 2009. *Using the Analytic Hierarchy Process to Structure the Supplier Selection Procedure*. [on-line] www.77.homepage.villanova.edu/robert.nydick/documents/Vendor%20Selection.pdf
Oreský, M. 2011. *Nákup v podniku*. Bratislava: Ekonóm.
Pekarčíková, M. & Trebuňa, P. 2011. *Zásobovacia a distribučná logistika*. Košice: Typopress.
Pružinský, M. & Mihalčová, B. 2016. Material flow in logistics management. In *Proceedings International Conference on Engineering Science and Production Management (ESPM), 16-17.04.2015, Tatranská Štrba, Slovakia*. Balkema: CRC Press.
STN EN ISO 9001: 2008: Systémy manažérstva kvality.
Seňová, A. et al. 2011. Influence of economical and marketing environment to the effectiveness of investment of mining business. In *Proceedings 11th International Multidisciplinary Scientific Geoconference and EXPO, SGEM 2011, 20.-25.6.2011, Varna, Bulgaria*. Bulgaria: STEF92 Technology Ltd.
Seňová, A. et al. 2015. Model evaluation of the economic efficiency of Slovak enterprises by European standards at the time of Euro-integration tendencies. In *Proceedings 2nd International Multidisciplinary Scientific Conference on Social Sciences and Arts (SGEM 2015), Albena, Bulgaria*. Bulgaria: STEF92 Technology Ltd.
Weiss, E. et al. 2013. Comparison of methods, using for evaluation of mining firms in conditions of Slovak Republic. *Metalurgia International*, 18(10): 105–107.

New Trends in Process Control and Production Management – Štofová & Szaryszová (Eds)
© *2018 Taylor & Francis Group, London, ISBN 978-1-138-05885-9*

Flight simulators use efficiency in flight training

V. Socha & L. Hanakova
Department of Air Transport, Czech Technical University in Prague, Prague, Czech Republic

L. Socha
Department of Air Transport Management, Technical University of Kosice, Kosice, Slovakia

S. Vlcek
Masaryk Institute of Advanced Studies, Czech Technical University in Prague, Prague, Czech Republic

ABSTRACT: Flight simulators are recently becoming an important part of pilot training. The integration of flight simulators into aviators' trainings marked a significant shift in pilot training methods. Nevertheless, flight simulators are still rarely used in initial pilot trainings, e.g. when obtaining personal pilot licence or ultralight pilot licence. The incorporation of simulation technologies into this kind of trainings has the potential of enhancing the acquisition of correct skills and procedures while increasing the number of flight hours, and potentially reducing training costs when compared to real flights. The major part of the submitted article therefore aims at the evaluation of efficiency of the use of flight simulators in practicing basic piloting techniques. Training progress of a sample of pilots undergoing a training at a flight simulator was examined in order to set an optimum training duration. The examined parameters were piloting precision and psychophysiological condition of pilots. In regard to the estimation of the optimum training duration, financial efficiency of the use of flight simulators was evaluated with the main emphasis on the so called 3E principle. The results show that an effective duration of pilot trainings using flight simulators is 11 training hours, with the major objective of acquiring basic piloting techniques. The results also suggest that increasing the requirements for initial pilot training by a mandatory flight simulator exercises as a part of the training would create business opportunity of operating such training devices. Based on the performed calculations it was concluded that pursuit of business of operating flight simulators may prove beneficial considering the economy, safety as well as the actual education.

1 INTRODUCTION

Currently, flight simulators are becoming an important part in pilot training. Proceeding from the integration of flight simulators into training programs for pilots, major changes in training methods are emerging (Boril et al. 2016). Compared to conventional training procedures during actual flights, not only risks are reduced and the quality of training is enhanced, furthermore, flight simulators have the potential of reducing costs for pilot trainings. Integrating flight simulators for type trainings and checkup testing of flight personnel is becoming a common practice. Nevertheless, flight simulators are scarcely used in the initial pilot trainings, for example in trainings for personal pilot licence or ultralight pilot licence. It is the incorporating of simulation technologies into this kind of training that may enhance the acquisition of correct skills and procedures while increasing the number of completed flight hours, and potentially reducing the training costs compared to real flights.

Flight simulator as a training tools should represent a number one choice of every aviation school. It however remains to estimate the accurate extent of the training on flight simulator for aviators as well as to estimate the right time to switch to real aircraft.

It is apparent that the acquisition of a skill might come to a stage when one is overly confident about their actions and does not pay due attention to the activity performed (Kozuba & Pila 2015). This unwanted phenomenon may reduce precision which might have grim consequences on safety.

The economics of pilot training is not to be dismissed as well due to the recent trend of reducing costs while complying with safety standards, which points to the economic efficiency of flight simulators. It appears that the use of flight simulators in pilot trainings would become a less expensive option compared to flights in real aircraft.

Despite the field examined in this work is wide and complicated as to its framework, the major part of this article focuses on the evaluation of training and economical efficiency of the use of flight simulators in practicing basic piloting techniques.

2 MATERIALS AND METHODS

2.1 *Setting an optimum training extent*

A study was carried out at Faculty of Aeronautics of Technical University of Košice in order to set the extent of an effective flight simulator training. The selection of the participants followed a completion of theory tests, psychology tests and a medical check. The objective of this selection process was to ensure an even level of knowledge among the subjects and to exclude those subjects who shown higher emotional instability, impaired concentration and decreased ability to multitask when exposed to increased pressure, in order to enhance the reliability of the experiment results. All selected participants, except those excluded based on the above, met the valid requirements for medical fitness of flight staff, and had no previous experience with flying. A total of 30 subjects participated in the experiment.

The participants were divided into three groups (Group A, Group B, Group C). Group A consisted of 8 males and 2 females with the average age of 22 ± 5 years, Group B consisted of 9 males and 1 female with the average age of 23 ± 3 years, and Group C consisted of 5 males and 5 females with the average age of 21 ± 2 years. All three groups underwent a basic theory preparation to acquire basic piloting technique (BPT) in the duration of 2 hours. Its purpose was to introduce students to the cockpit ergonomics, arrangement of respective controls and displays, their functionality and purpose during flight. Practical pilot training was carried out in accordance with the designed training methodology (see Fig. 1) and the set flight schedule under the supervision of an instructor.

The first flight hour was dedicated to the introduction of subjects to BPT directly at the flight simulator with analog display. The first training at the simulator was followed by the first measuring (M1) of heart rate (for the purposes of stress factor evaluation) and piloting precision (for the purposes of evaluating effectiveness) in all subjects. Then the experiment proceeded with basic training of piloting technique on flight simulator with analog display of flight, navigation and motor data (in the duration of 8 flight hours) focusing on mastering BPT with the objective of following the required flight parameters while performing straight and level flight (HPL), 360° horizontal turn (H360) with a 30° bank angle, 180° climbing turn (C180) and 180° descending turn (D180) at the climbing/descending vertical speed of 500 ft/min and a 15° bank angle. This series of four maneuvers was repeated threes times during each flight. The maneuvers were arranged according to their difficulty, from the least to the most difficult. The easiest maneuver was the straight and level flight and the most difficult were C180 and D180. The sequence of the described flight maneuvers was strictly set in order to ensure the training and measuring uniformity. Once the initial pilot training program on the flight simulator with analog display was completed, measurement (M3) of heart rate and piloting precision was performed during the 11th flight hour. The described methodology was applied to all three groups and was uniform and complete for Group A and Group B. There was a different number of performed measurements in Group C, in which the subjects were measured also during simulated flights in 6th flight hour (M2).

The most significant difference is that participants in Group C completed a training in the total of 16 flight hours, and therefore the psychophysiological condition and performance was measured during the last, 16th flight hour (M4). The inter group difference in training duration is demonstrated in Figure 1.

Measurements of physiological data (heart rate) were performed by the FlexiGuard system (Schlenker et al. 2015). The entire measurement concept follows the practices of previous studies (Socha et al. 2016, Regula et al. 2014).

2.2 *Data processing and statistical analysis*

For the purposes of evaluating the activity of the autonomic nervous system based on the outside stimuli, a spectral analysis of the measured R-R intervals by the FlexiGuard system was used. A custom designed MATLAB software was used to determine power spectral density. The principle of processing the signal relied on removing the linear trend and subsequent calculation of power spectral density using Welch's method, which devides the signal into overlapping segments which are

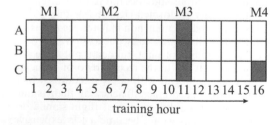

Figure 1. Training schedule for Group A, B and C along with the highlighted measurements.

then averaged. The Hanning window was used to eliminate spectral leakage (Regula et al. 2014). The accuracy of the results was verified by the commercial Kubios software (Niskanen et al. 2004) designed for this purpose.

The extent of stress can be quantified by comparing the LF/HF ratio which directly reflects the sympathovagal activity. The LF/HF ratio was calculated for each subject and each measurement individually. The data obtained from each group and each measurement was tested for normal data distribution by Kolmogorov-Smirnov test. The hypothesis of normal (Gaussian) distribution was rejected in each dataset, the hypothesis of normal distribution of data was rejected at the level of significance $p = 0.05$. Due to the failure to verify normal data distribution, non-parametric methods were used for statistical testing.

Median, the first and the third quartile, minimum and maximum were calculated for each dataset. These calculations were used to create box plots, which demonstrate the distribution of the measured data. Wilcox on test for two independent samples was used to evaluate match between two independent samples. The tests were carried out at the level of significance $p = 0.05$.

The analysis of piloting precision was based on the deviations from the required values of the tested parameters in the form of absolute error. When evaluating HPL, the possible absolute error applied to magnetic course (MC) and altitude (H), in the case of H360 the absolute error followed the failure to proceed with the required 30° bank angle (B) and altitude (H). The C180 and D180 observed failures to follow the prescribed 15° bank angle and the vertical speed (VS) of 500 ft/min. The assessment of the normality of these data concluded that the normal distribution hypothesis was rejected at the level of significance $p = 0.05$. The subsequent statistical analysis was carried out in a similar manner as the stress level analysis. Due to the limited space within the paper, for the presentation of the results shown in Figure 3, the medians of errors for all groups during each measurement were calculated. Subsequently, calculated medians were rescaled according to the maximum values of the parameter. This normalization set vary in a range (0,1) for each measured parameter.

2.3 Evaluation of the proposed changes in pilot training based on the 3E principle

Following the results of setting the effective extent of pilot training on flight simulators, it was possible to set our aim on creating a model which considers the integration of flight simulators into pilot training based on the 3E principle (efficiency, effectiveness, economy). In order to preserve the effectiveness of training and pilot testing, the overall benefit is to exceed the costs. In this case, the criteria for evaluation of the efficiency of invested funds (in the form of more expensive trainings) may be considered the reduced number of flight accidents and deaths resulting from flight accidents. In the year to year comparison of accidents caused in general air traffic by aircraft with Maximum Take-Off Mass (MTOM) up to 2 250 kg, the number of accidents was reduced by nearly 40% and the number of deaths was reduced by 46% from 2009 to 2013. Despite this decrease corresponds to the long term trend of reducing air traffic accidents and related deaths, the fatal accidents ratio is improving only in several percent, and the number of fatalities per accident remains the same, representing roughly 2 fatalities per accident. If the number of fatal flight accidents caused by human factor in the EASA member states dropped by 50% in the following 5 years compared to 2013, it might save over 40 lives a year. This figure can be reached by extending the pilot training aiming to enhance the training of take-off, landing and solving crisis situations. As a result of incorporating such extension of requirements for obtaining a pilot licence, pilots in real flight traffic would not experience excessive stress levels, and would be able to resolve consequences of arising situations easier, faster and more accurately, therefore eliminating risks of possible accidents.

By extending pilot trainings by flights on simulators, further conditions of the 3E principle would be met—effectiveness and economy. Operating costs of flight simulators are significantly lower compared to operating costs of real aircraft (see Table 1) and by completing test flights pilots would

Table 1. Comparison of costs, fuel/energy costs and operating costs per 1 flight hour for selected single (SEP) and multi engine (MEP) piston aircraft compared to an hour of operation of FNTP.

Class	Type	Cost (10^3€)	Cph* (l or kWh)	OC**
2*SEP	Cessna 152	37.42***	30	69.76
	Cessna 172	355.47	50	116.27
MEP	Piper PA-34	>975	95	220.91
2*FNPT	TRD40	405.06	11.5	2.18
	Piper PA-34	–	14	2.63

*Consumptoin per hour. This are approximated values which depends on various factors (particular type of aircraft, operating power of aircraft, weather situation etc.).
**Operating costs for 1 hour. At the price of 2.32€l of fuel including excise duty, or 0.19€kWh.
***The production of the Cessna 150 and/or Cessna 152 was ceased, therefore we consider a second hand price.

not be exposed to excessively elevated costs. Due to the fact that this would result in enhanced quality of pilot training while minimizing extra costs, the condition of efficiency of extending the pilot training is met. From the point of view of economy, it is necessary to set such duration of the extended pilot training on simulators, that will ensure reaching the objectives, i.e. sufficient reduction of stress levels when resolving situations which arise during respective stages of flight, while minimizing the extra costs. This condition is met by extending pilot trainings for obtaining pilot licence for aircraft below 2 250 kg MTOM, therefore especially PPL(A) a CPL(A), by the requirement of completing the set number of flight hours on a flight simulator as described above.

3 RESULTS

3.1 Evaluation of psychophysiological condition

Analysis of heart rate through the LF/HF ratio as the indicator of stress level showed that the greates stress in all examined groups was measured during M1. This result corresponds to the initial assumption. During M3 measurements, a significant drop in the LF/HF median was observed compared to to M1 at the level of significance $p = 0.05$.

The M4 measurement shows that the stress level during simulated flights in the 16th flight hour did not show significant difference from M3 compared to all three previous measurements in the three groups. Similarly, significant difference was not concluded either when comparing results of the M2 measurement against M1 and M3.

The estimated course of the stress levels depending on the course duration is demonstrated by a decreasing function in the Figure 2. From the above it follows that a statistically significant decrease in pilots' stress levels during trainings on flight simulator takes place under the described conditions after 11 hours of training. These findings will be

further used in the evaluation of the effectiveness of the use of flight simulators in pilot trainings.

3.2 Evaluation of the piloting precision

The results of the analysis of absolute errors in respective simulated flights are unified and illustrated using polynomial expression of the medians of the data from respective measurements from all three groups in Figure 3. The descending turn (D180) was not assessed as over the course of individual measurements, several complications took place during the data collection. From the illustrated it follows that the function of pilots' error rates in dependance on the training duration is decreasing. Based on this result, an assumption is verified that extending the duration of the training results in the decrease in pilots' error rates, and at the same time results in enhanced piloting performance.

From the point of view of optimum extent of the use of flight simulators, it is necessary to define the balance between the decreasing error rates and increasing piloting precision. Individual statistically significant inter-group differences calculated using Wilcoxon Two Sample Test are summed up in Table 2.

The presented analysis of the precision of the individual performed flight maneuvers during simulated flights implies that a sufficient increase in piloting precision was observed between M1 and M3. It can be thus assumed, that sufficient increase in piloting performance took place during the 11 hour training. It can be concluded that the

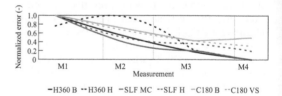

Figure 3. Presentation of absolute errors in respective simulated flights using polynomial expression of normed medians.

Table 2. Statistically significant differences between piloting precision measurements.

	M2*	M3	M4*
M1	SLF MC H360 B	SLF H SLF MC H360 H H360 B C180 B* C180 VS*	SLF H SLF MC H360 H H360 B C180 B C180 VS

*Valid for Group C only.

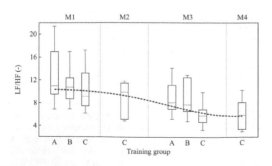

Figure 2. Boxplot representation of LF/HF distribution in Group A, B and C.

optimum duration of pilot training on flight simulators is 11 flight hours.

No further statistically significant decrease in error rates was observed in Group C between M3 and M4 measurements, therefore we can assume that incorporating any additional simulation flights into the pilot training would not result in any further significant increase in piloting precision in pilots in training.

4 DEFINITION OF THE BUSINESS OPPORTUNITY

Extending the pilot training by the necessity to complete one of its parts on a flight simulator in the Czech Republic would create an interesting business opportunity based on operation of such devices. The Brno-Tuřany airport appears as another suitable location for accommodating another flight simulator. Exactly such location of the flight simulator would be geographically favourable due to the integrated regions of Jižní Morava and Zlín, south parts of the Olomouc and Pardubice region as well as the eastern part of the Vysočina region.

4.1 Calculation of the annual costs and returns

Recently, there are six ATOs running directly at the Brno-Tuřany airport, out of which (as suggested above) none is operating a flight simulator. The entire calculation in this study follows the assumption that one of these facilities would decide to include a flight simulator in their portfolio. All costs, whether personal or operating, which will be considered in the calculation, will be considered as allocated costs related to the operation of a given facility.

The initial investment required to purchase a TRD40 flight simulator replacing the Cessna 172 aircraft represents 404 110.37€ with guaranteed warranty of 15 years. This investment shall be covered by a bank loan. The loan interest totals 7% (Doganis 2010) and is to be repaid in 8 years. The annual investment S due by the end of a calendar year is defined by the following formula:

$$S = D \frac{r}{12} \frac{(1 + \frac{r}{12})^{12n}}{(1 + \frac{r}{12})^{12n} - 1},$$ (1)

where D is the initial size of the loan, r is the interes rate in% and n is the duration of the loan in years. In this case, the amount of the annual installment is 67 675.47€. The overview of the interest and amortization rates as well as the amount of capital in the individual years is presented in Table 3.

Table 3. Overview of installment amount, interest rate, amortization and capital of the loan in individual years of repayment.

Year	Installment (€)	Interest (€)	Amortization (€)	Capital (€)
1	67 675.47	28 287.73	39 387.74	36 4722.64
2	67 675.47	25 530.57	42 144.86	32 2577.78
3	67 675.47	22 580.45	45 095.01	27 7482.76
4	67 675.47	19 423.79	48 251.68	22 9231.09
5	67 675.47	16 046.18	61 629.29	17 7601.80
6	67 675.47	12 432.12	55 243.35	12 2358.45
7	67 675.47	85 65.08	59 110.38	63 248.10
8	67 675.47	44 27.37	63 248.10	0

The planned extent of utilization of the flight simulator is 6 flight hours a workday and 10 flight hours a day during weekend, i.e. 50 flight hours a week. The annual operation time totals 50 weeks a year, it therefore follows that the estimated annual extent of utilization is 2500 flight hours. The final sum per flight hour on the flight simulator for the end user including the instructor fee thus totals 82.32€ excluding VAT. The estimated annual returns would thus represent 205 796.95€.

Among the operating costs allocated to the operation of the simulator, there are costs for electrical energy, rent and operation of non-residential space to accommodate the flight simulator, and the flight simulator insurance. The power consumption, i.e. the simulator's consumption of the electrical energy per hour is 11.5 kW, amounting to the total annual power consumption of 28750 kW. In the presented calculation we considered the first year of the operation of the flight simulator per 0.19€ for 1 kWh with 5% annual increase in the price. Other operating costs are considered to remain constant over the entire 15 year period used in the calculation. The rent and operating costs for the non-residential space was estimated at 6 735.17€ per year, and the flight simulator maintenance was estimated at 4 490.12€ per year, with the annual insurance fees of 1 571.54€.

The calculation of the allocated labour costs follows the personnel requirements for ensuring the operation of the simulator. Therefore, a flight instructor is required for the entire period of the estimated operation of the simulator, as well as a receptionist, whose estimated labour costs would equal 0.5 labour hour per each operating hour of the flight simulator. The instructor's gross wage per hour is 8.23€ and the receptionist gross wage per hour is 4.86€ with the planned 5% annual rise. The 25% wage bonus for work on Saturdays and 100% wage bonus for work on Sundays also needs to be considered. The annual labour costs would therefore total 44 655.62€.

Among the costs, there are also the above mentioned interest for the loan as well as depreciation. The flight simulator falls into group 2 of depreciation assets (Pelc 2011) and its depreciation is distributed into 5 years. In the case of straight line depreciation, which is assumed to be the case, the depreciation rate in year 1 is 11%, and 22.25% for the remaining 4 years. After subtracting the total annual costs (operating and labour costs, interest and depreciation), from the total returns from the operation of the flight simulator, earnings before tax are calculated. Based on this amount, the 19% corporate income tax is calculated, which is due in the following fiscal period.

4.2 Economic assessment of the investment

In order to assess the investment, it is necessary to estimate the cash flow for individual fiscal periods. In order to calculate the cash flow, the earnings before tax is adjusted by the taxes paid for the previous fiscal period, amortization which is paid to the bank as a part of the loan repayment, and by depreciation, which are a cost, but are not an expense. Cash flow is then discounted by (in this case required) 10% discount rate. The cummulated and discounted cash flow sets the net output value of the investment (NPV), which corresponds to the final value of the cummulated discounted cash flow for the evaluated period and determines the amount of the profit from the investment:

$$NPV = -INV + \sum_{i=1}^{n} \frac{CF_r}{(1 + IRR)^r}, \quad (2)$$

where INV is the initial investment (CF_0), CF_r is the cash flow in year r, n total number of years and i is discount rate (Veber 2012).

The critical criteria for making the investment is the positive value of NPV. The period, in which the cummulated discounted cash flow shifts from negative to positive value of NPV represents the payback period. The discount rate, by which the final value of the cummulated discounted cash flow was equal to 0, corresponds to the internal ratio of return (IRR), which determines the relative return of the investment for the evaluated period (Veber 2012):

$$0 = -NPV = INV + \sum_{i=1}^{n} \frac{CF_r}{(1 + i)^r}. \quad (3)$$

The calculated figures imply, that by the discount rate of 10% and the above estimations of the distribution of costs and returns, while considering the financing by a loan and the set period of 15 years, the value of NPV totals 149 467.89€, and

IRR of 15.5%. The return of the investment then represents 10 years. Considering the above data, it appears profitable to proceed with the investment and extend the existing services pilot trainings at the Brno-Tuřany with the training on the FNTP TRD40 flight simulator.

4.3 Risks

The most significant risk for the project of operating flight simulator at the Brno-Tuřany airport poses particularly an insufficient interest in the provided service. Nevertheless, due to the long term trend in the increase of the issued licenses, this risk may be considered low.

5 CONCLUSION

The objective of this paper was the evaluation of the current state of the use of flight simulators in the training of pilots of light aircraft, and the design of a modification which would result in the enhanced efficiency of training procedures for obtaining the PPL (A) licences. It would further result in enhanced safety of air traffic, which is directly proportional to the skills of the pilots. Sufficiently trained pilots are capable of better handling of stress situation which may arise during individual flights.

Based on the data collected within the scope of a study carried out at Faculty of Aeronautics of Technical University of Košice, an optimum extent of the use of flight simulators in pilot trainings was set. The designed solution expanding the current procedures for pilot training was discussed from the point of view of effectiveness, efficiency and economy with the objective of reducing fatalities caused by air traffic accidents.

Extending the requirements for pilot training with compulsory completion of a part of training at flight simulators creates a business opportunity for operating certified flight simulators. Within the scope of this work, this opportunity was analysed for ATO operating at the Brno-Tuřany airport. Based on the performed calculations it can be implied that operating a flight simulator is economically profitable, and the investment for the purchase of a flight simulator may be made in the presented extent.

Incorporating flight simulators into the lifelong educational process of pilots (also pilots of light aircraft) might represent a stimuli for further utilization of flight simulators. Currently, pilots are obtaining their qualification documents with a lifelong validity, which increases risks of air traffic accidents especially with uneven or infrequent events of actual flying. Establishing the requirement of mandatory check up tests for the light aircraft pilot licence holders might have beneficial effects on enhancing the air traffic safety and should have positive impact on other fields dealing with air-safety management (Vittek et al. 2016, Kraus et al. 2016).

ACKNOWLEDGEMENTS

This work was done in the frame of operational program No. ITMS 25110320049 co-financed from EU funds and grant no. SGS17/150/OHK2/2T/16 co-financed form Grant Agency of the Czech Technical University in Prague. The authors would also like to thank Andrej Madoran, B.A., for translation of this work.

REFERENCES

Boril, J., Jirgl, M. & Jalovecky, R., 2016. Use of flight simulators in analyzing pilot behavior. In *IFIP Advances in Information and Communication Technology*, 255–263. Springer International Publishing.

Doganis, R., 2010. *Flying off course: airline economics and marketing*. New York: Routledge.

Kozuba, J. & Pila, J., 2015. Chosen aspects of pilots situational awareness. *Naše more* 62(SI): 175–180.

Kraus, J., Vittek, P. & Plos, V., 2016. Comprehensive emergency management for airport operator documentation. In *Production Management and Engineering Sciences—Scientific Publication of the International Conference on Engineering Science and Production Management, ESPM 2015*, 139–144. CRC Press/Balkema.

Niskanen, J.P., Tarvainen, M.P., Ranta-aho, P.O. & Karjalainen, P.A., 2004. Software for advanced HRV analysis. *Computer Methods and Programs in Biomedicine* 76(1): 73–81.

Pelc, V., 2011. *Daňové odpisy: strategie pro podnikatelskou praxi firem a podnikatelu*. Prague: C.H. Beck. (In Czech).

Regula, M., Socha, V., Kutilek, P., Socha, L., Hana, K., Hanakova, L. & Szabo, S., 2014. Study of heart rate as the main stress indicator in aircraft pilots. In *Proceedings of the 16th International Conference on Mechatronics—Mechatronika 2014*. IEEE.

Schlenker, J., Socha, V., Smrcka, P., Hana, K., Begera, V., Kutilek, P., Hon, Z., Kaspar, J., Kucera, L., Muzik, J., Vesely, T. & Viteznik, M., 2015. FlexiGuard: Modular biotelemetry system for military applications. In *International Conference on Military Technologies (ICMT) 2015*. IEEE.

Socha, V., Socha, L., Hanakova, L., Lalis, A., Koblen, I., Kusmirek, S., Mrazek, P., Sousek, R. & Schlenker, J., 2016. Basic piloting technique error rate as an indicator of flight simulators usability for pilot training. *International Review of Aerospace Engineering (IREASE)* 9(5): 162–172.

Veber, J., 2012. *Podnikání malé a střední firmy*. Prague: Grada. (In Czech).

Vittek, P., Lališ, A., Stojić, S. & Plos, V., 2016. Challenges of implementation and practical deployment of aviation safety knowledge management software. In *Communications in Computer and Information Science*, 316–327. Springer International Publishing.

New Trends in Process Control and Production Management – Štofová & Szaryszová (Eds)
© 2018 Taylor & Francis Group, London, ISBN 978-1-138-05885-9

Critical elements in the area of possible improvement and development of flight training effectiveness

L. Socha & P. Kalavsky
Faculty of Aeronautics, Technical University of Kosice, Kosice, Slovakia

S. Kusmirek & V. Socha
Faculty of Transportation Sciences, Czech Technical University in Prague, Prague, Czech Republic

ABSTRACT: Study of aviation accidents in general came to conclusion that aviation accidents are always caused by more than just a single factor. They are often caused by teh combination of mistakes, whiche singe failure itself doesnt have to neccesarly cause the critical situation. As long as the crew is not able to break teh chain of smaller mistakes by correct procedure, it can lead to exhaustion of physical resources able to solve more compliacted mistakes, which in a long term can lead to a cathastrophe, human as well as financial losses. In these cases we talk about the failure of crew—incorrect management of human resources on board of a plane. Listed facts certainly show the decrease in number of incidents and increase in safety of flights could be achieved by better training of pilots. There are many areas which could lead to the improvement of this. They mainly include teh increase in the technical knowledge among the crew, improvment of trained processes and procedures and in gaining more experience during the practial traning of pilots. The improvement of practical training of pilots could be the key factor to the overall improvement in safety onboard. In a relation to this it is very necessary to specify basic critical elemnts of the practical training. For this purpouse we created a survey presented to the target group. The result of this survey shows over 66% of particiapnts havent noticed any effort from authorities to improve traning processes of pilots (60 respondcnts) At the same time all participants of this survey agreed on the fact that its of esence to pay a lot of attention to this issue. Based on results its easy to say that introduction of new technologies and new processes during training of pilots and personal on board would increase teh effectivnes of the training and the safety of flights, leading to the decrease in number of aviation incidents, meaning less financial and especially human losses.

1 INTRODUCTION

Devclopment of air tranportation focuses on overall increasing of safety in this industry. With the introduction of new tehcnologies, pressure on the teoretical and tehcnical knowledge of pilots increases. Safety on board is the main goal of the International Civil Aviation Organization (ICAO). Since the begining of the existance of the civil aviation, the safety has been increasing, however there is still plenty of space for improvement. Statistics show that three out of four incidents happen as a result of a human error. Because this fact has a huge influence on safety of flights, it is necessary to achieve improvement in this area of human factors/errors.

Study regarding avition accidents lists as one of the main reason fo avition accidents is the human error (Wiegmann 2003). Human eror occurs in situations when skills of the pilots are insufficient or when the pilot is faced with the suprising or unknown situation (Kozuba & Pila 2015). Even though results vary, its clear to say that human error is responsiblc for 70–85% of all aviation accidents (Weigmann & Shappell 1997). The most common human erros among pilots are related to his flying experience (as many as 80% of cases). Out of these erros about half of them lead to the chain reaction of events. Pilots with most flying experience are less likely to be involved in the aviation accident caused by teh human error, hower due to the fact of them flying more chalanges flies the probability of them being in the fatal accident is higher (Bazargan & Guzhva 2011).

Nowadays we can watch the effort to increase the level of safety in aviation industry (Vittek et al. 2016). People search for more complex approach to the elimination of human and technological errors in their mutual connection to safety on board of planes. Crews are required the knowledge of tehcnological limits. At the same time technologies are reuquired to help eliminate human weaknesses and to strengthen the human strenghts. It it very important to improve the proccesses of training of crews, their development and mantaining the usage in real life. Economic factor of air transportation pressures companies to achieve the lowest maintance costs (Szabo et al. 2015, Hospodka 2014).

One of the areas which would improve the safety, effectivness and keep costs low at the same time, is the improvement of practical training of pilots (Socha et al. 2016). In this relation it is imporant to specify basic critical elements of practical training. That is why this article is focused on critical elements which could support the improvement of effectivity of the training of pilots in the industry. The reserached performed via survey was created for this purpouse.

2 MATERIALS AND METHODS

Research done by survey asked people employeed as crew members no matter of their nationality or work position. Specialists like these should be able relevant people to express their opinions regarding the topic of this study. They themselves see the problems, issues and hazards in the avtiation. All of them had to pass the teoretical and technical part of the training. They are also involved in develomplent and implementation of new tehcnologies, proccesses etc. Thats why they should be able to point outmistakes and defecencies in the training of crews, as well as a possible improvements. The survey was also available to wide public through the web application created for this purpouse.

The survey was divided into two parts. First part consisted of nine questions and three subquestions where particiapnts could express their opinion freely. Another five questions were pilot oriented. Of course, all particiapnts could answer these questions in their own free form, even if the answer wasnt completely relevant to the question.

Questions in the survey were focused on the area of usage of the flight simulators and verifycation of its processes in traning. Another area regarded the oportunities for improving working, and training processes and syllabuses. The third and the last area was focused on the misleading displaying of information on board and the impact it can have on the mood in the cabin, work pressure and other aspects related to it.

The goal of this survey was the identification of possible opportunities of more frequesnt and effective usage of flight simulators (Boril et al. 2016) in training and verification of usefullness of these technologies as well as the verification of its beneficiality of new technologies, and proving our theories presented in this study.

The survey was published online for these purpouses and it was done annonymously. Overall we recorded 138 attemps for an answer. Out of this there 91 answers evaluated as solid and relevant answers. 31 answeres were an attempt for an unwanted commercial via sharing other links to other websites and 16 answers were not completed.

3 RESULTS AND DISSCUSION

Details regarding each question are listed bellow. Number of answers regarding question No. 1 "What trainings does you company use flight simulators for?", are displayed in Table 1.

As shown in Table 1 two participants said that their airline companies use flight simulators for different kinds of trainings. In both cases it regarded the so called Airport qualifications, which is gaining the permission to land at certain airport, which requires this permission (Kraus 2016), e.g. the airport in Insbruck.

As we can see, all four kinds of trainings represent approximatelly the same number. It is between 60–70%.

About usage of flight simulators for verification of routine procedures testify answers for question No. 2 of our survey "Have you ever used a flight simulator to verify the accuracy (effectivness) of routine procedures?". The positive answer was marked by 27 respondents, the negative by 64.

The replies show that only 27 participants of survey have met with situation, when the air company was veryfying correctness or effectivity of routine procedures. This represents only 30% of the total. This result may indicate that survey participants were mostly young and less experienced pilots (crew members), or they work in airlines where such a test aren't performed. This could indicate space for wider use of simulation technologies or space for improvement activities of airlines. This could have a direct impact on increasing the safety culture of the airline and therefore on the safety o fair transport.

The previous question directly develops into question No. 3. of survey. "If Yes, what procedures were involved?". The question was supposed to indetify additional space for using of flight simulators. The answers of participants of survey are shown in the Table 2.

Table 1. Answers to question No. 1 of survey.

Answers	Number of answers	
	Yes	No
Type rating	61	30
Recurrent Training	55	36
Low visibility procedure	56	35
Emergency situations	62	29
Other	2	89

Table 2. Answers to question No. 3 of survey.

Answer	Number of answers
Use of check-lists, standard, substandard and emergency procedures	1
Visual approach at night	1
Standard operation procedures	1
CRM	2
Verification of procedures in QRH issued by manufacturer —errors found	1
Middle and final phase of approach	1

Table 3. Answers to question No. 5 of survey.

Answer	Number of answers
Change of normal procedures	2
Check-lists	1
Substandard and emergency procedures	1

Table 4. Answers to question No. 7 of survey.

	Number of answers	
Answers	Yes	No
Initial training	81	10
Type rating	78	13
Recurrent training	77	14
Transition training	73	18
Briefings	57	34
Taxiing	30	61
Take off phase	77	14
Phase of initial ascend	68	23
Phase of flying at the normal alltitude	65	26
Phase of initial descend	70	21
Landing phase	79	12
Communication	73	18
Emergency situations	78	13
Other	1	90

From Table 2 flows, that 7 participants of survey indicated in which case of verification procedures was flight simulator used. It's only a very small number (27%) of the total participants who claimed to have met with verification of accuracy (effectivity) of used procedures. Along with it is clear that only in two cases it is the same sphere. In our case CRM – crew resources management—on the board of plane. This suggest that there is a lot of space for using flight simulator for mentioned verification. It also confirm an opinon, that the participants of survey meet very rarely with such a use of flight simulator.

The question No. 4 "Have you performed attempts to improve routines in the traing (e.g. for improving safety)?" have survey participants elected affirmitive answer 31 times and negative answer 60 times.

This result points to the fact that 31 respondents met with attempts for improvement of routine procedures in training. This represents only 34% of involved participants. This result indicates that only a small part of airlines deals with quality of flight training. This result may be influenced by age of survey participants who don't receive procedures offten. In our opinion it is a clear evidence that this issue have not enough attention by airlines, from which were the survey parcitipants.

Previous question directly develops question No. 5 of survey "If so, which procedures it were? What changes in the procedures you performed?". The answers are along with the frequency shown in Table 3.

Those replies confirm our view, that airlines don not pay much attention to this issue. This shows the type of airlines that operate in central Europe. These are mainly small companies with low amount of aircraft. Mainly deal with non-regular commercial air transport (e.g. charter flights or renting of aircraft including crew to another airlines). Such companies are suffering mainly to the lack of human resources that could adress the issue of systematic improvement of training and operational procedures in the company.

The importance of the issue of improving the training procedures evidenced by results obtained by survey. To question No. 6 "Do you think it is appropriate to pay attention to the improvement of procedures and shcemes of training?" responded positively by all participants. All survey participants agreed on the fact that appropriate (necessery) pay attention to the improvement of practices and schemes in training.

To question No. 7 of survey "If so, which areas of training you think is necessery to pay more attention to?", were gained the results presented in Table 4.

Table 4 shows how it is necessery to pay more attention to individual phases of flight. In addition to phase briefing and taxiing, wherein the participants expressed the requirment only on 63% resp. 33%, all other phases expressed an increased requirment for 72% to 89%. Only in one case (it is 1%) they know to identify necessity of paying more attention to another phase of of flight training as specified in the questionaire survey. This is the danger arising from the change in the display of flight and navigation data in the cockpit of aircraft.

473

In Table 5 are collected answers to subquestion "What problem/troubled process in the sphere it is?".

Despite a number of positive responses, survey participants were able to identify a specific problem or troublesome procedure only in 27 cases. This represents only 3% of the total number of positive answers. This situation can be explained by deeper unwilingness to deal with the issue, the low level of theoretical knowledge of participants or eventual fear to express their opinion. On the basis of such factor is necessery data in Table 5 take only informative.

Question No. 8 of survey "Do you think that such research is aimed at analyzing training methods of pilots can help to?" be an effort to verify hesitance research in various fields. In the opinion of survey participants, the answers given in Table 6 shows a clear view that the research makes sense. In 72% of the participants agreed that the analysis of training methods can help to gain practical experience in 99% to increase safety in air traffic, in 59% to

increase efficiency in air transport and in 60% for pilot pressure relief. In one case (1%) said survey participant other. He said that by analysis of training methods is possible to reduce the risk resulting from the change in altitude.

To question No. 9 of survey "It has your company aircraft equipped with traditional cockpit?" were recorded 84 positive answers.

This question in dirrect response to the following question No. 10 monitors the number of aircraft used to the traditional cockpit equipped with analog devices to display fly and navigation information. Out of the positive answers to question No. 9 it is clear that the monitored airlines has used a large number of aircraft with analog display of flight data.

To question No. 10 of the survey "Does your comany dispose with aircraft equipped glasscockpit?" responded positively by all participants of survey. This question in direct response to the previous question and monitors the number of aircraft used in the airlines of the survey participants, with modern cockpit with a digital display screen to display flight and navigation information (ie. Glasscockpit). As is clear from the responses, every monitored airline uses aircraft with glasscockpit.

From answers to question No. 9 and 10 it is clear that at present are simultaneously used aircraft with two ways to view the flight data. Thus aircraft cockpit with analog display data and glasscockpit with digital display data. This situation is likely to last for the life of current aircraft.

Since question No. 11 of survey "Have you experienced pilots in difficulty in the transition from analogue cockpit to glasscockpit?" were questions focused on the issue of display of flight data and problems associated with the transition from one cockpit type to another. In 86% of survey participants responded to a question No. 11 positively. Thus they confirmed that there occurs difficulties with pilots who experienced transition from analogue view to glasscockpit.

To question No. 12 of our survey "Do you think it would be appropriate to introduce in the shceme of training retraining aimed at glasscockpit controlling?" 99% (90 respondents) of survey particpants expressed a clear need to introduce into scheme of training a part, which would be aimed on glasscockpit controling.

To question No. 13 of survey "Have you ever done flights on aircraft, which was equipped with an analogue cockpit?" answered all of the participants in the survey positively. It follows that all the participants in the survey have practical experience to perform flights in the traditional cockpit with analog display of flight and navigation data.

To question No. 14 of survey "have you ever done flights on aircraft which was equipped with

Table 5. Specific answers to No. 7 subquestion.

Answer	Number of answers
Initial training	4
Type rating	2
Recurrent training	2
Transition training	0
Briefings	4
Taxiing	1
Take-off phase	2
Phase of ascending	2
Phase of flying at the normal alltitude	1
Phase of descending	2
Landing phase	4
Communication	0
Emergency situations	2
Others	1

Table 6. Answers to question No. 8 of survey.

Answers	Number of answers	
	Yes	No
Gain of practical experience	66	25
Increase safety in air trafic	90	1
Increase efficiency in air transport	54	37
Pilot pressure relief	55	36
Would have minimal asset	0	91
Would have none asset	0	91
Other (reduce the risk resulting from the change in altitude)	1	90

glasscockpit?" the 83 survey participants answered positively. It follows that 91% of survey participants have experienced some flight with an aircraft equipped with glasscockpit. This is less than issue No. 13 but the difference is very small (less than 10%). This reflects the fact that the develompent of aviation is almost at halfway transition from the era of analog display of flight and navigation data to digital (glasscockpit). When the life of the aircraft is about 25 years, we can expect that this journey will last for about 13 years.

Question No. 15 "If the answers for questions No. 13 and No. 14 are YES, in which cockpit have you experienced better orientation during the first contact?", directly completes the answers to question No. 13 and 14. In this issue was an effort to find out what kind of cockpit pilots feel better orientation during first contact. 77% of respondetns (70 respondents) argued that thay feel better in the cockpit using an analogue display of flight and navigation data during they first contact. This information may appear to be against the logic of introducing glasscockpit. This result may be affected by the fact that the pilots were on they initial training on aircraft with an analog display. But it can also tell, that it is easier and faster to orient in an enviroment with an analog display data. Here arises space for scientific research. As one of the unverified but plausible explanation may be that person easily and quickly perceives basic geometrical patterns as intricate pattern, which has the most rather spread out easier and you then fold in a resulting data (DiCarlo et al. 2012). Just analog display provides simpler geometric shapes. Pilots admit that in the cockpit with analog view is not always perceive the display value but only position of arrows on dashboard, which represent the actual values. If they learn where these arrows have to point (if the value is correct), so they just have to chceck their location. For simplicity are almost always "correct" positions shaded with green color, and bordered by a distinct color (yellow to increase focus close by limits and red to highlight the limits). Conversely, glasscockpit displays flight and navigational data as complex data screen. This means that a small viewing area is indicated by the large amount of data to be read, the values assigned to certain parameters and all these information then join in of the resulting pattern which pilot finaly works with. The advantage of this view is that the relatively small area (less than would be required with analog display) you can display larger amounts of data (Socha et al. 2015). Of course the implemetation of elements of microelectronic has reduced the weight of the display units, which is a basic parameter of efficiency of air transport.

The question No. 16 of this survey "In case you answere questioned number 13 and 14 Yes, which cabin would it be easier to work in after getting familiar with both of them?" it completes questions number 13, 14 and 15. All these answers prove the fact that after getting to know both cabins with different kinds of informational systems, its easier to navigate and work in the digital system cabin. 91% of corespondents (83 people) confirmed this fact. The explanation is simple and is partialy included in the analysis of results to answers in quesiton No. 15. To process any kind of visual informatin, our brain has to differenciate visual information between simpler one (a line, a ring, a square) and more difficult one (number, letter, sign). These are decoded and assigned different meaning (for example some sign, number, letter etc.). This information is then connected into one unit which creates the whole picture presented to individual (DiCarlo et al. 2012). If we manage to teach brain this formula and force it to function according to it, we will achieve the result when a human can evaluate rationally more information. In that case the main factor influencing evaluating of information, for example distance. It is natural that pilot encounter both kinds of these brain functions and they often mistaken then for each other. Thats why the last question was oriented towards gaining information about which change is easicr for pilot.

The questiont No. 17 "Which transition is the easiest according to you?" all participants (100%) agreed that the transition from analog to digital information system was the best. We think that this is caused by the fackt that its easier to find requested information in digital (glasscockpit) than analog one. Basically, the final conclusion is that digital (glasscockpit) displaying of information is ideal for pilots who know what kind of information they need during the flight.

We can also look at it from other side. If the results of our survey are true than it also shows the fact that its should be easier to teach pilot to fly in the analog cockpit and then switch him over to glasscockpit. This thought cannot be closed though and it should be the subject of other studies.

4 CONCLUSION

According to results from this survey we can conclude that approximatelly two thirds of companies employeeing participants of this survey use flight simulators for basic and advanced training of their employees. Although only one third of companies from this survey provide flight simulators to their pilot while practicing specific critical flight simulations. Absolute majority of respondends agree that current processes and trainings need to be reviewed and renewed in order to improve the safety in civil aviation. Participants of this survey were younger and less experienced pilots proving that airline companies dont pay enough attention to training their pilots and dont use flight simulators and training processes as efficiently and effectivelly as possible.

Over 66% of respondents havent noticed any effort by airline companies to improve or even change training processes. At the same time all participants agreed that their is a need to pay more attention to this issue. They even decisively answered that this change would bring a lot of advantage in gaining more experience, increasing the level of safety and effectiveness among pilots while decreasing the work pressure on them. There are lot of deficiences in this area.

Nowadays, ariline companies use both kinds of onboard information systems. Meaning they use both analog and digital displaying of on board information. Most of respondets expressed a lot issues and inconvinience when transfering from one information system to another. In case of an analog system it was 86% of participants. Introduction or expansion of usage of simulators for this kind of issue would be very helpful. 77% of asked people said they felt more comfortable in analog sysem cabin while 91% ofpeople said the same about digital system cabin. As mentioned above, the trasnfer from one to another cabin can be very complicated for most crews due to lack of training (Socha et al. 2015).

As mentioned already, pilots of large ariline companies who hold certificates like CPL(A) or ATPL(A) were asked to participate in this survey. If the same questions were presented to pilots with PPL(A) certification, it would be very likely that those pilots have never used flight simulator or never were presented with an option of using one in the training. The advantage of the training without a flight simulator is strictly financial. This creates the window of oportunity to expend the training experince of pilots with PPL(A) certificate.

ACKNOWLEDGEMENTS

This study was funded by Ministry of Education, Science, Research and Sport of the Slovak Republic within execution of the project No. ITMS 26220220161 (Research of pilots training methods using flight simulator) co-financed by EU funds. This work was also supported by the Czech Technical University in Prague, junior research grant number No. SGS17/150/OHK2/2T/16. Autors would like to thank Pavol Langer, for translation and language check of this article.

REFERENCES

Bazargan, M. & Guzhva, V.S., 2011. Impact of gender, age and experience of pilots on general aviation accidents. *Accident Analysis & Prevention* 43(3): 962–970.

Boril, J., Jirgl, M. & Jalovecky, R., 2016. Use of flight simulators in analyzing pilot behavior. In IFIP *Advances in Information and Communication Technology*, 255–263. Springer International Publishing.

DiCarlo, J., Zoccolan, D. & Rust, N., 2012. How does the brain solve visual object recognition? *Neuron* 73(3): 415–434.

Hospodka, J., 2014. Cost-benefit analysis of electric taxi systems for aircraft. *Journal of Air Transport Management* 39: 81–88.

Kozuba, J. & Pila, J., 2015. Chosen aspects of pilots situational awareness. *Naše more* 62(SI): 175–180.

Kraus, J., 2016. Determining acceptable level of safety of approach to landing. In *Proceedings of 20th International Conference Transport Means 2016*, 230–235. Kauno Technologijos Universitetas.

Socha, V., Schlenker, J., Kalavsky, P., Kutilek, P., Socha, L., Szabo, S. & Smrcka, P., 2015. Effect of the change of flight, navigation and motor data visualization on psychophysiological state of pilots. In *2015 IEEE 13th International Symposium on Applied Machine Intelligence and Informatics (SAMI)*, 339–344. IEEE.

Socha, V., Socha, L., Hanakova, L., Lalis, A., Koblen, I., Kusmirek, S., Mrazek, P., Sousek, R. & Schlenker, J., 2016. Basic piloting technique error rate as an indicator of flight simulators usability for pilot training. *International Review of Aerospace Engineering (IREASE)* 9(5): 162–172.

Szabo, S., Liptáková, D. & Vajdova, I., 2015. Robustness as a method of airline pro-active disruption management. *International Review of Aerospace Engineering (IREASE)* 8(4): 151–156.

Vittek, P., Lalis, A., Stojic, S. & Plos, V., 2016. Runway incursion and methods for safety performance measurement. In *Production Management and Engineering Sciences—Scientific Publication of the International Conference on Engineering Science and Production Management, ESPM 2015*, 321–326. CRC Press/Balkema.

Weigmann, D. & Shappell, S., 1997. Human factors analysis of postaccident data: Applying theoretical taxonomies of human error. *The International Journal of Aviation Psychology* 7(1): 67–81.

Wiegmann, D., 2003. *A human error approach to aviation accident analysis: The human factors analysis and classification system*. Burlington: Ashgate.

New Trends in Process Control and Production Management – Štofová & Szaryszová (Eds)
© 2018 Taylor & Francis Group, London, ISBN 978-1-138-05885-9

Incremental innovation and implementation of concept industry 4.0

A. Sorokač & B. Mišota
Institute of Management, Slovak University of Technology in Bratislava, Bratislava, Slovak Republic

ABSTRACT: The starting point for implementation of concept Industry 4.0 is input information, not only with enough quantity but also with required quality. In other words, if we want to take correct decisions, we need an image of monitored manufacturing process or of the complete production. To ensure requirements specified above, it is important to create an infrastructure of information-communication technologies with support of automation means. Creation of this infrastructure can be expensive. One way is to implement innovation in individual functional increments. This article is deal with incremental innovation within the material flow of the production process, namely information flow, which describes the material flow. For successful implementation of investment projects will be described 3I model (innovative interdisciplinary incremental) with methodology of incremental innovation. This means, explain individual methods, which we use for efficient implementation of new technology. We need verified methods from differently specialism of given production.

1 INTRODUCTION

Nowadays, it is necessary to find a solution that innovations of technological process will go hand in hand with eco-innovations (Spirko Spirkova Caganova & Bawa 2016). New information and communication technologies (only ICT) give us new opportunities for increasing the quantity and quality of information for fully or partially automated production. The aim of the innovation of the manufacturing processes (Zajko et al. 2014) in the industry is improvement of manufacturing plant performance (Grell & Hyránek 2012). The significant contribution to control of specific production or technological operations is ability to manage material flow. So, that we can properly manage material flow, we need logistical information system (only LIS), which provides information flow. Currently, they are deployed contactless technologies (such as RFID, 3D cameras etc.) to gather information about movement of elements in production (Clausen et al. 2015).

Production control and logistics production process is provided by control systems based I/O (input/output) systems. Industry 4.0 (Lee et al. 2013) was supported by the software PLC that working with database concept. In other words, there is a direct link between control systems and databases for increasing quality and quantity of collected information.

Rational concept in terms of synergies of ICT, databases system and control system reduces directly own investment for their implementation (Vieira et al. 2013). If we want to achieve the progress of production process, we need to implement new ICT or new control systems. Among others, implementation new ICT or control system can increase safety of production (according to EN 61508).

One of the possibility for permanently increase the efficiency of LIS is incremental innovation (Ritala & Hurmelinna-Laukkanen, 2013). Either, we divide one bigger investment into several smaller individual function increments, or we analyze bottlenecks and determine increments to maintain a steady progress within ICT trends (Mejjaouli & Babiceanu 2014).

For this purpose, has been created 3I model—incremental innovation interdisciplinary model of increasing the efficiency of information flow production process.

2 RESEARCH GOAL

According to consideration describe above, we determined goals:

A. Creation of innovation model for implementation ICT together with automation means into Logistic information system.
B. Creation of innovation model for implementation of incremental innovations.

3 METHODOLOGY OF CREATING INTERDISCIPLINARY INCREMENTAL INNOVATION MODEL—3I MODEL

Starting point is innovation model Stage-Gate® Xpress (Cooper, 2008). Innovation model

Stage-Gate® Xpress has been modified for individual processes in terms of control access methodologies of disciplines in innovation process (Ettlie & Elsenbach 2007). Methodologies are chosen as a cross-section through whole production process or part of this process, which require transfer of material or persons.

Generally, 3I model is open access structure for access of other methods for identification of constrain material flow or methods for creating current model of logistic. Basic rule: We can use diverse methods with criteria creating by us, but we won't change it during the innovation process.

3.1 Superior methods as a control function of innovation process

Control function puts decision step into practice, which it applies strict criteria for continuation of subsequent events. Decision criteria may be changed only at beginning of innovation process, which is defined in initialization phase, and then it will not change and it must be strictly observe. Criteria should be set, so that it can stop the innovation process when the criteria doesn't satisfy. The criteria are creating by closed questions, or by bounded real value based on empiricism of real projects. It is important to set achievable criteria.

Control function in the form of innovation methodologies should support parallel course of operation. It is an important feature of the methodology, because it minimizes time of one increment. The idea of concept is to set short investment, but with given frequency of increment.

3.2 Methods with control access to innovation process

The concept of incremental innovation model, in general, in the first phase make current production model and search bottleneck of production through the selected methodology or empiricism bottleneck of production. The second phase will include change in the new model. Methodology for creating model of production process must be:

- Clear
- Easy
- Modular
- Implementable (not necessary) to enterprise software

3.3 Methods for bottleneck identification

The methods should allow for specificities of shortcomings and specify of time frame established control methodology. The starting point choice of method is acquisition of added value by manufacturing part, which means method of production:

- Piece production
- Series production

It is not excluded use several methods as follows:

- Cascade follow-up. One method narrow down and follow methods describe the problem in detail
- Cascade supplementary. One method can determine the problem alone, but in case of a more detail parameters is used supplementary second methods

4 DISCUSION

According to methodology has been formulated 3I model—Incremental interdisciplinary innovation model (Fig. 1). Model reads from up to down in 4 parts:

I. Initialization
II. Analyses
III. Research and realization
IV. Evaluation of investments

Time framework is secure by control function of innovation model Stage-Gate® Xpress. Access of methodic is from left to right. 3I model creates matrix structure in Cartesian coordinate system, when axis y is time axis and axis × describe processes line.

4.1 Realization innovation via 3I model

3I model have been created for logistic of manufacture process, namely logistic information system (LIS). So, individually step and rules have been dimensioned according to require of implementation ICT. For creation of simulation model, which it describes current performance of logistic, 3I model propose support of simulation software. Condition for testing simulation model with new ICT is verification results with experts of manufacture process of current simulation model. We should assess results from current model not only from perspective of influence on performance of logistic or manufacture process, but also from perspective of influence on costs of manufacture. Simulation model allow complex assessment of analyzed process respectively material flow inside of this process. More complex evaluation of all aspects, then more credible will be results of simulation model with innovation.

3I model is linear, but it is open protentional changes according to type of manufacture and innovation's increments. It is possible to create loop, but is necessary respect gate, which decide of continuous of innovation process. Model enables to modify processes according to actual material flow, so logistic of manufacture process.

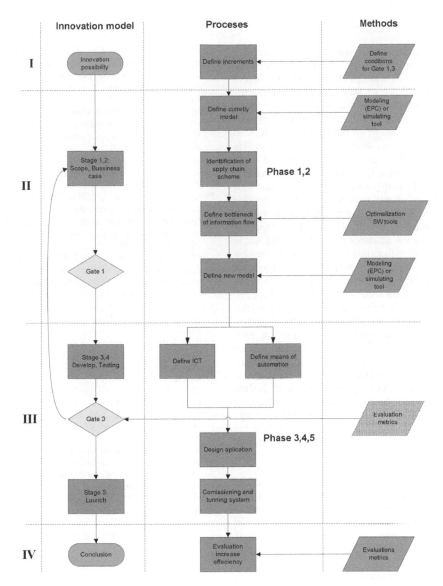

Figure 1.   3I model.

4.2   *Realization Industry 4.0 via 3I model*

Creation of synergic decentral system in the meaning of concept Industry 4.0 is demanding in terms of time and costs. Incremental innovations split the hole system on several individually function blocks via needs of LIS and ICT. We can divide synergic cooperation of decentral systems, so that every increment partial increase effectivity of LIS. On start, we need to know the whole system, milestone, and frequency of implementation of individually blocks.

Modular control system has advantage, control system supports idea of incremental innovation and Industry 4.0. Control system is "heart" of whatever manufacture technology. It collects information, evaluate and control technology process in our case—logistic system. These control systems are usually modular (we can extend control system on input or output) and they can work decentral.

Although, we split the whole system on smaller function blocks, it is important select method in steps from the side of time, frequency of implementation and financial source. Frequency of implementation incremental innovations shouldn't be strictly given, but we can't stop incomplete whole system. We should set financing to realization of every increment for creation complex synergic system. Base is real critically view on milestone and total time.

5 CONCLUSION

Trend, digitalization of production to increase level of automation of production processes is one of the key benefits of the concept of Industry 4.0. There are many solutions to increase production efficiency via digitization, various solutions are different in time and financial demands. Article present way which it is possible to implement new ICT to increase rate of traceability of material flow via innovation model. 3I model used incremental innovation for distribution time and financial demands.

In case of Industry 4.0, decentral systems with interoperability enable split a new LIS on smaller function blocks. 3I model consider these attributes in implementation a new means of automation and ICT. If we are discussing about split the whole system to smaller function block, then we are discussing about decentral systems. If we are speaking about connecting realized incremental innovation, then we are speaking about interoperability. Interoperability allow cooperation between function blocks. Availability enough quantity and quality of collecting information and used database solution with computerized databased provide to user prediction of servicing and minimalization of failure, generally.

Digitalization of manufacture processes in the direction of future factory in the spirit of Industry 4.0 can discourage companies from investing of new technology, because it means not only significant financial costs. Rationality of project for implementation of new technology with support of simulation model enable to reduce input costs, as well divide them to tolerable financial parts.

Rising portfolio of ICT and control systems products provides opportunities for creation new variation of final solution. If we harmonize opportunity of testing variations in simulation model of material flow with created function blocks of complex logistic information system, we can optimize input costs.

Goal of article, presented 3I model is minimalize input costs with maximum increase of effectivity of LIS, which indirectly affects to performance of logistic of manufacture process, nay affects to performance of manufacture process.

REFERENCES

Clausen, U. et al. 2015. Logistics trends, challenges, and needs for further research and innovation. In *Sustainable Logistics and Supply Chains: Innovations and Integral Approaches*. Breda: Springer International Publishing, pp. 1–13.

Cooper, R.G. 2008. Perspective: The stage-gates® idea-to-launch process—Update, what's new, and NexGen systems. In *Journal of Product Innovation Management*, 25(3), pp. 213–232.

Ettlie, J.E. & Elsenbach, J.M. 2007. Modified Stage-Gates-Regimes in New Product Development. In *Journal of Product Innovation Management*, 24(1), pp. 20–33.

Forés, B. & Camisón, C. 2016. Does incremental and radical innovation performance depend on different types of knowledge accumulation capabilities and organizational size? In *Journal of Business Research*, 69(2), pp. 831–848.

Grell, M. & Hyránek, E. 2012. Matrix models for production systems efficiency's measurement/Maticove modely na meranie vykonnosti produkcnych systemov. In *E+ M Ekonomie a Management*, 15(5), pp. 73–88.

Lee, J. et al. 2013. A Cyber-Physical Systems architecture for Industry 4.0-based manufacturing systems. In *Manufacturing Letters*, January, Zväzok 3, pp. 18–23.

Mejjaouli, S. & Babiceanu, R.F. 2014. *Integrated monitoring and control system for production, supply chain and logistics operations*. San Antonio, DEStech Publications Inc., pp. 29–36.

Ritala, P. & Hurmelinna-Laukkanen, P. 2013. Incremental and radical innovation in coopetition-the role of absorptive capacity and appropriability. In *Journal of Product Innovation Management*, January, 30(1), pp. 154–169.

Spirko, M. et al. 2016. Eco-innovation in manufacturing process in automotive industry. In *Lecture Notes of the Institute for Computer Sciences, Social-Informatics and Telecommunications Engineering*, LNICST 166, pp. 533–540,

Vieira, C.L.D.S. et al. 2013. ICT implementation process model for logistics service providers. In *Industrial Management & Data Systems*, 19 April, 113(4), pp. 484–505.

Zajko, M. et. al. 2014. *Inovačné procesy a konkurenčná schopnosť malých a stredných podnikov v SR*. Brno, Knowler.

New Trends in Process Control and Production Management – Štofová & Szaryszová (Eds)
© *2018 Taylor & Francis Group, London, ISBN 978-1-138-05885-9*

Succession and generational change in family businesses

P. Srovnalíková & E. Šúbertová
Faculty of Business Management, University of Economics in Bratislava, Bratislava, Slovak Republic

ABSTRACT: Involving family members into a business is closely related to preparation of young people for succession in family business. The importance of training in the field of succession intensifies its meaning with increasing age of founders. Slovak family businesses are mostly in a period where the first generational change is in progress or is planned for. According to surveys in only one third of Slovak companies in which generational change takes place, the process is successfully mastered. Underestimating the preparation of succession may mean that the successor does not share the mindset and has not built up sufficient interest to maintain and advance the business. It is necessary to deal with aspects of succession and generational change in family businesses. Subject of our investigation and analysis will be primarily economic, organizational, and legal conditions, training and education that allow family businesses to implement succession and thus ensure their effective development.

1 INTRODUCTION

The article deals with the analysis and examination of succession and generational change in family businesses. In addition, attention is paid to the education and training of young generations and raising family business members. Family businesses have a significant presence in the structure of production of gross domestic product, in employment of people, at balancing regional development, in introducing innovations into economic practice and significantly affect other spheres of social life such as family, education, and social relations. Traditionally the definition of family business in developed countries, especially the US, in most cases the principle of ownership is established. This means that the main accent in the family business is placed on the fact that one family owns the total or majority equity of the company, although multi-generationally speaking. Success of family firms abroad is measured not only by the amount of profit made, but also the number of generations that the company passed over. From this aspect it is not the goal of a family enterprise to make instant profit, but also the continuity of existence.

For example, as evident by the statistics, three quarters of European companies are family businesses. Overall, in the world the family businesses generate 60% to 90% of GDP per year (FFI 2016). In economically developed countries, the designation "family company" stands for quality, prestige,

vetted and stable company to which the whole family proudly stands. It goes without saying, that in the family business you own, you spend all the time in it trying to build it for upcoming generations and their children. Especially when you give your name to the company, you take better care of it than the anonymous joint stock company where there are commonly estranged owners. A family business usually takes place in the form of small and medium-sized enterprises, although in the world market we may find not inconsiderably big number of large companies that are also family businesses (Mura et al. 2015). Therefore, those small businesses have a greater potential to handle emergencies and survive in times of economic fluctuations (Belás et al. 2015; Kozubíková et al. 2016). This form of family business has a certain potential in developed countries, where some companies gave the fourth generational change in motion, where the companies in Slovakia in most cases only prepare for their first. Despite or perhaps because of the results are the surveys of PwC and different professional communities interesting and can be instructive for the succession process of education. Habánik et al. (2016) argue that here are various forms of institutionalized support for this form of business worldwide, in the European Union and also Slovakia. Kordoš et al. (2016) state hat within the member states of the European Union, considerable attention is paid to the family business succession, not only at a national level but also at the level of education of the family business.

2 THEORETICAL QUESTIONS EXAMINING THE ISSUE OF GENERATIONAL CHANGE IN FAMILY BUSINESS

Status and trends of succession in a family business, its importance and evaluation of the various member states of the European Union and the world are different. With he problem of family business and defining a succession is dealt on level of several economist and expert entities. Where, for example, the Institute for Research in Bonn defines business family business as "at least two related individuals directly involved in the management of the family business and in possession of at least half the capital" (Hesková & Vojtko 2008). Dutch economists define a family business as one meeting at least two of the following four requirements: 1). One family owns more than 50% of the business; 2). one family has an impact on the decisions and strategies; 3). at least 2 members of the statutory body that leads the company come from the same family; 4). if a company founded less than 10 years ago, then at least one family member is a member of managing entity or owns shares in the company (Overview 2013). For example, in the UK, two thirds of small businesses are in the hands of families. Federation of Small Businesses (FSB) show in their statistics that 6% of them inherited and 9% draw funds from family members. However only 21% of them state, that their future looks positive (De Massis et al. 2013). On the contrary, the family businesses in France develop slowly and rarely can be found on the list of companies with high turnovers. On the other hand, family businesses in Italy have a long tradition. Even the "Analysis of Hyperion Corporate Finance claims that the generational change in 60% of Italian companies is expected in next five years. In today's world, there may be other forms of succession in family businesses. For example, when two or more generational family firm is divided because of different production processes and at the same time it connects them to a particular unit, the so-called family cluster (Haviernikova & Strunz 2014). The demographic crisis caused that many companies are led by people over 70 years, with many at their 50 still waiting to take over the business from the older generation" (Európske 2014).

Intergenerational relationships in a family business are part of the object of scientific interest (Astrachan 2010, Chrisman et al. 2010, Zellweger et al. 2012). Also for reasons as the actual economic and financial security of families, relations and coherence of family members is strengthened. An important aspect of the family business at the same time is more enthusiasm of family members, use of their abilities, skills and transfer valuable knowledge as well as the mutual trust (Arregle et al. 2007,

Chrisman et al. 2010). Generally, family businesses are not among the group of entrepreneurs, who aim to maximize profits. On the contrary, focus on long-term operation of their family companies with a vision for the future that their descendants will take over management of the family business as inheritance (Astrachan 2010). This motivates the owners to build a solid and stable corporate base that will enable it to implement the descendants and build on the already strong foundations. Between family members, there is also greater solidarity and thus have a longer-term stability of family businesses, and therefore sometimes survive for several generations.

While statistics suggest that only 5–15% of the family businesses live up to only third, study of the American Institute of family firms shows that only 30% of companies will survive after the transfer of the second generation, 12% in the third generation and only 3% of companies get to the fourth or next generation (Astrachan & Shanker 2003). Family businesses are characterized by the force of culture and values that are reflected in mutual trust in the field of finance and control. Apart from the economic and financial security of families in family businesses it has an important meaning in consolidating relations and overall coherence of family members. Such a way of doing business is often more effective than business relations outside of family. Each family has in fact a natural desire to survive and if existence depends on the results of its activities then develops generally greater joint efforts for a successful outcome as foreign workers (SFA 2016).

Based on our analysis, the objective is to define and characterize the processes, mechanisms, and methods of generational change in family businesses, forming in terms of the development of family business in the current conditions of economic and social development in Slovakia. Of course, some elements of these mechanisms and modes of generational exchange, forming in the process of generational change first enter into a stable and dominant significance. Some will be sufficient for a generation change, some will be valid in the future. It is important to recognize and identify the mechanisms and conditions of their formation in terms of the characteristics of their own family businesses, as well as the characteristics of the external business environment.

3 SUCCESSION PLANNING AND PREPARATION OF GENERATIONAL CHANGE IN FAMILY BUSINESSES

Business development planning in the process of transfer of property between generations brings a number of challenges in today's complex business environment. Slovak family businesses and their owners are to recognize this situation as evidenced by our survey, which was focused on family business in Slovakia and takes the form of a structured interview in late 2016 and early 2017. We asked the owner and managers of family businesses a question: How many family members are currently working in your family' business? In the survey, we found out that almost 60%, which is more than half of the correspondents, the management and ownership in the company is in hands of the family. Roughly, a third of family businesses confirmed that some of the management is employed outside of the family. In 16%, the corporate ownership shares have also people who are not family members. Survey suggest that the family business in Slovakia takes place primarily within the family, where we can see a graphic representation of the percentages, how many family members are working within the family business.

In family businesses works up to five family members most often—the percentage of owners who said they employ two to five family members,

amounts in aggregate about 90%. Interesting is also the view of the number of generations that are active in the family business. 64% of family businesses employ two generations and 26% one. Three active generations operating in a company in Slovakia are not yet isolated.

As suggested by the data of our survey, only 10% of family firms operate within three generations. Based on this we can assume that they are grandparents, parents and children. It can be concluded that in family businesses in Slovakia there is certain level preparation, training, and education of the young generation of the family business. Up to five family members work in family businesses most often and the percentage of owners who said they employ one (self-employment of the owner) to five family members, amounts in aggregate about 90%. Interesting is also the view of the number of generations that are active in the family business. Approximately 64% of family businesses employ two generations and 26% employ only one. Three generations actively working in a family business in Slovakia are rare and according to our survey, there make about 10%. Owners of family businesses with a vision for the future and the idea of succession to the younger generation are not too optimistic. It may be that their descendants will not be interested in management and will want to go their own way. To the question asked in the survey—what owners of family businesses contemplate for the future, to do with their company; 4% responded that they want to sell the company to management, 1% is considering to enter the stock market, 5% would sell the family business to investors, 15% to another company, 6% would vote for a merger with another company, 24% considered property to be passed on within family and the employment of management and 45% of

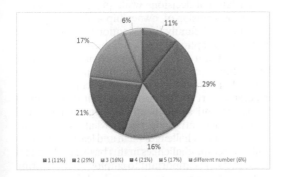

Figure 1. How many members of the family are working currently in the family business.
Source: Own research.

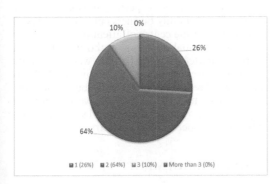

Figure 2. How many generations are currently working in the family business.
Source: Own research.

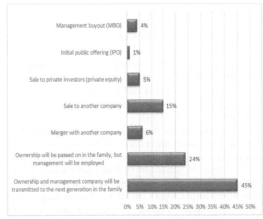

Figure 3. Foreseeable future of current owners.
Source: Own research.

Figure 4.   Transition process stages.
Source: Own processing.

respondents want the ownership and management of family firms to advance to the next generation in the family.

Therefore, the most common procedure in Slovakia, is passing on the family business into the hands of the next generation. First, it means need of owners for systematic and timely engaging family members in running the company, giving them the necessary insight and understanding, guiding them often and across all areas of business and last but not least also gradually entrusting power. The succession divided into a number of phases, which are shown in (see Fig. 4).

Successful intergenerational transfer of business is usually associated with a hard and long-term work on both sides, skills and confidence of individuals involved and other family members, employees and business partners. The owner must have an interest in realizing the changes in the family business. His personal involvement is the key in the whole process, as well as support for actions of the new successor. Secondary factors such as formal agreements in case of sick leave, death, divorce, or dispute in the family, which are important to have set up so that in such cases the company is not in danger.

4 EDUCATING AND TRAINING THE DESCENDANTS OF THE OWNERS OF FAMILY BUSINESSES

The situation of family companies in the world, such as America and Germany is different from those in Slovakia. In Western European countries, some firms are resolving the fourth generational exchange. Since entrepreneurship in Slovakia as it is has just little over 20 year history, the family business in Slovakia did undergo single generational change at most or the departure of founder only awaits (PwC 2014).

Education in the family business is organized by several European universities through economic or management study programs. Some offer specialized courses of economic or business studies. Training programs are most often oriented to the future managers and owners of family businesses, focusing on problems of family firms as a business versus family, legal and financial aspects, growth, innovation and internationalization, conflict resolution, and transfer of business in particular. In Europe, other approaches took place as well—for example by organizing visits to family businesses (so called "In-house visits"), mainly older members of the family and the rising generation to exchange experiences, approaches, practical solutions. Inspirational approach has been chosen in England at the London Business School. Their leadership research initiative in family businesses suggests that family businesses have unique abilities to implement joint activities (PwC 2014). Explained by the family system and its functioning, structure, personalities and culture. This unit is unique and unrepeatable. Slovakia, unfortunately, has not similarly developed educational infrastructure and institutional elements. Except for the Faculty of Commerce at the University of Economics in Bratislava, which addresses the issue of family business within the scope of a family business program.

5 CONCLUSION

Family businesses have to cope with a number of organizational decisions while planning and managing the transfer of ownership and passing various stages of family life. Mentioned problems are not only exceptional for family businesses and are more related to the predictable changes that occur in the family, as well as corporate life and adolescence. Family businesses are an essential and irreplaceable part of the economy, their owners must recognize and subsequently manage, nurture and educate their offspring to do what their unique opportunities, challenges them and learn to successfully overcome challenges in this business. Based on these facts, it is important to employ their offspring in family businesses and so they lead the way in the management and overall cycles of the family business. Processes of education, training, and succession planning in family businesses in Slovakia are subjects of awareness, expertise, and experience of their very own family business entrepreneurs. And it should be at the center of attention for both the state and regional and local authorities. It is not a private matter of those families who run family businesses even when we talk about family business. The development of family businesses, the effectiveness, and efficiency has a very important impact on the overall economic and social development and prosperity of the country.

ACKNOWLEDGEMENT

Article was created within the project VEGA no. 1/0918/16 50% and project VEGA no. 1/0709/15 50%.

REFERENCES

Arrègle, J.L. et al. 2007. The development of organizational social capital: Attributes of family firms. *Journal of Management Studies*, 44: 73–95.

Astrachan, J.H. 2010. Strategy in family business: Toward a multidimensional research agenda. *Journal of Family Business Strategy*, 1: 6–14.

Astrachan, J.H. & Shanker, M.C. 2003. Family Businesses' Contribution to the U.S. Economy: A Closer Look, *Family Business Review*, 9: 27–41.

Chrisman, J.J. et al. 2010. Intellectual foundations of current research in family business: An identification and review of 25 influential articles. *Family Business Review*, 23: 9–26.

Belás, J. et al. 2015. Approach of the SME Entrepreneurs to Financial Risk Management in Relation to Gender and Level of Education, *Economics and Sociology*, 8(4): 32–42.

De Massis, A. et al. 2013. Research on Technological Innovation in Family Firms: Present Debates and Future Directions, *Family Business Review*, 22: 10–31.

Európske partnerstvo pre verejné stratégie a Slovak Business Agency. *Rodinné podnikanie alebo (ne)zostane to v rodine*. 2014. Výskumná správa. Family Business. Bratislava: EPPP. www.eppp.sk/docs/eppp_studia_fb.pdf

Európske rodinné podniky to majú ťažké. http://www.euractiv.sk/podnikanie-v-eu/clanok/europske-rodinne-podniky-to-maju-tazke-013695

Family firm institute. 2016. Global Data Points. http://www.ffi.org/?page = globaldatapoints

Habánik, J. 2016. Competitiveness of Slovak economy and regional development policies, *Journal of International Studies*, 9(1): 144–155.

Haviernikova, K. & Strunz, H. 2014. The Comparison of Selected Methods Used for Identification of Cluster Potential in the Regions of the Slovak Republic. In *International Multidisciplinary Scientific Conferences on Social Sciences and Arts (SGEM 2014), Proceedings, Vol. IV.* Sofia: STEF92 Technology.

Hesková, M. & Vojtko, V. 2008. *Rodinné firmy zdroj regionálneho rozvoja*. Zeleneč: Profess Consulting. https://www.pwc.com/sk/sk/sukromnivlastnici/assets/prieskum_medzi_podnikatelmi_a_rodinnymi_firmami_2014.pdf

Kordos, M. et al. 2016. Cluster policies implementation in Slovakia, *Actual Problems of Economics*, 181(7): 90–96.

Kozubíková, L. et al. 2016. The Role of Entrepreneur´s Gender, Age and Firm´s Age in Autonomy. The Case Study from the Czech Republic. *Economics and Sociology*, 9(2): 168–182.

Mura, L. et al. 2015. Quantitative Financial Analysis of Small and Medium Food Enterprises in a Developing Country. *Transformation in Business & Economics*, 14(1): 212–224.

Overview of Family Business Relevant Issues. http://ec.europa.eu/enterprise/policies/sme/files/craft/family_business/doc/familybusiness_study_en.pdf

Poradenská spoločnosť PwC 2014. Prieskum medzi slovenskými podnikateľmi a rodinnými firmami. *Výskumná správa.* https://www.pwc.com/sk/sk/sukromnivlastnici/assets/prieskum_medzi_podnikatelmi_a_rodinnymi_firmami_2014.pdf

Slovenská Franchisingová Asociácia. *Rodinné podnikanie potrebuje aj na Slovensku novú modernú definíciu*. 2016. Bratislava: SFA. https://www.sfa.sk/sk/novinky/detail/rodinne-podnikanie-potrebuje-aj-na-slovensku-novu-modernu-definiciu

Zellweger, T. et al. 2012. From longevity of firms to transgenerational entrepreneurship of families: Introducing family entrepreneurial orientation. *Family Business Review*, 25: 136–155.

New Trends in Process Control and Production Management – Štofová & Szaryszová (Eds)
© 2018 Taylor & Francis Group, London, ISBN 978-1-138-05885-9

The improvement of the bank infrastructure of Ukraine

V. Stoika
Faculty of Economics, Uzhhorod National University, Uzhhorod, Ukraine

P. Maciaszczyk
Faculty of Social and Humanities Sciences, State Higher Vocational School Memorial of Prof. Stanislaw Tarnowski in Tarnobrzeg, Tarnobrzeg, Poland

I. Britchenko
Faculty of Technical and Economic Sciences, State Higher Vocational School Memorial of Prof. Stanislaw Tarnowski in Tarnobrzeg, Tarnobrzeg, Poland

ABSTRACT: The article is devoted to the prospects of creation and functioning of cooperative banks and development banks in Ukraine. The article deals with the peculiarities of cooperative banks and development banks. Scientists analyzed the views on the activities of banks and their impact on economic development. The article deals with the peculiarities of development and operation of the cooperative banks of developed countries: Germany and Poland. A three-tier model of national cooperative bank system has been suggested. The influences of developments banks operations on the national economy are defined. The main grounded differences between development banks and commercial banks are defined. The problems of creating development banks in Ukraine are analyzed.

1 INTRODUCTION

The article considers the ways for improving the current Ukrainian bank infrastructure through the creation of cooperative banks and development banks. International experience confirms that such banks are stable and remain financially stable in times of crisis. In addition, the direction of their activity and functions are of great importance for the further development of Ukraine. Cooperative banks have a significant impact on the stimulation of the development of the agrarian sector, rural areas and farmers, they also receive a high level of trust from the population. Development banks also have a positive impact on social and economic development of the country. During the period of Ukraine's independence, several financial institutions, referred to as development banks, have been created but the principles of their activity did not differ from commercial banking. The authors have considered the main problems in the process of creating both cooperative banks and development banks in Ukraine. The need for developing an appropriate legislation in order to regulate the activities proposed for the creation of banking institutions has been concluded by the authors.

Over the last decade, the Ukrainian economy along with the banking sector have been affected by the most profound crisis since gaining independence. This was due to armed conflicts in the eastern part of the country and internal economic imbalances. Last year, the banking system of Ukraine was under the influence of negative transformations, whose result was a decline in the number of banks and equities, outflow of deposits, high level of bad debts, and decreased financial stability, among others. This situation requires the development and implementation of effective measures aimed at eliminating such negative effects not only for the financial sector but also for the economy as a whole.

Building a competitive advantage on the EU market in cooperation with Ukraine is possible mainly through the development of the agricultural sector, as well as through the development of science and technology in the area of innovation. This strategy requires modernization of the banking sector as the main supplier of financial and credit resources. In modern economic conditions, a possible solution to this problem lies in the creation of cooperative banks and the so-called development banks in Ukraine.

Cooperative banks occupy a significant place in the banking system in developed countries, especially in Western Europe. During the financial crisis, their activities were of great importance because these institutions contributed to maintaining the financial stability of the sector and its economic effectiveness as well as strengthened their position on the market.

The first cooperative banks in Western Europe were established in Germany. Throughout almost 150 years of development history, they have evolved into a powerful financial system, that plays an important role in the banking sector in Germany. Nowadays, local cooperative banks are independent financial institutions whose activities are governed by cooperative and banking law. At the beginning of their operation, their main members were craftsmen and farmers. These days, most of the more than 13 million members belong to the middle class society. Approximately 75% of all enterprises, 80% of farmers and 60% of craftsmen in the country are members of cooperative banks (http://minfin.kmu.gov.ua/control/uk/publish/article.htm).

The basis for creating the cooperative credit sector in Germany was the idea of self-help through mutual help, which originated more than 125 years ago and was based on the principles of self-management and co-liability. At present, credit cooperatives offer a wide range of financial services due to the fact that their activities are focused on accepting savings deposits and providing short-term and medium-term loans to their members. In this case, it is worth noting that the volume of long-term loans is constantly increasing. The business activity of Volksbank and Reiffeisenbank is to carry out traditional banking services and other operations such as: product and service operations, acquiring savings deposits, involvement of other types of deposits, providing various types of loans, membership guarantee, payment transactions and international operations, including the purchase and sale of cash and non-cash means of payment, counseling, mediation, assistance in managing the family budget, protection and management of securities and other assets, mediation or sale of construction contracts, insurance, etc., collective purchase of essential goods, aggregate sales of agricultural products, and delivery of goods to members. In addition, cooperative banks in Germany lead active pro-social charitable activities. On average, every bank spends around 5% of its profit on charitable purposes, as well as support for culture and sport.

It should be noted that due to the increasing competition in the banking market, the number of regional cooperative banks in Germany (and other countries) is gradually decreasing through combining them in order to increase competitiveness. There were about 60 regional banks in Germany after the Second World War, 8 in the 80s and 3 in the 90s (SGZ-Bank in Frankfurt, WGZ-Bank in Düsseldorf and GZB-Bank in Stuttgart) and now – 1 (WGZ-Bank) (Boychuk 2009). This trend does not mean refining the role of regional cooperative banks. On the contrary, thanks to them, the German cooperative system has acquired significant stability and financial strength. Regional cooperative banks play a key role in creating a cooperative credit system. In Germany (as opposed to other countries—and this is also a feature of the German system), they have fulfilled their historic mission—strengthening bottom-up capitalization. Credit cooperatives have been transformed into large independent financial institutions, while cooperative financial structures at regional level have ceased to be important.

Cooperative banks in Poland are important for stimulating rural and agricultural development. They were founded in 1861, and their main customers were farmers and members of rural communities. The activity of cooperative banks was limited geographically (within the territorial community where the bank was located) and in terms of the scope of services (opening and running of accounts, deposits, loans and guarantees, acceptance of securities, and execution of term deposits in subsidiaries). By the end of 2010, there were 646 banks in Poland, of which 576 were cooperative banks (Oliynyk 2012). In modern conditions, cooperative banks in Poland operate as universal banks but place a strong emphasis on maintaining the agricultural sector and rural areas. In years 1997–2007, the share of agricultural loans was 74% in the structure of investment loans and 59% in the structure of current loans (Siudek 2010).

The development of cooperative banks in Poland has led to the provision of preferential loans for agriculture by these institutions, and additionally, through these institutions, payments are made to farmers from the European Union resources and the state budget. Some researchers noted that privileged loans have played an important role in creating a stable, well-functioning system of cooperative banks that provide financial services to rural entrepreneurs. The granting of preferential loans to agricultural enterprises in Poland in 1990 was carried out on a large scale. From the beginning of 1995, the volume of preferential loans was about $1 billion a year, and the share of credit debt of rural households accounted for around 80% (Oliynyk 2012).

The experience of EU countries confirms that the presence of cooperative banks and their service activities have become an effective tool for updating and developing socio-economic development. It is undeniable that these financial institutions are able to function effectively not only in the territory of the given country but also to establish a strong network of institutions in other countries and, in addition, they may be one of the largest banks in the world. Cooperative banks benefit from a high level of public trust and have a high share in the financial services market. These institutions operate as universal ones but what is characteristic is that they are oriented toward supporting agriculture and rural areas.

Cooperative banks have several advantages in comparison with commercial ones and are characterized by the following features:

1. Adequate level of capitalization and financial stability.
2. Specificity of a business model that consists in creating ownership of the membership contributions in the bank, thus they are less sensitive to changes in the economy.
3. Emergence of counter-cyclical behavior and national deposit guarantee schemes.
4. Effectiveness of the control mechanism (in line with the "bottom up" principle).
5. Democratic governance, which is based on the active participation of their members in making management decisions.

During the independence of Ukraine, cooperative banks were not formed. In this regard, W. Goncharenko (2012) emphasizes that such a situation affects the national economy and that the banking system is very sensitive to internal and external financial crises. The reason for this was the lack of understanding of the place and role of cooperative banks in the national economy by the National Bank of Ukraine and politicians, which is reflected in the imperfection of the legal regulation of the formation and operation of cooperative banks.

In Ukraine, the procedure for the establishment of cooperative banks is regulated by the Law "On Banks and Banking Activity" (The Law of Ukraine). Article 8 of this law provides for the procedure for the formation of cooperative banks. In particular, it is assumed that cooperative banks are created on a territorial basis and will also be divided into local and central cooperative banks. Minimum number of members of a local cooperative bank is at least 50 people. The members of the Central Cooperative Bank are local cooperative banks.

It should be noted that the Bill on Amendments to several Legislative Acts of Ukraine regarding Operation of the Credit Cooperation System was drafted in Ukraine in 2008 (The Law of Ukraine). It envisages, among others, the scope, powers and functioning of cooperative banks, but these amendments have not been accepted so far.

An important issue, which also requires legal regulation, is the organizational structure of cooperative banks in Ukraine. Some researchers (Marchuk 2016, Goncharenko 2012, Stoyko 2014) believe that this model should be based on a "bottom up" principle and consist of three levels, each of which has corresponding functions. At the first, local level of the system of credit cooperatives should be local cooperative banks, which will provide services to their members and will have a savings and credit nature. They will participate in proposing appropriate changes in legislation. They will be sovereign in creating financial policies at the bank level and will be able to provide other financial and non-financial services to small businesses, farms and small farmers. The second, regional level of the cooperative system should be created by regional cooperative banks, aimed at providing effective support to the functioning of local cooperative banks. The main task of these organizations should be to support first level institutions in the provision of additional services (in the area of liquidity, development of new financial products, internal payments, currency, payment, investment and other operation collaterals). The third, national level consists of a central cooperative bank, which complements the structure of a full system of cooperative banks. This institution cooperates and is composed of the second level organizations and aims to provide them with additional financial services (improving performance in the context of international banking operations in the area of equities, international payments, foreign exchange and exchange markets, leasing, etc.). Thus, a central cooperative bank will combine all cooperative banks into one system that will give them a competitive advantage in relation to large system banks without losing their independence and autonomy.

Contemporary conditions for the development of the banking sector in Ukraine and the observed qualitative changes indicate the need to establish cooperative banks. The formation of a cooperative banking system is linked to the provision of financial and credit support to small businesses, the agricultural sector, the financing of investment projects aimed at stimulating the development of the regional economy as well as financial and credit services in rural areas. Periods of economic crisis require greater state influence in the investment process, by supporting the involvement of credit institutions. In the structure of the economic system of almost every developed country in the modern world, a special financial institution, for instance, development bank, can be found. Often these banks are part of a large network of branches in the country and have a special tool for implementing state economic policies in different sectors or directions of development. Such a bank performs additional functions to ensure long-term financing of social projects whose implementation should contribute to economic growth, economic modernization and more. W. Diamond (1957) draws attention to the fact that in many countries development banks have been created as a tool for financing specific projects. They played a key role in the postwar reconstruction of Western European economies, and also helped in the "big leap forward", which led to the development of newly industrialized countries. The author states that the adopted bank's development objectives may differ significantly from country to country in terms of ownership, activity, policies and the institution's activities. W. Diamond generalizes that government-backed financial institutions primarily aim at providing long-term loans in the context of stimulating the economy.

At the end of the 20th century, there were about 750 development banks, 32 of them regional and about 700 national, including special export development banks and small business support banks in 185 countries. In summary, one country has three Development Banks on average: 95.8% of them are national development banks and 4.2% are international and regional development banks. In developing countries there are on average 3 or more banks, whereas in developed countries there are much fewer of them (Kovalev 2016). The trend is that development banks are mostly created in emerging economies because they are designed to "revive" the economy of a country, which is supposed to contribute to its long-term economic growth.

In emerging countries, unconditional governmental economic development occurred through the creation of mechanisms to mobilize investment capital, in particular in the Republic of Korea, where the creation of the banking system resulted from the need for economic growth. In order to finance economic development projects in the industrial and agricultural sectors, the Korean Development Bank was established as a public credit institution. Subsequently, the entire system of specialized banks was created to provide financial support for both the strategic and industrial sectors (Korean Bank of long-term loans, Export-Import Bank of Korea, commercial and banking corporations that provide medium- and long-term loans to non-financial corporations and activate the inflow of foreign capital).

The main differences between development banks and commercial banks are as follows (Emelyanov 2009):

– Firstly, development banks use a special mechanism for mobilizing credit resources, they tend not to perform deposit operations, because the source of their funding is primarily the state budget, and partly government loans received from international financial organizations. This is the main difference between development banks and commercial banks and other financial institutions,
– Secondly, development banks in their business activity are guided by the state's interests in the socio-economic development of the country, and receiving profits is not their goal. One of the most significant development banks' operations, with the exception of the refinancing of other banks, is the provision of guarantees from financial institutions on the repayment of the granted loans. Of course, development banks implement commercial principles, striving for positive performance, but in the case of losses, their coverage is at the expense of public funds,
– Thirdly, development banks do not create competition for commercial banks, because they have their own spheres of business—investing and lending to core industries that have a long return on investment and high risk,
– Fourthly, operations of development banks, as a trusted agent, are carried out by commercial banks.

During the period of Ukrainian independence, there were several financial institutions, known as development banks. One of them is the Ukrainian Bank for Reconstruction and Development (UBRD). It was created to develop innovative infrastructure, as well as strengthen the position of financial support for innovative activities, in accordance with the Regulation of the Cabinet of Ministers of Ukraine № 655 of 05.05.2003 (The Law of Ukraine). The mission of the bank is to create favorable conditions for the development of the Ukrainian economy by supporting domestic producers and the development of investment activity (The Mission, Strategy...). The bank should specialize in medium- and long-term financing of legal persons with a focus on reconstruction and production development loans. However, within 12 years of its existence, UBRD has failed to become a significant financial institution. The reason for this was the lack of a distinct specialization of the bank, which functioned as a counterpart to a traditional commercial bank. The UBRD's activity, as opposed to other countries, is governed by general banking laws, so the conditions for loans are not different from those of other commercial banks. The scale of activity and the main UBRD indicators show that it is one of the small banks in Ukraine apart from the head office in Kiev.

UBRD has no other branches and, at the end of 2015, it had assets of PLN 120 million. (0.0069% to GDP). Actual lack of macroeconomic efficiency and ineffective management led UBRD to take a decision in 2016 on the privatization of the bank.

Another private Ukrainian Development Bank is also now in liquidation. This decision was taken in December 2015. This financial institution was initially actually an analogue to a commercial bank and provided a wide range of banking services for both natural and legal persons. From 2014, the existence of the bank was largely linked to the political process and its owner was a representative of the ruling elite (iPress.ua). During its activities, the bank failed to become a financial institution for development as its activities were based on the general principles of commercial banking. Another attempt to create a Ukrainian Development Bank was made in 2013, when the Verkhovna Rada of Ukraine was introduced to the bill on the establishment of a new Ukrainian Development Bank (The Law of Ukraine). It was supposed to be created on the basis of UBRD with the transfer of all its assets under control. The purpose of establishing this bank was planned credit support for structural changes in the economy, the development of the banking system and long-term domestic and foreign investment in priority sectors for the Ukrainian economy. The bank was also to promote the development of public-private partnership and serve as an agent of the Ukrainian government. The bill received the support of the Cabinet of Ministers of Ukraine and the National Bank of Ukraine, but it was subjected to a series of critical remarks of the chief scientific and expert leadership of the Verkhovna Rada of Ukraine, and as a result was sent back to the revision, which has never been realized. As a result, on November 27, 2014, the bill was withdrawn from consideration.

One of the comments concerns the concept of the need to adopt such a law. In Ukraine, all the relationships resulting from the registration, servicing, reorganization, and liquidation of banks are already regulated by the Law "On Banks and Banking Activity". It should be noted that, because of the lack of a special law determining all aspects of the development bank, as a special Institute in the structure of the Ukrainian economy, the activity of these banks previously failed and was ineffective. World practice proves that the effectiveness and scope of banking activities can be achieved by adopting a special law that exists in many countries (Canada, India, Korea, and Brazil, among others). Ukrainian scientists Kindzersky (2015), Aborci & Mitusin (2016) also point out the need to adopt a special law that will define all the characteristics of the future Ukrainian Development Bank. Another problem that needs to be addressed is shaping of the development bank's liabilities. Among the researchers (forumkyiv.org), there is a conviction that the capital of development bank created in Ukraine should be at the level of 3–4 billion dollars; moreover, it is necessary to raise funds in the same amount from external sources. Banks in their activities attract various sources of funding, in particular state budget funding, loans from other financial institutions, attracting funds on local and international financial markets, and attracting funds from international and regional development banks. In particular, 89% of development banks attract international funds, financial institutions and issues securities (Kovalev 2016).

The area in which the development bank's activity may be concentrated should be "... based on the neo-industrial paradigm of the need to issue high added value products and the creation of innovative infrastructure. On this basis, it is possible to draw up an indicative list of strategic industries for the Ukrainian economy, where the most needed resources of the bank are" (Matyushin & Abborchi 2016, p. 41). The promising areas for the development of Ukraine are: machinery, transport, energy, electronics, environment, nanotechnology, agriculture, and also the military-industrial complex.

Therefore, the formation of a development bank in Ukraine can not only improve the infrastructure of the banking system, but also solve important economic development problems and long-term financing of priority sectors, public investment, infrastructure projects, and small and medium enterprises. To do this, first of all, it is necessary to meticulously draw up the relevant laws regulating the activity of such a bank and carry out an in-depth study of international experience in the functioning of development banks as well as the main problems of the Ukrainian economy such as bureaucracy and corruption. The solution to these problems requires the involvement of representatives of the government, the National Bank of Ukraine, experienced bankers-practitioners, and leading academic economists in the study of this problem.

2 CONCLUSIONS

On the basis of the study of experience of developed European countries concerning the formation of cooperative banking sector the main lines of development of cooperative banks in Ukraine have been suggested.

The expediency of creation of cooperative banks in Ukraine has been substantiated. The attention is focused on the problems of legislative and organizational nature regarding the operation of such financial institutions in Ukraine. A three-tier model of national cooperative bank system has been suggested. It should be based on a "bottom-up" principle and consist of three levels, each of which must perform the relevant function. The first level of credit cooperative system should be represented by local cooperative banks. The second level should be formed by regional cooperative banks. At the third, national level should be the central cooperative bank, which logically completes the construction of a full-fledged cooperative system. The result of the activity of these institutions should be a rational financing of small businesses and a rapid response to their needs, as well as the reduction of loan costs to farmers and the promotion of agriculture in general.

In the structure of the economic system of almost every developed country in the modern world, a special financial institution, for instance, development bank, can be found. Often these banks are part of a large network of branches in the country and have a special tool for implementing state economic policies in different sectors or directions of development. The main differences between development banks and commercial banks are as follows: 1) development banks use a special mechanism for mobilizing credit resources; 2) development banks in their business activity are guided by the state's interests in the socio-economic development of the country, and receiving profits is not their goal; 3) development banks do not create competition for commercial banks; 4) operations of development banks, as a trusted agent, are carried out by commercial banks. The formation of a development bank in Ukraine can not only improve the infrastructure of the banking system, but also solve important economic development problems and long-term financing of priority sectors, public investment, infrastructure projects, and small and medium enterprises.

The experience of developed countries clearly suggests that the solution to the strategic tasks of the national economy, by increasing the investment activity of credit institutions, requires the effective participation of the state. The effectiveness of close cooperation between the government and banks, as well as non-financial corporations, is essential for the realization of common national interests, which is particularly important at the stage of exiting the economic crisis and also in the process of European integration of Ukraine.

REFERENCES

Boychuk, V. 2009. Overview of the system of cooperative banks in Germany. *Bulletin of Credit Co-operation* (6): 19.

Diamond, W. 1957. *Development banks*, Baltimore: Johns Hopkins Press

Emelyanov Yu. 2009. Development Banks and the State in the Contemporary World Development Architecture. *Economics and Politics* (5): 13.

forumkyiv.org/uk/press-center/press-releases/chipotriben-ukrayini-bank-rozvitku

Goncharenko, V. 2012. Conceptual approaches to solving the problem of lending to the agrarian sector of the national economy. Poltava: PUETT.

Goncharenko, V. 2012. The Importance of the Revival of Cooperative Banks in Ukraine in Crisis Deplores in the World Economy, International Co-operative Movement: Genesis and Trends in Contemporary Development. Poltava: PUETT.

http://minfin.kmu.gov.ua/control/uk/publish/article.htm; retrieved: 10.02.2017

ipress.ua/ru/news/nbu_reshyl_lykvydyrovat_bank_aleksandra_yanukovycha_149209.html

Kindzersky, Yu. 2015. *State policy of structural and technological modernization.* Kyiv: Institute for forecasting economics of the National Academy of Sciences of Ukraine.

Kovalev, M. 2016. *Development Banks: A New Role in the 21st Century.* Minsk: Publishing Center of BSU.

Marchuk, V. 2016, Cooperative banks: prospects of creation in Ukraine. Moscow: Economics.

Matyushin, A. & Abborchi, A. 2016. National Development Banks as a Tool for Providing Non-Industrial Growth. *Industry Economics* (1): 37.

Oliynyk, O. O. (2012). Functioning of cooperative banks: experience of Poland. *Actual problems of economic development of the region.* 8 (1), 49–54.

Siudek, T. 2010. Polish cooperative banking sector in the face of systemic transformation and European integration in agriculture. *Agricultural Econonomics—Czech*, nr. 56 (3): 132.

Stoyko O. Ya. 2014. Necessity and Prospects for the Revival of Cooperative Banks in Ukraine in: O. Ya. Stoiko, Productivity of agro-industrial production. Economic sciences. Vip. 25. 31–38.

The Law of Ukraine, http://zakon2.rada.gov.ua/laws/show/2121–14; retrieved: 10.02.2017

The mission, strategy and objectives of the Bank, http://ubrr.com.ua/node/63; retrieved: 10.02.2017

New Trends in Process Control and Production Management – Štofová & Szaryszová (Eds)
© 2018 Taylor & Francis Group, London, ISBN 978-1-138-05885-9

Application of graduates in the Slovak labour market

M. Stričík & M. Bačová

Faculty of Business Economics with seat in Košice, University of Economics in Bratislava, Košice, Slovak Republic

ABSTRACT: It is important to monitor the application of different population within the analysis of labour market. Our contribution is focused on current state of graduates' application in labour market and the suggestion of possible changes in education regards to their application. The major factor of functional each modern and developed society and also economy is the share of high educated population. The number of high educated population has increased in last twenty years by 100%. The graduates are not registered in unemployed register very long and nearly 97% of all unemployed graduates are employed during half a year since their registration at the Local Labour Office, Social Affairs and Family. In summary the most important is the language preparation, ICT skills and also connection with practice during study (training and professional experiences) according to the practical requirements.

1 INTRODUCTION

The use of human resources including the graduates influences the economic growth of every country. The negative of high unemployment of graduates is their motivation to work abroad. The level of education of labour, level of business environment and economic growth has important effect on the employment of the graduates (Buchtová et al. 2013). Managing human resources in companies is related to the application of graduates in the labour market (Gallo & Mihalčová 2016). The important factor of the functioning of each modern and developed society and economy is the share of the population with university degree, higher education. The number of higher educated people has increased for last twenty years more than 100%. The whole level is very closely connected with work productivity, capacity of innovation and technological maturity and it also identifies the level of knowledge development. It is necessary to say that the number of university students in Slovakia are still low in comparison with the developed countries of EU and their employment is influenced the overall economic situation of every country. On the other side the university students hold position with lower qualification requirement in the countries with the high share of graduates. University degree starts to be not enough for professional opportunity in labour market. This happens when the structure of core curriculum and the field of study do not react on the requirements of labour market and economy flexibly. EU creates the framework and suggests the pro-

cedures which support the employment growth of young and calls on the Members States (Grandtnerová 2014). One of the five targets for the EU in 2020 is to achieve at least 40% of 30–34 – year-olds completing third level education (Europe 2020).Slovakia is fully committed to this objective. This share of the population was 28.4% in Slovakia in 2015, which was deeply under the EU average. The share of population with completed higher education is on average 38.7% in EU (2017 European Semester: Country Reports European Commission, 2017). The similar issue has other countries outside Europe such as China where this share was more than 24% (Ren et al. 2011). These authors concluded that the graduates have increasing difficulties when they look for job.

2 MATERIAL AND METHODS

The object of our investigation is the market labour in Slovakia from the education structure point of view. We focused mainly on the market labour of graduates in years from 2006 to 2016. The main aim of the contribution is to assess the present state of application graduates in labour market and suggest the possible changes in their education according to the application in labour market.

This study analyses the issue of development of the graduates' number and their application in labour market on the basis of available statistical data (from school statistical data and statistical data about employment and unemployment). We use the median, arithmetic average and standard deviation for statistical analysis in our work.

3 RESULTS AND DISCUSSION

Study at university at the present is very popular and attractive. The number of graduates is still problem, because not every graduate is able to find a job immediately after leaving school (Potužáková & Mildeová 2015). The absence of work—experiences and habits is the consequence of missing practical experiences. The employers are unwilling to prepare young qualified workers and prefer "finished workers". Deterioration of the status of graduates in labor market is the consequence of economic crisis. Graduates without working habits have the competitive disadvantage when they look for a job. The negative of high graduates unemployment is going to work abroad. The economic growth, level of education and level of business environment have important effect on graduates' employment. It is necessary to know the conditions which are offered to students coming to schools. Insufficient preparedness for the profession is the reason of failure after entry into employment. Young people with qualification have not secure employment automatically. The reason is high unemployment rate, low economic ability to support new job opportunities and school system which is inadequately adaptable for labour market requests. The Local Labour Offices, Social Affairs and Familyare very helpful according to graduates. Their main role is to find and offer job for them. Negative experiences with employers regards to the work experiences and communication are the most frequent reasons for disappointment. Graduates without work experiences and practice are handicapped. Although they studied hard, there is not demand for them in labour market. Economic and psychological problems, dependent on parents and higher risk of anti—social behavior are the consequences of this fact.Success of graduate integration and their professional adaptation into the job depends on the level of preparation on this profession. Although young people are qualified the integration in labour market is not guaranteed automatically.

Figure 1 shows the share of unemployed graduates to whole number of unemployed who are registered with the Central Office of Labour, Social Affairs and Family.

We can conclude that although the unemployment in Slovakia is falling, unemployment of graduates is growing. The percentage of graduate's unemployment is low. It was the lowest in 2006 (0.94%) and grew in 2010 (1.71%) and was till until 2013 (Fig. 1). The increase of unemployment was observed since 2014. The highest percentage was in 2015 (2.89%), it is more than 0.42% than in 2014 and 1.95% than in 2006. The unemployment fell in 2016; it was 2.47% for graduates unemployment.

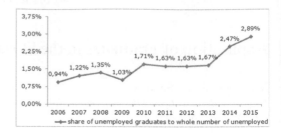

Figure 1. Share of unemployed graduates to whole number registered unemployed people (in%). Source: Own data processing. www.upsvar.sk.

The number of graduates (Bachelor and Master degree), who got diploma from 2006 to 2010 grew. There were 43,872 graduates in internal form of study and 26,509 graduates in external form in 2010. The difference is 34,455 graduates, that is 96%. There is a decrease in number of graduates since 2010, average 6% per year. The number of graduates who left the school in 2015 was 55,239 students and 53,851 in 2016. The difference between the year 2010 and 2015 is 15,142 and 16,530 graduates between 2010 and 2016. The number of graduates in external form (2008–2013) was higher than the average (21,647), five and six times higher value than median – 23,945. Therefore is median more appropriate position characteristic for statistical file evaluation than the average. Median is more appropriate indicator of prevailing tendency than the arithmetic mean because in most years the value was higher than the average value. The number of unemployed graduates increased from 2006 (2,715 unemployed graduates) to 2015 (9,611). The difference in the number of unemployed graduates is 6,896 (354%).The highest increase of unemployed graduates we can observe in late 2009 and early 2010. The number rose by 2,630 and the number of registered unemployed graduates rose by 2,568 than in 2013. Economic growth in Slovakia caused the decrease of unemployment. The consequence of this fact was that graduates unemployment decreased to 6,814, annual decline was 2,797 graduates (Table 2).

Focusing on regions we can see, that the number of registered unemployed graduates in the highest in Presov region and Kosice region and the lowest in Bratislava region. There was the sharpest rise of the number of unemployed graduates in Bratislava region from 2006 to 2015, where the number of unemployed rose by 460% over 2006. In spite of this fact the number of unemployed graduates was still in absolute figure the lowest in comparison with other regions. The growth of the number of unemployed was double and triple from 2005 to 2015 in every region. We can observe the decline

Table 1. Graduates (Bachelor and Master degree).

Year	Graduates (Bachelor and Master degree)		
	full-time form	part-time form	Together
2006	21,105	14,821	35,926
2007	24,433	17,345	41,778
2008	35,402	24,432	59,834
2009	42,508	27,301	69,809
2010	43,872	26,509	70,381
2011	42,653	26,642	69,295
2012	42,493	24,201	66,694
2013	40,699	23,689	64,388
2014	39,953	20,255	60,208
2015	38,271	16,968	55,239
2016	37,896	15,955	53,851
Median	40,326	23,945	62,298
Average	37,208	21,647	58,855
Standard deviation	8,004.52	4,530.11	11,947.2

Source: Own data processing. www.uips.sk

Table 2. Unemployed graduates (Bachelor and Master degree) according to regions.

Year	SR	BA	TT	TN	NR	ZA	BB	PO	KE
2006	2,715	154	242	264	422	315	331	473	514
2007	2,921	163	262	293	428	391	349	521	464
2008	3,346	179	300	326	404	459	413	667	562
2009	3,897	285	311	448	501	541	422	762	627
2010	6,527	488	576	746	821	880	673	1,311	1,032
2011	6,515	488	600	729	789	891	665	1,265	1,088
2012	6,931	542	623	685	865	935	682	1,387	1,210
2013	6,647	544	604	693	850	850	683	1,293	1,054
2014	9,215	837	796	984	1,115	1,162	934	1,858	1,398
2015	9,611	861	820	1,029	1,153	1,238	977	2,005	1,531
2016	6,814	576	534	723	773	922	694	1,557	1,035

Source: own data processing www.upsvar.sk

of the number of unemployed graduates in every region since 2016.

We can conclude that the number of unemployed graduates (Bachelor and Master degree) in Kosice region rose by 836 (200%) registered unemployed from 2006 to 2015. The highest growth of unemployed graduates was observed in 2008 compared with the year 2009 when the number of unemployed graduates rose by 98%. The significant growth was not recorded from 2009, even there was a decrease recorded in Kosice region in 2012 (by 12%), 2014 (by 0.13%) and 2015 (by 15%). The highest number of registered unemployed graduates in 2016 in

Table 3. Unemployed graduates (Bachelor and Master degree) - Kosice region.

Year	Kosice region	District							
		GL	KE I-IV	KE okolie	MI	RV	SO	SN	TV
2006	474	23	153	62	61	49	18	49	59
2007	511	16	177	72	71	28	18	53	76
2008	495	16	174	57	61	27	13	67	70
2009	978	34	300	133	115	91	24	132	149
2010	1,060	39	329	143	141	105	27	125	151
2011	1,247	41	385	195	169	98	39	143	177
2012	1,095	38	338	157	142	98	30	134	159
2013	1,510	46	474	171	204	103	55	208	158
2014	1,508	57	547	187	182	115	48	203	178
2015	1,310	33	399	152	225	91	61	212	137
2016	1,035	26	334	140	148	63	37	167	120

Source: Own data processing. www.upsvar.sk

Kosice district was in districts Kosice I to IV, then Michalovce, Spisska Nova Ves and Kosice area.

The best situation was in districts Roznava, Sobrance and Gelnica. The sharpest increase of the number of unemployed graduates (Bachelor and Master degree) we can observe after crisis. The number of registered unemployed rose by 100% in every district in 2009 compared with 2008.

We consider that the following suggestions are important for possible changes in higher education and graduates application in labour market:

– Increase collaboration between labour market and universities;
– Focus on the relationship between school system and graduates competences; and
– Ensure synergy between education and practice regarding to long-term sustainable development of graduates.

The collaboration between organizations and universities is very important according to graduates' application in labour market. Special orientation programs for new employees who are graduates could be helpful, with the emphasis on preparing simulation games (Olexová & Gajdoš, 2016) applicable not only in organizations but also in universities. The result of this collaboration is higher employment of graduates and the higher of their income.

4 CONCLUSION

The percentage of unemployed graduates to whole number of registered unemployed is low. There were more students who studied on a full-time form than part-time form in last ten years.

If present trend are not quickly changed, today's level of unemployed graduates will endanger their future prospects regards to their occupation. This can have serious consequences on the growth and social cohesion. The most important is the collaboration between labour market and universities. Universities should prepare graduates who will be able to apply in working environment after leaving university. Therefore they should adapt core curriculum to labour market conditions.

REFERENCES

2017 European Semester. 2017. *Country Reports | European Commission*. [online]. Retrieved 20th March 2017 from: <https://ec.europa.eu/info/publications/2017-european-semester-country-reports_en>.

Buchtová, B. et al. 2013. *Nezaměstnanost*, Praha: Grada Publishing, a.s.

Európa 2020. Retrieved 18th March 2017 from: <http://ec.europa.eu/europe2020/europe-2020-in-a-nutshell/targets/index_sk.htm>.

Gallo, P. & Mihalčová, B. 2016. Models of evaluation of managing people in companies. In: *Quality – Access to Success.* 155(17): 116–119.

Grandtnerová, L. 2014. *Podporazamestnanosti pre mladých ľudí* [online]. Retrieved 18th March 2017 from: <www.ceit./IVPR/NSZ/nsz_10.pdf>.

Olexová, C. & Gajdoš, J. 2016. Logistics simulation game proposal—A tool for employees' induction. In:*Quality Innovation Prosperity.* 20(2): 53–68.

Potužáková, Z. & Mildeová, S. 2015. Analysis of causes and consequences of the youth unemployment in the European Union. In: *Politická ekonomie* 63(7): 877–894.

Profesia SK. Retrieved 18th March 2017 from: <http://www.profesia.sk/>.

Ren et al. 2011. Human resources, higher education reform and employment opportunities for university graduates in the People's Republic of China. In: *International Journal of Human Resource Management.* Oct. 2011, Vol. 22 Issue 16, pp. 3429–3446. 18 p. DOI: 10.1080/09585192.2011.586871, Databáza: Business Source Complete. Retrieved 19th March 2017 from:<http://web.a.ebscohost.com/ehost/pdfviewer/pdfviewer?vid=1&sid=7616b3e2-76fe-4d5a-8ba0-45ac8e0321dd%40 sessionmgr4008&hid=4112>.

Štatistický úrad Slovenskej republiky Retrieved 20th March 2017 from: www.statistics.sk.

Tkáč, M. 2001. *Štatistické riadenie kvality.* Bratislava: Ekonóm.

Úrad práce, sociálnych vecí a rodiny Retrieved 18th March 2017 from: <www.upsvar.sk>.

Ústav informácií a prognóz školstva. Retrieved 18th March 2017 from: <www.uips.sk>.

Vysoké školy v SR. Retrieved 20th March 2017 from: <https://www.minedu.sk/vysoke-skoly-v-sr/>.

Zákon č. 131/2002 Z. z. Zákon o vysokých školách.

New Trends in Process Control and Production Management – Štofová & Szaryszová (Eds)
© *2018 Taylor & Francis Group, London, ISBN 978-1-138-05885-9*

Road design by taking into account the analysis of stress-strain state of bare boards

T. Suleimenov, T. Sultanov, G. Tlepiyeva & Y. Sovet
L.N. Gumilyjov Eurasian National University, Astana, Kazakhstan

ABSTRACT: Stress-strain condition of three nearby lying non isolated plates. The calculation problem of construction, working commonly with a resilient base, has an extremely big practical importance, since during the projection of many constructions, decision comes down to these problems. Construction on an elastic foundation in the form of the concrete bases are widely used in the construction of residential, industrial, administrative, agricultural, cultural and welfare and other objects of the national economy. These structures may have different shapes and sizes, operate under the influence of combinations of different types of loads and difficult operating conditions and have different rigidity characteristics that significantly affect to the course of calculation.

1 INTRODUCTION

Calculations are considered as the most important stage in the design of base plates. The important thing here is to choose a suitable method of calculation. In calculations there are used calculation schemes, which possibly more accurately reflect properties of real fundamental constructions and layered base. The concept of the calculation scheme in the theory of elasticity includes the idealization of materials, components, assemblies of structures, the structure and composition of the base, as well as loads acting on the buildings. Selection of the calculations scheme should provide high levels of efficiency and effectiveness. The level of mechanics' development, computer science and mathematics removes many traditional limitations on the choice of calculations scheme and allows the calculation of structures with the maximum approximation in terms of their actual work. At the same time simple calculation schemes allow to analyze the work of some constructions and commonly consider the work of structure and loads acting on it (Pasternak 1954, Aytaliyev et al. 2000).

2 FINITE ELEMENT MODELING OF MECHANICAL STATE OF LAMINATED BASES

2.1 *The basic equations of equilibrium and motion of the laminate base*

Numerous theoretical and experimental data (Barnshtein 1984) suggest that the grounds are sloping, layered and anisotropic. The angle varies for different areas in the range of 0001 / to 150.

Therefore, in the calculation and design of structures and foundations of buildings should take into account the anisotropic elastic foundation. The equation of state of such arrays is described according to (Aytaliyev 1996), i.e.

$$
\begin{cases}
\varepsilon_x = a_{11}\sigma_x + a_{12}\sigma_y + a_{13}\sigma_z \\
\varepsilon_y = a_{21}\sigma_x + a_{22}\sigma_y + a_{23}\sigma_z \\
\varepsilon_z = a_{31}\sigma_x + a_{32}\sigma_y + a_{33}\sigma_z \\
\quad \gamma_{yz} = a_{44}\tau_{yz} \\
\quad \gamma_{xz} = a_{55}\tau_{xz} \\
\quad \gamma_{xy} = a_{66}\tau_{xy}
\end{cases}
\tag{1}
$$

Expressing the voltage through the strain, these formulas can be rewritten as

$$
\begin{cases}
\sigma_x = A_{11}\varepsilon_x + A_{12}\varepsilon_y + A_{13}\varepsilon_z \\
\sigma_y = A_{21}\varepsilon_x + A_{22}\varepsilon_y + A_{23}\varepsilon_z \\
\sigma_z = A_{31}\varepsilon_x + A_{32}\varepsilon_y + A_{33}\varepsilon_z \\
\quad \tau_{yz} = A_{44}\gamma_{yz} \\
\quad \tau_{xz} = A_{55}\gamma_{xz} \\
\quad \tau_{xy} = A_{66}\gamma_{xy}
\end{cases}
\tag{2}
$$

entering technical constants: E1, E2 – Young's modules in the direction of plane isotropy and normal there to, v_1, v_2 Poisson's ratio characterizing cross compression of isotropic plane on tensile and normal there to isotropic plane, G1, G2 – modules of shift for isotropic plane and any perpendicular to it, can be rewritten in the form (Aytaliyev 1994, Aytaliyev 1996).

$$\begin{cases}
\sigma_x = \dfrac{1}{R}\Big[\big({}_1-{}_2\nu_2^2\big)\varepsilon_+ +\big({}_1\nu_1+{}_2\nu_2^2\big)\varepsilon, \\
\qquad +\big({}_2\nu_1\nu_2-{}_2\nu_2\big)\varepsilon_k\Big] \\[6pt]
\sigma_y = \dfrac{1}{A}\Big[\big(E_1\nu_1+E_2\nu_2^2\big)\varepsilon_x +\big(E_1+E_2\nu_2^2\big)\varepsilon_y \\
\qquad +\big(E_2\nu_2-E_2\nu_1\nu_2\big)\varepsilon_z\Big] \\[6pt]
\sigma_z = \dfrac{1}{A}\Big[\big(E_2\nu_1\nu_2-E_2\nu_2\big)\varepsilon_x +\big(E_2\nu_2-E_2\nu_1\nu_2\big)\varepsilon_y \\
\qquad +\big(E_2-E_2\nu_1^2\big)\varepsilon_z\Big] \\[6pt]
\tau_{yz} = \dfrac{1}{G_1}\gamma_{yz} \\[6pt]
\tau_{xz} = \dfrac{1}{G_1}\gamma_{xz} \\[6pt]
\tau_{xy} = \dfrac{1}{G_2}\gamma_{xy}
\end{cases} \qquad (3)$$

where: $A = \dfrac{E_1}{E_1-2\nu_2^2 E_2-\nu_1^2 E_1-2\nu_1\nu_2^2 E_2}$

If the laminate transport array lying at an angle to the horizontal is considered and the coordinate axis is drawn then the above formula is a description of transport array excluding the inclination angle of this array on horizon. To take into account the isotropic plane's inclination angle coordinates to the normal view according to the Figure 1 compose a table of the cosines directs

a) general form

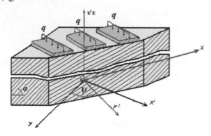

b) estimated cross-sectional plane

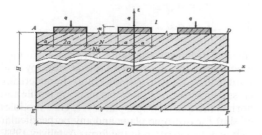

Figure 1. Elastic anisotropic system «base plane».

(Table 1) of transition from the eldest coordinates system to the new by rotating axis XOZ to φ angle.

On the basis of Table 1 composes table of transformation of coefficients (Table 2) according to the (Aytaliyev et al. 2000) and by using the formula of transformation $a_{ij} = \sum_{m=1}^{6}\sum_{n=1}^{6} a_{mn}q_{im}q_{jn}$, and find the constants in new coordinates:

$$\begin{cases}
\sigma_x = A_{11}'\varepsilon_x + A_{12}'\varepsilon_y + A_{13}'\varepsilon_z + A_{15}'\gamma_{xz} \\
\sigma_y = A_{21}'\varepsilon_x + A_{22}'\varepsilon_y + A_{23}'\varepsilon_z + A_{25}'\gamma_{xz} \\
\sigma_z = A_{31}'\varepsilon_x + A_{32}'\varepsilon_y + A_{33}'\varepsilon_z + A_{35}'\gamma_{xz} \\
\tau_{yz} = \qquad\quad A_{44}'\gamma_{yz} + A_{46}'\gamma_{xy} \\
\tau_{xz} = A_{51}'\varepsilon_x + A_{52}'\varepsilon_y + A_{53}'\varepsilon_z + A_{55}'\gamma_{xz} \\
\tau_{xy} = \qquad\quad A_{64}'\gamma_{yz} + A_{66}'\gamma_{xy}
\end{cases} \qquad (4)$$

After putting values of inclination and transformation angles we get:

$A_{11}' = A_{11}\cos^4\varphi + 2(A_{13}+2A_{55})\sin^2\varphi\cos^2\varphi + A_{33}\sin^4\varphi$

$A_{12}' = A_{12}\cos^2\varphi + A_{32}\sin^2\varphi$

$A_{13}' = A_{13}\cos^4\varphi + (A_{11}+A_{33}-4A_{55})\sin^2\varphi\cos^2\varphi + A_{31}\sin^4\varphi$

$A_{15}' = (A_{11}-A_{13}-2A_{55})\sin\varphi\cos^3\varphi + (A_{13}-A_{33}+2A_{55})\sin^3\varphi\cos\varphi$

$A_{21}' = A_{12}$

Table 1. Table of cosines' directives on rotation of Y axle.

Eldest	X	Y	Z
New			
X_1	$\cos\varphi$	0	$-\sin\varphi$
Y_1	0	1	0
Z_1	$\sin\varphi$	0	$\cos\varphi$

Table 2. Transformation table of coefficients in rotation of Y axis.

i,j	1	2	3	4	5	6
1	$\cos^2\varphi$	0	$\sin^2\varphi$	0	$-2\sin\varphi\cos\varphi$	0
2	0	1	0	0	0	0
3	$\sin^2\varphi$	0	$\cos^2\varphi$	0	$\sin\varphi\cos\varphi$	0
4	0	0	0	0	0	$\sin\varphi$
5	$\sin\varphi\cos\varphi$	0	$-\sin\varphi\cos\varphi$	0	$\cos^2\varphi-\sin^2\varphi$	0
6	0	0	0	$-\sin\varphi$	0	$\cos\varphi$

$$A_{22}' = A_{22}$$

$$A_{23}' = A_{21}\sin^2\varphi + A_{23}\cos^2\varphi$$

$$A_{25}' = (A_{11} - A_{13})\sin\varphi\cos\varphi$$

$$A_{31}' = A_{13}$$

$$A_{32}' = A_{23}$$

$$A_{33}' = A_{33}\cos^4\varphi + 2(A_{13} + 2A_{55})\sin^2\varphi\cos^2\varphi + A_{11}\sin^4\varphi$$

$$A_{35}' = (A_{11} - A_{13} - A_{55})\sin^3\varphi\cos\varphi$$

$$A_{44}' = A_{44}\cos^2\varphi + A_{66}\sin^2\varphi$$

$$A_{46}' = (A_{66} - A_{44})\sin\varphi\cos\varphi$$

$$A_{55}' = A_{55}\cos^4\varphi + (A_{11} - 2A_{13} + 2A_{55} - A_{33})\cos^2\varphi\sin^2\varphi + A_{55}\sin^4\varphi$$

$$\acute{R}_{66} = \acute{R}_{44}\sin^2\varphi + \acute{R}_{66}\cos^2\varphi$$

Other parts of the equation of state are zero.

The perpendicular axis lies at an angle φ, longitudinal axis, which divides operating loads, is located at angle ψ. For the direction of this axis strictly to the center between the pillar we make second rotation of axis coordinates relative to z' axis.

Similar to the fist situation we make a Table 3 of cosines' directors and also Table 4 of symbols of coefficients converting.

Finally we get

$$\begin{cases}
\sigma_x' = d_{11}\varepsilon_x + d_{12}\varepsilon_y + d_{13}\varepsilon_z + d_{14}\gamma_{yz} + d_{15}\gamma_{xz} + d_{16}\gamma_{xy} \\
\sigma_{yx}' = d_{21}\varepsilon_x + d_{22}\varepsilon_y + d_{23}\varepsilon_z + d_{24}\gamma_{yz} + d_{25}\gamma_{xz} + d_{26}\gamma_{xy} \\
\sigma_z' = d_{31}\varepsilon_x + d_{32}\varepsilon_y + d_{33}\varepsilon_z + d_{34}\gamma_{yz} + d_{35}\gamma_{xz} + d_{36}\gamma_{xy} \\
\tau_{yz}' = d_{41}\varepsilon_x + d_{42}\varepsilon_y + d_{43}\varepsilon_z + d_{44}\gamma_{yz} + d_{45}\gamma_{xz} + d_{46}\gamma_{xy} \\
\tau_{xz}' = d_{51}\varepsilon_x + d_{52}\varepsilon_y + d_{53}\varepsilon_z + d_{54}\gamma_{yz} + d_{55}\gamma_{xz} + d_{56}\gamma_{xy} \\
\tau_{xy}' = d_{61}\varepsilon_x + d_{62}\varepsilon_y + d_{63}\varepsilon_z + d_{64}\gamma_{yz} + d_{65}\gamma_{xz} + d_{66}\gamma_{xy}
\end{cases}$$

$$(5)$$

where coefficients

Table 3. Table of the cosines directors on Z axis rotation.

Eldest	X_1	Y_1	Z_1
New			
X_2	cosψ	−sinψ	0
Y_2	sinψ	cosψ	0
Z_2	0	0	1

Table 4. Table of coefficients' converting on rotation of Z axis.

i,j	1	2	3	4	5	6
1	cos²φ	0	sin²φ	0	−2sinφcosφ	0
2	0	1	0	0	0	0
3	sin²φ	0	cos²φ	0	2sinφcosφ	0
4	0	0	0	0	0	sinφ
5	sinφcosφ	0	−sinφcosφ	0	cos²φ−sin²φ	0
6	0	0	0	−sinφ	0	cosφ

$$d_{11} = A_{11}'\cos^4\psi + 2(2A_{66}' - A_{12}')\sin^2\psi\cos^2\psi + A_{22}'\sin^4\psi$$

$$d_{12} = 2A_{12}'(\cos^4\psi + \sin^4\psi) + (A_{11}' + A_{22}')\sin^2\psi\cos^2\psi$$

$$d_{13} = A_{13}'\cos^2\psi + A_{23}'\sin^2\psi$$

$$d_{14} = A_{15}'\cos^2\psi + A_{25}'\sin^2\psi$$

$$d_{15} = A_{15}'\cos^3\psi + A_{25}'\sin^2\psi\cos\psi - 2A_{64}'\sin\psi\cos^3\psi$$

$$d_{16} = (A_{11}' - A_{12}' - 2A_{66}')\cos^2\psi\sin\psi$$

$$d_{22} = A_{22}'\cos^2\psi + 2(A_{12}' + 4A_{66}')\sin^2\psi\cos^2\psi$$

$$d_{23} = A_{23}'\sin^2\psi + A_{23}'\cos^2\psi$$

$$d_{24} = A_{15}'\sin^3\psi + (A_{13}' + 2A_{64}')\sin\psi\cos^2\psi$$

$$d_{25} = (A_{15}' - 2A_{64}')\sin^2\psi\cos\psi + A_{25}'\cos^2\psi$$

$$d_{26} = (A_{11}' + A_{12}')\sin^3\psi\cos\psi + (A_{21}' - A_{22}' - 2A_{66}')\cos^3\psi\sin\psi$$

$$d_{33} = A_{33}'$$

$$d_{34} = A_{35}'\sin\psi$$

$$d_{35} = A_{35}'\cos\psi$$

$$d_{36} = (A_{31}' - A_{32}')\cos\psi\sin\psi$$

$$d_{44} = A_{44}'\cos^2\psi + A_{55}'\sin\psi\cos\psi$$

$$d_{45} = (A_{55}' - A_{44}')\cos\psi\sin\psi$$

$$d_{46} = A_{46}'\cos^3\psi + (A_{51}' - A_{64}')\sin^2\psi\cos\psi$$

$$d_{55} = A_{55}'\cos^2\psi - A_{44}'\sin^2\psi$$

$$d_{56} = A_{46}'\sin^3\psi + (A_{51}' - A_{52}' - A_{46}')\sin\psi\cos^2\psi$$

$$d_{66} = A_{66}'\cos^4\psi + (A_{11}' - 2A_{12}' + A_{22}')\sin^2\psi\cos^2\psi$$

Equation derives us the formula of intensely deformed substance of anisotropic obliquely layered array. However analytical solution of this task is harder. Besides of this fact we can solve this task by the final elements way (FEW).

2.2 Finite element partition of layered base

It is needed to mentally divide considered area into a bunch of elements in the form of cuboids, which contact in node. For a given finite element approximation functions we introduce the condition that the movement is distributed linearly and does not depend on each other (Aytaliyev et al. 1997).

$$\begin{cases}
u(x,y,z) = \alpha_1 + \alpha_2 x_i + \alpha_3 y_i + \alpha_4 z_i \\
+ \alpha_5 x_i y_i + \alpha_6 x_i z_i + \alpha_7 y_i z_i + \alpha_8 x_i y_i z_i
\end{cases}$$

$$(6)$$

Here α_1, α_2, ...α_{24}- constants, which are subject to determination through displacement and components of nodes displacement.

When calculating the massive bodies in the FEM dependences for three-dimensional-stress state are used. Construction of element's stiffness matrix (MSE) is one of the main stages of the calculation. Composition of MSE can be made as following:

– Displacement field over the area of finite element (FE) is assigned as a function of the displacement units;
– Deformation over TBE area is expressed in terms of its displacement (the ratio of Coshi);
– Strain over CE area is expressed in terms of its deformation (generalized Hooke's law);
– Based on the principle of possible displacements defined stiffness characteristics.

The matrix formulation is given by the expression (6)

$$[K] = \int_V \{\varepsilon\}^T [\sigma] dV = \int_V \{u\}^T [D]\{\varepsilon\} dV \qquad (7)$$

where $\{u\}$ – displacement matrix,
$[D]$ – matrix of elasticity, which expresses dependence of strain from deformation.

Displacement law of any point considered KE from its nodal displacements is given

$$u(x,y,z) = \frac{1}{abc} \begin{bmatrix} (a-x)(b-y)(c-z)u_i + \\ +x(b-y)(c-z)u_j + \\ +y(a-x)(c-z)u_k + \\ +xy(c-z)u_l + \\ +z(a-x)(b-y)u_m + \\ +xz(b-y)u_n + \\ +yz(a-x)u_p + xyzu_q \end{bmatrix} \qquad (8)$$

Using the Cauchy relation, we obtain expressions for the strain of TBE. In the form of matrix can be written as $\{\varepsilon\} = [B]\{u\}$

$$\begin{cases} \varepsilon_x = \dfrac{\partial U(x,y,z)}{\partial x} \\ \varepsilon_y = \dfrac{\partial V(x,y,z)}{\partial y} \\ \varepsilon_y = \dfrac{\partial W(x,y,z)}{\partial z} \\ \gamma_{xz} = \dfrac{\partial U(x,y,z)}{\partial z} + \dfrac{\partial W(x,y,z)}{\partial x} \\ \gamma_{yz} = \dfrac{\partial V(x,y,z)}{\partial z} + \dfrac{\partial W(x,y,z)}{\partial y} \\ \gamma_{xy} = \dfrac{\partial U(x,y,z)}{\partial y} + \dfrac{\partial V(x,y,z)}{\partial x} \end{cases} \qquad (9)$$

Applying the generalized Hooke's law, we obtain for strain the field of CE $\{\sigma\} = [D][B]\{u\}$. To determine the stiffness characteristics apply expression to the three-dimensional $k_{ij} = \int_V \varepsilon_i \sigma_j dV$ finite element takes the form.

$$k_{i,j} = \int_0^a \int_0^b \int_0^c (\sigma_{xj}\varepsilon_{xi} + \sigma_{yj}\varepsilon_{yi} + \sigma_{zj}\varepsilon_{zi} + \tau_{xyj}\gamma_{xzi} + \tau_{yzj}\gamma_{yzi}$$
$$+ \tau_{xyj}\gamma_{xyi}) dxdydz \qquad (10)$$

By using this expression we get a value of subintegral for different combinations of polynomials. By solution obtained symmetrical MSE sized 24×24, and besides matrix blocks are symmetrical to each other. Form MSE in $\{f\}^e = [k]\{u\}^e$ type. For this we pick up elementary parallelepiped from array (Fig. 2).

All values of angles are expressed through i, then for

$$j = i - NP,$$
$$k = i + NN,$$
$$l = i + NN - NP, \qquad (11)$$
$$m = i + 1,$$
$$n = i - np - 1,$$
$$p = i + NN + 1,$$
$$q = i + NN - NP + 1.$$

Based on the element's stiffness matrix, we construct matrix of system's stiffness. In general substance matrix of system's stiffness has size $3 \, N \times 3 \, N$. As is known all final elements are connected to each other accordingly each node comes as vertex "stars", consisted from 8 elements. To reduce the dimensions of the equation's system we use redirection of elements' stiffness matrix parts to matrix of system's stiffness. For example, I has a form

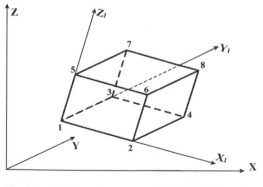

Figure 2.   Final element in form of parallelepiped.

500

Table 5. Table of redirection.

i	i	k	l	m	n	p	q
14	15	17	18	5	6	8	9
13	14	16	16	4	5	7	8
11	12	14	15	2	3	5	6
10	11	13	14	1	2	4	5
23	24	26	27	14	15	17	18
22	23	25	26	13	14	16	17
20	21	23	24	11	12	14	15
19	20	22	23	10	11	13	14

$$
\begin{aligned}
F_{X_i} = & K_{i,1}U_{i-NP-NN-1} + K_{i,2}U_{i-NN-1} + K_{i,3}U_{i+NP-NN+1} \\
& + K_{i,4}U_{i-NP+1} + K_{i,5}U_{+1} + K_{i,6}U_{+-NP+1} \\
& + K_{i,7}U_{i-NP+NN-11} + K_{i,8}U_{i-NN+1} + K_{i,9}U_{i+NP+NN+1} \\
& + K_{i,10}U_{i-NP-NN} + K_{i,11}U_{i-NN} + K_{i,12}U_{i-NP+NN} \\
& + K_{i,13}U_{i-NP1} + K_{i,14}U_i + K_{i,15}U_{i-NP} \\
& + K_{i,16}U_{i-NP+NN} + K_{i,17}U_{i+NN} + K_{i,18}U_{i+NP+NN} \\
& + K_{i,19}U_{i-NP-NN-1} \\
& + K_{i,20}U_{i-NN-1} + ... + K_{i,81}W_{++NP+NN+1}
\end{aligned}
$$

Finally we get a Table 5 of redirection.

System of linear algebraic equations has $\{F\} = [K]\{U\}$ form, where $\{F\}$ – vector – column of acting load; $[K]$ quadratic matrix of system's stiffness; $\{U\}$ – vector column of unknown displacements. Search of solution is began with random values $x_i^{(0)}$, $i = 1, 2,..., n$ (null iteration) and form consistent approximations $x_i^{(j+1)} = x_i^{(j)} + v_i^{(j)} x_i^{(0)}$, $i = 1, 2,..., n$; $j = 0,1,2,...$, which converge to the exact solution $\bar{x}$ at $j \to \infty$. Calculation ratios for widespread iteration method of Zeidel has form

$$
x_i^{(j+1)} = x_i^{(j)} - \frac{1}{a_{ij}}\left(\sum_{k=1}^{i-1} a_{ik}x_k^{(j+1)} + \sum_{k=1}^{n} a_{ik}x_k^{(j)} - f_i\right),
$$
$$
i = 1,2,..., n; j = 0,1,2,... \tag{12}
$$

In programming Zeidel's method it's necessary to wonder about the condition of iteration process $\overset{max}{\underset{i}{}}|x_i^{(j+1)} - x_i^{(j)}| \leq \varepsilon$, where ε – adjusted small value. Moreover, it's necessary to consider that the method of Zeidel gives convergence to exact solution only when main diagonal dominance condition is executed: $\sum_{\substack{j=1 \\ j \neq i}}^{n}|a_{ij}| \prec |a_{ii}|$, $i = 1, 2,..., n$. By solving system of linear algebraic equation we find displacement of arrays' points.

2.3 Algorithm for calculating the stress-strain state of the base

The preceding paragraphs set out the general concepts, methods and formulas required to calculate the stress-strain state of the stationary constructions fundamental slabs' base exposed to the concentrated, distributed load. In the next calculations following assumptions are accepted:

Base is submitted in the form of finite element system in parallelepiped type.

Base is exposed to only concentrated loads from the fundamental slabs since they are fixed firmly.

The scheme of base calculation is accepted:

1. Input design data;
2. Construction of elastic matrix;
3. Construction of element's stiffness matrix;
4. Construction of system's stiffness matrix;
5. Input values of vertical loads;
6. Determination of displacements in nodal points;
7. Determination of strain;
8. Determination of stress.

The followings are concrete explanation to the scheme of base's stress strain state calculation under the influence from fundamental slabs.

1. Input design data involves the introduction of elastic modules and shear modules for each layer of the array, Poisson's ratio for each layer of the array, the coordinates in the global coordinate system, limitations for all directions, numbers of elements' eight nodal points.
2. The elastic matrix is constructed on the basis of the above recommendations on the assumption that the array of layers has an inclination angle in relation to the horizon.
3. Based on the above algorithm, a program for calculating the base to the action of concentrated loads.

3 FINITE ELEMENT MODELING OF THE MECHANICAL STATE OF THREE NEARLYING NON-ISOLATED PLATES UNDER THE INFLUENCE OF THE WEIGHT OF THE BOILER AGGREGATE ON THE LAMINATE BASE

3.1 Finite element partitioning of three plates on the layered basis

Modern spatial structures in working process are exposed to the various force impact thus their construct elements are in significantly strain stressed state and separate nodes pass in to the stage of plastic deformation which significantly influence to exactness of calculations. Determination of kinematic and power factors in node elements of structure depending on the physics – mechanical and geometrical characteristics of materials and also depending on value and type of the external forces and moments with an aim

501

of appointment optimal parameters of constructions is the problematic issue of wrought solid body mechanics. In theoretical solution of real model tasks by construction and calculating the spatial structures of complex structure that are interconnected under the influence of variable forces difficulties in mathematical sphere are arose. Known variation and analytical method to deal with specific three dimensional tasks have limited usage: they are successfully used in investigation of construction elements state of structure. The advent of PC greatly expanded the class of tasks on the calculation and projection of three dimensional topology structures. It comes up need for the common and universal method of calculation on spatially inhomogeneous systems strength. For this the most effective was widely spread FEM.

The main advantage of FEM in comparison with the other is as following:

– representation of the spatial structure in the foreseeable geometric form;
– possibility of taking into account the complex physical and mechanical properties of structural elements and their forms;
– relatively easy to take into account the limited conditions for both power and for displacements;
– possibility of consideration multi-element constructions with different dimensions;
– solution of nonlinear FEM problems;
– easiness of algorithm and program formation in algorithmic high-level language.

In general, usage of finite element method to the study of stress-strain state of spatial structures is impossible without the involvement of modern computers that allow you to automate the entire process of calculation using matrix operations. With regard to the static and dynamic problems of FEM it makes it possible to consider the structural features of the structural elements, heterogeneous structure of the material, features of interaction of individual elements and allows the simultaneous determination of internal forces under the action of external force fields.

To solve and calculate such problems, it's necessary to determine the elastic base core that allows us to solve many problems in determining the stress-strain state of structures and fundaments contacting with an elastic base.

As an elastic base is considered an elastic transport pan-layered base.

We solve the problem of stress-strain state transport array which is located under the influence of a distributed load on the y-axis (Figure 1).

Write the equations for the generalized Hooke's law for transport array in Cartesian coordinates (x /, y /, z /) which's axis z / is perpendicular to the

plane of isotropy and let the stress state of the body is characterized by generalized plane deformation, i.e., all components of stresses and displacements are independent of y, $\left(\frac{\partial \sigma}{\partial y} = 0\right)$.

$$
\begin{cases}
\varepsilon'_x = a'_{11}\sigma'_x + a'_{12}a'_y + a'_{13}\sigma'_z, \\
\varepsilon'_y = a'_{12}\sigma'_x + a'_{11}\sigma'_y + a'_{13}\sigma'_z, \\
\varepsilon'_z = a'_{12}\sigma'_x + a'_{13}\sigma'_y + a'_{33}\sigma'_z, \\
\gamma'_{yz} = a'_{44}\tau'_{yz}, \\
\gamma'_{xz} = a'_{44}\tau'_{xz}, \\
\gamma'_{xy} = 2(a'_{11} - a'_{12}).\tau'_{xy},
\end{cases}
\tag{13}
$$

Draw coordinates (x, y, z) by the rotation of system method (x', y', z') for φ angle around axis y'. Then (13) equation is rewritten as:

$$
\begin{cases}
\varepsilon_x = a_{11}\sigma_x + a_{12}\sigma_y + a_{13}\sigma_z + a_{15}\tau_{xz} \\
\varepsilon_y = a_{12}\sigma_x + a_{22}\sigma_y + a_{23}\sigma_z + a_{25}\tau_{xz} \\
\varepsilon_z = a_{13}\sigma_x + a_{23}\sigma_y + a_{33}\sigma_z + a_{35}\tau_{xz} \\
\gamma_{yz} = a_{44}\tau_{yz} + a_{46}\tau_{xy} \\
\gamma_{xz} = a_{15}\tau_x + a_{25}\sigma_y + a_{35}\sigma_z + a_{55}\tau_{xz} \\
\gamma_{xy} = a_{46}\tau_{yz} + a_{66}\tau_{xy}
\end{cases}
\tag{14}
$$

Coefficient a_{ij} of (14) equation is expressed in coefficient a'_{ij} equation (13) and angle of rotation φ.

The equations of equilibrium of an elastic body in the absence of body forces in considered situation has a form:

$$
\begin{cases}
\dfrac{\partial \sigma_x}{\partial x} + \dfrac{\partial \tau_{xz}}{\partial z} = 0, \\
\dfrac{\partial \tau_{xy}}{\partial x} + \dfrac{\partial \tau_{yz}}{\partial z} = 0, \\
\dfrac{\partial \tau_{xz}}{\partial x} + \dfrac{\partial \sigma_z}{\partial z} = 0.
\end{cases}
\tag{15}
$$

If body has small deformations connected with components of displacement points of body ratio:

$$
\begin{cases}
\varepsilon_x = \dfrac{\partial u}{\partial x}, & \gamma_{yz} = \dfrac{\partial v}{\partial z} + \dfrac{\partial w}{\partial y} \\
\varepsilon_y = \dfrac{\partial v}{\partial y}, & \gamma_{xz} = \dfrac{\partial w}{\partial x} + \dfrac{\partial u}{\partial z} \\
\varepsilon_z = \dfrac{\partial w}{\partial z}, & \gamma_{xy} = \dfrac{\partial v}{\partial x} + \dfrac{\partial u}{\partial y}
\end{cases}
\tag{16}
$$

Defined displacement from the (14) equation integrating 2, 4 and 6 of them; since $\varepsilon_y = 0$, then put $v = v_0(x, z)$, then from $\gamma_{yz} = \frac{\partial v}{\partial z} + \frac{\partial w}{\partial y} = a_{44}\tau_{yz} + a_{46}\tau_{xy}$ we have, as $\dfrac{\partial v}{\partial z}=\dfrac{\partial v_0}{\partial z}$, $w=y\left(a_{44}\tau_{yz}+a_{46}\tau_{xy}-\dfrac{\partial v_0}{\partial z}\right)+W_0(x,z)$

similarly from $\gamma_{xy}=\dfrac{\partial v}{\partial x}+\dfrac{\partial u}{\partial y}=a_{46}\tau_{yz}+a_{66}\tau_{xy}$ we

have $u=y\left(a_{46}\tau_{yz}+a_{66}\tau_{xy}-\dfrac{\partial v_0}{\partial x}\right)+U_0(x,z)$ and by

reducing in system:

$$\begin{cases} v=v_0(x,z) \\ w=y\left(a_{44}\tau_{yz}+a_{46}\tau_{xy}-\dfrac{\partial v_0}{\partial z}\right)+W_0(x,z) \\ u=y\left(a_{46}\tau_{yz}+a_{66}\tau_{xy}-\dfrac{\partial v_0}{\partial x}\right)+U_0(x,z) \end{cases} \tag{17}$$

Substituting u, v, w values in 1,3 and 5 equality for (14) equation and equating members in similar level of y, then we get:

$$\begin{cases} \dfrac{\partial}{\partial x}\left\{a_{46}\tau_{yz}+a_{66}\tau_{xy}-\dfrac{\partial V_0}{\partial x}\right\}=0 \\ \dfrac{\partial}{\partial z}\left\{a_{44}\tau_{yz}+a_{46}\tau_{xy}-\dfrac{\partial V_0}{\partial z}\right\}=0 \\ \dfrac{\partial}{xz}\left\{a_{46}\tau_{yz}+a_{66}\tau_{xy}-\dfrac{\partial V_0}{\partial x}\right\}+ \\ +\dfrac{\partial}{\partial x}\left\{a_{44}\tau_{yz}+a_{46}\tau_{xy}-\dfrac{\partial V_0}{\partial z}\right\}=0 \end{cases} \tag{18}$$

$$\begin{cases} \dfrac{\partial V_0}{\partial x}=a_{11}\sigma_x+a_{12}\sigma_y+a_{13}\sigma_z+a_{15}\tau_{xz} \\ \dfrac{\partial W_0}{\partial z}=a_{13}\sigma_x+a_{23}\sigma_y+a_{33}\sigma_z+a_{35}\tau_{xz} \\ \dfrac{\partial U_0}{\partial z}+\dfrac{\partial W_0}{\partial x}=a_{15}\sigma_x+a_{25}\sigma_y+a_{35}\sigma_z+a_{55}\tau_{xz} \end{cases} \tag{19}$$

From the 2nd ratio (14) with consideration to $\varepsilon_y=0$ we express the strain σ_y:

$$\sigma_y=-\dfrac{1}{a_{22}}(a_{12}\sigma_x+a_{23}\sigma_z+a_{25}\tau_{xz}) \tag{20}$$

By integrating first two ratio (18) with consideration to third we get:

$$\begin{cases} a_{46}\tau_{yz}+a_{66}\tau_{xy}-\dfrac{\partial v_0}{\partial x}=-\theta z+V_1 \\ a_{44}\tau_{yz}+a_{46}\tau_{xy}-\dfrac{\partial v_0}{\partial z}=\theta x-V_2 \end{cases} \tag{21}$$

where: θ, V_1 and V_2 – constant.
Imagine U_0, V_0 and W_0 – in form of:

$$\begin{cases} V_0=V+V_1z-V_2x+v_0 \\ U_0=U-V_3z+u_0 \\ W_0=W+V_3x+w_0 \end{cases} \tag{22}$$

where: V_3, u_0, v_0, w_0 – arbitrary constant.

Then from (17) equation with consideration (16) we get common expression for displacements, that corresponds for strains depending on only to coordinates:

$$\begin{cases} u=-\theta_{yz}+V_2y+U-V_3z+u_0 \\ w=\theta_{yx}-V_1y+W+V_3x+w_0 \\ v=V+V_1z-V_2x+v_0 \end{cases} \tag{23}$$

V_1, V_2, V_3, u_0, v_0, w_0 constants characterize rigid body motion that not accompanied by deformation, θ relative angle of torsion around y axis.

Consider the properties of the model as an example for the following initial data E_1, E_2, v_1, v_2, ϕ, G_2, under the action of a concentrated load $P=1$. Figure 3 shows diagrams of deflections, the normal and tangent strains.

3.2 Basic equation of equilibrium and motion of bare plates on a layered base

Constructions with spatial elastic elements, interconnected rigidly are subjected to dynamic force actions. Depending on the magnitude of external forces and inertia forces of the elastic elements of the system design plate—on-base is in a complex three-dimensional stress-strain state. Spatial construction from elastic elements is divided into eight node elements connected rigidly at the nodal points. They then enumerated sequentially. They then enumerated sequentially. In addition to a fixed global coordinate system is considered a local coordinate system of each plate. Plates of coordinates are given. It's admitted that the external forces are known and the action of external forces is applied to the nodes.

Displacement of any element is determined by formulas of function – forms. Equation of the generalized Hooke's law for isotropic material of element "e" without any limitations and assumptions are written in form of:

$$\{f_i^s\}=[D]\{e\} \tag{24}$$

3.3 The algorithm for calculating the stress-strain state of the broken at the end of three elements lying next to the bare plates in the finite element method

The preceding paragraphs set out the general concepts, methods and formulas required to calculate

503

a) deflection

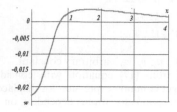

b) strain σ_x

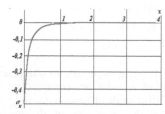

c) voltage σ_y

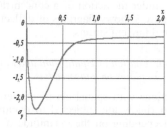

d) equal lines of strains

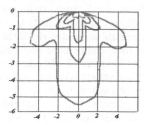

Figure 3. Diagrams of deflections and stresses under the action of a concentrated load.

the stress-strain state of fundamental plates, exposing impact the weight of the boiler unit. The further calculation takes the following assumptions:

1. Plate represents in form of final element system as parallelepiped.
2. Plate is exposed only to concentrated loads from the weight of boiler aggregate, as they are fixed firmly.

The calculation scheme of plate is taken as:

1. Input design data;
2. The construction of the elastic matrix;
3. The construction of the matrix of the stiffener;
4. The construction of the system stiffness matrix;
5. Input values of vertical loads;
6. The determination of displacements at the nodal points;
7. The definition of deformation;
8. The definition of strain.

The followings are specific explaining to the scheme of calculation of the stress-strain state of the plate when exposed to loads from the weight of the boiler aggregate.

1. Input of design data involves the introduction of elasticity and displacement modules, plates of Poisson's coefficients, the coordinates in the global coordinate system, limitations in all directions, numbers of eight nodal elements' points.
2. The elastic matrix is constructed on the basis of the above recommendations in assumption that the plate has an inclination angle in relation to the horizon.
3. Based on the above algorithm is made a program for calculating the plate by the action of concentrated loads.

4 CONCLUSION

Existing methods of determining forces in ferro-concrete fundament provide opportunity to assess its strength, but do not allow considering the stage of normal exploitation. In this regard, urgent tasks are associated with the development of methods, algorithms and computer programs to determine the efforts of the fundamental structures with the case of deformation features at all stages of loading. The calculation results depend on proper accepted models of the base, which reflect to the groundwork. One of the most important tasks of engineers and researchers is to ensure reliability by conducting appropriate calculations.

REFERENCES

Aytaliyev Sh. M. 1994. Scientific potential of Western Kazakhstan. "Kazakhstan science" N 4. (15–28.02.1994), p.1.
Aytaliyev Sh. M. 1996. Some regional problems of Western Kazakhstan in the mechanics of subsoil use. "Problems and prospects of development of science

and technology in the field of mechanics, geophysics of oil and gas, energy and chemistry of Kazakhstan". (Abstracts of the report of the International Scientific and Technical Conference 22–24.05.1996.), Aktau, p.146–147.

Aytaliyev Sh. M., Kudaykulov A. N. & Mardonov B. M. 1997. Mechanics of rocks in new problems of oil production, Almaty, p.6–7.

Aytaliyev Sh. M., Masanov Zh. K. & Nigmetov M. Zh. 2000. Finite element modeling of the mechanical state of offshore drilling platforms on a layered foundation. Materials of the International Conference dedicated to the 70th anniversary of Turksib. Almaty.

Barnshtein M. F. 1982. Calculation of spatial deep-water structures for wave, wind and seismic impacts. Construction mechanics and calculation of constructions. N1 p.47–54.

Barnshtein M. F. 1984. Dynamic calculation of high structures for the effect of wind. In book: "Dynamic calculation of buildings and constructions" under supervision of Korenev B.G. and Rabinovich I.M. M: Governmental construction publishment, p. 169–197.

Brebbya K. & Wocker S. 1983. Dynamics of offshore structures. Translation from english. L: Shipbuilding, p. 230.

Davenport A.G. 1967. Gust loading factors//J. Af the structural Division. Proc.ASCE.-1967, von.93.- *No.* 3 - p.11–34.

Gorodetskyi A. S. & Moyanskyi V. V. 1973. Construction of a stiffness matrix for the End Element of a three-dimensional continuum. Calculation of spatial structures. Kuybishev: 3 edition, p. 108–119.

Guide to the calculation of buildings and structures to the effect of wind. 1978. M: Governmental construction publishment, p.216.

Halfin I. Sh. 1979. Marine oilfield structures and wave influences on them. M: Shipbuilding.

Lehnitskyi S. G. 1977. Theory of elasticity of an anisotropic body. M: Science, p. 25–29.

Mechanics of rocks with regarding oil exploration and production problems. 1994. M: World, El Akiten.

Pasternak P. L. 1954. Basics of a new method for calculating foundations on an elastic foundation using two bed coefficients. M: Governmental construction publishment.

Pirson W. & Mockowitz L. 1964. A proposed spectral form for fully developed wind seas based on a similarity theory of Kitaigorodsky // J. Geophys. Res.- Vol. 69 - *No.* 24.- p. 5151–5190.

Postnov V. N. & Harhurim I. E. 1983. Finite element method in calculations of ship structures. L: Shipbuilding, 1978, p.235.

Standard of rules and regulations 2.01.01.82 Building Climatology and Geophysics. M: Governmental construction,

Yerzhanov Zh. S., Aytaliyev Sh. M. & Masanov Zh. K. 1971. Stability of horizontal workings in an oblique-layered massif. Almaty: Science, p. 160.

Yerzhanov Zh. S., Aytaliyev Sh. M. & Masanov Zh. K. 1980. Seismic stress of underground structures in an anisotropic massif. Almaty: Science, p. 213.

Zenkevich O. K. 1975. Method of end elements. M., Worlds, p. 541.

New Trends in Process Control and Production Management – Štofová & Szaryszová (Eds)
© *2018 Taylor & Francis Group, London, ISBN 978-1-138-05885-9*

Cost-benefit analysis of VFR to IFR airport change

K. Svoboda, J. Kraus & M. Štumper
Faculty of Transportation Sciences, Czech Technical University in Prague, Prague, Czech Republic

ABSTRACT: This article focuses on the economic evaluation of rebuilding airport for IFR traffic to support the development of aviation nodes, which are important economic elements in their regions and often overlooked. In total, three variants of rebuilding are analyzed. The cheapest one is here only to show the cheapest possible way of creating an IFR airport while following the regulations. In a real life situation, such option would make no sense and is not considered in the cost-benefit analysis. Therefore, cost benefit analysis is applied only for Variant I and II. The results show that the mid-option is the best with return of investment period 5.04 years. It is shown, that it is economically viable to transform a VFR airport to IFR airport.

1 INTRODUCTION

Travelling via air is still for many people considered as a luxurious mean of travel and that makes it very sensitive to economic development and its eventual crisis. Nevertheless, one can observe increasing demand for air travel both in Europe and all over the world. In addition, together with the expansion of air travel comes the importance of airports as part of infrastructure of every state. Especially controlled airports capable of accepting IFR flights can significantly affect economic prosperity of a region and attract investment of large companies, job opportunities and passenger comfort. When considering the predicted doubling of transported passengers over the next 20 years, it is obvious that some airports will hit their capacity cap. Construction of new terminals and runways solves the problem only partially. It will be necessary to invest into building and development of new IFR airports, which will serve individual regions.

Construction of completely new airports is, however, very expensive and has to deal with number of obstacles, such as ecological and political ones. (Riha et al. 2014) Therefore, solution that is more economical might be rebuilding of suitable VFR airports to IFR airports. This paper discusses the issues of such rebuilding.

2 PARAMETERS OF VFR AND IFR AIRPORTS

The following paragraphs discuss some differences in VFR and IFR airports. Although exact numerical values are important when designing an airport, they are not presented in this article as one can easily find them in ICAO Annex 14 (or national regulation derived from this annex) (MoT 2016, ICAO 2016) and they would glut the article with very detailed data.

2.1 VFR airport

VFR airport is an airport equipped only with runways without any radio navigation equipment enabling instrument approach. This kind of airport can be used only in Visual Meteorological conditions (VMC), i.e. visibility of five kilometers and a cloud base at an altitude enabling safe flight over obstacles. VFR airports are, logically, significantly simpler than airports capable of handling IFR traffic. However, when planning construction of VFR airport, it should be considered whether there is a possibility of a future development and maybe even transformation to IFR airport. If such possibility exists, it is appropriate to design the airport with instrument approach obstacle limitation surfaces in mind. It is important to think in a timeframe of decades, instead of just years.

The critical aircraft defines movement areas and their dimensions and properties are described in ICAO Annex 14. VFR airports are mostly equipped only with unpaved runways and taxiways. Only a handful of VFR airports—mostly old military airports—have paved runway and taxiways. Although unpaved RWY and TWY are cheaper, they are also more prone to damages caused by meteorological conditions, such as heavy rain or melting snow. On the other hand, short-term closure is usually not a big deal.

Due to the nature of VFR airports, it is not necessary to describe the parameters of VFR airports in more detail. It is more convenient to talk about IFR airports, as they are the desired outcome of potential transformation.

2.2 *IFR airport*

IFR airport is an airport that is equipped with at least one runway enabling instrument approaches. According to the type of equipment used, the approach minima are defined. There are two types of instrument approach, 2D and 3D. The 2D approach only provides an information about horizontal position relatively to the runway centerline. The 3D approach provides the information about horizontal position as well, but on top of that also information about position relative to glide slope.

Obstacle limitation surfaces are important "part of" infrastructure of IFR airports. That is because the aircraft descend to low altitudes with any visual contact with ground or obstacles and therefore completely rely on properly defined and maintained obstacle limitation surfaces. Their detailed description is provided in ICAO Annex 14, however it is worth mentioning that their dimensions depend on type of instrument approach. The lower the aircraft is allowed to descend without visual contact with runway, the stricter are the obstacle limitation surfaces.

The movement areas are usually paved, especially runways. Although, the regulations do not prohibit unpaved runway with instrument approach, it does not make economic sense. Unpaved runway is capable of receiving only aircraft with maximal take-off weight of 5700 kg, therefore the income from landing fees would not cover the cost of the necessary infrastructure.

Visual navigational equipment is important part of IFR airport and there are various types of such equipment. They are important for both locating the airport (and acquiring visual contact) and navigating around the airport. This is mainly achieved by the use of lighting equipment in the form of approach lights, runway lights and taxiway lights.

Because IFR airports are aiming at attracting larger jet airplanes, they are also required to provide proper services. One of the most important ones is the presence of firefighting unit and a required number of vehicles. For example, if an airport wants to handle typical Boeing 737–800 it is required to have a unit with two firefighting vehicles. Furthermore, every IFR airport must have a plan for removing an airplane unable to move, however they are not required to own an equipment capable of doing so.

There is also a requirement for system of surveillance and management of movement on movement areas. The airport must also have a perimeter security in place, which is mostly ensured by a perimeter fence. The last requirement is the one for appropriate airport maintenance, mainly in the form of keeping the movement areas clear of foreign objects, snow and ice.

3 MODEL AIRPORT

At the beginning, it is important to select a suitable airport for transformation. There are several airports in the Czech Republic, that have similar parameters, have paved runway, but are only equipped for VFR operations. For the purpose of the cost benefit analysis to be usable at more than just one airport, pre-selection of suitable VFR airports was made based on the following parameters: paved runway, runway length at least 1800 meters, runway width at least 45 meters. If will be these parameters kept during transformation, the airport would be able to handle large business jets and also Boeing 737 and Airbus A320, which are the most common type of airplanes for intra-European flying. With this capability, the transformation seems to already make sense, as the larger airplanes would bring substantial income for the airport (Doganis et al. 1995).

In the Czech Republic, there are four airports with mentioned parameters: České Budějovice, Hradec Králové, Přerov and Plzeň/Líně (ŘLP 2017). Furthermore, all these airports are close to main regional cities with approx. 100 000 inhabitants, and as ex-military airports they have quite good facilities, which could be used as for example airport offices.

The desired final parameters of the "new" IFR airport are following:

– Runway with the dimensions of 2500 × 45 meters.
– 1 taxiway parallel with the runway, 4 taxiways connecting the parallel TWY and RWY
– 2 aprons with dimensions of 300 × 100 meters each.

4 VARIANTS OF VFR TO IFR AIRPORT TRANSFORMATIONS

Three transformation variants were designed, of which two represent standard solution in more expensive and cheaper way. The third variant is mostly used as an example of the cheapest possible way of transforming VFR airport to IFR. It also serves as a proof, that the cheapest does not mean the best and that the low costs are greatly offset by many negatives (Endrizalová & Němec 2014). The following paragraphs describe each of the variant.

4.1 *Variant I*

The first variant represents the most expensive (complex) solution. It consists of Precision approach system, runway surface adjustments and most of the services provided by the airport. (Horonjeff et al. 2010) Annual terminal capacity shall be approximately 250 000 passengers.

One-time investments:

- Movement areas – RWY and TWY strengthening, modifications of surroundings
- Navigation – precision approach system, full approach light system, ground signs, hardware and software for radar surveillance
- Property – terminal, control tower, firefighting station, fuel station, other facilities
- Vehicles – adhesion measuring vehicle, snow plow
- Ground handling – vehicles (pushback, tugs, luggage trolleys, fuel tank truck, de-ice vehicle, sewage and clean water vehicle, a car), system for luggage processing, ground power unit, stairs, fork-lift, luggage loading belt
- Meteorological station
- Removal of obstacles
- Other – perimeter fence, apron lights
- Periodical investments:
- Navigation equipment maintenance
- Movement areas maintenance
- Radar data

4.2 Variant II

The second variant is different from the first one in several significant aspects. First, it does not include precision approach system, which reduces costs by not buying the system but also by not acquiring the full approach light system or obstacle removal. Further, an external company will provide ground handling and security, thus reducing financial risks for the airport operator. Annual terminal capacity stays the same at 250 000 passengers.

One-time investments:

- Movement areas – RWY and TWY strengthening, modifications of surroundings
- Navigation – non-precision approach system, intermediate approach light system, ground signs, hardware and software for radar surveillance
- Property – terminal, control tower, firefighting station
- Vehicles – adhesion measuring vehicle, snow plow
- Ground handling – external company
- Meteorological station
- Removal of obstacles
- Other – perimeter fence, apron lights

Periodical investments:

- Navigation equipment maintenance
- Movement areas maintenance
- Radar data

4.3 Variant III

This variant differs in a way, that it does not expect building an airport capable of handling an airplane in the size of Boeing 737. Its goal is to figure out the minimal amount for which a VFR airport can be transformed to IFR airport. It mainly uses the fact, that regulations do not require paved RWY and TWY for IFR airport. However, the probability of such airport becoming and IFR airport is very close to zero, as the allowed aircraft would never be able to pay for the needed investments (Phang 2002).

One-time investments:

- Movement areas – unpaved RWY and TWY, modifications of surroundings
- Navigation – non-precision approach system, lighting system, ground signs
- AFIS station
- Firefighting station (if needed)
- Security and handling – external company
- Meteorological station
- Removal of obstacles

Periodical investments:

- Navigation equipment maintenance
- Movement areas maintenance

5 FINANCIAL COSTS OF VFR TO IFR TRANSFORMATION

This chapter presents final costs for all the needed equipment in order to fulfil the requirements of above-mentioned variants. The values are derived from financial results of regional airports in Pardubice, Brno and Ostrava (Financni sprava 2014). Other data were acquired from consultations with people at Airport Vodochody, Air Navigation Services CR and from company ADB Airfield Solutions.

5.1 Cost of Variant I

The one-time investment costs are 888 290 000 CZK (Czech crowns) and the regular annual costs are 123 356 600 CZK. The equipment includes 2 pieces of GPU, 6 baggage trolleys, 2 firefighting vehicles and as employees 150 people. The airport would be equipped with ILS CAT I and CVOR (Pleninger 2014, Hospodka 2014).

5.2 Cost of Variant II

The one-time investment costs are 849 621 000 CZK (Czech crowns) and the regular annual costs are 113 483 000 CZK. There are 2 firefighting vehicles and 132 employees included (employees of external company are excluded). The airport would be equipped with CVOR and NDB.

5.3 Cost of Variant III

The one-time investment costs are 40 840 000 CZK and the regular annual costs are 17 898 600 CZK. The navigational equipment of this airport would include two NDBs, one DME and two PAPIs and approach lighting system for non-precision approach. There is only an AFIS station located in an already existing building, which would also serve as a little terminal with annual capacity of 20 000 passengers.

6 COST-BENEFIT ANALYSIS (CBA)

CBA is a proper method for evaluating large investment projects. Its advantage lies in the ability to evaluate not only direct financial costs and benefits, but also those of non-financial nature by assigning them a value. All the positives, benefits and profits are compared to all the negatives, costs and damages. The basis of a project is to bring more positives than negatives.

Transforming a VFR airport to a general model of IFR airport is a project with a goal to build an international airport of regional importance. The aim is to fulfil future demand and offer people and business an option to distant airports, therefore providing more comfortable and safe way of reaching other countries (markets) (Vittek et al. 2016). There are in total three variants, but the cheapest one was there only to show the cheapest possible way of creating an IFR airport while following the regulations. In a real life situation, such variant would make no sense and therefore is not considered in the cost-benefit analysis. Only Variants I and II will be dealt with further.

The following lists show benefits and costs which will be used in the analysis (for both variants).

Benefits:

1. Increase in number of job opportunities
2. Income from renting out space in unused buildings
3. Income from renting out commercial space inside the terminal
4. Income from fees
5. Saved time from not travelling to larger airport (Prague)
6. Saved money from not travelling to larger airport (Prague)
7. Benefit from new investors coming to region

Costs:

8. Costs for rebuilding the airport
9. Costs for airport maintenance

6.1 Assigned values to benefits and costs

Table 1. Margin settings for A4 size paper and letter size paper.

Benefits/ Costs	Variant I [CZK]	Variant II [CZK]
1	343,170	343,170
2	25,920,000	25,920,000
3	8,400,000	10,500,000
4	64,000,000	45,036,800
5	100,800,000	96,132,960
6	95,400,000	90,982,548
7	4,829,800	4,829,800
8	888,290,000	849,621,000
9	123,356,600	11,348,3000

6.2 Discount rate determination and conversion of costs and benefits to actual value

The project of transforming an airport has a lifetime of 20 years. Following the expert literature, the discount rate is determined to be 3.5% (Boardman 2006).

Present Value (PV)
Present Value is a sum of all future cash flows acquired from the investment, transformed to their actual value (Brent 2007). The calculation is according to this formula:

$$PVCF_t = CF_t * discount\ factor \tag{1}$$

$$discount\ factor = \frac{1}{(1+r)^t} \tag{2}$$

where $PVCF_t$ = present value of cash flow in year t; CF_t = cash flow in year t; r = discount rate.

Calculation of criterial indicator of project present value:

$$PV_t = \sum_{t=1}^{n} \frac{CF_t}{(1+r)^t} \tag{3}$$

The present values of each variant are following:

Variant I
Discounted benefits: 4 259 536 182 *CZK*
Discounted costs: 1753 267 355 CZK
Project present value: $PV_I = 2\ 506\ 268\ 827\ CZK$

Variant II
Discounted benefits: 3 822 095 688 *CZK*
Discounted costs: 1 612 933 879 *CZK*
Project present value: $PV_{II} = 2\ 209\ 161\ 809\ CZK$

Net Present Value (NPV)
Net Present Value is amount of net profit generated over time by the project (Sieber 2004).

$$NPV = \sum_{t=0}^{n} \frac{CF_t}{(1+r)^t} \qquad (4)$$

$$NPV = CF_0 + \sum_{t=1}^{n} \frac{CF_t}{(1+r)^t} = CF_0 + PV = PV - 1 \quad (5)$$

where NPV = net present value; PV = present value; I = amount of investment in year zero; CF_t = cash flow in year t; r = discount rate; t = time frame from year 0 to n.

NPV for both variants is following:

$$\begin{aligned} NPV_I &= 2\,506\,268\,827 - 888\,290\,000 \\ &= 1\,617\,978\,827\,CZK \end{aligned} \qquad (6)$$

$$\begin{aligned} NPV_{II} &= 2\,209\,161\,809 - 849\,621\,000 \\ &= 1\,359\,540\,809\,CZK \end{aligned} \qquad (7)$$

Return on investment period
Return on investment period describes the number of years required for the income cash flows to even out the initial investment. (Sieber 2004) For the purpose of this article, non-discounted values are used in order to be able to use the following simple equation:

$$return\ on\ investment\ period = \frac{CF_0}{CF_t} \qquad (8)$$

where CF_0 – initial investment; CF_t = cash flow constant for all t (from 1 to n);

Variant I
Return on investment period: 5.04 years

Variant II
Return on investment period: 5.46 years

Benefits to costs ratio (BCR)

$$BCR = \frac{discounted\ value\ of\ benefits}{discounted\ value\ of\ costs} \qquad (9)$$

Variant I
$$BCR_1 = \frac{4\,259\,536\,182}{2\,641\,557\,355} = 1{,}61 \qquad (10)$$

Variant II
$$BCR_{II} = \frac{3\,822\,095\,688}{2\,462\,554\,879} = 1{,}55 \qquad (11)$$

7 EVALUATION OF THE ANALYSIS

Authors are well aware of omitting costs that the transformation would have on the environment. However, due to the lack of expertise in this area, it was better to leave it out than to try to come up with some "random" numbers.

When looking at the criterial indicators, one can see that the present value is higher than the initial value of investment and that the benefit to cost ratio is higher than one. Both variants are acceptable. The parameter "return on investment period" is significantly lower than the lifetime of the project, therefore confirming the acceptability of the project. The last step would be to choose a variant to build. From the numbers above, Variant I appears to be slightly better.

8 CONCLUSION

The purpose of this article was to give readers a brief insight into possibility of transforming VFR airport to an IFR one and evaluating its pros and cons. Airports are usually understood as a simple economic subjects, evaluated based on their income and profit. However, they are more than that, they serve as a gateway for investors and people from the region to travel abroad while tourists can visit the region.

It was shown on a simplified model, that it is economically viable to transform a VFR airport to IFR airport. However, the VFR airport must already have some predispositions, such as paved runway. If the VFR airport had unpaved runway, the cost of building a completely new runway would probably make the project unviable. In the Czech Republic, there are already some real ideas about conducting transformation in some ways similar to the one presented here.

ACKNOWLEDGEMENT

This paper was supported by the Grant Agency of the Czech Technical University in Prague, grant No. SGS17/154/OHK2/2T/16.

REFERENCES

Boardman, A. et al. 2006. *Cost—Benefit Analysis Concepts and Practice*. Third Edition, Pearson Education, Inc., ISBN: 0-13-143583-3.
Brent, R., 2007. *Applied Cost—benefit Analysis*. Second Edition, Edward Elgar Publishing.
Doganis, R. et al. 1995. The Economic Performance of European Airports. *Research Report 3*, Department of Air Transport. Cranfield University. Bedford.

Endrizalová, E. & Němec, V. 2014. The Cost of Airline Service, *MAD - Magazine of Aviation Development* 2(8).

Financni sprava. 2014. *Daňové příjmy rozpočtů krajů a obcí dle zákona o rozpočtovém určení daní.*

Horonjeff, R. et al. 2010, *Planning & Design of Airports,* Fifth Edition, The McGraw-Hill Companies, Inc.

Hospodka, J. 2014. Electric taxiing—Taxibot system, *MAD—Magazine of Aviation Development* 2(10) ICAO, 2016, Annex 14. MoT of the CR, 2016, Aviation regulation L14.

Phang, S-Y. 2002. *Strategic development of airport and rail infrastructure: the case of Singapore, Transport Policy.*

Pleninger, S. 2014. The Testing of MLAT Method Application by means of Usage low—cost ADS-B Receivers, *MAD—Magazine of Aviation Development* 2(7).

Riha, Z. et al. 2014. Transportation and environment-economic research, *WMSCI 2014–18th World Multi-Conference on Systemics, Cybernetics and Informatics,* Proceedings, pp. 212. ŘLP, 2017, AIP of the CR: AD 2 - Letiště.

Sieber, P. 2004. *Analýza nákladů a přínosů—metodická příručka,* Ministerstvo pro místní rozvoj, 2004.

Vittek, P. et al. 2016. Runway incursion and methods for safety performance measurement. *Production Management and Engineering Sciences—Scientific Publication of the International Conference on Engineering Science and Production Management,* ESPM 2015, pp. 321.

New Trends in Process Control and Production Management – Štofová & Szaryszová (Eds)
© 2018 Taylor & Francis Group, London, ISBN 978-1-138-05885-9

Risk-based indicators and economic efficiency

S. Szabo, V. Plos, Š. Hulínská & P. Vittek
Faculty of Transportation Sciences, Czech Technical University in Prague, Czech Republic

ABSTRACT: The paper discusses the introduction of risk-based indicators and their subsequent use as a support tool for evaluating the economic efficiency of accepted safety measures. The first part is devoted to describing the principle of risk-based indicators, the second section describes usage of economic analysis and usage of risk-based evaluation indicators with regard to economic efficiency. The goal of this article is to declare new possibilities how to evaluate risks and how to assess cost and benefits for some safety measures. At the end there is a discussion about usage of risk-based indicators in a new economic way.

1 INTRODUCTION

Human activity is limited only by our imagination and every part always has its own specifics. Across all the parts it is possible to identify similar dependencies or similar need to decide between the same elements. It is no different in terms of trade-offs for financial benefits. However, it is most preferred to quantify the benefits in financial terms. Similarly, this is a trade-off in the number of safety for how much money. Safety Management Manual (SMM) in the area dedicated to risk management refers to the management system called ALARP—As Low, As Reasonably Practicable (ICAO 2013). This concept is related to the efficiency of resource allocation to safety risk management to a level that it is acceptable for the system (Lališ & Vittek 2015). In other words, we manage risk to a level at which costs are incurred for the management of risk is less than the revenues associated with the operation of the risky activities. If the cost of mitigating measures outweigh the possible income from the operation of the activity, activities should be canceled for the economic inefficiency. To analyze how effective is the risk management system, if the risks are managed to a sufficient level for a given costs, can serve several tools. One of them are risk-based indicators.

2 INTRODUCTION OF RISK-BASED INDICATORS

In order to well define, what do Risk-based indicators (RBI) means, it is necessary to explain two basic concepts. The first is a "risk". Under the term "risk" it is necessary to imagine a complete set of scenarios, probabilities and consequences of the implementation of undesirable events (Štumper & Kraus 2016). The indicator is measurable variable reflecting the current real state. The second is a Risk Influencing Factor (RIF). RIF is the aspect of the system, respectively activities affecting the level of risk systems/operations. This is a theoretical variable, which means it does not need to be determined how to measure this variable. Examples RIF may be a fluid leak.

RBI is then measured variable of the RIF which may be a number of fluid leaks during the period. (Øien 2001) RBI is also an indicator that monitors the performance of the corrective measures to reduce the risk in the context of monitoring the frequency of occurrences which deviates from the of corrective measures, together with the frequency of cracking defenses created by corrective measures associated with the direct implementation of risk. Under this you can imagine the realization of top events despite the corrective measures that were based on analyzes carried out in order to prevent the implementation of top events.

An indispensable condition for the safety assessment by RBI is the proper function of the safety management system. Two main processes of safety management are hazard identification and risk assessment.

Risk-based indicators implementation process can be divided into eight basic steps:

- The first step is to select those events that contribute most to the risk of the operation. The peak event must have a serious potential threat to safety.
- The second step is to identify risk influence factors. The entire risk analysis must be examined in detail and found all the elements that influence the occurrence of the event. This can sometimes be very difficult. In some cases it is very difficult to determine all the circumstances

that may lead to the realization of events with an impact on safety. This step may be represent a significant economic burden to the company in the form of working hours required to build complex accident scenarios.

- Based on this information it is possible to construct models of so called accident models.
- In the fourth step, there is an assessment of potential RIF's changes—whether it is even possible that factor will change over time, or whether it is a constant characteristic of the system, and if changes occur, how big those changes may be.
- After evaluating the possibility for changing the RIF follows the evaluation of the impact of changes to individual RIF on the size of the overall risk of the operational process/activity.
- Another, in order sixth step is the selection of Risk Influence Factors which change is causing serious change in the overall risk process. To these RIF are then selected indicators detecting a change individual parameters. This creating of indicators is based on discussions with operational staff, to set correct measurable quantities, characteristics of individual operations that need to be addressed, and the change can be serious for safe operation. It is also necessary to determine in what form will be collected data from operational staff.
- The penultimate step is to finalize the set of indicators and testing their application in real operations. Testing is focused not only on the indicators themselves, but also tested the data collection process for the evaluation indicators, evaluation and subsequent report creation. Test (tracking) period is about a few months—the normal time is 3 to 6 months. After this testing, there are some analysis and possible changes in the monitored variables to adapt to the actual operating conditions, etc.
- The last step is the routine use of indicators in normal operation. The most common approach to the use of indicators is to focus on changing parameter indicators for the given time period. When comparing we can get a percentage change of the parameters and on the condition that we know how we given parameter affects the overall risk, we can easily identify us as a given parameter change in a given time period, change the value of the overall risk of business operations.

Risk-based indicators can also be obtained by reforming the process of already used indicators. This can be illustrated by looking at where we have divided work into sub-processes. Anomalies in these processes are measured by process indicators. If we make evaluation of these indicators in terms of frequency and severity, respectively thus we evaluate the occurrence of events that these indica-

tors monitored and further evaluate their impact on the overall risk of the operation, we can already speak about Risk-based indicators. (Øien 2001)

3 RBI AS AN INPUT FOR COST-BENEFIT ANALYSIS

Cost—Benefit Analysis (CBA) is a methodology that their course progressively responsible fundamental question: "What whom the realization of the project brings and what to whom takes?" Thus defined impact events are then aggregated and converted to cash flows and are included in the calculation of key indicators on the basis of which to decide whether the project is ultimately beneficial for the company or not. When comparing two or more investment, then allow the calculated indicators to determine their order, or specify preference for one project over another. (Layard & Glaister 1994)

Cost-Benefit Analysis is an English expression that can create misleading, since in this case, not a cost in the accounting sense, but rather a sort of "detriment", or rather, any negative impacts of the project.

CBA advantage is the fact that this is a systematic process successfully applicable to every project and the ability to almost full use of theoretical and methodological apparatus of corporate finance, because after all the definition of the effects of actions are treated as cash flows.

3.1 CBA procedure

The first point is to define the very essence of the project. So what is the objective and by what means will be achieved. If we are clear about what we want to implement by the project, we set a group of entities that will be affected by the implementation and realization of its gain some benefit. These persons are called beneficiaries. If there is no right more projects, among which we choose—we compile the ranking, there is always the option not to implement the investment. It is therefore necessary to describe what the situation will be when the project will be implemented and the situation when not be implemented.

Furthermore, we come to the stage where we are trying to write down all the costs and benefits of the project, in all its life stages. The first phase of the project life is pre-investment phase costs attaching to exclusively pre-investment stage of the project must be accounted for in the analysis, because it is a cost that will be incurred regardless of whether the investment will or will not be realized. Most of these are costs associated with the preparation of project documentation, administrative costs associated with the preparation and implementation of projects, the costs of drafting the analysis itself. These costs are not relevant in the assessment. These are called sunk cost. (Riha et al. 2014)

After the pre-investment phase followed by a phase of investment. This is the period between the start of the investment and the commencement of its operation. At this stage, the costs far outweigh the benefits. The post-operational phase, which occurs after the project start up and lasts until the end. In this period should outweigh the benefits outweigh the costs and expenses incurred in the investment phase of the project. At the moment, the project is closed occurs liquidation phase. It is a time when the already completed project still influences the costs and benefits of some stakeholders. Unlike the costs incurred in the pre-investment phase it is necessary, these costs and benefits are offset because there is a direct link to the fact that the project is implemented or not. (Nas 2016)

3.2 Usage of RBI

If the RBI will be used for the evaluation of current operating conditions is guaranteed relative accuracy by the fact, that RBI eliminate the relative subjectivity of risk assessment, but the results of the evaluation RBI is the numerical expression. This evaluation and analysis of non-safety/risk getting a quality field in quantitative.

3.3 Economics of safety

In an environment of air transport is a big emphasis on safety, but it must be maintained ALARP principle, respectively the problem must be solved 2P—protection vs. production goals. On the basis of improper management of safety management in a dangerous situation when allocating resources essential business functions, organizations that directly and indirectly support the provision of services. Deciding on the allocation of resources can lead to a problem that is being called the dilemma of the two Ps. It is the allocation of resources to production targets (Production goals—Ps) and safety objectives (Protection goals—Ps). It is therefore important that investments in security were balanced with respect to the production goals of the company. Investment in safety therefore not exceed the revenue that the risk of surgery entails (Endrizalová & Němec 2014). For this reason, all investments must be evaluated in terms of impact on risk mitigation, whether the investment will be spent on alleviating effective, that adequately mitigate the risk and profit from such activities will still be sufficient.

For this purpose, the cost-benefit analysis, which assesses the cost-effectiveness. The input for cost benefit analysis may thus also data from experience with risk-based indicators, respectively data containing real change in risk after the application of corrective measures. Based on data from the past can therefore proceed to analyze the effectiveness of the management of resources, when we know how the risk has changed after the application of corrective measures (Skálová & Kraus 2015). Another aspect that must be taken into account when evaluating risk arising indirectly profit from the operation, which derive profit from operations as such, but is generated from the subsequent respectively related operations.

3.4 Risk-based indicators and economics

Risk-based indicators find its importance not only as a tool to assess the level of safety, where they provide a quantitative assessment, but the outputs from the risk-based indicators usage may very well serve to evaluate the economic acceptability of the remedial measures if the risk comes too high and there is a need to mitigated these risks when we need to continue of the processes (Jeřábek & Kraus, 2015). RBI thus serves as an effective tool for addressing ALARP, thus mitigating risks from unacceptable area to a level where it is possible to run the operation – (yellow area) shown in the following Figure 1.

So if RBI is used for deciding whether a given activity to operate or not. The second use of RBI may be a requirement to further reduce the level of risk down to an acceptable level, i.e. in the green area. This activity is not explicitly required by legislation, it is not absolutely necessary for the operation, but the operator can provide some example. Competitive advantage. At that time, RBI will be used to assess whether the money spent will reduce the risk of such a level to which society demands, therefore, that the money spent will return, or is

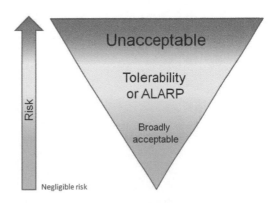

Figure 1. ALARP method—level of risks (Leathley 2016).

it better to stay on the level of risk in the yellow area, as has been the one to provide a sufficient level of safety, if the risks constantly monitored and controlled.

4 CONCLUSION

Use RBI for the economic evaluation of remedial measures appear to be very useful due to its quantified assessing the level of risk primarily by analyzing risks change after application of corrective measures. With the thus obtained information about the change in the level of risk to the resources spent can operate in various economic analyzes, cost-benefit analysis and the like. There are currently conducted by air carriers such measure, when introducing data collection systems that after running-up mode will providing data bases for risk assessment using RBI.

So despite the original intention that should RBI serve mainly safety manager for the evaluation of the need for remedial measures, RBI could also serve as a basis for the economic department in terms of the appropriateness of investing in safety.

ACKNOWLEDGMENT

This paper was supported by the Grant Agency of the Czech Technical University in Prague, grant No. SGS17/151/OHK2/2T/16.

REFERENCES

Endrizalová, E. & Němec, V. 2014. The Costs of Airline Service. *MAD—Magazine of Aviation Development*. 2(8): 14–16.
ICAO, 2013, *Doc 9859. Safety Management Manual (SMM), Third Edition*, Montreal: International Civil Aviation Organisation.
Jeřábek, K. & Kraus, J. 2015, Helicopter Approach to Offshore Objects. *Naše More* 62(2): 74–77.
Lališ, A. & Vittek, P. 2015. Safety Key Performance Indicators system for Air Navigation Services of the Czech Republic. *Transport Means—Proceedings of the International Conference*: 537–542
Layard, R., & Glaister, S. 1994. *Cost-benefit analysis*. Cambridge: Cambridge University Press.
Leathley, B. 2016. A is for ALARP, *IOSH Magazine*.
Nas, T.F. 2016. *Cost-benefit analysis: Theory and application*. Lexington: Lexington Books.
Øien, K. 2001, Risk indicators as a tool for risk control. *Reliability Engineering*. 74(2): 129–145.
Riha, Z. et al. 2014, Transportation and environment-economic research, WMSCI 2014–18th World Multi-Conference on Systemics, Cybernetics and Informatics, Proceedings: 212–217
Skálová, M., & Kraus, J. 2015. Thermal imaging cameras for aerodrome surveillance system. *Transport Means—Proceedings of the International Conference*: 678–682
Štumper, M. & Kraus, J. 2016. Safety Study in Aviation. *MAD—Magazine of Aviation Development*, 4(19): 19–22.

New Trends in Process Control and Production Management – Štofová & Szaryszová (Eds)
© 2018 Taylor & Francis Group, London, ISBN 978-1-138-05885-9

The "Family 500 plus" programme and the expenditure structure of households

A. Szromnik
Cracow University of Economics, Cracow, Poland

E. Wolanin-Jarosz
The State Higher School of Technology and Economics in Jarosław, Jarosław, Poland

ABSTRACT: The main rules of the "Family 500 plus" Government Programme have been presented in the article. Its influence on household budgets have also been discussed. The publication has theoretical and empirical character. The essence of the 500 plus Government Programme has been explained in the theoretical part, mainly its influence on bringing up children by giving financial benefits. However, the basic part of the publication—the empirical one, is based on the direct research results conducted among the inhabitants of Podkarpackie (pilot trial – 61 people) making use of the "Family 500 Plus" Government Programme. The respondents assessed the influence of the programme on the changes of expenses in large families.

1 INTRODUCTION

The government demographic "Family 500 plus" is a family policy programme realized with the use of financial instruments such as money transfers. The aim of this programme is to improve financial situation of Polish families by increasing the households disposable income. Within the last years, work efficiency has grown faster than salaries in Poland, thus, the government was going to increase consumption ability of Polish people by introducing this programme. As part of the "Family 500 plus" programme, households, in which the children are brought up, receive regular financial support of 500 zł every month for the second and the next child until they become adults. The means are given regardless of the family's income or parents' marital status. The money is received also by single parents or legal guardians irrespective of maintenance payments. Families with low income receive support also for the first and an only child if they fulfil the required income criteria.

According to the market analytics, the indirect aim of the "Family 500 plus" programme is also support of Polish economy. The Standard & Poor's agency reports that social programmes, including mainly the one "Family 500 plus", they increase domestic demand (S&P: "Family 500 plus" programme, 2017), and in turn it is a development impulse increasing national PKB (GDP—Gross Domestic Product) (consumption is one of the ingredients of the PKB account).

In the present article an attempt was made to analyse and assess the influence of the "Family 500+" on the changes of expenses in households. The basic part of the publication—the empirical one, is based on the direct research results conducted among the Podkarpackie inhabitants.

2 THE RULES, CONDITIONS AND LEGAL BASIS OF THE "FAMILY 500 PLUS" PROGRAMME FUNCTIONING

The Act of 11 February 2016 about the help of the country in bringing up children enforces the "Family 500+" programme (Journal of Laws from 2016 item 195). The essence of this programme is extending child-support benefits on the biggest number of people possible, who bring up children. According to the act, parents and legal guardians of children below the age of 18 will receive the child-support benefit 500 zł every month for the second and the next child, regardless of the income. Financial support is also given to families with one child with low income—oscillating less than 800 zł net per person. However, in households, in which there is a disabled child, the assumption income is higher and is 1200 zł net (Journal of Laws 2016 item 195).

The "Family 500 plus" programme's aim is to realize three fundamental tasks: minimization of poverty (especially as far as the youngest are concerned) by financial support of households bringing up offspring, birth rate improvement and investments in human capital.

Within the programme, financial means are directed to families who bring up children in order to secure the expenses concerning fulfilling every day and upbringing needs of the offspring. Educating children and preparing them to life is connected to sizable financial burdens of people, who support them, especially in families where there are many children. That is why these households very often have economic barriers dependent on the income. It was necessary to introduce such solutions, which would help to exclude or minimize them, especially among young people who decide to have a baby, especially the second or the next one.

As it has already been mentioned, the "Family 500 Plus" programme concerns a very important demographic issue as extension of birth rate. Basic trends which shape the demographic situation of Poland are, like in many other countries in Eastern Europe: decline of fertility rate (showing low birth index) and longer life expectancy of society. Both factors lead to ageing of Polish society. At the same time, by analysing present and predicted population age structure and international migrations, the decrease of number of people should be expected. It is worth mentioning at this point that Poland is one of the countries in the world, where there are fewest births. Birth rate for our country has been diminishing regularly from 1989 (GUS (the Main Statistical Office), 2017). In 1997 r. it was1.5–which means that 1.5 child fell on one woman in reproductive age. However in years 2001–2015 this rate fluctuated around the value of 1.3. It should also be stressed that the value of fertility rate, which guarantees substitution of generations is 2.1. Then, next generations are numerous enough to provide social and economic stability of a given country.

In connection to the above, the "Family 500+" programme is supposed to be a stimulus, which should encourage families to make decisions to have more offspring. This aim is to be realized by the guarantee of supportive, long-term financial help from the country, regardless of the households income. This programme is an investment in human capital as well as it finishes with belief that family policy equals social policy.

Realization of a child-support benefit within the "Family 500 Plus" programme is the duty of a proper local government unit (a village mayor's, a mayor's or a city's president), where a person, who applies for 500+ financial means, lives. An appropriate organ can authorize an employee of a given organizational entity (it is usually a worker of a social welfare centre) to conduct procedures in cases of the child-support benefits range. It should be stressed that the realization of a child-support benefit in a social welfare centre cannot cause the decrease of efficiency in realizing the

tasks of social assistance. It cannot also violate employment norms of social workers appointed in the regulations about social service (The Journal of Laws 2016 item 195).

In case of a child-support benefit realized in connection to coordination of social security systems (for example temporary stay of one of the parents abroad) a proper organ is voivodeship marshal. This organ makes decisions in cases concerning child-support benefits. The benefit, however, is paid by the commune.

The child-support benefit and the costs of its service are financed in form of designated subsidy from the country budget. It should be stressed that the benefit itself and the costs of service, which will be paid by the proper organs, come from the same designated subsidy. In case of communes and counties, that is units, which adequately pay child-support benefit and educational supplement, the costs have been defined as percent from the paid means, however in case of a voivodeship marshal, who makes decisions in cases connected to coordination of social security systems, but do not pays the benefits, the costs are defined individually. Decision in this range is made by the Minister of Finance, on the basis of conducted by the Minister of Family, Work and Social Policy analysis of the data, which comes from the demand for funds collected from marshals. Legal regulations determine that in 2016 2% of the dotation sum was dedicated to the service costs to communes and in consecutive years 1.5% (The Journal of Laws 2016 item.195). Counties, however, receive 1% of the dotation sum dedicated to payments realization for child-support benefits for Foster families and family children's homes for service. The service costs for counties are lowers in percentage taking into consideration the lack of necessity to verify the income of families who apply for a child-support benefit. At the same time, the child-support benefit is connected to realization of the main benefit for foster families. It is worth to stress that costs return of unduly received child-support benefits during a given budget year do not have influence on the height of operation costs. It should also be mentioned that the applications for child-support benefits form the "Family 500 plus" programme can be submitted in paper form—by personal intercession of the interested person in one of the proper offices or online, that is by electronic banking. Being the owner of the account serving "Family 500+" programme, required forms can be submitted and also choose a bank account to receive benefits.

From the data passed by the Ministry of Administration and Digitization results that in 2017 95% of applications were filled in electronically (Applications—"Family 500 plus", 2017).

As far as reporting commitments are concerned, a proper organ and the voivodeship marshal are obliged to prepare a material and financial report from realization of tasks concerning child-support benefit and provide it to the appropriate voivode. The voivode prepares a collective material and financial report from realization of tasks concerning child-support benefit and provides it to a proper minister who deals with family affairs (at present: the Minister of Family, Labour and Social Policy). The proper minister of family specifies formulas and the ways how to prepare material and financial reports from realization of the tasks concerning child-support benefits, in a regulation. He also specifies the time and the way of handing them in, taking into account the need to guarantee completeness and uniformity of information.

To sum up the above considerations, it should be stressed that the "Family 500 plus" programme is an enormous financial support mainly for large families (bringing up more than 3 children). These families mainly live in southern and eastern Poland. According to the research conducted by Ipsos Ltd. large families income in Poland, before introduction of the "Family 500 plus" programme, did not allow to meet every day needs of household members ("Large families in Poland—report" 2016). Only extra financial means in the sum of 500 zł per month covered current expenses of the family's needs and enabled them to pay for: "extra lessons for children, for example "English classes", tickets to the cinema and to the swimming pool, school trips or summer holidays for children. Moreover, about 30% of the respondents claim that the positive impulse to make a decision about having another child can only be improvement of financial and housing situation ("Large families in Poland" 2016).

3 THE AIM AND THE RESEARCH METHODOLOGY

The aim of the research was assessment of the influence of the financial measures from the "Family 500 plus" government programme on the changes of households' expenses.

In the conducted analyses, the main source material was achieved thanks to direct research (pilot studies) done in Podkarpackievoivodeship. The research sample in the number 61 people, were Rzeszów inhabitants using the "Family 500 plus" programme. The research method was a questionnaire. A survey questionnaire was used as a measuring instrument. It consisted of 12 questions adequate to the researched subject. The survey was conducted by the author of the article and trained to this purpose interviewers. The research period was January 2017.

The collected research material, after previous reduction, was used to create the data base. The analysis of the survey was done with the use of *Excel* programme. It depended on comparison of percentage distribution of the answers in the compared groups for most of the questions (Aczel 2000).

4 THE INFLUENCE OF INCOME FROM THE 500+ GOVERNMENT PROGRAMME ON THE CHANGES OF FINANCIAL MANAGEMENT IN HOUSEHOLDS— THE RESEARCH RESULTS IN PODKARPACKIE VOIVODESHIP

As far as the characteristics of the respondents are concerned, it should be mentioned that exactly 61 people took part in the pilotage survey and 55% of them were interviewees with higher education. However, 35% of the respondents had secondary and post-matura education. The smallest group included people with elementary and vocational education—only 10%. The vast majority of the respondents was in age range from 31 to 40–69%, people who were19 up to 30 years old belonged to the group of 20%. However, the least numerous group consisted of the respondents aged 41 to 50–11%. Moreover 55% of the people questioned claimed that they had 2 children to maintain, 35% of the respondents 3 children and 5% – "maintained 1 child" or "maintained 4 children".

By analysing the survey results, it should be noticed that according to the interviewees financial means received from the 500+ programme play a crucial role in the households budgets because they improve financial situation of a family to a great extent and provide higher standard of living. As it can be seen in the Figure 1 most of the respondents

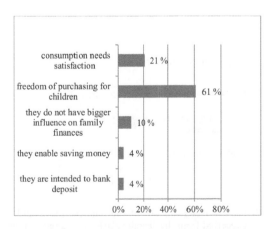

Figure 1. The role of financial means received from the "Family 500+" programme in the household budget.

(61% of the answers) declared that the financial means received from the 500+ programme give freedom of decision while doing shopping for children. The answer "they fully allow to satisfy everyday consumption needs" appeared on the second place (21% of the answers—Fig. 1). While the third place, indicating 10%, appeared the answer "they do not have bigger influence on the family's finances"

Moreover, more than half of the respondents (54%) claimed that the income from the 500+ programme influenced financial situation of the family in a very significant and noticeable way (Fig. 2). About 30% of the Podkarpackie inhabitants think that financial situation of their families changed only to some extent, but only 16% claim, that "it has not changed in significant way".

The respondents also specified what they were going to do with the money received from the "500+". The greatest interest of the received answers (32%) showed that the people who took part in the survey were going to buy products only for children for this extra income (Fig. 3). The following places were taken by "everyday shopping for the whole family" (20%) and "financial means (savings) saved to satisfy the future needs of the children" (19%).

It is worth mentioning that a great percentage (28%) of the answers were the following— "financial means from the 500+ are distributed to different aims".

It also results from the conducted survey that within the amount of money dedicated to shopping only for children, the most is spent on buying food (26% of the answers—Fig. 4). The second place took ex aequo per 16% – books purchase and expenses on learning foreign languages. About 25% of the finances from the Family 500

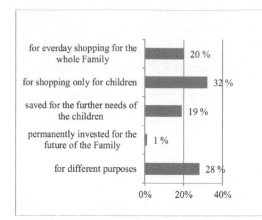

Figure 3. Allocation of the financial means received from the "Family 500+" programme.
The source: Own elaboration.

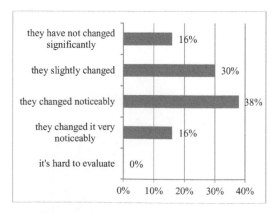

Figure 2. Determination of the financial means influence received from the "Family 500+" programme on the financial situation of the family.
The source: Own elaboration.

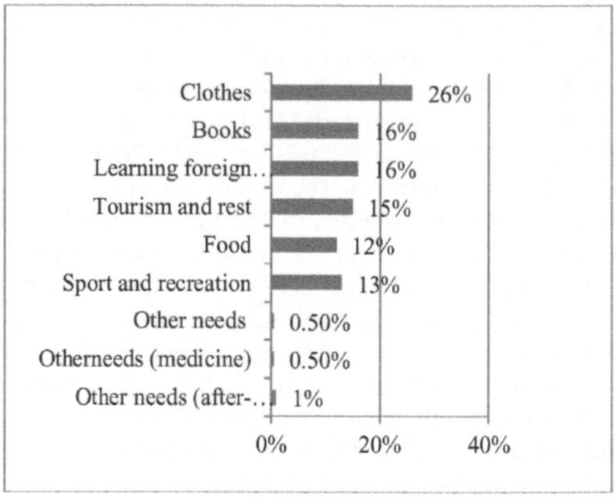

Figure 4. The structure of shopping dedicated only for children from the "Family 500+" programme income.
The source: Own elaboration.

plus programme is spent on tourism, leisure, sport and recreation and on purchasing food only 12%. The respondents also use extra income to buy electronic equipment, medicine or pay for extra lessons for children (for example singing lessons or playing a musical instrument lessons).

The aim of the conducted analyses was also to define if families who use the government educational programme "Family 500 plus" would accept other form of (than it is functioning at present) receiving this benefit Fig. 5). The respondents who would approve specified in the questionnaire forms of benefits wrote "1 – yes", however those who did not agree with the presented option marked "0 – no".

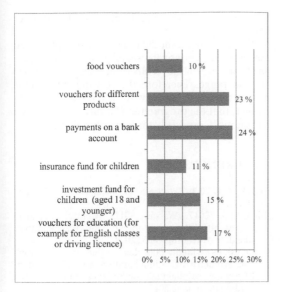

Figure 5. The respondents' opinion concerning different forms of benefits, which could be offered by the government instead of paying 500 zł within the "Family 500+" programme.
The source: Own elaboration.

The biggest group of the respondents (24%) claimed that the financial means achieved from the "Family 500+" programme could be paid on the children's account. 1% fewer people who took part in the survey would accept the form of vouchers to do any shopping. In turn, according to 17% of the inhabitants of Podkarpackie the benefits within the "Family 500+" programme could be paid in educational vouchers for children, which would be used to pay for extra education for children: private lessons, driving licence, learning foreign languages or other useful vocational skills and qualifications. The considerable percentage of the respondents –15% claims, that it would also be justified to pay on the investment fund for children, to use when they turn 18. However, 11% the inhabitants would agree to pay the money as investment fund for children and 10% would willingly receive food vouchers.

The direct research also concerned getting information about: "Who should not receive financial help from the country within the" Family 500+ "programme, even if they have adolescent children" – Figure 6.

According to 71% of the respondents, child-support benefits from the "500+" government programme should not be received families considered to be rich (wealthy). Significantly lower percentage of the people, who took part in the survey –14%, claim that it is inappropriate extra monthly income in the form of 500 zł is received by parents who

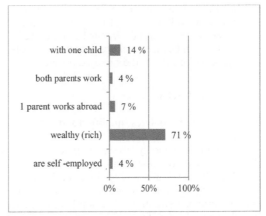

Figure 6. The respondents' opinions concerning the problem—"Who should not receive financial benefits from the "Family 500+" programme".
The source: Own elaboration.

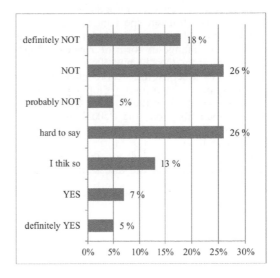

Figure 7. The respondents' opinions—"Will financial means from the "Family 500+" programme influence people to plan another baby in your family?".
The source: Own elaboration.

have one child. The respondents also selected the households in which one parent works abroad (7%), where both parents work abroad (4%) and the ones, which are self-employed (4%) as the ones that should not use the child-support benefits.

The respondent also defined their plans in the scope of enlarging a family due to additional income influencing their household budgets (Figure 7). It results from the conducted analyses that about 50%

521

of the respondents are not going to plan to have another baby and 26% do not have opinion in that matter. However, financial benefits from the "500+" programme can be an economical impulse to procreation process for 25% of the surveyed people.

5 CONCLUSIONS

The conducted analyses qualify to draw some interesting conclusions.

It should be generally stressed that according to the respondents the financial means received from the "Family 500+" play a crucial role in the households' budgets. They improve, to considerable extent, financial situation of a family and ensure it higher standard of living. They are mainly spent on purchasing for children they but also allow fully satisfy current consumption needs of the whole family. As far as shopping for children is concerned, clothes and books are bought the most frequently. Moreover, a considerable part of income from the "Family 500 Plus" is spent on paying for extra lessons for children (especially foreign languages) and on tourism, sport and recreation.

The surveyed people expressed their approval towards the change of form of receiving financial means from the "Family 500+" programme. They mainly accepted transfers to children's bank accounts and vouchers for any shopping.

It also results from the research thatthe child-support benefits from the "500+" programme should not be received by families considered to be rich (wealthy).

What is more, for half of the respondents, income received from the discussed educational programme is not enough impulse to plan another child.

Summing up, it should be stressed that presented results of the empirical research in the present publication confirm the results of market observations in this range. Therefore the research hypothesis for further, representative direct research can be drawn, that the programme "Family 500 plus" is a big support, especially for large poor families.

REFERENCES

Aczel, A.D.2000. *Statistics in management.* Warsaw: PWN.

Birth rate. 2017. Warsaw: GUS (Central Statistical office). http://stat.gov.pl/obszary-tematyczne/ludnosc/ [the date of reading: 3.03.2017].

Conclusions—"family 500 plus" online. 2017. Warsaw: The Ministry of Administration and Digitization. https://mc.gov.pl/aktualnosci/rodzina500plus [the reading date 10.02.2017].

Large families in Poland—report. 2016. Warsaw: Ipsos Polska. http://www.3plus.pl/wielodzietni-w-polsce-raport-2016 [data odczytu: 28.02.2017].

Polish family in the countries of European Union—conclusions for Poland. 2009. Edited by Balcerzak-Paradowska, B. & Szymborski, J. Warszawa: Biuletyn RPO—materials number 67.

2017. S&P. 17mld zł for the Family 500 plus programme. 2017. http://www.bankier.pl/wiadomosc/Peklo-17-mld-zl-na-program-Rodzina-500-plus-w-2016-r [the reading date: 15.02.2017].

The Act of 11 February 2016 about the help of the country in bringing up children enforces the "Family 500+" programme [Journal of Laws from 2016 item 195].

New Trends in Process Control and Production Management – Štofová & Szaryszová (Eds)
© 2018 Taylor & Francis Group, London, ISBN 978-1-138-05885-9

Improving the quality of production by the eight disciplines problem solving method

M. Šolc, L. Girmanová & J. Kliment
Faculty of Metallurgy, Technical University of Košice, Košice, Slovak Republic

A. Divoková
Faculty of Mechanical Engineering, Technical University of Košice, Košice, Slovak Republic

ABSTRACT: The article deals with the issue of process improvement and quality in practice. As this instrument is primarily used in organizations, which deal with automotive industry, article describes the quality assurance process in the automotive industry. The next part of the article deals with IATF Characteristics Standard 16949:2016, which is currently devoted to the quality management system in the automotive industry. Subsequently, the article describes the tools GLOBAL8D, preparation before implementation in the organizations, requirements for progressive implementation and the individual steps implemented in practice. Conclusion the article describes the experience in implementing tool GLOBAL8D in the practice, its benefits respectively benefits to organizations, process of the improvements of the implementation tool in organizations based on customer feedback after some time.

1 INTRODUCTION

Automotive industry is related to the secondary sector, which is engaged in motor vehicles developing, manufacturing, marketing and sale. In the automotive industry, all automotive producers, relevant service part organizations and sub-suppliers are included. The whole industry is under intense competitive pressure, there are high demands and requirements for safety and quality. That's why the automotive industry is among the bearers of modern methods of management. Problems may arise in any sector of the organization, regardless of the overall success of the organization and can take any form and shape. Commonly, the problems are often the result of multiple causes at different levels. To solve the problem, it is necessary to identify the possible causes of the problem and to find ways for the causes elimination in order to prevent the recurrence. GLOBAL8D is one of the tools that can be used to solve the quality problems. The process of problem solving by the GLOBAL8D tool is used to identify, correct and eliminate the problem. Global8D tool is based on a standardized procedure, which puts emphasis on facts in order to contribute to the products and processes improvement.

2 QUALITY ASSURANCE PROCESS IN THE AUTOMOTIVE INDUSTRY

2.1 *Preventive versus reactive approach*

All areas of modern industry have adopted a standardized set of tools and methods used in designing processes and communicating their performance (Evers 2011). Quality control stories and 8D methods have their share of success in the industries. These techniques have proven again and again in many industries across the world with varying degree of success (Lanke & Ghodrati 2013).

In order to assure the quality in automotive industry, it is necessary to meet the IATF 16949:2016 requirements, as well as the specific customer requirements in connection with the internal requirements of each organization, including legislative requirements. Already, quality has to be ensured in the product and process development phase. Then, consistent high quality product delivered on time, in the required amount is the output of the mass production process (Kliment & Šolc 2016). The use of core tools is considered as important element of quality management throughout the "life cycle" of the automotive industry products, such as APQP (Advanced Product

Quality Planning), FMEA (Failure Mode and Effect Analysis), SPC (Statistical Process Control), MSA (Measurement System Analysis) a PPAP (Production Part Approval Process) (Kužma et al. 2016). The use of these methods is not only an appropriate element of management but it is also a requirement of IATF 16949:2016 standard and customers, as well. Without the use of documentation, application of these methods is hardly feasible in practice. (Mičietová & Šulgan 2010).

It should be emphasized, the approach to quality management and system documentation updating can be reactive (caused by external complaint or customer audit), but also preventive (e.g. regular FMEA team meetings, processes improvement). (Nagyová & Markulik 2014). "What is the difference between reactive and preventive approach to quality?"

Reactive approach: Generally, the reactive update of such documents may be caused by internal or external customer complaints. GLOBAL8D (8D Report) is one of the methods for similar problems solving. It consists of several steps designed to identify the problem, create the solving team, to find the root cause of a problem, implement the corrective and preventive actions and verify effectiveness of the actions. Revising the system documentation is one of the significant activities associated with the problem solving (if it is required by the nature and type of the problem). If there are changes in the actual production process, it is necessary to revise the set of following documents: flow chart of a process, FMEA process, management plan, work instructions. In practice, this fact is often forgotten. Reactive approach is clearly seen in this example (the need for changes in the process, completion of a list of possible errors, review/completion of existing methods for product/process control have been caused by customer complaint) (Šurinová & Kliment 2014).

Preventive approach: the counterpart to the reactive approach. Reverse FMEA (known as "Go and See" in English) is one alternative to this approach. It is a proactive approach tool for identification of risk points in the process in order to prevent both the occurrence of nonconformities and subsequent updating of relevant documentation (Weber 2010).

2.2 Sustained success of an organization according to IATF 16949:2016

To meet the customer requirements for quality, reliability, safety as well as automotive service, system actions have to be carried out by the automobile manufacturers and automotive industry suppliers (Pauliková et al. 2016). Requirements for quality are even increased by both competition between automobile manufacturers who transfer their requirements to suppliers and competition between automotive industry suppliers

(Teplická 2015). Quality management system built on the quality management principles is able to help to meet these demanding requirements. Everyone who wants to be a supplier to the automotive industry is forced to unconditionally meet the requirements of generally accepted standards. In the field of automotive industry, IATF 16949:2016 Quality management system for organizations in the automotive industry is one of the most widely used standard for quality management (Šolc & Kliment 2016).

According to this standard, the goal of the certification is to demonstrate by verifying that the quality management system is functional and effective and customers can confidently expect a desired constant level of quality of production. Such verification is also a guarantee that the organization is likely to meet additional specific requirements of the automotive industry and customers. (Kliment & Šolc 2015).

Certificate IATF 16949:2016 issued by certification body confirms, that the organization in a competitive environment meets the compliance with globally recognized quality standards applied in the automotive industry. After successful IATF 16949:2016 certification, organization proves its capability and the competitiveness of the organization is increased. The confidence of interested parties and public is also increased. Ultimately, the number of second-party audits and customer visits could be reduced.

3 GLOBAL8D METHOD

3.1 Characterization of the GLOBAL8D method

In many areas, it is difficult to define the term quality. It may be related to products or services, can be focused both on customer satisfaction and on the added value creation (Šolc & Kliment 2016). Generally, there are two approaches to quality improvement (Kliment & Šolc 2015):

Reactive approach – solution of the problem that has already occurred and corrective actions implementing (Global 8D);

Proactive approach – preventing the problems (FMEA, control plan, Poka-yoke, ...).

It is possible to use a large number of support tools, which deal with different partial processes increased the quality level (Kotus et al. 2011).

The Global8D method is a basic troubleshooting methodology, which is often used when dealing with complaints, particularly in the automotive industry (Kaplík et al. 2013). The method is based on the PDCA or Plan-Do-Check-Act improvement cycle, also known as the Deming or Stewart cycle (Purdy et al. 2010). Global8D is very effective standardized method of saving time and money. It is usually applied, when the problem or complaint with unknown causes is suddenly appeared. The problem has to be efficiently solved as soon as possible and particularly, the customer has to be protected from adverse consequences (Heck & Smith 2005). Global8D process involves the use of teamwork and team judgment, use of the work with data from related processes and similar problems or data derived from the problematic process. Integration of different quality management methods and tools can be also applied. By introducing the interim containment and corrective actions (e.g. use of repeated measuring or complete retesting on a different test equipment), Global 8D prevents the nonconforming items to get further to the customer. These steps are used for temporary quality assurance and enable us to continue in the production process. Global8D is designed as a practical simple worksheet, but its completion is usually not quite easy.

3.2 The sequence of GLOBAL8D method steps

The term "process troubleshooting" means the procedure for the systematic implementation of improvements and the elimination of problems and failures. In general it can be described as a sequence of steps to be taken if the problem is obvious (Fig. 1). It helps to find the correct, timely

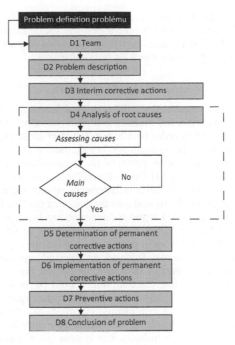

Figure 1.   Process scheme GLOBAL8D.

Table 1.   The sequence of GLOBAL8D method steps.

Step	The characteristic of steps
D1	Team approach: Establish a team of people (2 to 10 people) with the product/process knowledge, who will solve the problem
D2	Problem description: Specify the problem to provide as much detail regarding the current issue and deal with the analysis of interested parties
D3	Interim containment actions to prevent the damages: Stop the process, propose the containment actions to prevent damage, propose a plan for the implementation of containment actions and verify its functionality
D4	Root causes analysis: Determinate the possible causes of the problem, analyze the root cause and verify its accuracy
D5	Design of permanent corrective actions: Choose and verify the permanent corrective actions and make the risk analysis
D6	Implementation and validation of permanent corrective actions: Validate the effectiveness of corrective actions and monitor the process
D7	Prevent recurrence: Identify and implement the practices and procedures to prevent recurrence of this and similar problems
D8	Close the issues and recognize the efforts of the team: Recognize the collective efforts of the team, the team needs to be formally thanked by the organization

and complete solution of the problem. Global8D method can be used for improvement of both the product and the process. Its use is primarily focused on identifying problems related to product (customer complaints, as well as the internal complaints within the organization). Global 8D is the standardized method based on the following principles:

– Facts-oriented system – based on real data and controlled data collection.
– Root cause identification – to address the root cause of the problem, not only the results of the problem.

4   CHARACTERISTICS OF THE ORGANIZATION

4.1   Final products of the organization

In recent years, organizational production program is going through a change, from the electrotechnical industry to the production of plastic parts for the automotive industry. There is still a broad product range in the organization. The core products of the organization are: electronic products (telephones, home intercom, cash registers, electric installation material and tubes, connecting forks, lamp sockets, horns and bells, building locks and hinges), plastic products (parts for gardening equipment, e.g. mowing-machine decks, parts for their own final products). An important part of the production is focused on components for the automotive industry: plastic components for connectors, metal and plastic components for automotive locks, interior parts (dashboard, car center consoles, car door trims), piezo ceramics for parking sensors.

4.2 The initial GLOBAL8D template

In the organization, there is established a quality management system, that shall be continually improved in order to enhance customer satisfaction and also to bring maximum benefit itself with regard to the future development of the organization. Problem solving by GLOBAL8D method is an attempt to meet the customer requirements, resolve the customer complaints and gain the customer confidence.

Using the 8D report is due to the involvement of organization in the automotive industry, in which this method is most commonly applied. In organization, 8D report is issued on demand and most often it is a part of customer complaints. Based on this process, internal complaints could be resolved, what can lead to improvement within the organization (process of continuous improvement). As the organization has long been active as a supplier to the automotive industry, the system troubleshooting using the GLOBAL8D method has already been set. The established 8D template as well as the 8D method steps had to be changed in order to reduce the amount of internal complaints and retain a competitive advantage. To save the time and energy, the template shall be extended and integrated. The aim of this process is to regain the customer satisfaction and minimize the negative consequences.

4.3 Design for a new GLOBAL8D template

The structure of the new 8D template has been designed with respect to the original template form. The optimization of 8D template has been carried out with the intention to work with the template more effectively. The whole template form has been changed. All steps have been adjusted in accordance with current standards for better clarity for the customer and for the worker using the template. To identify the problem as accurately as possible, the pre-defined sorting have been added to the template. Instead of the original three-language, the template has been modified to dual-language (Slovak and English).

In the first section of the template, there is a specification of the team, the names of researchers and the team member contacts.

In the second step, the user shall specify the problem to provide as much detail regarding the current issue and then describe the problem as accurately as possible. The problem description is very important for the controlling of other nonconformity. Therefore, the picture documentation, problem identification and the specific questions

for better and more broadly problem description have been added in this section.

In the third step, pre-defined sorting and also specific questions to select the correct interim containment actions have been added. In this point, the boxes for the recording the responsible person and the date when the interim actions were introduced have been added.

The fourth point is one of the most important points. It is necessary to complete this section on the basis of data obtained from the previously implemented analysis. In this section, the most appropriate and generally recommended method 5x Why has been well established in organization. There is another separated template for 5 Why method. On its basis, the fourth section of the 8D GLOBAL template can be completed. Therefore, this point has not undergone any larger changes. Only one pre-defined question about the importance of the problem has been added. 5WHY technique of asking questions is appropriate to use the quick solution to a particular problem, where the results of analyses based on data collection from the past are not, whether any reason not available, or because the advanced statistical tools could not be used (Folta 2012).

In the fifth, sixth and seventh point, as well as the third point, it is necessary to record the responsible person and the date of introduction, completion and verification of corrective actions. Validation of the corrective actions effectiveness has been introduced in the sixth point.

The final point is determined by customer feedback. If the customer is satisfied with the resolution of the complaint, it is possible to close the issues and recognize the efforts of the team.

4.4 Case study: application of the new GLOBAL8D template in the process of complaint

Complaint process is usually caused by deviation from the defined requirements. The deviation has been notified by the customer to the supplier. Plastic connector (8-pole connector) was the subject of the complaint. In the complaint process, the new 8D template was applied.

Plastic component Plastic connector (Fig. 2 and Fig. 3) is used to contact the rear view mirror. Plastic connector is pressed from the plastic granulate of PBT type on the ENGEL VC 200/80 press (5.2 t). The forming unit for this component pressing is 8-fold. Pressed pieces removing runs automatically. The pressed pieces are controlled in 4-hour intervals.

At the beginning of the complaint process, a team of people involved in the root cause identification and corrective and preventive actions determination was established. Since the departments that are important in the process of complaint are pre-defined in the template, so the task was to find

Figure 2.   8-pole plastic connector – side view.

Figure 3.   8-pole plastic connector – a front view.

a people with the relevant knowledge of product/ process, as well as the knowledge in the techniques needed to solve the problem and to implement the corrective actions.

In the second step, the problem has to be described. For the complaint resolving, the main task was to specify the problem to provide as much detail regarding the current issue and then describe the problem as accurately as possible. In this complaint, the burr inside the connector was the problem. For better clarity, the defect was photographically documented and added to the template. In order to analyse the problem, it was necessary to gather all relevant information. The analysis showed, that the problem was in the hole no. 3. There was the burr and thereby the contact knife did not snap in the hole and this could lead to withdrawal of the contact knife. By pre-defined questions, the problem was identified in more detail, the deviation was determined. It was found, where and when the problem occurs, what is the extent of the problem and who is affected.

In the third step, the interim containment actions have been proposed. This step is optional, but in this case it was necessary to introduce the interim actions in order to prevent that the customer receive more nonconforming parts. As the first, the sorting of parts in the warehouse of the organization and the customer's warehouse was determined. On the road, was not necessary

to separate the parts. Furthermore, three other interim containment actions have been designed and implemented: the immediate expedition stop (the first day the complaint was received in order to prevent the customer damage); sending the nonconforming parts to customers for analysis; request the photo label of the box, in which the non-conforming parts were found. The effectiveness of the provisional measures was 100%.

The root cause was identified in the fourth step. This section was completed on the basis of data obtained from the previously implemented analysis. For analysis, the 5 Why method was applied. After this method performing, it was found that the main cause of the problem is the burr inside the connector, because of the punch damage in cavity no. 5, in hole no. 3 see. red arrow in Figure 4.

After the root cause determination, the permanent corrective actions were proposed:

1. Form repairing—changing the punch in the cavity no. 5
2. Checking the insertion of the contact knife to the hole no. 3 in the cavity no. 5, after correction of the form.

Both actions have been verified. Subsequently, the permanent corrective actions have been implemented and validated. In the next step in this complaint process, the practices and procedures to prevent recurrence of these nonconformities have been implemented. There have been three preventive actions: Visualization of the defect (nonconformity) and location of the image with conforming and nonconforming parts in the workplace, control plan updating and FMEA P updating.

The difference between the initial and new designed template is considerable. The new template was designed according to current standards. Compared with the initial template, working with the new one is more simple and understandable both for employees and customers. One of the main benefits for the organization is modification from the original three-language template to dual-language (Slovak and English).

Figure 4.   8-pole plastic connector/error – a front view.

Already, the first step showed that defining the departments from which people should be involved in problem solving is effective. Filling this point in the process the complaint took place quickly and with no problems. Since the departments have been pre-defined, it was necessary only to choose people from different departments. That resulted in the time saving.

In describing the problem in the second step, this step also proved to be effective. In the process of complaint it was beneficial that the description of the problem was defined in more detail. The pre-defined questions have been helpful in determining the root cause of nonconformity as well as in determining the interim containment actions and permanent corrective actions.

In the third step, define the sorting proved to be very effective. More precise data on how many parts are nonconforming and where they occur have been available. Sorting as well as increasing the efficiency of the actions has proved effective, because the functionality of all actions was verified.

Adding the validation in the sixth step has been the other change made in the template. It helped ensure that the selected corrective actions were right and functional.

From the employees point of view, new template structure is clear and transparent, and in completing some parts of the template was also saved the time.

5 CONCLUSION

Global8D realization leads to a rapid and permanent removal of nonconformities. It seems that by following of 8D method step, each received complaint could be reliably solved. The entire process of Global8D application leads to the desired result, but also depends on other important aspects. This approach ensures that problem solving, decision making and planning are based on the data in order to be properly solved a real problem and not just the consequences that this problem mask. The effectiveness of this method depends on the cooperation between the employees. This method requires open communication, due to difficulties that can arise in in dealing with problems of different nature.

ACKNOWLEDGEMENTS

This contribution is the result of the project implementation VEGA 1/0904/16: MINIMAX-3E The utilization of processes capability and performance and products dimensional tolerances in the management of material consumption and related economic, energy and environmental consequences.

REFERENCES

Evers, C. T. 2011. An application-based graduate course in advanced quality tools. *ASEE Annual Conference and Exposition, Vancouver, 26 June 2011.* Vancouver Convention Centre.

Folta, M. 2012. Problems solving model in supply chains. CLC 2012: *Carpathian Logistics Congress, Priessnitz SpaJeseník, 7–9 November 2012.* Jeseník: TANGER Ltd.

Heck, D. & Smith, L. 2005. What abour Ford's Global8D tool? *Quality progress* 38(5): 8–10.

Kaplík, P. et al. 2013. Use of 8D method to solve problems. *Advanced Materials Research* 801: 95–101.

Kliment, J. & Šolc, M. 2016. IATF 16949—issue a new standard of quality management in the automotive industry is coming (in slovak). *Quality* 24(3): 26–29.

Kliment, J. & Šolc, M. 2015. Correlation of quality and risks in the automotive industry (in slovak). *New trends in safety and health, High Tatras, 18–20 November 2015.* High Tatras: TU Kosice.

Kliment, J. & Šolc, M. 2015. Global 8D—the tool of improvement quality in the automotive industry (in slovak). *Kosice safety review* 5(2): 156–161.

Kotus, M. et al. 2011. *Quality technological improvement.* Trnava: Tripsoft.

Kužma, D. et al. 2016. Application of PLM Software NX in practice (in slovak). *Transfer of innovation* 33: 62–67.

Lanke, A. & Ghodrati, B. 2013. Reducing defects and achieving business profitability using innovative and lean thinking. *IEEE International Conference on Industrial Engineering and Engineering Management, Bangkok, 10–13 December 2013.* Bangkok: IEEE Computer Society.

Mičietová, M. & Šulgan, M. 2010. Use of innovative technologies in automotive industry (in slovak). *Transport and Communications* 5(3): 1–8.

Nagyová, A. & Markulik, Š. 2014. Application of Balanced Scorecard methodology in automotive industry. *QMOD-ICQSS: Quality Management and Organizational Development, Prague, 3–5 September 2014.* Prague: Lund University Library Press.

Pauliková, A. et al. 2016. QFD—support to higher efficiency of industrial automotive production. *Production Engineering Archives* 10(1): 21–24.

Purdy, K. et al. 2010. Defect reduction through Lean methodology. *SPIE—The International Society for Optical Engineering, Monterey, 13 September 2010.* Monterey: Photomask Technology.

Šolc, M. & Kliment, J. 2016. Development of the automotive industry and process quality assurance in the automotive industry (in slovak). *Quality, technologies, diagnostics of technical systems, Nitra, 24–25 May 2016.* Nitra: SPU Nitra.

Šolc, M. & Kliment, J. 2016. The process of identification security risks in the automotive industry. *SGEM, Sofia, 28 June–06 July 2016.* Sofia: STEF92 Technology Ltd.

Šurinová, Y. & Kliment, J. 2014. Reactive versus preventive approach to quality in the automotive industry (in slovak). *Quality* 22(1): 14–16.

Teplická, K. 2015. *Optimization methods and their application in practice (in slovak).* Kosice: TU Kosice.

Weber, J. 2010. Towards an aspect driven approach for the analysis, evaluation and optimization of safety within the automotive industry. *SAE Technical Paper.*

New Trends in Process Control and Production Management – Štofová & Szaryszová (Eds)
© 2018 Taylor & Francis Group, London, ISBN 978-1-138-05885-9

Assessing the impact of current trends in payment services

L. Štofová & P. Szaryszová
Department of Management, Faculty of Business Economics with seat in Košice, University of Economics in Bratislava, Slovak Republic

V. Serzhanov
Faculty of Economics, Uzhhorod National University, Uzhhorod, Ukraine

ABSTRACT: Significant trends in banking include the development of electronic banking and the expansion of services offered in this way. The basic feature of electronic banking is the delivery of services anywhere and in any circumstances. It basically changes customers' access to the use of bank products. For clients and banks, it provides an immense amount of opportunities and is a determining factor influencing customer satisfaction. Its content, status and development depend on demand, legislation and technological progress. The authors of the paper are focused on evaluating and comparing the business conditions of selected providers of payment services for businesses. The main aim of the paper is to analyze the impact of current payment services on business users by means of a nonparametric method of measuring the effectiveness of information, increasing the protection, availability and especially the prices for the use of payment services.

1 INTRODUCTION

Sustainable development of information technologies and their accessibility in everyday life extends the quantity and quality of information channels between financial institutions and its clients. For bank customers, the introduction of electronic banking means the possibility of using new attractive banking products. The client can communicate with the bank quickly, comfortably and without obligation of official time (Olexová, 2014). That is why clients are increasingly beginning to require the introduction of e-banking, which is another reason for the banks to introduce and further expand this service.

New data confirm that the European Commission is well on track to achieve digital goals, which also means considerable progress for the banking sector. Citizens and businesses in the EU are increasingly using the internet, buying more online, and increasingly trusting information and communication technologies that are also much more in control.

2 CURRENT TRENDS IN THE FINANCIAL MARKET

2.1 Safety of the functioning of the financial market

The credibility and stability of the financial market and its institutions is one of the basic conditions for a proper functioning of the economy. This trustworthiness and stability can not be ensured by market mechanisms alone, and therefore the activity of financial market operators and, therefore, payment service providers and electronic money issuers are regulated by a number of restrictive and prescriptive rules, in particular in the form of legislation (Sylvie & Xiaoyan 2005).

For the proper functioning of modern financial markets, it is also necessary to ensure their security. The security of business in the financial market is one of the problems that every time solves according to existing options (Ferencz & Dugas 2012). The question of the safety of the functioning of financial markets has many different assumptions and dimensions, with the aforementioned problem being decomposed, for example, to the following elements (Roztocka & Weistroffer 2015):

– The technical aspect of the functioning of the financial markets, when it is necessary to base the functioning of the market mechanism on reliable technical equipment. The proper functioning of the financial market is in many ways conditional on the sound operation of their infrastructure. Especially nowadays, as more and more subscribers make use of internet banking or make online transactions via online payments, the risk of attacking the software of a payer or a payment institution increases every day.

– The economic aspect of the functioning of the financial markets, where the efficient functioning of markets is essential. Here is the need to look for cost savings, introduce technical, economic and organizational innovation, etc.
– A legal aspect ensuring the functioning of financial markets by quality legislation, their maintenance and timely renewal, which also entails the need to adapt legislation at the current level of technology and the requirements of economic efficiency.
– The psychological side of financial market participants, ensuring the appropriate behavior of financial market users. The psychological side includes the necessary level of financial literacy of financial market participants and is associated not only with the functioning of its own financial market trading but also with the functioning of the elements of the wider financial market infrastructure (rating, media, research, osvetta). In order to ensure the appropriate behavior of financial market users, it is necessary to improve the financial literacy of individual participants, in particular by promoting financial education, which is increasingly being paid attention to in the Slovak Republic as well as in the other European Union countries. The safety of the functioning of modern financial markets can not be ensured without adequate financial regulation.

The approach to regulation of financial markets, of payments has in the past been characterized by the micro-prudent nature of regulation, while the current approach emphasizes the macro-prudential nature of regulation, and this change is already institutionally embedded and is crucial for the functioning of the financial markets and their security (Gallo & Mihalčová 2016). Whilst considerable emphasis has been placed on repression, today there is a strong emphasis on prevention, which also plays a major role in self-governance and progress in the management and governance of financial organizations (Antošová et al. 2014).

2.2 Trends in development of payment services

In the world economy, radical changes can be observed recently. It is the gradual globalization of financial markets, the expansion of capital markets, securitization, the progressive development of information and telecommunication technologies and social change, which are also driven by current trends in banking (Table 1).

These banking trends indicate that banks are going through a turbulent period. After a long period of essentially unchanged classic banking, current banks are exposed to almost continuous radical changes.

Table 1. Current trends in the banking sector.

The transition from the classical model to the transition model banking	In the classical model, the bank performed a an intermediary between creditors and borrowers, providing a single service within a single entity. On the contrary, the transitional model is based on the fact that individual activities are no longer carried out within one entity but through specialized institutions.
Consolidation	As a result of the slackening of the banking sector and increased competition, the number of banks is decreasing, their trading network is reduced and the number of employees decreases.
Mergers and acquisitions	Banks are now characterized by a largewave of mergers and acquisitions, resulting in larger and stronger entities.
Reducing costs	It is in the strategic intentions of each of the world's most important banks, because it is the least risky way to increase the profitability of banks of their competitiveness.
Income diversi-fication	Expresses the desire for banks to maximize revenue sources, diversify them, especially product and geographic. Banks enter the product dimension into other activities such as banking and insurance interconnection, client asset management, and greater investment banking investment. Geographical diversification means an effort to diversify incomes with regard to the place where these profits are generated.
Gradual globalization	Gradual globalization means that the place and time for the conduct of banking business is becoming less important, which is also supported by the gradual convergence of regulatory and legal regulations between countries.
Using new information technologies	It represents a change in client access and a change in traditional ways of delivering products to clients. The consequence is the growth of self-service zones and e-banking products.
Social change	These are changes in social security models, demographic impacts, increasing population mobility, rising living standards of the population and reduced loyalty of clients, which also act on current trends in banking.
Raising value for shareholders	The current trend is an increase in this figure, which is expressed as the ability of the bank to generate an inflow of money in the long run to exceed the cost of capital. It is a comprehensive expression of the need for a high-quality financial management system for the bank.

3 METHODS AND MATERIAL

The field of electronic banking evolves very quickly, depending largely on the latest technologies. It is therefore generally difficult to estimate its current level. What is once introduced as a new bank is already part of the offer of another bank. Banks are aware of the importance of electronic distribution channels and the need to orient their strategy primarily to clients in order to satisfy their wishes and needs. The establishment of more personal relationships with clients is also acknowledged by European banks. The level of service offered is also significantly higher.

3.1 Research methodology

In this study, we focused on internet banking, which represents a modern, complex management variant as a current account, Credit card. Charges and options for individual services can be obtained from the bank pricelist available on the websites of individual banks, or by verifying the bank's infoline. As a research method for analyzing clients' satisfaction with the electronic banking of individual banks, we chose an informally managed interview, in which respondents (bank clients) had to point out the individual service criteria of selected banks. We conducted the survey on a sample of 54 respondents. They were business managers and managers as a client without a difference in the length of Internet banking. The information obtained is without mentioning the business name, as the respondents respond much more sincerely and spontaneously. The survey sample consisted of production agricultural holdings, car industry, assembly electronics enterprises, pharmaceutical companies, food businesses, tourism businesses and small businesses.

In addition to selecting banks, it is important to determine how the quality of selected banking services will be assessed. Quality can be perceived by each client, in the following part, it is characterized by criteria that should reflect the quality of the service while maximizing it. Based on the study of domestic and foreign literature, it is unclear what these criteria should be. In this research, based on the analysis of empirical studies, criteria were used: functions, security, fees, user interface, which also focused on questions in the informal interview with clients of banks.

Based on the obtained results, authors compare the selected payment services between banks from the point of view of the payment policy. In the analysis, quantitative scientific research methods as well as informally managed debates in the banking environment are used to map attitudes and opinions on the use of electronic services by selected banks. Consequently, it is intended to identify the effectiveness of Slovak commercial banks using the Data Envelopment Analysis (DEA) method.

DEA assumes no random error and performs relatively simple programming partial formulations. Each consumes varying amounts of inputs and produces different outputs, Decision Making Unit (DMU) consumes xii input quantities for output y output quantity. It is assumed that these inputs (x_{ij}) and outputs (y_{ij}) are non-negative. The DMU efficiency can then be written as:

$$h_j = \frac{\sum_{r=1}^{s} u_r y_{rj}}{\sum_{i=1}^{m} v_i x_{ij}} \tag{1}$$

where h = efficiency rate; j = indicator of n different DMUs; r = indicator of s different outputs; i = indicator of m different inputs; u and v = weights assigned to each input and output; x = inputs; y = outputs.

By using mathematical programming techniques, DEA assigns optimum scales under the given conditions. The scales for each DMU are assigned under the conditions that no other DMU has efficiency greater than one if it uses the same weight, which means that the effective DMU will have a ratio of one.

Additional data from the annual reports of selected commercial banks were also used to estimate the efficiency of the banking sector. When estimating efficiency, inputs and outputs are first defined. As suggested by Berger & Humphrey (1997), there is no consistent agreement in the literature on the subject of inputs and outputs to be used in the bank efficiency analysis.

With regard to the range of the set that were identified in the study by Stavárek (2005), it indicates the significance of the number of factors included in the analysis, which significantly affect the results of efficiency using non-parametric techniques. The excessive number of variables artificially increases the number of effective units and reduces the discriminatory power and the ability to analyze the ability to report. The empirical literature states that the number of units to be evaluated should be at least two or three times greater than the sum of the input and output variables in the model.

4 RESULTS OF THE MEASUREMENTS

When using electronic banking, security is the most important aspect. Empiric research can be said that one of the main factors in choosing a bank is its credibility. Users do not have to worry about losing their funds in these overbought banks. Compared to security, we focused on the form of signing in to the system. As well as the level of security, the price associated with the use of electronic banking is an important factor in the decision-making of the users that the bank chooses. In these criteria, we have focused on all fees associated with the use of electronic banking.

4.1 System aspects of electronic banking

Since each criterion has a different weight for potential clients and each criterion has a different number of categories, the individual scores are converted by weight into successive features, security, fees, and user experience. The recalculation of points has been chosen to take into account the number of categories that are different for each of the criteria. The greatest emphasis is placed on the security of internet banking, but since a higher form of security has been chosen for all services, this weight is now assigned to the same weight as functions and fees. On the other hand, the user environment interface criteria have a lower weight, since they are additional (supporting) elements when using Internet banking services compared to other monitored elements. In the following Table 2, we calculate the points weighted by the scales.

The winning BusinessNET service is from UniCredit Bank Slovakia, but given that the bank offers a retirement allowance for other businesses through another company. It does not offer the possibility to have an overview of the contract and the pension insurance account in Internet banking. Which was one of the preconditions set by potential clients. However, all conditions are met by ČSOB Inter-netbanking, which was ranked as the second and therefore this service should be recommended to clients.

4.2 Assessing the effectiveness of selected commercial banks

Several studies can be found in empirical literature to estimate the effectiveness of banks in the Slovak banking sector. Some empirical studies (Bems & Sorsa 2008, Matoušek et al. 2008, Mamatzakis et al. 2008) examined the efficiency of banks in several European countries where the Slovak banking sector included the panel analysis. Bonin et al. (2005) or Fries & Taci (2005) estimate efficacy in the banking sector in the 1990 and examined the impact of privatization on the efficiency of banks. The results of the studies have shown that the privatized banks are more efficient than the state-owned banks, but also there are significant differences in efficiency. The privatized banks with majority foreign participation were more efficient than banks with domestic ownership. Rossi et al. (2005) estimated average cost efficiency averaging 0.67 between 1995 and 2002, while profit efficiency was 0.47. The Slovak banking sector has experienced a significant level of cost and profit inefficiency, which means that the average banks worked well above the cost-effective boundary and beneficial profit margins. It was also found that cost efficiency increased between 1995 and 2002. Stavárek & Polouček (2004) estimated efficiency and profitability in selected banking sectors including Slovakia. They found that the banking sectors of the Central European countries were less efficient than in other EU Member States. In the conclusions of the survey results the conventional claim of higher efficiency of foreign-owned banks than conventional banks is contradictory. They found that the size of banks is one of the factors that determine efficiency. In order to achieve high efficiency the bank should have a broad portfolio. It should be well known and offer a wide range of products and services. In other words, small banks have to focus on specific market segments and offer special products. Any other bank structure leads to a reduction in relative efficiency (2005) identified the increased value of the Slovak banking sector's efficiency during the period 1999 to 2003 and also found that the Slovak banking sector was less efficient than the banking sectors of other Vysehrad quartets. Iršová & Havránek (2011) estimated banking efficiency in five countries of Central and Eastern Europeincluding Slovakia. The results showed that between 1995 and 2006 average cost efficiency was 51.8% and profitability was 43.2%.

These studies looked at efficiency in several banking sectors. On the other hand, Stavárek & Šulganová (2009) estimated efficiency only in the Slovak banking market. They applied the parametric Stochastic Frontier Approach and Cobb-Douglas production function in the period 2001–2005 and found that the average efficiency of banks increased and their results point to the better ability of Slovak banks to use inputs in the production process.

From the review of empirical literature it can be stated that only a few studies were examined by the individual Slovak banking sector. Most of the empirical studies were examined by several bank

Table 2. Quality services assessment of e-banking.

	Max. number of points	Scale	ČSOB	SLSP	Unicredit Bank
Function	40/30	0.74	11.1	9.25	11.84
Safety	25/30	1.2	14.4	13.2	12
Fees	15/30	2	16	16	22
User interface	25/10	0.4	5.6	4	5.2
Total points	–	–	47.1	42.45	51.04

sections, which also included Slovak republic. Further findings from the literature review are that most studies looked at banking efficiency during the 1990s. That is why we used the DEA method in this chapter the results of which could fill a gap in the time axis of the empirical literature. Contributions examined by the Slovak banking sector separately applied a parametric approach, while the final work uses a nonparametric approach to information efficiency in the area of information, increased protection accessibility and in particular the cost of using the payment services of selected commercial banks.

4.3 Estimation of the effectiveness of the identified aspects

The DEA method can be used to estimate the effectiveness according to the assumptions of constant and variable yields from the range. The DEA method is appropriate to use in banking as it can easily process multiple producers such as banks and does not require explicit specification of functional form for production boundaries or explicit statistical distribution for inefficiency as an econometric method.

Banking efficiency is estimated using DEA models an input-oriented model with constant yields on the scale and an input-oriented model with variable yields is used. The reason for using both methods is that the assumption of constant yields on a scale is only allowed if all production units operate at the optimum size. However, this assumption can not be met in practice and is therefore also calculated with variable yields on the scale (Tkáč, et al. 2013).

The results of DEA on the basis of constant yields on a range (CCR model) for each bank are shown in Table 3.

The average efficiency of the Slovak banking sector estimated using the constant yield model ranged from 59% to 66%. As a cause of inefficiency SLSP bank can be marked as a non-existent redirection from http to https (address with increased

Table 3. Efficiency of selected commercial banks in the CCR model.

| | Inputs | | Outputs | | |
DMU	Fees	Func-tion	Safety	User inter-face	Effectiveness (Average)
ČSOB	0.43	0.35	0.27	0.33	0.6550
SLSP	0.34	0.31	0.38	0.39	0.645
Unicredit Bank	0.44	0.52	0.21	0.46	0.5925

Table 4. Efficiency of selected commercial banks in the BCC model.

| | Inputs | | Outputs | | |
DMU	Fees	Func-tion	Safety	User inter-face	Effectiveness (Average)
ČSOB	0.26	0.29	0.28	0.39	0.6975
SLSP	0.28	0.34	0.35	0.47	0.64
Unicredit Bank	0.37	0.33	0.21	0.25	0.71

security). Table 3 shows that the most effective Slovak bank on average is bank ČSOB followed by Unicredit bank Slovensko, a.s. and SLSP. If banks had an efficiency of 100% for most of the period under review, it means that these banks produced their outputs within the period of maximum efficiency.

Table 4 captures the effectiveness of commercial banks in the variable yield model. The average profitability of commercial banks estimated using the DEA model with variable yields ranged from 64% to 71%. The Slovak banking sector can see increasing efficiency.

In the BCC model, banks are effective in the Unicredit bank Slovakia, a.s. ČSOB bank and SLSP bank operated throughout the analyzed period at the efficiency limit in the variable yield model. It can be seen that in the variable yield model the banks have achieved more efficiency than in the model with constant yields on the scale. It can be concluded that banks that have achieved lower efficiency in the CCR model do not have the optimal size and do not produce a sufficient amount of output using those inputs. It is also important to recall the fact that in order for the results to be reliable. It is necessary to include a corresponding amount of inputs and outputs in relation to the number of banks included in the data file. The fact that the Slovak banking sector is relatively small and consists of a limited number of banks automatically limits the completeness and optimality of the model.

5 CONCLUSIONS

CCR models are the most efficient large and medium-sized Slovak commercial banks. On the other hand, the small bank is the most efficient in the BCC model followed by the banks from a group of large Slovak banks. The average efficiency in all groups of banks is significantly higher in the model with variable yields confirming the above-mentioned findings regarding the inadequate size of the Slovak commercial banks. Obtaining different results within both model specifications is an interesting finding common to many e-banking efficiency studies. While smaller banks are typically more efficient in the CCR model, but under VRS assumptions. The efficiency values are much higher. All major banks included in the analysis are more effective in terms of gross yields. This means that these banks have chosen an inappropriate range of operations and use too many inputs and produce too few outputs.

From the Efficiency Estimates it was found that the average commercial bank efficiency estimated in the CCR model range ranged from 59% to 66% and the average commercial banks' effectiveness estimated using the DEA model with variable yields from the range was in an interval of 64%–71%. As a result of the inefficiency of individual banks, it is possible to identify technological influences in particular. Electronic banking combines both the latest data transfer techniques and security and security features. Therefore, one of the most significant influences in this respect is the discovery of new techniques and practices. Banks compete with each other and each of them is trying to offer the client something new and thus distinguish it from other banks. An important consideration is also the application of the latest security features that can help convince clients who are still hesitant about using e-banking for lack of confidence in their security.

Banking has undergone, especially during the 21st century dramatic changes that have brought significant innovations particularly in the area of e-banking. Almost every bank client has already been confronted with some of the modern electronic distribution channels. Not only are these ways of providing electronic banking services comfortable but also fast. These and many other non-profit benefits have helped to consolidate the position of these electronic channels in the provision of banking services. thereby also increasing the efficiency of banking services.

REFERENCES

Antošová, M. et al. 2014. Assessement of the balanced scorecard system functionality in Slovak companies. In *Journal of Applied Economic Sciences*, 9(1), 15–25.

Bems. R. & Sorsa. P. 2008. Efficiency of the Slovene Banking Sector in the EU context. In *Journal for Money and Banking* (Bančni Vestnik). 57(11). ISSN 0005-4631.

Bonin. J.P. et al. 2005. Privatization matters: Bank efficiency in transition countries. In *Journal of Banking and Finance*. 29. 2155–2178. ISSN 0378-4266.

Ferencz, V. & Dugas, J. 2012. *Management of innovation: scientific monograph*. 1st ed. Brusel: EuroScientia, 2012. 129 s. ISBN 978-90-818529-8-2.

Fries, S. & Taci, A. 2005. Cost Efficiency of Banks in Transition: Evidence from 289 Banks in 15 Post-communist Countries. In *Journal of Banking and Finance*. 29(1). 55–81. ISSN 0378-4266.

Gallo, P. & Mihalčová, B. 2016. Knowledge and Use of the Balanced Scorecard Concept in Slovakia related to Company Proprietorship. In *Calitatea*. 17(151), 64.

Iršová. Z. & Havránek, T. 2011. Bank Efficiency in Transitional Countries: Sensitivity to Stochastic Frontier Design. In *Transition Studies Review*. 18(2). 230–270. ISSN 1614-4015.

Mamatzakis. E. et al. 2008. Bank efficiency in the new European Union member states: Is there convergence? In *International Review of Financial Analysis*. 17(5). 1156–1172. ISSN 1057-5219.

Matoušek. R. 2008. Efficiency and scale economies in banking in new EU countries. In *International Journal of Monetary Economics and Finance*. 1(3). 235–249. ISSN 1752-0487.

Olexová, C. 2014. Business intelligence adoption: a case study in the retail chain. *WSEAS Trans Bus Econ.* 11: 95–106. E-ISSN: 2224–2899.

Rossi, S.P.S. et al. 2005. *Managerial behavior and cost/ profit efficiency in the banking sectors of Central and Eastern European countries*. Working paper [online]. No. 96 [vid. 25. dubna 2012]. Wien: Oesterreichische Nationalbank. Dostupné z: http://www.oenb.at/de/img/wp96_tcm14-27319.pdf.

Roztocki, N. & Weistroffer, H.R. 2015. Information and communication technology in transition economies: an assessment of research trends. In *Information Technology for Development*. 21(3). 330–364.

Stavárek. D. & Polouček, S. 2004. Efficiency and Profitability in the Banking Sector. In: *Reforming the Financial Sector in Central European Countries*. Hampshire: Palgrave Macmillan Publishers. s. 74–135. ISBN 1-4039-1546-6.

Stavárek. D. & Šulganová, J. 2009. Analýza efektívnosti slovenských bank využitím Stochastic Frontier Approach. In *Ekonomická revue* – Central European Review of Economic Issues. 12(1). 27–33. ISSN 1212-3951.

Stavárek. D. 2005. *Restrukturalizace bankovních sektorů aefektivnost bank v zemích Visegrádské skupiny.* Karviná: SU OPF. ISBN 80-7248-319-6.

Sylvie, L. & Xiaoyan, L. 2005. Consumers' attitudes towards online and mobile banking in China. In *International Journal of Bank Marketing*. Vol. 23 Issue: 5. pp. 362–380. doi: 10.1108/02652320510629250.

Tkáč, M. et al. 2013. Modern computation methods for business applications. Reviewers: Adrian Olaru, Jozef Mihok. 1. vyd. Vaterstetten: Adoram, 2013. 276 s. [13,85 AH]. ISBN 978-3-00-044092-2.

New Trends in Process Control and Production Management – Štofová & Szaryszová (Eds)
© 2018 Taylor & Francis Group, London, ISBN 978-1-138-05885-9

University education as a motivation towards one's own business

E. Šúbertová & D. Halašová
Faculty of Business Management, University of Economics, Bratislava, Slovak Republic

ABSTRACT: The economically active population of the Slovak Republic comprised 2,725,838 people to 28 February 2017. While the number of job applicants being disposable was 228,665. Although the number of unemployed people is decreasing gradually according to the long term statistics, the unemployment of young people has been high for a long time. In the given period, there were 12,601 graduates unemployed, out of them 2,939 were university graduates and 9,662 graduates from secondary schools. Therefore we carried out a survey on a chosen sample of respondents in the Faculty of Business Management of the University of Economics in Bratislava. We wanted to know whether the undergraduate students in the third year of bachelor study are interested in starting their own business, their attitude and motivation towards business in connection with their parents' entrepreneurial activities and possibilities to raise finance.

1 INTRODUCTION

1.1 *The important aim for business schools*

The requirements in the field of education to entrepreneurship in order to gain quality knowledge and develop entrepreneurial skills are becoming a high priority (Block 2009).

To start a business for young people and then continue successfully is not easy at all. In February 2017, the unemployment rate in Slovakia according to the Central Office of Labour, Social Affairs and Family reached 8.39 per cent. The highest number of 3,050 unemployed was in Prešov region while Bratislava region recorded only 663 unemployed graduates. Graduates lack the practical experiences and skills that would meet the labour market requirements. Setting up their own business can be a good starting point for them. Therefore, also from a social point of view, it is very important to prepare young people for entrepreneurship (Havierniková et al. 2016).

The training should aim at acquisition of entrepreneurial knowledge, models of evaluation and managerial skills, as well as completion of adequate practice (Gallo & Mihalčová 2016). Schools face an urgent task of preparing undergraduates during their studies in order that students can be able to set up and run their businesses immediately after they have graduated.

1.2 *The main objective and partial objectives*

The main objective of this contribution is to analyse prerequisites, preparedness and attitudes of graduating bachelor students towards entrepreneurship in Slovakia and their mutual comparison. To meet the given objective we defined four partial objectives.

The first partial objective was to describe the students' experience with a family business. We evaluated mainly hitherto students' participation in the family business.

The second partial objective aimed at the area of planned participation in business in view of students' family backgrounds where we analysed the influence of entrepreneurs from students' families, their consideration to engage in business and interest in active participation in business in the future.

The third partial objective was to find out what would motivate young people to run their business and then we should try to activate those factors gradually.

The fourth partial objective was to identify which sources of finance they would use to start a business.

1.3 *The methodology*

We adapted the preparation of this scientific article to the methodology of both primary and secondary surveys. In the secondary survey we oriented to the study of literature and collection of information (data) necessary for the analysis. We analysed the prerequisites and preparedness as well as attitudes of students towards the current business in their families and towards their own prospective businesses as a form of self-employment.

The questionnaire survey itself was carried out within three academic years: from 2011/2012 to 2015/2016 among the full-time students of the University of Economics in Bratislava at the Faculty of Business Management with the aim to find out their experiences and attitudes towards business. In total, 140 full-time students filled in our questionnaires in the academic year of 2011/2012, and 118 students participated in the last survey of the academic year 2015/2016.

According to the profile of the Faculty of Business Management of the University of Economics in Bratislava, the graduates from the faculty should launch careers as entrepreneurs or economists, managers in middle or senior positions, in manufacturing, export, import and finance organizations, or can find jobs in manufacturing companies, in organizing production strategies of the company. They can also work in consultancy for small and middle-sized enterprises, or in economic departments of enterprises regardless their size, ownership or legal forms. In our survey we aimed at their opinions, experiences, as well as plans in business and entrepreneurship.

2 THE SURVEY RESULTS

2.1 Results of the first objective

In the first part of the survey we identified what entrepreneurial experiences students gained in their families. We evaluated their previous participation in a family business.

We raised questions about a family business to all respondents included in the survey. The results showed that in the academic year of 2011/2012 almost 44 percent of respondents were from families in which at least one member ran a business. In most cases it was the father, less frequently the mother, and least frequently it was a sibling. There were almost 11 percent of cases where two family members were entrepreneurs, in most cases those were both parents.

In the academic year of 2015/2016 only 43.2 percent of respondents came from families in which at least one member was an entrepreneur. In most cases it was the father, i.e. in 64.3 percent of families, less frequently, in 17.9 percent it was the mother or a sibling.

The above findings show that a positive, or negative attitude towards entrepreneurship in three quarters of entrepreneurial families is formed by fathers. Gradually, the significant role of fathers was also supported by that of siblings, which can be described as a positive trend. In the studied group, the number of mothers—business women gradually decreases which seems to be logical to a certain extent due to a high degree of business risk which self-employed people encounter nowadays.

Table 1. Active participation of students in a family business in the academic years 2011/2012 and 2015/2016.

Indicator	Year 2011/2012		Year 2015/2016	
	number	%	number	%
Regularly	68	57.6	11	9.3
Occasionally	0	0	19	16.1
Never	50	42.4	88	74.6

Source: own calculations

The sharp decline in interest of students in business results from their assistance to parents in their businesses.

In the academic year of 2011/2012 we asked our respondents about their active participation in a family business and we found out that 57.6 percent of students from entrepreneurial families helped out in their parents 'businesses. Forty-two percent of students from entrepreneurial families never participated in a family business.

Out of them, 15 percent students helped out regularly and 43 percent occasionally, but not in family business.

In the academic year of 2015/2016 we found out that only 9.3 percent of students from entrepreneurial families regularly helped out their parents. In our sample, 16.1 percent of students occasionally helped out and 16.1 percent of students never participated in their family businesses. The remaining students did not come from entrepreneurial families.

2.2 Results of the second objective

The second partial objective focused on the planned participation in business in connection with the family background of students. We also analysed the influence of occurrence of entrepreneurs in the students 'families on their decisions whether to engage in business in the future. Considering that the percentage of students who were not involved in their family businesses increases we were interested in their plans in setting up a family business.

Only 13 students in the academic year of 2011/2012 planned to participate actively in running their family members' businesses. The main reason why they want to participate in running their family business is the vision of self-realization in taking over the family business, its expansion and, of course, the help to their families.

The interest in opportunity to run a business is still very low. In the academic year of 2015/2016 only 4.2 percent of students were interested in opportunity to participate in their family

Table 2. Interest in doing business in the academic years 2011/2012 and 2015/2016.

Indicator	Year 2011/2012		Year 2015/2016	
	number	%	number	%
Clear interest in business	91	77.1	14	11.8
Were nor interested in business	25	21.2	87	73.8
Were not able to decide	2	1.7	17	14.4

Source: Own calculations.

Table 3. Motivations towards business from the environment in the academic years 2011/2012 and 2015/2016.

Indicator	Year 2011/2012		Year 2015/2016	
	number	%	number	%
Yes	60	50.8	39	33.1
No	49	41.6	18	15.2
Yes or not	9	7.6	61	51.7

Source: Own calculations.

businesses. Most students from entrepreneurial families did not plan to participate in businesses of their family members.

The main reason of such decisions is the fact that their parents are not going to expand their businesses. It is an interesting fact because theoretically in the next years there should be the first generational change of the entrepreneurs who started their businesses after 1989 (Tóth 2016). However, their children are not interested in their parents' businesses and prefer to do business independently, in another field (Molly et al. 2010). Of course, this may be connected with the advent of new information and communication technologies.

Among other reasons were: different interests of the respondents, or they do not like the region where their families run a business. The respondents sometimes gave other reasons such as insecurity and high risk of business, many problems in the Slovak business environment, complicated legislation, insolvency of customers, as well as yearning of the respondents to set up their own company. Therefore we asked the respondents about their own intentions to do business and their interest in becoming entrepreneurs in the future.

In the academic year of 2011/2012 we found out that as many as 91 students, i.e. 77 percent of respondents were interested in doing business in the future while 21 percent of respondents did not want to run a business in the future and less than 2 percent were not able to decide.

We obtained some interesting results in the academic year of 2015/2016. Only 11.8 percent of respondents had a clear interest in doing business.

On the other hand the number of those who were motivated to do business increased: 22.9 percent gave as an incentive of doing business to be independent, 20.3 percent declared to have inner motivation and 10.2 percent expected to have a higher income, 21.2 percent of students did not plan to run a business in the future. The large group of students was not able to decide. We think

that the constantly changing legislation in Slovakia and often controversial opinions about entrepreneurship in the society influence the changing considerations of young people.

2.3 Results of the third objective

The third partial objective was to find out what motivate young people to do business and activate more these factors.

In our survey we tried to find out the motivation towards business from the environment of the students. We found out that almost 50.8 percent of students in the academic year of 2011/2012 are motivated by their environment. For more than 41.6 percent the environment was not motivating and 7.6 percent answered that the environment either motivated them or not, i.e. they were not able to determine it clearly.

In the academic year of 2014/2015 the motivation for doing business under 50 percent, what was caused by media coverage of various corrupt activities, or inappropriate and constantly changing legislation.

We differentiated the evaluation of causes and factors motivating to do business. In the academic year of 2011/2012, the students most frequently gave as motivation factors achievements of their friends and demanding work with employees.

The largest motivating and demotivating factors were in the academic year of 2011/2012 according to the respondents a high risk and poor support of the government for start-ups.

The evaluation of the reasons and factors that would motivate students to do business was very different. In the academic year of 2015/2016, the students most frequently gave as motivators mainly their own knowledge from the given field in the given environment and better salaries. They do not afraid of negative results in any legal persons forms (Srovnalíková 2015).

We are glad that the respondents—students are more confident about their knowledge and abilities than their predecessors.

Table 4. The sources for starting a business in the academic years 2011/2012 and 2015/2016.

Indicator	Year 2011/2012		Year 2015/2016	
	number	%	number	%
Bank loans	53	44.9	10	8.5
Own savings	26	22.0	59	50.0
Deposits or borrowings from family and friends	18	15.3	26	22.0
Investors	15	12.7	17	14.4
State aid or EU funds	6	5.1	6	5.1

Source: Own calculations.

2.4 Results of the fourth objective

In the fourth part of the survey we focused on a source for starting their business. Over 44 percent of students would most often use bank loans in the academic year of 2011/2012, 22 percent of respondents would use their own savings as a source of finance their business.

The third source of finance in the form of deposits and borrowings from family members and friends would be used by 15 percent of respondents.

Only 13 percent of students would try to raise finance from investors, e.g. from business angels. Five percent of the students in the both study periods would apply for a certain form of state aid, or EU funds (Šúbertová 2015).

In the academic year of 2015/2016 the situation changed, the students had a different approach to raising finance.

Their own savings as a source of finance would be used by 50 percent of students. The other source of finance were above 8.5 percent, for example the students would use to start their own businesses either bank loans or they would use the source from external investors, or they would apply for a state support in innovate activities. Similar results had authors (Machová et al. 2015).

3 DISCUSSION

3.1 The role of positive case studies from life

The macroeconomic situation in the form of a high rate of unemployment hit also students, in particular the students of humanities, in the past ten years. We are glad that some students do want to do business in the future. They see the opportunity of being independent as the major motivator. However, the lack of finance and changing legislation as well as fear of failure are reasonable. For this reason, it is necessary to use suitable positive case studies from the life of real and successful entrepreneurs. Economic and managerial challenges are for business important (Kajanová 2014).

3.2 The role of support organizations

We think that examples of successful entrepreneurs in the educational process can improve the relation of students to entrepreneurship what was verified by our partial research results and they were also a part of open questions.

In the direct and indirect support of beginning entrepreneurs should play a more significant role in addition to school, family and also various organization founded for development of entrepreneurship, e.g. Slovak Business Agency, Slovak Association of Small and Medium Enterprises and Entrepreneurs, etc. (Bondareva 2013).

Unfortunately, students rather search for information on the Internet than go to these organizations in person to consult the problem and use the advisory service that they provide.

The number of various business support institutions remains still a problem, which may be confusing for young people (Šebestová 2015).

Obtaining bank loans is another problem as many banks are cautious about lending money to young people without "history of their enterprises", however, many students have already become more realistic about raising finance and having their business plans approved by banks.

4 CONCLUSIONS

At present traditional forms of education are not sufficient as they do not reflect labour market needs any more. At the beginning they should have not only great entrepreneurial activity ideas, courage to do business and be able to make a radical decision, but they also have to work out their business plans or submit various grant applications. The expertise in the chosen sector of business, as well as some concrete knowledge on how to do business including cash flow management of an enterprise is a prerequisite for raising finance. The marketing communication using information and modern technologies, management of human resources, etc. are also very important. The contents of each academic subject and application of new modern teaching methods are essential. For teachers' preparation, the feedback in the educational process is also a very important aspect. It makes them to think about whether their activity is efficient as regards the better activation of students and their interest in the latest knowledge. Therefore we did the survey about preparedness of our graduates for successful practice.

The survey showed that young people want to walk their own path of independence and do not want to repeat "a business path of their parents". The students believe that their own direction will be new and more suitable for them. They are more interested in the area of business services than in traditional manufacture of products. On one hand, this may be a positive signal for the society that young people keep abreast of the latest developments in science and technology but on the other hand, overvaluation of various irrelevant information from the Internet sources may cause them some problems. The students could use their professional and language skills in business in combination with their desire for self-realization and independence. In such a case they would be employers and employees at the same time which would have a positive impact not only on economic situation in the family but also on economic and social development of both individual regions and Slovakia as a whole.

It is debatable whether it is positive or negative that the students have a low degree of confidence in their environment, either in their business partners or in the business environment as a whole. According to our survey each student believes only in themselves. It is clear that cooperation either within an enterprise or among enterprises carries a great risk. However, no enterprise can manage everything on its own. Therefore it is important to teach students how to find trustworthy business partners, or managers with necessary knowledge, capable workers as well as reliable business partners. For the above reason the quality university education is a necessary prerequisite for development of human capital.

ACKNOWLEDGEMENT

The contribution was prepared within the VEGA project No 1/0709/15, share 100 percent.

REFERENCES

Block, J.H. 2009. *Long-term orientation of family firms. An investigation of R&D investments, downsizing practices, and executive pay.* Wiesbaden, Germany: Gabler.

Bondareva, I.A. 2013. Tools to support the sustainable development of small and medium sized enterprises of the Slovak Republic. In *Proceedings of the international research and practice conference—Science, technology and higher education,* Westwood, Oct. 16, 2013. Westwood: Accent graphic communications.

Gallo, P. & Mihalčová, B. 2016. Models of evaluation of managing people in companies. *Quality—access to success,* 17(155): 116–119.

Haverníková, K., Okręglicka, M., & Klučka, J. 2016. *Theoretical and methodological issues of risk management in small and medium-sized enterprises.* Berlin: Merkur.

Kajanová, J. 2014. *Economic and managerial challenges of business environment.* Bratislava: Comenius University.

Machová, R. et al. 2015. *Inovačné podnikanie a hodnotenie inovačného potenciálu podnikateľských sietí.* Brno: Tribun.

Molly, V. et al. 2010. Family business succession and its impact on financial structure and performance. *Family Business Review,* 23: 131–147.

Šebestová, J. 2015. Regional Business Environment and Business Behaviour of SME's In Moravian-Silesian Region. In *Proceedings of the 12th International Conference Liberec Economic Forum 2015.* Liberec: Technical University of Liberec.

Srovnalíková, P. 2015. The Impact of tax license on small and medium enterprises. In *Proceedings Management and related disciplines,* University of Economics in Bratislava, Slovak Republic, 21–23 october 2015. Bratislava: Ekonóm.

Šúbertová, E. 2015. Characteristics and types of financial resources to support the development of small and medium – sized enterprises in Slovakia. In *Proceedings Financial management of firms and financial institutions,* 10th international scientific conference, 7th–8th september 2015, Ostrava: VŠB—Technical university of Ostrava.

Tóth, M. 2016. Vývoj vzniku mikro podnikov v Slovenskej republike v rokoch 2008 až 2014. In Zhodnotenie efektívnosti financovania projektov na podporu rozvoja podnikania a faktory zvyšovania výkonnosti poľno-hospodárskych podnikov. Bratislava: EKONÓM.

New Trends in Process Control and Production Management – Štofová & Szaryszová (Eds)
© 2018 Taylor & Francis Group, London, ISBN 978-1-138-05885-9

Options of increasing the flexibility of air traffic management

Z. Šusterová, D. Čekanová, L. Socha & S. Hurná
Faculty of Aeronautics, Technical University of Košice, Košice, Slovak Republic

ABSTRACT: The article focuses on the issue of increasing the resilience of the system of air traffic operation in case of negative external impacts such as maybe weather, strikes, illegal acts etc. In case of minimizing the negative impacts of these external factors it is difficult for the management to reorganize the air traffic generally so as to be the least disturbed, and airlines would minimize the eventual loss. The article, to evaluate the resistance of air network, aims at using advanced techniques and adaptive and permanent strategies. Both methods have been applied to the network of European routes. The analysis showed that the adaptive strategy is preferable because it increases the connectivity of the European airline network. Moreover, if all airports had a small excess capacity of 5 percent at least, which could be used in case of threat and danger, up to 18 percent fewer flights would be cancelled.

1 INTRODUCTION

Transport systems alongside with other infrastructures such as energy networks or communication networks are the core element of our society and economy. They guarantee a high level of mobility that we all use and which is vitally important for the cohesion of the market and the quality of life. Moreover, transportation systems allow for socio—economic growth and job creation. If such basic systems and networks suffer accidental failure or become the target of terrorist attacks, the whole society is greatly affected by this sudden disruption. Air transport is not an exception in this case, e.g. due to accidents and incidents (Socha et al. 2015, Socha et al. 2017). In 2008, this sector of industry employed 32 million people worldwide (Kolesár 2015). On the other hand, its vulnerability is obvious with the implications for the mobility of citizens, for example in case of strikes or the volcano eruption we can notice disruption of normal system behaviour.

Planning the scheduled flights should be finished and published almost a year before the actual date of departure, but the air transport system is very far from calling it the static system. Conversely, airlines constantly adapt their planning according to the needs of travellers, in the short and the long term.

Near future is a major challenge for the aviation system and the theory of complex networks. Air transport is growing around the world and individual countries are aware of the fact that the current transport system reaches its limits in the horizon of several years due to steadily growing demand for transportation and due to new business challenges (Rozenberg 2014). Precisely for this reason major investment programs, such as SingleSky,

have been launched. It is also important to require a higher degree of integration between the different modes of transport.

2 RESISTANCE OF AIR NETWORKS

For creation of the EATN network model we used data from Openflights (2010), which cover 525 airports and 3,886 airlines routes (Bongiorno 2015).

The article focuses on the use of two strategies to evaluate the possibility of increasing the resilience of the European aviation network when it is exposed to spatial danger. Because we want to assess the potential worst variant disruption, we will take into account the spatial danger that hits the centre of the network (the place where the great density of airports). The area around the geographic centre of the network in the European Region includes all the London airports (United Kingdom), Charles de Gaulle (France) and Schipol Airport (Netherlands), and is considered as the weakest points of the European aviation network (Dunn 2013). This spatial danger grows towards the outside of the geographical centre of the network. Airports are closed and connecting routes abolished as soon as you experience a threat. In both strategies of resistance several assumptions were made to be convinced that they are real and they can be used as a basis for sophisticated methods (social and economic elements, as well as the movements of individual aircraft). It is believed that the location of airports cannot be changed and that even existing airports cannot be removed or added to other airports (Sabo 2015). Similarly, it is assumed that the network cannot be added to other routes. In the most extreme cases of adding tracks to create a homogeneous network where each airport connected with every other airport surely creates the most flexible network, but on the other hand, it would be the least economically advantageous. It is also expected that a certain number of airports have surplus capacity which can be used even in the short term, especially in imminent danger (e.g. if the diverted routes, while some aircraft still in the air).

3 INCREASING FLEXIBILITY OF AIR TRAFFIC

The first strategy of resistance take into consideration the proposals for a constantly changing network topology, and when it is considering a compromise between optimal social and economic factors EATN current, and resistance networks. In this strategy, we use the generating algorithm for the network design by Wilkinson et al. (2012) to generate artificial representatives of the European air network, but we limit the maximum number of air routes that can be connected to any airport (Jin 2004). Reducing the number of air routes connected to hub airports in case of closure of the airport the impact on the rest of the network should be reduced (less air routes will be affected). In their algorithm Wilkinson, et al. (2012) begins with a series of starting nodes, m0, which are all connected via the routes and are assigned a spatial position. In this work, we have assigned a position to this node corresponding to one of the current airport in the European air network (Burghowt 2001). At each step, the network a new node is added and then assigned a specific position corresponding to the existing airports of European air network. Then we create between 1 and m0 interconnections and calculate the probability of the connection node to node in an existing network based on the degree of all nodes in the vicinity of the area of a circle with a diameter r around the airport, which we have assigned to the network where the size of the radius r represents the distance from which people are ready and willing to be transported to the airport in order to travel from it. According to Wilkinson et al. (2012) we also allow the creation of links between a pair of existing nodes (simulation of new air routes between the existing airports). The likelihood of connection (direct route) to the existing airport drops to zero for airports which have reached their maximum allowed capacity.

We have generated a series of networks with different maximum values of air routes for one airport (20, 50 and 100), in order to assess the impact on resistance. For comparison—the airport with the most connections in the network model has been London Stansted 133 having a connecting air routes (Džunda 2013). By limiting the number of

airport air routes, caused by disruption when the airport is closed because of the danger, should be reduced in comparison to untreated European air network which increases resilience of the network. It is known that many airports, including Heathrow, are not willing to reduce the number of air routes they operate (for financial reasons), and it can significantly reduce the possibility to implement this strategy; and so, we have developed and evaluated the permanent strategy for quantifying the potential to increase the resilience network EATN compared with adaptive strategy (Petruf 2015). Fig. 1 shows (a) the level of distribution and on the part (b) there is a spatial distribution level for the European aviation network and three nets adjusted by the permanent method. A degree of the network distribution $P(k)$ indicates the cumulative probability of whether a selected node number of connections k, or is greater. $P(k)$ is calculated by the sum of the node $k = 1, 2, ...$connections divided by the total number of links in the network graph (O' Kelly 1994). The spatial distribution of the degree is obtained by calculating the weighted geographical centre of the network first, and then counting the degrees of all points (i.e. the number of connections that have a node) in a given circuit.

In this second strategy of resistance the air routes are adaptively diverted from airports which are surrounded by increasing spatial threat to their nearest operating airport (i.e. the nearest airport out of danger), provided that there is sufficient excess capacity (Fig. 2). The amount of air traffic

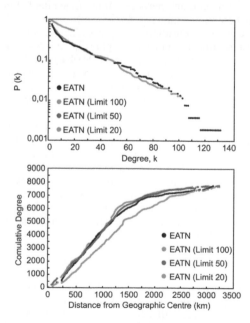

Figure 1. (a) the level of distribution and (b) the degree of spatial distribution for the European aviation network.

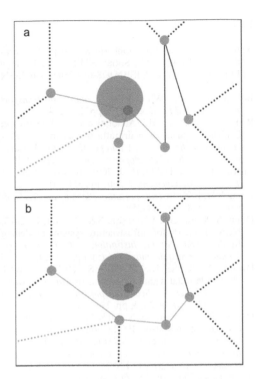

Figure 2. An adaptive strategy of air routes diverting in an emergency.

that arrives and departs from the airport depends on the given time of day, the day of the week, or the season (Li 2004). So, the vast majority of airports do not operate continuously at maximum power; for example, in the summer of 2012 it was reported that London Gatwick airport was 12% of the unused roller seats, while London Stansted and Luton were 47% and 51% of the unused roller seats (Transport Committee 2013b). We provided that each airport has a share of overcapacity, which may take diverted routes, given the proportion of air routes to be under normal conditions (Berechman 1996). For example, if we assume that each airport has a 10% spare capacity, an airport with 20 air routes (in normal operating conditions) will be able to accommodate two routes from airports closed (i.e. a total of 22 air routes).

In Figure 2 we can see an illustration of an adaptive strategy of diverting air routes in case of emergency. In both parts of the image, for better understanding, only a part of European aviation network is displayed; airports (nodes) are displayed by dots and air routes (links) by interconnecting lines (solid line shows a connection between two nodes and the dotted line between two nodes where one is out of the picture). The numbers next to each node indicate the degree of a node, which is equals the number of air routes connected to it. In part (a) we can see one airport (a red dot), which is

surrounded by spatial danger (red circle). The blue air routes are diverted to the nearest airport—the result can be seen in the image (b).

To estimate the impact that it has on the resilience of the European aviation network uses 5 different levels of excess capacity (5%, 10%, 20%, 50% and 100%). We are aware that in the real air network, it will be much less, if there is an airport at all, that would have spare capacity at 50% or 100%. We also use these values to assess the borders for extra capacity to improve the resilience of the system (Wang 2007).

If we want first to assess and quantify the impact that will have a permanent adaptive strategy for resistance of European air network, we map the share of cancellations and share of closed airports and closed (or affected) areas. Adaptive strategies of resistance increase the resistance of European air network, so the network is resistant to all sizes of space hazards, as compared with both comparative networks in the evaluation of the results in terms of cancelled air routes and airports closed (Antoško 2015). Redirecting of air routes for 5% of the excess capacity has resulted in 18% less air routes cancelled when 10% airports closed because of the danger. Using 20% excess capacity has resulted in the abolition of 29% of less air routes under threat of the same extend. Comparing these results with the results of ongoing strategies of resistance shows that adaptive strategy is better for this example. The permanent strategy has not increased the resistance of European aviation network significantly (reducing the proportion of cancelled air routes) unless major airports limited to 20 air routes. This is due to the placement of highly interconnected airports (hubs), which are located in the geographical centre of the network. Removing these hub airports has a disproportionate impact on the remaining network, which causes that the network is vulnerable to all sizes of space threats (Chon 1990). Networks created using permanent resilience strategies retain about the same effect as the unmodified European air network until the 40% of airports are closed.

If we want first to assess and quantify the impact that will have an adaptive and permanent strategy for resistance of EATN, we map the share of cancellations and share of closed airport air routes and closed (or affected) areas. Adaptive strategies of resistance increase the resistance of European air network, so the network is resistant to all sizes of space hazards, as compared with both comparative networks in the evaluation of the results in terms of cancelled air routes and airports closed (Kraus 2014). Comparing the results of two strategies of resistance we again show that an adaptive strategy is better in quality enhancing the resilience (elasticity) of the European air network, it ensures that connectivity is preserved even when the of the hazards has increased.

4 CONCLUSION

When creating air routes the airlines take into account dozens of factors from costs to continuity with other modes of transport. Network diagrams are undoubtedly one of the methods by which we can determine congestion of networks and their overall distribution in the area in which we are planning a new route. We have investigated two ways, in which it would be possible to increase the resilience of air transport networks with different disturbance of the flight operation such as extreme weather conditions. One strategy is adaptively changing the network topology, moves air routes in case that an airport is connected to the network, surrounded by impending danger while the second one permanently changes the network topology. Both of these strategies of resistance have been vividly applied to the European aviation network. We exposed a network to the spatial growing threat in the network geographic centre. The aim was to choose this location and affect the biggest disruption of the network. The network resistance was quantified by plotting the proportion of cancelled air routes, and the share of closed airports and air regions. It turned out that the permanent strategy has a little benefit for the resilience of the European aviation network. This strategy is very ineffective until the maximum number of air routes at every airport is not drastically reduced, and this results in the maximum reduction in the number of cancelled air routes by only 15%. It was found that the air transport network in the United States and China have a similar topological character as the European air network, and it is likely that the same conclusions can be applied for them. The adaptive strategy is better for increasing the resilience of the European aviation network and therefore increasing the connectivity of the network (enables all open airports to be available), but also preserves the effectiveness of European networks (which means it is still relatively easy to navigate flights to all open airports).

We have also previewed to the problems of overcapacity which is necessary for a positive impact on the resilience of the European aviation network with adaptive strategy. If all airports had little excess capacity of 5%, which could be used in an emergency, then air routes would be cancelled by about 18% less due to possible small spatial danger (covering 10% of the area network) compared to the unmodified European air network. It is therefore clear that a relatively small capacity growth of about 5% can dramatically increase the resilience of the European aviation network. Airports generally do not work constantly at their maximum capacity, thus redirecting the air routes in dangerous situations is a viable option.

REFERENCES

Antoško, M. et al. 2015. One runway airport separations. In*SGEM 2015*. Sofia: STEF92 Technology Ltd., 2015 p. 241–248. ISBN 978-619-7105-34-6. ISSN 1314–2704.

Berechman, J. & de Wit, J. 1996. In *Journal of Transport Economics and Politics*. 3, 1996 pp. 251–274.

Bongiorno, C. et al. 2015. Adaptative air traffic network: Statistical regularities in air traffic management. In *Proceedings of the 11th USA/Europe Air Traffic Management Research and Development Seminar*, ATM 2015.

Burghowt, G. & Hakfoort, J. 2001. In *Journal of Air Transport Management*. 7, 2001 pp. 311–318.

Chou, Y.H. 1990. In *Transportation Planning and Technology*. 14, 1990, pp. 243–258.

Dunn, S. & Fu, G.& Wilkinson, S.& Dawson, R. 2013. Network theory for infrastructure systems modelling In *Proceedings of the Institution of Civil Engineers: Engineering Sustainability*. 166 (5), pp. 281–292.

Džunda, M. & Cséfalvay, Z. 2013. Selected methods of ultra-wide radar signal processing In *Marine Navigation and Safety of Sea Transportation: Advances in Marine Navigation*, 2013, pp. 239–242.

Džunda, M. & Kotianová, N. 2015. The accuracy of relative navigation system. Production Management and Engineering Sciences In *Scientific Publication of the International Conference on Engineering Science and Production Management*, ESPM 2015, pp. 369–376.

Jin, F. et al. 2004. In *The Professional Geographer*. 56, 2004, pp. 471–487.

Kolesár, J. et al. 2015. Application of forecasting methods in aviation In Production Management and Engineering Sciences—Scientific Publication of the International Conference on Engineering Science and Production Management, ESPM 2015, pp. 419–424.

Kraus, J. et al. 2015. Comprehensive emergency management for airport operator documentation. In *International Conference on Engineering Science and Production Management*, ESPM 2015, pp. 139–144.

Li, W. & Cai, X. 2004. In *Physical Review*. E69, 2004, pp. 046–106.

O'Kelly, M.E. & Miller, H.J. 1994. In *Journal of Transport Geography*. 2, 1994, pp. 31–40.

Petruf, M. et al. 2015. Roles of logistics in air transportation [Ulogelogistike u zračnomprometu] In *Nase More*. Volume 62, 2015, pp. 215–218.

Rozenberg, R. et al. 2014. Comparison of FSC and LCC and Their Market Share in Aviation. In*International Review of Aerospace Engineering (IREASE)*. Vol. 7, no. 5 (2014), pp. 149–154. ISSN 1973-7459.

Sabo, J et al. 2015. Flight planning and its impact on the environment. In *Transport Means - Proceedings of the International Conference 2015-January*, pp. 632–636.

Socha, V. et al. 2015. Effect of the change of flight, navigation and motor data visuazation on psychophysiological state of pilots. In *Proc. of IEEE 13th International Symposium on Applied Machine Intelligence and Informatics, 22–24 January 2015*.Herlany: IEEE.

Socha, V. et al. 2017. Basic piloting technique error rate as an indicator of flight simulators usability for pilot training. In *International Review of Aerospace Engineering* 9(5), pp. 162–172.

Wang, J. & Jin, F. 2007. In *Euroasian Geography and Economics 48*, 2007, pp. 469–480.

New Trends in Process Control and Production Management – Štofová & Szaryszová (Eds)
© 2018 Taylor & Francis Group, London, ISBN 978-1-138-05885-9

Evaluation of processes focused on efficiency and functionality

K. Teplická
Faculty of Mining, Ecology, Process Control and Geotechnology, Technical University of Košice,
Košice, Slovak Republic

J. Kádárová
Faculty of Engineering, Technical University of Košice, Košice, Slovak Republic

S. Hurná
Faculty of Aeronautics, Technical University of Košice, Košice, Slovak Republic

ABSTRACT: The paper deals with the problem of assessing the effectiveness and functionality of the process in terms of quality management system. The entire evaluation process is based on the philosophy of the Balanced Scorecard, which aims to assess the performance of the organization. Post gives key insights into the application of quantitative methods in quality evaluation processes. Evaluation of processes we provide on based of processes maps in the company. The emphasis in the evaluation we place on the main processes and processes related to the maintenance and continuous improvement of quality management system. Evaluation results indicate malfunctioning processes: internal audits, mold, development and laboratory testing, promotion and communication, production of materials. Evaluation processes is the foundation of quality management system in practice.

1 INTRODUCTION

Global business strategy is shaping its future development and places high demands on its capacity for adjustment. Enterprise "world class" can maintain high performance solely based on continuous study of new processes and the crucial points of comparison with competitors. Innovative companies use the Balanced Scorecard as a strategic management system to manage its long-term strategy and use controlling for evidence all indicators for BSC Potkány & Hitka (2005). Quantified properties using BSC businesses: critical to the implementation of management processes, to clarify and convert the vision and strategy into specific objectives, to communication and alignment of the strategic plans and the ratios, to planning and establishing objectives and harmonization of strategic initiatives, the improvement of strategic feedback and learning process. We present these principles in this paper in evaluation of business processes. On these principles, it is also based evaluation of the effectiveness and functionality of processes. Preventive maintenance policy can improve the operation efficiency of the product and functionality of processes (Chen et al. 2016). Maintenance planning is tool how to improve the efficiency and functionality of equipment to satisfy the production requirements and to minimize the total production cost Erfanian & Pirayesh (2016).

2 EVALUATION OF PROCESSES

BSC approach was evaluated in a company that manufactures refractory ceramic materials used for the linings of heat aggregates. Basic areas of use of those products is metallurgy, ceramics, and cement and lime industry, chemical industry, construction. On a global scale assessment of refractory materials leads to a significant reduction in their production. The largest proportion of refractory materials e.g. 60% is used in the production of iron and steel. Decline in the production of refractories is now attached to the steel industry, as the impact of technological changes in steel production to reduce specific consumption of these materials. The main objective of the company was to evaluate process performance by applying the Balanced Scorecard approach (Antošová et al. 2014). This instrument BSC is used in production companies and BSC perspectives are monitoring indicators of processes in the company. BSC approach is very important tool for outsourcing and its using for some activities in the company (Potkany et al. 2016). As the trend of the world production of refractory materials decreased to 2010, it is for the focus of a strategy, which enables the company to continue to be competitive. The first step was to develop a SWOT analysis of the company so that we know what direction the business will move in and what its strategic goal. Performance evaluation in the company we have made based on the application of methods of assessing the effectiveness of processes and functionality of operational processes per the methodology of quality management system ISO 9001:2016 (Seňová & Antošová 2015). We use indicator efficiency of process. This is indicator of efficiency of process where R = reality of process; P = plan of process.

$$K = R/P \qquad (1)$$

The second indicator is index of functionality where K = indicator efficiency of process; w = weight of process.

$$I = K/w \qquad (2)$$

Evaluation of efficiency and functionality is based on SWOT analysis (Teplická 2003). These indicators classify Strength and Weakness in the company and indicators determine strategic goals of the company. Significant of indicators present processes: Weaknesses—processes in which the overall outcome of the evaluation largely effective, ineffective, respectively, mostly functional, nonfunctional. Strengths—the processes by which the overall outcome of the efficiency and functioning of an effective and functional (Teplická 2008). In this part, we analyses strength and weakness of company front the point of view process management. Weakness is essential to improve. Strength is possible to use as a competitive advantage. Each perspective shows to strength and weakness of financial, consumer, internal processes, and education (Fig. 1). The strength built new instruments of modern

Table 1. Valuation of process—efficiency.

Type of efficiency	Indicator
Effective process	$K \geq 0,85$
Mostly effective process	$0,85 > K \geq 0,70$
Ineffective process	$K < 0,70$
Maximum	$K = 1,2$

Table 2. Valuation of process—functionality.

Type of functionality	Indicator
Functional	$I \geq 1$
Mostly functional	$1 > I \geq 0,90$
Nonfunctional	$I < 0,90$

SWOT		STRENGTH	WEAKNESS
	Financial	Support for the construction of the new research center	A high proportion of products of low added value
	Consumer	The existence of customer service Marketing support and technical support	Insufficient promotion and external communication with customers
	Internal processes	Free production capacities	Obsolete technology equipment A high proportion of manual work The lack of own raw material resources The occurrence of non-conforming production
	Education	Highly qualified workforce	

Figure 1. BSC—SWOT analysis in the company.

management. One of these instruments is outsourcing. Outsourcing is primarily used as a strategic tool for resources optimizing and its significant is very important for evaluation efficiency and functionality of processes. Enterprises decide to remove some supporting activities to external units, which can be provided with ensuring a higher quality and lower costs and processes are more functionality (Potkány et al. 2016). Addressing weaknesses of the company can be realized through outsourcing. Outsourcing is an opportunities for no effective and non-function processes in the company. The current tough competition prevalent within the market economy is forcing business entities to achieve efficiency in their processes (Němec et al. 2015).

Taking stock of the SWOT analysis, we started evaluating each business process to find out fundamental weaknesses e.g. weaknesses in the evaluation of processes and take advantage of opportunities in terms of the external environment (Fig. 2). The second part of SWOT analysis was monitoring opportunities and threats for company in business environment.

Performance evaluation in the company we have made based on the application of evaluation methods. In the company, we evaluated the effectiveness and functionality processes the following approach. Out of business processes elaborated in detail the process of filling the production plan as the algorithm for the solution of all business processes. We got the data that can measure and evaluate business processes in the company. We analyzed production process and summarized data of plan of production.

This company produces 17 products and plan for year 2016 was followed (Table 3). Ordinary brick, Komag 97, Hight earth material built base of production. We monitored all products of company.

SWOT		OPPORTUNITIES	THREATS
Financial perspective		The potential growth of supply of products to lime and cement and construction Industries.	The stagnation of the European steel industry because of overproduction in China. The decrease in the consumption of refractory materials per ton of metal.
Consumer perspective		Strategic focus on metallurgy industry. A large market for manufactured products. The possibility of applying the aluminum industry.	Entry of competitors on the market with customer-oriented prices
Perspective of internal processes			Changing production technology and the associated investment costs.
Perspective of education			Shift production to China, the loss of skilled workers.

Figure 2.   BSC—SWOT analysis in the company.

Table 3.   Plan of production for year 2016.

	Year 2016/(tone)
Fulfillment of production plan	Plan
Ordinary brick	6772
Hard brick	2394
Decorating brick	3000
Hight earth material	5617
Special ceramic	13
Concrete	593
Hight earth concrete	2274
Special concrete	771
Brick refractory	770
Billets	246
Basic product	289
Material	180
Komag 96	1760
Komag 97	4141
Komag 98	1636
Komag 99	192
Basic material	944
Summary:	31,592

Real production was changed, some products decreasing and some products increasing. Production of these products depends on demand at market. Real production in tone is presented in (Table 4).

Production of some products was decreasing for example Decorating brick, Concrete, Hight earth concrete, Billets, Komag 98, Komag 99, and production of some products was increasing for example hard brick. Plan production was exceeded for products Ordinary brick, Hight earth material, Komag 97 very markedly. We compared plan of production and reality of production (Table 5),

Table 4.   Reality of production for year 2016.

	Year 2016/(tone)
Fulfillment of production plan	Reality
Ordinary brick	7361
Hard brick	4767
Decorating brick	893
Hight earth material	5726
Special ceramic	18
Concrete	467
Hight earth concrete	1508
Special concrete	404
Brick refractory	291
Billets	227
Basic product	236
Material	88
Komag 96	1366
Komag 97	4553
Komag 98	1455
Komag 99	89
Basic material	654
Summary:	33,453

Table 5.   Variance analysis of products.

	Year 2016/(tone)
Fulfillment of production plan	Variance
Ordinary brick	589
Hard brick	2373
Decorating brick	−2107
Hight earth material	109
Special ceramic	5
Concrete	−126
Hight earth concrete	−766
Special concrete	367
Brick refractory	479
Billets	−19
Basic product	53
Material	92
Komag 96	394
Komag 97	412
Komag 98	−181
Komag 99	−103
Basic material	290
Summary:	1861

because indicators of efficiency and functionality influence production.

Complete change presented 1861 tone of products—it is positive trend for company. This positive trend influenced efficiency and functionality of processes in the company because limit of indicators was observed only in several processes. We compare plan of production and reality of production and result was positive because plan was fulfilled. Production increases about 1861 tone (Fig. 3).

Increase in production will be pursued about efficiency and functionality in business processes. Next, we calculated indicators of process efficiency and operational processes functionality by equations (1, 2) and we came to the following conclusions.

We monitored all processes in the company and we calculated indicator of efficiency of processes (Fig. 4). All processes have indicator (0.81–1.07). Only one process has indicator of efficiency 0.59 for process Communication and it means ineffective process. Processes in production are effective. This result is confirmed by the result of comparison of the plan and reality of production and its growth.

We monitored all processes and we calculated indicator of functionality of processes (Fig. 5). This indicator reaches values (0.43–1.09). All processes under 0.90 are nonfunctional. It means negative situation in company. Between nonfunctional processes belong Production of material, Communication, Development and testing, Production of mold, Internal auditing. These processes are related to production and thus their malfunction could influence the reduction of production in some products. Reduction of production is 3302 tone. Very positive process for functionality is Requirements of customer.

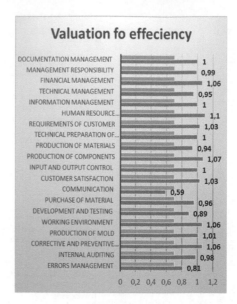

Figure 4.   Valuation of efficiency of processes.

Figure 5.   Valuation of functionality of processes.

3   CONCLUSIONS

BSC method is a qualitative tool of performance management, which is especially suitable for the derivation of the objectives of the vision of the company, their measurement by business characteristics and to determine the actions that lead to these objectives. The benefit of this method is

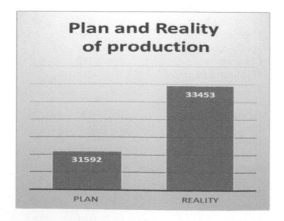

Figure 3.   Comparison analysis—plan/reality.

properly balanced and strategic monitoring of critical success factors and their quantification at different levels of governance. Continual improvement and reduction of costs is very important factor in all area of industry. Cost reduction is connected with efficiency and functionality of all business processes. High level of efficiency and functionality built base of continual improvement (Dolinayova et al. 2016). Functionality and efficiency of the processes be an important factor in the quality and the decisive criterion for the future development of the company. Therefore, it is essential to use the strategic tools of global economics and management, enabling businesses to move forward and approaching the level of the enterprise "world class". Knowledge economy means increasing of knowledge operating intensity, business process improvement framework is developed through an integrative adaptation of the concepts of knowledge intensity and knowledge management to the principles of process redesign and re-engineering reported (Sallos et al. 2016).

ACKNOWLEDGEMENTS

This contribution is part of project VEGA 1/0741/16 Innovation controlling of industry companies for maintenance and improvement of competiveness.

REFERENCES

Antošová, M. et al. 2014. Assessment of the Balanced scorecard system functionality in Slovak companies. *Journal of Applied Economic Sciences,* 9(1): 15–25.

Chen, Z. et al. 2016. Joint optimization of degradation-based burn-in, quality, and preventive maintenance. In *Proceedings International Conference on Industrial Engineering and Engineering Management.* Bali: Indonesia.

Dolinayova, A. et al. 2016. Social and Economic Efficiency of Operation Dependent and Independent Traction in Rail Freight. *Procedia Engieneering,* 134: 187–195.

Erfanian, M. & Pirayesh, M. 2016. Integration aggregate production planning and maintenance using mixed integer linear programming. In *Proceedings International Conference on Industrial Engineering and Engineering Management.* Bali: Indonesia.

Němec, F. et al. 2015. A proposal for the optimization of storage areas in a selected enterprise. *Nase More,* 62: 101–108.

Potkány, M. et al. 2016. Outsourcing in conditions of SMEs the potential for cost savings, *Polish Journal of Management studies,* 13 (1): 145–156.

Potkány, M. & Hitka, M. 2005. Controllingová koncepcia integrovaného manažmentu v spojitosti s motivačným programom organizácie. *Manažment v teórii a v praxi,* 1(3): 42–48.

Sallos, M. P. et al. 2016. A usiness process improvement framework for knowledge-intensive entrepreneurial ventures. *Journal of Technology Transfer,* 1–20 (in press).

Seňová, A. & Antošová, M. 2015. Business performance assessment and the EFQM excellence model 2010 (Case study). *Management,* 20(1): 183–190.

Teplická, K. 2003. *Ekonomika environmentálne orientovanej kvality—teoretické možnosti a ich aplikácie v praxi.* Dizertačná práca: TU v Košiciach, F BERG.

Teplická, K. 2008. *Uplatnenie moderných trendov nákladového riadenia vo výrobných podnikoch.* Habilitačná práca: TU v Košiciach, F BERG.

New Trends in Process Control and Production Management – Štofová & Szaryszová (Eds)
© 2018 Taylor & Francis Group, London, ISBN 978-1-138-05885-9

Energy clusters as a tool for enhancing regional competitiveness

A. Tokarčík & M. Rovňák
Faculty of Management, The University of Prešov, Prešov, Slovak Republic

G. Wisz
Subcarpathian Renewable Energy Cluster, Rzeszow, Poland

ABSTRACT: Increasing demands for efficient management of energy in enterprises and public institutions forces them to seek solutions that are appropriate for their structure and market segment. Optimal energy policy of enterprises and public institutions and its implementation in practice has become a benchmark for quality assessment of a company and its brand, as well as the criterion for the optimal management of public funds in municipalities. The complexity of solved problems presents a professional framework for defining the tasks and objectives that cannot be solved by their internal expert groups or employees. Optimal networking is an important element in the development of professional backgrounds in the market and linking between theory and practice in solving specific tasks in the optimal management of energy while reducing negative impacts on the environment.

1 INTRODUCTION

In the first phase of globalization, a source of competitive advantage was globality itself, i.e. the company's ability to activate and mobilize inputs and assets across national borders. Enterprises with international cooperation thus gained an advantage over companies that operated only in national level. We are now at the stage that is much less intuitive because globality is now taken for granted. The company must now seek input source in cheaper locations, seek internal sources of capital and must produce with low labour costs. And therefore what can now get business from remote locations is no longer a competitive advantage, because now they all have access to it.

If the competitiveness of the region will be seen as the ability of companies continuously and profitably produce goods and services that are successful in open markets (Korecet et al. 2011), then building energy clusters appears to be the best solution. Energy Cluster in essence represents the institution in the field of management of energy efficiency, which builds a framework for the participation and cooperation of companies and other institutions of various types and sizes that contribute to increasing the performance of the region. Clusters focused on optimal management of energy represent a group of regional actors who create harmony between scientific knowledge in regions and environmental requirements as well as offer comprehensive solutions to problems in the field of energy efficiency and renewable energy sources implementation. Grouping of public and private sector leads to the formation of stable background in the creation and management of regional policy aimed at dealing with energy, taking into account the historical and philosophical relations of the population and characteristics of the territory in different regions. It turns out that the cluster members represent specific entity characterized by high competitiveness and expansionism. And this fact should be considered for the role of clusters in the economy of dynamic development of regions. (Plawgo 2014, p. 9). The common aim of the actors of the energy cluster is to define, characterize and ensure the sustainability of energy resources in selected regions, with long-term increase in energy efficiency. The role of the energy cluster in the region is the creation of a human society that meets the needs of the region, without compromising the ability of future generations and their ability to meet their energy needs. Create an institution in the field of energy efficiency means building a healthy competitive environment for the involvement and cooperation of companies in line with those of other institutions of different types and sizes that can better contribute to the performance of the region in meeting the energy efficiency targets. (Majerník et al. 2015)

2 THE ROLE AND IMPORTANCE OF ENERGY CLUSTERS

Energy clusters play a special role in creating links between economic subjects and research institutions whose relations are key elements of modern models of innovation processes. Cluster policy in terms of optimal management of energy, due to its horizontal nature is an important part of several areas of economic policy, especially innovation, regional and industrial policy. Energy clusters form an effective mechanism at concentrating resources and measures, which form one of the best ways to develop economically possible mutual cooperation between market actors. They have the ability to influence, create and accelerate development and play an important role in the growth of poles throughout the region and even the country. (Bakowski & Dworzycki 2010). Energy clusters represent a set of related companies and regional universities and research institutions whose bonds have the potential to consolidate and increase their competitiveness. Energy Cluster is a representation of interconnected companies such as specific suppliers, service providers and associated institutions in the field of renewable energy and energy efficiency that are competing, but also cooperate closely and complement each other. The vision of local energy cluster is to create a modern cross-sectional structure of legal entities on the principle of mutual assistance and cooperation to ensure the competitiveness of the region in the best possible utilization of energy sources of the region with regard to the available technologies. Energy cluster is a catalyst for a more efficient and dynamic performance of their own energy efficiency activities of cluster members in the region, while supporting innovation strategies in the deployment of renewable energy, reducing carbon footprint and environmental management in order to reduce the long-term impact on climate change. Active activity of energy cluster in the region ensures growth of skilled labor, increases employment in the region and expanding international cooperation. In terms of ensuring the competitiveness of the region, energy cluster reduces the limitations of small companies and increasing specialization in the field of energy management. Energy cluster in the region brings together companies from different value chain and allows smaller companies to specialize and cooperate in competition against larger, vertically integrated companies. Cooperation with larger companies provides a mechanism for smaller companies to access the international networks of larger companies in the energy cluster.

3 THE PROCESS OF ENERGY CLUSTER FORMATION

The creation of an energy cluster in the region, which has its position and is accepted not only in scientific circles but also in general public, is demanding. In principle, it is a long process that has two phases. In the phase of its occurs institutional establishment of the cluster and the development phase is focused to the stabilization of the cluster in the region. The second phase consists of establishing cooperation with other clusters at home and abroad. In the phase of the cluster formation, cluster goes through three stages (Szajna & Kamycki 2011, p. 40). The first stage is informal and focuses on the needs of the region. The result is the definition of the newly built energy cluster activities and setting goals. The second stage is an association of members of the energy cluster in the region and its institutional establishment in the form of clearly identifiable cluster. The aim of the second phase is the creation of internal structures. In the third stage is defined sphere of development and professional orientation of the cluster in the following period. The objective of this stage is to provide an analysis of the activities and needs of cluster members.

During the implementation of the activities of the different phases, the facilitator helps in the process. He develops activities leading to the start up and development of cooperation among potential members of the energy cluster. The selection of the facilitator is a key moment, which is necessary for that person to meet not only professional and organizational skills, but also operates in the region in which the cluster was created. The deployment stage of Szajna cluster (2011, p. 39) is made up of four stages. Forming stage is characterized by establishing cooperation, exchange of technology implementation plans and innovative solutions. Growth stage is represented by the growing number of cluster initiatives, implementation of new technologies, building and developing new strategies. The stabilization stage is represented by the implementation of original ideas and innovative practices; it balances the differences between the cluster parties, develops the structure of the cluster, using the energy potential of the region. The final stage is the stage of decline. In fact a period when the structure and actors are too big, dynamic development has stagnated, the requirements of the region are changing, changing the environmental due to changes in various market segments, and it is necessary to seek new opportunities. Based on the analyzes it can be stated that regional clusters in Slovakia are mostly based on competences, which is a focus on a particular area of energy management as main initiative of the cluster. It is therefore a win-win partnership companies, universities and regional institutions, which brings a number of benefits to all its member organizations (Duman et al. 2009)

4 CHARACTER OF THE ACTIVITIES OF ENERGY CLUSTERS

Based on the experience of existing regional clusters with an energy focus, whether as a main activity or one of the activities carried out, it is possible to create a group of basic scenarios for the operation of regional energy clusters. Possible scenarios initiatives which can be combined depending on the built energy cluster are as follows:

- Clusters aimed at fostering interdisciplinary approaches in research related to the efficient use of energy and diversification of energy,
- Clusters aimed at measuring and analyzing energy loss and environmental impacts of the energy factors,
- Clusters aimed at bringing together the human potential in the management of energy,
- Clusters aimed at energy sources and energy services,
- Clusters focused on the economy in the implementation, development and sustainable energy policies.

Energy clusters designed to promote interdisciplinary approaches in research related to energy, are seeking to establish research cooperation between universities and research institutions at home and abroad. Activity of the cluster is focused on issues of implementation of energy efficient equipment, procedures and methods leading to continuous improvement. The proposed solutions are not only scientific and technical, but also social, economic, cultural, political and security related, including direct impact on the real environment. The result of such cluster can be seen in receiving requests for scientific research tasks and solve them within the cluster. The disadvantage may be a break with the practice, which is a risk that achieved results of scientific research will not be feasible in the current economic, and legislative conditions.

Energy clusters aimed at measuring and analyzing energy and environmental impacts will focus on exploring the potential for improving the existing situation. The main activity is the creation of functional groups and solutions for optimal management of energy and diversification of energy resources for the promotion of renewable energy sources. The advantage of these clusters are methodologies, processes and solutions created on the basis of good experience from practice, taking into account the ability of partners to implement the results in the region. These clusters are characterized by a good knowledge of the environment, the important role played by cross-linking the public and private sector.

The role of energy clusters aimed at linking human potential is the creation of new ideas toward business opportunities in order to ensure energy-acceptable solutions with customers. It is grouping of specialized companies, researchers, students and organizations focused on "energy-efficient" solutions. Great advantage are solution unique to the market, while some individuals have the opportunity to gain a dominant position in the market. The proposed solutions are unique, but the disadvantage can be solutions only for specific conditions.

Energy clusters focused on energy services and products are characterized by seeking local opportunities to promote local ties, create new innovations with a new business opportunities. Partnerships are built on a very specific level. The task is to gain a dominant position in the market segment with an added value of exclusivity. The disadvantage of these clusters is their dependence on the lifecycle of products and services in the relevant market segment.

Based on the Triple Helix model (Fig. 1), the optimal solution appears to be close cooperation of energy clusters in the region presented by the following composition (Bakowski & Dworzyckij 2010):

- The business environment defines the requirements of practice and requires higher educated individuals suitable to their competitive environment. It promotes research and creates conditions for further business activities.
- Universities and research—development centers constitute new tools for business and create innovative tools for the competitive environment in the market. Results of research and development are part of their own know-how.
- Energy and environment define requirements appropriate for the environment on the one hand and on the other hand they offer energy-efficient products, services and design solutions. The benefit of the business environment create a raw material base (biomass, renewable energy, photovoltaics, low-energy boilers, LED technology, etc.), with expectation of energy and environmentally sound practices and solutions affecting the reduction of carbon footprint.

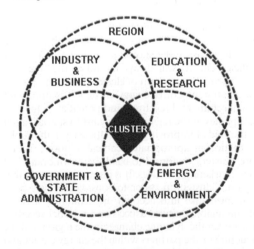

Figure 1. Institutional cooperation of the cluster based on the Triple Helix model (Bakowski & Dworzyckij 2010).

- The Government creates support mechanisms with regard to the long-term sustainability and sets limits in handling energy instruments in the area. Government funds research, supports research, and develops support mechanisms for the benefit of clusters using tools for their development and stabilization in the region. The dominant position is in the application of knowledge of clusters in strategic documents. It uses the expertise of clusters in decision-making at long-term planning.

A distinctive feature of the energy clusters operating in the regions is the diversion structures, focusing on the energy potential of the region in which it operates. Danger may be the creation of similar structures in the region with different economic support. In this case, the key becomes the establishment of cooperation with other clusters and creating attractive conditions for business in the regions in order to provide long-term benefits. Factors limiting the scope of the region are then binding on the regional authorities, government agencies and corporate structures in the regions. Built cross-border cooperation is consequently seen as an innovative element of ensuring the development of relevant regional clusters. Possible stabilizing factor seems to be the attraction of foreign investment.

Cluster management should be recognized as a new professional qualification that requires high quality standards and professionalism, so companies and institutions co-operating in clusters provide efficient service and are able to fully enjoy the benefits of relations between universities, industry and state administration. The Commission encourages such efforts, which are already developing in some regions of the EU. (European Commission 2008, p. 9)

5 CONCLUSION

There is no single universal process how to create energy cluster or the rules of evaluation that can be used to measure the success of a cluster and cluster policies. Quality will always depend on the objectives of a particular program or cluster policy. In order to maintain competitive energy clusters it is important to collect internal resources and potential of the best companies in the region with regard to specialization in the field of improving energy efficiency in the implementation of appropriate technical solutions. Therefore, clusters are forced to build their own brand and own marketing practices. It is important to note that regional clusters in order to consolidate their position must take into account the economic strength of the region, cultural aspects and regional specificities. Given the dynamic changes in the region and the structure of the partners within the energy cluster and international environmental policy it is necessary to constantly review marketing tools of clusters. Active clustering is the basis for building regional prosperity.

Clustering, however, includes not only networking but also confidence in the actors. Energy clusters provide a forum for dialogue among key actors in the region, with a focus on growth and solution to the specific region in an energy-efficient management of energy, as well as the deployment of a variety of energy sources, including building energy mix. Energy clusters are key components of the innovation potential of the regions. They stimulate interest in investment by the private sector, job creation and other economic and social values. They also contribute to reducing the energy intensity of government and public sector as a whole. Pilot verification procedures and solutions in the energy cluster environment for future innovative project partnerships and initiatives are useful in a particular environment. Creating of know-how for members of clusters and networking of partners helps to reduce energy dependence of governments and the private sector and thus the dependence on imported energy sources from unstable environment for the benefit of the developing region.

ACKNOWLEDGEMENT

The paper was elaborated in the framework of the project "V4EaP Visegrad University Studies Grant No. 61500079."

REFERENCES

Bakowski, G. & Dworzycki,J. 2010. *Klastre*. Ministerstvohospodárstva. Dostupné z: http://issuu.com/kpr_europa_2020/docs/klastry_pl/1?e = 0.

Duman, P. et al. 2009. *Klastre na podporu rozvoja inovácií: analytická štúdia*. SIEA. Dostupné z: http://www.siea.sk/materials/files/inovacie/slovenske_klastre/Klastre-SIEA.pdf.

European Commission, 2008, *Towards world-class clusters in the European Union: Implementing the broad-based innovation strategy*. [cit. 2014-04-29]. Dostupné z: http://ec.europa.eu/enterprise/policies/innovation/policy/clusters/index_en.htm.

Korec, P. et al. 2011. Regionálna konkurencieschopnosť v kontexte globalizácie, novej ekonomickej geografie a inovačných procesov In: *Geographia Cassoviensis* V. Roč. 2, s. 57–66. Dostupné z: http://geografia.science.upjs.sk/ images/ geographia_cassoviensis/articles/GC-2011–5-2/Korec.pdf.

Majerník, M. et al. 2015. Innovative model of integrated energy management in companies. *Quality, Innovation, Prosperity* 19 (1): 22–32.

Örjan, S. et al. 2006. *Zelená kniha klastrových iniciatív*. Czechinvest, ISBN 91-974783-3-4.

Plawgo, B. 2014, Benchmarking klastrów w Polsce–edycja 2014. Warszawa: Polska Agencja Rozwoju Przedsiebiorczości. ISBN 978-83-7633-289-5.

Szajna,W. & Kamycki, J. 2011. Tvorenie klastrov. In: *Szajna, W.*, Ako vytvoriť klaster: sprievodca. Rzeszów, Vysoká škola informatiky a manažmentu a Ústav ekonomiky, s.38–53 ISBN 978-83-60583-81-4.

New Trends in Process Control and Production Management – Štofová & Szaryszová (Eds)
© 2018 Taylor & Francis Group, London, ISBN 978-1-138-05885-9

Risk analysis of firefighting lifts and safety rules

M. Tomašková & M. Nagyová
Faculty of Mechanical Engineering, Technical University of Košice, Košice, Slovak Republic

ABSTRACT: A fire fighting lift is designed to transport firefighting services and equipment with unobstructed access to all floors of a building in a firefighting operation. Buildings of a more than 22, 5 m fireheight above ground must be equipped with firefighting lift sifat that height there are situated business premises classified under group 6 or 7 and more than 60 m. Firefighting lifts and their safety components have to meet general health and safety regulations. A firefighting lift can be used as a normal passenger lift at any time other than in the even to fire. Reliability of power supplies and circuitry is essential to the operation of a firefighting lift. The purpose of the contribution was to evaluate the calculation of the target residual risk.

1 INTRODUCTION

Pursuant to par. 2 sec. 4 of the Act of the National Council of the SR No 314/2001 Coll on fire prevention as amended "Firefighting devices include firefighting equipment, fire extinguishers, fire valves, devices for extinguishing of sparks in pneumatic conveying systems, equipment for the supply of water for firefighting, equipment for permanent energy supply during a fire, firefighting lifts, evacuation lifts, emergency lighting and other devices for evacuation of people and firefighting operation". Sec. 5 (a) of Law No 324/2001 Coll. stipulates that for the purposes of providing conditions for effective extinguishing of a fire (apart from other obligations) both a legal entity and a natural person—entrepreneur are obliged to: provide and install the appropriate types of firefighting equipment, firefighting materials, firefighting devices, adequate means for protection against firein buildings, facilities and premises, keep firefighting equipment in proper working condition, secure and carry out their regular inspection and maintenance by a qualified professional person if stipulated by the law, maintain and preserve records containing information of their operation.

1.1 *Firefighting lift*

A lift for transportation of firefighting services and equipment that is accessible to all floors of the building where a firefighting action is required. A firefighting lift can also be used as an evacuation lift installed in a building in the protected escape route of type B or C. The lift is designed with additional protection and operating elements which allow its operation directly by the firefighting service when fighting a fire (Chrebet et al. 2013).

A firefighting lift provides quick access for firefighting services and firefighting equipment to all floors of the building; its operation must be safe during the course of the fire fighting.

A liftof a special fire proofed encased structure installed either inside a building, or on the outside of a building used solely for the operational needs of firefighting services.

1.2 *Evacuation lift*

A lift used for evacuation of persons; it should not be used as an escape route and its capacity is not included in the total capacity of escape routes; its operation must be safe for the specified period of time during a fire.

The lift used for evacuation of people is accessible from the protected escape route or is located directly in this area being operated from two independent power supply sources.

A firefighting lift unlike a normal lift must be designed to ensure its operation as long as it is necessary in the event of a fire in a building. This lift may be used as a passenger lift when there is nofire.

The reliability of power supply sources and circuits is important for the operation of a firefighting lift.

The Decree of the Ministry of the Interior of the SR No 94/2004 Coll. stipulates technical requirements on fire safety during erection and use of structures.

A firefighting lift is a lift used for transport of firefighting teams and firefighting equipment with access to all floors of the building in the event of a fire. An evacuation lift is also used

Figure 1. Pictogram for a firefighting lift. (STN EN 81-72:2003).

as a firefighting lift according to (Regulation of the Ministry of the Interior no. 94/2004 Coll.), which must be installed in the protected escape route of type B or C. Two evacuation lifts can be installed in one shaft. Pursuant to par. 85 sec. 1 of the Decree No 94/2004 Coll. a building must be equipped with a lift if its firefighting height above ground level is:

– More than 22.5 m, if at that height there are located premises included in groups 6 or 7,
– More than 60 m.
– Par. 85 sec. 2 of the Decree No 94/2004 Coll.
– A firefighting lift is installed in the protected escape route of type B or C and has sufficient supply of power for at least
– 45 minutes, if it is part of the protected escape route of type B,
– 90 minutes, if it is part of the protected escape route of type C.

Terms and definitions according to STN EN 81-72 (Safety rules for construction and installation of lifts) part 72 Firefighting lifts:

– Control system: a system which responds to input signals and generates output signals making the equipment operate in the required manner.
– Evacuation: an organised and controlled movement of persons in a building from a dangerous area to a safe area. Evacuation can be from one floor to another floor and not necessarily from the building. (Folwarczny & Pokorný 2006).
– Evacuation level: the level with exits for evacuation of people from a building. It is not necessarily at the same level as the level with the access for the fire service.
– Fire compartment: part of a building separated by walls and/or ceilings in order to prevent spreading of fire and hot gasses within the area.

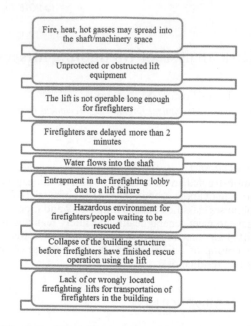

Figure 2. Hazards, hazardous situations and events (STN EN 81-72:2003).

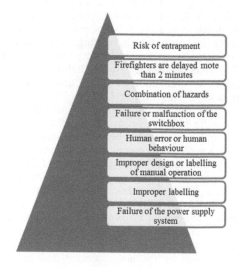

Figure 3. List of significant hazards and dangerous situations (STN EN 81-72:2003).

– Firefighting lift: a lift commonly installed for the transport of passengers that has additional protection, operation and signalling which enable it to be used under the direct control of the fire service.
– Fire protection: measures aimed at preventing the outbreak and spread of fire in all cases into

the protected escape routes and effective management of fire including determination of fire resistance, fire load and behaviour of building materials and structures during a fire.
- Firefighting lobby: a fire protected environment providing protected access from the area in the building to firefighting lifts.

A firefighting lift is installed in the shaft with firefighting lobbies in front of every shaft door of the lift. The area of each firefighting lobby is given by the requirements for the transportation of stretchers and the location of the doors in every single case (Pokorný & Gondek 2016).

If there is another lift in the same shaft, then the shaft as a whole shall meet the fire resistance requirements for firefighting lift shafts.

This level of fire resistance must also apply to the door of the firefighting lobby and the machine room.

If there is no firefighting wall to separatea firefighting lift from other lifts in the shaft, then all the lifts and their electrical equipment must have the same fire protection as the firefighting lift to ensure the proper functioning of the firefighting lift (Fig. 5).

The lift must be so designed as to function correctly under the following conditions:

A firefighting lift, compared to a normal lift, is designed to operate as long as it is required when a fire breaks out in a building. At any time other than in the event of fire in the building the lift can be used as a normal passenger lift. To reduce the risk of the entrance being obstructed when the lift is required by the firefighting service during

Table 1. Verification table (STN EN 81-72:2003).

Sub-clause	Visual inspec-tions[a]	Compli-ance with the lift design[b]	Measure-ments,[c]	Design documen-tation check[d]	Func-tional test,[e]
5.2.1	(STN EN 81-1, STN EN 81-2, STN EN 81-5, pr STN EN 81-6, pr EN 81-7)				
5.2.2	x				
5.2.3			x	x	
5.2.4			x		
5.3.1	x		x		
5.3.2	x		x		
5.3.3	x			x	
5.3.4	x	x	x		
5.3.5	x	x	x		
5.4	x	x	x	x	
5.6	x				
5.7		x		x	
5.8.1	x	x	x	x	
5.8.2	x	x		x	
5.8.3		x			
5.8.4		x			
5.8.5		x		x	
5.8.6		x	x		
5.8.7		x			x
5.8.8 a,b, c,d,e,f	x	x	x		x
5.8.8 g		x	x		x
5.8.8 h	x	x			x
5.8.8 i, j, k, l, m		x			x
5.8.9	x	x			x
5.10		x			x
5.11.1				x	
5.11.2	x			x	
5.11.3	x	x		x	
5.11.4	x				
5.12		x			x
7	x				

Note: Where the installer uses a product subject to testing, the tests and inspections are carried out as specified in the product documentation
a. The results of visual inspections are only to prove that the required labelling meets the requirement and that the content of the documents delivered to the owner is in compliance with the requirements
b. The results of the compliance with the lift design are to show that the lift is installed consistent with the design and that its components are in compliance with the design documents
c. The measurement results are to prove that the specified measurable parameters are met
d. The results of checking the design documentation have to show that the design requirements set forth by the standard are met in the design documentation (e.g. spatial arrangement, specifications)
e. The result of the functional test has to show that the lift functions properly as intended including the safety devices

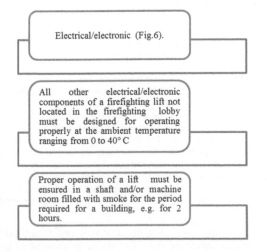

Electrical/electronic (Fig.6).

All other electrical/electronic components of a firefighting lift not located in the firefighting lobby must be designed for operating properly at the ambient temperature ranging from 0 to 40° C

Proper operation of a lift must be ensured in a shaft and/or machine room filled with smoke for the period required for a building, e.g. for 2 hours.

Figure 4. Verification of safety requirements and/or protective measures (STN EN 81-72:2003).

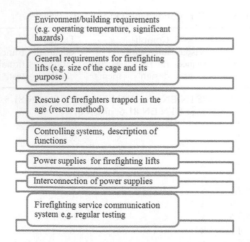

Figure 5. Rescue operations from outside (STN EN 81-72:2003).

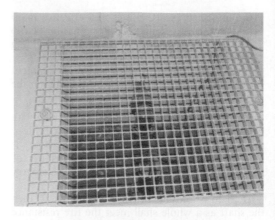

Figure 6. Sump with a pump for drawing water used to extinguish a fire.

the firefighting operation, its use for transportation of goods should be restricted. (STN EN 81-72:2003)

Rescue operations from outside:

2 FIREFIGHTING CONCEPT FOR HIGH RISE BUILDINGS

An example of potential hazards:

- National building regulations apply to: building structures, checking for smoke, alarm systems, installation of fire extinguishing equipment, water hydrants etc.
- According to firefighters: the term high-rise building means any building with floors above the level of fire department vehicle access.
- With the developments in construction of high rise buildings architects and firefighters came up with two suggestions, the first being to design buildings that resist fire, stop the spread of smoke and provide a high level of safety for people in the building. The other was to provide these buildings with appropriate firefighting equipment and rescue techniques, which were both practical and efficient.
- Firefighting lifts, their number and location within the building are determined by national regulations and are an important means of extinguishing fire, transportation of firefighters and fire escape equipment under supervision of firefighters (Baron et al. 2016).

2.1 Description of risks in closed spaces

Different methods are currently used for risk analysis. The combined method of risk consists of two mutually supporting processes—a method

Figure 7. Electrical safety switch.

Table 2. Determining risk using the combined method.

Categories of risk:

I.   Unacceptable risk (the activities cannot continue or start if the risk is not reduced)

II. Undesired risk (safety measures and checks of adherence to them are necessary, and without this the activities cannot continue)

III. Acceptable risk (the risk is acceptable but monitoring and control of the risk is necessary)

IV. Acceptable risk (the risk is acceptable but monitoring and control of the risk is necessary)

for identification of sources of risk and a method for assessment of risk. The entire process of risk assessment can be expressed by one method. The resulting value of risk represents a combination of the parameters D and TP, where the total probability TP is given by the sum of the parameters

Table 3. Determining risk using the combined method 1. (Salvová 2016).

Danger/Risks	D	F	P	O	TP
Unprotected or stopping of elevator equipment	3	5	1	3	H
Water running into shaft	2	5	3	3	H
Trapped person in the fire vestibule due to elevator breakdown	2	5	3	3	H
Collapsing of building construction before fire-fighters can finish freeing persons	4	4	3	1	H
Insufficient or incorrect placement of the fire lift for transport of fire-fighters in the building	0	3	1	1	L
Defect or error in fuse box function	1	4	1	3	L
Human error	4	5	3	3	H
Unsuitable labelling of manual controlling	2	3	3	1	L
Power outage from electricity network	4	3	2	1	H
Presence of hazardous materials	3	4	1	1	M
Lacking or limited access to persons with physical disabilities	4	5	2	3	H
Movement with poor precision when stopping	2	4	4	5	H
Breakdown or stopping of the lift cage outside of opening area	4	2	3	3	H

Table 4. Determining risk using the combined method 2. (Salvová 2016).

Danger/Risks	D	F	P	O	TP
An elevator accident upon failure of any mechanical part	4	3	2	5	H
Fall of cage upon tearing of load-bearing ropes	4	3	2	3	H
Shocking of a service worker by electrical current	4	5	2	1	H
Activation of the shaft doors	3	4	3	3	H
Missing or unsuitable lighting by shaft doors	0	5	4	1	L
Insufficient enclosure of the elevator shaft	3	5	3	1	H
Slippery floor in the machine due to unwanted substances on the floor	2	5	3	1	M
Insufficient ventilation of the cage	2	3	2	3	M
Lacking or unsuitable control during fires	3	5	1	1	M
Unsuitable securing of entry doors to shaft or depth	2	5	2	3	M
Dangerous access to depths	2	5	3	1	M
Unsuitable lifting equipment	2	4	1	1	L
Unsuitable glass in doors	1	4	1	1	L
Lacking or unsuitable protection against catching of fingers in sliding cage or shaft doors	2	3	2	3	M

(*Continued*)

Table 4. (*Continued*).

Danger/Risks	D	F	P	O	TP
Lacking self-closing device with sliding doors	2	2	2	3	L
Non-compliant fire resistance of shaft doors	3	2	2	1	M
Possibility of movement of cage doors with opened shaft doors	2	3	3	3	M
Cage without doors	3	3	2	1	M
Insufficient barriers with cage	2	4	2	3	M
Lacking of a speed limiter	4	3	2	3	H
Lacking or unsuitable buffers	4	3	3	3	H
Excessive distance between cage and shaft doors	3	3	3	1	M
Lacking closing valve	1	3	2	1	L
Lacking controlling device for freeing of rope/chains	2	3	2	3	M
Lacking control of time of running	0	4	2	1	L
Lacking or unsuitable control of cage loading	3	3	2	3	H

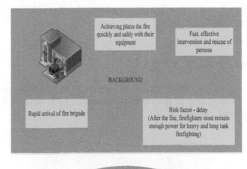

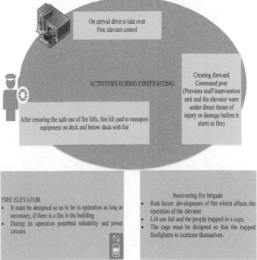

Figure 8. Fire concept in high-rise buildings.

frequency F, probability P and the possibility of prevention O (Salvová 2016).

3 CONCLUSION

After the assessment of hazards using the combined method given in Table 3 it follows that evacuation lifts and fire lifts belong to the group where high risk (40.5%) predominates; the second group shows moderate risk with a share of 31%, and the last one is with a low risk at 28.6%. Therefore, it is necessary to implement measures in order to minimize risks (Bosák 2016).

ACKNOWLEDGEMENT

The paper was prepared within the project: *The contribution was prepared in the scope of the project: VEGA no. 1/0150/15 Development of methods of implementation and verification of integrated safety systems for machines, machine systems and industrial technology and APVV -15-0351 Development and application of risk-management models in terms of technological systems in line with the industry (industry) 4.0.*

REFERENCES

Salvová, V. 2016. *The intervention of fire brigades and evacuation of people from vulnerable areas through fire and evacuation lifts,* Thesis, TU, SjF, KBaKP.

Folwarczny, L. & Pokorný, J. 2006: *Evacuation,* SPBI.

STN EN 81-72:2003 *Safety regulations for the construction and assembly of elevators, fire lifts.*

Ružinská, E. et al. 2014. The study of selected fire-technical characteristics of special wood products surface treatment by environmentally problematic coatings, *Advanced Materials Research Volume 1001*, Pages 373–378.

Chrebet, T. et al. 2013. Moment of lignocellulosic materials ignition defined by critical mass flow rate, *Applied Mechanics and Materials*, 291–294, pp. 1985–1988.

Baron, P. et al. 2016. M. Research and application of methods of technical diagnostics for the verification of the design node, Measurement: *Journal of the International Measurement Confederation.*

Baron, P. et al. 2016. The correlation of parameters measured on rotary machine after reparation of disrepair state, *MM Science Journal.*

Pokorný, J. & Gondek, H. 2016. Comparison of theoretical method of the gas flow in corridors with experimental measurement in real scale. *Acta Montanistica Slovaca*, 21 (2), Košice: Technická univerzita v Košiciach. p. 146–153.

Bosák, M. et al. 2016. Research an development of new Technologies of decontamination of ash mixture tailings, *Production Management and Engineering Sciences*—Scientific Publication of the International Conference on Engineering Science and Production Management, ESPM 2015.

New Trends in Process Control and Production Management – Štofová & Szaryszová (Eds)
© *2018 Taylor & Francis Group, London, ISBN 978-1-138-05885-9*

Trends in corporate environmentally related taxes

K. Vavrová, P. Badura, M. Bikár & M. Kmeťko
Faculty of Business Management, University of Economics in Bratislava, Bratislava, Slovak Republic

ABSTRACT: Environmental taxes are one of the most frequently discussed topics. The European Commission considers environmental taxes to be one of the most effective environmental policy objectives. The aim of this paper is to present the most important information on the impact of the various sectors of economic activity—transport, energy, industry, agriculture, forestry and tourism on the environment. The impact is measured by indicators and reports. The assessment is based on the methodology of the Organization for Economic Cooperation and Development (OECD) and the European Environment Agency (EEA). The methodology for assessing the indicators explores the causal link between the actual status and environmental trends. International and national legislation and documents, valid for different environmental organizations, have also been analysed. The introduction of environmental taxes can ultimately lead to many desirable effects within companies.

1 INTRODUCTION

Growing concern about climate change has brought environmental issues to the forefront of the policy agenda in many European countries. Global warming is one of the most important challenges currently facing the world (Suwirman 2016). Taxes, charges, tradable permits and other economic instruments can play an important role in achieving cost-effective control of greenhouse gas emissions, but their potential scale and revenue contribution raise many wider economic and fiscal policy implications. A number of European countries introduced carbon taxes during the 1990s, though a proposal for an EU-wide carbon energy tax was ultimately unsuccessful. More recently, attention has shifted to emissions trading. In the European, union a number of tax measures have been implemented primarily with environmental objectives. They included national environmental taxes and there are environmental tax measures. The increasing use of environmental taxes, emissions trading and other economic instruments has been partly driven by recognition of the limitations of conventional environmental regulation. They can initiate the necessary changes to the economies, resource utilization, behaviour, and general approach to nature (Aydin 2010). Extensive and far reaching changes to existing patterns of production and consumption will be needed, and these changes will inevitably entail substantial economic costs. The search for instruments capable of minimizing these costs, and of achieving behavioural changes across all sectors, has led policy makers to pay much closer attention to the potential for incentive based environmental regulation, that is, through economic instruments. This paper provides an overview of key economic issues in the use of taxation as environmental policy. The relationship between environmental knowledge, attitudes and behaviour is complex (Hines et al. 1987). Furthermore, residential electricity consumption in the six Gulf Cooperation Council countries has increased rapidly over recent decades amid a steep increase in population and relatively fast economic growth (Squalli 2007, Reiche 2010). Energy is in a market economy an indispensable aspect of our daily lives in company. We need it for heating, cooling, lighting fixture and moving around; it is essential for the functioning of our offices, work places and the entire economy. Its importance makes its accessibility a politically sensitive topic. This is one of the reasons why the Commission has proposed its Energy Union Strategy. The price of energy is also sensitive. On the one hand, low prices can be beneficial because they raise our purchasing power and standard of living and they reduce costs for our companies and so increase their competitiveness. At the same time, since energy is delivered through markets, energy suppliers need prices to cover their costs and to finance investment to ensure the future delivery of energy. High prices send signals to reduce the use of high carbon energy or to encourage energy efficiency and the use of innovative eco designed products and clean technologies.

The history of energy prices and costs shows major changes and major impacts. In the 1970s and 1980s, restrictions by oil suppliers drove up prices and triggered economic shocks. More recently, new energy supplies and growing use of alternative energy sources have boosted supply, while energy efficiency measures and weak growth have reduced demand and brought wholesale prices down. The EU has found that the more competitive and liquid the energy market is, the more diverse and numerous our energy supplies and suppliers are, the less exposed we are to such volatility.

2 AIM AND METHODOLOGY

The aim of this paper is to assess the economic instruments suitable for assessing the environmental policy.

2.1 Economic instruments

2.1.1 Advantages of environmental taxes and other economic instruments

Regulatory policies which stipulate that polluters must use particular technologies or maintain emissions below a specified limit may achieve compliance but do not encourage polluters to make further reductions below this specified limit. Indeed, where regulations are negotiated basis, polluters may fear that any willingness to exceed requirements may simply lead the regulator to assign the firm a tougher limit in future. By contrast, environmental taxes provide an ongoing incentive for polluters to seek to reduce emissions, even below the current cost-effective level, since the tax applies to each unit of residual emissions, creating an incentive to develop new technologies that have marginal cost below the tax rate (Fischer et al. 2003).

2.1.2 Disadvantages of environmental taxes and other economic instruments

Economic instruments such as environmental taxes have, however, a number of identifiable drawbacks and limitations that may be sufficiently important to rule out their use in particular applications. If pollution damage varies with the source of emissions, then a uniform pollution tax is liable to result in inefficiency, and source-by-source regulation may be needed to achieve a more efficient outcome (Helfand et al. 2003). In principle, an environmental tax does not need to be constrained to apply the same rate to all sources, and could thus achieve the efficient outcome through appropriately differentiated tax rates. However, once the tax rate has to be set individually for each source, the tax may become exposed to lobbying influence from the regulated firms. Also, some forms of environmental tax may have to apply at uniform rates, even where damage is known to differ between locations. Thus, for example, environmental taxes on pollution related inputs may be

unable to differentiate between sources, because of the difficulty of preventing resale of inputs to firms with more damaging emissions. Collective wisdom of humanity for conservation of biodiversity, embodied both in formal science as well as local systems of knowledge, therefore, is the key to pursue our progress towards sustainability (Sushmita 2014).

2.1.3 The balance between costs and benefits of using environmental taxes within companies

The considerations above imply that environmental taxes are likely to be particularly valuable where wide-ranging changes in behaviour are needed across a large number of production and consumption activities. The costs of direct regulation in these cases are large, and in some cases prohibitive. In addition, where the activities to be regulated are highly diverse, society may gain substantially from changing these damaging activities in the most cost-effective manner. In other areas, market instruments may work less well. In the next section we discuss the high costs be incurred in operating well-targeted environmental taxes. In other cases, an outright ban might be substantially easier to implement and enforce than a tax rate that requires fine measurement, or where avoidance activities are costly or dangerous.

2.2 Indicators

Stochastic indicator is composed of two curves: the curve labelling of % K and % D curve. Mathematical calculation is apparent from the attached formulas:

$$\%K = 100 \times \left[\left(C - Lx \right) \div \left(Hx - Lx \right) \right] \qquad (1)$$

where C = the last closing price (close) at the relevant time period (timeframe); Lx = the lowest point (low) for a selected period of time; and Hx = the highest point (high) for the selected time period.

$$\%D = 100 \times \left[Hn \div Ln \right] \qquad (2)$$

where Hn = an n-day sum (C-Lx); Ln = an n-day sum (Hx-Lx) (In case of daily chart).

The results of these two mathematical expressions are two lines that oscillate between 0 and 100. Curve %K while is usually denoted by the solid line and the slow curve %D dashed or dotted line. The curve shows the %K (expressed as a percentage), which is the closing price towards range x days (or period). %D line then is nothing else than the n-day moving version of %K.

3 ANALYTIC SYSTEM

Indicators are measurable quantities, providing information on the development process in quantitative and qualitative terms. They should be evaluated on a set of sectoral indicators, indicators green, indicators of resource. Sectoral indicators are a means of assessing progress in the implementation of sectoral policies in relation to the environment. Based on analysis of indicators and regular evaluation of the European Environment Agency (EEA), the Organisation for Economic Development and Economic Cooperation (OECD), Office of the European Union (EURO-STAT), were evaluated the possibilities of evaluation indicators in Slovak Republic.

3.1 Sectoral indicators—indicator of production, consumption and price of electricity

The indicator describes the development of production and consumption of electricity, electricity generation by source and balances its imports and exports.

In 2014 the Slovakia produced 27,254 GWh of electricity. Compared to the previous 2013 production there was a decrease by 1336 GWh (down about 4.7%). In the long term years 1990–2014, there was an increase of electricity production by about 13.3%. In the period 2000–2014 there was a decrease of electricity by about 11.7%, which was affected by the shutdown of V1 Jaslovske Bohunice. The highest share in electricity production in 2014 had the nuclear power plants (56.9%). They were followed by hydropower (16.8%), thermal electricity-plant (12.8%) and other power plants (11.8%). The share of photovoltaic power plants accounted for 1.7% of total production. In 2014 the total consumption of the electricity in Slovak republic was in the amount of 28,355 GWh. Compared to 2013 there was a drop by 325 GWh. Consumption of electricity has in a long term (1990–2014) the same course.

With the development of the internal market, wholesale electricity markets in Europe have undergone major changes in recent years. Wholesale electricity exchange markets have been established in almost all Member States, to provide day ahead, forward and intraday trading. Flexible and liquid markets can allow a more efficient matching of supply and demand that lowers generation costs and therefore prices. Such exchanges should also steer bilateral 'over-the-counter' contract prices in the most mature markets.

Gradually, these separate national wholesale markets are being coupled with neighbouring markets, which together with more transmission grid inter-connections create more efficient markets. Prices are driven by various factors, including fuel mix, cross-border interconnections, market-

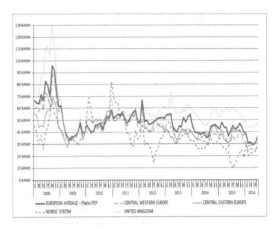

Figure 1. Trends in the EU electricity prices. *Source*: Platts and European power exchanges.

coupling and market supplier concentration and weather conditions. Similarly, consumer and industry demand, demand management, energy efficiency and the weather influence the prices, as well.

European wholesale electricity prices peaked in the third quarter of 2008 and, apart from a slight recovery in 2011, have been falling ever since. Prices have fallen by almost 70% since 2008 and by 55% since 2011 and in 2016 reached levels not experienced for 12 years (Fig. 1).

The passes through of reduced coal and gas prices, together with other factors, have been key drivers of electricity prices. Econometric analysis suggests that a 1% increase in the share of fossil fuels (coal, gas and oil) in the power generation mix results in an increase of € 0.2–1.3/MWh in the wholesale electricity price, depending on the regional market, in several markets, the rise of low marginal cost solar and wind-powered electricity decreases wholesale prices. Econometric analysis suggests too, that every percentage point increase in renewable share reduces the wholesale electricity price by € 0.4/MWh in the EU, on average; the actual reduction depends on the regional market and the fuel source being replaced by renewables. The impact of renewables is even greater (€ 0.6–0.8/MWh) in north-western Europe, the Baltics and the Central and Eastern Europe.

3.2 Environmental taxes on energy

Under the Kyoto Protocol the EU is committed to reduce greenhouse gas emissions by 8% by 2008–2012, measured against a baseline of the 1990 emissions level. Within this overall EU target, the UK is required to achieve a 12.5% emissions reduction. The 2003 Energy White Paper stated a further ambition to achieve a 60% cut in CO_2 emissions by 2050, "with significant progress by 2020".

Figure 2 shows the development of the weighted average EU electricity price for households, broken down into its three main components (energy, network, and taxes and levies).

However, as can be seen from Figure 2, other components were subject to bigger changes. On average, the network component increased annually by 3.3%. The taxes and levies component also grew significantly, with its share of the average price increasing from 28% to 38%.

The figures for electricity prices for industry show smaller increases, with the EU average increasing between 0.8% and 3.1% a year from 2008 to 2015. Figure 3 uses a representative industrial consumption band (2000–20,000 MWh/year). Large energy consumers, including more electricity intensive industries, may produce their own power, have long-term contracts for energy supply or often pay lower network tariffs, taxes and levies which can result in prices 50% lower than for other industrial consumers in the same country.

The energy component of average industrial electricity prices fell by 2.8% a year between 2008 and 2015 (Fig. 4). The difference in this component across Member States also shrank, by 12%.

This partial price convergence implies that EU energy policies promoting increased competition, which results from market coupling and cross border trade, are having a positive impact. That said, seven Member States actually saw a rising energy component over the period which, in some cases, could be an indication of inadequate price competition at retail level, allowing suppliers to avoid passing on lower wholesale prices.

The network component of the industry price increased annually by 3.2% over the period and the share of the taxes and levies component increased significantly, from 12% to 32% of the price. On average, almost two thirds of the network price component is attributed to distribution networks, but data remains weak due to calculation methods differing across Member States. As with household prices, the taxes and levies component of industry data was broken down into subcomponents. As some sub-components (VAT and certain other taxes) are reimbursed to industry, this component remains significantly lower than for households: Industry pay € 34/MWh and households pay € 79/MWh.

The energy price developments in recent years have not increased the energy cost share of production costs for European businesses, which lies at less than 2%. However, to reduce the cost of energy to industry, most Member State governments provide subsidies through exemptions and reductions in energy taxes and levies (e.g. renewable energy or energy efficiency levies, or network tariffs). Depending on the characteristics of a business and the Member State in which it is based, it may enjoy energy prices 50% lower than another company in the same sector.

That said, for some industries in which energy costs are more significant and exposure to international competition is high, there is a need to assess energy costs more closely. These are energy-intensive industries of a certain economic significance and trade exposure (where energy costs

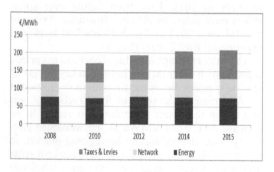

Figure 2. Components of average EU household retail electricity prices.
Source: Eurostat, Commission data collection.

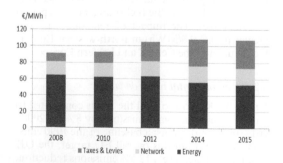

Figure 3. Components of average EU industry retail (company) electricity prices.
Source: Eurostat.

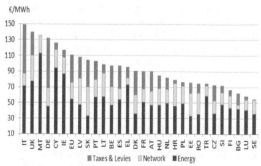

Figure 4. Industry retail electricity prices in 2015.
Source: Eurostat.

account for at least 3% of total production costs and for up to 40% on average or more in some cases). The analysis undertaken for the Commission shows that, for the 14 sectors selected, energy cost shares and absolute energy costs fell in most cases between 2008 and 2013. This is the result of the fall in energy prices, tax exemptions and reductions, lower energy consumption related to reduced production levels, a shift in production to less energy-intensive products, the uptake of energy efficiency measures and slower reductions in other production cost factors.

The European economy overall is not a highly energy-intensive economy. For decades, the EU has been restructuring its economy in the face of changing domestic and global markets and demand for different goods and, increasingly, services. Restructuring also results from changing resource availability, price signals and technology developments. However, as noted above, certain energy intensive industries face international competition. For this reason, it is important to assess how energy prices and costs developments in the EU compare with international developments.

As noted in earlier sections, average EU electricity and gas prices for industry rose comparatively moderately from 2008 to 2015 and those in Asian countries (China, Korea and Japan in particular) rose significantly faster.

The (limited) data available for comparing the industry's energy costs and energy intensities around the world suggests that Chinese energy intensive industries are far more energy intensive than US and EU industries. In contrast, some EU industry sectors appear to be more energy intensive than its US counterparts. Despite this, in most cases, energy appears to account for lower proportions of the production costs of energy intensive industries in the EU than of those in the USA and higher proportions than those in Japan. However, energy cost shares have fallen faster in the USA than in the EU since 2008, which signals, beyond the lower energy prices in the US, that US energy intensive industry could also be 'catching up'. As from 2008, evidence on energy efficiency improvements in some European energy intensive industries signals that these seem to have slowed down or even stopped. This could be explained by factors such as possibly limited scope for technical improvement, reduced capacity utilisation rates but also insufficient availability of capital to invest.

3.3 *An international comparison of industry energy costs and results*

A competitive and properly functioning energy market is expected to deliver the energy that households and industry need in the most cost effective manner. The clearer the price signals and the closer the align-

ment between prices and production costs, the more efficient energy production and consumption will be. However, in various respects, the energy market is not functioning properly. A range of market and regulatory failures have led governments to intervene in a wide variety of ways over the years to steer the development of the energy sector. Regulatory or financial measures affecting energy producers or consumers have been introduced to achieve policy goals such as reducing pollution and greenhouse gas emissions, improving security of energy supply or reducing the cost burden of energy on poor households or vulnerable businesses. Such measures often subsidise energy production and consumption and may correct price signals to reflect market failures. They include explicit measures reflected in prices, as illustrated in the energy taxes and levies subcomponents discussed above. However, some (e.g. some energy demand measures or below cost regulated prices) may also blunt price signals that would otherwise steer energy consumption and production, energy efficiency and investment. Fossil-fuels subsidies are particularly problematic, as they make the clean energy less advantageous and hamper the transition to a low-carbon economy.

In 2014, energy taxes collected by EU Member States totalled € 263 billion, equivalent to 1.88% of EU GDP. Excise duties make up the largest part of energy taxes. In 2015, excise duty revenues alone amounted to € 227 billion. Reduced consumption of energy products would have resulted in a fall in excise duty revenues, but Member States often increased the rate of excise duty. Thus energy consumption continues to provide an important tax base for public revenues, helping Member States to consolidate difficult fiscal positions. More generally, energy taxation can have a positive impact on growth compared to taxation on labour and investment. Energy prices have a further impact on broader, macroeconomic aspects of the EU economy, through inflation (Fig. 5). Energy plays

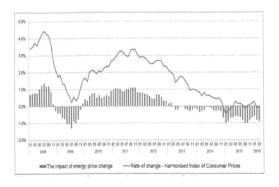

Figure 5. Impact of energy prices on EU inflation. *Source*: Eurostat

a clear role in household expenditure and industry costs, as well as a role through petroleum prices, in the transport sector in particular. As such, energy prices are a significant factor affecting inflation. Energy price peaks in 2008 and even 2011 contributed 1% to EU inflation at the time, just as lower prices are now having a deflationary impact on the EU economy.

3.4 *Conclusions*

The rapid decline in energy commodity prices in the last years, notably for oil but also for gas, stems from technological change as well as market and geopolitical developments. It has changed the energy landscape dramatically. In Europe, a major energy importer, it has brought temporary relief for households and businesses in difficult economic conditions and given a 'one-off' boost to the economy. It shows the importance of developing global markets for energy, especially for an energy-importing region such as the EU, and it reduces price differences vis-à-vis other regions. This is especially relevant for gas, where the global development of LNG markets and new sources of supply are creating opportunities for Europe: it can help narrow the gas price differential with other parts of the world such as the USA and, from a decarbonisation perspective, it improves the competitive situation of gas versus coal. However, lower prices may distract attention from the energy challenges we are facing of energy security, competitiveness and climate change (Fischer et. al. 2003). These have not disappeared. Indeed, the data and analysis in this paper show a much more nuanced picture and can help us to identify the right approaches and policies as we develop the Energy Union within this changing energy landscape. Low prices should be not taken for granted. The data in this paper shows how supply and demand can shift abruptly. This is even more the case as new resources such as shale gas and oil wells are rapidly exhausted and need constant replacement investment to maintain production. Environmental taxes may be introduced to correct market failures, for general fiscal reasons or directly to finance for example the investments in electricity.

REFERENCES

Aydin, F. 2010. Secondary school students' perceptions towards global warming: a phenomenographic analysis. In *Natural and Social Sciences Res.* Essays 5 (12), 1566–1570.

Fischer, C. et al. 2003. In *Journal of Environmental Economics and Management.* Volume 45, Issue 3, May 2003, Pages 523–545.

Helfand, G. et al. 2003. 'The Theory of Pollution Policy', in Mäler, K.-G., and Vincent, J. R. (eds.), Handbook of Environmental Economics, 1, Amsterdam: North Holland Elsevier.

Hines, J.M. et al. 2010. Analysis and Synthesis of Research on Responsible Environmental Behavior: A Meta-Analysis, In *The Journal of Environmental Education,* Volume 18, 1987—Issue 2, Pages 1–8. Published online: 15 July 2010.

Reiche, D. 2010. Energy policies of Gulf Cooperation Council (GCC) countries possibilities and limitations of ecological modernization in rentier states. In *Energy Policy.* Volume 38 (5), 2395–2403.

Squalli, J. 2007. Electricity consumption and economic growth: bounds and causality analyses of OPEC members. In *Energy Economics.* Volume29 (6), 1192–1205.

Sushmita, J.G. 2014. Environmental conservative and management systems in ancient India. In *International Journal of Research in Humanites, Arts and Literature*, Volume 2, Issue 4, Pages 49–62.

Suwirman, N. 2016. The Perceptions of Environmental Knowledge. In *Natural and Social Sciences* (IMPACT: IJRANSS) ISSN (P): 2347–4580; ISSN (E): 2321-8851 Vol. 4, Issue 11, Nov 2016. 113–122.

New Trends in Process Control and Production Management – Štofová & Szaryszová (Eds)
© 2018 Taylor & Francis Group, London, ISBN 978-1-138-05885-9

The potential of using event marketing

J. Wyrwisz
Faculty of Management, Lublin Technical University, Lublin, Poland

ABSTRACT: The aim of the paper is pointing out event marketing as the effective and acceptable tool in the communication of the organization with market environment. The research approach based on preliminary research of the writing and internet was applied in the article. An overall description of event marketing showing the specify and possibilities of its application was included.

1 INTRODUCTION

Creating appropriate communication with the environment is an important task of an organization determining its position on the market. The dynamically changing market, rising competition and rigid market rules enforce more effective marketing management on organizations. Marketing communication encompasses many aspects, undergoes changes dictated by consumer behaviors and technology. Effective ways of communication with the entities of internal and external environment are sought for. The typical promotion instruments are losing their significance and strength of impact, and thus the message targeted at specific customers should be visible, different from others and should strongly expose the brand, creating its image.

Considering these conditions, event marketing can be viewed as a new and prospective tool of marketing promotion. It focuses on building contact between the consumer and the brand as well as commitment. It gives an opportunity to directly commune with a brand, and experience it with all the senses. It is also a chance to intuitively position the brand. (Jaworowicz 2016) Marketing events cover a very broad spectrum of events, such as concerts, picnics, congresses, jubilees, festivals, shows, sports competitions or ambient actions. Organization of events targeted at different groups of market environment becomes an integral part of marketing strategies of many companies.

2 EVENT MARKETING IN MARKETING COMMUNICATION

Marketing communication involves a whole range of instruments and actions through which information about the company and its products is transferred. Communication is aimed at shaping the needs and preferences of buyers and stimulating demand (Wiktor 2013).

In today's market conditions, entities typically manage a relatively complex marketing communication system, covering identification and evaluation of key message recipients, communication actions planning, tool selection, and control of quality and effects of communication processes. Due to the increasing degree of customer preference differentiation, there is a need to create and manage multiple marketing communications programs tailored to specific market segments. The company integrates and coordinates numerous tools and communication channels to create a clear, consistent and effective message for the environment. Integrated marketing communication means conveying the value, identity, and characteristic features of the company distinguishing it from its competitors. At the same time, it is a process of receiving information from the environment and reacting to it on a partnership basis. (Rosa 2005) Table 1 lists the classic and modern marketing promotion instruments.

Due to the ever-increasing need for creativity of undertaken marketing activities, non-standard promotional solutions, including event marketing, are gaining in importance.

Event marketing is defined as a promotional tool consisting in organizing a variety of group or mass events to meet the organization's objectives in relation to the external and internal environment.

Table 1. Traditional and modern tools of marketing communication.

Tools of marketing communication	
Classic	Modern
Advertising	Event marketing
Public relations and publicity	Word of mouth marketing/ buzz marketing
Personal sale	Mobile marketing
Direct marketing	e-marketing
Sales promotion and POS	Sponsoring product placement

(Jaworowicz http) It is a complementary set of marketing solutions that coordinates and supports their effectiveness. Selected event marketing elements are also presented as public relations components and a sales promotion tool with consumer and business applications. (Kalinowska-Żeleźnik 2009) Organization of various types of events aims at achieving specific and desired indicators related to image or sales benefits (Mazurkiewicz 2015).

Events are identified as cultural, sporting, scientific or entertainment events, presenting the company and its products in an original and unconventional way. They are a tool of non-personal communication reaching target groups through the use of multiple tool combinations, supported by direct brand information. (Grzegorczyk 2009).

Although the term event marketing permanently entered the marketing dictionary, there also function several other terms: promotion event, sponsorship marketing, or event organization. (Jaworowicz 2017).

Assigning event marketing to a specific promotion-mix instrument, an attempt to classify it as a type of advertising or sales promotion may be encountered. Event marketing is sometimes considered a public relations element. (Żukowska 2015) However, it is difficult to agree with such order because in practice these tools are combined and often used interchangeably within the framework of the event itself. Event marketing should be considered an overriding activity, using various marketing tools in accordance with the outlined elements of the adopted marketing strategy. (Jaworowicz 2017) The event can combine essential functions of: (Olejniczak 2013).

– Communication (information),
– Persuasion,
– Cultural studiem,
– Education,
– Consolidation,
– Company and product image creation.

It is very difficult to determine the scope of the concept of event marketing and its clear boundaries. The scale of activities within this issue is very wide and varied. Organizers of such events may be any market entities undertaking marketing activities to promote their products. Thus, these are manufacturing and commercial companies, as well as non-profit organizations and individuals. It has been observed that event marketing has become one of the most important instruments of territorial marketing in recent times, and local government units willingly take advantage of this tool to promote cities and regions. (Stępowski 2016) It should be stressed that the challenges placed on event organizers by their participants require undertaking many well planned and coordinated actions. High attractiveness and entertainment,

usefulness of time and place, and security provision are expected from the events. Event marketing entails certain benefits for the participant, that is the opportunity to meet a famous person, rivalry, artistic experience, or participation in a show or feast. It is desirable from the point of view of the customer to achieve material benefits, that is product samples, awards in competitions, promotional gadgets, or refreshments. The event should, however, primarily provide non-material values based on emotions, new stimuli, and experiences (Walkiewicz 2016).

The growing popularity of events is primarily associated with the fact that they are a place of extraordinary meetings of consumers and brands, during which unique opportunities to present products in action are created. Events have also become a unique communication platform with particular participation of the media and their ability to produce attractive content. (Jaworowicz 2016) The most important features of event marketing are: (Jaworowicz 2016).

– Multi-level communication tool implementing complex marketing objectives,
– Alternative to traditional promotion tools,
– Emotionally engaging branding tool,
– Means of reaching the media and content creation moderator,
– Communication platform integrating other tools and brands,
– Live campaign and touchpoint—a place of meeting with the brand,
– Environment in which B2B and B2C relationships are built.

3 TYPE OF EVENTS

The approach of the subject literature to the classification of events is very varied, indicating numerous divisions of marketing events. The classification takes into account the factors of various importance influencing the nature of the event. (Grzegorczyk 2009) Typical classification in terms of importance distinguishes: (Florek 2013).

– Mega-events, one-off large format events with international publicity, attracting many participants and generating great economic benefits,
– Special events, most often performances or celebrations, which take place to mark an occasion,
– Events characteristic of a given place, inseparably linked and often automatically identified with it,
– Significant events, similar to mega-events but of a smaller scale,
– Festivals celebrating a specific theme, identified with a given city,
– Small-scale local events.

Another division indicates four groups of events: (Grzegorczyk 2009).

– Institutional events,
– Business events,
– Incentive events,
– Special events.

Taking into consideration the criterion of the place, an attached event, that is unambiguously associated with the place, a traveling event and stationary event can be distinguished. In terms of the duration, one-day, multi-day and stage events can be mentioned. In turn, using the repeatability criterion, one-off, several-time and cyclical events are observed. From the point of view of marketing event attendees, there exist internal, external and mass events. The scale criterion in turn allows for distinguishing chamber and group events, as well as mass and global events. Taking into account the availability, events can be closed ticketed, by invitation only or open to anyone. The effectiveness of the event is seen among others in the publicity, therefore, media publicity or event aimed only at direct participants can be distinguished. In terms of activity area, events are divided into online/multimedia or offline events. Event objectives determine the distinction of events into image, political, sales, nonprofit (charity, religious, ecological) and product events (e.g. involving a product whose advertising is legally restricted) as well as guerilla marketing events. The criterion of integration with participants distinguishes active and passive events. In turn, the customer's attachment to the event concerns fixed and casual events. Distinction of marketing events can also be made on the basis of the company's engagement criterion; sponsorship or the organizer's own contribution can be distinguished here (Grzegorczyk 2009, Jaworowicz 2016).

Taking into account the theme of the event, there is a broad spectrum of events: public/local government, jubilee, cultural, musical, theatrical, film, other fields of art and culture, historical, political, religious, sporting, economic, scientific, fair and exhibition, picnic and feast, gastronomic, awarding, record breaking, carnival, parade and gala, extreme, happenings, multimedia, light and sound spectacles, pyrotechnics, and children's. (Florek 2013).

4 IDENTIFICATION OF AREAS AND CONDITIONS OF EVENT MARKETING APPLICATION

The specific features of event marketing provide a wide range of its application possibilities. They allow this form of promotion to be original and surprising for the audience. They allow for conveying, apart from the brand logo, content and intangibles as added value. (Gruszka 2016) This is due to the following conditions: (Szkotak http and Kalinowska-Żeleźnik 2009).

– Direct and individual contact with customers,
– Easy access to specific stakeholders and possibility of quick feedback from customers,
– Diverse and arbitrary impact on participants,
– Participants' emotional involvement with the brand,
– Interaction with the brand, product, and organization representatives,
– Possibility of using participant/consumer behavior in a group,
– Creation of direct contact with the brand in all its forms and elements,
– Unique opportunity to directly observe the reaction to the brand.

Event preparation is based on a typical plan of marketing activities that includes identifying image and economic goals, specifying target audiences, developing content, planning operational activities, and budgeting and auditing. The model course of this process does not imply its simplicity. For the event to be effective, it is necessary to develop a number of issues that will ultimately result in an attractive event—from the idea of the event, place, creative approach to its individual elements, precise scenario, visual identification system, to the technical and security issues.

A stage that requires a special commitment and to a great extent determines the success of the event is its promotion. Here classic and nonstandard promotion tools are also applied. In addition to the media coverage on television and radio, outdoor, word of mouth or the so-called ambush marketing is used. Among the events best suited for this type of promotion are tournament, international and cyclical events (Waśkowski 2009).

Modern technologies, including mobile technologies, and popularization of their access made the implementation of events move to virtual space. Special event services serve this purpose. This makes it easier for companies to build relationships with their environment. These activities are often accompanied by the introduction of a new product to the market, an important event related to the brand and the company, but also film festivals and fairs (Jando 2017).

More and more often the implementation of events is taking into account the use of a number of modern technologies. They raise the level of the event and make it more attractive to the participants. Applications which provide all the needed information in one place are being prepared. (Gruszka 2016) Social media is an indispensable attribute of the event, which takes on the role of information, promotion and entertainment.

5 CONCLUSIONS

Effective communication with the environment is one of the most important aspects of the company's functioning in the changing conditions of today's market. Active competition, the flood of information and consumer's dominant position can make even the most attractive products go unnoticed. In order to prevent the negative effects of such a situation, a special role is attributed to the process of marketing communication with the broadly understood environment. The forms of impact on the customer, which allow direct contact and the possibility to interact with them are gaining in importance. In the area of solutions to this problem, there is the use of event marketing. It is distinguished by high effectiveness, whose source is attractiveness to the audience, high noticeability and branding in a particular context. Event is a complementary instrument that perfectly complements other elements of the promotion. It perfectly integrates into wider marketing activities. As such, the event is becoming an increasingly important tool for product marketing communication. The event industry is undergoing noticeable professionalization and emphasizes its autonomy.

REFERENCES

Florek, M. 2013. Podstawy marketingu terytorialnego, Poznań: Wydawnictwo Uniwersytetu Ekonomicznego w Poznaniu.

Gruszka, A. 2016. Event szczęśliwych marketerów. Marketing w Praktyce 223(9): 66–68.

Grzegorczyk, A. (ed). 2009. *Event marketing jako nowa forma organizacji procesów komunikacyjnych*. Warsaw: Wyższa Szkoła Promocji.

Jando, O. 2017. Kilka słów o event marketing, http://www.sferamenedzera.pl/marketing/kilka-slow-o-event-marketingu/ (25.03.2017).

Jaworowicz, P. & Jaworowicz, M. 2016. Event marketing w zintegrowanej komunikacji marketingowej. Warsaw: Difin 2016.

Jaworowicz, P. 2017. Event marketing znany i nieznany. https://marketingowe.wordpress.com/ 2014/04/06/event-marketing-znany-i-nieznany/ (25.03.2017).

Kalinowska-Żeleźnik, A. 2009. Event marketing jako forma komunikacji marketingowej. Zeszyty Naukowe Uniwersytetu Szczecińskiego 559, Ekonomiczne Problemy Usług 42: 429–436.

Mazurkiewicz, B. 2015. Gry miejskie oparte na lokalizacji jako sposób promocji miasta. *Handel wewnętrzny* 4(357): 328–336.

Olejniczak, A. 2013. Event marketing jako jedna z form innowacji marketingowych w instytucjach naukowych i badawczych. *Marketing i Rynek* 10: 9–15.

Rosa, G. 2005. Komunikacja marketingowa. Szczecin: Wydawnictwo. Naukowe Uniwersytetu Szczecińskiego.

Szkotak, M. Event marketing. http://nowymarketing.pl/a/9166,e-event-marketing-encyklopedia-marketingu (25.03.2017).

Walkiewicz, S. 2016. Event bardziej kulturalny. *Marketing w praktyce* 219(5): 64–65.

Waśkowski, Z. 2009. Wykorzystanie ambush marketingu w promocji przedsiębiorstwa. *Zeszyty Naukowe Uniwersytetu Szczecińskiego 559, Ekonomiczne Problemy Usług* 42: 482–489.

Wiktor, J.W. 2013. Komunikacja marketingowa. Warsaw: PWE.

Żukowska, J. 2015. Marketing communication. Warsaw: Warsaw School of Economic.

New Trends in Process Control and Production Management – Štofová & Szaryszová (Eds)
© 2018 Taylor & Francis Group, London, ISBN 978-1-138-05885-9

Evaluation of financial situation in tourism company through summary index

J. Zuzik, K. Čulková, M. Janošková & A. Seňová
Faculty of Mining, Ecology, Process Control and Geotechnologies, Technical University of Košice,
Košice, Slovak Republic

ABSTRACT: One of the main goals of any company is to maintain long term financial stability, since it must permanently evaluate results of its activities. Individual indexes of financial analysis confirm financial stability of the company as a tool for deep knowing of economical processes and events. This analysis help to find problems in given company, it presents also useful source of information about real situation for its managers. Contribution analyses financial situation of chosen tourism company in Slovakia. Authors used calculation of summary indexes for evaluation of present state of financial situation in subject, mainly by two methods: Altman model Z-score, and bankruptcy model that by the way of indexes file enables to determine probability, if the company is prospering or having problems, as well as Kraliček Quick test as bonity model, enabling to evaluate bonity of company. According achieved values of four indexes there was calculated index of bonity.

1 INTRODUCTION

Idea of financial analysis can be defined in narrow sense as method of knowing, in broader sense as tool for deep knowing of economic events and processes that create assumption for their effective influencing. Financial analysis is used in all phases of management cycle with goal to come to characteristics of events, processes and mainly to know causation and relations among events that are not known at first glance and that create economy of the company as a subject of analysis. It enables qualitative and reliable knowing of basic situation and its development tendencies. It creates reliable base for effective management, and it is part of any serious decision (Zalai 2013).

In the contribution there is analysed financial situation of chosen tourism company in Slovakia. Financial analysis had been done in time horizon 2013–2015, according data, obtained from balance sheet and loss and profit statement. Since in summary indexes of evaluation there are also data from cash flow, such information had been used from cash flow statement.

2 LITERATURE REWIEV

Financial analysis of the company serves for managers of the company, as well as for its owners and businessmen that must permanently follow up performance of their business. But it can serve also for brokers, financial analytics, investment advisors and experts during providing of finances for business making (Kislingerová & Hnilica 2005).

The importance of financial analysis can be seen also from the view of global economic crisis, when the value of a number of companies decreased. The very fact proves financial analysis can be helpful for industries consolidation as well (Mixtaj et al. 2013). Traditional criteria for analysis of financial situation and performance of the company had been extended by modern methods, mentioned for example by Zalai (2013). Beside other industries possible using of financial analysis can be made in mining companies. Weiss R. and Weiss E. (2013) presented analyses of selected economic indicators of mining companies that in 2011 and 2012 made significant acquisitions. Zhu et al. (2016) analyses the financial behaviour of crisis. More importantly, a new approach is provided for testing the financial effect.

Other useful area for financial analysis application is in area of tourism companies. Weiss et al. (2012) dealt during their research with analysis of financial income in the Slovak economy in tourism sector with aim to access the impact of the economic crisis on the sector of tourism services. The results speak about negative impact of the economic crisis, involved also Slovak tourism industry as the worldwide trend. One of the reasons that most influenced the recession in tourism is less business activities of companies. The result of this factor is also reduced employment and a decline in disposable income of the population.

Results and information from financial analysis can be used for decreasing of business risk, connected with specific activity of the company (Weiss et al. 2012), helping easing of proper management. Financial analysis is an important subject also in the field of international finance and investigating of the influence of country risks, including economic, financial, and political risks (Lee 2016). Higher economic, financial and political risks generally lead to higher inequality.

3 METODOLOGY

In the contribution there is evaluated financial situation of International Hotel Dukla***, characterized by attractiveness of its history, modern equipment and single facilities. It is situated in historical center in Prešov immediately besides historical building of Jonáš Záborský theatre. Hotel provides accommodation in 60 over standard rooms with capacity of 101 beds. Rooms are divided to basic categories, mainly: double bed and single bed rooms with possibility to have extra bed, luxury apartments, classical apartments, representation apartments and business apartments. On the top of the hotel there is oasis of Wellness center with panoramic view to the city with complex offer of services. Hotel offers in wellness center massage, sauna, Jacuzzi, solarium, fitness center, relax room and bar. Part of the hotel is also congress hall that is situated at 2nd floor. Hall is equipped with modern technique with capacity for 200 persons. General using is determined for congresses, symposiums, conferences, as well as celebration of jubilee, weddings, graduation ceremonies and other social and cultural events. Congress hall has own bar, summer observatory terrace, parquet, small lounge for 20 persons and small hall for negotiations for 15 persons that properly fill services of main hall and create possibility for uninterrupted conversation of top managers.

Basic source of data for financial analysis presents financial reports, mainly statements from financial accounting, providing real state of property and liabilities (balance sheet), real revenues and costs (loss and profit statement), real flow of incomes and expenses (cash flow statement). Cash flow statement is in some cases part of annex of financial statements. Such reports are filled by annual report in companies that must have audit according law about accounting.

Authors used calculation of summary indexes for evaluation of present state of financial situation in subject, mainly by two methods: Altman model Z-score, and bankruptcy model that by the way of indexes file enables to determine probability, if the company is prospering or having problems, as well as Kraliček Quick test as bonity model, enabling to evaluate bonity of the company according points classification.

Hotel Dukla*** Prešov has activities of number of provided services, review of sale from accommodation, restauration, and other services is illustrated by Table 1.

4 RESULTS OF FINANCIAL SITUATION ANALYSIS TROUGH SUMMARY INDEXES

4.1 *Altman methodology*

Actual financial situation of the hotel had been analysed by Altman methodology, which speaks about its future development. Altman calculated multi variable discrimination function for analysis of 66 companies, while half of the group was prospering, and half not prospering. Statistical analysis found out that the best statement of the financial situation and its future development are given by following indexes.

Table 2. and Figure 1. illustrate single calculation of Z-score in chosen hotel.

In Hotel Dukla*** Prešov during time horizon 2013–2015 financial situation was analysed as bad, hotel is threatened by bankruptcy, since Z-score achieved negative values. Consequence of bad result can be mainly negative or minus result of economy that configures in various indexes of

Table 1. Structure of sale from offered services.

	2014	2015
Accommodation	341.092	300.204
Restaurants	384.268	362.181
Other	126.577	130.872
Summary	851.937	793.357

Table 2. Z-score calculation in chosen hotel.

	2013	2014	2015
X_1 = net working capital/total assets	0.08	−0.02	−0.02
X_2 = not divided profit/total assets	−0.01	−0.02	−0.04
X_3 = EBIT/total assets	−0.23	−0.20	−0.02
X_4 = own capital/foreign capital	0.48	0.44	0.36
X_5 = sales/total assets	0.25	0.24	0.23
$Z = 0{,}717x_1 + 0{,}847x_2 + 3{,}107x_3 + 0{,}42x_4 + 0{,}998x_5$	−1.05	−0.52	0.27

the evaluation. Therefore there is necessary the hotel would increase value of its economic result and by these way also individual indexes, which could increase value of final result. Table 2. and Figure 1. illustrate single calculation of Z-score in chosen hotel.

4.2 Kraliček quick test

From the group of bonity models of financial analysis we used Kraliček test (called Quick test), which enables to evaluate analysed company very quickly and its informative value is rather high. In afford to provide balance of the index; there are used indexes from four basic areas of analysis. Quick test is using following indexes: repayment of debts from cash flow, quota of own capital, cash flow in percentage from sales, return on assets (ROA). Calculation of the model is made according following table by pointing of individual results.

During calculation according the model we resulted from cash flow statement in 2014 and 2015. It is necessary to know also situation of the company before 2012 with aim to compare the results of the index from long term view. Since the test is

created according point methodology, the result is number of allocated points. Result of the test for 2014 is value 9, which speaks about creditworthy company. In 2015 hotel achieved value 4, which is at the limit of bonity and area of unclear future. During comparing of the values there is rather big decrease in spite value 4 is yet in part of good result. Due to the cash flow increasing there is necessary hotel would elaborate redeemable and targeted financial plan as well as to observe the plan. There is also necessary the hotel would observe its payment discipline and relations with important consumers and suppliers.

Development of the index is given also by Figure 2.

4.3 Bonity index

Result of complex economic situation of the company is given also by bonity index. Besides it presents indicator of bonity, it is based on multi variable discrimination analysis.

During calculation for analysed hotel Dukla*** Prešov for 2013 and 2014, it was calculated according cash flow index, since cash flow had been recorded only in this period. Results, qualified bonity index B in 2013 was at value 0.03, which means certain problems in economic situation of the company. In 2014 the value decreased to –0.64, which means bad financial situation. Values of this index in individual

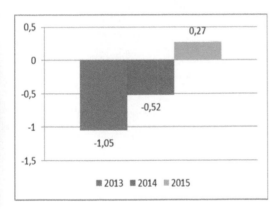

Figure 1. Graph for Z-score according calculation in hotel.

Table 3. Kraliček Quick test.

	2014	Points	2015	Points
R1 = equity/total assets	0.25	3	0.22	3
R2 = (foreign sources— short term financial property)/cash flow	8.09	2	58.81	0
R3 = (net profit + interest)/ total assets	–0.29	0	–2.40	0
R4 = cash flow/sales	0.30	4	0.05	1
Summary		9		4

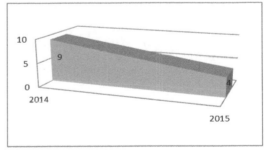

Figure 2. Development of Kraliček index.

Table 4. Bonity index in hotel Dukla***.

	2014	2015
X_1 = cash flow/foreign capital	0.12	0.17
X_2 = total assets/foreign capital	1.73	1.62
X_3 = EBIT /total capital	–0.02	–0.04
X_4 = EBIT /total performance	–0.08	–0.18
X_5 = stocks /total assets	0.94	0.85
X_6 = sales /total assets	0.24	0.23
$B = 1.5x_1 + 0.08x_2 + 10x_3 + 5x_4 + 0.3x_5 + 0.1 \times 6$	0.03	–0.64

years were rather low and they yet decreased mainly due to the cash flow and economic result decreasing. It is necessary hotel would increase not only index of economic result, but also its cash flow.

5 CONCLUSION

Financial analysis is important part of financial management of the company. Since it acts as backward information about situation. To which company got to in individual areas of financial management. In which areas company succeeded to fill the goals and vice versa. In which areas company backslides behind expectations. etc. The hotel should follow up development of individual indexes also in the future. To improve their values. which influence single financial situation.

ACKNOWLEDGMENT

Contribution is partial result of project solving VEGA No 1/0310/16.

REFERENCES

Čulková. K.. et al. 2015. Development of established and cancelled companies in Slovakia. *Journal of Applied Economic Sciences.* 10(5): 644–653.

Kislingerová. E. & Hnilica. J. 2005. *Finanční analýza—krok za krokem.* Praha: C.H. Beck.

Lee. C.C. & Lee. C.C. 2016. The impact of country risk on income inequality: A multilevel analysis. *Social Indicators Research.* 1–24.

Ivaničková. M. et al. 2016. Assessment of companies' financial health: Comparison of the selected prediction models. *Actual Problems of Economics.* 180(6): 393–391.

Mixtaj. L. et al. 2013. Impact of acquisition on financial indicators of mining. In Proceedings SGEM 2013. 13th International Multidisciplinary Scientific Geoconference Science and Technologies in Geology. Exploration and Mining: Vol. 1. Albena. Bulgaria. 16–22 June 2013. Albena: STEF92 Technology Ltd.

Mižíková. I. & Csikósová. A. 2009. Insurance as an important factor reducing the risk in industry. *Acta Montanistica Slovaca.* 14(3): 260–267.

Weiss. R. & Weiss. E. 2013. *Tourism and its risk.* Eger: Líceum Kiadó.

Weiss. E. et al. 2012. The options of minimalization of financial risks in a travel agency. *Acta Geoturistica.* 3(2): 36–41.

Zalai. K. a kol. 2013. *Finančno-ekonomická analýza podniku.* Bratislava: Sprint 2.

Zhu. Y. et al. 2016. In press. Financial contagion behaviour analysis based on complex network approach. *Annals of operations research.* 1–19.

New Trends in Process Control and Production Management – Štofová & Szaryszová (Eds)
© 2018 Taylor & Francis Group, London, ISBN 978-1-138-05885-9

Financial resources of innovative activity in small and medium-sized enterprises

A. Zych

State Higher Vocational School Memorial of Prof. Stanislaw Tarnowski in Tarnobrzeg, Tarnobrzeg, Poland

ABSTRACT: The force for the long—term development is knocking economics off its balance by business—oriented entrepreneurs. Therefore, it is very important to make the optimal conditions for innovative business. Innovative activity is basically based on all branches of economic life. Its functioning is determined not by one, individual phenomenon but the whole its data set. It is necessary to take into account economic, informational, financial, market-based, social, institutional, scientific, legal aspects and so on. In the market economy which is characterized by its globalization and prominent competition where the entrepreneurs have to function nowadays, one of the crucial factors of creation, introducing the improvement of different innovations are financial dimensions. That is why; the aim of this article is the analysis of financial resources of innovative activity in small and medium-sized enterprises of agri-food industry which have been functioning in Rzeszow County.

1 INTRODUCTION

A lot of research concerning the management of innovations shows that they are firmly connected with raising productivity or making an enterprise more competitive on the market. It seems that the ability to create and absorb innovations is one of the most important challenges of the 21st century. Although the analysis of innovation is rather frequent in other countries, they are very rare on the regional and local scale.

Small and medium sized enterprises are more often perceived as an important source of innovations for the whole economy. It involves rising an issue of ways in which small and medium sized enterprises realize and implement innovations and the question of methods of evaluation of innovation (Hoffman ET AL. 1998, Suku 2009, Varis & Littunen 2010, Wang Y.L. et al. 2010).

Innovation processes might be also observed in agri-food industry which because of the great importance for national economy and its specific of production gets its support. This sector in Rzeszow County is concentrated mainly on trades such as: meat, grain and miller's, fruit and vegetables, dairy and confectionary and it is one of the best developed fields of economy of the province. Agriculture is the area linking four functions: agricultural, industrial, touristic and recreation.

Gaining, using and introducing innovations is connected with appearing barriers which might be defined as circumstances delaying, stopping or modifying a process of generating and implementing innovations (Mirow et al. 2008).

The innovative activity of enterprises depends on internal and external conditions mutually related with each other. Even though the range of capabilities of affecting on innovative activity of an enterprise changes depending on its external conditions, the activity in this field is mostly determined with factors depending on an enterprise which create and form the immediate environment of an innovator, facilitating and stimulating its attitude to introducing innovative solutions. The proper recognition of these determinants allows increasing innovation of enterprises (Zych 2015).

Innovations are the peculiar instrument of enterprise, introduced by businessmen; therefore, they should be a priority in the innovative activity of enterprise, even at the time of present crisis, as they can help companies in development. In order to do that, companies are forced to evaluate their innovations

The innovations are the strategic stimula for the economic development and enterprise and being innovative are the main powers causing the economic development. The force for the long—term development is knocking economics off its balance by business—oriented entrepreneurs (Cash et al. 2011). Therefore, it is very important to make the optimal conditions for innovative business in the business reality. It makes enterprises for achievement of definite purpose taking activity innovative. About biggest purpose and development enterprise is most important meaning. However, financial factors and credit policy are these ones which have the most negative effects on the innovative activity of the enterprises. Among them are high prices of resources essential for their activity, taxes, inflation and indicator of table rates. The introduction of innovations generally means the improvement of their activity and functioning. Innovative activity is considered as a reason of increase of competition and the development but it is depended on financial resources which are connected with the possibility of the particular enterprise to benefit from inner and also external financial resources (Zych & Maciejewska 2016).

The basic foundation of gaining competitive advantage, generating an attractive market offer or effectiveness in creating innovations is having their own financial potential. Each company should ask itself what influences its innovation. The question should raise the issue of methods of suitable, internal and external environment of a company which allows it to improve its effectiveness, efficiency in raising a level of innovation. To answer this question is an aim of this article in which internal and external determinants of financial potential need to be discussed. These are the ones which allow companies to adjust to appearing dynamic changes or predict and overtake changes that will come.

2 METHODOLOGICAL ASSUMPTIONS AND PROBABILITY SAMPLING FOR MARKET-BASED RESEARCH

The major units of sampling were small and medium-sized enterprises of agri-food industry in Rzeszow County in accordance with the classification of Polish Central Statistical Office. On the other hand, the units of market-based research were the owners and the managers of these enterprises. The enterprises were treated as certain identical repertory, concurrently the micro-enterprises were excluded from this market research. According to Oslo handbook, it is indicated to conceptualize the extent of enterprises on the grounds of the number of employees. This recommendation is in accordance with analogical proposals formulated in various handbooks from 'Frascati family'. In order to obtain comparability of the data, the handbooks recommended applying the following classes of extent of enterprises: from 10 to 49 of the employees-small enterprises; from 50 to 249 – medium-sized enterprises.

The attention was focused on the group of these enterprises for in the structure of MSP, they have had dominant involvement in innovative activity.

The criterion of sampling was determined by the following factors taken into account inclusively:

– The enterprise of business activity within, Rzeszow country,
– The enterprise which employs from 10 to 249 people,
– The enterprise of public or private sector,
– The agriculture based on agri-food industry.

The market research was carried out within those enterprises which fulfill above criterion on the basis of written compilation. This compilation was attained for the order from the Provincional Statistic Office in Rzeszow in accordance with the public register of national economy. For the reason of the information included in the REGON registry this written compilation is reliable in the case of enterprises and institutions as far as different aspects are concerned. Oslo handbook recommends applying different official repertories of enterprises in order to establish the absolute written compilation. These repertories are used for statistical purposes such as national and provincial governor's office.

The written compilation was made within 49 enterprises. The market research was made in January 2017. As the result of this research 43 questionnaires were obtained, from which one was rejected after the verification. For the ultimate analysis 42 questionnaires were taken into consideration.

The research was carried out through a method of direct inquiry, a postal survey, an e-mail and a phone.

3 THE CHARACTERISTIC OF MARKET RESEARCH ENTERPRISES

Among the enterprises where the market research was carried out, 83.3% were small economic subjects. On the other hand, 16.7% were medium-sized enterprises (Fig. 1).

Almost 54.8% of the enterprises have been on the market for 10 years, nearly 35.7% of them have been existing on the market from 6 to 10 years. Those enterprises which were established from 2008 and 2013, therefore which have been running their business from 1 year to 5 years, constituted 7.1% of the registered traders. Those enterprises which were established in 2012, therefore which have been functioning on the economic market for a year-just 2.4%. Then, it can be assumed that the enterprises

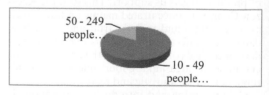

Figure 1. The structure of conducted market research of enterprises according to the extent of enterprises (in %). The source: Personal study on the basis of market research.

in which the market research was made, can be characterized as those enterprises which have been functioning for a long time on the market. The logical conclusion can be made that the opinions of these enterprises can be treated as entirely reliable and based on long established experience.

The majority of enterprises where the market research was done are those which are run by the legal persons on the basis of business activity register (52.4%). It is the simplest form of business activity as far as the legal aspects are concerned. Once, there are those economic subjects of the certain character. The most characteristic elements of such business activity are: low financial investment needed for running the business (very often by the owner of the business who makes all the decisions personally by himself/herself), narrow range of business, focused on family-oriented business, making the decisions basically on the local level. Almost 28.6% are civil partnership, 9.5% are limited company corporations. The smallest group was established by enterprises in the form of registered partnership (7.1%) and also joint-stock company (2.4%).

The market research of particular enterprise was also analyzed on account of the sort of economic activity. The majority of the enterprises where the research work was carried out are those economic subjects which run productive activity. As far as the area of influence is concerned in the field of supply and sales, the respondents clearly indicated local, regional and domestic range.

4 RESULTS AFTER CONDUCTING RESEARCH

The enterprises make business expenses for innovative activity. In the recent years within small and medium-sized enterprises based on agri-food industry of the Rzeszow County, their structure has not changed essentially. Despite the growing number of financial service on the market and the increasing interest of MSP sector of new forms of financial support, still, the most popular form of funding innovative solutions are internal funds and also earned profit (financial surplus). In the case of external sources of finance, the most popular form of funding are commercial credit, family and friends' financial support, banking credit and leasing. For small and medium-sized enterprises based on agri-food industry in Rzeszow county, the basic form of funding (Table 1), also in the case of innovative activity, are internal funds (85.7% among the tested) and also financial surplus (95.2%). The remaining forms of funding have been applied only subsidiary. The costs and the risk of running the innovative activity within the market research enterprises are the financial burden of their own-

Table 1. The source of funding the innovative activity within the market research enterprises (in %)*.

The source of funding	Market-based research enterprises (%)
Internal funds	85.7
Financial surplus	95.2
Commercial credit	38.1
Family and friends financial support	35.7
Bank credit	30.1
Leasing	28.6

*The possibility of few range of indication.
The source: Personal study on the basis of market research.

ers. Moreover, those enterprises which have been newly established and those which are the smallest ones on the market are not treated favorably by the financial institutions when they try to make use of external sources of finance, quite on the contrary.

Despite the permanent development of innovative activities within small and medium-sized enterprises in Rzeszow County they encounter a lot of obstacles and barriers difficult to overcome effectively and efficiently. The result of that is the economic slowdown or even the cessation of their innovative activities. As the market research shows, for the development of innovative activities, the most important are financial barriers and economic factors.

The enterprises of small and medium-sized of agri-food industry in the Rzeszow County found that the most important external barriers of innovative activities (Table 2) are economic and financial factors. It means extensive costs of innovation, high costs of launching a new product on the market, tax burden, high bank rate of credits and also too high costs of compilation of innovative solutions.

The barriers of innovative activity of the enterprises can also be the result of internal weaknesses (Table 3). Similarly, as it is in the case of external barriers, the most significant internal barriers for the market research enterprises, are financial factors. In that way, it means the low level of profit, the lack of internal funds for the innovative activity and also bad financial position of the enterprise.

As the market based research of the author shows, among the inner dimensions of innovative activity, the financial factors have had the most significant importance for creating, introducing and diffusion of innovation.

Summing up, it is also worth citing the general results for the particular inner groups, the

Table 2. The external financial factors in the opinions of market-based research enterprises (in %)*.

The external financial factors	Market-based research enterprises (%)
Extensive costs of innovation	95.2
High costs of launching a new product on the market	85.7
Tax burden	83.3
High bank rate of credits	64.3
Too high costs of compilation of innovative solutions	50.0

*The possibility of few range of indication.
The source: Personal study on the basis of market research.

Table 3. The internal financial factors in the opinions of market-based research enterprises (in %)*.

The internal financial factors	Market-based research enterprises (%)
The low level of profit	92.8
The lack of internal funds for the innovative activity	88.1
Bad financial position of the enterprise	85.7
The lack of financial reserves for risky operations	85.7
The low financial potential of the enterprise	69.0
The low self expenditure for research and development	45.2

*The possibility of few range of indication.
The source: Personal study on the basis of market research.

dimensions of innovative activity of small and medium-sized enterprises in Rzeszow County based on agro-food industry. They show that in the opinion of the market-based research, they hold resources which help them to come to the fruition and implementation of the innovations. Recapping-the most significant factors for innovative activity within the respondents are financial factors (92.8%), human capital (83.3%), market factors (81.0%), information factors (64.3%) and organizational factors (47.6%). For less important however also positively influencing were considered, generative factors (38.1%).

5 CONCLUSIONS

Conclusions after the research carried out by the author show that among all the tested enterprises in Rzeszow County, there are entities which are truly innovative, mainly these of medium size. For tested respondents, an innovation is an introduction of brand new or improved products or processes. The innovative activity of small and medium sized enterprises is connected with the introduction of new products rather than processes. Projected innovative research show that nearly 88.1 per cent of the tested enterprises is going to introduce some innovations within the next two years. They are going to introduce new or improved products, gain new market share and introduce new technologies. The main reason for introducing innovations on the market by small and medium-sized enterprises in the county are reactions to the changes in the environment. It is environment pressure which is the main driving force for innovative activity of tested enterprises (the improvement of competition, the need for surviving on the market, increasing or maintaining the market, the need for introducing innovations because of customers' expectations, fear of not being competitive enough on the market). The predominant financial reason of introducing innovations is the hope for increasing sales and profits of an enterprise and decreasing the costs of its activity (Zych 2016).

The effect of innovative activity which is expected by tested enterprises is the improvement of both customer service and the quality of produced goods or services. Therefore, small and medium-sized enterprises of the Province cooperate with their customers. For the respondents, the introduction of innovations generally means the improvement of their activity and functioning. According to the owners of the surveyed enterprises, innovative activity is perceived as a reason for increase of competition and development of their companies. However, it depends on their financial resources. In the group of the enterprises which were tested, an innovation is an action being forced by the market, their environment and economic necessities. The basic source of financing of innovative activity are own resources of an enterprises. In tested enterprises, innovations which dominate are these which are mainly developed by their own companies and its source are their employees, managers.

The most significant restrictions for innovative activity and also the most serious weaknesses of small and medium-sized enterprises based on agri-food industry in Rzeszow County are connected with the high costs of innovation and also with insufficient financial potential. Indeed, the financial potential of the enterprise plays very often the crucial role in estimating how many unsuccessful trials the particular enterprise can afford. Moreover, according to the respondents' innovative activity is carried with the high risk that the expenditure made, will not be retrieved. What is more, the respondents also indicated the difficulties with external funding.

The factors which influence innovative activity are also those connected with central authorities, the increase of expenditures on developmental and research activity. They stimulate cooperation between research-expansional institutions, universities and enterprises. Improving and supporting the innovative potential in the area by endorsing financial support, creating social and organizational conditions, the funds simplifying enterprises the availability to credits by giving them bank and credit guarantees, are also crucial factors in innovative activities of the enterprises. It may also contribute to economic development of the area because the expenditures from the budget of local authorities can be retrieved in the form of local taxes which are paid by well-functioning, dynamically developing enterprises.

The strong points, according to the respondents, are good opinion of customers, an ability to react quickly because of the market changes and their flexibility. The weak point, according to the respondents is high risk concerning innovations. The main limits concerning innovations are the high cost of introducing them and insufficient financial resources.

According to the respondents, economic factors and credit policy are these ones which have the most negative effects on the innovative activity of the tested enterprises. Among them are high prices of resources essential for their activity, taxes, inflation and indicator of table rates. As far as the law and political conditions are concerned, the respondents mentioned indolence in passing new bills, expenditures, the lack of cooperation between academic institutions and economy as the main factors restraining their innovative activity. Another factor limiting their innovative activity is the system of education in Poland.

According to the respondents, the main factors having a positive impact on their innovative activity are taking advantages of grants, subsidizations and the access to external sources of financing and increasing of opportunities of using modern technologies and human resources. The determinants which have the greatest negative impact on limiting the innovative activity of the enterprises is competition. Therefore, the tested enterprises should pay more attention to structural competition and creation of new needs of customers which they will be able to satisfy later.

REFERENCES

Cash,, I. et al. 2011. *Jak jednocześnie zwiększyć innowacyjność i spójność działań w organizacji*: 146. Harvard Business Reiew: Polska.

Hoffman, K. et al. 1998. Small firms, R&D, Technology and Innovation in the UK: a Literature Review, *Technovation* vol. 18, no. 1/1998.

Mirow C. et al. 2008. *The Ambidextrous Organization in Practice: Barriers to Innovation within Research and Development*, referat z konferencji Academy of Management, Anaheim 2008.

Suku B. 2009. *Incremental Innovation and Business Performance: Small and Medium-Size Food Enterprises in a Concentrated Industry Environment*, Journal of Small Business Management vol. 44, Issue 1/2009.

Varis M. & Littunen H. 2010. *Types of Innovation, Sources of Information and Performance In Entrepreneurial SMEs*, European Journal of Innovation Management vol. 13, no. 2/2010.

Wang Y.I.. et al. 2010. *Learning and Innovation in Small and Medium Enterprises*, Industrial Management & Data Systems vol. 110, no. 2/2010.

Wiśniewska, J. & Janasz, K. 2013. *Innowacje i jakość w zarządzaniu organizacjami*: 34–98. CeDeWu: Warszawa.

Zych, A. 2015. *Wybrane czynniki innowacyjności w małych i średnich przedsiębiorstwach na przykładzie branży rolno-spożywczej*: 7,133. Wydawnictwo Państwowej Wyższej Szkoły Zawodowej im. prof. Stanisława Tarnowskiego w Tarnobrzegu: Tarnobrzeg.

Zych, A. & Maciejewska, Ż. 2016. Innovations as the factor of competitive advantage of the enterprises-patterning of innovative activity: 83. In *A. Dudziak, M. Stomy, L. Rydzak (eds.): The selected problems from the range of management-theory and practice.* Libropolis: Lublin.

Zych, A. 2016. *Istota i uwarunkowania innowacyjności przedsiębiorstw*: 45. Państwowa Wyższa Szkoła Zawodowa im. prof. Stanisława Tarnowskiego w Tarnobrzegu: Tarnobrzeg.

Author index